ADVANCED TOPICS
IN SCIENCE AND TECHNOLOGY IN CHINA

ADVANCED TOPICS
IN SCIENCE AND TECHNOLOGY IN CHINA

Zhejiang University is one of the leading universities in China. In Advanced Topics in Science and Technology in China, Zhejiang University Press and Springer jointly publish monographs by Chinese scholars and professors, as well as invited authors and editors from abroad who are outstanding experts and scholars in their fields. This series will be of interest to researchers, lecturers, and graduate students alike.

Advanced Topics in Science and Technology in China aims to present the latest and most cutting-edge theories, techniques, and methodologies in various research areas in China. It covers all disciplines in the fields of natural science and technology, including but not limited to, computer science, materials science, life sciences, engineering, environmental sciences, mathematics, and physics.

Chengzao Jia

Characteristics of Chinese Petroleum Geology

Geological Features and Exploration Cases of Stratigraphic, Foreland and Deep Formation Traps

With 161 figures, 119 of them in color

ZHEJIANG UNIVERSITY PRESS
浙江大学出版社

Springer

图书在版编目(CIP)数据

中国石油地质学特征：岩性地层油气藏、前陆冲断带油气藏和深部油气藏的地质学特征与勘探实例 = Characteristics of Chinese Petroleum Geology: Geological Features and Exploration Cases of Stratigraphic, Foreland and Deep Formation Traps: 英文 / 贾承造著. —杭州：浙江大学出版社，2011.12
（中国科技进展丛书）
ISBN 978-7-308-08270-9

Ⅰ. ①中… Ⅱ. ①贾… Ⅲ. ①石油天然气地质—中国—英文 Ⅳ. ①P618.130.2

中国版本图书馆CIP数据核字(2011)第162382号

浙测地图审[2011]142号

Not for sale outside Mainland of China
此书仅限中国大陆地区销售

中国石油地质学特征：岩性地层油气藏、前陆冲断带油气藏和深部油气藏的地质学特征与勘探实例
贾承造　著

责任编辑	伍秀芳　张月红	
封面设计	俞亚彤	
出版发行	浙江大学出版社	
	网址：http://www.zjupress.com	
	Springer-Verlag GmbH	
	网址：http://www.Springer.com	
排　　版	杭州中大图文设计有限公司	
印　　刷	浙江印刷集团有限公司	
开　　本	710mm×1000mm　1/16	
印　　张	38.5	
字　　数	977 千	
版 印 次	2011 年 12 月第 1 版　2011 年 12 月第 1 次印刷	
书　　号	ISBN 978-7-308-08270-9 (浙江大学出版社)	
	ISBN 978-3-642-23871-0 (Springer-Verlag GmbH)	
定　　价	268.00 元	

Preface

The Chinese petroleum industry has a glorious and long history. The Chinese petroleum explorationists and geologists have made important contributions throughout the history of petroleum exploration. During the development of the petroleum industry, advancements in the understanding of Chinese petroleum geology were always accompanied by huge growth in the petroleum reserves. At the end of the 1950s, the Daqing oil field was discovered in the Songliao basin; along with the development of the geological theory of the terrigenous basin generating oil, the old geological theory that only marine facies could generate oil was revised. At the end of the 1970s, because of the exploration breakthrough in the Bohai Bay basin, the geological theory of composite petroleum accumulation in the faulted basin was developed and annual crude oil production reached 100 million tons. Entering the 21st century, the Chinese petroleum industry has developed into a new stage. The rapid development of the domestic economy brought a serious challenge to domestic oil and gas supplies. Chinese petroleum companies invested a huge amount of capital in domestic exploration. Oil and gas exploration was carried out in eastern and western China on a large scale. As a result, there were major advances in natural gas exploration. At present, the new circumstances that we encounter are increasingly diversified exploration targets; the exploration focal points and hot spots may include several types of basin, complicated oil and gas reservoirs, various regions with different exploration levels and old and new exploration domains. In accordance with oil and gas exploration in the near future, petroleum explorationists and geologists are now facing the major need to develop new geological theories to be applied to oil and gas exploration so as to achieve new breakthroughs and maintain and increase domestic oil and gas supplies. This issue can also be viewed as a challenge and an opportunity facing Chinese petroleum geologists and Chinese petroleum exploration engineers. The Chinese petroleum industry needs a forward-thinking geological theory and a suitable exploration strategy.

In recent years, new geological theories have been developed, which were motivated by the following: to challenge existing geological theories that were being applied during oil and gas exploration by the petroleum industry; to examine the use of new geological theories from overseas and from within China (such as the introduction of the theory of stratigraphic sequences); to make use of the progress in technologies that specialized in seismic technology and drilling technology. The author believes that the application of new geological theories are in the course of development, to face the latest challenges facing oil and gas exploration at the beginning of the 21st century, incorporating the newest discoveries and advances in knowledge that were uncovered during the petroleum exploration of recent years, and which absorb the newest study results in fundamental disciplines of petroleum geology as well. These new geological theories could be called "Chinese Exploration Theories for Petroleum Geology at the beginning of the 21st Century", which include applying new theories to petroleum geology and exploration of the stratigraphic reservoir; to a study of the tectonics of the petroleum basins of western China; to thrust belt exploration in the terrigenous foreland basin; to exploration of the lower and middle petroleum systems (at a deep level) in the superimposed basin; to natural gas geology and related gas accumulative theory; to the geology of non-conventional oil and gas resources and to related technology for exploration and development and to petroleum resource evaluation and forecasting of the remaining petroleum resources.

At the same time, considering the circumstances that the Chinese onshore petroleum exploration is confronting (which include the situation of the petroleum resources, the economic circumstances and the technical conditions), the author believes that the strategy for onshore petroleum exploration at the beginning of the 21st century should include the following four measures: firstly, the major exploration domains should include the stratigraphic reservoir, the thrust belt in the foreland basin, the middle and lower petroleum systems (at a deep level) in the superimposed basin and comprehensive exploration in previously explored regions. Furthermore, the seven large petroleum basins should be the major targets, including the Songliao basin, Tarim basin and other large basins; secondly, we should not only accentuate oil exploration and accelerate natural gas exploration, but should also positively carry out risky exploration in new basins and new domains; thirdly, while developing applied seismic technology and drilling technology, we should also highlight comprehensive geological research and geoscience studies in fundamental disciplines; lastly, we should pay close attention to non-conventional petroleum resources and new forms of energy.

The author is not only a geologist, but also an explorationist. Almost all of his working years have been spent either at the site of petroleum exploration and development or in geological field studies. At present, the author still concentrates on petroleum exploration research. With many years of working experience (either in scientific study or in exploration), the author profoundly understands that new scientific discoveries come from first hand scientific data collected at the work site and, in addition, that new exploration ideas come from petroleum exploration practices. These new scientific discoveries and new exploration ideas not only motivate the creation of new geological theory and new exploration theory, but also direct the development of the petroleum industry. In recent years, during geological research and exploration, the author has achieved some success by carefully studying and evaluating the data at first hand. Some research results have been tested and proven by exploration practices, which also have made a number of contributions to Chinese onshore petroleum exploration. The author is happy to see a hard working petroleum industry and the important progress that has been achieved during oil and gas exploration in recent years.

This book includes four parts. Part I contains three chapters that fully introduce the plate tectonics, the sedimentology, the petroleum geology and the distribution pattern of oil and gas resources in the Chinese petroleum basins. Chapter 1 discusses the tectonic background and related developments in the Chinese petroleum basins. Chapter 2 discusses the distribution features of petroleum resources and exploration potential in the Chinese petroleum basins. Chapter 3 includes the evaluation of the Chinese petroleum resources, the current situation of petroleum exploration and development. In particular, the author discusses distribution patterns, the undiscovered oil and gas resources and the exploration domains.

Part II introduces the geological theory of Chinese stratigraphic reservoirs, the technology and exploration practices. Part II contains five chapters, from Chapters 4 to 8. Chapter 4 discusses the types of reservoir, the characteristics of a trap and the controlling factors in the creation of traps in a Chinese stratigraphic petroleum reservoir. Chapter 5 discusses the major types and related characteristics of petroleum accumulative regions to be found in stratigraphic reservoirs. In this chapter the author presents the concept of the accumulative petroleum combination of structure–stratigraphy; in addition, the author also discusses the classification of accumulative combinations and their features. In establishing three types of petroleum accumulative combination (which include the type of source rock inside, above and beneath the reservoir) and due to the controlling factors of the accumulative petroleum combination, Chapter 6 presents the categories of accumulative combinations and related creative mechanisms. Chapter 7 discusses the petroleum accumulative features and distribution patterns of stratigraphic reservoirs in four different basin prototypes. Chapter 8 presents successful exploration examples that include the lithologic type of oil reservoir in the Bayindulan depression in the Erlian basin, the lithologic type of oil reservoir in the Changling depression in the southern part of the Songliao basin, the lithologic type of oil reservoir in the Mesozoic low permeable sandstone in the Eerduosi basin and the stratigraphic type of oil and gas reservoir in the carbonate reef flat in the Tazhong area.

Part III not only introduces the characteristics of petroleum geology in the foreland thrust belt which contains extremely complicated structural geology, but it also discusses the related exploration practices. Part III contains Chapters 9 and 10. Chapter 9 discusses the characteristics of petroleum geology in the Chinese foreland thrust belt. Chapter 10 discusses successful examples of petroleum exploration in the Chinese foreland thrust belt, which not only include oil and gas exploration in the Kuche foreland thrust belt and the discovery of the Kela 2 gas field, but also oil and gas exploration in the Jiuquan foreland thrust belt and the discovery of the Qingxi oil field.

Part IV introduces the geological characteristics of oil and gas reservoirs that not only utilize carbonate rocks and volcanic rocks as the reservoir sequences, but that are also positioned at a deep level in the superimposed basin. In addition, Part IV also discusses the potential of the oil and gas resources and exploration practices. Part IV includes four chapters, from Chapters 11 to 14. Chapter 11 proposes methods of exploration in searching for a petroleum system at a deep level in the onshore superimposed basins. Chapter 12 discusses the geological features of oil and gas reservoirs in carbonate rocks in China. Chapter 13 introduces the geological features of volcanic reservoirs in China. It also utilizes the geochemistry of non-organic carbon dioxide and its resource as an example of how to carry out an in-depth investigation. Non-organic carbon dioxide occurs in the Cretaceous volcanic petroleum system in the Songliao basin. Chapter 14 discusses successful examples of petroleum exploration at a deep level in the superimposed basins, in particular emphasizing the Lunnan Ordovician oil and gas reservoir in the Tarim basin, the reef flat gas reservoir in the Changxing–Feixianguan formation in the Longgang area in the Sichuan basin, the Cretaceous volcanic gas reservoir in the Xujiaweizi faulted basin in the Songliao basin and the Carboniferous volcanic gas field in the Zhungeer basin.

The author is full of gratitude to the enormous support from PetroChina, the Research Institute of Petroleum Exploration and Development (RIPED), as well as the Science and Technology Division of PetroChina. The author would like to show particular appreciation to Professor Kai Qian, Professor Guoqi Wei, Professor Xuanjun Yuan, Dr. Benliang Li, Dr. Hui Qu and Dr. Yongfeng Zhang; they contributed to this book by editing and writing some chapters. Last, the author would like to thank Dechin Wang and Ian McIntosh for their contribution in improving the English of the manuscript.

Chengzao Jia
Beijing, China
Oct. 2011

Contents

Part I

Characteristics of Chinese Petroleum Geology and Evaluation of Oil and Gas Resources

1

Tectonic Background and Development of Chinese Basin

From a geological point of view, most of the present-day Chinese continental landmass is located along the southern margin of the Eurasian plate and along the northern margin of the Indian plate whereas, to the east, it borders on the trench-arc-basin system formed by the subduction of the Pacific plate. In earlier geological times, the principal elements of the Chinese landmass were located between the Gondwanaland and Laurasia continents. Throughout geological history, the Chinese landmass has experienced many phases of divergence and convergence. Besides three small principle plates (that were the Tarim plate, Huabei plate and Yangtze plate) that served as basic building blocks, the Chinese landmass also converged with more than twenty micro terrains that include Zhungeer terrain, Chaidamu terrain, Qiangtang terrain and others (Zhang, 1991; Li, 1982). During the Paleozoic era, most of the Chinese landmass was located on the northern margin of Gondwanaland and it was positioned to the south of the Paleo-Asian ocean. During the Mesozoic era, the Chinese landmass was located on the southern margin of the Laurasia continent and it was positioned to the north of the Tethys ocean (Ren and Xiao, 2002; Xiao et al., 1991; Jia, 1997; Yang, 2002). Destruction of paleo-oceanic crusts and the convergence of ancient plates determined that Chinese sedimentary basins were predominately developed on top of a series of small size paleo-plates. After experiencing the Paleozoic marine depositional phase and the Mesozoic terrestrial depositional phase, Chinese sedimentary basins are at present surrounded by orogenies that represent the nature of the paleo-oceanic crust. The orogenies include Tianshan–Yinshan orogeny, Qilian mountain–Qinling orogeny, Kunlun mountain orogeny, Longmen mountain orogeny and Helan mountain orogeny. The geological age of orogenies is diverse (Chen and Wang, 1997; Lu et al., 1994; Hendrix et al., 1994; Graham et al., 1993; Huang et al., 1977; Shu and Zhou, 2002; Jia et al., 2001; Chen et al., 1986; Hsu, 1988; Zhu et al., 2007; Luo, 1991; Wu et al., 1997; Zhang, 1996) (Fig. 1.1). There are 373 sedimentary basins in China, which offer a total basin area of 548×10^4 km^2. Among the basins, fifteen are large size basins; one large size basin can cover an area larger than 10×10^4 km^2. Together, large size basins provide a

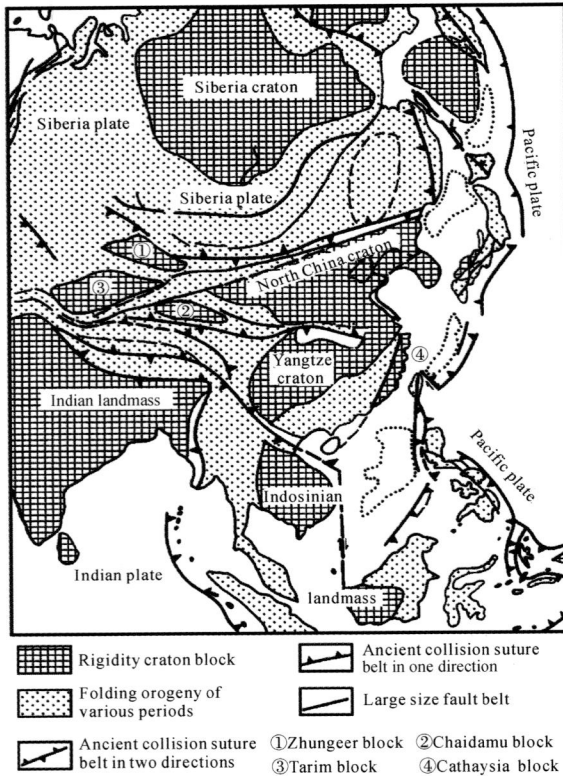

Fig. 1.1.   Tectonic outline of China and its neighboring areas

total basin area of 257×10^4 km^2. Large basins include the Songliao basin, Bohai bay, Eerduosi basin, Sichuan basin, Zhungeer basin, Chaidamu basin and Tarim basin. Furthermore, there are 39 medium size basins; one of them can cover an area of $(1 - 10) \times 10^4$ km^2. Jointly, medium size basins present a total basin area of 100×10^4 km^2. Chinese sedimentary basins were developed on top of the basements of small plates. The basements include pre-Sinian acidic and intermediate-basic metamorphic rock series and the Paleozoic fold basement. The tectonic origin and development of Chinese basins were decided by multi-phase reduction of the oceanic basin, the convergence of several micro plates, divergence and convergence of ancient plates and the growth and destruction of the paleo-oceanic crust. Tectonic evolvement and development of Chinese basins can be divided into four phases: (1) the form of the Precambrian basement; (2) the drafting of small continental crusts and deposition in the marine basins during the Paleozoic era; (3) the convergence of continental crusts and deposition in terrestrial basins during the Mesozoic era, and (4) orogenic movements and the form of basins on the inner continent during the Himalayan tectonic period. The orogenic movements and the development of basins in the fourth phase were controlled by dual tectonic mechanisms: one was the collision and compression between India plate and Eurasia plate; the other one was the back-arc extension caused by the subduction of the western Pacific plate.

1.1 Characteristics of Tectonic Framework and Distribution of Geophysical Field

In China, the surface structures are very well developed and the outline of the last tectonic movement is evident. At a deep level in the earth, because the deformation dynamics process caused the exchange of substance and energy, and because of interaction between different layers of the sphere, the surface structures correspond with multiplicity, intricacy and activity. Over the past two decades, after several generations of efforts by geologists and with investment, using geophysics exploration methods (that include seismic, geothermal, gravity, conductivity and magnetic methods) and analytical technologies of tomography and geoscience transect mapping, remarkable results have been achieved in the fields of regional lithosphere structure, seismic sounding and geodynamics. The outcome of research work initially reveals the deeply buried structural background and its characteristics in the Chinese landmass (Yuan, 1996; Xu and Shi, 2003).

The depth distribution of Moho in the eastern Asian continent and adjacent waters reveals that, in western-central China, the crustal thicknesses vary horizontally with distinct thickness variation and irregular contour lines. Overall, the crustal thickness decreases inside the basin and it gradually increases towards surrounding orogeny belts. According to the crustal thickness, the crust can be divided into different blocks (Teng et al., 2002).

Based on thirty years of work, using nearly 50,000 km of seismic sounding data that include 2,000 km of seismic reflection profiles and DSS data from neighboring countries and adjacent territorial waters, Teng et al. (2002) published the depth distribution of Moho in the eastern Asian continent and its adjacent waters. The crustal thicknesses vary horizontally with distinct thickness variation and irregular contour lines. Overall, the crustal thickness is thinner on the edge of the landmass and it gradually increases towards the inner continent. According to the crustal thickness, the crust can be divided into different blocks (Fig. 1.2). The boundary of the crustal block is based on a steep gradient zone with a strongly altered contour line of crustal thickness. Basically, the crustal thickness is stable inside blocks. However, an alteration of crustal thickness might occur in an isolated block. Utilizing teleseismic data and seismic tomography technology, Xu et al. (2001) obtained details of the velocity structure of the lithosphere and information about the deep geodynamics in the Tarim basin, Tianshan mountain and Zhungeer basin. In the Tarim basin, the thickness of the lithosphere is about 200 – 250 km; however, in the Zhungeer basin, the thickness of the lithosphere is 170 km; in the Tianshan orogeny, the thickness of the lithosphere is close to 150 km. The structure of the lithosphere also has a distinctive variation horizontally.

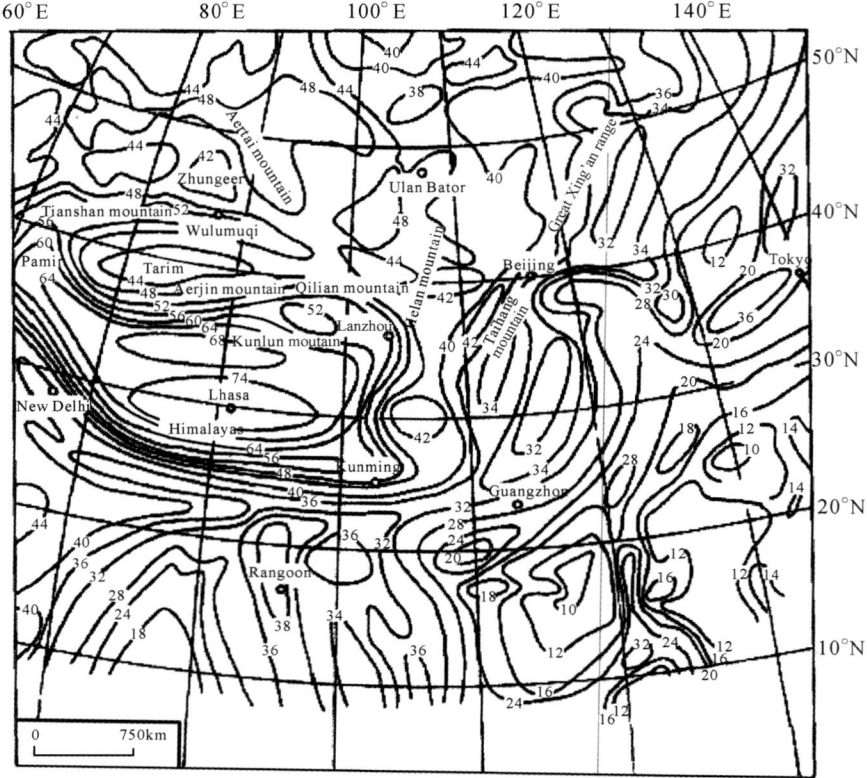

Fig. 1.2. Distribution map of crustal thickness (km) in eastern Asian continent and adjacent waters (Teng et al., 2002) (with permission from Journal of Science in China)

It is well known that characteristics of the geophysical field represent the deep buried structures. Therefore, via geophysical inversion analysis, the details of the deep structural framework and structural information will be revealed. Due to the ambiguity of geophysical inversion analysis, the joint inversion analysis combines the data of geothermal, gravity and magnetic methods and lithosphere structure, in order to obtain information about deep structures and the lithosphere inside basins. Detailed comprehensive geophysical interpretations were carried out for the Tarim basin, Zhungeer basin, Chaidamu basin, Eerduosi basin and Sichuan basin. The gravity data and the aero-magnetic data demonstrate a dissimilarity between different sections of a basin. On the edge of a basin, a closely packed gradient zone of gravitational anomaly and high magnetic anomaly is predominant. Inside a basin, the changes are gentle; the magnetic anomaly is lower, and gravitational anomalies alternate between high and low (Jia, 1997; Jia et al., 2001).

The sedimentary basin has lithospheric heterogeneity; vertically, the lithosphere can be divided into different layers; horizontally, the lithosphere may be separated as blocks. Vertically, the upper crust of the lithosphere is a brittle layer that contains fractural deformation; the middle and lower portions of the crust are ductile layers that contain plastic deformation and elongation flow; the upper mantle of the lithosphere is a high strength brittle layer. The characteristics of deformation in different vertical layers control the faults of inner basins; along the décollement horizon in the crust, the large size deep fault converges at the middle portion of the crust. Because of variations in the physical property and litho-mechanical property, there are some décollement surfaces and ductile layers inside the sedimentary cover where numerous fault-propagated folds have developed. Horizontally, the entirety of the inner basin is good, with weak deformation; every layer of the lithosphere is basically parallel to each other and parallel with the sea level. However, at the conjunction area between basin and range, the transitional zone is strongly deformed because here is the weak zone of the lithosphere thermal rheology. The lithosphere is easily deformed by structural stress. The characteristics of the lithosphere of a basin are clearly exhibited in the gravity field, aero-magnetic field and geothermal field.

1.2 Basement of the Craton Basin Formed During the Precambrian

In this section, the basement of the sedimentary basin refers to the basement of the craton basin. Wang (1985) strictly defined the terminology of a platform (craton) to a continental crust that had been consolidated and matured and that is tectonically stable. The terminological usage of platform (craton) does not include the Phanerozoic fold zone. In addition, Wang further divided the development of a platform into three stages that are the formation of a continental shield, the formation of a proto-platform and the formation of a platform.

1.2.1 Continental Shield Formed Earlier than 2,800 Ma

The development of the continental shield refers to the creation of a large scale, tectonically stable, matured crust of the earth. The continental shield is a crystalline basement that is predominantly made of gray color gneiss with a geological age greater than 2,800 Ma. The study of the Precambrian geology shows that giant cratons were made of many pieces of landmass of various sizes, such as the Siberia craton, North American continent, Indian landmass, African continent and Australian continent. Approximately 3 billion years ago, some continental shields (ancient gneiss areas) were created on the surface of the earth. When granite and greenstone belts were welded onto the borders (or on top) of the shields, continental accretions occurred and giant cratons were formed. The end of the early Proterozoic era and the end of the middle Proterozoic era were the two most important periods for continental accretion, which were also the main periods when the Laurasia continent and Gondwana landmass were formed and converged into a supercontinent.

Numerous studies indicated that major Chinese platforms were created by divergence and convergence of continental shields; for example, the North China platform contains the Jiliao shield (3,720 – 3,650 Ma) and Shanbei shield; the Yangtze platform contains the Kangding shield (3,100 – 1,700 Ma) and Huangling shield (2,850 Ma); the Tarim platform contains the Tabei shield (3,040 Ma) and Tanan shield (2,785 Ma). These continental shields served as a center for landmass convergence. In China, the oldest granite occurred in the Huabei region and Tarim basin (Ren, 1980; Wang, 1990). In the eastern Hebei province, the isotope stratigraphic age is 3,850 – 3,550 Ma; near the Anshan area, the isotope stratigraphic age is 3,840 Ma (Liu, 1991). In the Hongliu gap of Gansu province, the geologic age datum is 3,488 Ma; in the Luketage area of the Tarim basin, the geologic age datum is 3,263 Ma (Hu et al., 1992). These chronological data indicate that in the Huabei region and Tarim basin, continental shields occurred even in the early Archean Eon. In many locations, the volcanic sedimentary rock series of the late Archean era has been discovered on the margin of continental shields, which include various types of volcanic rocks, aluminum-rich terrigenous clastic rock and calcium-magnesium carbonate rock. The greenstone formation predominately occupied the area between the continental shields; the age of granite, migmatite and other metamorphic rocks is in the range of 3,000 – 2,500 Ma (Zhang, 1997).

According to distribution of the lithologic and stratagraphic unit of the Archean-Proterozoic, isotopic chronology data, rock type and its correlation, pattern of structural deformation and aero-magnetic anomalies, Zhai et al. (2002) divided the Chinese landmass of the Archean Eon into several continental blocks, which are the Zhungeer-Kazakhstan block, North China–South Tarim–Yangtze super block, Aertai block, Songliao block, Cathaysia block, Gangdise block and Himalayan block. In addition, Zhai et al. pointed out that during the middle and late Archean Eon, the North China terrane, South Tarim terrane and Yangtze terrane were individual continental shields. During the Fuping movement at the end of the Neoarchean era, controlled by a uniform stress field, three terranes converged into a super block that was oriented in an NW-SE direction or almost in an EW direction.

1.2.2   Basement of the Craton Basin Formed During 2,800 – 800 Ma

In China, the continental landmass primarily experienced the Luliang tectonic cycle and the Jinning tectonic cycle. Through continental accretion and convergence, a proto-platform and a platform were formed. The major platforms are the North China platform (around 1,800 Ma), Yangtze platform (around 1,000

Ma) and Tarim platform (around 800 Ma). Similar to the accretion pattern of the North American platform-shield that expands outwards concentrically, the Chinese cratons are small in size. Because of divergence during the late period, the broken pieces of a craton drifted away in a vast ocean. Finally, they became the small cratons that were inlaid between the giant folded belts. Dissimilar to the North American giant craton and bordered folded belts, Chinese cratons were tectonically unstable; therefore, Chinese cratons are also called paraplatforms. Small Chinese cratons can better reflect the correlations between continental margin and inner craton; the comparisons are not only between sedimentary sequences and deformed patterns, but also between the tectonic cycle and stratigraphic cycles.

After the structural-thermal event, the main portion of the North China craton and several large landmasses in the Tarim basin were formed. The structural-thermal event happened during the Statherian period of the Paleoproterozoic era (around 1,850 Ma) and the North China craton was finally consolidated during the same geological period. The Changchengian system of the Mesoproterozoic erathem was the sedimentary cover for the North China craton, or it was the base litho-formation of the inner continent and continental margin rift (Wu et al., 1991). In southern China (that includes the Yangtze craton and Cathaysia continental block), the continental shields were formed later than the one in the North China craton and Tarim craton. The isotopic data of the earliest metamorphic rocks of amphibolite facies show a geological age range of 3,200 – 3,000 Ma. These metamorphic rocks were discovered in the following areas: the Kangdian area, the Huangling-Shennongjia area in western Hubei Province, Zhanggong mountain in Jiangxi Province and the Jianning area in Fujian Province (Lin, 1987, Ma, 1993, Zhao, 1993; Shen, 1993). In southern China, the Paleoproterozoic (2,000 – 1,800 Ma) is the most important period for the development of the crust (Shen, 1993); the Cathaysia continental block had been formed during this period. Furthermore, in the upper Yangtze region (from central Sichuan to western Hubei) and the region from northern Jiangsu Province to the southern Yellow Sea, two relatively large size continental blocks were formed (Zhang, 1997). During the Mesoproterozoic era, terrigenous clastic rock (quartz sandstone and carbon-rich, aluminum-rich argillaceous rocks) and magnesian carbonate rock were deposited on these continental blocks. The rifted volcanic-sedimentary rocks were predominately deposited in the areas between the continental blocks, such as the Sichuan–Yunnan area and the Motianling area.

During the Mesoproterozoic era, after the accretions from the Archean shields, the North China craton, Yangtze craton and Tarim craton were continually solidified with the continental lithosphere; the region of re-merged continental landmass of the North China–Southern Tarim–Yangtze continually increased and this joint continental landmass gradually became more stable and more rigid. During the early period of the Mesoproterozoic era, divergences occurred on rigid continental landmasses along the pre-existing ductile shear zone, the active zone or folding zone. Limited divergence occurred on the North China craton, southern

Tarim craton, Chaidamu continental block and Yangtze craton; for example, a three-armed rift system appeared on the North China craton (Ma et al., 1987). From the middle and late Mesoproterozoic–early Neoproterozoic, on the North China craton, the rifting event tended to stop with stable sedimentary deposits. The depositional thickness is about 1,000 m; the sedimentary sequences are divided into three systems: the Changchengian system, the Jinxian system and the Qingbaikouan system. Based on continental margin rifting, the active type of continental margin deposits appeared on southern Tarim craton and Yangtze craton. For example, on the western margin of Yangtze craton, Yanbian group, Huili group and Kunyang group represent a typical sedimentary combination of trench–island   arc–back   arc   basin   (Zhai,   2002).   Thus,   from   the Mesoproterozoic–early Neoproterozoic, the North China craton is much more stable than the Tarim craton and Yangtze craton. During the middle and late Neoproterozoic era, the accretion of terranes was lateral; and the Tarim terrane and Yangtze terrane converged due to the folding activities of sedimentary strata on the Mesoproterozoic active continental margins of the South Tarim terrane and Yangtze terrane. During the Proterozoic, the convergence of cratons formed a supercontinent–Rodinia, including convergence of the North China craton and Yangtze craton and convergence of the Tarim craton and Yangtze craton. Zhang (1997) called the joint Chinese cratons a "Proto-Chinese landmass". However, Zhu (1997) considered the joint Chinese cratons as a "Chinese portion of Rodinia". Evidence for the joint Chinese landmass includes the isotopic chronological age of granite, migmatite and metamorphic rocks that occur in a converged suture area (Yue, 1995). The most important event is the moraine deposit on Chinese cratons during the end of the Neoproterozoic era (the Sinian period). Hence, the basement of the Chinese landmass had been formed, which included the Archean shield and the metamorphic folded basement from the Mesoproterozoic and Neoproterozoic. The rock series and petrofabric of the basement are complicated.

1.3   Drifting of Small Paleozonic Plates and Marine Deposits

In China, the long duration of tectonic evolvement and tectonic intricacy determined the particularity of the Paleozoic marine basin and its oil and gas potential. Since the 1970s, many geologists and petroleum geologists have conducted research in a broad range of scientific disciplines (Wang et al., 2001; Zhu, 1986). Qiu and Gong (1999) believed that, since the Proterozoic Eon, Chinese continental plates had experienced multi-phases of divergence and convergence. Employing small tectonic plates as its core (such as the Tarim plate, North China plate and Yangtze plate), the Chinese continental plate was created from the convergence of more than twenty micro terranes; these microterranes include Zhungeer, Chaidamu and Qiangtang (Jia et al., 2001). Over the past dozen years, petroleum exploration in the Paleozoic marine basin (which was represented by the Tarim basin) has been expanded with much scientific data acquired. Via studies of the geological structure and petroleum geology in the Tarim basin and considering both the exploration status in the marine basins and the study results, Jia et al. (1997; 2001) demonstrated the characteristics of Chinese marine basins regarding their structural origin, evolution pattern and petroleum geology.

1.3.1 Drifting of Early Paleozonic Plates and Marine Deposits

During the late Neoproterozoic era, due to the divergence of the supercontinent Rodinia, the North China plate, Yangtze plate and Tarim plate were separated from Rodinia; these three plates were linked by three interconnected oceans that were the Paleo-Asian ocean, Paleo-China ocean and Proto-Tethys ocean (Wang et al., 2001; Zhi, 1986).

During the early Paleozoic era, these small cratons were located in middle-low areas of latitude in the southern hemisphere; the distances between the Tarim plate, North China plate and Yangtze plate were large (Table 1.1). For example, during the Ordovician period, the three plates were located south of the equator. However, they were positioned at different longitudes. The Tarim plate was located at longitude 181.5° E, the North China plate was located at longitude 28.8° E and the Yangtze plate was located at longitude –17.4° E. The distances between the three plates were great. Later, the plates drifted northward. During the end of the Paleozoic era, the separated plates tended to converge together to construct the embryonic form of the ancient Chinese continent (Hsu, 1988). According to Li (1982), during the early Paleozoic, the distance between the North China plate and Yangtze plate was 4,000 km at least. The Tarim plate, North China plate and Yangtze plate were floating in Kunlun ocean, Qinling-Qilian ocean and Tianshan ocean, respectively (Li and He, 2002) (Fig. 1.3).

Table 1.1 Reconstructed latitude and longitude data for three ancient Chinese plates based on ancient paleo-magnetic information (Jia, 1997) (with permission from Petroleum Industry Press)

Name of plates	Cambrian		Ordovician		Silurian	
	Longitude (° E)	Latitude (° N)	Longitude (° E)	Latitude (° N)	Longitude (° E)	Latitude (° N)
Tarim plate			181.5	–21.2	164.5	12.5
North China plate	21.9	–24.0	28.8	–21.3		
Yangtze plate	–5.2	–3.2	–17.4	–1.0	–4.9	–1.3

Note: Reference point of Tarim plate is 41.7° N, 80.5° E; Reference point of Yangtze plate is 29.6° N, 103.4° E; Reference point of North China plate is 37.8° N, 112.4° E

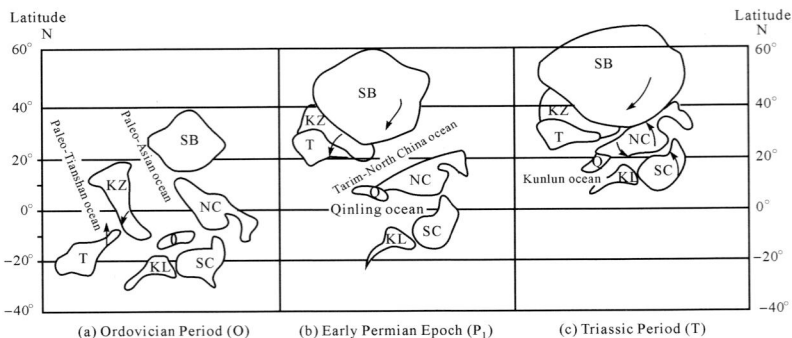

Fig. 1.3. Diagram of ancient tectonic plate movements in China and adjacent areas (Li and He, 2002) (with permission from Journal of Marine Origin Petroleum Geology)
SB, Siberian plate; KZ, Kazakhstan plate; T, Tarim plate; Q, Chaidamu (Qaidam) plate; KL, Middle Kunlun plate; SC, Yangtze (South China) plate; NC, North China plate

From the Sinian period to the early Paleozoic era, the main portion of tectonic plates merged below sea level. On the margin of cratons, sediments were predominately deposited in an aulacogen, the passive continental margin and marginal depression, for example the Yanliao area on the northern side of the North China plate, Kuluketage aulacogen on the eastern Tarim plate and the Helan mountain aulacogen in the northwestern Eerduosi basin (Chen et al., 1986). In the western Sichuan basin, a peripheral rift might have developed in the Longmen mountain area (Luo, 1991; Guo et al., 1996) and a suite of organic rich shale developed in the deep water environment. Occasionally, a deep depression might develop on the inner craton plate, which received calcareous mudstone or mudstone deposits that showed signs of hydrocarbon source rock. For example, in the central Tarim basin, the Ordovician calcareous mudstone belongs to the sequence of an inner craton depression. The cratons were limited by oceanic basins and received carbonate deposits of platform facies on the inner craton. During the early Paleozoic era, some craton plates steadily subsided, such as the Tarim plate, Sichuan plate and Eerduosi plate. These ancient tectonic plates subsided below sea level in the same geological period and they received epicontinental deposits. From the uplift area on the inner craton plate towards the oceanic basin, the horizontal facies consist of inner platform lagoon facies, dolomite platform facies, marginal platform facies and neritic facies. On the margin of a platform, the reef deposit and bio-clastic deposit on the tidal flat formed a massive layer of oil and gas reservoir sequences. During the Late Ordovician epoch, the North China plate was lifted above the water and it became an ancient landmass that experienced a long period of erosion. During the Early and Middle Silurian epochs, argillaceous clastic deposits of neritic facies occurred on the margins of the Tarim plate and Yangtze plate. During the Late Silurian epoch, both plates were lifted above the water and they became ancient landmasses (Chen et al., 1986). Because the ancient tectonic plates rose above the sea water, and because carbonate formations went through an erosion process or underwent fresh water leaching, weathering crust type karst reservoirs were formed. At same time, the marine deposits that were developed on ancient tectonic plates were compressed by surrounding oceanic crust subduction during the Caledonian movement. Ancient uplifts formed inside the tectonic plate, which became a beneficial place for petroleum migration and accumulation during later geological periods.

1.3.2 Late Paleozoic Plate Convergence and Marine–Terrigenous Basin

During the Late Carboniferous epoch, the Tarim plate, North China plate and Yangtze plate began to drift northward, in successive order. The oceans that separated these ancient plates were closed, including South Tianshan ocean, North Tianshan ocean, Helan sea trough, Kunlun ocean and Qinling ocean. On the southern margin of the Eurasian plate, the borders of three ancient plates were welded together to form the ancient Chinese landmass (Fig. 1.3).

These small Chinese craton plates were surrounded by the Hercynian orogenies and they developed into a stable landmass with widespread sedimentary deposits of marine–terrigenous facies. The Tarim-North China plate converged with the Kazakhstan-Siberian plate; Zhungeer basin was formed north of Tianshan mountain. Kunlun-Qinling ocean was subducted northward and it was closed. To the north of Kunlun-Qinling ocean, the Tarim plate, North China plate and Chaidamu plate received sedimentary deposits of marine–terrigenous facies and terrestrial facies (Huang et al., 1977; Shu and Zhou, 2002). During the Carboniferous period, based on the analysis of the Cathaysian flora and the distributions of gypsum and other minerals, the North China plate and Tarim plate were located in lower altitude tropical-subtropical regions which had a dry-mesic climate (Teng et al., 2002). During the end of the Early Permian epoch, they drifted to a higher altitude region and collided with the Siberian plate. Along with the closing of the Paleo-Asian ocean, the marine transgression had been stopped and Zhungeer basin was formed. The region that was located north of Kunlun-Qinling had become the North China continent. At this time, the ancient Yangtze plate still floated near the equatorial region and it was still submerged under the sea. Until the Middle Triassic epoch, due to the closing of the Paleo-Qinling oceanic basin, the ancient Yangtze plate was welded onto the southern margin of the North China plate. During the end of the Middle Triassic epoch, the Jinshajiang Tethys oceanic basin and the basins north of it were finally closed; the Yangtze plate collided with the North China plate and Upper Triassic terrestrial basins of Sichuan basin and Eerduosi basin were formed. Along with tectonic plates drifting northward to converge into a joint continent, rifting valleys and the oceanic basin were closed; the peripheral basins on the border of cratons were converted into orogenies; the Paleozoic marine sedimentary sequences were inlaid between the IndoSinian orogenies. During this geological period, affected by intermittent regional extension of tectonic events, on the edge of craton plates, rift basins or limited oceanic basins were formed. Since the Late Carboniferous epoch, on the margins of craton plates, limited oceanic basins or rift zones appeared at Tianshan mountain, Qilian mountain, the South Kunlun–South Qinling areas. The depressions on the inner craton and the rift basins on the margins of a craton received the Carboniferous–Permian sedimentary deposits of marine–terrigenous facies.

1.3.3   Creation and Development of Marine Craton Basin in China

The subsidence of a craton basin was primarily related to the elevation and subsidence of the mantle plume, or it was associated with the opening-closing movement of the tectonic plate. Along with the divergence of a supercontinent, the tectonic plate responded accordingly with a subsidence event, which formed a craton basin. Rifting or volcanism was often beneath a craton basin; for example, the subsidence of the Tarim basin occurred after the divergence of Tianshan ocean; the subsidence of the Eerduosi basin and the subsidence of the Sichuan basin happened after the divergence of Qinling ocean. Therefore, the craton cycle was related to the supercontinent cycle (Wilson cycle), because the development of a craton basin primarily followed the cycle of continental divergence and convergence. From divergence to convergence, all kinds of prototype basins were developed (Liu, 2005). Because the craton basin predominantly appeared during the drifting period of a plate, thus the Chinese craton basins were mainly developed during the oceanic developmental period in the Paleozoic era. From the Cambrian period–Early Ordovician epoch, thermal subsidences happened subsequent to divergence events in the Tarim basin, Eerduosi basin and Sichuan basin. Carbonate platforms were formed during the early subsidence phase and evaporite platforms were formed during the final subsidence phase, which jointly constituted a transgression-regression cycle. From the Middle and Late Ordovician epochs–the Silurian period, affected by the subduction and collision on the edge of a plate, an uplift event might happen at the inner craton basin and form an unconformity surface. On the other hand, the flexural subsidence might occur at the edge of a craton and form a submerged surface with a foreland basin superimposed on top, which became an important developmental stage of a Chinese craton basin. All marine basins that were formed on top of the cratons had experienced two developmental processes that were the Paleozoic stage and the Mesozoic (or the Cenozoic) stage. All these marine basins were distributed on ancient landmasses of the Tarim, Yangtze and Eerduosi.

The Paleozoic craton basins were stable, and which either occurred on an inner craton as a depression or a rift basin, or appeared on the edge of a craton as a peripheral basin.

The following are the developmental stages of the Tarim basin (Jia, 1997; Jia et al., 2001; 2005). After the crystalline basement of the Tarim plate converged in the Mesoproterozoic era, from the Sinian period–Ordovician period–middle Carboniferous epoch, the Tarim plate was positioned at a latitude of around twenty degrees south. These were the craton block stage and the marginal rifting stage with sedimentary deposits of marine carbonate rock and clastic rock. From the Late Carboniferous epoch–Permian period, the Tarim plate drifted northward to a latitude of twenty degrees north; the continental convergences happened and the oceanic basins around Tarim plate were closed (Fig. 1.3). During this interval, an extensive basalt eruption occurred during the Permian period. The clastic deposits of marine–terrigenous facies were superimposed on top of marine formations and they were interbedded with the Permian basalt. Beside the early Triassic compressive structural deformations (that were caused by a thrust fault) occurring on the edge of the Tarim plate, from the Triassic–Cretaceous, the intracontinental depressions were superimposed by the lacustrine basins throughout the Tarim plate. During the Cenozoic era, the Indian plate collided with the Tibetan plateau and it propagated a long-rang effect into the continent. Thus large scale intracontinental subduction happened and, due to the basin and range coupling on the edge of the Tarim plate, the fold-thrust belt and massive terrestrial molasse deposits were formed (Fig. 1.4). The molasse deposits were caused by flexure on the edge of the Tarim plate. For the region south of the Tarim plate, marine formations developed as well; they were under the control of the Tethys tectonic domain since the Mesozoic. The Tethys ocean was subject to subduction, convergence and collision northward, in successive order (Huang et al., 1977; Shu and Zhou, 2002). The marine basin on the southern margin of the Tarim plate was the first one that experienced tectonic alterations of compressive uplifting and magma activity in the extensional environment. The Tarim basin that was located at the far side of the tectonic event center received a relatively weak form of tectonic alteration. In addition, because the Tarim plate had a Precambrian crystalline basement, which had a strong resistance to structural deformation, the Tarim basin was preserved well. Overall, in the intra-basin, the structural framework was stable and the Paleozoic marine formations were favorably developed. However, on the edge of the basin, due to the fold-thrust belt and the compressive flexural subsidence (deposit) of the Indosinian tectonic period and the Himalayan tectonic period respectively, the possibility of petroleum migration and accumulation in the Paleozoic marine formations was limited (Fig. 1.4).

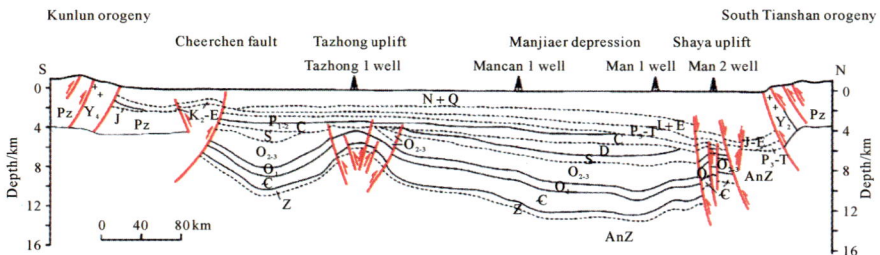

Fig. 1.4.   Structural section (N-S) of Tarim basin

The following are the developmental stages of the Yangtze plate and Sichuan craton basin (Chen et al., 1986; Luo, 1991). During the Sinian period, the Yangtze plate converged with a crystalline basement. From the Sinian–Ordovician–Silurian periods, the ancient Yangtze plate was positioned around $1° - 3°$ south of the equator and it experienced a craton block stage and a continental margin rifting stage. From the Sinian period, the Yangtze plate was in the state of a constantly subsiding craton basin and it mainly received marine carbonate deposits. Beginning with clastic sedimentary deposits from the shallow water domain, the Yangtze craton basin generally received dolomite deposits of the Dengying formation from the deep water domain. The dolomite is about $700 - 1,000$ m thick and it is one of the major gas reservoirs in the Sichuan basin. During the Cambrian period, a phosphorus condensed section with a slow depositional rate occurred on the Yangtze plate. Above the condensed section, it deposited carbonate rocks from the deep water domain and formed a widespread carbonate platform. On the other hand, turbidite fan was deposited on the slope zone on the southeastern edge of the Yangtze craton. Originating in the Middle Ordovician epoch, the continental margin was transformed into the Middle Ordovician–Silurian flysch foreland basin. From the Devonian–Permian periods, the ancient Yangtze plate was still located south of the equator. During this time interval, not only was the ancient landmass elevated and eroded to form the ancient Central Sichuan uplift but, also, on the northern margin of the Yangtze plate, the Mianlue oceanic basin was split open and it received massive marine carbonate deposits. Additionally, a basalt eruption happened during the Permian period. During the Triassic period, the Yangtze plate drifted to a latitude of $15°$ N and it converged with the North China plate. Because of compressive stress that was brought on by the convergence event, fold-thrust structures were well developed on the western and northern margins of the Sichuan basin; on the intra-continent, the depression and lacustrine basin were superimposed. The massive southern China carbonate platforms were commonly developed in most of the areas and they were transformed into evaporite platforms during the Middle Triassic epoch. The total thickness of the platform is about $1,000 - 2,000$ m; the sequences of a carbonate-evaporite platform in the Permian–Triassic systems are the major natural gas reservoirs in the Sichuan basin. During the Cenozoic era, influenced by the collision between the Indian plate and Tibetan plateau, on the continental margin the thrust faults were deformed. Because of basin and range coupling, the entire Sichuan basin was elevated and eroded (Fig. 1.5); the edges of the Sichuan basin were limited by the uplifted mountains, which shaped the present topography of the basin. In the entire marine facies deposition area that was controlled by the Yangtze plate, the similarity of geological factors include the tectonic development, sedimentary sequences, source rock of natural gas, reservoir formations, sequence combination of the reservoir and petroleum accumulation phases. The key dissimilarity is the inconsistency in the preservation of the marine prototype basin. Since the Paleozoic era, the continental crust of the middle and lower Yangtze plate was subducted, converged, collided and accreted from the southeast to the northwest, which was a processing of repeated structural complications of the crust (Li and He, 2002). From the Indosinian tectonic period–Yanshanian tectonic period, the Pacific plate was subducted and slid towards the eastern Asian continent. The compressive and extended tectonic movements from the southeast direction of various phases directly worked on the Yangtze plate. The marine basins on the middle and lower Yangtze plate were the first group that were uplifted and eroded, and that experienced fault cutting and magma activity. At present, we have only discovered some bitumen sands that indicate the existence of an ancient oil reservoir. On the upper Yangtze plate (Sichuan basin), that was far away from the tectonic event center beside the ancient Central Sichuan uplift, an inner basin was elevated and eroded during the Caledonian tectonic period. Beside the large scale fold-thrust activities and uplifting and erosion activities that happened around the edges of the Sichuan basin during the Indosinian and Himalayan tectonic periods, the tectonic deformations of the entire inner basin were relatively weak with gentle structures and fully developed formations. The marine sequences that were deposited during the Paleozoic era were favorably preserved and might have developed the Paleozoic petroleum system and natural gas accumulation (Fig. 1.5).

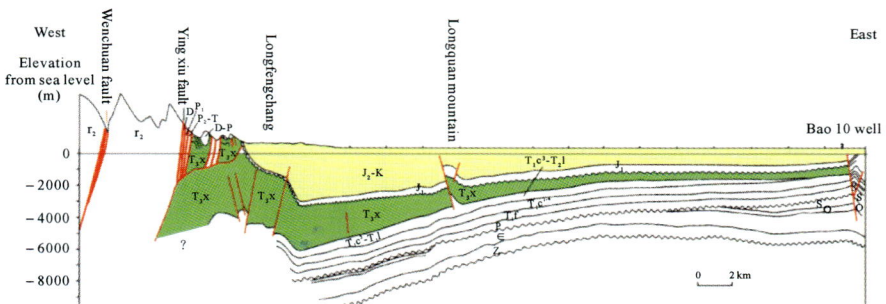

Fig. 1.5. Structural section (W-E) of Sichuan basin

The following are the developmental stages of the North China plate and Eerduosi basin (Li, 2002). The crystalline basement was fused during the Mesoproterozoic era. From the Sinian–Ordovician periods, the ancient North China plate was positioned near the equator and it was surrounded by the Mongolian ocean and Qinling ocean. The aulacogen and the rift basin were formed on the edge of the plate and they predominately received carbonate sediments. From the Silurian–Devonian periods, the North China plate was elevated above the water; the early Paleozoic marine carbonate rocks were eroded away and a sedimentary hiatus commonly occurred. During the Paleozoic era, a relatively intact inner craton basin and continental peripheral basin developed on the North China plate. Because of rifting activities, a rift valley, rift trough and aulacogen trough were formed. During the Cambrian–Ordovician periods, in the North China craton basin, shallow water carbonate was deposited on the platform that was transformed into an evaporite platform during the late stage of this time interval. Later on, a further uplift brought a dissolution type of erosion to the platforms. In the Eerduosi basin, this type of platform is the major gas reservoir.

However, in central Hebei province, carbonate rocks are the sequences of the oil reservoir in the Cambrian–Ordovician systems. On the southern margin of the North China craton, in Helan aulacogen trough and on the margin of the ancient Liupan landmass, gravity-flow deposits occurred in deep water and semi-deep water, which were slump breccia and turbidite with a thickness of 4,000 – 5,000 m. Along with the subduction of North Qilian ocean and North Qinling ocean, the North China platform was elevated. However, the Silurian–Devonian foreland basin coarse grain, clastic sediments were only deposited on the southern continental margin of the Alashan terrane. During the Carboniferous–Permian periods, the ancient North China plate drifted to a latitude of 12° N (Fig. 1.3). Because the surrounding oceanic basins were closed, the North China plate received massive coal measures deposits of marine–terrigenous facies. During the Triassic–Cretaceous periods, the North China plate was transformed into the developmental stage of an intra-continental faulted lacustrine basin. During the Cenozoic era, under the influence of dual geo-dynamic systems, which were the collision between Indian plate and Tibetan plateau and the subduction of the Pacific plate, the North China plate was tilted and an extensional faulted basin occurred on the continental margin. Although the Eerduosi basin, that was a portion of the western margin of the North China plate, experienced some tectonic development and had the same formations when compared to the North China plate, since the Indosinian tectonic period, under the influence of the subduction of the eastern Kula–Pacific plate, the North China plate experienced a back-arc extension in a direction from west to east of gradually increasing tectonic strength. On eastern side of the ancient North China craton, the marine basin was altered into a shattered rift basin by extension of the faulted block during the Indosinian and Himalayan tectonic periods. At present, even the Paleozoic formations are preserved inside the North China basin. Conversely, the prototype basin was cut into pieces by deep extended large faults. Because the extensional stresses gradually decreased westwards, on the western margin of the North China plate, the Eerduosi basin of the Paleozoic era had been totally preserved. The existing marine basin has a relatively large size block and complete overlying formations, which are the fundamental conditions for petroleum accumulation in the Paleozoic marine formations in the Eerduosi basin (Fig. 1.6).

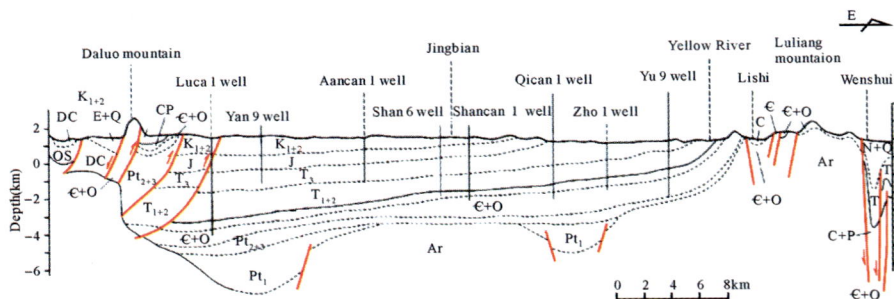

Fig. 1.6.   Structural section (W-E) of Eerduosi basin

1.4 Mesozoic Plates Convergence and Terrestrial Basin Deposits

During the late Paleozoic era, ancient oceanic basins were closed. Chinese terrestrial basins began to develop during the Mesozoic and Cenozoic eras; the developmental progress of a terrestrial basin can be divided into the following stages.

(1) From the Early–Middle Triassic epochs, northern China terrestrial basins started to develop north of the Kunlun–Qinling mountains. After the Permian period, the Paleo-Asian ocean was totally closed and the northern China terrestrial basins were formed north of the Kunlun–Qinling mountains. In particular, because of the convergence between the Tarim–North China plate and Kazakhstan–Siberia plate, terrestrial basins were initially formed on both the northern and southern sides of Tianshan mountain, which included the widespread Tarim basin and Zhungeer basin (Xiao et al., 1991; Jia, 1997). (2) During the Late Triassic epoch, Chinese basins progressed into a diversity of marine basins and terrestrial basins in eastern and western China. During the Late Triassic epoch, both the Jinshajiang Tethyse oceanic basin and the ancient Qinling–Dabie oceanic basin were closed; the Yangtze plate converged on the southern margin of the North China plate; a joint continent was formed in northwestern China and inter-continental subsidence occurred with related sediment deposits. During this time, Sichuan basin and Eerduosi basin entered the end phase of marine deposits. From the Triassic–Jurassic periods, the terrestrial deposits were commonly distributed over the entire region of northwestern China (Jia, 1997). (3) From the Jurassic–Cretaceous periods, coal measures developed in the joint continent. Since the Jurassic period, because of the subduction of the Pacific plate, a back-arc extension happened in northeastern China. Songliao basin, that was located on the eastern margin of the North China plate, experienced this extension event; the ancient plate was shattered into pieces or was cut by faults; lacustrine basins were developed in the rifting region (Yang et al., 2002). (4) The characteristics of the Chinese terrestrial basin are the following: the northern region progressed to the terrestrial basin stage earlier than the southern region, and the western region moved into the terrestrial basin stage earlier than the eastern region. For example, Zhungeer basin, that was located in northwestern China, received terrestrial basin deposits during the Permian period. However, Songliao basin was formed during the Cretaceous period in eastern China; moreover, not until the Tertiary period, was Bohai basin formed. With the closing of the Tianshan–Xingmeng ocean during the Permian period, the region that was located north of the Kunlun–Qinling mountains advanced into the terrestrial basin developmental stage. In company with the closing of Jinshajiang ocean and Mianlue ocean during the Late Triassic epoch, the entire Yangtze area started to receive terrestrial deposits. Until the New Tethys ocean was closed during the Cenozoic era, the entire southwestern area and Tibetan plateau were uplifted and became a continental landmass when the marine deposits ended. The primary developmental stage of the Chinese terrestrial basin was initiated after the Permian period. During the Early Triassic epoch, a landmass appeared in northern China and an ancient ocean still occupied southern China. During the Late Triassic epoch, on the southeastern side of southern China,

seawater retreated, a landmass formed and a terrestrial basin started to develop; an embryonic form of basinal type diversity was created. During the Jurassic period, the area of sea water continually decreased and the region of terrestrial basins expanded in western-central China. The eastern portion of Pacific plate compressed the Asian continent along the island arc belt, which brought large deposits of limnetic facies and strong volcanic activity. During the Cretaceous period, dry or semi-dry regions expanded in central China and most of the deposits were red colored sediments with gypsum. In northeastern China, the climate was humid. Because the Great Xing'an Range was uplifted to the west of Songliao basin, a large scale depression basin was created based on multiple small rift basins, which contained large size, deep water lakes. For example, during the period when the Qingshankou formation was deposited, the basin covered an area of 26×10^4 km^2 and the lake covered an area of 8.7×10^4 km^2, which is the reason why a large scale oil field developed inside the Songliao basin.

1.4.1 *Terrestrial Basins were Formed During the Early–Middle Triassic Epochs in Northern China*

During the Late Carboniferous–Early Permian epochs, in company with the creation of the Eurasian continent and Pangea continent, the Paleo-Asian ocean was closed. The small craton terrains, such as Zhungeer, Tarim and North China, collided and converged with the Siberian plate and Kazakhstan plate, in successive order (Li, 1982; Ren and Xiao, 2002; Jia, 1997; Hendrix et al., 1994; Hsu, 1988). Because of accretion that happened on the southern margin of the Eurasian continent, the Tianshan–Xingmeng orogeny was formed. This tectonic movement achieved convergence, so that the Tarim and North China plates were welded onto the Eurasian plate. Since then, the topography has changed on the Chinese landmass; terrestrial basins occurred in northern China and marine basins occupied southern China (Fig. 1.7a). Due to the closing of oceanic basins, the sea water retreated and terrestrial basins were commonly developed. During the same time interval, on the edge of small craton plates, intermittent regional extensional movements created rift basins that received deposits of marine–terrigenous facies. After the Permian period, there was no oceanic basin north of the Kunlun–Qinling mountains; these regions were either uplifted above sea level or were developed into terrestrial basins. On both sides of the Tianshan orogeny, the Late Permian–Triassic sedimentary sequences of fluviolacustrine facies and depositional characteristics are similar in the Tarim basin, Zhungeer basin, Yili basin, Tulufan–Hami basin and Santanghu basin. The framework of these basins was a multi-faulted graben basin. In the North China region, marine–terrigenous facies and terrestrial facies were deposited. Conversely, coal measures of marine-terrigenous facies were deposited in the southern portion of the North China region. In Songliao basin in northeastern China, sedimentary deposits of fluviolacustrine facies commonly occurred, in company with terrestrial volcanic rock series (Zhu et al., 2007) during the Late Permian epoch. South of the

Kunlun–Qinling mountains, it was the territory of the marginal sea or oceanic basin that related to the Tethys ocean (Luo, 1991). Therefore, during the Late Permian–Middle Triassic epochs, the terrestrial depositional system was basically limited to the region north of the Kunlun–Qinling mountains. Utilizing the Kunlun–Qinling mountains as a boundary, the structural framework can be clearly divided into two sub-divisions: a marine depositional environment in the southern region and a terrestrial depositional environment in the northern region. In addition, a terrestrial basin was distributed from the western region to the eastern region (Fig. 1.7a).

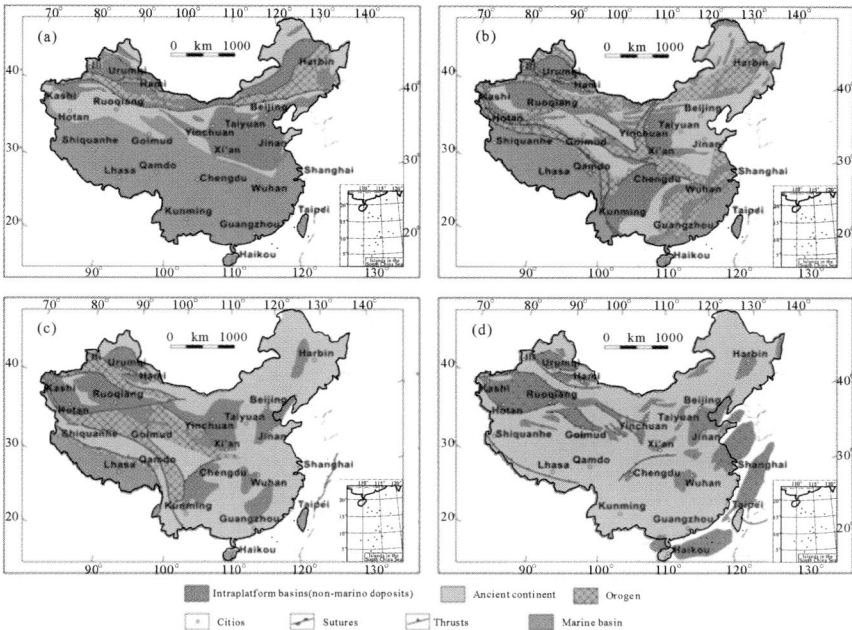

Fig. 1.7. Distribution map of Chinese terrestrial basins
(a) Distribution map of terrestrial basins during the Permian period; (b) Distribution map of terrestrial basins during the Late Triassic epoch; (c) Distribution map of terrestrial basins during the Jurassic–Cretaceous periods; (d) Distribution map of terrestrial basins during the Early Tertiary epoch

1.4.2   Diversity of Marine Basins and Terrestrial Basins Originated During the Late Triassic Epoch

The time interval of the Late Triassic–beginning of the Jurassic was an important period during which Tethys ocean and Paleo-Pacific ocean were subducted towards the Paleo-Asian continent. Because the Yangtze terrain drifted northward and the Paleo-Pacific plate moved in a northwestern direction, accretions occurred on the eastern and southern margins of the Paleo-Eurasian continent (Wu et al., 1997). During the Late Triassic epoch, because ancient oceans were closed, which included Mianlue ocean, Jinshajiang ocean, Lancangjiang ocean and Qinling sea trough (Huang et al., 1977; Shu and Zhou, 2002), theYangtze plate and North China plate converged. To the north of the Tethys ocean, a group of basins joined together to form a joint continent that was adjacent to the Siberian plate in the north. Influenced by subduction, convergence and collision events of the Tethys ocean on its southern margin (Huang et al., 1977; Shu and Zhou, 2002; Xu et al., 2001), intra-continental subsidence occurred with related sedimentary deposits on the joint continent (Fig. 1.7b). Large scale marine regression occurred in southern China, which completed the paleo-geographic settlement of the ocean on the southern side and landmass on the northern side. From the point of view of altitude, the new topographic diversity revealed that the eastern side was higher than the western side and the northern side was higher than the southern side (Fig. 1.7b). While the Tibetan plateau still remained in the domain of the Tethys ocean, the rest of the Chinese landmass became a terrigenous continent north of the Qiangtang terrain, which consisted of the main portion of the Paleo-Eurasian continent containing terrestrial basins. During the Late Triassic, the Chinese paleo-geographic settings could be divided into three sections. The first section was the marine depositional region of the Paleo-Tethys ocean in southwestern China (the Qinghai–Tibetan area). The second section was the terrestrial basin region in western-central China, which includes the Eerduosi basin, Sichuan–Yunnan basin, Zhungeer–Tuha basin and Tarim basin, and which were depression basins. The third section was the uplifting and erosion region in eastern China (which included the Northeastern Highland, North China Highland and Yunnan–Guizhou–Guangxi Highland) (Zhang, 1996). In western-central China, the river-lacustrine depositional system included terrestrial basins that were near the ocean and the intermountain basins. In the upper Yangtze area, the foreland basin predominately contains the Sichuan basin; the lithology includes thick layer sandstone with argillaceous shale and siltstone interbedded with a coal layer. The boundary of the basin was limited by the fold-thrust belts; furthermore, the basin was contracted towards the inner basin. The entire North China basin initiated a contraction event towards the west to form the Eerduosi depression basin that received sedimentary deposits of fluviolacustrine facies. On the western margin, diluvial sandstone and diluvial conglomerate were deposited in front of Liupan mountain, with a thickness of 2,000 m. The Triassic system was commonly missing in the North China basin, which infers a tectonic uplift in the eastern region. The Eerduosi basin was formed in the western region of the North China basin, which received a deposit of black color, argillaceous shale of shallow–deep lacustrine facies with a thickness of a few thousands meters and a deposit of delta facies in the fluviolacustrine system. In the Zhungeer–Tuha basin, the sedimentary rocks cover an area of 10×10^4 km^2 with a thickness of $500 - 1,000$ m. In the Tarim basin, large scale lacustrine deposits and fluviolacustrine delta deposits occurred. In the region that was located south of the Kunlun–Qinling mountains, during the Late Triassic epoch, large scale, inner continental, terrestrial lakes were developed in the Sichuan basin; the deposits include thick layer sandstone with argillaceous shale and siltstone interbedded with a coal layer. The boundary of the basin was limited by the fold-thrust belts; in addition, the basin was contracted towards the inner basin (Yuan, 1996).

The Sichuan basin covers an area of 18×10^4 km^2 and it has a Proterozoic metamorphic crystalline basement. During the Late Triassic epoch, because oceanic basins were closed along the western and northern margin of the Sichuan basin, the marine depositional interval ended and the Sichuan basin progressed into the terrestrial foreland basin development stage. Foreland sequences of the Xujiahe formation in the Upper Triassic series were deposited on the front of Longmen mountain with a thickness of 4,000 m. After a short uplift event, the Sichuan basin continually received deposits of fluviolacustrine facies during the Jurassic–Cretaceous periods. Using the Xujiahe formation of the Upper Triassic series as an example, the following discussions focus on the characteristics of asymmetrical deposits in a foreland basin. During the time that the lower member of the Xujiahe formation was deposited, overthrust movements were very active in the Longmen orogeny; a strong subsidence event occurred in the western Sichuan basin, which caused asymmetrical sedimentary deposits in the basin; the formations were thicker on the western side of the basin and they were thinner on the eastern side of the basin (Fig. 1.8). The sedimentary deposits of the Xujiahe formation gradually moved forward from the western mountain range and the central Sichuan uplift towards the center of the Sichuan basin. The sedimentary system consists of (fluvial fan) braided river–braided delta–lowstand lacustrine. In the Sichuan basin, the region of the lowstand lake was like a "∩" shape (an upside down "U" shape) and the lake region was tilted in a southwest direction; in addition, small sandbars formed in the lowstand lake. Transitional sedimentary facies can be divided into a coarse–medium grain clastic depositional belt of fluvial fan (river) facies and (fan) delta facies and a fine grain clastic–argillaceous depositional belt of lacustrine facies (Fig. 1.8). The fluvial fan and fan delta were located on the western margin of the Sichuan basin and on the central Sichuan uplift. On the western margin of the Sichuan basin, all kinds of fans were linked laterally to form a fringe of alluvial fans, which were oriented in a southwest-northeast direction and which paralleled Longmen mountain structural belt. For example, in Fig. 1.8, the second member of the Xujiahe formation was deposited during the beginning of tectonic activity; the thickness of the second member of the Xujiahe formation was in the range of 115.0 – 560.0 m in a mountain front depression basin that was located on the front margin of Longmen mountain. The thicknesses of sediments were various and unstable; the alterations in thickness and the shift of facies occurred horizontally; the lithologies were coarse grain clastic rocks of conglomerate and sandstone. In the central Sichuan area, because it was influenced by a lateral movement of structural uplift and under supply of sediments, the slope area of the foreland uplift did not generally have the deposits of a lowstand system tract that should be developed during the early phase of a base level uplift. On the other hand, the lake transgressive systems tract developed a sedimentary overlap in ascending order on the slope of the foreland uplift; only the largest pan-lake surface could overlap and submerge the foreland uplift.

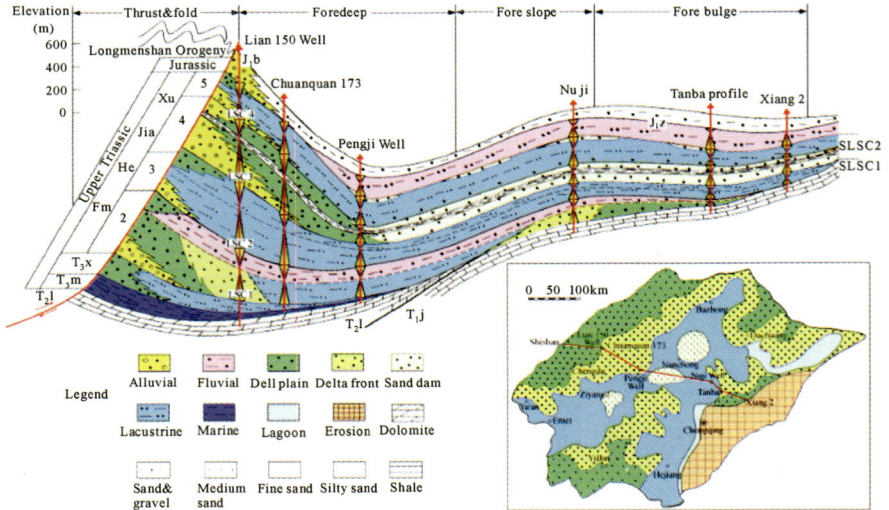

Fig. 1.8. Distribution map of sedimentary facies and stratigraphic column of the western Sichuan basin during the Late Triassic epoch

1.4.3   Intra-Continental Faulted Basin Developed During the Jurassic–Cretaceous Periods

From the Jurassic–Cretaceous periods, the Paleo-Pacific ocean was closed along the eastern margin of the Paleo-Asian continent and a collision occurred between the Paleo-Asian continent and Western Pacific plate. During the same time interval, Bangong-Nujiang ocean disappeared along the southern margin of the Paleo-Asian continent and the Lhasa terrane collided with the Paleo-Asian continent (Fig. 1.7c). In addition, the Paleo-Tethys oceanic domain still had marine deposits in the region south of the Tibetan plateau while the rest of the Chinese basins received terrestrial deposits. In western-central China, terrestrial coal measures were deposited in large scale depression basins. In northeastern China, a combination of clastic rocks and coal measures were deposited in a group of rift basins. In North China–South China, a group of small basins received red color sediments.

In western-central China, a series of large size terrestrial basins developed near the coast, which demonstrated the characteristics of an inner-craton depression with stable sedimentary deposits and less thickness (Fig. 1.7c). Tarim basin, Zhungeer basin, Eerduosi basin and Sichuan basin developed into large size intra-continental depressions. On both sides of the ancient Kunlun mountain, Tianshan mountain and Qilian mountain, medium–small size rifted lacustrine basins developed, for example on the northern margin of Chaidamu basin, Yili basin, Yanqi basin, Santanghu basin, Ejinaqi basin and Gansu corridor. Also, many rift basins formed north of the Kunlun–Qilian mountains. In particular, during the Jurassic period, the intra-continental basins that were located near the coast were accompanied by gentle topography and a wide-open water body; the humid environment benefited the development of coal measures strata of limnetic facies; the lithology of fluviolacustrine facies includes sandstone, mudstone and coal measures. Influenced by the Pacific plate, large scale, northeast oriented complex anticlines and synclinoriums formed in eastern China. In the same time interval, a series of rifted graben basins were developed, of various sizes. The rift basins in the Greater Xing'an Range and on the southeastern coast are examples of a volcanic-sedimentary basin group. Erlian basin is an example of a pull-apart rift basin group (Fig. 1.7c). During the Jurassic–Cretaceous periods, the coal measures and red beds of some depositional facies occurred together in terrestrial basins in both northern China and southern China. The continental crust was very active with uplift events in eastern China; conversely, the continental crust was stable with slow subsidence in western-central China.

In eastern China, the Mesozoic Songliao basin is a typical rift basin. The following discussion uses the Songliao basin as an example to demonstrate the depositional characteristics of a rift basin. Songliao basin is a Mesozoic back-arc rift basin that covered a region of 26×10^4 km^2 with a folded basement Carboniferous–Permian in age and 4,000 m thick in deposits of Cretaceous fluviolacustrine facies. Songliao basin experienced five developmental phases, which were early rift phase, rift-expansion phase, post-rift depression phase, structural reversal phase and continental rift phase (Chen, 2000; Zhang et al., 1996; Chen and Chen, 1996). Every developmental phase had its own signature deposit (Fig. 1.9). The volcanic rock deposit developed during the rift phase. During the rift-expansion phase, a series of divided rift basins and single rift basins were distributed from the edge to the center of Songliao basin. Depositional facies were transformed from alluvial fan → river → fan delta → palustrine → delta → lake shore → lacustrine. At the same time, the sedimentary types varied either between different grabens or, in different development phases, within a graben. Vertically, lake water was absent to the present day; the aquatic-region varied from small to large; the grain size of sediments went from coarse grain to fine grain and was then transformed back to coarse grain again (overall, the grain size had a tendency to be coarse). The post-rift depression phase can be divided into two sub-phases. The early sub-phase was the transitional phase from rift basin to depression or was the initial phase for a depression. During this time interval, the climate was dry and hot and the topographic relief was gentle; the aquadynamic condition was weak. A joint lacustrine basin was gradually formed. From the edge to the center of the basin, depositional facies were transformed from alluvial fan → river → lake shore sandy mudflat→ river end delta → delta → shallow lacustrine. Along with facies alteration, the lake region gradually expanded. The late sub-phase was the developmental phase of a depression. The lake had transgression and regression twice and the region of Songliao basin expanded on a significant scale. From the edge to the center of the basin, the depositional facies were transformed from alluvial fan → river → delta → lake shore →deep lacustrine. During the lake transgression period, the lake shore facies developed very well. However, during the lake regression period, the delta facies expanded wonderfully. The deposits of river facies were primarily formed during the structural reversal phase and the continental rift phase.

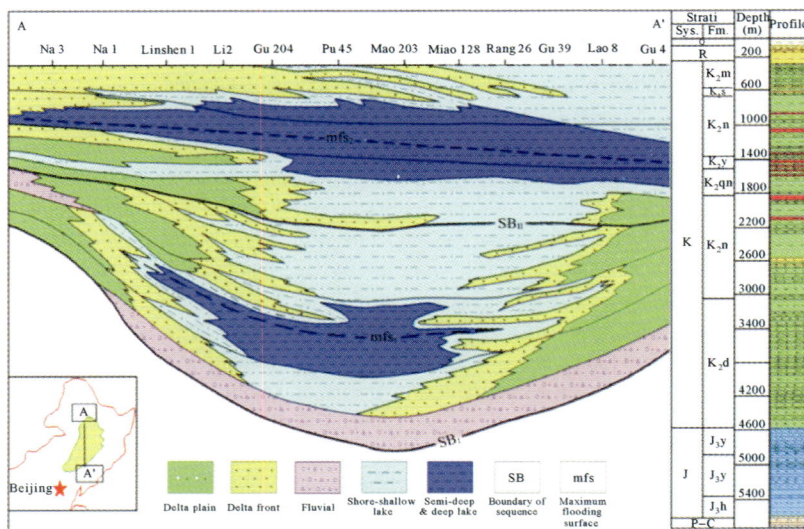

Fig. 1.9.   Cress section of depositional system and stratigraphic column of the Upper Cretaceous series in Songliao basin (see Fig. 1.8 for map legend)

1.5   Development of Intra-Continental Orogenies and Basins During the Himalayan Tectonic Period

The inner continental deformations and the development of the rift basin happened during the Palaeogene. The New Tethys ocean collided with the Eurasian plate along the suture zone at Yaluzangbu (Brahmaputra) river, which reactivated orogenic belts in western central China, and which superimposed regenerated foreland basins (or thrust belts) in an asymmetrical pattern (half-graben like) in the direction of a craton (Chen and Wang, 1997; Lu et al., 1994; Hendrix et al., 1994). In eastern China, influenced by persistent subduction of the Pacific plate and back-arc extension, Bohai bay progressed to the developmental phase of a rift basin and a depression basin was formed by geothermal cooling effect during the late period (Yang et al., 2002). During this tectonic period, tectonic events formed a rift valley in eastern China and developed inner continental deformation in western China. Since the Neogene, the rapid uplifting of the Tibetan plateau settled the topographic relief of present day China, in which the western terrains are higher than the eastern terrains (Harrison et al., 1992; Allégre et al., 1984; Tapponnier et al., 2001). Furthermore, the gigantic basin and range system was formed surrounding Tibetan plateau in western-central China and the rift basin group of the West Pacific Rim was developed in eastern China.

1.5.1 Collision Between Indian Plate and Eurasian Plate and Development of Regenerated Foreland Basin

Prior to the geological time interval of 55 – 40 Ma, the Tethys ocean initiated a subduction event along the Yaluzangbu River and it was closed. During the Palaeogene period, the transgressions of the Tethys ocean occurred many times from the western region to the eastern region. In the Tarim basin, massive sequences of marine–terrigenous facies developed in southwestern Tarim and the Kuche area; these sequences included coal measures of terrestrial facies, reef limestone of neritic facies, a gypsum-salt layer of lagoon facies and sandy mudstones of lacustrine delta facies and river facies. During the Palaeogene period, in Chaidamu (Qaidam) basin, fine grain clastic sediments with stabilized minerals were deposited in a low energy environment; the sedimentary basin was a sliding-extensional, disk like depression. The marine transgressions were stronger in the western region than in the eastern region. However, the Eerduosi (Erdos) basin and Sichuan basin that were located on the eastern side of the Helan–Longmen mountains were slightly influenced by the marine transgression.

Since the Neogene period, marine deposits stopped on the Chinese landmass; the creation of inner continental orogenies and terrigenous basins was in progress. The Indian plate carried the Himalayan terrain and Gangdise terrain to collide with the Eurasian plate along the suture zone of the Yaluzangbu River. In western-central China, in the direction from orogeny belts to cratons, regenerated foreland basins (or thrust belts) were superimposed in an asymmetrical pattern (half-graben like) (Lu et al., 1994; Hendrix et al., 1994; Graham et al., 1993). Under a persistent compressive stress field, the lithosphere in the basin area was deformed and flexed because of its rigid physical properties; the rigid lithosphere was subducted under the orogenies and, as a chain reaction, the orogenies were thrust towards the basins. This phenomenon has been proved by seismic sounding data (Kao et al., 2001). In particular, deformation is represented by the center uplift in a basin and a mountain front depression that is located at the junction area between a basin and an orogeny; the depression subsided on account of structural movement and sediment overloading. For example, the Kuche depression is located at the junction area between the northern margin of Tarim basin and the southern margin of South Tianshan mountain; the southwestern Tarim depression is located at the junction area between the Tarim basin and West Kunlun mountain.

In addition, the mountain front depression in the southern Zhungeer basin and the western Sichuan basin all belong to this type of foreland basin. Because the basement of a basin was a rigid terrain previously, the structural deformations were mainly concentrated on the areas of accretion and convergence, which were located on the margin of the basement terrain. The convergence zone was a vulnerable area for structural deformation; this area was reactivated by the collision between the Indian and the Eurasian continents and a series of regenerated foreland basins were formed in this vulnerable area (Lu et al., 1994; Hendrix et al., 1994; Graham et al., 1993). Massive molasse deposits settled in the foreland depression with a maximum thickness of 4,000 – 7,000 m; the thicknesses of molasse deposits gradually reduced when approaching the center uplift of a craton. Because it had tough rigidity and great strength, the lithosphere was plunged under the orogeny. Accordingly, the orogeny was thrust in the direction of the basin (Kao et al., 2001). The deformation was symbolized by the central flexural uplift in a terrain and depression on the margin of a terrain. Under structural loading and sediments loading, the margin of a terrain was not only depressed, but it also received sedimentary deposits. During the late Cenozoic era, because surrounding mountains were rapidly uplifted, the mountain front depression zone should display a rapid subsidence, which formed a series of thrust belts that surrounded the eastern side and northern side of the Tibetan plateau. Due to the differentiation in the physical properties of the lithology in sedimentary strata, some décollement surfaces were formed inside strata, which offered a varied type of structural deformation. Also, this helped develop the combination of fault-related folds, salt-related folds and the combination of strike-slip fault and thrust fault.

1.5.2   Subduction of West Pacific Plate and Development of Back-Arc Rift Basin in Eastern China

In eastern China, the tectonic development mainly reflected a back-arc extension that was triggered by the subduction of the Pacific plate. Along the eastern margin of the North China plate, Songliao basin and Bohai bay are examples of a rift basin. Under the extension force, the Paleozoic craton basins were broken into pieces or were cut by faults; the rift basin received lacustrine deposits and the uplifted region became the buried hill. The back-arc extension happened in eastern China, which was brought on by the subduction of the Pacific plate. On the eastern margin of the North China plate, extensional basins developed at Songliao basin and Bohai bay (Fig. 1.7d). During the early–middle Mesozoic, the eastern region was in an uplifting state. In the eastern North China terrane and the lower Yangtze area, some small, separated, terrigenous basins developed on the uplifted background with limited dimensions. Since the late Mesozoic, because the subduction direction of Pacific plate changed, and because of the slip line field that derived from the Indian plate being pushed northward, an extensional tectonic settlement took place in China and even in the whole eastern Asia continent, which formed a series of rift basins with a few aspects of a pull-apart nature. Among these rift basins, Bohai bay, the southern North China basin, Jianghan basin and North Jiangsu–South Yellow Sea basin were superimposed on top of the Paleozoic craton basins. These rift basins were all developed in a rift-depression phase during the early Palaeogene, which shows that the Paleozoic craton was not only altered by the Cenozoic rift valleys, but was also superimposed by rift basins.

In conclusion, the continental shield of the Chinese landmass and the Paleozoic craton basins were formed by the convergence of three drifting plates that were the Tarim plate, North China plate and Yangtze plate. During the Mesozoic era, terrigenous basins developed on small terrain basements. In China, these terrigenous basins were broadly distributed from the northern region to the southern region, from the eastern region to the western region. Also, these terrigenous basins were divided by multiple orogenies that were formed during the Hercynian–Indosinian tectonic periods. During the Cenozoic era, western China was controlled by a gigantic basin and range system that was around the Tibetan plateau. On the edge of the ancient craton, regenerated foreland basins were formed. Influenced by the subduction of the Pacific plate and controlled by a back-arc extension event, the rift basin group developed in eastern China. The distribution of an oil and gas basin is shown in Fig. 1.10. These oil and gas basins include superimposing the basins of different periods and the combination of different types of basins.

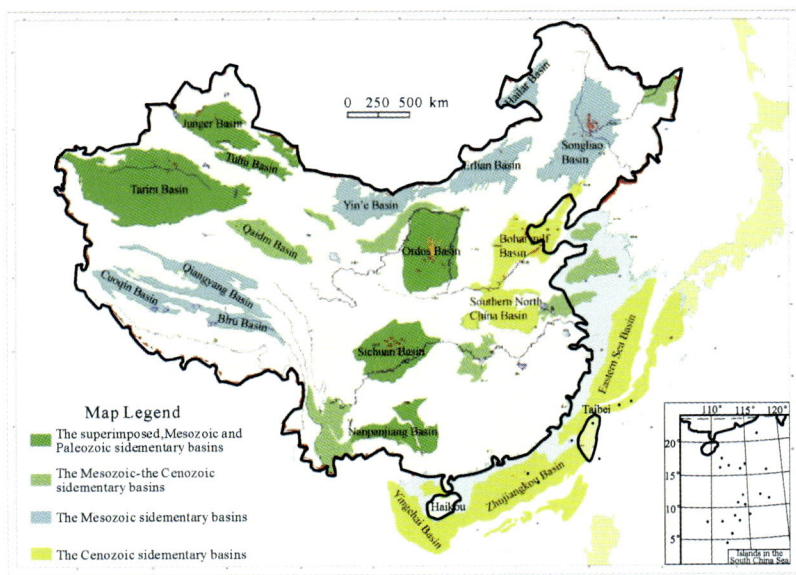

Fig. 1.10. Distribution map of oil and gas basins in China

References

Allégre, C.J., Courtillot, V., Tapponnier, P., et al., 1984. Structure and evolution of Himalaya-Tibet orogenic belt. Nature, 307:17-22.

Chen, J.W., 2000. Depositional filling pattern of a large backarc rift basin—a case study of the Songliao basin. Experimental Petroleum Geology, 22(1):50-54 (in Chinese with English abstract).

Chen, Z.N., Chen, F.J., 1996. Kinematic characteristics of inversion structures in Songliao basin. Geoscience—Journal of Graduate School, China University of Geosciences, 10(3):390-396.

Chen, F.J., Wang, X.W., 1997. Genetic types, tectonic systems and geodynamic models of Mesozoic and Cenozoic oil and gas bearing basins in China. Geoscience, 5:409-424.

Chen, H.J., Sun, Z.C., Zhang, Y.C., 1986. Framework of Chinese petroliferous basins. Petroleum Geology & Experiment, 8(2):97-106.

Graham, S.A, Hendrix, M.S., Wang, T.B., 1993. Collision successor basins of western China: impact of tectonic inheritance on sand composition. Geo. Soc. Amer. Bull., 105:323-344.

Guo, Z.W., Deng, K.L., Han, Y.H., et al., 1996. Development and Evolvement of the Sichuan Basin. Geological Publishing House, Beijing.

Harrison, T.M., Copeland, P., Kidd, W.S.F., Yin, A., 1992. Raising Tibet. Science, 255:1663-1670.

Hendrix, M.S., Dumitru, T.A., Graham, S.A., 1994. Late Oligocone-Early Miocene unroofing in the Chinese Tian Shan: An early effect of the India-Asia collision. Geology, 22(6):487-490.

Hsu, K.J., 1988. Relict back-arc basins: principles of recognition and possible new example from China. In: New Perspectives in Basin Analysis. Kleinpell, K.L., Paola C. (eds.), Springer-Verlag, New York, pp. 245-263.

Huang, T.K., Ren, J.S., Jiang, C.F., Zhang, Z.M., Xu, Z.Q., 1977. An outline of the tectonic characteristics of China. Acta Geologica Sinica, 2:117-135.

Jia, C.Z., 1997. Structure Features and Petroleum in Tarim Basin of China. Petroleum Industry Press, Beijing, pp. 1-438.

Jia, C.Z., Yang, S.F., Chen, H.L., Wei, G.Q., 2001. Structural Geology and Natural Gas of Basin Group in North Fringe of Tethys. Publishing House of Petroleum Industry, Beijing, pp. 1-162.

Jia, C.Z., Wei, G.Q., Li, B.L., 2005. Superimposed-composite characteristics of micro-craton basins and their bearing petroleum systems, Central-Western China. Geological Journal of China Universities, 11(4):479-492 (in Chinese with English abstract).

Kao, H., Gao, R., Rau, J., et al., 2001. Seismic image of Tarim basin and its collision with Tibet. Geology, 29:575-578.

Li, C.Y., 1982. Manual of Asian Geotectonic Map. Geology Press, Beijing.

Li, D.S., He, D.F., 2002. Petroleum geology of sedimentary basins in northwest China. Marine Origin Petroleum Geology, 7(1):1-6 (in Chinese with English abstract).

Liu, H.P., 1995. Types of foreland basin and patterns of fold-thrust fault. Earth Science Frontiers, 2(3):59-68.

Lu, H., Howell, D.G., Jia, D., 1994. Rejuvenation of the Kuqa foreland basin, northern flank of the Tarim basin, Northwest China. International Goel. Rev., 36:1151-1158.

Luo, Z.L., 1991. Dynamic model of crust evolution in Longmenshan Orogenic belt. Journal of Chengdu College of Geology, 18(1):1-7.

Qiu, Z.J., Gong, Z.S. (eds.), 1999. Oil and Gas Exploration in China (Vol. 1). Petroleum Industry Press, Beijing.

Ren, J.S., Xiao, L.W., 2002. Tectonic settings of petroliferous basins in continental China. Episodes, 25(4):227-235.

Shu, L.S., Zhou, X.M., 2002. Late Mesozoic tectonism of southeast China. Geological Rev., 48(3):249-260.

Tapponnier, P., Xu, Z.Q., Roger, F., et al., 2001. Oblique stepwise rise and growth of the Tibet Plateau. Science, 294:1671-1677.

Teng, J.W., Zeng, R.S., Yan, Y.F., et al., 2002. Depth distribution of Moho and tectonic framework in eastern Asian continent and its adjacent ocean areas. Since in China, Series D, 32(2):89-100.

Tian, Z.Y., Zhang, Q.C., 1997. Lithofacies and Paleogeography with Related Oil and Gas in Chinese Petroleum Basin. Geological Publishing House, Beijing, pp. 65-149.

Wang, S.Y., Hearm, T.M., Xu, Z.Y., et al., 2001. Pn velocity structure of uppermost mantle beneath Chinese continent. Since in China, Series D, 31(6):449-454.

Wu, Z.H., Wu, G.G., Wang, J.P., 1997. Constraints of the Meso-Cenozoic global velocity field of lithosphere on the tectonic evolution of China and its adjacent areas. Episodes, 20(2):117-121.

Xiao, X.C., Tang, Y.Q., Li, J.Y., et al., 1991. Tectonic Evolution of the Southern Margin of the Paleo-Asian Composite Gigantic Suture. Beijing Science and Technology Press, Beijing, pp.1-29.

Xu, Z.H., Shi, Y.L., 2003. Lithosphere structure and geodynamics of Chinese continent. Acta Seismologica Sinica, 25(5):453-464.

Xu, Y., Liu, F.T., Liu, J.H., et al., 2001. Deep features of continental collision belts in northwestern China and their dynamic implications. Chinese Journal of Geophysics, 44 (1):40-47.

Yang, S.F., Jia, C.Z., Chen, H.L., et al., 2002. Evolution of Tethyan tectonic belt and development of basin groups in northern fringe, and prospect of natural gas exploring in Tarim. Chin. Sci. Bull., 47(suppl.):36-43.

Yuan, X.C. (ed.), 1996. Atlas of Geophysics in China. Geological Publishing House, Beijing.

Zhang, K., 1991. On the disintegration, displacement, collision and convergence of pan-China plate and evolution of its oil and gas bearing basins. Xinjiang Petroleum Geology, 12(2):91-106.

Zhang, W.Z., 1996. Formation and evolution of terrestrial hydrocarbon basins in

China. Explorationist, 1(1):10-17.

Zhang, G.C., Zhu, D.F., Zhou, Z.B., 1996. Extensional and inversion structural styles of Songliao basin. Petroleum Exploration and Development, 23(2):16-20.

Zhu, X., 1986. Structure of Petroliferous Basins in China. Petroleum Industry Press, Beijing (in Chinese).

Zhu, R.K., Xu, H.X., Deng, S.H., Guo, H.L., 2007. Lithofacies palaeogeography of the Permian in northern China. Journal of Palaeogeography, 9(2):133-142.

2

Sedimentology of Chinese Petroleum Basins

The Chinese petroleum industry is famous for producing oil and gas from terrigenous sedimentary basins. As a result, geological studies of the terrigenous petroleum basin and the use of related technologies for petroleum exploration and development are at an advanced stage in the world. If the idea that terrigenous stratigraphy can generate oil is the central idea behind the geological theory, then the characteristics of reservoir sequences in the terrigenous basins are the significant foundation of this geological theory. This chapter will discuss the distribution of the stratigraphy, the major types of depositional facies and the characteristics of the petroleum system in the Chinese petroleum basins. In addition, the nature of the depositional evolution in the major petroleum basins will be briefly introduced.

2.1 Distribution of Stratigraphy and Types of Depositional Facies

First, we will introduce the distribution of stratigraphy.

2.1.1 Developmental Characteristics of the Stratigraphy and Their Distribution Pattern

In the geological history of the earth, Chinese geological developments were placed in an important and crucial position. China contains a vast land with well developed and various types of stratigraphy. Therefore, the Chinese strata contained a full range of fossils that have a systematic entirety and consist of complete specimens in various collections. The entirety of this range of fossils is remarkable in the world. In October 2002, the "The Instruction of the Regional Chronostratigraphic (Geological Chronology) Table of China" was published by the Commission for Stratigraphy in China. This book represents the most recent results of regional chronostratigraphy (Geological Chronology) in China. Table 2.1 introduces the chronostratigraphic system in the marine stratigraphic region of China. The establishment of a stratigraphic stage in the marine stratigraphic region only occurred up until the Triassic period. From the Jurassic–Quaternary period, the distribution of marine strata was limited on the Chinese landmass. Therefore, there is no clear impact unless the marine stratigraphy was correlated over a broad range. As a consequence, the stratigraphic stages are also very hard to introduce. Table 2.2 presents the stratigraphic system in the terrigenous stratigraphic region of China. In the terrigenous stratigraphic region, the stratigraphic stages mainly show the stratigraphy above the Triassic system, which highlight the widely developed Mesozoic and Cenozoic terrigenous strata. Even though the Carboniferous and Permian systems contained stratigraphic stages, in the Carboniferous system the stratigraphic stages mainly show the characteristics of marine–terrigenous facies, which could not reflect the real terrigenous stages. On the other hand, because the time line of stratigraphic stages in the Permian system could not be accurately defined, the stratigraphic stages of the Permian period are not applicable at present.

In China, since the Phanerozoic era, during the multi-cycles of geological development, the strata of marine facies and marine–terrigenous facies were widely developed to form the Pre-Paleozoic erathem and Paleozoic erathem. The terrigenous strata developed to form the Mesozoic erathem and Cenozoic erathem. The stratigraphy was well developed and contained many hydrocarbon source sequences and reservoir sequences. However, variations were present between different petroleum basins (Table 2.3). The following are some brief descriptions of stratigraphic distributions in the major petroleum basins (Zhai, 1987; Zhai et al., 2005).

● **Archean Eonothem–Paleoproterozoic Erathem (Ar-Pt$_1$)**

In China, the oldest stratigraphic strata were positioned in the Archean eonothem–Paleoproterozoic erathem. The lithology consisted of rocks from the metamorphic series with assorted degrees of alteration. In the North China region, commonly distributed lithologic series included the Taishan group, Qianxi group, Fuping group and Wutaishan group. These groups belonged to the Archean eonothem. The isotopic age was greater than 2,500 Ma; the longest isotopic age was 3,500 Ma. The lithology was a set of rocks from the high metamorphic series, which included granulite, gneiss, marble, hornblende schist and pegmatite. The Hutuo group of the Paleoproterozoic erathem had an isotopic age of 2,500 – 1,800 Ma, and was made of a set of rocks from the epimetamorphic rock series which included phyllite, slate, feldspar quartzite and marble. This set of strata functioned as the basement for the Bohai bay basin, Eerduosi basin and Tarim basin. In the Bohai bay basin, the oil and gas reservoirs were discovered in the weathering crust on the top of the metamorphic basement.

Table 2.1   Regional stratigraphic table of marine facies and the geological time scale for China

Eonthem (Eon)	Erathem (Era)	System (Period)	Series (Epoch)	Stage (Phase)		Ma
Phanerozoic Eonothem (Eon) PH	Cenozoic Erathem (Era) Cz	Quaternary System (Period) Q	Holocene Series (Epoch) Q_h			0.01
			Pleistocene Series (Epoch) Q_p			2.6
		Neogene System (Period) N	Pliocene Series (Epoch) N_2			5.3
			Miocene Series (Epoch) N_1			23.3
		Palaeogene System (Period) E	Oligocene Series (Epoch) E_3			32
			Eocene Series (Epoch) E_2			56.5
			Palaeocene Series (Epoch) E_1			65
	Mesozoic Erathem (Era) Mz	Cretaceous System (Period) K	Upper (Late) Cretaceous Series (Epoch) K_2			96
			Lower (Early) Cretaceous Series (Epoch) K_1			137
		Jurassic System (Period) J	Upper (Late) Jurassic Series (Epoch) J_3			
			Middle Jurassic Series (Epoch) J_2			
			Lower (Early) Jurassic Series (Epoch) J_1			205
		Triassic System (Period) T	Upper (Late) Triassic Series (Epoch) T_3	Tulongian Stage (Phase) T_3^2		
				Yazhiliangian Stage (Phase) T_3^1		227
			Middle Triassic Series (Epoch) T_2	Wait to be set up T_2^2		
				Qingyanian Stage (Phase) T_2^1		241
			Lower (Early) Triassic Series (Epoch) T_1	Caohuan Stage (Phase) T_1^2		
				Yinkengian Stage (Phase) T_1^1		250
	Paleozoic Erathem (Era) Pz	Permian System (Period) P	Upper (Late) Permian Series (Epoch) P_3	Meishan Sub-Stage (Sub-Phase) P_3^2	Changxingian Stage (Phase)	
					Baoqing Sub-stage (Sub-phase)	
				Wujiapingian Stage (Phase) P_3^1	Laoshanian Sub-stage (Sub-phase)	
					Laibinian Sub-stage (Sub-phase)	
			Middle Permian Series (Epoch) P_2	Lengwuan Stage (Phase) P_2^4		257
				Maokouan Stage (Phase) P_2^3		
				Xiangboan Stage (Phase) P_2^2		
				Qixian Stage (Phase) P_2^1		277
			Lower (Early) Permian Series (Epoch) P_1	Longlinian Stage (Phase) P_1^2		
				Zisongian Stage (Phase) P_1^1		295

(To be continued)

(Table 2.1)

Eonothem (Eon)	Erathem (Era)	System (Period)	Series (Epoch)	Stage (Phase)	Ma
Phanerozoic Eonothem (Eon) PH	Paleozoic Erathem (Era) Pz	Carboniferous System (Period) C	Upper (Late) Carboniferous Series (Epoch) C_2	Xiaoyiaoan Stage (Phase) C_2^4	
				Dalan Stage (Phase) C_2^3	
				Huashibanian Stage (Phase) C_2^2	
				Luosuan Stage (Phase) C_2^1	320
			Lower (Early) Carboniferous (Epoch) C_1	Dewuan Stage (Phase) C_1^3	
				Datangian Stage (Phase) C_1^2	
				Yanguanian Stage (Phase) C_1^1	354
		Devonian System (Period) D	Upper (Late) Devonian Series (Epoch) D_3	Shaodongian Stage (Phase) D_3^4	
				Wait to be set up	
				Xikuangshanian Stage (Phase) D_3^2	
				Shetianqiaoan Stage (Phase) D_3^1	372
			Middle Devonian Series (Epoch) D_2	Dongganglingian Stage (Phase) D_2^2	
				Yingtangian Stage (Phase) D_2^1	386
			Lower (Early) Devonian Series (Epoch) D_1	Sipaian Stage (Phase) D_1^4	
				Yujiangian Stage (Phase) D_1^3	
				Nagaolingian Stage (Phase) D_1^2	
				Wait to be set up	410
		Silurian System (Period) S	Top (End of) Silurian Series (Epoch) S_4		
			Upper (Late) Silurian Series (Epoch) S_3		
			Middle Silurian Series (Epoch) S_2	Ankangian Stage (Phase) S_2^1	
			Lower (Early) Silurian Series (Epoch) S_1	Ziyangian Stage (Phase) S_1^3 — Nantaliangian Sub-stage (Sub-phase)	
				Ziyangian Stage (Phase) S_1^3 — Matiwanian Sub-stage (Sub-phase)	
				Dazhongban Stage (Phase) S_1^2	
				Longmaxian Stage (Phase) S_1^1	438
		Ordovician System (Period) O	Upper (Late) Ordovician Series (Epoch) O_3	Qiantangjiangian Stage (Phase) O_3^2	
				Aijiashanian Stage (Phase) O_3^1	
			Middle Ordovician Series (Epoch) O_2	Darriwilian Stage (Phase) O_2^2	
				Dawanian Stage (Phase) O_2^1	
			Lower (Early) Ordovician Series (Epoch) O_1	Daobaowanian Stage (Phase) O_1^2	
				Xinchangian Stage (Phase) O_1^1	
					490

(To be continued)

(Table 2.1)

Eonothem (Eon)	Erathem (Era)	System (Period)	Series (Epoch)	Stage (Phase)	Ma
Phanerozoic Eonothem (Eon) PH	Paleozoic Erathem (Era) Pz	Cambrian System (Period) ϵ	Upper (Late) Cambrian Series (Epoch) ϵ_3	Fengshanian Stage (Phase) ϵ_3^3	
				Changshanian Stage (Phase) ϵ_3^2	
				Gushanian Stage (Phase) ϵ_3^1	500
			Middle Cambrian Series (Epoch) ϵ_2	Zhangxian Stage (Phase) ϵ_2^3	
				Xuzhuangian Stage (Phase) ϵ_2^2	
				Maozhuangian Stage (Phase) ϵ_2^1	513
			Lower (Early) Cambrian Series (Epoch) ϵ_1	Longwangmiaoan Stage (Phase) ϵ_1^4	
				Canglangpuan Stage (Phase) ϵ_1^3	
				Qiongzhusian Stage (Phase) ϵ_1^2	
				Maishucunian Stage (Phase) ϵ_1^1	543
Proterozoic Eonothem (Eon) PT	Neoproterozoic Erathem (Era) Pt₃	Sinian System (Period) Z	Upper (Late) Sinian Series (Epoch) Z_2	Dengyingxian Stage (Phase) Z_2^1	630
			Lower (Early) Sinian Series (Epoch) Z_1	Doushantuoan Stage (Phase) Z_1^1	680
		Nanhuan System (Period) Nh	Upper (Late) Nanhuan Series (Epoch) Nh_2		
			Lower (Early) Nanhuan Series (Epoch) Nh_1		800
		Qingbaikouan System (Period) Qb	Upper (Late) Qingbaikouan Series (Epoch) Qb_2		900
			Lower (Early) Qingbaikouan Series (Epoch) Qb_1		1000
	Mesoproterozoic Erathem (Era) Pt₂	Jixianian System (Period) Jx	Upper (Late) Jixianian Series (Epoch) Jx_2		1200
			Lower (Early) Jixianian Series (Epoch) Jx_1		1400
		Changchengian System (Period) Ch	Upper (Late) Changchengian Series (Epoch) Ch_2		1600
			Lower (Early) Changchengian Series (Epoch) Ch_1		1800
	Paleoproterozoic Erathem (Era) Pt₁	Hutuoan System (Period) Ht			2300
Archean Eonothem (Eon) Ar	Neoarchean Erathem (Era) Ar₄				2500
	Mesoarchean Erathem (Era) Ar₃				2800
	Paleoarchean Erathem (Era) Ar₂				3200
	Eoarchean Erathem (Era) Ar₁				3600

Table 2.2   Regional stratigraphic table of terrigenous facies and the geological time scale for China

Eonothem (Eon)	Erathem (Era)	System (Period)	Series (Epoch)	Stage (Phase)	Ma
Phanerozoic Eonothem (Eon) PH	Cenozoic Erathem (Era) Cz	Quaternary System (Period) Q	Holocene Series (Epoch) Q_h	Haven't set up the Stage	0.01
			Pleistocene Series (Epoch) Q_p	Salawusuan Stage (Phase) Malanian Stage (Phase)] Q_p^3	
				Zhoukoudianian Stage (Phase) Lishian Stage (Phase)] Q_p^2	
				Nihewanian Stage (Phase) Wuchengian Stage (Phase)] Q_p^1	2.60
		Neogene System (Period) N	Pliocene Series (Epoch) N_2	Mazegouan Stage (Phase) N_2^2	
				Gaozhuangian Stage (Phase) N_2^1	5.30
			Miocene Series (Epoch) N_1	Baodean Stage (Phase) N_1^4	
				Tonggurian Stage (Phase) N_1^3	
				Shanwangian Stage (Phase) N_1^2	
				Xiejian Stage (Phase) N_1^1	23.3
		Palaeogene System (Period) E	Oligocene Series (Epoch) E_3	Tabenbulukian Stage (Phase) E_3^2	
				Wulanbulagean Stage (Phase) E_3^1	32
			Eocene Series (Epoch) E_2	Caijiachongian Stage (Phase) E_2^4	
				Yuanquan Stage (Phase) E_2^3	
				Lushian Stage (Phase) E_2^2	
				Lingchan Stage (Phase) E_2^1	56.5
			Palaeocene Series (Epoch) E_1	Chijiangian Stage (Phase) E_1^2	
				Shanghuan Stage (Phase) E_1^1	65
	Mesozoic Erathem (Era) Mz	Cretaceous System (Period) K	Upper (Late) Cretaceous Series (Epoch) K_2	Furaoan Stage (Phase) K_2^6	
				Mingshuian Stage (Phase) K_2^5	
				Sifangtaian Stage (Phase) K_2^4	
				Nenjiangian Stage (Phase) K_2^3	
				Yaojian Stage (Phase) K_2^2	
				Qingshankouan Stage (Phase) K_2^1	96
			Lower (Early) Cretaceous Series (Epoch) K_1	Quantouan Stage (Phase) K_1^6	
				Sunjiawanian Stage (Phase) K_1^5	
				Fuxinian Stage (Phase) K_1^4	
				Shahaian Stage (Phase) K_1^3	
				Jiufotangian Stage (Phase) K_1^2	
				Yixianian Stage (Phase) K_1^1	137
		Jurassic System (Period) J	Upper (Late) Jurassic Series (Epoch) J_3	Dabeigouan Stage (Phase) J_3^3	
				Wait to be set up	
				Tuchengzian Stage (Phase) J_3^1	
			Middle Jurassic Series (Epoch) J_2	Toutunhean Stage (Phase) J_2^2	
				Xishanyaoan Stage (Phase) J_2^1	
			Lower (Early) Jurassic Series (Epoch) J_1	Sangonghean Stage (Phase) J_1^2	
				Badaowanian Stage (Phase) J_1^1	205
		Triassic System (Period) T	Upper (Late) Triassic Series (Epoch) T_3	Wayaopuan Stage (Phase) T_3^3	
				Yongpingian Stage (Phase) T_3^2	
				Hujiacunian Stage (Phase) T_3^1	
			Middle Triassic Series (Epoch) T_2	Tongchuanian Stage (Phase) T_2^2	
				Ermayingian Stage (Phase) T_2^1	241
			Lower (Early) Triassic Series (Epoch) T_1	Heshanggouan Stage (Phase) T_1^2	
				Dalongkouan Stage (Phase) T_1^1	250

(To be continued)

(Table 2.2)

Eonothem (Eon)	Erathem (Era)	System (Period)	Series (Epoch)	Stage (Phase)	Ma
Phanerozoic Eonothem (Eon) PH	Paleozoic Erathem (Era) Pz	Permian System (Period) P	Upper (Late) Permian Series (Epoch) P_3	Sunjiagouan Stage (Phase)	
				Wait to be set up	
			Middle Permian Series (Epoch) P_2	Xiashihezian Stage (Phase)	
				Wait to be set up	
			Lower (Early) Permian Series (Epoch) P_1	Taiyuanian Stage (Phase)	
		Carboniferous System (Period) C	Upper (Late) Carboniferous Series (Epoch) C_2	Jincian Stage (Phase) C_2^4	295
				Benxian Stage (Phase) C_2^3	
				Yanghugouan Stage (Phase) C_2^2	
				Hongtuaoan Stage (Phase) C_2^1	320
			Lower (Early) Carboniferous (Epoch) C_1	Yushuliangian Stage (Phase) C_1^3	
				Chouniugouan Stage (Phase) C_1^2	
				Qianheishangouan Stage (Phase) C_1^1	354

● **Mesoproterozoic Erathem (Pt_2)**

The Mesoproterozoic erathem was widely distributed in the North China region and the following provinces: Hebei Province, Liaoning Province, Shanxi Province, Ningxia Province, Henan Province and Anhui Province. In addition, the Mesoproterozoic erathem was commonly distributed in the basement of petroleum basins that included the Bohai bay basin, Eerduosi basin and Subei basin. It also occurred in a peripheral area of the Tarim basin. The Mesoproterozoic erathem contains the Changchengian system and Jixian system. The best cross section is located in Jixian County in Hebei Province. The Changchengian system contained strata of terrigenous facies and strata of littoral–shallow sea facies. Both carbonate rocks and clastic rocks were well developed. Additionally, there were small amounts of volcanic rocks. The Jixian system mainly consisted of carbonate rocks and grayish black color shale of platform facies in a restricted sea area with well-developed stromatolite. Among the Mesoproterozoic strata, the Wumishan formation has the most important reservoir sequences in the Renqiu buried hill type of oil reservoir in the basement of the Jizhong depression in the Bohai bay basin, with a geological oil reserve of 400 million tons.

Table 2.3 Statistical table of drilling encountered strata that include stratigraphic formations, hydrocarbon source sequences and petroleum reservoir sequences in major terrigenous petroleum basins in China

System	Series	Songliao			Erlian			Bohai Bay			Sichuan			Eerduosi (Erdos)			Jiuquan			Chaidamu			Tuha			Zhungeer			Tarim		
		I	II	III	I	II	III	I	II	III	I	II	III	I	II	III	I	II	III	I	II	III	I	II	III	I	II	III	I	II	III
Q	Q	gray			gray			gray			gray			gray			gray			gray			gray			gray			gray		
N	N₂	gray			gray			gray			gray			gray			gray			gray	blue	yellow	gray			gray			gray		
N	N₁	gray			gray			gray		orange	gray			gray			gray		orange	gray	blue	orange	gray			gray		orange	gray		yellow
E	E₃	gray			gray			gray		orange	gray			gray			gray			gray	blue	orange	gray			gray	blue	orange	gray	blue	yellow
E	E₂	gray			gray			gray	blue	orange	gray			gray			gray			gray	blue	orange	gray			gray	blue		gray	blue	yellow
E	E₁	gray			gray			gray	blue	orange	gray			gray			gray						gray			gray	blue		gray	blue	yellow
K	K₂	gray	blue	orange																						gray					
K	K₁	gray	blue	orange		blue	orange											blue	orange							gray		orange			
J	J₃	gray		yellow		blue																									
J	J₂	gray											orange	gray		orange	gray			gray		orange	gray	blue	orange			orange		blue	orange
J	J₁											blue	orange	gray	blue	orange	gray			gray	blue	orange	gray	blue	orange			orange		blue	orange
T	T₃											blue	yellow	gray			gray			gray	blue		gray		orange		blue			blue	orange
T	T₂												yellow				gray										blue			blue	
T	T₁											blue																		blue	
P	P₃							gray					yellow	gray		yellow							gray	blue	orange		blue	orange			orange
P	P₂	gray						gray				blue	yellow	gray	blue	yellow							gray	blue			blue	orange		blue	
P	P₁	gray						gray					yellow	gray	blue	yellow											blue	orange			
C	C₂				gray																										
C	C₁				gray																									blue	

(To be continued)

(Table 2.3)

System	Series	Songliao			Erlian			Bohai Bay			Sichuan			Eerduosi (Erdos)			Jiuquan			Chaidamu			Tuha			Zhungeer			Tarim		
		I	II	III	I	II	III	I	II	III	I	II	III	I	II	III	I	II	III	I	II	III	I	II	III	I	II	III	I	II	III
D	D_3																															
	D_2																															
	D_1																															
S	S_3																															
	S_2																															
	S_1																															
O	O_3																															
	O_2																															
	O_1																															
Є	$Є_3$																															
	$Є_2$																															
	$Є_1$																															
Pt	Pt_3																															
	Pt_2																															
	Pt_1																															
AR																															

Note: I – Formation; II – Hydrocarbon source sequence; III – Oil and gas reservoir sequences (Red color represents oil sequence and yellow color represents gas sequence)

● **Neoproterozoic Era (Pt₃)**

The Neoproterozoic erathem included the Qingbaikouan system and the Sinian system. The Qingbaikouan system was mainly developed in the North China (Huabei) region. The Sinian system was mainly developed in the South China (Huanan) region. However, both the Qingbaikouan system and the Sinian system were developed in the Tarim basin. In the South China region, the greatest thickness of Neoproterozoic strata could reach 5,000 m. On the other hand, the thickness of the same strata was only 1,000 m in the Tarim basin. In the Bohai bay basin, the Qingbaikouan system mainly contained clastic rocks and claystone that went through a light metamorphic process; drilling exploration obtained oil and gas shows. In the Tarim basin, the Qingbaikouan system contained clastic rocks of littoral–shallow sea facies and carbonate rocks; in the upper portion, carbonate rocks contained plentiful stromatolite and oncolite. The Sinian system was widely distributed in China. The standard section of the Sinian system is located at the Three Gorges area in the Yangtze river valley. During the early Sinian period, the Yangtze region mainly received clastic deposited that filled the rifted valley; at the same time, moraine rock and volcanic rocks were also commonly developed. During the late Sinian period, the peripheral area of the Yangtze region had been transformed into a passive continental margin with well developed clastic rocks and carbonate rocks. A medium size gas field was discovered in the Weiyuan area in the Sichuan basin. The Tarim basin mainly contained clastic rocks of marine facies that were interbedded with volcanic rocks and carbonate rocks.

● **Paleozoic Erathem (Pz)**

The Paleozoic erathem was widely distributed in China. It not only contained carbonate rocks, but it also contained clastic rocks of marine facies and marine–terrigenous facies. The Paleozoic strata are the primary target for petroleum exploration in the basins of centra-western China, such as in the Sichuan basin, Eerduosi basin, Tarim basin and Zhungeer basin.

The Cambrian system and the Ordovician system were completely developed in the Yangtze plate, Tarim plate and North China plate. The lithology was a set of carbonate rocks of platform facies. At present, these strata are important reservoir sequences for the Tarim basin and Bohai bay basin. Large and medium size oil and gas fields were discovered in the Bohai bay basin (the Jizhong depression and Bozhong depression) and the Tarim basin (the Tazhong uplift and Tabei uplift).

During the Silurian period, the marine territory of the platform was reduced relatively, but the terrigenous source area continually increased. In most areas, the marine strata were absent or terrigenous clastic rocks were deposited. In the Yangtze region, the thickness of strata was 1,000 m approximately. However, in the Tarim region, the thickness of the same strata could reach as much as 4,000 m.

Influenced by the Caledonian tectonic movement, during the Devonian period the North China plate and the Tarim plate were connected to create a landmass. In the North China region, the Devonian system was absent. However, in the Tarim basin, the thickness of Devonian strata was in the range of 500 – 1,500 m. These strata were mainly distributed in the northern depressions and the lithology consisted of clastic rocks of shallow sea facies and of marine–terrigenous facies.

Compared with the Devonian system, the Permian system was widely distributed in the Chinese landmass with reduced thickness (usually around 1,000 m). The lithology consisted of a set of deposits of shallow sea facies and marine–terrigenous facies. On the North China plate, coal measures of marine–terrigenous facies were widely distributed. These strata were the important reservoir sequences in the Eerduosi basin, Bohai bay basin and Zhungeer basin. For example, in the Eerduosi basin, a large size gas field was discovered in the delta system in the Sulige area with a natural gas reserve of a thousand billion cubic meters.

● **Mesozoic Era (Mz)**

Compared with the Paleozoic erathem, the distribution and lithologic combination of the Mesozoic erathem were very different. The Mesozoic strata of marine facies were mainly distributed in southern China and the Tibetan highland. However, the good combination of petroleum generation–storage–seal was absent. The favorable exploration domains were only to be found in the Feixianguan formation of the Lower Triassic series and the Mesozoic marine strata in the Qiangtang basin. In northern China, most of the areas received terrigenous deposits. The Mesozoic terrigenous lacustrine basins of the Songliao basin, Eerduosi basin, Zhungeer basin and Sichuan basin are the important petroleum basins in China. The discovered large size oil and gas fields include Daqing and Xifeng.

The Triassic system has the important reservoir sequences in the Sichuan basin, Eerduosi basin and Zhungeer basin. In the Sichuan basin, the deposits of marine facies occurred in the Feixianguan formation, the Jialingjiang formation and the Leikoupo formation of the Middle–Lower Triassic series. The lithology consisted of limestone and dolomite of bio-reef flat facies. Terrigenous clastic rocks in the foreland basin and coal measures appeared in the Xujiahe formation in the Upper Triassic series. At present, these formations (either marine facies or terrigenous facies) are the important exploration targets. In the Eerduosi basin and the Zhungeer basin, terrigenous clastic rocks were developed in large scale depressed basins. These terrigenous strata are the major exploration domains.

The Jurassic system was widely developed in northern China. Here were the important coal producing sequences in the Mesozoic erathem. The Mesozoic strata were commonly developed in large size petroleum basins in central and western China. The lithology was a set of terrigenous coal measures that formed the important reservoir sequences in the Tarim basin, Zhungeer basin, Tuha basin, Chaidamu basin, Eerduosi basin and Sichuan basin. Several coal bed methane gas fields were discovered in the Jurassic coal measures in the Tuha basin and the

Chaidamu basin. In eastern China, the Jurassic strata were mainly developed in some faulted depressed type of terrigenous basins with a set of volcanic rocks that were interbedded with coal measures.

In Chinese petroleum regions, the Cretaceous system was distributed over a broader area when compared with the Jurassic system. These strata were well developed in northeastern China, the North China region and northwestern China, such as in the Songliao basin, Erlian basin, Eerduosi basin, Jiuquan basin, Zhungeer basin and Tarim basin. In every Chinese petroleum basin, almost all of the Cretaceous system consisted of terrigenous clastic deposits. Marine deposits only occurred in part of the Tibetan highland and the southwestern depression in the Tarim basin. The Cretaceous system is one of the major oil producing sequences in China, such as the Songliao basin. Since the 1970s, the annual oil production in the Songliao basin has been 500 million tons approximately.

● Cenozoic Era (Cz)

The Cenozoic erathem was widely distributed in many Chinese petroleum basins and it mainly contained terrigenous clastic rocks. The Palaeogene system and the Neogene system are the important reservoir sequences for eastern China and its marine territories; the typical example is the Bohai bay basin.

The Palaeogene system is one of the most important reservoir sequences in China. The oil and gas layers were widely distributed in the Bohai bay basin, Jianghan basin, Subei basin, Chaidamu basin, Zhungeer basin, Tarim basin and the marine territories of the Pearl River estuary and Beibu Gulf. Among these basins, the rifting depressed basins in eastern China and in the marine territories on the continental shelf contained terrigenous lacustrine clastic deposits of nearly ten thousand meters thickness. These clastic strata not only functioned as high quality hydrocarbon source rocks, but they also worked as primary reservoir sequences.

The Neogene system was distributed in most of the Chinese petroleum basins. It was present in particular in the petroleum basins of the Bohai bay basin, Subei basin and the continental shelf of eastern China. The lithology consisted of a set of deposits of river–lacustrine facies of great thickness. In the Neogene system, the hydrocarbon source rocks did not develop well. However, the oil and gas reservoirs (that were generated in the lower strata with storage in the upper strata) were commonly distributed, such as in the Bohai bay basin. Dozens of oil fields were discovered in the Neogene system with a reserve magnitude of hundreds of millions of tons.

The Quaternary system was the cap rock for most of the Chinese petroleum basins. In the Cenozoic foreland basin of western China, the Quaternary terrigenous clastic rocks were of great thickness, such as in the Chaidamu basin and Tarim basin. The greatest thickness of the strata could be more than 3,000 m. Conversely, in the lowland region of eastern China, the thickness of the Quaternary strata was less (usually less than 500 m); the lithology consisted of marine deposits and terrigenous deposits alternatively. Bio-gas reservoirs were discovered in the Quaternary system in the Chaidamu basin and the middle and lower reaches of the Yangtze river.

2.1.2 Characteristics of Depositional Evolution

In Chinese petroleum basins, the nature of the depositional environment experienced fundamental alteration from marine facies to terrigenous facies. During the Proterozoic and Paleozoic era, the primary depositions occurred in marine faces. During the Mesozoic era and Cenozoic era, most of the deposits appeared in the terrigenous facies. Thus, most of Chinese petroleum basins were superimposed basins that were constructed by the strata of marine facies and terrigenous facies. According to (Wang, 1985; Hu et al., 1991; Zhai et al., 2005), the following is a brief discussion of depositional evolution in Chinese petroleum basins.

The Mesoproterozoic era and the Neoproterozoic era can also be called the pan-oceanic era. During this geological time interval, oceans existed all over China. The Xinmeng ocean was located on the border of China and Mongolia, the North China ocean stretched from Tarim to North China, the Qinkun ocean appeared south of the North China ocean, the Himalayan ocean was positioned in the Tibetan region and the South China ocean was located in the southern region. The paleo-landmasses were small and few in number. Most of them drifted in the North China ocean or on its margin. These small landmasses included (from west to east) the Tarim terrane, Alashan terrane, Inner Mongolian terrane, Luxi terrane and other small terranes.

During the Early–Middle Neoproterozoic era, after sediments were deposited and influenced by the Jinning movement (Tarim movement), sea water retreated from the Chinese landmass. Up until the Early Sinian epoch, when marine transgression occurred, large areas were still available. In northern China, from Tianshan mountain to the North China region, there was a whole area of land that was oriented in a W-E direction that was called the Northern Terrane. In southern China, the upper Yangtze region did not receive deposits as well. Thus it was called the Upper Yangtze terrane. During the Late Sinian epoch, because the marine territory expanded, the land of the Northern terrane was only preserved in the North China region that was called the North China terrane. In southern China, the Upper Yangtze Terrane was also reduced to islands that were located in the Dali–Gejiu area. During the Sinian period, there were four major marine territories. Xinmeng ocean was located in northern China, the Himalayan ocean was located in the Tibetan region, the Yangtze–Tarim ocean stretched from the Xinjiang–Sichuan–Hubei area, and the South China ocean was located in southern China.

During the Cambrian period, the marine territory continually expanded. Obviously, the North China Terrane was submerged under water. At the same time, North China, the Yangtze and tarim were separated by deep water territories. In northern China, there were the Alashan Terrane, Dongsheng terrane and

Yingshan–Yanbei island; in southern China, there were the Songpan terrane, Kangdian terrane, Dabie terrane, Jiangnan island and Minzhe island group. The marine territories included Xinmeng ocean in the northern region, the Himalayan ocean in the southwestern region, the Tarim ocean in the western region, the Qilian–Qinling ocean in the central region, the North China ocean in the eastern region, the Yangtze ocean and South China ocean in South China.

During the Ordovician period, the distribution of oceans and terranes was similar to that in the Cambrian period; only the size of an individual terrane decreased. After the Early Ordovician epoch, the North China region (that included the major portion of Eerduosi) was uplifted entirely. During the Ordovician period, the marine territories included the Xinmeng ocean, Himalayan ocean, Tarim ocean, the Qilian–Qinling ocean, the North China ocean, the Yangtze ocean and the South China ocean. The major terranes included Chaidamu, Alashan, Yinshan–Yanbei, Dabie and Minzhe (Fig. 2.1).

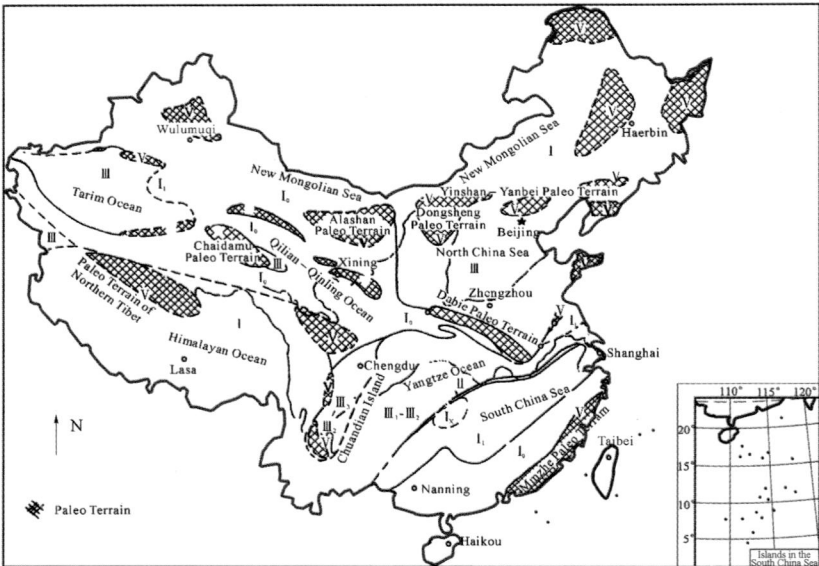

Fig. 2.1. Settings of paleo-oceans and paleo-terrains during the Ordovician period and the distribution of depositional facies (Wang, 1985) (with permission from SinoMaps Press)

During the Silurian period, the distribution of marine territories and terranes was the same basically as in the Late Ordovician epoch. The marine territories included the Xinmeng ocean, Himalayan ocean, Tarim ocean, Qilian–Qinling ocean, Yangtze ocean and South China ocean. The terranes included the North China Terrane, Northern Tibetan Terrane, Chaidamu Terrane, Southwestern Terrane and Minzhe Terrane.

After the Silurian system was deposited, a large scale marine regression occurred in China. During the Devonian period, the marine territory was the smallest one in the Paleozoic era. The oceans included the Xinmeng ocean, Himalayan ocean, Qinling ocean and South China ocean.

During the Early Carboniferous epoch, the distribution of the oceans and terranes was similar to that in the Late Devonian epoch. During the Late Carboniferous epoch, the largest marine transgression occurred since the Late Ordovician epoch; large size connected paleo-terranes disappeared. In particular, the Northern Terrane (that stretched from North China to Tarim) was submerged under water. During the Carboniferous period, the marine territories included the Xinmeng ocean, Northern ocean, Himalayan ocean and South China ocean. The terranes included the North Zhungeer Terrane, Aerjin Terrane, Qinling Terrane and the Yangtze terrane.

During the Permian period, the distribution of marine territories and terrane was clearly different when compared with the Late Carboniferous epoch. However, this spreading pattern was similar to that in the Late Devonian epoch or in the Early Carboniferous epoch. From Tarim to North China, the terranes had been connected to create a united landmass. On the other hand, the climax of marine transgression took place in southern China and most of the landmasses were submerged under water. During the Permian period, the major oceans included the Xinmeng ocean, Qilian–Qinling ocean, Himalayan ocean and Southern Sea; the Northern Terrain was the major landmass at that time. Inside the territory of the Northern Terrain, there were three relatively large terrigenous basins, which were the North China terrigenous basin, the Zhungeer terrigenous basin and the Tarim terrigenous basin. The Lower Permian coal deposits in the North China region and the Upper Permian mudstone in the northwestern region were the oldest hydrocarbon source rocks among the Chinese terrigenous deposits. Thus, in China, the Permian period was the first developmental phase of hydrocarbon source rocks in terrigenous facies.

During the Early and the Middle Triassic epochs, southern China was under water and north China was above water. Northern China was in the course of being uplifted. During these early times, because of the dry climate, red color clastic deposits were commonly developed. The large scale basins included the Zhungeer basin, the northern part of the Tarim basin, and the North China region. The sedimentary type of deposits belonged to the dry climate deposition in the intra-continental depressions. Influenced by the Indosinian movement, seawater also withdrew from southern China gradually. During the Later Triassic epoch, because the climate was humid, coal deposits were generally developed in southern China. In some large scale basins, lacustrine basins were broad and in a deep water environment. In China, the Late Triassic epoch was the second developmental phase of hydrocarbon source rock in terrigenous facies. The reservoir sequences were mainly developed in the Zhungeer basin, Tarim basin, Tuha basin, Eerduosi basin and Sichuan basin. During the end of the Triassic period, the Post-Indosinian movement influenced broad regions, the territory of marine deposits sharply decreased, the strata that were deposited prior to the Jurassic were commonly eroded.

During the Early Jurassic epoch, another developmental stage of lacustrine basins was initiated (Fig. 2.2); the climate was humid and most of the sediments contained coal. The Early and Middle Jurassic epochs were important phases for coal making. This was also the third developmental phase of hydrocarbon source rock in terrigenous facies. The reservoir sequences were mainly distributed in the Zhungeer basin, Tuha basin, Tarim basin, Eerduosi basin, Chaidamu basin and Sichuan basin. In western China, the sediments were deposited in a large scale intra-continental humid climate type of depression or intra-continental humid climate type of faulted depression. In eastern China, the sediments were deposited in faulted depressions that were either located on the intra-continent or were positioned near the ocean. During this time, the Sichuan basin was surrounded by mountains or highlands. It received deposits of sand and mud in the littoral–shallow lacustrine environment or it received fresh water carbonate deposits. The sediments did not contain coal. This was an intra-continental semi-arid zone and semi-humid zone climate. During the Middle Jurassic epoch, the framework of the basins had not changed; the paleo climate had gradually become more of a dry climate. Coal deposits mainly occurred in the Xinjiang region and northeastern China. During the Late Jurassic epoch, because the Yanshanian movement was at its most energetic stage, volcanic eruptions occurred widely in eastern China.

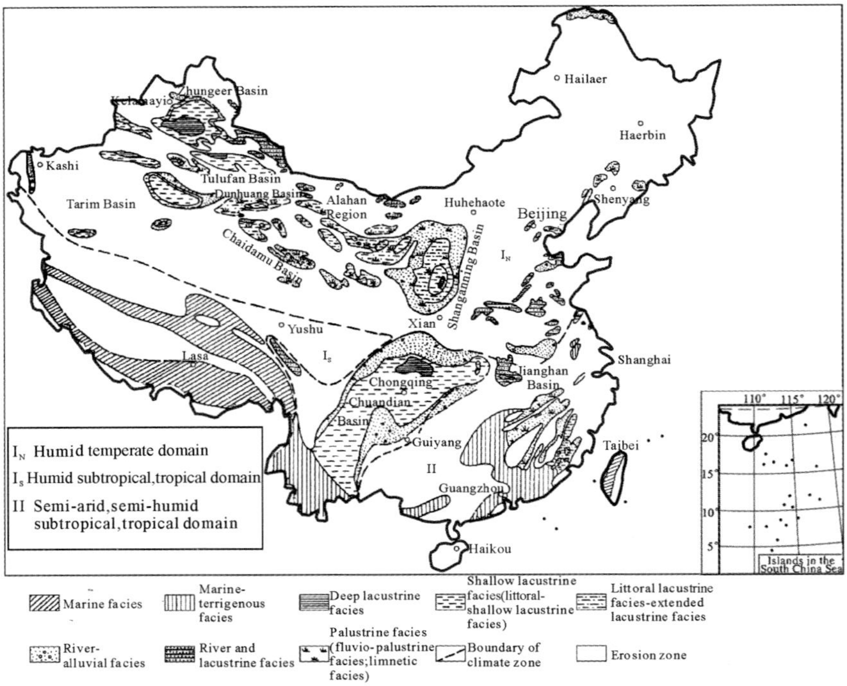

Fig. 2.2.   Map of sedimentary basins and depositional facies during the Early Jurassic epoch in China (Hu et al., 1991) (with permission from Petroleum Industry Press)

The Early Cretaceous epoch was the most important developmental phase for the Mesozoic erathem. In northern China, the large scale basins subsided one after another. In southern China, uplifting events were in progress. During the early–middle phases of the Early Cretaceous epoch, the deposits of limnetic facies and the deposits of relatively deep water lacustrine facies represented the fourth developmental phase of hydrocarbon source rock. The diversity of the paleo climate zone was clear. Northeastern China and Inner Mongolia belonged to the humid climate zone, the area from North China to the northern part of the Xinjiang region belonged to the semi-arid and semi-humid subtropical climate zone, the area from south China to the southern part of the Xinjiang region belonged to the semi-arid subtropical climate zone and the Tibetan region belonged to the humid subtropical climate zone. The discovered Cretaceous oil fields and forecast petroleum regions were distributed in the humid climate zone and transitional climate zone in northeastern China. During the Late Cretaceous epoch, the Yanshanian movement brought very different influences to northern China and southern China. In northern China, the large scale depositional basins were either reduced in size or disappeared. In southern China, the basins were continually developing. During the Late Cretaceous epoch, the arid climate zone expanded and red color deposits were common. Because the humid climate zone shrunk in size, the sediments were widely distributed either in the intra-continental arid climate type of faulted depression or in the intra-continental arid climate type of depression. Even the Songliao basin was still in the humid climate zone of northeastern China. However, its depositional region had clearly decreased. During the Late Cretaceous epoch, the humid climate zone of northern China had migrated towards the north. The Upper Cretaceous series was mainly distributed in the Songliao basin and the Sanjiang area. The Songliao basin was the most important depositional region. Because it had an extensive lake region and relatively deeper water, a set of petroleum sequences was developed here. The Late Cretaceous epoch was the fifth developmental phase of hydrocarbon source rock in terrigenous facies in China.

The Palaeogene was not only the most important depositional period of the Himalayan tectonic period, but it was also the period when petroleum basins were widely distributed over the Chinese landmass. Because the Pacific plate was subducted toward the Eurasian plate, a series of extensional type of faulted depressions occurred in eastern China. Conversely, because the Indian plate pushed toward the north, many large scale depressions appeared in western China. The Palaeogene was also the period when paleo climate zones were diversified and were of a similar nature to that in the Early Cretaceous epoch. Marine deposits occurred from Taiwan to the southern part of the East China Sea, in the southern part of Tibet, and the southwestern part of the Tarim basin. The Palaeogene was not only the period when the paleo-structure, paleo-climate and paleo-geography were obviously diversified, but it was also the period that contained most of the depositional types. In the humid climate zone of northern China, there were the intra-continental humid type of faulted depressions (such as the Yilan–Yitong graben) and the intra-continental humid type of depressions (such as the Sanjiang basin). In the transitional climate zone from North China to the Zhungeer basin, there were the intra-continental type of depressions (such as the Zhungeer basin), the intra-continental type of faulted depressions (such as the grabens in the Great Bend of the Yellow River, in the Yinchuan area and in the Fengwei area), and near the ocean type of faulted depressions (such as the Bohai bay basin). In the arid climate zone from central China to the Tarim basin, there were near the ocean arid type of depressions (such as the southwestern part of the Tarim basin), the intra-continental arid type of depressions (such as the Chaidamu basin) and the intra-continental arid type of faulted depressions (such as the Jianghan basin). In the humid climate zone of southern China, there were the intra-continental humid type of faulted basins (such as the Lunpola basin and the Baise basin) and near the ocean the humid type of faulted depressions (such as the Sanshui basin, the Pearl River Estuary basin, Beibu Gulf basin, Yingge Sea basin and Southeastern Hainan basin). The Palaeogene was the developmental period for reservoir sequences in the basins that were located in the continental rift valley system and on the continental margin. Thus, these regions became the major oil producing regions in the Palaeogene system.

During the Neogene period, because of the uplift of the Himalayas, sea water finally retreated from Tibet (Fig. 2.3). On the other hand, because of the extension of the South China Sea, marine transgression occurred on the continental shelf in the South China Sea. From southern China to Tibet, the uplifting was the major tectonic event and the isolated small faulted basins were scattered around. However, because northern China was subsiding, many large scale depressions were developed. For the paleo climate zones, because northwestern China was located at the intra-continent, it was in the arid climate zone. The rest of China basically belonged to the humid climate zone. In the arid climate zone of northwestern China, the types of basin included the depressions (such as the Tarim basin, Zhungeer basin, Chaidamu basin and Jiuxi basin) and the faulted depressions (such as the grabens in the Great Bend of the Yellow river, the Yinchuan area and the Weihe area). In the humid climate zone, the types of basin included the intra-continental humid type of depressions (such as the Songliao basin), the intra-continental humid type of faulted depressions (such as the small basins in southwestern China), and near the ocean the humid type of depressions. Based on the framework of the Neogene period, the paleo-geography of the Chinese landmass underwent new developments during the Pleistocene epoch. Gradually, the stair-like topography was developed on the Chinese landmass, in the highlands on the western side and the lowlands on the eastern side. In addition, the territorial division between marine terranes and above water terranes, the Yangtze river, Yellow river and other major water systems of the present day, were developed successively.

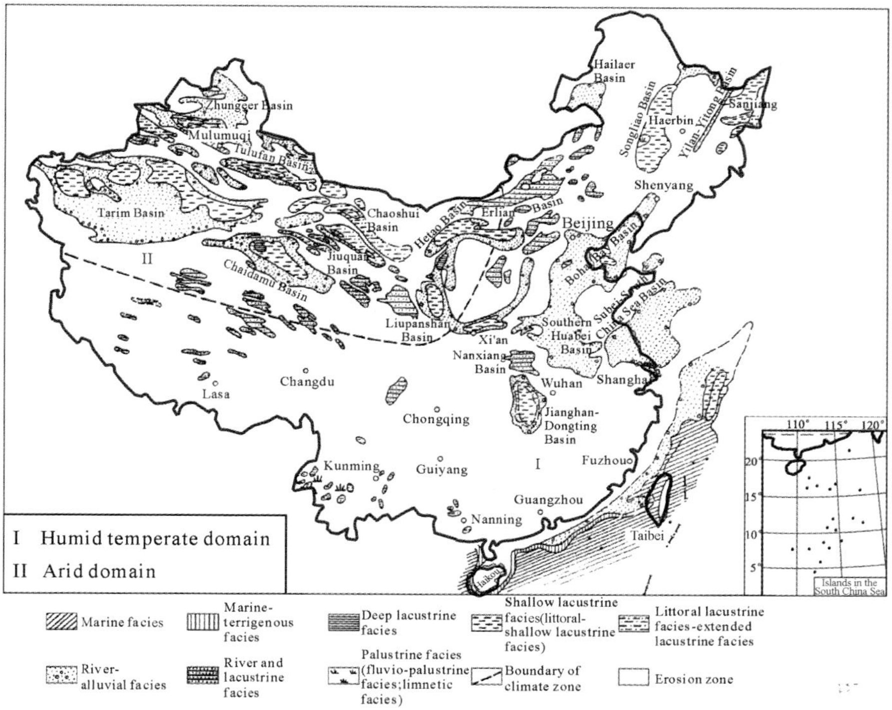

Fig. 2.3. Map of sedimentary basin and depositional facies during the Neogene period in China (Hu et al., 1991) (with permission from Petroleum Industry Press)

2.1.3   Major Depositional Facies and Their Attributes

● **Terrigenous Depositional System**

The development of Chinese terrigenous lacustrine basins was controlled by the evolution of regional tectonic movements. The development of terrigenous lacustrine basins and related depositional features were manipulated by several factors that included the regional tectonic background and structural development, regional paleo geography, paleo climate, paleo topography and paleo aquatic system. Generally speaking, because terrigenous lacustrine basins were surrounded by uplifts or mountains, many streams carried a huge amount of clastic sediments and chemical substances into the basins. There were abundant supplies of terrigenous clastic material with plentiful sand and pebbles. Thus, inside a lacustrine basin, the sand bodies (that included sandstone and conglomerate) were well developed. Among the lacustrine strata, the proportion of the sand body was larger than in the marine strata.

Using an analysis of depositional characteristics of the paleo lacustrine basins and a comparative study of lake basins of the present day (Wu et al., 1992; Feng et al., 1994; Gu et al., 1994; Qiu, 1992; Xue et al., 2002), geologists confirmed that in a terrigenous lacustrine basin there were six depositional systems of alluvial fan, river, delta, under water fan, lacustrine and palustrine (Table 2.4).

Table 2.4 Classification of depositional systems and sedimentary facies for a terrigenous lacustrine basin

	Depositional system	Sedimentary facies	Subfacies	Micro facies and framework of sand body	Principle sedimentation
I	Alluvial fan system	Dry land fan Wetland fan	Fan root Mid-fan Distal fan	Main channel, lateral channel, slot beach, sheet wash zone, braided river trench, sheet flood (between channel)	Mudslide Tractional current
II	River system	Meandering river Braided river Anastomosed stream	River channel Between channel	Channel-lag deposit, marginal bank, channel bar, natural levee, crevasse fan, flood plain, oxbow lake	Tractional current
III	Delta system	Meandering delta Braided river delta Fan delta	Plain, Front margin: inner zone, outer zone Delta front	Distributary channel, inter-distributary channel, underwater (inter-)distributary channel, estuary dam, sand sheet	Tractional current is principle flow; Gravity flow is secondary flow
IV	Underwater fan system	Near shore underwater fan Sub-lacustrine fan Fluxoturbidite	Supply channel: Inner fan Mid-fan Outer fan	Main channel, natural levee, braided channel, inter-channel, sand sheet without channel, slump lens	Gravity flow is principle flow; Tractional current is secondary flow
V	Lacustrine system	Fresh water lake Brackish water lake Salt water lake	Lake shore Shallow stand lake (Medium) Deep stand lake Lake bay	Clastic beach-bar Carbonate beach-bar Bio-reef	Lake current, wave, chemistry, biology
VI	Palustrine system	Lake swamp River flood plain swamp Delta plain swamp			Biology

The lacustrine sand bodies were mainly developed in the former four systems. The lacustrine system and the palustrine system primarily received deposits of mudstone, coal measures and gypsum–salt. There were sand bodies of beach bar facies and turbiditic sand bodies in small amounts. Therefore, in a terrigenous lacustrine basin, there were essentially four kinds of sand bodies, which included the sand bodies of alluvial fan facies, of river facies, of delta facies and of underwater fan facies. The sand bodies of beach bar facies were developed in some lacustrine basins. For the sand bodies of alluvial fan facies and river facies, their depositional mechanisms did not relate to the presence of a lacustrine system; these sand bodies were produced by the tractional current in the typical continental environment. For the sand bodies of delta facies, of underwater fan facies and of beach bar facies, their depositional mechanisms were definitely related to the presence of a lacustrine system. The sedimentations of these sand bodies were various, which not only included tractional current and gravity flow, but also included lake current and waves. At present, the name lacustrine sand body is given to sand bodies if their depositional mechanism is related to a lacustrine system. In a broad sense, the lacustrine sand body was extensively distributed in every subfacies of a lacustrine system of various types from middle– deep water to littoral–shallow water. The lacustrine sand body was the first choice for oil and gas accumulation. The oil and gas were generated by the hydrocarbon source rocks of argillaceous rock and coal measures in the lacustrine basin. In the Mesozoic and Cenozoic oil and gas fields, most of the reservoir sequences consisted of various sand bodies that were deposited in the lacustrine environment.

Exploration practice and study showed that, in the terrigenous lacustrine basin, all types of sand body could contain oil and gas. Whether the lacustrine sand body could or could not contain oil and gas was mainly decided by the distance between the sand body and petroleum generating center (or petroleum generating layer) and the correlated relationship of generation–storage–seal. In addition, the size of a sand body, the thickness of a sand layer and its connectivity, and also the quality of the physical properties could influence the productivity. Because the terrigenous lacustrine basins varied in their geological framework, basin development, depositional environment, depositional model and petroleum accumulation, all kinds of sand body were developed of various sizes, dissimilar shapes and different buried depth. For example, in the depressed type of terrigenous lacustrine basin, the primary type of petroleum sand body was of the large size river delta. Secondly, the sand bodies of the fan delta and of the braided river delta could also contain oil. In addition, the oil sand bodies could be developed in diluvial (alluvial) fans and rivers. On the other hand, in the faulted depressed type of terrigenous lacustrine basin, many types of sand body might contain oil and gas. The types of sand body could occur in the following places: rivers, all kinds of delta, near shore underwater fans, base fans of lacustrine, beach bars and alluvial fans.

According to the statistical study of sand body related oil reserves in the Chinese terrigenous lacustrine basin, the sand bodies from all kinds of delta systems were the major type of oil bearing sand body in the terrigenous lacustrine basin, which represented 55.3% of the proven oil reserve. Among these oil reserves, the sand bodies of river facies contributed 13%, the sand bodies of underwater fan facies contributed 12.6%, the sand bodies of alluvial fan contributed 6.5%, the sand bodies of beach bar facies contributed 5% and the basement rock of the basin contributed 7.6% (Fig. 2.4).

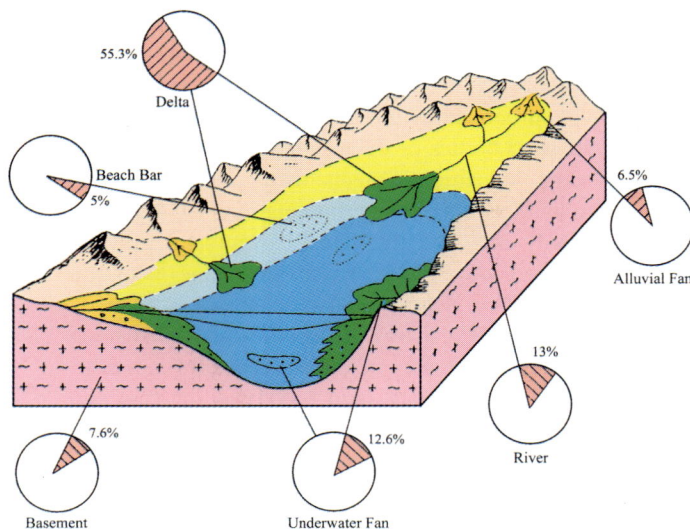

Fig. 2.4. Depositional system of terrigenous lacustrine basin and distribution of petroleum reserve within the system (Qiu et al., 1997) (with permission from Petroleum Industry Press)

The sand body of alluvial facies was either developed on the margin of a terrigenous lacustrine basin or developed in most parts of the basin during the early phase of sediment infill. The dry fan was well developed in the Chinese petroleum basin, which commonly formed a course grain clastic sandy conglomerate entity. Because the grain sizes in the mid-fan subfacies were ideally sorted, thus a favorable petroleum accumulative zone was developed here. Because the alluvial fan was close to the source of the substance and because it was far away from the hydrocarbon source center in deep water, there were very few reports that the alluvial fan functioned as reservoir sequences around the world. However, in the Chinese terrigenous lacustrine basins, the sandy conglomerate that functioned as reservoir sequences was not only discovered in the alluvial fan facies frequently, but also contained a large size oil field. For example, in the Zhungeer basin, the Kelamayi oil field contained a proven oil reserve of 8×10^8 t. The Triassic reservoir sequences in the Kelamayi oil field consisted of seven alluvial cones; the connected sandy conglomerate was developed on the river bed in the alluvial fans with the best oil bearing properties. Furthermore, the discovered oil fields that utilized the sand body of alluvial fan facies as their major reservoir sequences were located in the following basins: the Zaoyuan oil field in the Huanghua depression, the Caoqiao oil field in the Jiyang depression in the Bohai bay basin and the M3 oil layer in the Miocene series in the Laojunmiao oil field in the Jiuquan basin.

All types of river sand body could be developed in the terrigenous basins.

However, the best developed type was the braided river sand body. The braided river was widely developed in the horizontal depositional system in the faulted depressed basin, the small intra-mountain basin and the mountain front basin. In these depositional systems, the braided river usually ran directly into the lacustrine basin or formed a braided river delta instead of being transformed into a meandering river. Thus, the sand body of braided river facies was the most important type of sand body in the river system in terrigenous basins. Except in the vertical depositional system of a large depressed type of lacustrine basin, a large scale meandering river system was very hard to develop. Therefore, generally speaking, the sand body of a highly curved meandering river did not hold a significant position in volumetric matter. In China, the braided river type of restricted river channel deposits was discovered in the vertical depositional system that was developed during the tectonic active phase of the terrigenous basin, such as in the Jurassic system in the Eerduosi basin.

All types of river sand body were developed in terrigenous lacustrine basins with broad distribution and on a large scale. In China, the oil fields that utilized the river sand body as major reservoir sequences were discovered in the Mesozoic and Cenozoic terrigenous petroleum basins. In the Bohai bay basin, the large and medium size oil fields (such as the Gudao, Gudong, Chengdao and Beidagang) utilized the Neogene sand body of braided river facies as major reservoir sequences. In the Songliao basin, the major reservoir sequences in the Fuyu oil field consisted of meandering river deposits in the Quantou formation. In the Eerduosi basin, the major reservoir sequences in the Maling oil field were the sand bodies of braided river facies in the Jurassic Yanan formation. This phenomenon was one of the special characteristics of petroleum geology in the terrigenous petroleum basin. Because the river deposited reservoir sand bodies were developed very well in the terrigenous basin, during different developmental phases of the basins the river sand bodies were deposited in every tectonic position with various levels of accumulation.

In terrigenous basins, all kinds of delta sand bodies were developed well, which was the most important component for terrigenous clastic rocks. Because they were close to the source rocks of terrigenous facies, delta sand bodies became the most important reservoir sequences in the lacustrine basins. Recently, the proven reserves from the reservoir sequences of delta sand bodies represented more than 50% of the total proven reserves in China. Oil and gas reservoirs generated from delta sand bodies have been discovered in almost every Chinese terrigenous petroleum basin.

In the Mesozoic and Cenozoic terrigenous petroleum basins of China, the distribution pattern of oil and gas fields indicated that, in a petroleum basin, the largest clastic rock type of major oil field was always developed in the largest river–delta (or alluvial fan–fan delta) system of the basin. Beside the unique preconditions of abundant hydrocarbon sources and the thickest reservoir sand bodies, there was another significant and interesting phenomenon that was the perfect correlation between the sand body belt and structural trap. The depositional system of the delta brought abundant sandy clastic substances into the lacustrine basin. These clastic sediments were enveloped by nearby mudstone of lacustrine facies. During the depositional process, a variation in compaction between the composite sand body and surrounding mudstone occurred and it produced an embryonic structural format. Under the influence of this embryonic structural format, the structural activity of the late period usually carried on the deformation and finally created a structural trap. This was one of the important reasons why the composite sand body of a delta generally correlated very well with the structural trap in terrigenous petroleum basins. The Daqing placanticline that was the largest oil structure in China was the product of a perfect correlation between the large size placanticline and the largest depositional system of the river–delta. In the lacustrine basin, another important characteristic of the delta deposits was that the constructive delta was in a dominant position due to the lower energy level of the water body. When a river ran into a lacustrine basin, it created a constrictive river controlled delta; when an alluvial fan directly went into a lacustrine basin, it developed a constructive fan delta. Among the paleo geographical factors that decided the type of lacustrine delta and the nature of reservoir sequences in the sand body, the most critical factors were the slope gradient and the distance to the source center of clastic substances.

During the major depositional period, along the long axis of a large size depressed basin, a constructive delta of bird foot shape was commonly developed. The supply center of clastic material was far away from the depositional center, the slope gradient was small, the length of the river was long with a broad river valley. The river was gradually transformed into a meandering river with few bends or with many bends. Finally, as distributary channels, the river ran into a lacustrine basin to construct a bird foot shape delta that mainly consisted of the sand body in the distributary channels and digit shaped sand dams. This type of delta sand body contained good reservoir conditions on a large scale, which was the primary target when exploring for a large oil field. For example, in the Daqing oil field in the Songliao basin, the oil layers of Sha, Pu and Gao were made of sand bodies in a typical meandering river delta in the Qingshankou formation– Yaojia formation in the Lower Cretaceous series. During the middle–late phase of the Early Cretaceous epoch (the depositional period of the Quantou formation– Nenjiang formation), the growth of the Songliao basin progressed into a high peak phase of development of a large depression. In particular, when the Qingshankou formation was deposited, a unified central depressed region was developed in the basin to create a large scale paleo lacustrine basin in the Songliao. During this time, the fundamental depositional characteristics of the lacustrine basin included utilizing the central depression as the subsidence and depositional center. Several depositional systems brought the sediments from several material source centers on the peripheral areas into the lacustrine basin to create a huge delta depositional system with five large rivers. Among these depositional systems, the largest one was the Northern Depositional System that was developed along the long axis of the basin and that came from the north. Starting at the marginal alluvial fan and passing through the broad alluvial plain, the rivers were transformed from braided rivers into highly curved meandering rivers. After flowing through the distributary delta plain, these rivers not only ran into the lacustrine basin to construct a large scale river delta system, but they also deposited a set of sandy mudstone sequences with a thickness of more than 1,000 m. The Daqing oil field was developed downstream of this depositional system; it contained sand bodies as reservoir sequences that included part of the alluvial plain river system and the whole delta system. The Daqing oil field is the largest oil field in China with an initially proven oil reserve of more then 40×10^8 t.

In a terrigenous basin, especially on the steep slope side of a faulted basin and a lacustrine foreland basin, if the horizontal depositional system of the alluvial fan directly ran into the lacustrine basin, a constructive fan delta or braided river delta might usually be developed. Along the short axis of the lacustrine basin, on the steep slope side, the depositional center was near the supply center of clastic material; the slope gradient was very steep. Sometimes, the distance between the supply center of clastic material and the depositional center of the lacustrine basin was only a few thousands meters with a steep rate of incline. With this kind of paleo geographic background, because the alluvial fan of the mountain was directly connected with the lacustrine environment, the alluvial plain could not be developed. In addition, a huge amount of course grain clastic material was directly deposited in the lacustrine basin through the alluvial fan to form sandy conglomerate entities in the constructive fan delta or in the braided river delta. The fan delta and braided river delta were usually developed during the depression and expansion of the lacustrine basin. Because they were next to the hydrocarbon generating depression, the supply of oil was plentiful. With the perfect combination of storage and seal, the oil and gas reservoirs were discovered in the sand bodies of the fan delta facies and braided river facies in several terrigenous basins. For example, in the Liaohe area, the Xinglongtai oil reservoir of fan delta facies was located on the steep slope zone of the Western Depression and the Shuguang oil reservoir of braided river delta facies was located on the gentle slope. In the Kuche foreland basin in the Tarim basin, there were oil and gas reservoirs of braided river delta facies (such as the Kela 2 gas field).

Aiming at petroleum accumulation and reservoir distribution patterns in the terrigenous delta sand bodies, geologists accomplished a lot of research work and gained a series of important research results. Among these, the most important one was that the depositional facies belt and the micro facies of the sand body not only controlled the petroleum accumulation and distribution in the delta sand body, but they also controlled the type of oil reservoir. Generally speaking, oil and gas were highly accumulated in the inner belt of the delta front with the structural–lithologic type of oil reservoir. On the other hand, the lithologic type of oil reservoir was developed in the outer belt of the delta front and the structural type of oil reservoir was developed on the delta plain. For example, in the Bohai bay basin, the Xinglongtai oil field was positioned in the fan delta sand bodies that were mainly distributed in the subfacies of the front margin. The detailed micro facies included the underwater distributary channels, the estuary dam and the shallow flats of intra-distributary channels. However, the fan front of the delta and fan delta plain did not contain any oil layer.

The sand body of the underwater fan was an important type of reservoir sand body in the rifted basin. The Bohai bay basin is a typical example. Oil and gas reservoirs have been discovered in the underwater fan type of sand bodies in every petroleum depression in the Bohai bay basin in succession. These include, for example, the Bonan oil field of underwater fan type in the Zhanhua depression, the Liangjialou underwater fan type of oil field in the Dongying depression, the nearshore underwater fan type of oil field in the northern belt of the Dongying area, the nearshore underwater fan type of oil field in the western depression in the Liaohe area, the nearshore underwater fan type of oil field on the foot wall of the Daxing fault in the Langgu depression and the Niuzhuang slumped turbidite type of oil field in the Dongying depression. The underwater fan type of reservoir sand body was distributed in the Palaeogene mudstone of deep water–semi-deep water lacustrine facies in the rifted and faulted basin in Bohai bay. Because they were either directly in contact with source rocks or they were enveloped by the source rocks, the sand bodies not only experienced oil accumulation at an early stage, but they also experienced continual accumulation throughout geological time. At present, this kind of sand body is the important exploration domain in the Bohai bay basin, especially for discovering the lithologic type oil reservoir. Because of its special mechanism and distribution pattern, the underwater fan sand body had different accumulative conditions and patterns when compared with other types of sand body.

● Marine Depositional System

According to the geography, the marine basin can be subdivided into the marine environment and the transitional environment of sea to land. Using the characteristic analysis of deposition in the paleo marine basin and the comparative study of a modern marine basin, geologists learned that, in the marine environment, the Chinese petroleum basin contained the following depositional systems: littoral system, bio-reef (flat) system, shallow sea continental shelf system, semi-deep ocean system and deep ocean system (Table 2.5). On the other hand, in the transitional environment of sea to land, there are the delta system and the estuary system. Petroleum exploration practices and the study results showed that multi-type reservoir sequences were developed in the marine basin; the diversified types of reservoir sequences were developed in the delta facies, littoral facies, base fan facies in the ocean facies, carbonate facies and bio-reef facies. If compared with all kinds of sandstone reservoir sequences in the terrigenous lacustrine basin, the marine reservoir sequences are clearly different. At present, oil and gas exploration in the marine depositional system in China is focused on river controlled delta facies, bio-reef facies and shallow water carbonate facies.

Table 2.5 Primary depositional system and sedimentary facies of marine basin (modified from Liu, 1980) (with permission from the Geological Publishing House)

Depositional environment	Depositional system	Facies	Subfacies
Marine-terrigenous transitional environment	Delta	River-controlled delta Wave-controlled delta Tide-controlled delta	Delta plain Delta front Pro-delta
	Estuary		
Marine environment	Littoral sea (cost)	Barrier cost (Restricted sea)	Lagoon Tidal flat Barrier island (flat)
		Non-barrier cost (Open sea)	Costal sand dune Backshore Foreshore Nearshore
	Bio-reef, flat	Patch reef, atoll reef	
	Neritic shelf	Supratidal zone, intertidal zone	
	Semi-deep ocean	Slope fan, sub-marine fan	
	Deep ocean	Sub-marine fan	

In the Eerduosi basin, during the Late Carboniferous epoch and Early Permian epoch, the river controlled delta depositional system was developed in the marine–terrigenous facies, which had well developed subfacies of the delta plain and poorly developed subfacies of the front margin. The delta plain consisted of the distributary channel, crevasse fan and flood basin. The delta front mainly consisted of underwater distributary channel facies, estuary dam facies, distal bar facies and inter-distributary channel facies. However, the estuary dam was poorly developed. During the late Paleozoic era, the distribution of the depositional system of the Eerduosi basin was controlled by the substance source center in the north. Four river delta systems were developed to cover more than half of the basin. These river delta systems were located at Shenmu–Mizhi, Yulin, Sulige and Tiekesumiao. A large size gas field was discovered in the Sulige area.

The littoral depositional system can be divided into the barrier type and the non-barrier type of littoral facies. The typical beach sand body was developed in the Tarim basin during the Early Silurian epoch and the Late Devonian–Early Carboniferous epochs (when the Donghe sandstone was deposited), which was the non-barrier type of sandy littoral coast, which included all kinds of sand bodies that were developed in different facies of the beach sand dunes, the backshore, the foreshore and the shoreface. The discovered large and medium size oil and gas fields included the Tazhong oil field and others. The typical depositional system of tidal flats was developed in several formations in the Tarim basin. These formations included an upper member of bitumen sandstone of the Kepingtage formation, Tataeiertage formation, Yimugantawu formation and Kalashayi formation of the Carboniferous system. The deposits of the tidal flats mainly included the micro facies of tidal channels in the subtidal zone, a sandy flat in the intertidal zone, an intermixed flat of sand and mud, an upper mud flat in the intertidal zone and a mud flat in the supratidal zone. The oil and gas fields were discovered in the tidal channel of the subtidal zone and in the sandy flat of the intertidal zone.

The carbonate depositional system in the marine craton basin was mainly developed in the Tarim basin (in the Cambrian–Ordovician systems and the Carboniferous system), the Eerduosi basin (in the Cambrian–Ordovician systems and the Carboniferous–Permian systems), and in the Sichuan basin (the Cambrian–Ordovician systems, the Carboniferous–Permian systems and Triassic system). The depositional systems of the carbonate platform mainly included three types, which were located on the continental shelf, on the gentle slope and in the epicontinental sea. Among these depositional systems, the continental shelf type of carbonate platform could be further divided into a type of composite reef flat on the platform margin and a type of marginal flat on the platform.

In the Tarim basin, the Cambrian–Ordovician systems contained types of marginal reef and marginal flats. During the Middle and the Late Ordovician epochs, the sea level frequently fluctuated. The period when the sea level was slowly changing or was stable was the best time for the development of the bio-reef. In general, the bio-reef contained more than 30% of the reef-building organisms. These organisms included bryozoan, sponge, receptaculites, stromatoporoids, coral, blue-green algae, solenopora. The reef-attaching organisms included gastropoda, brachiopods, echinoderm, ostracoda, lamellibranch, encrinite, trilobite, cephalopods and all kinds of algae. The position of marginal facies on the carbonate platform was changeable. They could not only migrate constantly during the different geological periods, but could also migrate by leaping a long distance within a very short time interval (Fig. 2.5).

In the Sichuan basin, during the Early Triassic epoch, the oolitic beach on the platform margin was developed on the northeastern side of the basin with a depositional thickness of 300 – 500 m. The lithology included sparite bio-clastic limestone and grain clastic oolitic limestone, which were distributed in the pattern of a sheet with small thickness and in a single layer. The lithology was unstable vertically and appeared as an irregular pattern horizontally. The oolitic beach on the platform margin was developed in a high energy environment on the margin of the platform and acted as a barrier. This oolitic beach was located on the margin of the Kaijiang–Liangping sea trough; it was a mixed entity of oolite, sandy clastic and oncolite of great thickness and stable distribution. Because of dolomitization and dissolution, the secondary pores and fractures were developed. Therefore, the sequences became perfect reservoir sequences, such as the oolitic beaches in Dukouhe, Shatuo and Tieshan–Liangping.

During the Early Cambrian epoch and the Carboniferous period, the gentle slope type of carbonate platform of shallow water marine facies were developed in the Tarim basin. The subfacies included an evaporate platform, restrictive platform, semi-restrictive platform and wide open platform. The evaporate platform was of a relatively weak aquadynamic condition with arid climate and strong evaporation. Here, the dolomite flat (that mainly contained penecontemporaneous dolomite) was very easy to deposit. The combinations mainly included a combination of dolomite flat, a combination of dolomitic gypsum and limestone flat and a combination of beach–dolomite flat. However, the combination of calcareous dolomite flat and the combination of dolomitic limestone flat hardly ever occurred.

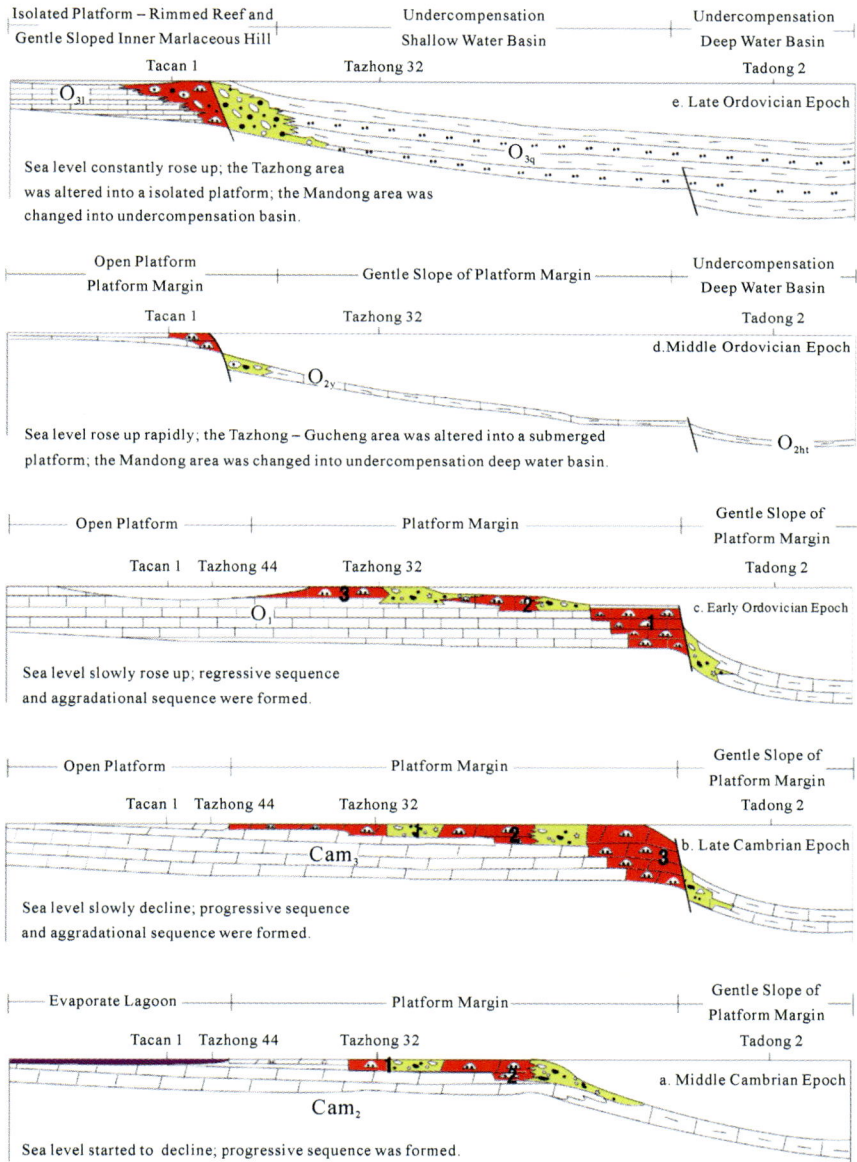

Fig. 2.5. Developmental model of carbonate platform margin from the Middle Cambrian–Late Ordovician in the Tarim basin

During the Ordovician period, because the Eerduosi basin and Bohai bay basin were parts of the North China epicontinental sea, the carbonate rocks of the epicontinental sea and the evaporate rocks of the restrictive sea were deposited alternatively in cycles. Among the formation, the members of Ma 1, Ma 3 and Ma 5 were deposited in the restrictive sea on the continental shelf in an evaporative environment. The lithology included a dolomite and gypsum–salt layer. On the other hand, the members of Ma 2, Ma 4 and Ma 6 were deposited in the epicontinental sea on the continental shelf. The lithology included micrite and dolomite. Now, several weathering crust types of buried hill oil and gas fields have been discovered in the Ordovician system in the Bohai bay basin and Eerduosi basin.

2.2   Characteristics of the Petroleum System

We will introduce the types of source rocks, reservoir sequences and cap rocks and their distributions, respectively.

2.2.1   Types of Source Rocks and Their Distributions

In China, the history of continental tectonic development and the history of infilling a basin determined that Chinese on-shore petroleum basins have two distinct characteristics. The first one was the long duration and multi-cycled history of plate tectonic evolution, which created the widely distributed superimposed basins; these superimposed basins consisted of basins that had different tectonic characteristics. In the major petroleum basins, multiple sets of hydrocarbon generating sequences were commonly developed vertically (Table 2.6). The large petroleum basins were developed on three paleo plates that included the North China plate, Yangtze plate and Tarim plate. For example, two sets of the most important petroleum systems that consisted of the Paleozoic marine sequences and the Mesozoic–Cenozoic terrigenous sequences were developed in the following basins: the Bohai bay basin, Eerduosi basin, Sichuan basin and Tarim basin. Secondly, because of the convergence of small terrains and because of strong tectonic activities at the intra-continent, the Chinese marine sequences had poor conditions for preserving oil and gas. As an alternative, the Mesozoic–Cenozoic widely distributed terrigenous basins produced most of the oil and gas in China. The terrigenous faulted basin, terrigenous depressed basin and foreland basin have been major exploration domains and research targets at all times.

Table 2.6 Hydrocarbon source sequences and their basic characteristics in the major Chinese petroleum basins

Basin name	Major hydrocarbon source sequence	Major developmental environment	Lithology of major hydrocarbon source rock	Type of kerogen	Organic carbon (%)	Type of hydrocarbon
Songliao	K_2	Terrigenous depression lake	Lacustrine mudstone	I, II	$1 - 2.2$	Oil
	K_1	Terrigenous rifting lake	Lacustrine mudstone, Coal measure	III, II	$1 - 2.6$	Gas
Bohai Bay	E_{2-3}	Terrigenous rifting lake	Lacustrine mudstone	II, I	$0.8 - 3.8$	Oil
	C-P	Marine– terrigenous facies	Coal measure	III, II	$1 - 3$	Gas
Eerduosi	T_3, J_1	Terrigenous depression lake	Lacustrine mudstone	II, I	$1.2 - 1.9$	Oil
	C-P	Marine– terrigenous facies	Coal measure	III, II	$1 - 6$	Gas
Sichuan	J_{1-2}	Terrigenous depression	Lacustrine mudstone	II, I	$0.3 - 1.0$	Oil
	T_3	Terrigenous foreland basin	Coal measure	III, II	$1.6 - 14.2$	Gas
	P	Craton	Coal measure, Marine mudstone	III, II	$0.4 - 8.0$	Gas
	Є, S	Craton	Marine mudstone	II, I	$0.4 - 3.1$	Gas
Chaidamu	Q_1	Terrigenous foreland lake	Lacustrine mudstone	III	$0.3 - 2.0$	Bio-gas
	E_3, N_1	Terrigenous rifting lake	Lacustrine mudstone	II, I	$0.2 - 0.6$	Oil
	J_{1-2}	Terrigenous depression lake	Coal measure, Lacustrine mudstone	III	$1 - 3$	Coal generated oil
Tuha	J_{1-2}	Terrigenous depression lake	Coal measure, Lacustrine mudstone	III	$1 - 3$	Coal generated oil
	P	Terrigenous foreland basin	Lacustrine mudstone	II, I	$1 - 2$	Oil
Zhungeer	J_{1-2}	Terrigenous depression lake	Coal measure, Lacustrine mudstone	III	$1 - 2$	Coal generated oil
	P_2-T_3	Terrigenous foreland basin	Lacustrine mudstone	II, I	2.5	Oil

(To be continued)

(Table 2.6)

Basin name	Major hydrocarbon source sequence	Major developmental environment	Lithology of major hydrocarbon source rock	Type of kerogen	Organic carbon (%)	Type of hydrocarbon
Zhungeer	C	Marine–terrigenous facies	Coal measure Marine mudstone	III, II	$1-3$	Gas
Tarim	J_{1-2}	Terrigenous depression lake	Coal measure Lacustrine mudstone	III, II	$0.5-5.5$	Gas
	T	Terrigenous foreland basin	Coal measure Lacustrine mudstone	III, II	$0.5-3.5$	Gas
	$\in$-O	Craton	Marine mudstone	II, I	$0.3-1.8$	Oil, Gas
Beibu Gulf	E_{2-3}	Terrigenous rifting lake	Lacustrine mudstone	II, I	$1.7-2.2$	Gas
Pearl River Estuary	N_1	Terrigenous rifting lake	Lacustrine mudstone	II, I	$1-5$	Oil
	E_{2-3}	Terrigenous rifting lake	Lacustrine mudstone	II, I	$1.1-1.4$	Oil
East China Sea	E_3-N_1	Terrigenous rifting basin	Lacustrine mudstone	II, I	1.1	Oil, Gas
South China Sea	E	Terrigenous rifting basin	Lacustrine mudstone	II, I	$0.7-1.7$	Oil, Gas

In China, the source rocks of marine facies were mainly developed during the Paleozoic era and they were commonly distributed in the lower petroleum system in the superimposed basin. The lower Paleozoic erathem contained several kinds of source rocks that included argillaceous rocks of marine facies, shale stone, micrite and algal limestone. In the marine argillaceous rocks, the content of organic carbon was in the range of 1% – 2% in general. Hydrocarbon source rocks of marine facies were mainly distributed in the Tarim basin, Eerduosi basin and Sichuan basin. In the upper Paleozoic erathem, the hydrocarbon source rocks consisted of coal measures that included coal and carbonaceous mudstone. These coal measures were distributed in the Eerduosi basin, Zhungeer basin, Bohai bay basin.

In China, hydrocarbon source rocks of terrigenous facies were widely developed, which included argillaceous rocks of lacustrine facies, coal measures of terrigenous facies and carbonate rocks of lacustrine facies. Dark colored mudstone, shale stone and oil shale were major hydrocarbon source rocks in the Chinese terrigenous lacustrine basin. Most of the proven oil and gas reserves came from the argillaceous hydrocarbon source rocks and the content of organic carbon was in the range of 1% – 3%. Chinese terrigenous hydrocarbon source rocks occurred in seven different geological period or epochs, which included the Late Permian epoch, Late Triassic epoch, Early–Middle Jurassic epochs, Cretaceous period, Palaeogene period, Neogene Period and Quaternary period. In particular, the terrigenous hydrocarbon source rocks were widely distributed in the Middle–Lower Jurassic series, Upper Cretaceous series and Eocene–Oligocene series.

In the Upper Permian series, the terrigenous hydrocarbon source rocks were mainly distributed in the Tarim basin, Zhungeer basin and Tuha basin in the Xinjiang region. The lithology was a set of black color lacustrine mudstone and shale that was interbedded with bitumen oil shale; the thickness of source rocks could reach 1,400 m. In the Upper Triassic series, the hydrocarbon source rocks were mainly distributed in the large scale terrigenous lacustrine basins in Zhungeer, Eerduosi and Sichuan in western-central China; the lithology included lacustrine mudstone and coal measures. In the Middle–Lower Jurassic series, the coal measures functioned as hydrocarbon source rocks that were widely distributed in many petroleum basins that included Tarim, Zhungeer, Tuha, Chaidamu, Eerduosi, Sichuan, Erlian and Songliao. The hydrocarbon source rocks in the Cretaceous system consisted of lacustrine mudstone, which was mainly distributed in the Songliao basin and Erlian basin in northeastern China; the source rocks in the Songliao basin played a major role in oil generation. The hydrocarbon source rocks in the Palaeogene system were distributed broadly. The discovered oil generating depressions were scattered across nearly all of China. In the Eocene–Oligocene series, the hydrocarbon source rocks of near ocean lacustrine mudstone worked as major source rocks in the Bohai bay basin, East China Sea basin and South China Sea basin. In the Neogene and Quaternary systems, the hydrocarbon source rocks were distributed across a small region. At present, these source rocks have only been discovered in some mountain front depressions or in inter-mountain basins in western China, in the southeastern costal region and in the basins of marine territories.

2.2.2   Types of Reservoir Sequences and Their Distributions

The Chinese reservoir sequences included various types of rocks, which were the clastic rocks of marine facies and terrigenous facies, volcanic rocks and metamorphic rocks. The distribution range of reservoir sequences included all Chinese stratigraphic sequences across the geological timespan. In addition, commercial grade oil and gas flows have been discovered in the strata from the crystalline metamorphic rock series in the Archean eonothem to the Quaternary deposits. However, as far as the geological oil reserves are concerned, most of the oil was accumulated in the Mesozoic and Cenozoic terrigenous strata that contained about 90% of the total proven oil reserves in China (Qiu et al., 1997).

In China, the terrigenous clastic rocks were the most important reservoir sequences in the Mesozoic and Cenozoic petroleum basins. They were not only widely developed and distributed in the Songliao basin, Bohai bay basin, Eerduosi basin, Zhungeer basin, Tuha basin, Tarim basin and Sichuan basin, but they also contained most of the discovered oil and gas fields with 80% of the total proven oil reserves. The terrigenous clastic rocks were mainly made of sand bodies of river facies and all kinds of delta facies. Secondly, they also contained sand bodies of underwater fan facies and beach bar facies. The reservoir sequences of sandstone mainly contained a porous type of storage space with a buried depth greater than 2,000 m. The porosity was usually less than 25%, the permeability was less than $1,000\times10^{-3}$ μm^2, which placed the reservoir sequences in the category of medium–low permeability. For the carbonate rocks of lacustrine facies, only a small number of medium–small size oil and gas fields were discovered in the Bohai bay basin and Jiuquan basin.

The marine carbonate reservoir sequences were very important worldwide. However, as a crude oil reservoir, it only contained a very small amount of proven oil reserves in China. On the other hand, the Chinese marine carbonate reservoir mainly contained natural gas. The distribution of Chinese carbonate reservoir sequences was limited, and mainly occurred in the middle and lower petroleum systems in the superimposed basins that were located either in central-western China (the Tarim basin, Sichuan basin and Eerduosi basin) or in eastern China (Bohai bay basin). At present, the targeted exploration domains include the Ordovician system of the Tarim basin and the Upper Permian–Lower Triassic series of the Sichuan basin.

In China, the sandstones of marine facies and marine–terrigenous facies were not only developed very well, but they also functioned as oil and gas reservoir sequences. The marine sandstone was mainly distributed in the Carboniferous system and Silurian system in the Tarim basin, which was made in general of sand bodies of littoral facies. The sandstone of marine–terrigenous facies was mainly distributed in the Carboniferous system–Permian system in the Eerduosi basin, and commonly consisted of large scale sand bodies of river controlled delta facies.

In Chinese petroleum basins, the volcanic reservoir sequences were not only well developed in general, but they were also the important exploration domain at a deep level in the Songliao basin, Bohai bay basin and Zhungeer basin. At present, several large gas fields have been discovered with 10×10^{10} m^3 of gas reserves in each of them. In China, there are four phases of volcanic rocks, which occurred during the Pre-Cambrian period, Paleozoic era, Mesozoic era and Cenozoic era. The volcanic rocks in the Sinian system belonged to the intermediate–basic type and they appeared in the Tarim basin. Most of the Paleozoic volcanic rocks occurred in the upper Paleozoic erathem in western China, such as in the Permian system in the Tarim basin, in the Carboniferous– Permian system in the Zhungeer basin, Tuha basin and Santanghu basin. The Mesozoic volcanic rocks belonged to the intermediate–basic type and the acidic type, which were distributed in the rifted basin of eastern China, such as in the Songliao basin, the Huoshiling formation of the Jurassic system and the Yingcheng formation of the Cretaceous system. The Cenozoic volcanic rocks mostly consisted of basalt and they were mainly distributed in the rifted basin of eastern China, such as in the Neogene system in the Bohai bay basin.

In the Chinese petroleum basins, the types of volcanic reservoir sequences were varied. The lava type of lithology included basal, andesite, dacite, rhyolite and trachyte; the pyroclastic type of lithology included agglomerate, volcanic breccia, tuff and fused pyroclast. In eastern China, the Mesozoic volcanic reservoir sequences were developed during the Late Jurassic–Early Cretaceous epochs. Although the types of lithology crossed the spectrum from the basic type to acidic type, the majority of volcanic rocks belonged to the acidic type. The Cenozoic volcanic reservoir sequences in eastern China were developed in the following locations: the Jiangling depression in the Jianghan basin, the depressions within the Jiyang depression in the Bohai bay basin and the eastern depression within the Lower Liaohe depression. Even the types of lithology included a broad range from the acidic type to the basic type, but the majority of volcanic rocks belong to the intermediate type. In western China, the Carboniferous–Permian volcanic reservoir sequences mainly consisted of the intermediate–basic type of volcanic rocks that were mainly distributed in the Zhungeer basin, Santanghu basin, Tuha basin and Tarim basin. At present, the discovered Chinese volcanic reservoir sequences mainly contain the fractural type of storage spaces.

The metamorphic reservoir sequences are mainly distributed in the Pre-Paleozoic basement in the rifted faulted basins (such as the Bohai bay basin and Erlian basin) in eastern China. Several buried hill type of oil and gas fields have been discovered in the basins, such as the Renqiu oil field in the Bohai bay basin, which utilized the Mesoproterozoic erathem and the Neoproterozoic erathem as the primary reservoir sequences.

2.2.3   Types of Cap Rocks and Their Distributions

The Chinese petroleum basins contain well developed cap rocks. If categorized by lithology, the types of cap rocks includ mudstone (shale), a gypsum–salt layer, salt layer and carbonate rocks. According to the statistical data from 14 natural gas reservoirs (the natural gas reserve of these reservoirs is larger than 100×10^8 m^3), among the immediate cap rocks, eleven of them are made of argillaceous rocks, which represent 79% of total cap rocks. Conversely, three cap rocks are made of the gypsum–salt layer, which represents 21% of the total cap rocks. The regional cap rocks in these 14 gas reservoirs total 13 argillaceous rocks, which represent 93% of the total gas reservoirs; only one gas reservoir utilized a gypsum–salt layer as its regional seal, representing 7% of the total gas reservoirs. Therefore, the argillaceous rocks are the major cap rocks for the Chinese gas reservoirs. These cap rocks are distributed in all of the stratigraphic sequences, excluding the Quaternary cap rocks. Other cap rocks are in the middle–late phases of the digenesis process with dense lithology; the displacement pressure is high and the sealing ability is good.

The argillaceous rocks are widely distributed in almost every depositional environment in huge quantities. In China, most of discovered oil and gas reservoirs are related to the mudstone that not only functioned as cap rock, but also had great thickness. For example, in the Qiongdongnan (Southeastern Qionghai) basin of China, the thickness of the mudstone seal for the Tertiary gas reservoir can reach 294 – 370 m; in the Weiyuan gas field in the Sichuan basin, the thickness of the Cambrian clay shale (that functioned as cap rock for the Sinian gas reservoir) is 230 m. However, if compared with the relatively thicker mudstone seal, the relatively thinner mudstone seal might have equally good (or even better) sealing ability. Along with the increase in buried depth, because of mudstone compaction, the porous water was discharged, the diameter of the pores was reduced and the capillary pressure (resistance) was increased. Therefore, mudstone of lesser thickness received stronger compaction than mudstone of greater thickness. In addition, compared with mudstone of greater thickness at the same buried depth, the mudstone of lesser thickness had a smaller throat diameter, larger displacement pressure and was easy to fracture. Thus, when evaluating the sealing ability of mudstone of lesser thickness, geologists should not only look at the laboratory data, but they should also consider the buried depth, regional stability and the severity of tectonic activity in the area. How thick was the mudstone in order to be a cap rock? Here, we only can offer some extreme examples. In the Liaohe depression, the middle portion of the Sha 3 member ($E^2s_3^{middle}$) in the Shuguang oil and gas field contained a massive conglomerate that was interbedded with mudstone of lesser thickness. Where the mudstone was thinnest (which was also the highest point of the gas reservoir), the thickness of the mudstone was only 2 m. The water layer appeared above the mudstone and a 23.8 m thick gas layer was located below this mudstone. In the Eerduosi basin, north of the Shengli well, the thickness of the cap rock was 4 – 5 m; this cap rock worked as a sealing layer for the gas reservoir that was positioned in the Shihezi formation in the Permian system. Additionally, in the Liujiazhuang area, the thickness of the cap rock for the gas reservoir was only 1.2 – 2.0 m; this gas reservoir was also located in the Shihezi formation in the Permian system. Another example is in the Gudao oil and gas field in the Jiyang depression. The shallow gas reservoir in the Minghuazhen formation in the Neogene system was covered by a direct cap rock of mudstone 0.6 m thick. However, the gas reservoirs that were sealed by thin layered cap rock had smaller gas reserves, lower productivity and lower remaining pressure. Thus, the preconditions for developing a large–middle size gas reservoir did not exist.

The salt layer and the gypsum–salt layer made high quality cap rocks; they had excellent sealing ability and could generally capture a very high gas column. Among the Chinese petroleum basins, the salt layer and the gypsum–salt layer were mainly distributed in the marine strata in the Tarim basin, Sichuan basin and Eerduosi basin. For example, in the Eerduosi basin, the Ordovician gypsum sealing layer commonly had a permeability of $(0.012 - 202) \times 10^{-3}$ μm^2 and a bursting pressure of 0.2 – 1.5 MPa.

Could carbonate rocks function as good cap rock? Geologists still have different opinions. Some researchers think that, because the rock was brittle, easy to be dissolved, easy to be re-crystallized, and because it fractured well, pure carbonate rock could not seal off natural gas. Others consider that because pure carbonate rocks were not only dense and homogeneous but because they also had great thickness and stability horizontally, they could become good cap rock only if they were not altered by strong tectonic movement and the paleo-karsts process. However, most researchers agree that, if it functioned as cap rock for a natural gas reservoir, the important precondition was that carbonate rock must contain argillaceous substances. With an increase in argillaceous content, it became more flexible, harder to dissolve, and less fractured. Thus, carbonate rock could become good cap rock. In the Suqiao–Wenan area in the Jizhong depression, the average thickness of the argillaceous limestone in the Feng 4 member in the Ordovician system (O_1f_4) was 42 m; the content of argillaceous materials was larger than 20% and the porosity was less than 0.01×10^{-3} μm^2. This argillaceous limestone was the perfect cap rock for the Ordovician buried hill type of gas reservoir in this area. In the Eerduosi basin, the test result indicated that, under the conditions of no fracture, no dissolutive pores and no karsts, the dense type of carbonate rocks and argillaceous carbonate rocks had good sealing ability. However, in the areas that had stronger tectonic movement, because they contained well developed fractures, carbonate rocks could only seal off a small gas reservoir. Actually, they were cap rocks with poor sealing ability.

2.2.4 *Development of Sequences and Nature of the Petroleum System*

In China, the combinations of petroleum generation–reservoir–seal are diverse. In western China, the superimposed basins in the Tarim basin and the Zhungeer basin contained oil and gas that were generated in the lower strata and were stored in the upper strata. However, in central-eastern China, the terrigenous petroleum basins mainly contained an in situ type of petroleum system. In different prototypes of basins, the sequence of developments controlled the nature of the petroleum system. The following sections will introduce the sequence developments and related petroleum systems in four major prototype petroleum basins.

● **Terrigenous Depressed Basin**

The terrigenous depressed basin usually had a wide open, gently sloped lacustrine basin and the depositional background of a stable paleo aqua-system for supply of sediments. Therefore, the distributions of the lithology, litho-facies and the depositional thickness were relatively stable. In addition, the depositional system was broadly distributed horizontally. However, the depositional rate was relatively low and the thickness of the third order sequences was commonly small. Therefore, in a depressed basin, the framework of second order sequences and its development directly controlled the distributions of major source rocks and cap rocks. Furthermore, this framework also controlled the combination of petroleum generation–reservoir–seal on a macro scale and the distribution of this petroleum system. In the basin, the framework of the third order sequences only controlled the petroleum system of the next order and related distribution. Via stratigraphic analysis of sequences for the strata that were deposited during the depressed stage in the Songliao basin, Eerduosi basin, Zhungeer basin, geologists discovered that the second order sequences (2 – 3 of them as the major depositional entities) controlled the petroleum system on a macro level in every basin. The major source rocks were developed in the second order sequences in the lower lacustrine transgressive system tract; the regional cap rock was developed in the second order sequences in the middle (or upper) lacustrine transgressive system tract; the favorable delta sand body that functioned as reservoir sequences was developed between these two lacustrine transgressive system tracts. Together, they constructed the middle level petroleum system that contained the most abundant oil and gas in the depressed basin.

Songliao Basin. From the base of the Quantou formation to the top of the Nenjiang formation, there were two complete second order sequences and eight third order sequences. The second order sequences on the lower level basically consisted of the strata from the Quantou formation and the Qingshankou formation, with a geological time interval of 27 Ma approximately; the second order sequences on the upper level were made of the strata from the Yaojia formation and the Nenjiang formation with a geologic time interval of 15 Ma approximately. The geological time interval of each of the third order sequences was in the range 5 – 2 Ma usually. In the Songliao basin, the development of second order sequences controlled the features of the related petroleum system (Fig. 2.6). The reservoir sequences that were made of delta sand bodies were developed in the middle phase of the lowstand and the highstand lacustrine transgressive system tract. The source rock and regional cap rocks were developed in the late phase of the lacustrine transgressive system tract.

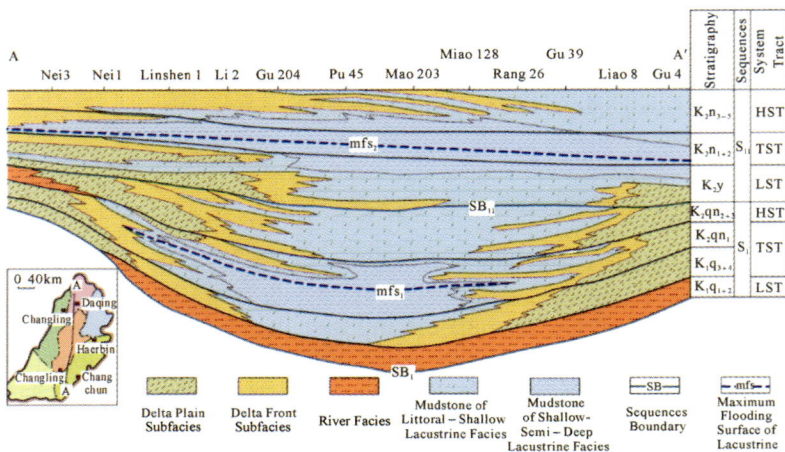

Fig. 2.6. The taxonomy of second order stratigraphic sequences in the Songliao basin during the depressed period (Jia et al., 2008) (with permission from Petroleum Industry Press)

The Qing 1 member, Nen 1 member and Nen 2 member were the second order sequences that contained sediments of the late phase of the lacustrine transgressive system tract. The lacustrine mudstone that was developed during the two phases of the largest flood contained abundant organic materials. This mudstone not only functioned as source rock in the Songliao basin, but also worked as an important regional cap rock. In the Songliao basin, the two sets of regional cap rocks separated three sets of petroleum systems vertically. The lacustrine mudstones in the Nen 1 member and Nen 2 member were deposited during the greatest flooding period and they were considered to be the most important regional cap rocks in the Songliao basin. According to the statistical data, the regional cap rocks in the Nen 1 and Nen 2 members sealed off the oil layers in the Yaojia formation beneath (which included grouped oil layers of Saertu and Putaohua (SP)) and the oil layer in the Qingshankou formation (which included the grouped oil layers of Gaotaizi (G)). The proven oil reserves from these two formations represented 78.1% and 14.8% of the total proven oil reserve in the Songliao basin respectively. Together, the grouped oil layers from these two formations developed the middle level petroleum system that contained 92.9% of the total proven oil reserve in the basin. Another regional cap rock was made of the strata from the Qing 1 member and it sealed off the oil in the lower level petroleum system (which included oil layers in Fuyu and Yangdachengzi (FY)). These oil layers contained 6.9% of the total proven oil reserve in the basin. The upper level petroleum was located above the Nen 1 member, which only contained 0.2% of the total proven reserve in the basin.

Between the two largest flooding periods of Qing 1 and Nen 1, the lacustrine basin expanded and contracted periodically. The delta of the highstand system tract in the Qingshankou formation, the delta of the lowstand system tract in the Yao 1 member and the delta sand bodies of the early stage of the lacustrine transgressive system tract in the Yao 2 and Yao 3 members were developed between the largest flooding periods. These sequences consisted of the high-frequency sequences of the next order that were developed during the next level of lacustrine transgression and regression. The frequent alterations to the Songliao basin and the nature of the high-frequency sequences controlled the combination of storage–seal at the next level. The frequent change in the water level in the lacustrine basin and the alteration in the depositional rate caused the frequent migration of delta lobes. Therefore, in a vertical direction, the sandstone and mudstone were interbedded with each other repeatedly (especially on the front margin of the delta) to create multiple sets of the combination of storage–seal in the next order. In addition, because the sand bodies in the underwater distributary channel on the delta front and the sand bodies in the estuary dam were pinched-out in a lateral position, these were the favorable places to develop the lithologic traps.

Eerduosi Basin. The Late Triassic Yanchang formation, the Early Jurassic Fuxian formation and the Yanan formation represented the structural sequences that were developed during the late Indosinian movement and during the early Yanshanian movement respectively. These strata reflected the completed progression of the creation, development and decrease in the two inner continental depressed basins separately. There were two second order sequences: the lower second order sequences (SSQ1) and the upper second order sequences (SSQ2) (Fig. 2.7). In the lower second order sequences (Yanchang formation, SSQ1), the lacustrine basin experienced a completed developmental process that included the initial growth period (the Chang 10 member), the lacustrine basin expansion period (the Chang 9 – Chang 7 members), and the lacustrine basin decreasing and disappearing period (the Chang 6 – Chang 1 members).

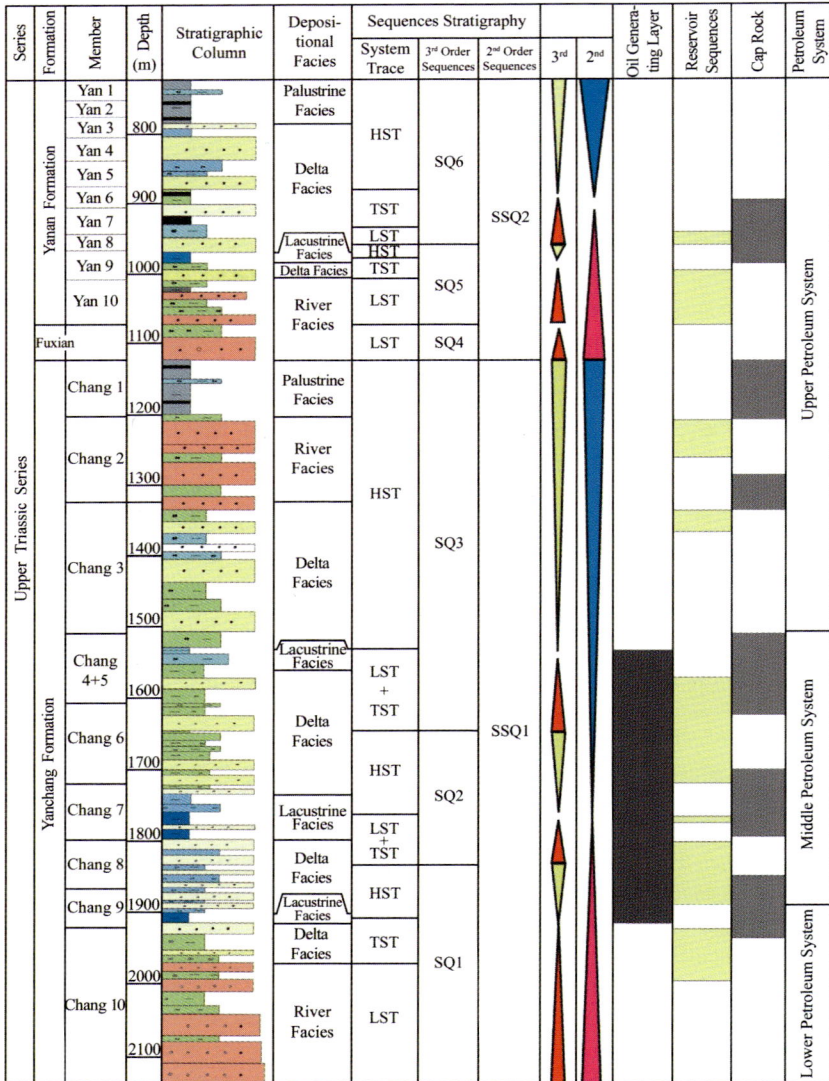

Fig. 2.7. Classification of sequences stratigraphy from the T_3 and related petroleum system in the Eerduosi basin

At present, geologists generally agree that "in the Eerduosi basin, the Mesozoic erathem only contained one set of source rocks". In the second order sequences of the Upper Triassic Yanchang formation, the dark colored mudstone and the oil shale of semi-deep water lacustrine facies were the most important hydrocarbon source rocks in the lacustrine transgressive system tract. In the second order sequences of the Lower Jurassic series, although the lacustrine transgressive system tract did not contain effective hydrocarbon source rocks, the mudstone of lacustrine facies and the marsh peat of delta plain facies in the Yan 9 member and the marsh peat in the Yan 6+7 members created effective cap rocks. Because of the large scale unconformity surface on the top of the Triassic system, the sandstone that filled the incised valley and the fault that connected the Triassic source rocks, the Mesozoic Yanan formation–Yanchang formation were joined together to form a united oil and gas system in the basin. In the second order sequences of the Upper Triassic series, the lacustrine transgressive system tract contained hydrocarbon source rocks; in the Lower Jurassic series, the lacustrine transgressive system tract contained the regional cap rocks. Between these two system tracts, all kinds of delta sand bodies and a river delta functioned as the major reservoir sequences. Together, they created a combination of oil generation–reservoir–seal in the second order sequences. Furthermore, this petroleum system controlled the macro-distribution of oil and gas in the Mesozoic erathem in the basin.

According to the characteristics of the Mesozoic petroleum system and exploration practice, the strata in the Yanchang–Yanan formation were divided into three petroleum systems (or three combinations of oil generation–reservoir–seal), which were the lower combination, middle combination and upper combination. In the lower combination, the reservoir sequences consisted of sand bodies of river facies and delta facies that were positioned in the Chang 10 member. The source rock and the cap rock were the mudstone of lacustrine facies in the Chang 9 member. This was a typical combination that generated oil in the upper strata and stored oil in the lower strata. The middle combination contained grouped oil layers in the Chang 9–Chang 4+5 members. The mudstone of lacustrine facies not only functioned as source rock, but it also worked as cap rock. Among the mudstones, the mudstones that were deposited during lacustrine transgression worked as regional cap rocks, which occurred in the Chang 9, Chang 7 and Chang 4+5 members. All kinds of delta sand bodies and the turbiditic sand body were the reservoir sequences. Together, the mudstone and sandstone created an in situ type of combination. The upper combination contained grouped oil layers in the Chang 3–Yan 1 members and the source rock came from the middle combination. The Chang 1 member, Yan 9–Yan 6 members functioned as regional cap rocks; the sand body of river facies in the Chang 2 and Yan 10 members and the sand body of delta facies in the Chang 3 and Yang 9 members were the reservoir sequences. This was the combination that generated oil in the lower strata and stored oil in the upper strata.

In these three combinations, the lower combination had poorly accumulated oil that only represented 1.2% of the total proven oil reserves in the Mesozoic erathem in the basin. The middle combination contained most of the oil that represented 59.9% of the total proven oil reserves in the Mesozoic erathem in the basin. The upper combination mainly had the structural–lithologic type of oil reservoirs that contained 39.9% of the total proven oil reserves in the Mesozoic erathem in the basin.

● **Terrigenous Faulted Basin**

In the terrigenous faulted basin, the creation of stratigraphic sequences and the characteristics of their components were controlled by various factors. The episodic tectonic movement was the most important control factor, which controlled the development of sequences and the overall characteristics of the petroleum system. Because the episodic movement of the basement fault caused periodic expansion and contraction of the storage space and the lacustrine region, the basin contained multi-cycled subsided events and depositional events, which created several sets of sequences, multiple sets of the combination of oil generation–storage–seal, and more than a few oil and gas layers. The multi-phases of episodic tectonic movement controlled the developments of the multi-ordered stratigraphic sequences and multi-classified petroleum systems. In the same basin, the variations in fault activity resulted in different combinations of oil generation–reservoir accumulation in various areas. In other words, fault activity controlled the variation in the major oil and gas layers not only in the structural type of oil reservoir, but also in the lithostratigraphic type of oil reservoir.

In eastern China, the Palaeogene system in the Bohai bay basin was a typical faulted basin. The episodic rifted movement of several orders controlled the development of stratigraphic sequences in multiple orders. The episodic subsidence movement of the basement fault was the primary control factor in the development of sequences in 1 – 4 orders in the terrigenous faulted basin (Chi et al., 1996). Correlated with tectonic episodic orders, in the Bohai bay basin, the depositional sequences and the petroleum systems of the Palaeogene System and Neogene System could be divided into four orders (Fig. 2.8).

During the Palaeogene period, the Bohai bay basin experienced three phases of second order episodic rifted movements; in correlation, three second order sequences were developed. Every second order sequence contained a completed depositional cycle from water transgression to water regression. A completed second order sequence contained four system tracts with distinct depositional features; these system tracts included the alluvial fan system tract, the water transgressive system tract, the water regressive system tract and the river flood plain system tract. However, some second order sequences might miss one or two system tracts. The following are the features of composition. (1) During the early stage of subsidence in a rifted basin, the alluvial fan system tract was commonly developed at the base of the first set of sequences. The meandering river–flood plain system tract was usually developed on the top of the last set of sequences at the rifted stage. These two types of system tract were poorly developed during the mid-term of the episodic rifted movement. (2) Because the terrigenous faulted basin had a higher depositional rate and a lower level of sediment sorting, the third order sequences normally contained a "half cycle" deposit. In most situations, the different types of system tract were hard to identify; therefore, we named the third order sequences "semi-sequences group". (3) The subsidence center of a large terrigenous rifted basin had clearly migrated. Thus, every set of sequences in each depression might have very different developmental features. For example, in the Bohai bay basin, the subsidence center of the Palaeogene period had migrated from onshore to marine territory. As a result, the developmental features of sequences in each depression were varied.

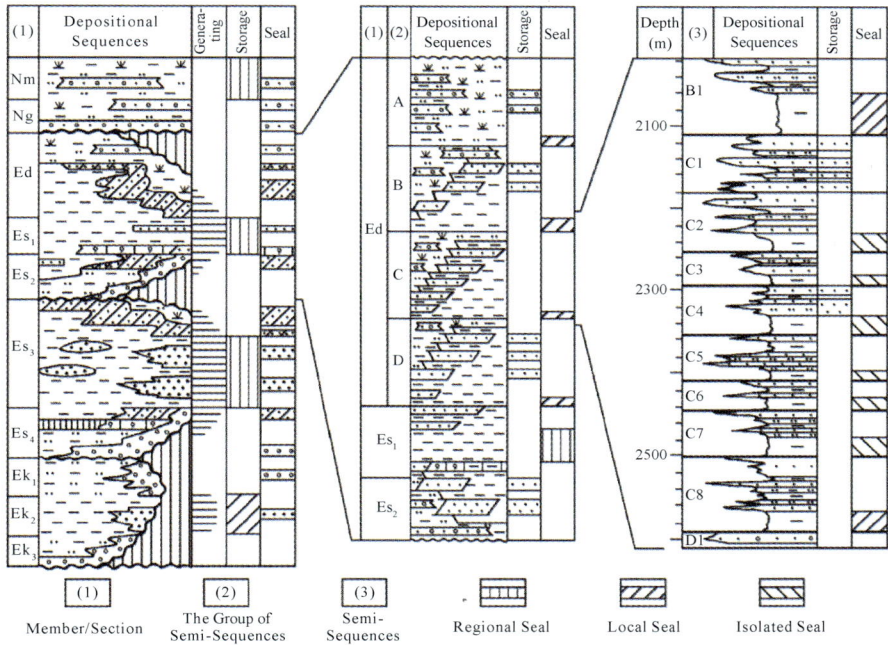

Fig. 2.8. Sequences stratigraphy of the Tertiary system and related petroleum system in the Bohai bay basin (Chi et al., 1996) (with permission from Acta Petrolei Sinica)

The Bohai bay basin was a rifted type of petroleum basin. The rifted basin was developed during the Palaeogene period; the lacustrine strata were deposited in the Kongdian formation–Dongying formation. The typical features of these sequences included well developed hydrocarbon source rocks, varied types of reservoir sand bodies and a combination of oil generation–storage. Therefore, each major depression commonly contained 3 – 4 sets (at least 2 – 3 sets) of oil generating and storage sequences that were superimposed vertically (Fig. 2.9).

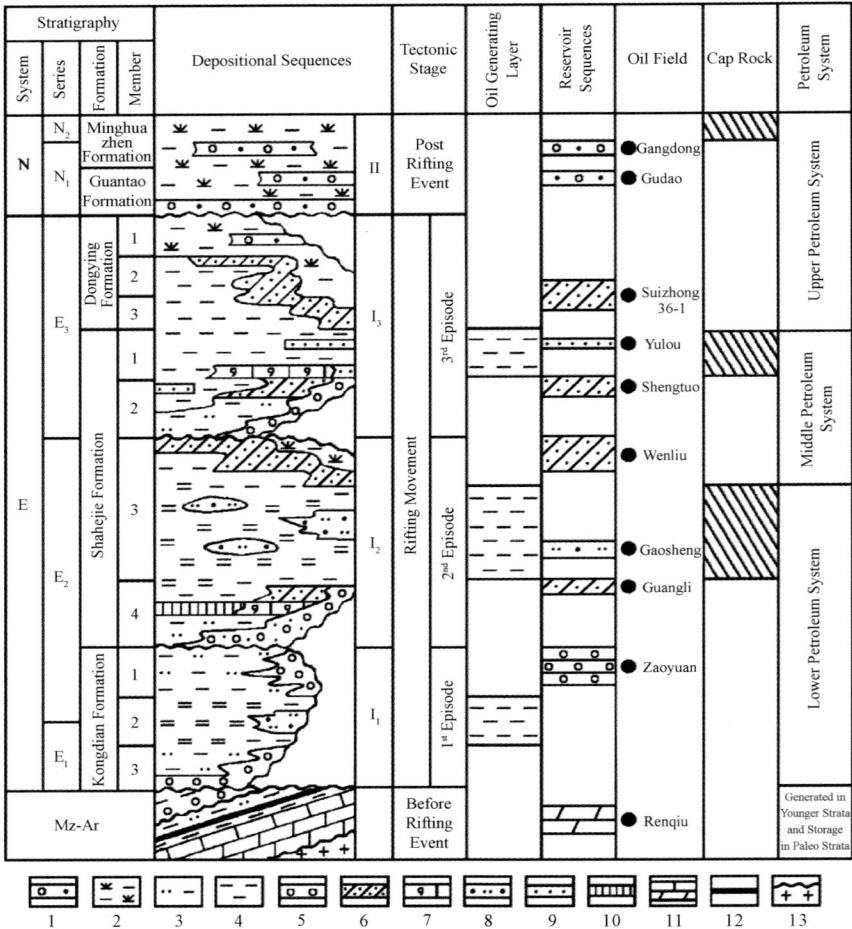

Fig. 2.9. Composite diagram of depositional sequences and related petroleum system in the Bohai bay basin

The results of basin modeling indicated that the Sha 3 member had the major source rocks for each petroliferous depression, which contained 54.02% of the total oil resource and 44.9% of the natural gas resource. Secondly, the Sha 4 member and the Kongdian formation also functioned as source rocks, which contained 24.74% of the total oil resource and 13.36% of the natural gas resource. Thirdly, the resource sequences on the list of source rocks were the Sha 1 and Sha 2 members. The hydrocarbon generating ability of the Dongying formation was poor; the Dongying formation was mainly distributed in the Liaohe area and in the beach flat areas of the Huanghua depression. Because the buried depth of source rocks was relatively shallow, at present the organic materials are still at an immature–low maturity stage. Because, in the Bohai bay basin, the depositional center migrated toward the marine territory, in the Liaohe area and the Huanghua depression, the highest oil generating sequence was the Dongying formation. However, in the Jiyang, Jizhong and Linqing depressions, the highest oil generating sequences were the Sha 1 and Sha 2 members. In most of the depressions, the Sha 3 member was the major oil generating sequence. In a few depressions, the Sha 4 member and the Kongdian formation functioned as major source rocks. Thus, in every depression, the in-depth exploration that targeted major oil generating sequences will be the determinating factor in the future.

According to the development of stratigraphic sequences and the geological settings of oil generation–storage–seal, in the Bohai bay basin, the Tertiary System was divided into three oil generating combinations at the upper, middle and lower levels. The Kong 2 member, the middle and the lower portions of the Sha 3 member and the Sha 1 member were the three sets of major source rocks in the Palaeogene system. The lacustrine mudstone in the Sha 3 and the Sha 1 members was widely distributed in the entire basin. Combined with the upper member of the Minghuazhen formation in the Neogene system, this created three major regional cap rocks in the basin. The Pre-Tertiary buried hill type of oil field was sealed by the Tertiary mudstone of lacustrine transgression, which was a unique combination that "generated oil in younger strata and stored oil in the older strata". The middle and lower oil generating combinations belonged to the in situ type with the most abundant oil and gas reserves, which contained 63% of the total proven reserves. The upper oil generating combination generated oil in the lower strata and stored oil in the upper strata, which contained 26% of the total proven reserves. The Pre-Tertiary oil generating combination generated oil in the upper strata and stored oil in the lower strata, which contained 11% of the total proven reserves. For every individual depression, because of the differentiations in the geological conditions and because of the influence of depositional center migration in the Bohai bay basin, the reservoir sequences were varied. Relatively speaking, in the beach flat area of the Bohai bay basin, reservoir sequences of a younger age appeared in the Huanghua, Jiyang and Bozhong depressions, which contained the upper combination of oil generation. Most of the proven reserves were in the Neogene system.

● **Terrigenous Foreland Basin**

In the foreland basin, the depositional sequences were mainly controlled by periodic activity of the thrusting orogenic belt. Because the foreland basins in different areas were created at different times, the combination of oil generation–reservoir–seal varied. According to the time of creation of the foreland basin and the characteristics of superimposed alteration in the late period, in central-western China the foreland basins could be divided into three types: the early period type, the late period type and the superimposed type. Comparing all types of foreland basin, there were differences in the geometry of the basin, the nature of the structure and depositional sequences. Therefore, different types of foreland basin contained different petroleum systems. In the same foreland basin, because the episodic thrust movement with multi-phases alternated with relatively quiet periods, several sets of petroleum combinations were developed vertically.

The early period type was represented by the Late Triassic foreland basin in western Sichuan (Fig. 2.10). According to the characteristics of the boundary of depositional sequences and the combinational relationship of the depositional cycle, in the western Sichuan foreland basin, the Upper Triassic series could be divided into four individual 3rd order sequences and twelve system tracts. Sequence I corresponded to the Xu 1 member; sequence II was compatible with the Xu 2 and Xu 3 members; sequence III matched with the Xu 4 member; sequence IV equalled the Xu 5 member, approximately. The lowstand system tract, the lacustrine transgression system tract and the highstand system tract were developed in each sequence. Usually, the lowstand system tract contained a sandy conglomerate and sandstone of alluvial–river facies; the lacustrine transgressive system tract contained mudstone or sandstone of lacustrine–delta facies; the highstand system tract contained delta deposits of thin layered sandstone that were interbedded with mudstone and coal layers. During the largest flood period, the black colored mudstone, coal layers and carbonaceous mudstone were commonly developed. In the western Sichuan foreland basin, each sequence in the Upper Triassic Xujiahe formation had a distinctive depositional cycle that was represented by grain size alteration, going from coarse grain to fine grain to coarse grain vertically.

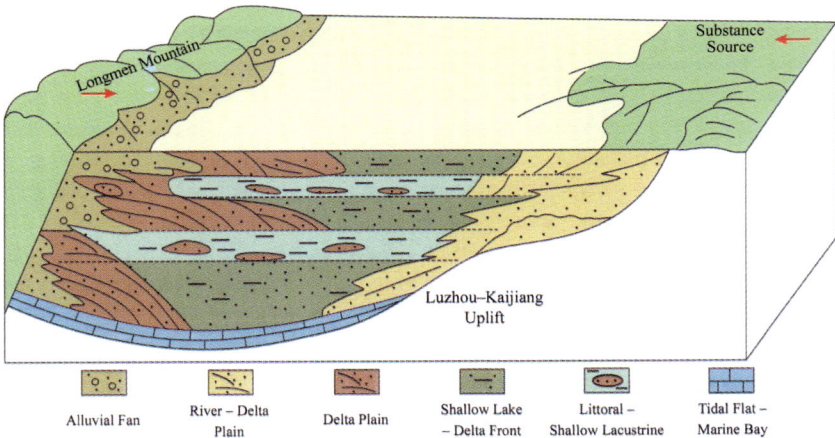

Fig. 2.10. Model of a foreland basin petroleum system during the Upper Triassic epoch in the western Sichuan basin

The major hydrocarbon source rock in the Upper Triassic series in the western Sichuan foreland basin included all sequences, from sequence I to sequence IV. In sequence I, the hydrocarbon source rocks included mudstone of marine–terrigenous facies in the lowstand and highstand system tracts. In sequences II, III, and IV, the source rocks included clay shale of littoral–shallow water lacustrine facies and semi-deep water lacustrine facies in the lacustrine transgressive system tract and in the highstand system tract. Vertically, in the western Sichuan foreland basin, the Upper Triassic series could be divided into three sets of the combination of oil generation–reservoir–seal. In the Xu 1–Xu 3 members, the discovered combinations of oil generation–reservoir–seal included the gas reservoirs in the Pingluoba, Baimamiao, Moxi, Zhongba and Xinchang areas. In the Xu 3–Xu 5 members, the discovered combinations of oil generation–reservoir–seal included the gas reservoirs in the Bajiaochang, Moxi, Guangan and Chongxi areas. In the Xu 5–Xu 6 members, the discovered combinations of oil generation–reservoir–seal included the gas reservoirs in the Moxi, Guangan and Longnusi areas.

The late period type was represented by the Tertiary System on the northern margin of the Chaidamu basin. Two individual second order sequences and fourteen third order sequences were developed in the Palaeogene and Neogene systems. In the Chaidamu foreland basin, during a period of strong tectonic movement, the storage space and water depth were continually increasing. The aggradational type or regressive type of semi-sequences group was developed during the early phase; the progressive type of semi-sequences group was developed during the late phase. In humid climate conditions, the coarse grain sand body of alluvial fan facies was deposited on the mountain front. On the other hand, relatively fine grain sediments were deposited in the depression; the lithology mainly included gray, grayish green colored mudstone, silty mudstone and sand bodies of delta facies and beach bar facies in the littoral–shallow water lacustrine basin. The sand body was distributed over a relatively large area. For example, the upper portion of the Lower Ganchaigou formation contained two third order sequences and all of them contained the deposits of strong tectonic movement in the humid climatic environment. In arid climatic conditions, the area of the water body was largely changed. On the mountain front, the coarse grain sand body of alluvial fan facies was developed. In the depression, the sand body was distributed over a relatively small area; the sand body of delta front facies was limited to the edge of the lacustrine basin. On the margin of the depression and the higher ground of the inner depression, thin bedded sandstone and a carbonate beach were developed in the shallow water zone. In Chaidamu, the Tertiary foreland basin contained an in situ type of petroleum system; the discovered Neogene oil reservoirs were located in the Nanbaxian, Lenhu #4 and Lenhu #5 areas.

The superimposed type of foreland basin was represented by the one on the western margin of the Zhungeer basin. During the Cenozoic era, because of the reactivation of the orogeny, a rejuvenated foreland basin was superimposed on top of the Permian foreland basin. Therefore, in a vertical direction, there were three extensional periods and three compressive periods, which occurred alternatively. Because each tectonic cycle (from extension to compression) controlled the development of one 1st order sequence, there were three 1st order sequences of the C-P; the T-J and the J-R. The subsidence–depositional centers of every sequence were varied. During the Permian period, the thrust activity on the northwestern margin was strong; the subsidence–depositional centers of sequence I were located at the Mahu depression; the subsidence–depositional centers of sequences II and III were located at the mountain front depressions of the northern and southern orogeny belts. In particular, during the Cenozoic era, because the southern margin experienced intensified thrust activity, in the Fukang depression on the southern margin the thickness of sediments was great, which caused most of the Zhungeer basin to incline towards the south. Every first order sequence would contain several second order sequences; from C-P to the Cenozoic erathem, there were nine second order sequences. Because each second order sequence was a completed depositional cycle of water transgression–regression, multiple sets of petroleum systems were identified.

The northwestern margin contained the most abundant oil and gas in the Zhungeer basin. On the northwestern margin, the multi-cycles of tectonic evolution and depositional development created several sets of petroleum systems. In descending order, three sets of regional cap rocks were developed in the Lower Cretaceous series, the Lower Jurassic series and the Upper Triassic series with stable distributions. In the Baijiantan formation in the Upper Triassic series, the thickness of the mudstone was 100 – 400 m. In most areas, the thickness of the mudstone was 70 – 110 m. That which occurred in the upper portion of the Sangonghe formation in the Lower Jurassic series and the Lower Wuerhe formation in the Middle Permian series not only functioned as source rock, but also worked as the regional cap rock. Therefore, on the northwestern margin, there were three petroleum systems at the upper, middle and lower levels (Fig. 2.11).

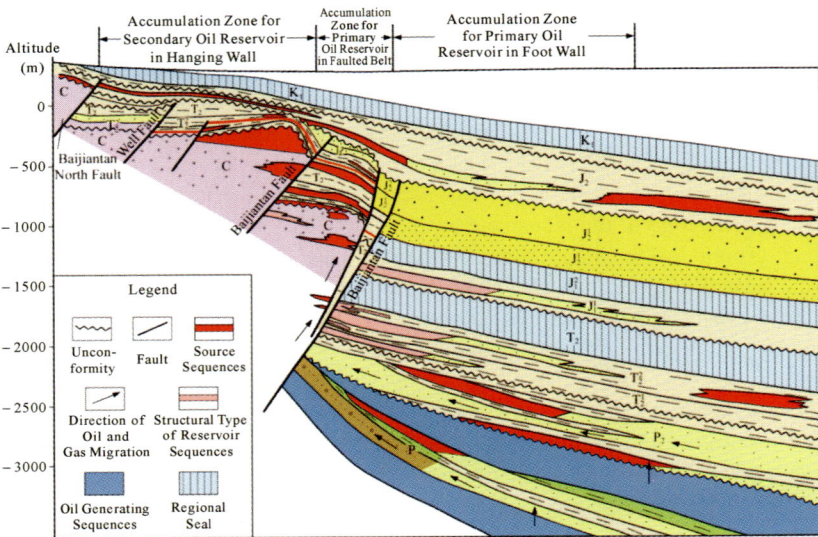

Fig. 2.11. Distribution model of petroleum reservoirs and related petroleum system on the northwestern margin of the Zhungeer basin

The most important mudstone occurred in the Baijiantan formation in the

Triassic system, which was deposited during the lacustrine transgression with stable distribution. The in situ type of lower petroleum system was the important exploration target in the region. On the northwestern margin, the discovered geological reserve of oil was 12×10^8 t; 74% of the oil reserve was in the Triassic and Permian systems.

● **Marine Craton Basin**

In China, three large marine craton petroleum basins were developed in the Tarim basin, Sichuan basin and Eerduosi basin. Against the background of two huge cycles of marine transgression and regression and against a regional tectonic background of extension and flexure, the Paleozoic erathem experienced two large stages of basin development, which were the development of the lower Paleozoic marine craton basin and the development of the upper Paleozoic marine–terrigenous basin (Zhao et al., 2002). Each stage not only contained a similar development of sequences, but it also had a comparable depositional combination. Among these basins, the Sichuan basin contained the best developed stratigraphic sequences. The marine deposits appeared in the strata from the Cambrian–Triassic systems and they could be divided into ten second order sequences (Fig. 2.12).

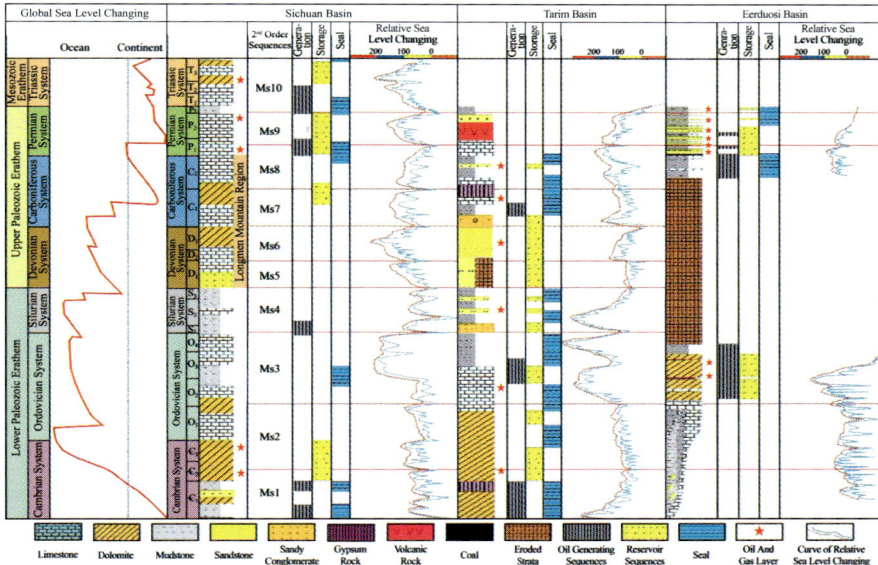

Fig. 2.12.   Development of sequences in craton basins from the Paleozoic era–Triassic period and related petroleum systems in the Sichuan basin, Tarim basin and Eerduosi basin

During the early Paleozoic era, China experienced a worldwide gigantic cycle of rises and falls in the sea level. The Sichuan basin experienced four cycles of rises and falls in sea level. These events happened during the C_1-C_2 epochs,

C_2-O_1 epochs, O_2-O_4 epochs and S_1-S_3 epochs. The Tarim basin experienced continual marine transgression from the Sinian–Middle Ordovician epoch and continual marine regression from the Late Ordovician epoch–Silurian–Devonian periods. The largest marine transgression occurred during the Middle Ordovician epoch. The Eerduosi basin experienced two cycles of sea level rises and falls. According to the cycles of sea level rises and falls, the early Paleozoic strata were divided into four second order sequences that occurred in the Lower Cambrian– Middle Cambrian series, the Middle and Upper Cambrian series–Low Ordovician series, the Middle and Upper Ordovician series and the Silurian system.

During the late Paleozoic era, most of the deposits belonged to the marine–terrigenous facies with distinct variations. During the middle of the Ordovician period, the Eerduosi basin was uplifted to become a terrane; the marine transgression did not happen until the Dala phase in the Late Carboniferous epoch. The direction of the marine transgression was from east to west. During the early phase of the Early Permian epoch, the sea water came from a southeast direction and the water submerged the entire basin. During the Middle Permian epoch, the sea water started to regress and the Eerduosi basin was gradually transformed into a terrigenous basin. In the Sichuan basin, during the Devonian period, because a faulted-depression occurred in the eastern section of the Paleo-Tethys ocean, the sea level was rising. This geological event continued until the Early Carboniferous epoch. From the Permian–Triassic periods, the overall developmental trend of paleo geography was the expansion of marine territory and submerging terranes. During the late Paleozoic era, the largest sea level uprising event happened in the Early Permian epoch. In the Tarim basin, the marine transgression started in the Late Devonian epoch. During the Early Permian epoch, the marine transgression was on the largest scale. During the Middle Permian epoch, the sea level was falling and volcanism occurred over broad areas. Most of the region was uplifted and the deposits were transformed rapidly into terrigenous facies. According to the sea level alterations, the strata of the late Paleozoic era could be divided into five second order sequences, which occurred in the Lower Devonian series, the Middle and Upper Devonian series, the Lower Carboniferous series, the Upper Carboniferous–lower portion of the Lower Permian and the upper portion of the Lower Permian–Middle Permian series. In the Sichuan basin, there was another second order sequence in the Upper Permian series–Lower and Middle Triassic series. The base boundary of this second order sequence could correlate with the Dongwu movement. In the Sichuan basin, the sea water totally regressed during the Middle Triassic epoch and it was gradually transformed into a terrigenous basin.

According to the similarity in sequence development and the parallel depositional combination, four sets of petroleum systems were developed in the Chinese marine craton basins.

Cambrian Combination. The Cambrian carbonate rocks were widely distributed in the Chinese marine craton basins that worked as hydrocarbon source rocks and reservoir sequences. Combined with the cap rocks of mudstone or the gypsum–salt layer above, the combination of petroleum generation–reservoir–seal was created.

For example, in the Tarim basin, the Lower Cambrian dolomite combined with the Middle Cambrian gypsum–salt layer to create a petroleum system. The Cambrian limestone (dolomite) combined with the Ordovician mudstone above to create another petroleum system. In the Sichuan basin, the Sinian algal dolomite (as a reservoir) combined with the Lower Cambrian clay shale (as a seal) to create a petroleum system. The Middle and Upper Cambrian dolomite (as a reservoir) combined with the Lower Ordovician clay shale (as a seal) to create another petroleum system. In the Sichuan basin, oil and gas were discovered in this kind of petroleum system.

Ordovician Combination. The Caledonian movement significantly altered the Chinese marine craton basin. The strata were uplifted and eroded. The weathering process and eluviation developed the dissolutive pores and holes in the Ordovician carbonate rocks that were widely distributed and that functioned as reservoir sequences. At the same time, the reef flat type of reservoir sequences was developed on the platform margin. These carbonate rocks combined with cap rocks (that included the dense argillaceous rocks or gypsum mudstone) above to create a favorable combination of reservoir–seal. For example, in the Eerduosi basin, the favorable reservoir sequences of dolomite (that contained dissolutive pores and holes and that occurred in the Ordovician karst type of gypsum dolomitic flat) combined with the cap rocks above to create a petroleum system. The cap rocks included the Middle–Upper Ordovician anhydrite argillaceous rocks and the Carboniferous bauxite in the Benxi formation. In addition, several sets of reservoir–seal combinations were developed in the upper Paleozoic erathem. In the Tarim basin, the karst type of reservoir sequences on the Ordovician weathering crust combined with the Carboniferous mudstone in the Bachu formation to create a reservoir–seal combination. Additionally, the Lower Ordovician reservoir sequences in the reef flat combined with the Middle–Upper Ordovician argillaceous limestone to create another combination of reservoir–seal. At present, large quantities of oil and gas are being discovered in this kind of reservoir–seal combination in the Chinese marine craton basins, such as in the Tazhong oil and gas field in the Tarim basin and the large gas field of Jingbian in the Eerduosi basin.

Carboniferous–Permian Combination. The Caledonian movement caused a large magnitude uplifting in craton basins. In three major marine craton basins, the strata of the Devonian–Carboniferous periods were absent in various degrees. The marine transgressions that happened during the Late Devonian epoch–Carboniferous period were initiated at different times in every basin. Because of long term exposure and erosion, peneplanation gradually took place. The clastic reservoir sequences and coal measures were developed during the early phase of the marine transgression. With the increase in transgressive magnitude, a carbonate platform was developed. As a result, widely distributed carbonate reservoir sequences were created. The Carboniferous–Permian systems contained three types of combination (reservoir–seal). The first one was the combination of carbonate rocks. For example, in the Sichuan basin, the Carboniferous dolomite (as a reservoir) combined with mudstone in the Liangshan formation (cap rock). In addition, there was an in situ type of combination in the

Middle Permian series. The Puguang and other gas fields were discovered in the first type of combination. The second one was the combination of coal measures and sandstone reservoir sequences. For example, in the Eerduosi basin, the in situ type of combination was developed in the coal measures of marine–terrigenous facies. The Sulige gas field was discovered in the second type of combination. The third one was the combination of marine sandstone, such as in the Tarim basin, the carboniferous Donghe sandstone. The Hadexun oil field was discovered in the third type of combination.

Triassic Combination. This combination was mainly developed in the Sichuan basin; the reservoir sequences included oolitic dolomite in the Feixianguan formation, the dolomite in the Jialingjiang formation, the dolomite in the Leikoupo formation. In correlation, the excellent cap rocks of gypsum in the Jialingjiang formation and the mudstone in the Upper Triassic series were developed. In the eastern Sichuan basin, several medium size gas fields were discovered in this kind of combination in the Feixianguan formation.

2.3   Features of Depositional Evolution in the Major Petroleum Basins

In China, the sedimentary rocks covered 670×10^4 km^2 of land; in total, there are 505 sedimentary basins that consisted of 424 terrigenous sedimentary basins, 12 marine sedimentary basins and 69 superimposed basins of marine and terrigenous facies. At present, drilling exploration is being carried out in 86 basins. As a result, 79 of them are petroleum basins. Among the petroleum basins, 32 of them have commercial grade oil and gas flows; 11 of them have low production rate oil and gas flows; 36 of them have oil and gas shows. The following is a basic introduction to the nature of depositional evolution in major petroleum basins in China (Zhai, 1987; Qiu et al., 1997; Jia et al., 2008).

2.3.1   Songliao Basin

The Songliao basin is located in northeastern China, which crosses over three provinces (Heilongjiang, Jilin and Liaoning) with a total area of 26×10^4 km^2. In China, the Songliao basin is one of the petroliferous onshore basins, which contains 144×10^8 t of oil resources and 2.4×10^{13} m^3 of natural gas resources. At the end of 2007, the proven oil reserve was 72.04×10^8 t and the proven gas reserve was $3,606.56 \times 10^8$ m^3.

The Songliao basin is a typical intra-craton terrigenous basin. The basement of the basin was made of Paleozoic–early Mesozoic metamorphic rocks, Hercynian and Yanshanian granites and intermediate–basic intrusive rocks. The sedimentary strata of the basin basically experienced four stages of tectonic movement that included thermal uplifting and a rifting stage (J_3), a rifting faulted stage (K_1), a depression stage (K_2) and a contractive folding stage (E-Q). The Cretaceous system was the major entity of sedimentary infill in the basin. During the Early Cretaceous, because the basin was in the rifting faulted stage, a series of faulted lacustrine basins were developed and they persisted in the deep level of the Songliao basin. Accordingly, a set of coal measures that worked as source rocks were deposited. In addition, the reservoir sequences that consisted of volcanic rocks and sandy conglomerate were also deposited. These sequences are important targets for natural gas exploration at present. The Late Cretaceous epoch was not only the depression phase of the post-rifting period for the Songliao basin, but it was also the major depositional phase for the basin. During the prime period for the development of the depression, the Qingshankou formation–Nenjiang formation were deposited over a broad distribution region and at a great rate of subsidence. The lacustrine hydrocarbon source rocks and large size sand bodies of river–delta facies were developed, which were the primary targets for petroleum exploration. Several large and medium size oil fields were discovered; the Daqing oil field was a typical example. At present, the synclinal region is the major exploration domain for the lithologic type of oil reservoir.

The Songliao basin was a large scale terrigenous depressed basin; the Upper Cretaceous series was the primary geological entity. In ascending order, the Upper Cretaceous series included the Denglouku formation, Quantou formation, Qingshankou formation, Yaojia formation, Nenjiang formation, Sifangtai formation and Mingshui formation. The Quantou formation–Nenjiang formation represented the major depositional entity of this time, which included two complete second order sequences and eight third order sequences (Fig. 2.13).

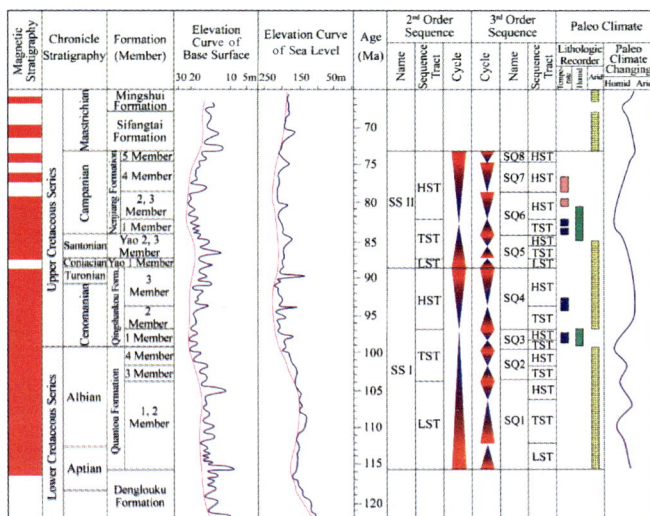

Fig. 2.13. Comprehensive proposal for classifying sequences stratigraphy in the Quantou formation–Nenjiang formation in the Songliao basin

The depositional system of the Denglouku formation and the Quan 1–Quan 2 members belonged to river alluvial facies; the function of these sequences was to fill up and to even out sags. The shallow water river–delta system was developed during the time that Quan 3–Quan 4 members were deposited. Together, they created the lower petroleum system in the Songliao basin. The reservoir sequences included the Fuyu oil layer, the Yangdachengzi oil layer and the Nongan oil layer. When the Qingshankou formation–Nenjiang formation was deposited, the depositional system was fully developed and was symbolized by the large river–delta depositional system. During this time, lacustrine transgressions occurred twice. Because of large scale lacustrine flooding, the deep water lacustrine basin was formed and the supply of sediment was inefficient. As a result, black color mudstone was deposited, which contained abundant organic materials and which became excellent hydrocarbon source rock. Also, during this time, perfect sandstone reservoir sequences were developed in the large river–delta system when the lacustrine deposits contracted. These sandstones formed the middle and upper petroleum systems in the Songliao basin; the reservoir sequences included the Gaotaizi oil layer, Putaohua oil layer, Saertu oil layer and Heidimiao oil layer. When the Sifangtai formation and the Mingshui formation were deposited, the depression stage came to the end. The basin was transformed from subsidence in a vertical direction to compression in a lateral direction. Furthermore, the depositional region decreased. The depressed center migrated toward the west, to the west of the Qijia–Gulong depression and the Changling depression.

When the Qingshankou formation–Nenjiang formation were deposited, the Songliao basin was a large scale intra-continental fresh water lacustrine basin. From the edge of the basin to the center, the depositional facies included alluvial facies, river facies, delta facies, littoral–shallow water lacustrine facies and semi-deep water–deep water lacustrine facies (Fig. 2.14). Surrounding the lacustrine basin, there were six major aquatic systems that supplied clastic materials to the basin; among these aquatic systems, the Northern aquatic system and the Baokang aquatic system were the major suppliers and they paralleled the long axis of the basin. On the western side, there were a series of aquatic systems that had a short axis. Horizontally, these six major aquatic systems created six depositional systems of alluvial fan–river–delta facies. The Northern delta system and the Baokang delta system on the southwestern side were long axial river delta systems that had great magnitude and a broad distribution range.

● Northern Depositional System

The Northern depositional system was the largest depositional system in the Songliao basin, which stretched along the long axis of the basin and extended into the lake. Two rivers that came from the Beian and Nahe areas brought sufficient clastic materials. Horizontally, this depositional system spread out far and wide and it formed the highly constructive delta deposits on the Daqing placanticline and on both sides of it. Under the efforts of difference compaction and tectonic movements, the Daqing placanticline structure was formed. This structure offered a highly favorable place for oil and gas to accumulate. In the Songliao basin, the Northern aquatic system created the most favorable depositional system for oil and gas accumulation. In the Daqing oil field, the major oil production sequences were developed in the sandstone of the Northern delta system. From north to south, the Northern depositional system included alluvial fan facies, river facies, delta facies, littoral–shallow water lacustrine facies and semi-deep water–deep water lacustrine facies.

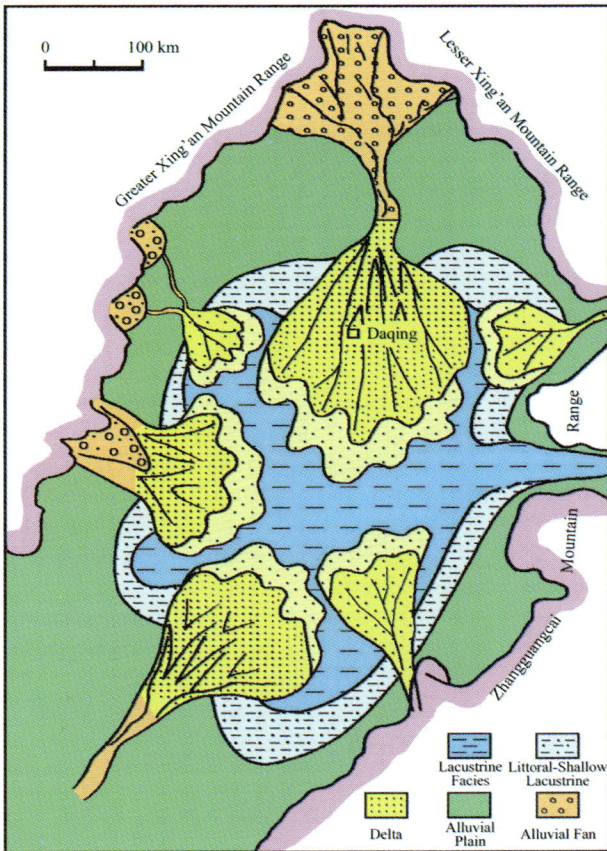

Fig. 2.14. Depositional systems in the Songliao basin during the depressed period

● **Yingtai Depositional System**

The Yingtai depositional system mainly concerns the aquatic system from the Tailai–Baicheng area, which extended into the lake along the short axis of the basin. This depositional system contained multiple orders of sub-depositional systems because of between four and six intermittent inter-mountain braided rivers that extended into the lake in parallel fashion. The Zhenlai–Yingtai sub-depositional system and the Baicheng–Honggang sub-depositional system were on a relatively large scale. The aquatic system in the Zhenlai–Yingtai area formed a fan delta system that was near the source center with a short transportation distance and was perpendicular to the long axis of the basin. The aquatic system in the Baicheng–Honggang area created a braided river delta system. The Yingtai depositional system included alluvial fan facies, braided river facies, braided river delta facies, littoral–shallow water lacustrine facies and semi-deep water–deep water lacustrine facies.

● Tongyu–Baokang Depositional System

The Tongyu–Baokang depositional system was located on the southern side of the Songliao basin and it extended into the lake along the long axis of the basin around the Tongliao and Baokang areas. This system crossed over four of the first order tectonic units that included the Kailu depression, the southwestern uplift, the western slope and the central depression and it was distributed in the Tongyu, Baokang, Qianan and Changling areas. In the Qingshankou formation, the geometry of the sand body, the axial direction of the sand body and the distribution range of the sand body were stable. The orientation of the sand body was nearly parallel to the long axis of the basin. The sand belt extended from the Seven 1 well toward the Qianan area and it was divided at the Changling area and at the Daqingzi well district. During the time that the Qing 1 and Qing 2 members were deposited, a high concentration sand belt reached an area near the Qian 110 well district. During the time that the Qing 3 member was deposited, the sand belt not only pushed in a northeast direction to reach the Cha 22 well district, but it also formed a widely distributed delta sand body in the Qianan area. From south to north, this depositional system included river facies, delta facies, littoral–shallow water lacustrine facies and semi-deep water–deepwater lacustrine facies. If compared with the Northern depositional system, this depositional system was on a smaller scale and had a smaller supply of sediments.

● Qiqihaer Depositional System

The Qiqihaer depositional system is located between the northern delta and the Yingtai delta and is on a small depositional scale and has a small distribution range. This depositional system was formed by a less important short axial aquatic system that came from the Fulaerji area on the northwestern side of the basin. Because the Northern delta system extended in a southwestern direction to create a sand dam, in the Taikang area the lacustrine basin was semi-restricted. In addition, compared with the wide-open lacustrine basin, the physical and chemical properties of the lacustrine water were clearly distinct. Therefore, the littoral–shallow water lacustrine facies contained the typical characteristics of a lacustrine bay and it was named the Taikang lacustrine bay. The sediments in the Taikang lacustrine bay were made of sandstone, argillaceous siltstone, silty mudstone, oolitic limestone, ostracoda limestone, algal stromatolite and a small amount of interbedded marlstone and dolomite.

• Southeastern Changchun–Huaide Depositional System

Two aquatic systems were developed on the southeastern side of the basin; one was started in the Huaide area, passing through the Shuangtuo, Dalaoyefu and Gudian areas, it arrived at the Qianguo area. Another one started in the Changchun area and passing through Nongan it also arrived at the Qianguo area. Then, the two aquatic systems were combined and they ran into the lake in the Qianguo area. When the Quantou formation was deposited, the aquatic system of Changchun–Huaide was in its prime. After that, it gradually disappeared. Thus, the shallow water delta of the Quan 4 member was widely developed on the southern side of the Songliao basin. The delta sand bodies of the Qing 1 and Qing 2 members would be extended to the Qian 133 well district. However, the sand bodies of the Qing 3 member retreated to the Shuangtuozi area and the area south of Nongan.

• Qinggang–Suihua Depositional System

The Qinggang–Suihua depositional system was small. At present, research work in this area is still in the preliminary stage. Compared with the Daqing placanticline, the Sanzhao area had a higher content of lithoclastic materials. Moreover, towards the east, the lithoclastic content increased. Therefore, it is reasonable to deduce that the depositional system in the Sanzhao area was influenced by this aquatic system.

In the Songliao basin, the distribution of the depositional system was clearly controlled by the paleo aqua-system and the developmental level of the lacustrine basin. The paleo aquatic system originated at the intersection of deeply extended faults, either in the inner basin or on the margin of the basin. Moreover, these aquatic systems flowed into the basin along the deeply extended large paleo faults. The deeply extended large faulted belts controlled the origination and distribution of the aquatic systems. The central faulted belt controlled two large axial aquatic systems that were the Beian–Nahe system on the northeastern side and the Baokang–Tongyu system on the southwestern side. The Siping–Haerbin faulted belt controlled the Huaide–Jiutai aquatic system and the Qinggang–Suihua aquatic system. The Second Songhua River–Deerbugan faulted belt controlled the Yingtai–Baicheng aquatic system and the Second Songhua River. The Binzhou faulted belt controlled the Qiqihaer aquatic system and the First Songhua River.

2.3.2   *Bohai Bay Basin*

The Bohai bay basin is located on the eastern side of the North China region and the southern side of the northeastern region of China. It consists of the marine territory of Bohai bay and its surrounding areas, including several sections or districts from surrounding cities and provinces. The attached areas are Tianjin, Liaoning, Beijing, Hebei, Henan and Shandong. The outline of the basin is of diamond shape; Bohai bay covers 20×10^4 km^2 of land approximately. The Bohai bay basin contains an abundant oil resource. The total resource is 275×10^8 t approximately. The proven oil reserve has reached 105×10^8 t. Since the 1980s, the Bohai bay basin has passed the Songliao basin to become the largest onshore crude oil production basin in China; the oil production rate has reached 60 million tons per year.

The Bohai bay basin is a superimposed basin that consists of the Mesozoic–Cenozoic terrigenous basin and the Proterozoic–Paleozoic marine basin. The stratigraphic sequences are completely developed. The crystalline basement of the basin is made of the Archean–lower Proterozoic metamorphic rock series. The Bohai bay basin experienced four developmental stages, which included the middle–late Proterozoic developmental stage, the Paleozoic platform stage, the late Jurassic–Palaeogene rifting stage and the Neogene depression stage.

During the Mesoproterozoic era, this region advanced into the developmental stage of the platform. On the northwestern side of the basin, the Mesoproterozoic–Neoproterozoic algal reef carbonate rocks were developed with a thickness of more than tens of thousands of meters. During the Neoproterozoic era, because this region was uplifted, the Sinian system was absent. During the Paleozoic era, the vertical movement was the primary tectonic act and the deposition was stable. Because the Caledonian movement caused uplifting in this region, the Silurian system, Devonian system and the Lower Carboniferous series were missing. The Ordovician system contained a set of carbonate deposits of stable thickness. The Carboniferous–Permian systems consisted of a set of river–delta deposits that contained a coal layer and that belong to marine–terrigenous facies. The Hercynian–Indosinian movements caused folding and uplifting in this region. The complex anticline and the synclinorium were developed. The strata experienced erosion.

After the Indosinian movement, under the influence of the subduction of the Pacific plate, the faults were well developed in this area. The activity of the faulted block and the activity of the magma were forceful. During the Early–Middle Jurassic epochs, the synclinorium of the Hercynian tectonic period was inherited. In addition, coal measures were deposited in the newer faulted basins. During the Late Jurassic epoch, this region moved into the developmental stage of a faulted basin. The activity of the faulted block was the major movement, which included three phases of vigorous movement in the Late Jurassic–Early Cretaceous epochs, the Eocene epoch and Oligocene epoch. On top of the faulted blocks, the dust-pan shaped depression and the cuesta type of uplift were created accordingly. Thus, the framework of the basin consisted of uplifts alternating with depressions. During the Neogene, the faulted activities in the basin were reduced, the surrounding mountain ranges were slowly lifted up and the basin subsided entirely. Therefore, stable stratigraphic overlapping was initiated and it finally covered all of the uplifts and bulges in the basin to form a united depositional basin in the Neogene.

The exploration practice indicated that, in the Bohai bay basin, the oil and gas were mainly accumulated in the Palaeogene and the Neogene systems. Moreover, there were multiple sets of petroleum systems (Fig. 2.16). During the Palaeogene, because the Bohai bay basin had the most powerful activities in the faulted block, the paleo geography consisted of multiple uplifts and multiple depressions. For example, in the Bohai bay basin, the Palaeogene system contained 54 depressions and 39 uplifts (Gao et al., 2001). Each depression could contain an isolated depositional system and petroleum system. The regional petroleum exploration was basically targeted at the level of an individual depression. Because of the faulted activity, the structural framework alternated with depression and uplift. The nature of the structural framework determined that each depression was not only near the sediment source center, but was also supplied by multiple source centers. Thus, various types of small scale reservoir sand bodies were developed (Fig. 2.15), such as the sand body of alluvial fan facies, fan delta facies, braided river delta facies, turbiditic fan facies and beach bar facies.

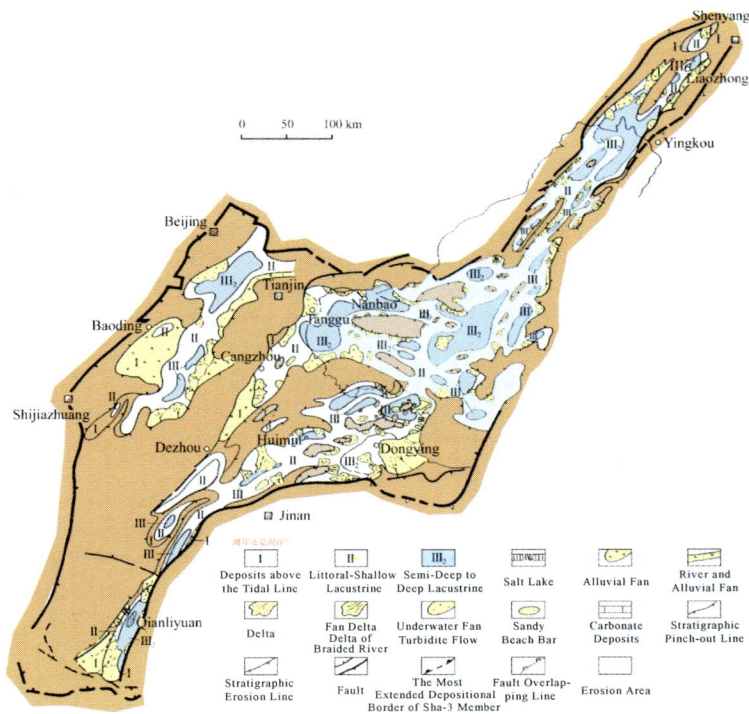

Fig. 2.15. Depositional system of the third member in the Shahejie formation, the Palaeogene system during the faulted depression period in the Bohai bay basin (Wu et al., 1992) (with permission from Petroleum Industry Press)

In the rifting basin, during the initial extension phase and during the decline phase, the primary sedimentation was for infilling the basin within the depositional facies of the alluvial fan and river facies. During the subsidence phase of a lacustrine basin, the elevations of the mountains were high and the water depth was great. Also, the slope gradient was substantial. This was a prime time for developing the deep water mudstone in the lacustrine basin and for developing deposits of gravitational flow. The near shore underwater fan was distributed in the steep slope zone. Conversely, the distal fan with supply channels was distributed in the gentle slope zone. During the contractive phase of a lacustrine basin and during the early phase of subsidence, because the lacustrine region was relatively small with shallow water depth, all types of delta were developed. The fan delta was distributed in the steep slope zone and the braided river delta was distributed in the gentle slope zone. During the phase of stable subsidence, because the lacustrine basin had been infilled previously, the topography was gentle and the water body was broad and shallow. As a result, the sand bodies of littoral–shallow water beach bar facies and carbonate rocks were well developed. Along with the development of the rifted basin, there were several phases of basic volcanic eruption, which formed unique volcanic reservoir sequences. The commonly developed volcanic reservoir sequences occurred in the initial extension phase.

2.3.3   *Eerduosi Basin*

The Eerduosi basin is located in the central-northern region of China; it crosses over five provinces (Shan'xi, Gansu, Ningxia, Inner Mongolia and Shanxi) and it covers 25×10^4 km^2 of land approximately. The tectonic position of the Eerduosi basin is on the western side of the Huabei platform and it is a depressed basin on the margin of a craton. The Eerduosi basin contains six 1st order tectonic units that include the Yimeng uplift, Weibei uplift, Jinxi flexural zone, Yishan slope, Tianhuan depression and the Western Margin thrust belt. The Eerduosi basin contains abundant oil and gas resources; the total oil resource is 85.9×10^8 t; the gas resource is 10.7×10^{12} m^3. At the end of 2007, the proven oil reserve was 16.8×10^8 t; the proven gas reserve was 1.53×10^{12} m^3. The sandstone reservoir sequences had lower porosity and lower permeability.

The Eerduosi basin is a large scale craton basin, which experienced uplifting and subsidence of the entire basin and the migration of the depression. In addition, it had a simple structure. The basement of the basin consists of the Archean–Paleoproterozoic metamorphic rock series. The thickness of sedimentary rocks (that include the Paleozoic erathem, Mesozoic erathem and Cenozoic erathem) is 1.1×10^4 m. The basin experienced five developmental stages. During the Mesoproterozoic–Neoproterozoic aulacogen stage, the North China craton was created, which was the embryonic form of the basin. During the early Paleozoic shallow sea platform stage, the major deposit was carbonate rock in the epicontinental sea. In addition, the buried hill type of gas reservoir was developed on the weathering crust. During the late Paleozoic littoral plain stage, the primary deposits included coal measures of marine–terrigenous facies and delta deposits. Here, a large scale lithologic type of gas field was developed with medium–low abundance. During the Mesozoic terrigenous depressed lacustrine basin stage, the primary deposits included large scale delta deposits of lacustrine facies and mudstone of deep water lacustrine facies. Here, the lithologic type of oil field was developed, which had low porosity and low permeability.

Because the basin was controlled by a paleo geological background of different periods, the Triassic and Jurassic systems contained different types of depositional system and different distribution patterns. In the Eerduosi basin, during the Late Triassic epoch, the paleo topography was higher on the northeastern side and lower on the southwestern side. When considering the slope gradient, the northeastern side was gentle and the southwestern side was steep. The major portion of the depression or subsidence center was a dustpan shaped depression that was oriented in a northwest direction. On the peripheral area of the basin, there were several paleo uplifts that functioned as sediment source centers. Among the sediment source centers, Yinshan mountain on the northern side of the basin and the Longxi paleo uplift on the southwestern side of the basin were the most important sediment source centers during the Late Triassic epoch. Therefore, the depositional framework consisted of river–delta depositional systems that mainly received sediments from the southern and the northern source centers (Fig. 2.16). These two depositional systems not only controlled the development of the lacustrine basin on a macro level, but they also offered important spaces in the Mesozoic and the Cenozoic erathem for oil accumulation in the Eerduosi basin.

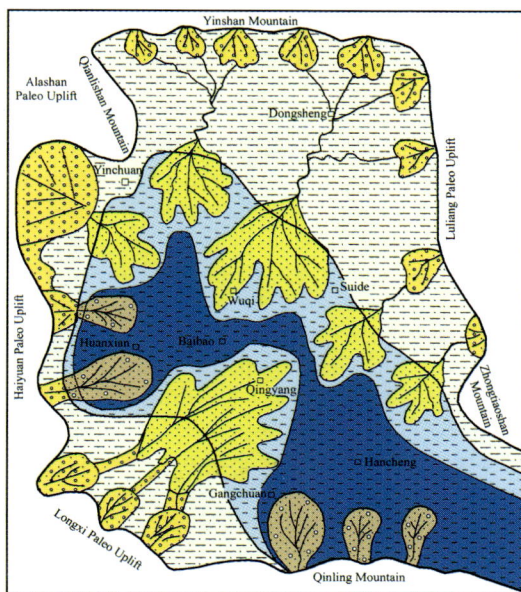

Fig. 2.16. Depositional system of the Eerduosi basin during the Late Triassic period

Precisely speaking, during the Late Triassic epoch, the Eerduosi basin was higher on the northeastern side and lower on the southwestern side; the broad gentle slope occurred on the northeastern side of the basin and the narrow steep slope appeared on the southwestern side of the basin. The main axis of the depression or the subsidence center was either extended or spread out in a northwest direction, which created an asymmetrical depression that had a dustpan like shape. Also, during the late Triassic epoch in the Eerduosi basin, the northeastern region and the southwestern region were the most important sediment supply centers. Influenced by the fault activities in the basement, which was oriented in a northeast direction and manipulated by the paleo topography of the basin in a northeast–southwest direction, the paleo aquatic systems carried the sediments into the lacustrine basin to create two of the most significant depositional systems. From the south facing mountain front of North Yinshan mountain to the center of the Eerduosi basin, a large scale stable river–delta system was developed, which had a small gradient, abundant sediment supply, a distant source center and a stream stretching a long way. The underwater fan and turbiditic depositional systems were developed on the extensional margin on the southern margin of Qinling mountain; the alluvial fan and braided river delta systems were developed at the edge of the strike-slip thrust belt on the southwestern margin of the basin. The alluvial fan and the underwater fan depositional systems were developed at the edge of the thrust belt on the northwestern margin of the basin; the depositional system of the meandering river delta was developed on the northeastern margin and the northern margin of the basin, where there were passive margins with broad gentle slopes.

At the end of the Triassic, influenced by the shear stress that was produced by the left rotation of the Eurasian plate, the Kula plate and the Pacific plate in a N-S direction, the North China craton was uplifted entirely and it suffered erosion. This tectonic movement caused a topographical change in the Eerduosi basin to form higher ground on the western side and lower land on the eastern side. In addition, at the top of the Yanchang formation, there were small hills, stepped grabens occurred frequently, gorges and canyons overlapped, and the slopes and sags were widely distributed. This kind of paleo geography was crucial to the Early Jurassic depositional systems and their distribution in the Eerduosi basin. The Early Jurassic depositional systems and their distribution were strictly controlled by the paleo geography. When the Fuxian formation and the Yan 10 member were deposited, from west to east, the paleo aquatic systems carried sediments to fill the paleo river channels. When the Yan 10 member was deposited, the Ganshan paleo river channel that was oriented in a W-E direction passed through the basin with a width of a few thousand meters. Other aquatic systems were to join it in the sharp angles. When the Yan 9–Yan 8 members were deposited, the lacustrine delta system was developed. The orientation of the depositional system was influenced by the paleo river channels. Five of the former river channels (such as the Ganshan, Ningshan and others) were transformed into delta systems of different sizes. After the Yan 7 member was deposited, the primary depositional systems were the anastomosing river system and the palustrine system; this was the important coal making phase in the Eerduosi basin. Overall, during the Early Jurassic epoch, the Eerduosi basin experienced three developmental stages. Accordingly, there were three depositional systems that included the alluvial fan–river depositional system in the Fuxian formation–Yan 10 member, the river–delta depositional system in the Yan 9–Yan 6 members and the anastomosing river–residual lacustrine depositional system in the Yan 4+5 members (this depositional system was developed by the uplifting of the basin and rejuvenated rivers).

2.3.4 Sichuan Basin

The Sichuan basin and surrounding mountains are located on the eastern side of Sichuan province. It is a structural basin that has a distinctive border frame that covers 200,000 km^2 of land. Daba mountain, Daliang mountain and Longmen mountain encircle the Sichuan basin. Above the basement of the basin, the stratigraphic sequences are fully developed with a total thickness of 6,000 – 12,000 m. The Middle Sinian–Middle Triassic strata contain marine deposits with a thickness of 4,000 – 7,000 m. Most of the marine strata are carbonate rocks. The Lower Sinian series was only developed in the northeastern and southeastern regions of Sichuan province; the Upper Sinian series was fully developed with stable distribution. The Cambrian system, Ordovician system and Silurian system contained the platform type of deposits that were widely distributed in the basin. The Carboniferous and Permian systems contained shallow sea platform deposits that covered the entire region. At the end of the Middle Triassic epoch, because the early phase of the Indosinian movement caused the uplifting of the entire Upper Yangtze region, erosion occurred in the Sichuan basin and the large scale marine transgression was finished. The Upper Triassic series contained a set of deposits of marine–terrigenous facies; the thickness of the strata was 250 – 3,000 m. The Jurassic–Quaternary systems contained terrigenous strata only, which mainly consisted of a set of clastic rocks; the thickness of the strata was 2,000 – 5,000 m. The Sichuan basin mainly contains a natural gas resource; the total gas resource is 7.2×10^{12} m^3. At the end of 2007, the proven natural gas reserve was 1.02×10^{12} m^3 and the proven oil reserve was 0.7509×10^8 t. At present, the proven petroleum reserves mainly occur in the reef flat type of reservoir sequences in the Upper Permian–Middle and Lower Triassic carbonate rocks and in the Upper Triassic sandstone reservoir sequences that had lower porosity and lower permeability.

The Upper Permian series mainly contained the Changxing formation that consisted of the slope type of platform carbonate rocks and bio-reef deposits. The Middle and Lower Triassic series contained the Feixianguan formation, Jialingjiang formation and Leikoupo formation. The Feixianguan formation was made of purple red color shale that was interbedded with limestone; the Jialingjiang formation and the Leikoupo formation were made of limestone and dolomite that was interbedded with a gypsum–salt layer. These three formations were developed on the carbonate platform of a shallow sea in hot and dry climate conditions.

When the Feixianguan formation was deposited, not only had the bio-reef facies belt been developed into oolitic beach facies on the platform margin, but the territory of the platform facies also had been enlarged. During this time, the marine territory of the Sichuan basin was a continental shelf that was influenced by the clastic debris of the continental margin, which belonged to the accretion model of a carbonate platform that appeared on the continental shelf in the shallow sea and that received clastic debris from the continental margin. The development of a carbonate platform on the eastern side of Sichuan divided the continental shelf into three sections, which included the platform in the semi-restricted sea on the western side, the carbonate platform in the middle section and the platform in the open sea on the eastern side. The clastic debris of the continental margin was mainly deposited in the semi-restricted sea on the western side. On the carbonate platform, because of the limitation of sea water circulation, the oolitic beach deposits were predominately developed in this kind of environment. On the eastern side and the southeastern side of the platform, because of stronger storm waves, grain dams were developed along the platform margin. Deposits of gravitational flow were developed on the slope of the open sea. The northern side of the platform was in the depositional environment of a tidal flat for a long duration.

The depositional period of the Jialingjiang formation was the same as the depositional period of the Feixianguan formation; several depositional cycles that consisted of limestone, dolomite, gypsum and salt were developed in the Jia 1–Jia 2 members, the Jia 3–Jia 4 members and the Jia 5 member.

During the Middle Triassic epoch, when the Leikoupo formation was deposited, the transportation direction of material was the opposite of that in the Early Triassic epoch. The marine basin was shallow on the western side and deep on the eastern side. Marine–terrigenous facies occurred in the eastern and western regions of Sichuan province. In the central and western regions, most areas still retained the depositional features of the Jialingjiang formation, which consisted of platform facies in the restricted sea. Along with the stabilization of the marine basin environment, the alternative deposits of lagoon and shallow sea were developed; the lithology was interbedded silt-crystalline limestone and calcareous dolomite. The largest marine transgressions occurred when the lower portions of the Lei 1 member, Lei 3 member and Lei 5 member were deposited; the primary lithology included algal dolomite and limestone. The marine regression happened when the upper portions of the Lei 1 member, Lei 2 member, Lei 4 member, and the middle portion of the Lei 3 member were deposited; the lithology included dolomite, gypsum dolomite and argillaceous dolomite. These marine transgressions and regressions developed the next level of reservoir–seal combinations that created excellent commercial grade gas reservoirs.

The Xujiahe formation in the Upper Triassic series contained a set of dark color coal measures containing sandstone and mudstone. When the Xu 1 member was deposited, the depositional environment was from the marine environment to the marine–terrigenous environment. The Xu 2 member and other members above it were deposited in the lacustrine, palustrine, delta and river environments. However, these depositional environments were often influenced by sea water. The carbonate platform in a shallow sea occurred on the peninsula of Longmen mountain and the area west of the Kangdian paleo terrain. The shallow water–semi-deep water lacustrine facies were distributed in the Jintang–Yaan area in the central-western region of the basin; the lithology consisted of clay shale that was interbedded with sandstone. The alternative facies belt (that contained delta and fan delta sandstones and lacustrine mudstone) was distributed on the western side of the basin. This alternative facies belt was located on the western slope of the mountain front depression of Longmen mountain, where there was also the steep slope of a depositional basin. The sandstones of progressive delta facies and the littoral–shallow lacustrine facies were distributed alternatively with the mudstone of littoral–shallow facies and palustrine facies in the Suining–Pujiang area on the southeastern side of the depression. The sandstones of delta plain facies and flood plain facies (that was interbedded with mudstone) were distributed in the Daxian, Chongqing, Zigong and southeastern region. In these areas, the predominant depositional environment was a braided river.

During the Late Triassic, when the Xujiahe formation was deposited, there were several sediment source centers and multiple depositional systems. The major sediment source centers included the Jiangnan paleo terrane, Longmen mountain paleo terrane, Qinling–Daba mountain paleo terrane and Kangdian paleo terrane. The eroded materials from these paleo terranes were carried by several rivers on the southeastern side of the basin, the river on the northeastern side of the basin and a series of short rivers from the Longmen mountain and Qinling mountain, to be deposited in the basin. Each river created an individual depositional system of river–delta.

2.3.5 Zhungeer Basin

The Zhungeer basin is located on the northern side of the Xinjiang region, which is a central region of the Asian continent. The basin is surrounded by folded mountains of the Hercynian tectonic period and it covers 13.4×10^4 km^2 of land. In western China, the Zhungeer basin not only contained the most abundant oil reserves, but it also produced the largest quantity of oil and gas. In this basin, the total resource of oil is 86.81×10^8 t; the total resource of gas is 2.51×10^{12} m^3. At the end of 2007, the proven reserve of oil was 19.83×10^8 t; the proven reserve of gas was 787.46×10^8 m^3.

The Zhungeer basin is a large scale multi-cycled superimposed basin; it endured three phases of tectonic stress, which included the late Paleozoic extension–compression movement, the Mesozoic depression procession and the Cenozoic powerful compression movement. The basin contained "dual layer structures", which consisted of the Pre-Cambrian crystalline rock series and the Pre-Carboniferous epimetamorphic rock series. In the Zhungeer basin, the stratigraphic sequences include the Middle and Upper Carboniferous series–Quaternary system; the greatest thickness of the sedimentary rock can reach 15,000 m.

Since the Carboniferous period, the Zhungeer basin has experienced seven stages of tectonic–depositional development, which include the development of the Early–Middle Carboniferous rifted basin, the development of the Late Carboniferous–Early Permian extension type of depressed basin, the development of the Middle Permian–Middle Triassic foreland basin, the development of the Late Triassic–Middle Jurassic insubstantial extension type of depressed basin, the development of the Middle Jurassic foreland basin, the development of the Late Cretaceous–Palaeogene inclined depressed basin and the development of the Neogene–Quaternary intra-continental foreland basin. The best developed terrigenous lacustrine basins occurred in the Permian period, the Triassic–Jurassic periods and the Cretaceous period. The area of the largest lacustrine region was greater than 10×10^4 km^2. The following description uses the Jurassic system as an example to show how the distribution of a depositional system and related depositional features in the Zhungeer basin occurred.

In the Zhungeer basin, the paleo tectonic background, the climatic conditions and the closeness of the sediments source center controlled the types of Jurassic depositional system and their distribution on a gentle slope in the shallow water region of a large scale depressed basin. Under the control of the surrounding mountain ranges, in the Zhungeer basin eight large depositional systems were developed in the Jurassic system (Fig. 2.17); these depositional systems were located at the Halaalate mountain, Zhayier mountain, Kelameili mountain, Wulungu, Delun mountain, Sikeshu, Yilinheibiergen mountain and Bogeda mountain. These eight large depositional systems controlled the Jurassic depositional structure of the Zhungeer basin.

(1) The Zhayier mountain front depositional system consisted of alluvial fan–river–delta–lacustrine facies. Zhayier mountain was the major sediments supplier. Overall, this depositional system was distributed in an NW-SE direction and it could be divided into two major branches. Passing the Guai 8 well district, one branch extended a long distance toward the east and it influenced the Mosuowan area; another branch was developed toward the Chepaizi area with an orientation in an N-S direction.

(2) The Halaalate mountain front depositional system consisted of alluvial fan–river–delta–lacustrine facies. The northwestern material source center of the Halaalate mountain was the major sediments supplier. This depositional system was distributed in an NW-SE direction with several branches. The delta lobes (that were spread in an E-W direction approximately toward the Luxi–Shinan area) ran into the eastern material source center in the Shixi area; the underwater distributary channels were well developed.

(3) The Delun mountain front depositional system consisted of braided river–braided river delta–lacustrine facies. Delun mountain was the major material

source center. This depositional system was oriented in an NW-SE direction; in addition, it influenced the western side of the Wulungu depression and the Luxi area.

(4) The Wulungu depositional system consisted of fan delta–lacustrine facies and alluvial fan–river–delta–lacustrine facies; this depositional system extended in a NE-SW direction. The Yilinheibiergen mountain front depositional system consisted of river–delta–lacustrine facies; the primary deposits were in the delta facies, which were oriented in an N-S direction.

(5) The Kelameili mountain front depositional system consisted of river–delta–lacustrine facies. The Kelameili mountain was the sediments supplier. This depositional system was oriented in a W-E direction approximately and it contained several branches. Among these branches, the one that was developed toward the west, from Kelameili mountain to the Dixi–Lunan area, could have stretched to the Shixi area when the Sangonghe formation was deposed. In the Shixi and Mobei areas, this depositional system met the Halaalate mountain front depositional system and the Zhayier mountain front depositional system.

(6) The Bogeda mountain front depositional system consisted of river–delta– lacustrine facies. The southeastern material source center of the Bogeda mountain was the sediments supplier; this depositional system was oriented in an NW-SE direction.

(7) On the southern side of the basin, both the Wulumuqi and Sikeshu depositional systems consisted of river–delta–lacustrine facies; they were small in scale and were oriented in an N-S direction.

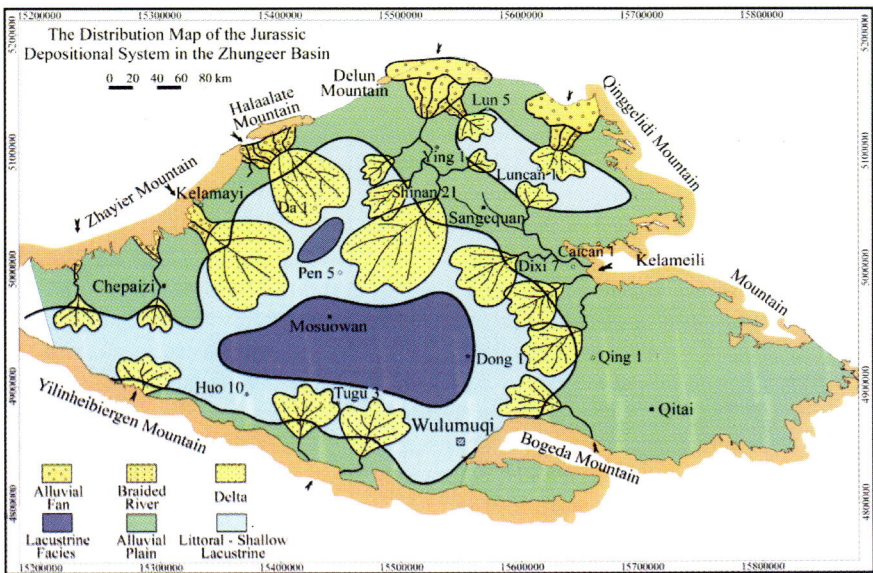

Fig. 2.17. Distribution map of the Jurassic depositional system during the depressed period of the Zhungeer basin

Under the control of the tectonic stages in the basin, and limited by the structural background of the tectonic formation, during the different developmental stages of the lacustrine basin, these eight large scale depositional systems differed in their developmental magnitude and their kind of distribution. Precisely speaking, two tectonic cycles created the two different types of tectonic frameworks and two different kinds of distribution patterns for these depositional systems.

2.3.6   Tarim Basin

The Tarim basin is located on the southern side of the Xinjiang region; it is one of the largest petroleum basins in China and it contains 56×10^4 km^2 of land approximately. In the Tarim basin, the total resource of oil is 86.81×10^8 t. The total resource of natural gas is 2.51×10^{12} m^3. At the end of 2007, the total proven reserve of oil was 19.83×10^8 t; the total proven reserve of gas was $8,600 \times 10^8$ m^3. The Tarim basin was stretched out in a diamond shape. The secondary tectonic units included the Kuche depression, Tabei depression, Northern depression, Tazhong uplift, Southwestern depression, Tanan uplift and Southeastern depression.

The Tarim basin is a large scale superimposed basin that consisted of the Paleozoic marine craton basin and the Mesozoic–Cenozoic terrigenous foreland basin. On top of the paleo basement of the continental crust, the stratigraphic sequences from the Sinian system–Quaternary system were well developed. Three major tectonic sequences (the upper, middle and lower tectonic sequences) were developed in the basin. The lower tectonic sequences contained two groups of sedimentary rocks; the first group occurred in the Sinian system, the Cambrian system and Ordovician system and the lithology included carbonate rocks (that were either deposited in the deep ocean craton basin and sea trough or deposited on a shallow sea platform) and clastic rocks. The second group included the clastic rocks of littoral–shallow sea facies in the Silurian system and Devonian system. The lower tectonic sequences were mainly distributed in the region of the platform and basin with a buried depth of 4,000 – 8,000 m. These sequences are the primary exploration targets for discovering the large scale lithostratigraphic type of oil and gas reservoirs in the region of the platform and basin. At present, the discovered large scale stratigraphic types of oil and gas fields include the Lunnan oil field and Tazhong oil field. The middle tectonic sequences also contained two groups of sedimentary rocks; the first group was developed in the Carboniferous–Permian systems, which not only contained the strata of shallow sea platform facies in the craton basin and the strata of marine–terrigenous facies, but which also contained volcanic rocks of the rifting stage. The second group consisted of Triassic terrigenous clastic rocks. These sequences were widely distributed in the Tarim basin. In the region of platform and basin, the buried depth of these strata was 3,000 – 4,000 m. The middle tectonic sequences are primary exploration targets for discovering the medium–small size lithostratigraphic type of oil and gas reservoirs in the region of platform and basin. At present, the discovered stratigraphic oil and gas fields include the Hadexun oil field and others. The upper tectonic sequences contained a massive layered strata of lacustrine and palustrine facies and red-bed molasses deposits; these strata not only occurred in the Jurassic system and the Cretaceous system, but they were also developed in the Palaeogene–Neogene rifted basin and foreland basin. In the Kuche depression, the upper tectonic sequences are the primary exploration targets for discovering the large and middle size structural types of oil and gas reservoirs. The discovered gas fields include the Kela 2 gas field and others.

In the Tarim basin, the carbonate platforms on the continental shelf were mainly developed in the Cambrian–Ordovician systems in the region of platform and basin. The reef and reef flat types of reservoir sequences were well developed on the platform margin (Fig. 2.18). These sequences are important exploration domains at present. According to the study of the paleo magnetic field in the Tarim plate, during the Ordovician period the Tarim plate was located between $20° – 30°$ south, in the southern temperate zone with an oceanic climate. The water temperature of the paleo-ocean was $13.49 – 25.53 °C$; the average water temperature of that time was $19.12 °C$. During the early phase of the Early Ordovician, the sea level was rising. During the middle–late phases of the Early Ordovician, the sea level was falling below the average. During the end of the

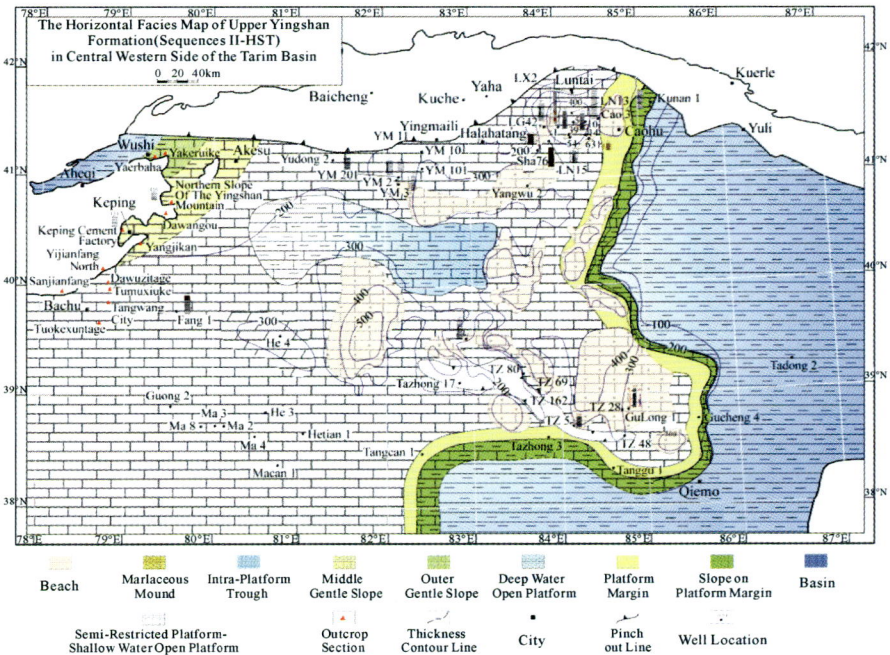

Fig. 2.18. Horizontal facies map of the upper Yingshan formation (sequences II-HST) in the Ordovician system on the central western side of Tarim basin

Early Ordovician–Middle and Late Ordovician, the sea level was rising again. During the Middle–Late Ordovician, the sea level was frequently changing. For bio-reef development, the favorable intervals were when the sea level was changing slowly or when the sea level was stable. In the bio-reef complex, the content of reef-building organisms was commonly larger than 30%. The position of the marginal facies belt on the carbonate platform was changeable. Throughout geological time, the marginal facies belts could either continual migrate or they could move a long distance within a very short duration. In the Tarim basin, the depositional system of a gently sloped carbonate platform was developed in the Carboniferous system, which included an evaporate platform, restrictive platform, semi-restrictive platform and wide-open platform.

In the Tarim basin, the nature of the Mesozoic–Cenozoic deposition included multiple marine transgressions on the southwestern side of the basin and a terrigenous environment in the remaining areas of the basin.

During the Triassic period, most of the Tarim basin endured weathering and erosion; the Triassic system only deposited in the Baicheng faulted basin and the Pulu faulted basin. Among the faulted basins, the Baicheng faulted basin was the largest one; during the Early–Middle Triassic, it covered 15,500 km^2 of land. In the Baicheng faulted basin, the sediments contained coarse grains on the northern side and fine grains on the southern side. Along the Tianshan mountain, the strata were deposited in diluvial–alluvial facies that were interbedded with shallow water lacustrine facies; southwards, the grain size became fine. All the sediments were deposited in the shallow water lacustrine basin. During the Late Triassic epoch, the faulted basin continually expanded and the lacustrine water became deep. The strata of the semi-deep lacustrine facies consisted of dark gray color, grayish black color mudstone with a thickness of 400 – 500 m. Because they contained abundant organic materials, these mudstones were important oil generating sequences.

During the Jurassic period, the paleo-geographic framework of the Late Triassic was inherited. However, because the climate was warmer and more humid, the precipitation increased and the rivers were more active. During the Early–Middle Jurassic, the lacustrine basin in the Baicheng faulted basin obviously expanded in a southeast direction. On the northern margin of the lacustrine basin, different sizes of delta sand bodies were developed. Together, they created a belt of delta sand bodies. During the Early Jurassic epoch, more than 2,000 m of coal containing a sandy conglomerate of diluvial–alluvial facies was deposited in the Qielieke faulted basin. The strata of lacustrine facies were developed during the Middle Jurassic epoch; the red-beds were developed during the late Jurassic.

During the Cretaceous, if drawing a line from Bachu to the Heshen 2 well district, the broad areas on both sides of this line were the weathering regions. A narrowed lacustrine basin (that was located in the Yingjisha–Zepu–Yecheng area, south of the weathering region) covered 20,000 km^2 of land; red color sandy mudstone of shallow water lacustrine facies was deposited in this lake.

2.3.7 Chaidamu Basin

The Chaidamu basin is located in the northwestern region of Qinghai province, Qilian mountain is to the northeast of the basin, Kunlun mountain is in the south of the basin, Aerjin mountain is to the northwest of the basin and it functions as the border between Chaidamu basin and Tarim basin. The basin contains in total 12.1×10^4 km^2 of land. The Mesozoic and Cenozoic sedimentary rocks were distributed over an area of 9.6×10^4 km^2 with the largest depositional thickness being 16,000 m. The Chaidamu basin contained 25.4×10^8 t of oil resources and 2.6×10^{12} m^3 of gas resources. At the end of 2007, the proven reserve of oil was 3.48×10^8 t; the proven reserve of gas was $3,056 \times 10^8$ m^3.

The Chaidamu basin is a terrigenous faulted and depressed type of composite basin that was developed during the Mesozoic and Cenozoic eras after the Indosinian movement on top of the Proterozoic metamorphic crystalline basement and the deformed by folding Paleozoic basement. It experienced four stages of development. Among the stratigraphic sequences, the Triassic system not only had the greatest thickness and broadest distribution range, but it also had the most important sequences for oil generation and storage.

Triassic–Middle Jurassic Faulted Basin Stage. During the Triassic period, since the seawater regressed from the eastern side of the Chaidamu basin, the isolated strata of the Late Triassic were deposited in the faulted lacustrine basins on the northwestern margin and on the northeastern side of the basin. During the Early Jurassic–Middle Jurassic, small scale, clearly separated subsided belts of diversified faulted basins were developed along the northern margin of the basin. In the humid climate, a set of coal measures was deposited. During the end of the Middle Jurassic, the early phase of the Yanshanian movement caused structural inversion and uplifted the faulted basins. Therefore, in these fault basins, the Middle–Lower Jurassic strata commonly endured weathering and erosion, which limited the overall distribution of the Middle and Lower Jurassic series.

Late Jurassic–Cretaceous Isolated Compressed Uplifting and Erosion Stage. After the Middle Jurassic, the mountains that surrounded the Chaidamu basin were reactive. In the arid climate, the Upper Jurassic and Cretaceous red color clastic sequences of river alluvial facies were rapidly deposited in the basin, which had a distribution range larger than that of the Middle and Lower Jurassic series. During the end of the Cretaceous period, because the late phase of the Yanshanian movement uplifted a part of the basin, an erosion process happened in some areas; the greatest thickness of eroded strata could reach 600 m.

Palaeogene Slip–Compressive Depression Stage. Controlled by the combined stresses of the N-S oriented regional compression and the NEE oriented Aerjin slip fault, a subsided depression with the characteristics of compression– distortion was developed on the southwestern side of the basin. At the same time, the surrounding mountains were uplifted. As a result, an embryonic form of the Chaidamu basin was created.

During the early phase of the Oligocene, influenced by the regional tectonic movement, the Chaidamu basin was transformed from a water discharge basin with an outflow into a river into a water collecting basin with inflowing streams. During this time, the alluvial facies were well developed. The alluvial facies combination that consisted of diluvial fan and river flood plain covered 90,000 km^2 of land, which represents more than 90% of the total depositional area. In the southern district of the western region, the sandy conglomerate in the Alaer alluvial fan is the important reservoir sequence of this basin.

During the middle and late phases of the Oligocene, the lacustrine basin was forcefully sunk, which created a depressed basin that utilized the subsidence center as its depositional center. Because the water depth of the central lake was deep and stable, dark color mudstone was continually deposited with a great thickness of 500 – 800 m. This was the major oil generating period for the basin. The sediments came from several source centers and the depositional pattern was asymmetrical; the aquatic system of eastern Qilian mountain still functioned as the primary depositional system.

Neogene–Quaternary Compressive Folding and Depression Stage. Influenced by the intra-continental subduction that was caused by the Himalayan movement, the basin was increasingly compressed by the stress from the south. Finally, the folded uplifting became the major tectonic event in the basin; accordingly, the region of the lacustrine basin was gradually reduced.

During the Miocene–early Pliocene, the depositional center migrated northeastward. The territory of the lacustrine basin expanded and the depth of water became shallow. This was the period when the second set of hydrocarbon source rocks were developed in the lacustrine basin; the stable lacustrine facies covered 7,630 – 16,100 km^2 of land. Most of the sediments were deposited in the shallow water region; the lithology included light gray, greenish gray and gray color mudstone, calcareous mudstone that was interbedded with argillaceous limestone, siltstone and a small amount of red color argillaceous rocks. In some areas, the fractural type of oil and gas reservoir had been developed in argillaceous rocks and argillaceous limestone.

During the middle and late phases of the Pliocene, because the climate became gradually more arid, the amount of inflowing water was reduced and the depth of water was shallow. Furthermore, salinization occurred in the water, which converted the fresh water lacustrine basin into a salt lake. Finally, the lacustrine basin disappeared. On the eastern side of the basin, shallow water palustrine facies were developed in the region that was close to the water, which was a favorable environment for natural gas generation. At present, gas shows are commonly discovered this region. After the middle Pleistocene, in the western region the uplift continued. However, in the eastern region the Sanhu area experienced diversified subsidence. Therefore, Quaternary lacustrine deposits were developed in relatively limited areas. At present, medium size bio-gas fields have been discovered in the Quaternary strata.

2.3.8 Basins in the Southeastern Coastal Region

In the South China Sea and the East China Sea, several petroleum basins were developed on the continental shelf. Most of them were Cenozoic basins. The large sedimentary basins included the North Yellow Sea basin, East China Sea basin, the basin of the Pearl River estuary, Yingqiong basin and the basin of Beibu Gulf. The southeast coast and the basins in the marine territories are future exploration domains for Chinese petroleum exploration. At present, important discoveries are being achieved in major petroleum basins that have great resource potential.

● North Yellow Sea Basin

The North Yellow Sea basin is located in the northern marine territory of the Yellow Sea. In the basin, the strata mainly consisted of Mesozoic–Cenozoic sequences. The geographical position of the North Yellow Sea basin is between the Liaodong peninsula, Shandong peninsula and Korean peninsula. The basin is in the shape of an elongated oval and it is 5×10 km^2 in area. The largest depositional thickness of the Mesozoic and Cenozoic erathem can reach 8,000 m. The development of the North Yellow Sea basin was initiated by regional uplifting during the Triassic. During the Late Jurassic, the initial rifting stage started. Manipulated by faults, many separated faulted basins were developed. The river sediments and pyroclastic were deposited in these faulted basins. In the semi-humid climate conditions, the region became a large size, deep water, wide open, fresh water structural lacustrine basin. During the Early Cretaceous, the second rifting stage began. The climate was arid and the supply of sediment was limited. The terrigenous red-beds were the major sediments in the basin. At the same time, in some isolated areas, seasonal shallow water salt lakes were developed. During the Eocene–Oligocene, the third rifting stage started. In the humid climate conditions, the subsidence was reactivated; the Palaeogene basin was developed with two kinds of depositional systems, a lacustrine system and river–delta system. Finally, during the Miocene–Pliocene, the regional subsidence created a shallow water basin that mainly received the sediments of delta facies. During the Pleistocene, sea water transgressed into the entire North Yellow Sea basin.

● Basin of the East China Sea

The East China Sea basin is a back arc rifting–depressed type of basin. It is located on the continental shelf in the East China Sea; the geographical position of the East China Sea basin is to the north of Taiwan and south of Jizhou Island. The basin is 270,000 km^2 in area approximately. The depositional types were varied on the northern and southern sides of the basin. In the Xihu (West Lake) depression on the northern side of the basin, in ascending order, the Pliocene series–Quaternary system contained the sediments of river plain facies–littoral shallow sea facies (these facies were transformed gradually); the Miocene series contained the sediments of river palustrine facies and lacustrine palustrine facies. Marine transgression occurred occasionally during this time. The Oligocene series mainly contained the sediments of shallow lacustrine facies and marine transgression happened intermittently during this time too. The middle–upper Eocene series contained the sediments of marine–terrigenous facies. In the Diaobei depression and Wendong depression on the southern side of the basin, the Palaeocene series contained the sediments of littoral–shallow sea facies, which were altered by palustrine facies. The Eocene series mainly contained shallow sea sandstone and mudstone, which were altered by bio-clastic limestone and the deposits of littoral facies and palustrine facies. The Miocene series–Quaternary system contained the sediments of littoral–shallow sea facies, which were altered by palustrine facies over a very short period.

- **Basin of the Pearl River Estuary**

The basin of the Pearl River Estuary is located on the continental shelf and continental slope, south of Guangdong, between Hainan and Taiwan. It is 17.5×10^4 km^2 in area. The basin mostly contained Cenozoic sequences with a thickness of more than ten thousand meters. The basin experienced rifting–pulling apart–depressed tectonic developments. In correlation, a depositional system of river and lacustrine–semi-restrictive sea–open sea was developed in the basin. The Palaeogene system contained the deposits of deep water lacustrine facies, which were the primary oil generating sequences. The lower Miocene series contained the deposits of marine facies, which were deposited in the newly formed semi-restrictive marine basin. Besides the depositional system of the delta, lobe-like sand bodies and sheet-like sand bodies were developed in the depositional system of the tidal, storm and ocean currents. During the early Miocene epoch, the marine transgression of the Zhuhai phase occurred. The development of a marine depositional system was initiated in the basin. The lower portion of the Zhujiang formation contained phase I carbonate rocks that were distributed over an area of 69,000 km^2; in the Dongsha uplifted area, the upper portion of the Zhujiang formation contained the superimposed and overthrust type of carbonate rocks of phases II to IV, with great thickness and with a well developed tower reef, "reef on the platform margin" and reef complex. This region has excellent potential for oil and gas exploration.

- **Yingqiong Basin**

The Yingqiong basin is located on the continental shelf in the western section of the northern territory of the South China Sea. The geographical location of the basin is southeast of Hainan Island and the western marine territory. The Yingqiong basin is 6.5×10^4 km² in area. The Palaeogene system contained the sediments of terrigenous facies and marine–terrigenous facies; the Neogene system contained the sediments of marine facies. During the rifting phase of the basin, the sediments revealed the features of a single faulted lacustrine basin. On the margin of the lacustrine basin, coarse grain sediments were deposited in the littoral–shallow water region of the lacustrine basin or in the delta. The sandstone of littoral–shallow water lacustrine facies was interbedded with several thin layered palustrine coal beds; the center of the rifted basin received the sediments of the semi-deep water–deep water lacustrine facies. A turbiditic fan might be developed here. During the faulted depression phase, the basin was in a marine–terrigenous environment. From the peripheral area to the center of the basin, the lithology was transformed from the massive sandstone of littoral facies to the interbedded thin layer mudstone and argillaceous limestone of shallow marine facies. During the depression phase, because the Northern Sea Trough in the Xisha region was extended, the basin received overlapping marine deposits over an extensive area. All kinds of sand bodies that were related to the littoral coast environment were developed during this time. When the Huangliu formation–Yinggehai formation were deposited, because the supply of sediments was abundant, a series of delta complexes were developed. Several thin layer coal beds of palustrine facies were deposited in the marine bay that was related to the development of the delta.

● **Basin of Beibu Gulf**

The basin of Beibu Gulf is a Mesozoic–Cenozoic rifting basin; it is located in the marine territory of Beibu Gulf. The geographical location is south of the coast line of the North China Sea and north of Hainan Island. The basin covers 12×10^4 km² in area approximately. The basin of Beibu Gulf experienced two large developmental stages, which were the Palaeogene rifting stage and the Neogene depression stage. The Palaeogene rifting stage contained five sub-stages. During the initial rifting sub-stage, the basin began to extend and the development of major faults was initiated. In intra-mountain basins, alluvial fan facies and alluvial plain facies were developed; the sediments mainly consisted of red color, coarse grain clastic rocks of multiple ingredients. During the extensional sub-stage, a water accumulating lacustrine basin was created; the sediments mainly included sandstone and mudstone of littoral–shallow water lacustrine facies and delta facies of river and lake. At the height of its development, the lacustrine basin was continually expanding. The semi-deep, water–deep water lacustrine facies were well developed; the sediments mainly consisted of dark color mudstone. During the contraction sub-stage, the basin decreased in size. The water depth became shallow and the region of littoral–shallow water increased. The sediments mainly contained interbedded gray color sandstone and mudstone. During the uplifting sub-stage, the region of the lacustrine basin decreased, the alluvial plain was well developed and the sediments mainly consisted of red color, multi-color sandstone and sandy conglomerate, which were interbedded with red color mudstone. During the end of the Palaeogene rifting stage, the entire region was uplifted and the lacustrine basin disappeared. During the Neogene, the entire region subsided; the marine transgression came from the southern side of the basin, which brought a marine environment into the basin. The sediments consisted of sandstone and mudstone of littoral–shallow sea facies. As a result, the entire region became a united depressed basin.

In addition, the Chinese marine territories also contain many marine basins that include the South Yellow Sea, Taiwan Strait, Chongsheng Sea Trough, Zengmu, Wenlai–Shaba, Zhongjiannan, Wanan, Beikang, Nanwei West, Nanwei East, Lile, Northwest Balawang, Bijia South, the Nansha Sea Trough, Andu North, Jiuzhang and Yongshu. Overall, the petroleum exploration levels in these basins are still at a preliminary stage. Limited by the content of this book, we will not discuss them in detail.

References

Chi, Y.L., Zhang, W.X., Zhang, H.F., et al., 1996. Genesis of stratigraphic sequence in continental rift basins. Acta Petrolei Sinica, 17(3):19-26.

Feng, Z.Z., Wang, Y.H., Sha, Q.A., et al., 1994. Sedimentology of China. Petroleum Industry Press, Beijing.

Gao, R.Q., Zhao, Z.Z. (Ed.), 2001. Chinese Petroleum Exploration in New Domains (Vols. 1 – 7). Petroleum Industry Press, Beijing.

Gu, J.Y., et al., 1994. Depositional Facies and Related Oil and Gas. Petroleum Industry Press, Beijing.

Hu, J.Y., Huang, D.P., et al., 1991. The Fundamental Petroleum Geology of Terrigenous Basins in China. Petroleum Industry Press, Beijing.

Jia, C.Z., Zhao, W.Z., Zhou, C.N., et al., 2008. The Geological Theory and Exploration Technology for Lithostratigraphic Type of Oil and Gas Reservoirs. Petroleum Industry Press, Beijing.

Liu, B.J., 1980. Sedimentary Petrology. The Geological Publishing House, Beijing.

Qiu, Y.N., 1992. The development of terrigenous clastic reservoir sequences in China. Acta Sedimentologica Sinica, 10:16-24.

Qiu, Y.N., Xue, S.H., Ying, F.X., 1997. The Terrigenous Reservoir Sequences in China. Petroleum Industry Press, Beijing.

Wang, H.Z., 1985. Chinese Paleogeographic Atlas. SinoMaps Press, Beijing.

Wu, C.J., Xue, S.H., et al., 1992. Sedimentology of Petroleum Basins in China. Petroleum Industry Press, Beijing.

Xue, S.H., Liu, W.L., Xue, L.Q., et al., 2002. Sedimentology and Petroleum Exploration in Lacustrine Basins. Petroleum Industry Press, Beijing.

Zhao, W.Z., Zhang, G.Y., et al., 2002. Marine Petroleum Geology and Superimposed Petroleum Basins in China. The Geological Publishing House, Beijing.

Zhai, G.M. (Ed.), 1987. The Collection of Chinese Petroleum Geology. Petroleum Industry Press, Beijing.

Zhai, G.M., Gao, W.L., et al., 2005. Petroleum Geology of China. Petroleum Industry Press, Beijing.

3

Oil and Gas Resources in China

China has relatively abundant petroleum resources. Because of the unique geological settings, there are two noticeable features of petroleum accumulation and distribution. Firstly, both marine sequences and terrigenous sequences contained sizable petroleum resources. However, the terrigenous sequences contained the major portion of the total petroleum resources. Secondly, despite a limited number of petroliferous rich depressions and a low abundance of widely spread petroleum reservoirs, the latter have increasingly become the main target for Chinese petroleum exploration at present and for the future. Although oil and gas exploration has been carried out for almost one hundred years in China, there are plenty of remaining oil and gas resources. In addition, the exploration in many areas is still at a preliminary stage and many new exploration domains need to be investigated.

3.1   Current Status of Petroleum Exploration and Production

Since the founding of the Chinese government over half century ago, thanks to the efforts of several generations of petroleum workers, the Chinese petroleum industry has achieved great success. Firstly, it took less than 20 years to get rid of the label "oil deficient country". Then, a few large scale oil and gas regions were continually being discovered and established. Three important crude oil production centers were set up, which included the eastern center, western center and marine territory. The eastern production center mainly contains Songliao and Bohai bay; the western production center contains Eerduosi, Chaidamu, Tarim, Zhungeer, Tuha and Yumen. The production center of marine territory includes Bohai bay marine territory and the Pearl River estuary. At the same time, four major natural gas production centers have been established onshore. Among the gas production centers, the Tarim basin is the major gas region for the project called "transportation of western gas to the east". The Chaidamu basin is the major gas region for the pipe line Ce–Ning–Lan; the Sichuan basin is the major gas resource for "transportation of Sichuan gas to the east"; the Eerduosi basin is the major gas region for the pipe line of Shan–Jing. In the Songliao basin, the fifth largest gas production center is under construction.

3.1.1   Development of Oil and Gas Fields and Production History in China

According to the 2008 domestic evaluation of proven reserves of oil and gas, there are 843 discovered oil and gas fields; 614 of them are oil fields; 229 of them are gas fields. Up to the end of 2009, the total domestic oil production has continually increased after passing the threshold of 100 millions tons in 1978 (Table 3.1). In 2005, crude oil production passed the threshold of 180 million tones for the first time. In 2007, Chinese crude oil production reached 186 million tons (Fig. 3.1). At the same time, the total oil production in the world was 3618 million tons. The Chinese oil production was ranked in the fifth position (Table 3.2). In 2009, total domestic crude oil production reached 189 million tons per year, which upgraded China to the fourth place internationally. Therefore, China is one of the largest oil producing countries in the world.

In recent years, the growth in natural gas production was fast. At the end of 2007, the remaining proven recoverable reserve of natural gas is 175.15×10^{12} m^3 in the world. Natural gas production in the world is 2.82×10^{12} m^3 per year. In China, the remaining economically recoverable reserve of natural gas is 2.56×10^{12} m^3. Natural gas production is 698.87×10^8 m^3 per year (Fig. 3.2). These numbers are ranked 13th and 9th in the world respectively.

Table 3.1   Annual oil and gas productions from 2002 – 2009

Year	Annual oil and gas productions	
	Oil ($\times 10^8$ t)	Gas ($\times 10^8$ m^3)
2002	1.69	326.30
2003	1.70	341.28
2004	1.75	409.80
2005	1.81	499.50
2006	1.84	593.78
2007	1.86	694.05
2008	1.89	774.73
2009	1.89	830.00

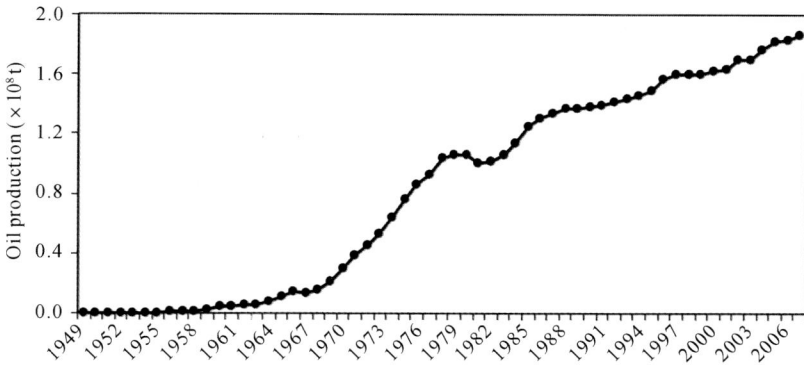

Fig. 3.1.   Curve of annual oil production in China

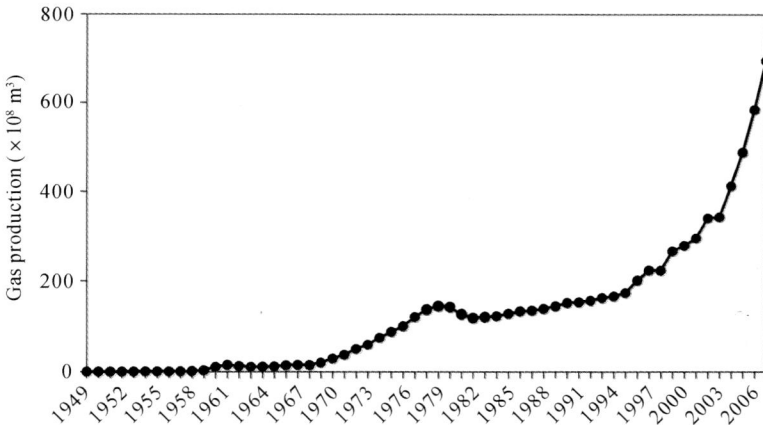

Fig. 3.2.   Graph of annual natural gas production in China

Table 3.2   Top 10 countries oil reserves and annual oil production in 2007 worldwide

Country	Oil reserves ($\times 10^8$ t)	Country	Annual oil production ($\times 10^8$ t)
Saudi Arabia	362.02	Russia	4.86
Canada	244.67	Saudi Arabia	4.21
Iran	189.61	United States	2.57
Iraq	157.55	Iran	1.96
Kuwait	139.06	China	1.87
United Arab Emirates	133.99	Mexico	1.57
Venezuela	119.24	Canada	1.32
Russia	82.20	United Arab Emirates	1.23
Libya	56.81	Venezuela	1.20
Nigeria	49.62	Norway	1.13

3.1.2   Production of Major Oil and Gas Fields

In recent years, domestic oil and gas production has been continually increasing. According to analysis of the development of oil and gas fields and the growth in petroleum production, Chinese oil and gas production has the following five noticeable features. First of all, the oil and gas production wells have been transformed from "a medium production rate in multiple wells" to "a lower production rate in multiple wells". Secondly, the number of horizontal wells has gradually increased annually. Thirdly, the oil and gas fields with low grade reserves (in China the low grade reserves represent 30%–40% of total proven reserves) have been effectively developed with the help of higher oil prices and advanced technology. Fourthly, the oil and gas recovery rates in the marine territory have been gradually increased every year. Lastly, the replacement rate of the oil reserve has increased a small amount. The replacement rate of the natural gas reserve has gradually progressed so as to be balanced. The major Chinese petroleum production regions will be discussed in the following sections (Table 3.3).

Table 3.3   Annual oil and gas production of major petroleum regions in China

Oil field	Annual oil production ($\times 10^4$ t)							
	2002	2003	2004	2005	2006	2007	2008	2009
Daqing	5013.10	4840.03	4640.03	4495.10	4338.10	4162.21	4020.01	4000
Shengli	2671.50	2665.51	2674.30	2694.54	2741.55	2770.08	2774.02	2783
Xinjiang	1005.02	1060.10	1111.06	1165.37	1191.66	1217.06	1222.49	1087
Changqing	610.12	701.56	811.00	940.00	1059.00	1213.02	1379.60	1572

● **Daqing Oil Field**

The Daqing oil field is the largest one in China. In 1976, the annual production of crude oil passed the threshold of 50 million tons for the first time. In 1994, the annual production reached the highest peak of 56 million tons. In 2002, the annual production was 50.13 million tons. The Daqing oil field remained at a high developmental level for 27 years with a stable annual production of 50 million tons. During 44 years of development, the Daqing oil field produced 17.7×10^8 t of crude oil in total, which represented more than 40% of the total domestic production during that time. In 2008, the annual production of the Daqing oil field was 40.2 million tons with a water rate of 91.4% and a recoverable rate of 82.24%. In recent years, the Daqing oil field has reduced its annual production. Compared with the annual production of 2003, the annual production in recent years has been reduced by 8.2 million tons. During five years, the annual reduction in crude oil production was about 1.6 million tons.

At present, the Daqing oil field has progressed to the high water cut stage. The secondary infill adjustment has been positioned. In the future, the tertiary infill adjustment can only be conducted in an isolated area that mainly contains an untabulated reservoir. Thus, oil field development will be more complicated and more difficult. Along with the diminution in infill adjustment in the oil field and the decrease in economic benefit, the stable production of an old oil field has turned to different technology for a third round of recovery that mainly depends on polymer flooding. During more than 40 years of oil field development, the Daqing oil field has solved the major problems at each development stage. Gradually, a series of oil field development and recovery techniques have been carried out, which include the technique of in-depth geological research through sub-dividing depositional facies in the reservoir sequences, the technique of water infilling and comprehensive adjustment that is centered on methodically detailed water infilling technology. Also, there has been the development of oil well design and adjustment at multiple stages and various phases, the technique of polymer flooding and the technique of water infilling for an oil field with low permeability in peripheral areas. As the technology gradually advanced, oil and gas was first recovered from good quality reservoirs. Then, exploration moved to poor quality reservoirs. Simple oil fields were developed, then complicated ones. Petroleum production progressed, recoverable reserves increased and finally there was an improvement in the recovery rate of oil fields. In conclusion, the oil field development model of water infilling to achieve a highly productive stable oil field over a long duration was used.

● **Shengli Oil Field**

Shengli oil field is located on the Yellow river delta, on the northern side of Shandong Province, on the coast of Bohai bay. The oil field region covers 8 cities and 28 counties. These cities are Dongying, Binzhou, Dezhou, Jinan, Huaifang, Zibo, Liaocheng and Yantai. In addition, the major working area covers 4.4×10^4 km^2 of land approximately. The Shengli oil region is the second largest oil region in China; the highest annual oil production has reached 33 million tons. In the middle of the 1990s, because of the shortage of investment for the newly developed area, the previous oil fields were forced to increase their production and the water content in the old oil fields gradually increased. Since then, the infrastructure of the oil fields has been adjusted. By utilizing oil recovery technology for the third round, the development conditions continually improved. Finally, the decreasing trend in annual production was turned around. Moreover, the production rate not only basically stabilized, but the oil production slightly increased. At the end of 2009, in the Shengli oil field, there were 69 discovered oil fields; the total oil bearing area covers 2,865.5 km^2 of land and the discovered geological reserve of oil is 49.33×10^8 t. The discovered geological reserve of gas is $2,367.96 \times 10^8$ m^3. In 2009, the annual production of oil was $2,783.5 \times 10^4$ t; the composite rate of water content was 90.84%. 80.64% of the recoverable reserve has been recovered. The remaining recoverable reserve of oil is $23,159.68 \times 10^4$ t.

● **Xinjiang Oil Field**

The development of the Xinjiang oil field began in the 1930s; the local government of Xinjiang joined with Soviet Union to develop the Dushanzi oil field. However, the production rate was very low. On October 29, 1955, a commercial grade oil flow was obtained in the Triassic Kexia formation in the Ke 1 well. This event led to the discovery of the Kelamayi oil field that was the first large oil field in the petroleum history of the New China. In 1957, the development of the Kelamayi oil field started. In 1958, a production of one million tons per year was established.

Since a joint stock company entered the market in 1999, according to the overall strategy of this company, the Xinjiang oil field was presented with an extraordinary opportunity to extensively develop western China and the developmental pace in the Xinjiang oil fields was increased. In 2002, the annual crude oil production in the Xinjiang oil field was 10.05 million tons. Since then, the Xinjiang oil field has become the first one in western China to have an annual production capacity of 10 million tons. In 2007, the annual crude oil production passed the threshold of 12 million tons. In 2009, the Xinjiang oil field produced 10.89 million tons of crude oil. Since 1981, the annual crude oil production of the Xinjiang oil field has been steadily and continually increasing.

● **Changqing Oil Field**

The Changqing oil field is located in the Eerduosi basin; it covers five provinces (or regions), which include Shanxi, Gansu, Ningxia, Inner Mongolia and Shaanxi. The total exploration area covers 37×10^4 km^2 of land. At the end of 2007, the Changqing oil field produced 1.06×10^8 tons of oil and 497×10^8 m^3 of gas in total.

From 1970 to 1989, the Changqing oil field was in an early developmental stage. The medium–low permeability oil reservoirs (> 50 mD) in the Jurassic Maling formation were developed. The production rate of a single well was 5 – 6 t; the annual production of crude oil was stable at 1 million tons approximately. From 1990 to 1999, the Changqing oil field progressed to a stable development stage; the super large, extremely low permeability oil reservoirs (2 mD) in the Ansai and Jinan areas (such as the Chang 6) were developed. The production rate of a single well was above 3.0 t. The additional annual crude oil production was 300,000 t. approximately. After 2000, along with the development of extremely low permeability and ultra low permeability oil reservoirs (<1.0 mD) in the Xifeng, Jiyuan and Huaqing areas, the Changqing oil field has moved into a fast development stage. The additional annual production rate of oil was more than 1 million tons. At present, the major oil fields are producing at speed, the oil recovery rate is relatively high and the effective exploitation of the oil reserve is the crucial factor that will keep the growth in production at a fast pace and will guarantee the stability of production as well.

At present, the growth in the Chinese oil reserves and oil production are still focused on the major basins that include Bohai bay, Songliao, Eerduosi, Tarim, Zhungeer and the Pearl River estuary. On the other hand, the growth in the Chinese gas reserves and gas production will be centered on Sichuan, Tarim, Eerduosi, Songliao, Zhungeer and the Pearl River estuary basins (RIPED, 1994).

The development of Chinese oil and gas fields is facing the following problems. Because most of the major oil fields are either at the stage of high water content or at a high recovery rate stage, a stable production rate is harder to achieve. Among the newly discovered oil and gas reserves, the percentage of low permeability reserves (that are harder to recover) is gradually increasing, which puts the development of oil fields in a very difficult situation. The recovery technique for heavy crude oil, the technique for the third round of recovery, and the recovery technique for low pressure, low permeability, and for gas fields of low abundance, are applied in limited areas only. In addition, the current recovery technology still needs to be further improved.

In 1990, the ratio of reserves to recovery was only 16. Then it continually increased. In 1998, the ratio of reserves to recovery for the first time passed the threshold of 40 (the precise figure was 42). In 2003, this ratio of reserves to recovery reached its peak value of 61 (Fig. 3.3). After that, because of an increase in the annual production, the ratio slightly decreased. In 2008, the ratio of reserves to recovery was 42. At PetroChina, the ratio of reserves to recovery increased from 10 (in 1990) to 67 (in 2003) to reach its peak value. After that, this ratio decreased to 39 (in 2008). In 2008, for Sinopec, the ratio of reserves to recovery was 75; for CNOOC, the ratio was 33. According to the ratio of reserves to recovery at present, the fast increase in Chinese gas production is based on excellent reserves.

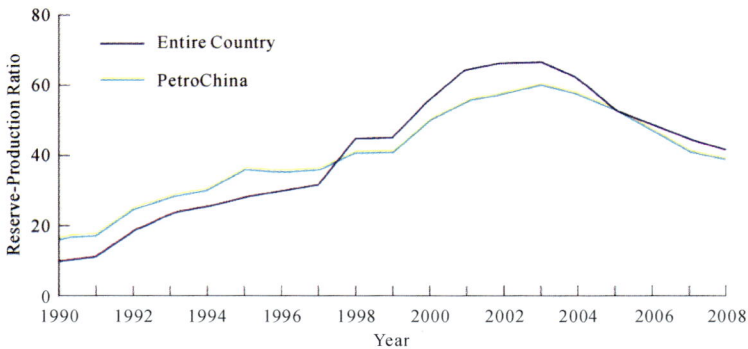

Fig. 3.3. Reserve-production ratio of natural gas in China

3.1.3 *Chinese Petroleum Reserves*

Chinese oil and gas exploration has been a great success. The proven reserves of oil and gas have increased in value. Following the policy of exploring for both oil and gas, every petroleum company focused their exploration activities on lithostratigraphic sequences, foreland basins and marine carbonate rocks. Many important petroleum reservoirs in the Bohai bay, Eerduosi, Tarim, Sichuan, Zhungeer and the Songliao basins were discovered in succession. At the end of 2008, PetroChina discovered 265.10×10^8 t of geological oil reserves in total. Among the oil reserves, 74.23×10^8 t were discovered to be technically recoverable reserves, and 67.04×10^8 t were discovered to be economically recoverable reserves. In 2008, the annual production of oil was 1.84×10^8 t. The accumulative production of oil was 47.19×10^8 t. The remaining technically recoverable reservoir reserves are 27.59×10^8 t. The remaining economically recoverable reserves are 20.40×10^8 t. At the end of 2008, PetroChina discovered 6.69×10^{12} m^3 of geological reserves of natural gas in total. Among the natural gas reserves, the technically recoverable reserves of gas were 3.78×10^{12} m^3. The discovered economically recoverable reserves of gas were 3.17×10^{12} m^3. In 2008, the annual production of gas was 584.46×10^8 m^3. The accumulative production of gas was 0.78×10^{12} m^3. The remaining technically recoverable reserves of gas are 3.01×10^{12} m^3. The remaining economically recoverable reserves of gas are 2.39×10^{12} m^3.

3.1.4   *Current Status of Petroleum Exploration and Exploration Phases*

● Current Status of Oil and Gas Exploration

After experiencing more than 50 years of ups and downs and after an arduous struggle, Chinese petroleum exploration has rapidly developed. Since the founding of the New China, Chinese geological workers have constantly studied advanced ideas and newer geological theories from around the world. At the same time, from their own exploration experiences, Chinese geological workers developed a set of geological theories based on the characteristics of Chinese petroleum geology, which included the theory of terrigenous oil generation, the theory of composite oil and gas accumulations and the theory of coal bed methane (Hu et al., 1991). These Chinese geological theories successfully advanced petroleum exploration in China. Accordingly, geologists also developed a set of unique exploration technologies (that included the technology for a surface geological survey, the technology for geophysical exploration, the technology for well boring, the technology for geological testing and the technology for petroleum resource evaluation for basin–region traps). The development of geological theories and the innovations in exploration technology made great contributions to the development of the Chinese petroleum industry. China has 505 sedimentary basins with a total depositional area of 670 km^2. Since the founding of the P.R. China, oil and gas exploration has been carried out in 105 basins. At the end of 2005, more than 600 oil fields in 79 large and medium size basins had been discovered. The total discovered geological reserve of oil was 256.08×10^8 t; the annual production of crude oil was more than 1.8×10^8 t. Therefore, with these achievements, China has become the fifth largest oil producing country in the world.

Since the "Ninth Five-Year Plan", Chinese petroleum exploration has emphasized the development of geological theories and the study of comprehensive geological configurations. At the same time, preparation prior to seismic surveys has been intensified. By widely utilizing advanced and practical mainstream exploration technology, geologists have aggressively explored new domains and realized multiple exploration achievements. In eastern China, in-depth exploration was continually practiced in mature basins. The discovered geological reserves of oil steadily increased. In western-central China, major exploration breakthroughs and great discoveries repeatedly occurred and were announced. The discovered geological reserves of oil increased greatly. In the nearshore marine territories, exploration frequently led to new discoveries; the discovered geological reserves of oil speedily increased. Chinese petroleum exploration moved into a new developmental period. The growth of discovered geological reserves attained a high value. At the same time, China not only highly valued geological research and innovative exploration technology, but also increasingly imported exploration technology. Important developments and innovations were achieved in multiple subjects. As a result, the new technology improved the possibility of discovery and made great breakthroughs. Moreover, the new technology expanded the petroleum exploration domain (Jia, 2007; Qu et al., 2006).

In China, PetroChina is not only the largest producer and supplier of crude oil and natural gas, but also the largest producer and supplier of chemical materials that are related to the oil refining business. Among the top 50 large oil companies around the world, PetroChina is positioned in the 7th place. At present, PetroChina has 412 blocks with exploration rights for oil and gas inside China. The exploration blocks cover 180.14×10^4 km^2 of land; these numbers represent 36% and 42% of total domestic numbers, respectively. PetroChina also has 384 blocks with mining rights. These mining blocks cover 6.36×10^4 km^2 of land; these numbers represent 62% and 82% of total domestic numbers, respectively. In total, there are 796 blocks that are either issued with exploration rights or mining rights. The total area of these blocks covers 186.53×10^4 km^2 of land; the numbers stand for 45% and 42% of total domestic numbers, respectively. At the same time, PetroChina absolutely dominates the mining rights in the new frontiers of coal bed methane, oil sands and oil shale.

Since the funding of the company in 1999, PetroChina has achieved important successes and made new developments in oil and gas exploration. The growth in oil and gas reserves and production have continually and steadily increased. At the end of 2004, the total domestic proven reserve of oil was 246.83×10^8 t; PetroChina controlled 156.13×10^8 t of the proven oil reserve, which represented 63% of the total domestic proven oil reserve. The total domestic proven reserve of natural gas was 4.38×10^{12} m^3; PetroChina controlled 3.24×10^{12} m^3 of the preserved gas reserve, which represented 74% of the total domestic proven gas reserve. The total domestic proven reserve of coal bed methane was $1,023.09 \times 10^8$ m^3. PetroChina controlled 754.45×10^8 m^3 of the proven reserve of coal bed methane (which included 402.19×10^8 m^3 of the reserve from the CNCM), which represented 74% of the total domestic proven reserve. During the 1st four years of the "Tenth Five-Year Plan", the growth in the oil and gas reserves equalled 13% and 300% of the oil and gas reserves of the "Eighth Five-Year Plan" respectively. The total discovered geological reserve of oil was 18.5×10^8 t, the average annual growth of the oil reserve was 4.63×10^8 t, crude oil production consistently increased and the average annual growth in oil production was 28×10^4 t. The total discovered geologic reserve of natural gas was 1.3×10^{12} m^3 and the average annual growth of the gas reserve was $3,238 \times 10^8$ m^3. Natural gas production rapidly increased, the average annual growth in gas production was 20×10^8 m^3, the growth rate of gas production was more than 10%. In 2004, PetroChina progressed into a new development period; the newly discovered proven reserve of oil was 5.255×10^8 t, which was the eighth time that the annual discovered oil reserve passed 5×10^8 t and which was a new highlight since the reform of the company. In the same year, the newly discovered natural gas reserve was $2,009 \times 10^8$ m^3, which continually keeps growing at a high pace. In 2004, the total domestic crude oil production was $17,499 \times 10^4$ t. PetroChina produced $10,446 \times 10^4$ t of crude oil, which represented 59.7% of the total domestic oil production. In the same year, the production of natural gas was 408×10^8 m^3. PetroChina produced 285.32×10^8 m^3 of natural gas, which represented 70% of total domestic gas production.

● **Exploration Phases**

According to the size of the exploration workload per unit area, the tendency of the annual growth in the reserves and the investigation level of the resources, the exploration history of a region could be divided into a preliminary phase, middle phase and later phase. The exploration forecast for a region is based on the analysis of the exploration phases of that region. The United States of America has a relatively longer petroleum exploration history and higher exploration level. Both oil and gas explorations have entered the later phases, the annual growth in the oil and gas reserves clearly show a decreasing trend, the explored reserve/resource rate of oil and gas is above 60% (Table 3.4).

According to the identification index of the exploration phase from the USA, compared with the statistical bar chart of the annual growth in the reserves and also by using the numbers from the newer round of domestic petroleum resource evaluation and the data of domestic proven reserves of oil and gas at the end of 2004, the explored reserve/resource rate of PetroChina is 32%. Because this rate revealed that the exploration level of the PetroChina just equals the exploration level of the 1950s – 1960s in the USA, PetroChina is in the middle phase of overall exploration. Based on these deductions, when considering that the explored reserve/resource rate of petroleum is 65% in eastern China and 45% – 50% in western-central China, for the next phase of petroleum exploration, the probability of an important discovery is great. Over a reasonably long duration, the annual growth in the proven reserve will retain its high base value and growth will be stable. According to the natural gas exploration history of the USA, when the explored reserve/resource rate reaches 10%, the proven reserve will enter the stage of fast growth. In China, the explored reserve/resource rate of natural gas (that does not include southern marine territory of the South China Sea) is 12.5%. This rate indicates that the natural gas exploration level in China equals the exploration level of the 1920s – 1930s in the USA, which was the time of fast growth in the reserves and which was an early exploration phase. Thus, the annual growth in the proven reserve will increase rapidly for a reasonably long duration.

Table 3.4 Growing steps of the petroleum reserves in the United States

Exploration stage	Major indexes		
	Proved resources (%)	Control area per single exploration well (km^2/single well)	Annual increase of reserves
Fast growing period	< 30	> 100	Fast increase
Stable growing period	30 – 60	5 – 100	High base value with small fluctuation
Slowly decreasing period	> 60	< 2 – 5	Slowly decreasing

• Characteristics of Petroleum Exploration and Growth of Oil and Gas Reserves

During almost one hundred years of petroleum exploration history, Chinese petroleum exploration and reserve growth have unmistakable Chinese characteristics and related pattern. The nature of oil and gas exploration has multiple phases; the nature of reserve growth has various peaks.

Most of the Chinese petroleum basins are superimposed basins since the Mesozoic and the Cenozoic terrigenous sequences were deposited on top of the Paleozoic marine sequences. The depositional and superimposed processes of many basins did not simply pile up several sets of depositional sequences. During the superimposed processes, the creation and distribution of depositional sequences in the deep and shallow levels had noticeable variation and distinctive alterations. Moreover, the source rocks, reservoir sequences and the petroleum reservoir that was located in the depositional sequences also displayed variation and alteration. In particular, for the superimposed basins that experienced large magnitude tectonic movement during the depositional interval between the early and late sequences, the petroleum accumulation procedure and the distribution patterns of the petroleum reservoirs displayed a clear distinction between the nearby depositional sequences or between the contiguous structural unit. Compared with the single cycle basins overseas, these characteristics of a Chinese superimposed basin were obviously different.

Differing from the basins overseas, the Chinese superimposed basins experienced multiple phases of tectonic movement throughout geological history. Some tectonic movements were quite strong. Therefore, there were various phases of oil and gas generation and accumulation. In addition, these different oil and gas phases were not only mixed up, but they also vanished in different measures due to the tectonic movements. Thus, in China, the procedure for oil and gas generation and distribution was very complicated. This kind of situation determined that the petroleum basin had abundant oil and gas resources, which offered a wonderful resource foundation for petroleum exploration. On the other hand, it also indicated the complicated nature of petroleum resources, which brought great intricacy and bigger obstacles to the exploration process. Thus, the exploration experience and acquired knowledge (obtained during the exploration in a set of sequences, an exploration domain, or a structural unit) would not be simply used for reorganizing the accumulation process and for advising an exploration project in another set of sequences, another domain, or another structural unit. The procedure of reorganization and the progress of the exploration should be intensified gradually with developmental phases (Zhao et al., 2007).

The exploration breakthrough from a sequence, a domain, or type of petroleum accumulation would not only encourage continual investigation on a large scale, but it would also lead to the peak in reserve growth. On the other hand, the exploration experiences and the acquired knowledge from a new sequence, a new domain, or new type of petroleum accumulation would prompt a newer understanding, a newer discovery, or newer achievements. Likewise, it also would lead to a new peak in reserve growth. For example, in the Eerduosi basin, the exploration of the 1970s mainly targeted the Jurassic sequences; petroleum exploration achieved a first peak in reserve growth. At the end of the 20th century, with the practice of in-depth exploration and the gaining of detailed knowledge, the exploration target was switched to the Chang 6 member in the Triassic system. Large oil fields in the Ansai and Jinan areas were discovered, which brought the second peak in reserve growth. Since 2000, along with the development of research work, geologists evaluated and explored the entire Triassic system. In the Longdong area, the Xifeng oil field was discovered in the Chang 8 member. In the Jiyuan area, future exploration targets were determined in the Chang 4+5 members and the Chang 6 member. These newest discoveries brought the third peak in reserve growth to the basin. Another example is found in the Jiuquan basin. After near 50 years without any discovery, in 2,000 geologists discovered the Qingxi oil field. The 3rd grade reserve has reached 1×10^8 t. This discovery changed the situation where it was said that "exploration would not find an oil reserve" in the Yumen oil field.

Therefore, compared with the single cycle basin found overseas, the progress in understanding and the exploration history of the Chinese petroleum basin took rather a lengthy time. In particular, the discovery of large and medium size oil and gas fields commonly contained multiple phases. The growth of the petroleum reserve clearly had several peaks.

Considering all previous domestic petroleum resource evaluations and petroleum resource evaluation worldwide, along with the developments in geological theory and the innovations in technology, together with the growth in newer exploration regions and domains, no matter if they are around the world or in China, the quantity of oil and gas resources will continually increase.

When considering the preconditions of the southern marine territory of the South China Sea, the new region in Qinghai and Tibet, and considering the developments in exploration theory and the innovation in technology, over the next 45 years, with a medium to high range in oil prices, a high peak in reserve growth could happen again. In the future, the total expected recoverable oil resource could reach 200×10^8 t and the total expected recoverable resource of natural gas could reach 22×10^{12} m^3.

3.2   Evaluation of Oil and Gas Resources

First we will introduce the history and the methods of resource evaluation.

3.2.1   History and Methods of Resource Evaluation

The evaluation of the Chinese petroleum resource could be divided into four phases. The first phase was the domestic evaluation of the oil and gas resource, which was carried out before the 1980s. The second phase was the first round of domestic petroleum resource evaluation that was conducted between 1981 and 1987. The third phase was the second round of domestic petroleum resource evaluation that was performed between 1991 and 1994. The fourth phase included the petroleum resource evaluations conducted by each of the large oil companies after the "Ninth Five-Year Plan" and the newest round of domestic petroleum resource evaluation that was carried out between 2002 and 2005.

- **Domestic Petroleum Resources Evaluation Before 1980**

Before the 1980s, both domestic and foreign oil companies, the government agency and scientists roughly estimated Chinese petroleum resources. Among these evaluations, typical estimations included the following: (1) The Mobil Petroleum Co. (1922) estimated that the total amount of the Chinese petroleum resource was 1.75×10^8 t; this estimation was based on the geological theory that the Chinese sedimentary basins basically did not have the conditions for oil and gas accumulation. (2) The Petroleum Ministry of China (1964) estimated that the total amount of the Chinese petroleum resource was 115.51×10^8 t; this estimation was based on the breakthrough in geological theory, whereby oil and gas were generated by terrigenous sequences. (3) The Petroleum Ministry of China (1971) estimated that the total amount of the Chinese petroleum resource was 217×10^8 t. (4) Shicong Guan (1979) estimated that the total amount of the Chinese petroleum resource was 450×10^8 t (Fig. 3.4).

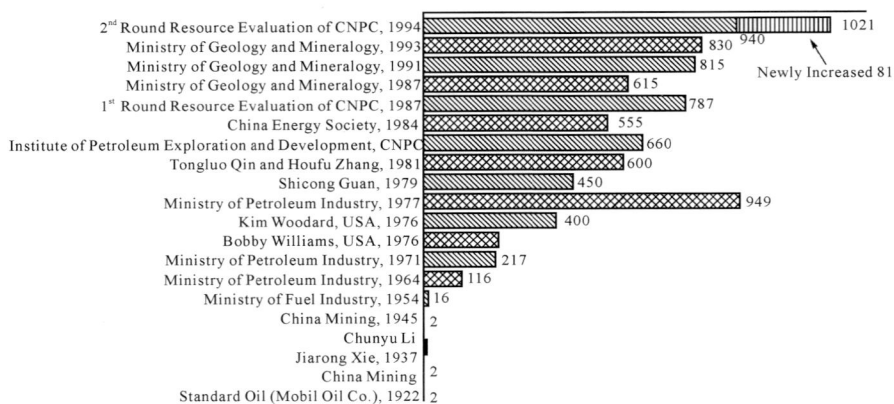

Fig. 3.4.   Change in Chinese petroleum resources (Unit: 10^8 t)

● **The First Round of Domestic Petroleum Resources Evaluation**

During 1981 – 1987, the Petroleum Ministry and the Geology Ministry worked together to carry out the first round of domestic petroleum resource evaluation. Compared with the previous estimations, this evaluation not only systematically summarized the characteristics of Chinese petroleum geology, but it also enhanced and developed the theory of petroleum geology. Based on the comprehensive analysis of evaluation theory and methods from foreign countries, analyzing and selecting from more than 50 kinds of evaluation methods that were used in China, considering Chinese geological conditions, considering the nature of evaluation techniques and the nature of laboratory analysis and testing, a system of evaluation of natural resources, which was the RES – 851 evaluation system, was developed. From the sedimentary regions in China, 283 sedimentary basins and regions that contained an area larger than 200×10^4 km^2 and that had a thickness greater than 600 m, were selected. By deduction, 143 of these basins (or regions) were then chosen as the evaluation targets. The basic unit of evaluation was a basin. By using the calculation method that mainly considered hydrocarbon generation–discharging–accumulation, and by using the RES–851 evaluation system, geologists projected the amount of the domestic oil resource to be 787×10^8 t and the amount of the domestic gas resource to be 33×10^{12} m^3.

● **The Second Round of Domestic Petroleum Resources Evaluation**

During 1991 – 1994, the CNPC conducted petroleum resource evaluation in mainland China. This round of evaluation also invited the CNOOC and calculated oil and gas resources in the near shore marine territories. The second round of resource evaluation not only used a huge amount of new data and greatly exceeded the last round of evaluation in the variety of items evaluated and in the details of the subject, but it also using the unified evaluation method and evaluation software (basin modeling was the primary method). According to the new exploration achievements, and according to the new theory of petroleum geology and the new knowledge (these new criteria were developed after the first round of petroleum resource evaluation), geologists systematically evaluated 150 petroleum basins, 618 regions and zones, and 7,792 traps. Finally, geologists proposed that the amount of the domestic oil resource was 940×10^8 t and the amount of the domestic gas resource was 38×10^{12} m^3 (this did not include the oil and gas resources in Taiwan and in the southern marine territory of the South China Sea). Furthermore, geologists analyzed the potential of the domestic petroleum resource and predicated the trend of future exploration in every region, which provided scientific evidence of the growth of the petroleum reserve and the growth of petroleum production in China.

- **Petroleum Resources Evaluation Conducted by Oil Companies after the "Ninth Five-Year Plan" and the Newest Round of Domestic Petroleum Resources Evaluation**

Entering the 21st century, the imbalance between petroleum supply and demand has been further intensified worldwide. For securing domestic supplies of oil and gas resources, for further understanding the inventory of the petroleum reserve and the potential of Chinese oil and gas resources, for better investigating and evaluating, preparing, managing, protecting and utilizing Chinese oil and gas resources, and for providing the scientific data to make the "Eleventh Five-Year Plan" and to create a medium and long term energy development plan, the Ministry of Land and Resources, the National Development and Reform Commission, and the Ministry of Finance worked together to carry out the newest round of domestic petroleum resource evaluation. They took into consideration the progress in the exploration and development of the Chinese oil and gas resources since the second round of domestic petroleum resource evaluation and the current situation based on recent evaluations from each oil company.

During the newest round of domestic petroleum resource evaluation, 129 basins from onshore and marine territories were selected as the evaluation targets; these basins covered in total 530×10^4 km^2 of land and water. There were fifteen companies and institutes that joined the evaluation, which included PetroChina, Sinopec, CNOOC, the China University of Petroleum (Beijing), the China University of Geosciences and the Chinese Academy of Geological Science (CAGS). In total, there were 726 scientists who attended the evaluation work.

The results of the newest round of domestic petroleum resource evaluation indicated that in 115 basins (either onshore or in the near shore marine territories, not including 14 basins in the southern marine territory of the South China Sea), the amount of the foreseeable oil resource was $1{,}086 \times 10^8$ t, the amount of the geological oil resource was 765×10^8 t, the amount of the recoverable oil resource was 212×10^8 t, the amount of the foreseeable resource of natural gas was 56×10^{12} m^3, the amount of the geological resource of natural gas was 35×10^{12} m^3, and the amount of the recoverable resource of natural gas was 22×10^{12} m^3.

The evaluation results showed that the overall exploration level of the Chinese petroleum resource was low. The amount of the undetermined geological resource of oil was 517×10^8 t, the amount of the undetermined recoverable resource of oil was 144×10^8 t, the amount of the undetermined geological resource of natural gas was 31×10^{12} m^3, and the amount of the undetermined recoverable resource of natural gas was 19×10^{12} m^3. The percentage of the explored oil resource is 33%; the percentage of the explored gas resource is 13%. The evaluation results also showed that China has abundant oil and gas resources. The amount of the recoverable oil resource has clearly increased. The amount of the natural gas resource has increased greatly. Overall, in China, oil exploration is in the middle phase and natural gas exploration is in the early phase.

Using prediction analysis, at present the Chinese explored reserve of oil is increasing steadily. Until 2020, the proven reserve of oil will show stable growth. The amount of the annual proven geological reserve of oil is $7 \times 10^8 - 9 \times 10^8$ t.The proven reserve of natural gas is rapidly increasing. The amount of the annual proven geological reserve of natural gas is $4{,}000 \times 10^8 - 5{,}000 \times 10^8$ m^3. The production of oil will increase steadily. In 2015, the amount of forecast oil production will pass the threshold of 2×10^8 t. In addition, the annual production rate of oil will keep at this level for the next 10 years. Natural gas production is increasing at a fast pace. In 2010, the amount of forecast gas production will pass the threshold of 800×10^8 m^3. In 2020, the amount of gas production will reach $1{,}500 \times 10^8$ m^3.

Through this round of petroleum resource evaluation, geologists not only acquired fresh knowledge of the hidden potential of the Chinese petroleum resource and the growth in the proven resource, but they also obtained a few new ideas for improving the Chinese petroleum industry in the future. First of all, the expansion in petroleum exploration and development should build on the foundation of intensifying the research in geological theory, increasing exploration activity and improving the exploration technology. Secondly, emphasizing scientific innovation in the sector of petroleum development will improve the oil and gas recovery rate on a large scale and will guarantee growth of the petroleum reserve and petroleum production. Thirdly, the completion of the energy structure calls for speeding up pipeline construction to promote the unification of the upper, middle and lower streams of natural gas production and to increase the proportion of natural gas in the primary energy consumption structure. Fourthly, stable growth of Chinese oil production requires a supportive policy with regard to older oil fields in eastern China to encourage the development of oil fields that have higher water content, heavier condensed oil and lower abundance. By determining the adequate production rate, these old oil fields could continue to produce oil.

Next, reducing the gap between Chinese technology and the most advanced technology in the world demands an intensified study of exploration and development technology in deep waters to improve petroleum exploration and development in marine territories. This strategy could not only increase the production of oil and gas, but it could also build a foundation for Chinese oil companies to develop oil and gas fields around the world. Lastly, through the annual dynamic evaluation of oil and gas resources and comprehensive system evaluation every five years, evaluations of special projects should be the focal point. These projects include the southern marine territory of the South China Sea, Qinghai and the Tibetan region. In addition, selected evaluation in the peripheral regions around China and in foreign countries should also be listed on the agenda. As a resource country, for advising on petroleum exploration, for making medium and long term energy development plans, for making energy policy and for providing important scientific data, the evaluation of petroleum resources requires systemization, institutionalization and dynamism.

- **Methods of Petroleum Resources Evaluation and the Current Technology**

For the evaluation of petroleum resources, the major foreign oil companies have developed some good methods. The major categories of evaluation methods and techniques include the area yield method and the system production rate method (combined with the geological analogy method) (Wu, 1994), the Delphi method or the method of subjective identification, the extrapolation method (based on the historical characteristics), the method of geochemical balance and the comprehensive method of geological modeling and statistical modeling. In recent years, based on a thorough analysis of the basic physical and geological factors that influenced resource evaluation and that controlled resource distribution, every major foreign oil company has introduced new evaluation methods and new techniques. In particular, during economic analysis, the foreign oil companies studied the supply curve and the information about short term energy demands. This study not only connected resource evaluation and market demand analysis, but it also utilized artificial intelligence in resource evaluation and in the study of the evaluation method, which repaid itself in the development of evaluation technology.

In China, the study of resource evaluation technology had a late start. Compared with the major foreign oil companies, there is a gap. Before 2000, during petroleum resource evaluation, although we used various methods, the genetic method was the only primary evaluation method. In general, the quantity of the petroleum resource was estimated by the genetic method. During individual evaluation, the statistical method was selectively used. Because the geological analogy method did not have a unified standard of comparison, when calculating the quantity of the resources, the factors determining the comparison were based on the relative correlation between the basins or between the depressions. In resource evaluation, most of the work focused on the geological evaluation. The evaluation of the economic value of the resources and the evaluation of the recoverable resources were relatively poor. After 2000, on account of the project "petroleum resource evaluation for major onshore petroleum basins" that was operated by the Research Institute of Petroleum Exploration and Development (RIPED), and especially on account of the newest round of domestic petroleum resource evaluation in 2005, Chinese petroleum revaluation technology has greatly improved in the areas of methods applied, parameters and software. At the evaluation level, the system of evaluation was based on two large evaluation levels of the basin and the target (block zone and block). For systemization of the evaluation method, the system of evaluation was set up using the genetic method, the statistical method and the geological analogy method.

Genetic Method.   Recently, by establishing the parameters of oil and gas migration and by establishing the relational model of geological factors, geologists solved the problem of how to scientifically collect the crucial parameters of oil and gas migration by the genetic method. Thus, the genetic method has progressed.

Geological Analogy Method.   The newly developed quantitative geological evaluation method solved the forecasting problem for oil and gas distributions in 3D space, which achieved the goal of integration with the international standard.

Statistical Method.   This method mainly includes statistical trend prediction, the oilfield scale serial method, the oilfield scale distribution method, the scale model between a large scale oilfield and a medium–small scale oilfield, the reserve growth model, the reserve variation model and regression analysis with the multiple geological factors model. In recent years, the petroleum resource distribution forecasting method has been the latest development.

3.2.2   *Measures of Oil and Gas Resources*

● **Result of the Newest Round of Domestic Petroleum Resources Evaluation**

Chinese oil and gas resources consist of conventional resources and unconventional resources. The conventional oil and gas resources are mainly distributed onshore and in marine territory. The unconventional oil and gas resources include oil shale, oil sands and coal bed methane.

According to the newest round of domestic petroleum resource evaluation, the quantity of the total conventional geological resource of oil is 895.1×10^8 t. The quantity of the recoverable resource of oil is 254.9×10^8 t. In the total conventional geological resource of oil, the onshore reservoirs contain 657.6×10^8 t of oil; 182.8×10^8 t are a recoverable oil resource, which represents approximately 73% of the total conventional oil resource. The marine territory contains 237.5×10^8 t of the conventional geological resource of oil; 72.1×10^8 t of this is a recoverable resource, which represents approximately 27% of the total conventional oil resource.

The total quantity of the conventional geological resource of natural gas is 43.87×10^{12} m^3; the quantity of the recoverable resource of natural gas is 27.48×10^{12} m^3. In the total conventional geologic resource of natural gas, the onshore reservoirs contain 26.93×10^{12} m^3 of gas; 16.78×10^{12} m^3 of this is a recoverable gas resource, which represents approximately 61% of the total conventional gas resource. The marine territory contains 16.94×10^{12} m^3 of the conventional geological resource of gas; 10.70×10^{12} m^3 of this is a recoverable resource, which represents approximately 39% of the total conventional gas resource.

China has abundant unconventional oil and gas resources, which include oil shale, oil sands and coal bed methane. Among these unconventional oil and gas resources, the quantity of the coal bed methane resource is 27.3×10^{12} m^3 with a buried depth of less than 1,500 m. The quantity of the oil sand resource is 71×10^8 t approximately (Chinese Academy of Engineering, 2005). The quantity of the oil shale resource is 62×10^8 t approximately. In addition, some scientists predicted that there are $600 \times 10^8 - 700 \times 10^8$ t oil equivalent of natural gas hydrate (NGH) deposited in an area of 1 million km^2 on the continental shelf and continental apron in the South China Sea. Furthermore, on the Tibetan highland, the permafrost region may contain a huge amount of natural gas hydrate. The exploration and development of unconventional oil and gas resources are still at a preliminary stage.

● **Distributions of Petroleum Resources**

According to the results of the newest round of domestic petroleum evaluation in 2006, the quantity of the foreseeable resource of oil is $1,086 \times 10^8$ t (not including the southern marine territory of the South China Sea). The quantity of the geological resource of oil is 765×10^8 t approximately. Among the geological resources of oil, the onshore reservoirs contain 658×10^8 t of oil approximately and the marine territory contains 107×10^8 t of oil approximately. The quantity of the foreseeable resource of natural gas is 55.89×10^{12} m^3 approximately (not including the southern marine territory of the South China Sea). The quantity of the geological resource of gas is 35.03×10^{12} m^3. In the geologic resource of natural gas, the onshore reservoirs contain 26.93×10^{12} m^3 approximately and the marine territory contains 8.10×10^{12} m^3 approximately.

The Chinese petroleum resource is mainly distributed in the large size petroleum basins that include the Songliao basin, Bohai bay basin, Eerduosi basin, Sichuan basin, Zhungeer basin, Tarim basin, the basin of the Pearl River estuary, Tuha basin and Chaidamu basin. The majority of the major petroleum basins are superimposed basins that contain marine sequences and terrigenous sequences. A huge amount of oil accumulated in the Mesozoic and Cenozoic erathem. On the other hand, a huge amount of natural gas accumulated in the Paleozoic erathem.

3.2.3   Exploration Domains of the Future

At the end of 2007, the total size of the discovered geological reserve of oil was 275.2×10^8 t; the size of the discovered recoverable reserve of oil was 76.1×10^8 t; the size of the discovered geological reserve of natural gas was 5.94×10^{12} m^3; the size of the discovered recoverable reserve of natural gas was 3.66×10^{12} m^3. The size of the remaining conventional geological resource of oil was 276.87×10^8 t. If we add the 50×10^8 t of oil from the southern marine territory of the South China Sea, the total size of the remaining resource of oil was 326.87×10^8 t. The remaining geological resource of natural gas was 16.05×10^{12} m^3. If we add the 5×10^{12} m^3 of natural gas from the southern marine territory of the South China Sea, the total size of the remaining resource of natural gas was 21.05×10^{12} m^3. Overall, the discovery level of the Chinese petroleum resource is low; the percentage of the explored geological oil resource is 32%; the percentage of the explored geological gas resource is only 12.5%. The remaining oil and gas resources are abundant, which provide the foundation for further development on a grand scale. The remaining oil and gas resources are mainly distributed in the following seven locations: the lithostratigraphic type of reservoirs, the foreland basin, the middle–lower petroleum system in a superimposed basin, the former area in a mature basin, the new area in a new basin, the southern marine territory of the South China Sea and unconventional exploration domains. These seven exploration domains contain the remaining conventional oil and gas. The size of the geological resources are 326.87×10^8 t and 21.05×10^{12} m^3 respectively (Table 3.5). Considering the distribution pattern of the remaining oil and gas resources, in future these seven exploration domains will be important exploration targets for PetroChina (Jia, 2005; Zhao et al., 2005).

Table 3.5   Statistical table showing remaining petroleum resources in China

Domain	Remaining oil resource		Remaining gas resource	
	Quantity ($\times 10^8$ t)	Percentage (%)	Quantity ($\times 10^8$ t)	Percentage (%)
Lithostratigraphic type of reservoir	111	24	2.31	6.5
Foreland basin	40	9	5.74	16.3
Middle and lower petroleum system in superimposed basin	27	6	4.65	13.2
Old petroleum region in matured basin	33	7	0.86	2.4
Newer petroleum region, new basin	66	14	2.48	7.0
Southern marine territory in the South China Şea	50	11	5.00	14.2
Non-regular resource	133 (Oil sands: 71; Oil shale: 62)	22	14.30 (Coal bed methane)	40.5
Total	460	100	35.30	100.0

● **Exploration Domains for the Lithostratigraphic Type of Petroleum Reservoir**

According to the unique nature of a terrigenous basin, a set of complete stratigraphic sequences for a terrigenous faulted basin and terrigenous depression was gradually researched in recent years, which proposed that different types of basin contained different types of exploration targets in the lithostratigraphic type of oil and gas reservoir. In addition, it was also suggested that the related development conditions and development configuration varied. In particular, in the exploration regions of eastern China, more detailed knowledge was obtained in the following areas: the depositional system, the distribution of the sand bodies, the characteristics of petroleum accumulation and the distribution pattern of oil and gas reservoirs (Jia et al., 2004). Geologists willingly changed their exploration philosophy. Geologists developed new exploration strategies and related technology for in-depth exploration, in order to discover the lithostratigraphic type of oil and gas reservoirs. Among the technological developments, high resolution 3D seismic technology and high resolution stratigraphic sequences analysis were two crucial technologies for exploring the lithostratigraphic type of oil and gas reservoirs. In some previous exploration areas, the biggest changes happened in geological identification and exploration philosophy. Therefore, exploration activities were transformed from discovering the structural type of oil and gas reservoirs to an in-depth exploration of the lithostratigraphic type of oil and gas reservoirs. By overturning a troublesome exploration situation, geologists discovered a huge amount of oil and gas in highly explored areas. For example, two mega size exploration regions were discovered for the exploration of the lithostratigraphic type of oil reservoir in the Xifeng and Jiyuan areas. At the end of 2004, in the Xifeng oil field, the size of the discovered geological reserve of oil was 1.99×10^8 t. The size of the controlled oil reserve was 0.59×10^8 t, the size of the foreseeable oil reserve was 2.56×10^8 t, the size of the third order oil reserve was 5.15×10^8 t. The established production capacity for crude oil was 95.5×10^4 t. The annual production of crude oil was 59×10^4 t in 2004. According to the geological conditions and the potential of the resources, the growth in the oil reserve is continually increasing in the Xifeng area. In 2004, three oil sand belts were explored in the Jiyuan area, which were located in Tiebiancheng, Baoziwan and Xiaojianzi East. The controlled oil reserve was 3×10^8 t. With additional exploration, the oil reserve could reach 5×10^8 t. In the Sichuan basin, the central Sichuan region is a former exploration region. During the past 50 years, the discovered structural types of oil and gas reserves were small in size. In recent years, geologists discovered a huge area for the exploration of the lithostratigraphic type of gas reservoir in Guangan. The size of the forecast natural gas reserve will be in the range of $3,000 \times 10^8 - 5,000 \times 10^8$ m^3. This was a new breakthrough in the history of natural gas exploration in the Sichuan basin (Qiu and Fang, 2009).

According to exploration experiences in recent years, in eastern China exploration activities have been carried out in most of the onshore exploration regions to discover the lithostratigraphic type of oil and gas reserves. In western–central China, the exploration has been conducted in the onshore exploration regions to discover the structural type of oil and gas reserves and the lithostratigraphic type of oil and gas reserves. The exploration domains for the lithostratigraphic type of oil and gas reservoirs have become important regions for the growth in the oil resources of PetroChina (Table 3.6). The annual growth in the discovered geological reserve represented more than 50% of the total annual growth of the oil reserve of PetroChina. During January to September 2005, the exploration of the lithostratigraphic type of oil and gas reservoirs maintained its pace and successful results kept coming in. PetroChina evaluated the potential of the remaining resource in the lithostratigraphic type of oil and gas reservoirs and evaluated favorable exploration areas or belts in nine exploration basins. The evaluation results indicated that these areas still contained great development potential. Along with the improvement in seismic technology and by widely utilizing sequence stratigraphy, these areas will be the important exploration regions for the growth in oil and gas reserves and the growth in the production of PetroChina (Jia et al., 2008).

Table 3.6 PetroChina evaluation of the lithostratigraphic type of oil and gas reservoirs in favorable Chinese petroleum basins

Basin	Remaining oil resource		Favorable domain	
	Quantity of remaining resource ($\times 10^8$ t)	Lithostratigraphic type of oil resource ($\times 10^8$ t)	Domain of 50 – 100 million tons	Domain of 30 – 50 million tons
Songliao	41.3	26.6	Honggang–Haituozi; Tahala-Changjiaweizi	Huaaopao Area
Bohai bay	32.7	12.7		
Eerduosi	33.7	27.6	Jiyuan Highland; Longdong Area	
Zhungeer	35.3	10.3	Fan on Northwestern Margin	Mosuowan–Mobei Area
Tarim	38.3	8.5	Gaimaiti Slope	Tazhong-I Slope Zone
Chaidamu	10.0	4.0		
Sichuan	2.9	1.5		
Tuha	4.5	2.4	Taipei Depression	
Erlian	5.6	3.1		

● **Exploration Domains in the Thrust Belt in a Foreland Basin**

In western–central China, the foreland basins could be divided into two phases and three types. In a foreland basin, because the oil and gas resources are abundant and because the exploration level is low, the exploration potential is great. In a foreland basin, the foreland thrust belt has the following characteristics; firstly, oil and gas accumulated during the late phase; secondly, hydrocarbons were generated by coal measures; thirdly, abnormally high pressure occurred in the region. Then, most of the accumulated petroleum was natural gas. Lastly, oil and gas accumulated on the large scale anticline that was located under the regional detachment layer. Thus, the foreland thrust belt is the important exploration domain in a foreland basin. Recently, following new geological theories (the theory of the fault-related fold in a foreland basin and the theory of late phase accumulation), and on account of persistent scientific innovation, PetroChina developed a series of exploration technologies that included seismic technology for a mountain range, drilling technology for a thrust belt on the mountain front and well logging technology for complicated geological conditions. These new technologies provided efficient technical support for oil and gas exploration in a foreland thrust belt. Oil and gas exploration has moved forward and a series of important discoveries of valuable strategic significance have been obtained (Jia et al., 2000).

In 2000, during natural gas exploration in the Kuche foreland basin, a large size, sandstone gas field, the Kela 2 gas field, was discovered. Among the Chinese onshore gas fields, the Kela 2 gas field has the highest reserve abundance, the best integrity, the highest production rate of a single well and good reservoir physical properties. This gas field covers 48.1 km^2 of land and it contains $2,840.29 \times 10^8$ m^3 of geological gas reserves. Next, the Dina 2 gas field was discovered on the eastern side of the Kuche depression. The total quantity of the discovered and controlled geological reserve of natural gas was $1,159.84 \times 10^8$ m^3. In the Tarim basin, the discoveries of the Kela 2 and the Dina 2 gas fields in the Kuche depression built the resource foundations for the project called "transporting western gas to eastern China". At the same time, in the entire western-central China, oil and gas exploration in the foreland thrust belt was motivated by new geological theory and by new exploration experiences. In 2003, on the southern margin of the Zhungeer basin, an important discovery was made in the Huoerguosi structure. The estimated quantity of controlled and foreseeable oil and gas equivalent reserves might exceed 2×10^8 t. As the founding region of the Chinese petroleum industry, the Jiuquan basin experienced more than 50 years of exploration and development. In 2000, the Qingxi oil field was discovered in the mountain front thrust belt at Qilian mountain. In 2004, the size of the explored reserve of oil was $5,750 \times 10^4$ t; the quantity of the 3rd order reserve had exceeded 1×10^8 t (the precise number was 1.02×10^8 t). In 2005, a new discovery was made in the area east of the Qingxi oil field, which brought the expectation of connecting the oil fields of Qingxi with the Yaxi oil field structure. The discovery of the Qingxi oil field increased crude oil production from 40×10^4 t (in 1998) to

75×10^4 t (in 2004) in the Yumen oil field.

According to the statistical data, in the foreland thrust belt, most of the oil and gas fields were distributed in the anticlinal structure. The nature of this kind of oil and gas field included mega size, high abundance and a high yield per single well. The economic efficiency of the exploration was very good. Thus, the foreland thrust belt offers a good chance of discovering a mega size oil and gas field. Even though exploration in the foreland thrust belt might encounter great difficulties, if geologists persist in using scientific innovations in seismic technology for the mountain range, improve the accuracy of resolution and aim for a mega size reservoir, the exploration will not only achieve a breakthrough, but it will also be a great discovery. According to resource analysis, in western China the overall exploration level in the foreland thrust belt is very low and there is an abundant remaining resource. In the PetroChina exploration blocks, the remaining geological resource of oil is 40×10^8 t and the remaining geological resource of natural gas is 5.74×10^{12} m^3. Recently, preliminary explorations prior to the seismic survey explored many target regions that have a large structural acreage and a better potential for a breakthrough. The target regions included the Kuche of the Tarim basin, the southern margin and the northwestern margin of the Zhungeer basin, the Taibei depression in the Tuha basin and the northwestern Sichuan basin. Using additional drilling exploration, several mega size oil and gas fields will be discovered with the aim of increasing the petroleum reserve and production of PetroChina.

- **Exploration Domains for the Middle and Lower Petroleum Systems in a Superimposed Basin**

Because of the distinctive geological developments, most of the Chinese sedimentary basins were large scale superimposed basins. In the past, oil and gas exploration mainly focused on the clastic reservoir sequences in the middle and shallow levels of a superimposed basin. Therefore, geologists always questioned the effectiveness of reservoir sequences and the economic viability of the resources in the middle and lower petroleum systems. Recently, geologists have discovered large size oil fields in the middle and lower petroleum systems in the Jidong beach flats in the Bohai bay basin, Hade, Tahe–Lunnan and the Tazhong I slope zone (the latter three areas are located in the Tarim basin). In addition, geologists have also discovered several natural gas fields (the quantity of the gas reserve is more than $300 \times 10^8 - 1,000 \times 10^8$ m^3) at the deep level of the Songliao basin, the Paleozoic erathem in the Eerduosi basin and the Cambrian system in the Sichuan basin. These discoveries not only provided detailed information of a geological nature, but they also expanded our understanding of the exploration potential in the middle and lower petroleum systems. Generally, two kinds of superimposed basins were developed in the Chinese landmass; one was made of a craton basin with either a foreland basin or rifted basin, the other one consisted of a faulted basin and a depressed basin. In these two kinds of superimposed basins, the petroleum systems included a lower craton petroleum system, an upper craton petroleum system and a petroleum system at the deep level of the Mesozoic erathem. By correlation, the nature of the petroleum geology and the petroleum distribution pattern varied. In the middle and lower petroleum systems, the reservoir sequences were mainly made of carbonate rocks and volcanic rocks and the lithology had distinct heterogeneity. At same time, during the exploration, considering the unique nature of the sedimentary layers in the middle and lower petroleum systems, geologists developed a correlated exploration technique. The exploration technologies for the weathering crust type of reservoir sequences in carbonate rocks and for volcanic reservoir sequences were introduced initially, which further promoted oil and gas exploration in the middle and lower petroleum systems (Jia, 2005).

Considering the potential of the remaining oil and gas resources in the middle and lower petroleum systems and reviewing the exploration practices of recent years, this exploration domain had been less investigated. However, it had an abundant remaining oil and gas resource and a good economic future. In fact, the exploration has obtained positive results. In the middle and lower petroleum systems in large size petroleum basins, the size of the remaining oil resource is 27×10^8 t and the size of the remaining gas resource is 4.65×10^{12} m^3. The remaining oil and gas is distributed either in the Tarim basin, Sichuan basin, Eerduosi basin and Chaidamu basin in western-central China, or in the Songliao basin and Bohai bay basin in eastern China. At present, the favorable exploration targets included the Paleozoic erathem that occurred in the Tarim basin (the Lunnan and Tazhong areas) and the Eerduosi basin, the Leshan–Longnusi paleo uplift in the Sichuan basin and the deeply buried sequences in the Songliao basin. The targets that would produce the exploration breakthrough included the middle and lower petroleum systems in the Zhungeer basin, the buried hill in Bohai bay and the Tadong and Yingmaili areas in the Tarim basin. The middle and lower petroleum systems will become one of the important onshore exploration domains for the growth of the petroleum reserve and production in the future.

- **In-Depth Exploration Domains in Previously Investigated Regions in the Matured Basin**

In the matured basin, although the previously explored regions had been investigated more fully, geologists never stop acquiring detailed geological knowledge. Also, because of the limitations imposed by the surface conditions and by earlier technology, many regions or belts (zones) are still at a lower stage of exploration and many newer sequences wait to be explored. Thus, the oil and gas resources are abundant and the exploration potential is great. Recently, using a study of basic geological conditions and through the application of new technology, geologists carried out a preliminary exploration. New discoveries have been repeatedly made in the previously explored regions (that had a relatively lower level of investigation) in the matured basin, in the new area (or zone) and in the new sequences. For example, in the Jidong exploration region in the Bohai bay basin, as a result of recollecting and reinterpreting the 3D seismic data twice, geologists enhanced their geological knowledge and discovered a high quality oil reservoir with a hundred million tons of reserves in the shallow buried sequences in a previously explored onshore region that had almost half a century of exploration history. In 2004, in the beach flats area (that had been less thoroughly investigated) of Bohai bay, important breakthroughs were made in the Nanbao 1 well and Laobao 1 well. The size of the discovered oil reserve reached 5×10^8 t, which was the most important discovery and the most exciting moment in nearly 30 years of exploration history for PetroChina. The northwestern margin of the Zhungeer basin has nearly 50 years of exploration history. A new discovery has been made every year. In 2005, comprehensive evaluation results indicted that, in this region, the remaining oil resource is more than 20×10^8 t. Therefore, the exploration potential is great. In the next 3 to 4 years, the growth in the discovered geological oil reserve will be $3 \times 10^8 - 4 \times 10^8$ t. In China, this is the only previously explored region that still shows growth in crude oil production. In addition, many low resistance oil layers have been discovered in previously explored regions in the Bohai bay basin, Zhungeer basin, and Tarim basin. These oil layers were not discovered during the previous oil exploration.

In the previously explored region in the matured basin, the targets of in-depth exploration were located in the Bohai bay basin (on the beach flats and in the onshore petroliferous depression), in the Songliao basin (in the placanticline, Sanzhao depression and the Fuxin uplift), on the northwestern margin of the Zhungeer basin and on the southwestern side of the Chaidamu basin. In these regions, the size of the remaining geological resource of oil is 33×10^8 t and the size of the remaining geological resource of natural gas is 0.86×10^{12} m^3. Certainly, in the previously explored region in the matured basin, the potential for in-depth exploration is great. However, the targets were well concealed in the previously explored region in the matured basin. The remaining resources were mainly distributed in the lithostratigraphic type, buried hill type and complicated faulted block type of targets. A thorough understanding of the characteristics of petroleum geology in the petroliferous depression is the critical step in the exploration. Additionally, fully exploiting the advanced technology will promote the exploration work at a more detailed level.

● **Exploration Domains in the New Region and New Basin**

Until now, PetroChina has registered mining rights in 35 sedimentary basins in China. Most oil and gas exploration has only been carried out in 8 large size basins. The rest of the medium and small basins (27 of them) have only had very little exploration work or no exploration at all. These remaining new regions and new basins will be important exploration domains for PetroChina in the future. In recent years, according to the strategy of focusing on economic efficiency, based on the evaluation of the basic conditions of petroleum geology, geologists conducted a system of unified queuing and optimization for some new regions and new basins inside PetroChina. For the new regions and new basins that are of a small size, poor efficiency and weak in of exploration value, economic efficiency evaluation will eliminate the one in the worst position. For basins that have a favorable exploration future, geologists will concentrate manpower and technology to evaluate them. New discoveries and new breakthroughs were obtained in the Wushi depression in the Tarim basin, Hailaer basin, Erlian basin (Bayindulan depression, Wuliyasitai depression and Saihantala depression), and the Hongze depression in the Subei basin. These regions become an important supplementary region for the growth in the petroleum reserve and production in PetroChina. In the mining blocks of PetroChina, the preliminary estimation indicated that the size of the remaining geological resource of oil is 66×10^8 t and the quantity of the remaining geologic resource is 2.48×10^{12} m^3. These remaining petroleum resources were distributed in the new region, new basins, the Luobupo depression in the Tarim basin, the Dajing depression in the Zhungeer basin, the Delingha depression in the Chaidamu basin, the Qiangtang basin group in the Tibet region, the northwestern side of the Sichuan basin and the deeply buried sequences on the southern side of the Songliao basin. In these regions, the petroleum resources are abundant. Especially in the Qiangtang basin group in the Tibet region, the basins are of a large size, have abundant resources and no exploration history.

In 2005, PetroChina specially designed a risky exploration project for the new regions and the new basins in order to increase investment. Among the 13 exploration targets, the exploration plan included 3,900 km of 2D seismic survey, 19 exploration wells with a total drilling extent of 6.96×10^4 m, and an investment of ¥10×10^8 (Renminbi). After the exploration work in 2005, geologists obtained noticeable results. The exploration wells dicovered good oil and gas shows, which included the Changshen 1 well in Jilin province, the Xinglong 1 well in northern China, the Wujia 1 well in southwestern China and the Ma 13 well in Santanghu in the Tuha basin. In particular, an important breakthrough was obtained in the volcanic rocks in the Yingcheng formation in the Changshen 1 well that was a deep exploration well, located in the Changling faulted depression on the southern side of the Songliao basin. The daily production rate of natural gas is 46.09×10^4 m^3. This discovery not only set the foundation for achieving an exploration strategy for both oil and gas in the Jilin exploration region, but it also offered a new exploration domain in the search for alternative resources of oil. In addition, this discovery also has great strategic significance for long term stable petroleum production in the Songliao basin and for establishing the Daqing oil field with its 100 year history.

● **Exploration Domain in the Southern Marine Territory of the South China Sea**

The southern portion of the South China Sea covers 350×10^4 km^2 of marine territory approximately; the traditional Chinese marine territory covers 198×10^4 km^2

of the territory. At present, in the southern portion of the South China Sea, there are 48 discovered sedimentary basins. Vietnam and other peripheral countries have discovered 172 oil and gas fields or structures. In 2002, the annual oil production was 1×10^8 t and the annual gas production was 740×10^8 m^3. In 2003, PetroChina started oil and gas exploration in the southern portion of the South China Sea. At present, PetroChina has the mining rights for 18 exploration blocks in 5 basins, which totally cover 12.7×10^4 km^2 of the region. From the preliminary evaluation and seismic survey, many large scale structural traps and seismic anomalies were discovered in the Wanan basin, Zengmu basin and Beikang basin. Among the discoveries, 16 of them are favorable structures that hold 1×10^8 t of resources.

According to the present data, in the regions where PetroChina has the mining rights, the estimated size of the oil resource is 50×10^8 t approximately and the estimated size of the gas resource is 5×10^{12} km^3. With an abundant petroleum resource, the potential for discovering a super size oil and gas field is great. The exploration future is bright. In order to promote oil and gas exploration in this region, PetroChina has set up a special department and financed a special fund. At present, the preliminary geological study and evaluation of targets have been finished. The drilling project is at the stage of platform selection.

● **Atypical Exploration Domains**

China has various kinds of abundant non-conventional oil and gas resources. To quickly get hold of these resources, PetroChina has conducted many preliminary comprehensive studies using modern exploration technology. Up until now, important results have been obtained in the exploration and development of coal bed methane and oil sands.

Coal Bed Methane: According to the newest results of resource evaluation for coal bed methane, the resource of Chinese coal bed methane is mainly distributed in 39 coal basins. At a buried depth of $300 - 1,500$ m (which is an adequate range for the exploration of coal bed methane), the size of the foreseeable resource is 27.3×10^{12} m^3, which is ranked third in the world. The areas that PetroChina registered for coal bed methane mining rights cover 6.93×10^4 km^2 of land and they contain 14.3×10^{12} m^3 of coal bed methane resources. These areas represent 56% of the total domestic registered area. The size of the resource represents 52% of the total resource. Since 1994, PetroChina has drilled 57 wells in 14 blocks. First of all, PetroChina discovered the first large size coal bed methane field – the Qinshui gas field; the size of the proven geological reserve is 352×10^8 m^3; the quantity of the controlled geological reserve is $1,100 \times 10^8$ m^3. Secondly, PetroChina also discovered and controlled one large size coal bed methane containing region, the Daning–Jixian coal bed methane region. The size of the controlled and forecast geological reserve is $1,500 \times 10^8$ m^3. Additionally, PetroChina discovered a gas containing basin, the Ningwu basin. In the Wushi 1 well, the daily production rate of water was 169 m^3; the daily production rate of gas was 3,112 m^3. The controlled geological reserve was 500×10^8 m^3.

Recently, in Chinese coal bed methane exploration and development, most of the wells have the characteristics of "four lows": low desorption pressure, low porosity, low permeability and a low single well production rate. Usually, the daily production rate is $1,500 - 2,500$ m^3. In order to increase the single well production rate and to complete the industrialization of coal bed methane production, the crucial step is to develop various kinds of technology for different types of coal bed methane. The USA has developed a complete set of drilling techniques. Air drilling and caving completion technology is mainly used for low rank coal bed methane. Vertical well fracturing technology is largely applied to middle rank coal bed methane and directional pinnate horizontal well technology is primarily used for high rank coal bed methane. Currently, PetroChina is increasing technical cooperation with the USA and is developing directional pinnate horizontal well technology in the hope of establishing more than 15×10^8 m^3 of production capacity of coal bed methane in 2010, so as to complete the industrialization of coal bed methane.

In China, there are many outcrops of oil sands (including bitumen), which are mainly distributed in Xinjiang, Qinghai, Inner Mongolian, Sichuan, Tibet, Guizhou, Guangxi and Zhejiang. In the past couple of years, PetroChina has made some progress in the business development plan for oil sands, the geological survey and evaluation, as well as the separation technology. In the next few years, PetroChina will intensify comprehensive evaluation of oil sands in order to understand the potential of Chinese oil sands and to determine the exploration direction and targets.

3.3 Distribution Patterns of Oil and Gas Reservoirs

3.3.1 Characteristics of Chinese Petroleum Geology

● **Plates that Made the Chinese Landmass Had a Multi-Cycled Developmental History; the Major Types of Basin Included the Terrigenous Basin and the Superimposed Basin**

The Chinese landmass is located between the Siberian plate, Indian plate and Pacific plate. It is a heterogeneous inconsistent landmass that was made of several small size craton terrains (that were of different size, from different geological ages and had distinctive variations) and the surrounding orogeny belts of different tectonic phases. The core of the Chinese landmass consisted of the North China plate, the Yangtze plate and the Tarim plate, and this was one of the most complicated tectonic regions in the world. These typical tectonic features and distinctive development processes created the unique characteristics of petroleum geology in China (see Section 1 for details) (Qiu et al., 1999).

- **Characteristics of the Chinese Petroleum Basin Included Multiple Stages of Basin Development, Several Phases of Hydrocarbon Generation, and Various Periods of Petroleum Accumulation, which Usually Created the Composite Type of Oil and Gas Reservoirs**

The Chinese petroleum basins were created and developed in a very special tectonic environment and they have three distinctive features. First of all, during the early Paleozoic era, because of the separation that was caused by the oceans, the consistency of depositional combinations was poor. Therefore, the hydrocarbon generation and accumulation were obviously regionalized. Secondly, during the late Paleozoic era, because the terrains were combined, the consistency of deposits had been improved. A good depositional combination of marine–terrigenous facies was developed, which commonly generated hydrocarbons. Thirdly, during the Mesozoic and Cenozoic eras, influenced by three mega dynamic systems, being the Paleo-Tethys, Pacific Rim and Paleo-Asian systems, the diversity of sedimentary basins was gradually clarified. In the area west of the Aerjin strike-slip fault, large size depressed basins were formed with good and consistent deposits. In the area east of the Aerjin strike-slip fault, because of the pull-apart–limited rifting movement, the faulted basin and the depressed basin were developed. As a result, the superimposed basin was created. The hydrocarbon depressions were clearly separated. Therefore, the regionalization of hydrocarbon generation and accumulation was distinct and the size of the reservoirs varied. The characteristics that were discussed above determined that, in the Chinese petroleum basins, under the Mesozoic and Cenozoic terrigenous sequences, Paleozoic sequences of marine facies and marine–terrigenous facies were commonly developed. This is a typical superimposed basin. Because the tectonic background, superimposed history, assorted superimposed patterns and different tectonic modifications that happened after the superimposed movement were all varied, there were many types of superimposed basins.

The superimposed Chinese petroleum basins had a complicated framework and they were formed from several monotype basins by stacking up the monotype basins in different directions. Therefore, these petroleum basins commonly had two sets or more of stratigraphic sequences that were deposited during different geological periods (or eras) and had different superimposed tectonic sequences as well. Because these basins had a longer developmental duration, a deeper depression and a higher geothermal gradient, they offered a beneficial place for organic materials to accumulate and form multiple sets of hydrocarbon source rocks. For example, in the Tarim basin, there are four sets of major hydrocarbon source rocks. The oil and gas accumulated in multiple sets of sequences during different geological periods to create various types of reservoirs. During their developmental process, the superimposed basins experienced different phases of tectonic movement, faulted activities, uplifting, erosion and re-subsidence. In addition, they also experienced relocation of the subsidence center and relocation of the depositional center. These geological events not only caused the migration of hydrocarbon source rock and the migration of the hydrocarbon depression over geological time, but they also caused an adjustment to the oil and gas accumulations. As a result, multi-sources, multi-geothermal centers and multi-phases of oil and gas reservoirs were developed. For oil and gas generation and distribution, no matter if the same set of hydrocarbon source rock was distributed in different depressions horizontally, or different sets of source rocks were superimposed in one depression vertically, the processes of oil and gas generation, migration and accumulation were not usually of an inherited nature and coherence. On the contrary, these processes clearly displayed variability.

In the Chinese petroleum basins, the petroliferous region not only inherited multiple phases of tectonic movement, but was also largely altered. In particular, the tectonic movement that occurred during the end of the Yanshanian tectonic period produced most of the oil and gas structures. Influenced by superimposed multiple phases of tectonic movement and by new tectonic movement, in the petroleum basins the hydrocarbon potential in the second order structure system had been largely changed. On the other hand, the basins that were formed during different tectonic movements had their own depositional mode, structural pattern, petroleum migration and accumulative patterns. Therefore, different types of oil and gas reservoirs were not only stacked up vertically, but they were also connected together horizontally to form a composite oil and gas accumulative belt.

3.3.2 Petroleum Distribution Pattern in China

In China, the intricate nature of the tectonic background, basin development, petroleum geology and geophysics greatly influenced the development of oil and gas reservoirs. Moreover, these reservoirs also developed a complicated configuration of petroleum geology, which will be discussed in the following eight sections.

- **Situations of a Few Large Basins and Plenty of Small Basins were Unfavorable for Developing Super Sized Oil and Gas Fields**

The analysis of geological conditions in large size oil and gas fields throughout the world indicates that the scale of plate tectonics was large, subsidence was stable and the large size basins were well developed. Usually, the large or super large size oil and gas fields developed under these kinds of conditions, for example the Paleozoic petroleum regions of marine facies on the North American plate and on the European plate or the oil and gas region in the Persian Gulf, which contained Mesozoic marine sequences that were deposited in the foreland basin in the Tethys tectonic domain. On the other hand, in China the total area of three paleo plates (the Tarim plate, the Zhongchao plate and the Yangtze plate) only represented 1/3 of the North American plate. More important, throughout geological history these three paleo plates had more vigorous activity if compared with the North American plate and the European plate. Therefore, the Chinese sedimentary basins did not have a large size and good stability like the North American plate and the European plate. During the Mesozoic era, the Tibet region belonged to the Tethys domain and it received Triassic–Jurassic deposits and Cretaceous deposits, which had good preconditions for oil and gas accumulation. However, due to the collision of the Indian plate, the foreland basin did not occur in this region. Instead, this region was uplifted to become the highland and, as a result, the preconditions for the development of a "Mid-Eastern style large size oil and gas field" were destroyed. There are 485 different sizes and different types of basin in China. Sedimentary rocks totally cover more than 670×10^4 km^2 of land. Among these basins, twelve of them have an area of more than 10×10^4 km^2 and fifty of them have an area of $1 \times 10^4 - 10 \times 10^4$ km^2. These 62 basins only represent 12.8% of the total number of the Chinese basins. However, they contain 97% of the total domestic oil resource; nine of the 62 basins contain 80% of the total natural gas resource. Obviously, Chinese oil and gas resources are mainly distributed in a few large and medium size basins. The Daqing oil field is a mega size oil field, which was developed in the unique geological conditions in the Songliao basin (a large size basin that covers 26×10^4 km^2 of land). Over a broad region of China, geologists have not discovered another mega size oil field.

- **Strong Tectonic Activities did not Benefit the Preservation of the Paleozoic Oil and Gas Fields of Marine Facies**

Although the original depositional regions of three paleo plates (the Zhongchao, Yangtze and Tarim) were relatively large during the Mesozoic era, influenced by tectonic movement on the stable continental margin and the prototype basin, the beneficial preconditions for oil and gas generation and storage were destroyed. Moreover, the basin was incised. For example, after the paleo plate of the Greater North China basin was incised, the Eerduosi basin was the only remaining part. Although the North China basin of the present day still contains the Paleozoic erathem, its basement has been cut into pieces, which is not beneficial for protecting the oil and gas reservoirs. During the Paleozoic era, the Upper Yangtze basin crossed the Sichuan, Yunnan, Guizhou, and Hubei provinces. After the Indosinian movement, because the Yungui highland was uplifted, the Upper Yangtze basin was reduced to become the Sichuan basin. Reducing the area of a basin and cutting by faults to expose a large area of Paleozoic strata were not beneficial for developing and preserving the large size oil and gas fields. The Tarim basin is the largest basin in China (it covers 56×10^4 km^2 of land). With well developed Paleozoic strata, the basin not only had good preconditions for hydrocarbon generation and storage, but it also experienced the development of an oil reservoir. It is reasonable to say that a super large size oil and gas field should have been distributed in the Tarim basin. However, during the Paleozoic era, the Caledonian movement and the early Hercynian movement caused the evaporation of some oil and gas. Accordingly, even though some large and medium size oil and gas fields were discovered, geologists still havenot discovered a mega size oil and gas field in the Tarim basin. Throughout geological history, Chinese plate tectonics were extremely active, although this was not beneficial for preserving an oil field in the Paleozoic marine craton basin, which saw the creation, however, of many terrigenous petroleum basins. Many large size terrigenous oil and gas fields were discovered among these terrigenous basins. This is a unique phenomena in the world.

- **Multiple Types of Basin and Varied Kinds of Depositional Formation Caused the Difference in Petroleum Distribution in the Geological Periods**

After the strong tectonic alteration during the Mesozoic and Cenozoic eras, among the prototype basins that developed on the Tarim, North China and Yangtze plates and that contained late Paleozoic marine strata, only three of them still kept their marine strata with good oil and gas accumulation conditions. These included the Tarim basin, Eerduosi basin and Sichuan basin. During the Mesozoic and Cenozoic eras, massive terrigenous sequences were widely deposited in China. In particular, they were overlaid on top of three craton basins. Therefore, if compared with Chinese marine strata, Chinese terrigenous strata were well developed. The marine sequences in these three craton basins had been deeply buried due to the massive terrigenous sequences above. As a result, the organic materials in the marine strata had progressed into the post mature stage. Under this kind of accumulative background, the characteristics of Chinese oil and gas distribution over geological time are as follows. The Mesozoic and Cenozoic terrigenous strata mainly contained oil and the total size of the oil resource was 760×10^8 t approximately, which represented more than 80% of the total oil resource. The Paleozoic marine strata mainly contained natural gas (if we do not consider the associated gas in the Tertiary oil field). The total size of the gas resource was 18×10^{12} m^3 approximately, which represented 47% of the total gas resource (Table 3.7).

Table 3.7 Characteristics of Chinese petroleum distribution over geological periods

Age of stratigraphy	Oil		Gas	
	Quantity ($\times 10^8$ t)	Percentage (%)	Quantity ($\times 10^{12}$ m^3)	Percentage (%)
Tertiary	487.8	51.9	11.02	29.5
Cretaceous	158.4	16.8	1.60	4.0
Jurassic	113.4	12.0	1.99	5.2
Permian	36.8	4.0	3.62	9.5
Carboniferous	48.5	5.0	8.03	21.1
Ordovician	19.6	2.0	5.09	13.4
Cambrian	8.0	0.8	1.18	3.1

● **Among Various Types of Hydrocarbon Source Rocks, the Mesozoic and the Cenozoic Terrigenous Source Rocks were the Most Important Ones**

Over the world, many petroleum basins contained various types of hydrocarbon source rocks that included marine source rocks, terrigenous source rocks and the source rocks of coal measures. However, most of the petroleum basins mainly contained only one type of source rock. For example, in the midwest region of the USA, the basins mainly contained Paleozoic marine source rock. In Western Europe, the North Sea Basin mainly contained coal measures as the source rock. Therefore, the recognition of the magnitude of hydrocarbon generation was relatively easy. On the other hand, the types of Chinese petroleum basin and the development of the sedimentary formation were more complicated. There was the large scale Paleozoic prototype basin of marine facies and there was the Mesozoic terrigenous rifting basin that was overlaid on top of the Paleozoic basin. The sedimentary formations included marine facies, marine–terrigenous facies and terrigenous facies. In addition, there were several types of hydrocarbon generating environments. Thus, the recognition of the hydrocarbon generation pattern of a basin is relatively difficult.

During the Late Permian epoch, Chinese terrigenous basins formed in the Tarim basin and Zhungeer basin. After that, many Chinese basins were transformed into terrigenous basins during the Mesozoic and Cenozoic eras. In a terrigenous basin, excellent hydrocarbon source rocks developed in the deep water lacustrine deposits and in lacustrine deposits. The latter were located near the ocean. There was not a systematic and complete theory of terrigenous sequences for the generation of oil in the world. With the efforts of two generations of Chinese geologists, at the end of the 1950s we developed the theory that the Chinese terrigenous sequences generated oil. The understanding that a petroleum generating center controlled the oil and gas distributions in a terrigenous basin was invaluable information for petroleum exploration in terrigenous basins in the following years. Coal measures were well developed in China, which were deposited in the late Paleozoic sedimentary formation of marine–terrigenous facies and in the Mesozoic–Cenozoic sedimentary formation of terrigenous facies. Many Chinese basins contained coal measures. Although, in the 1970s, we learned that, in foreign countries, coal measures could produced coal bed methane, it took us more than 20 years to understand that the source rocks of coal measures could produce commercial grade oil and gas. For example, the large scale gas fields in the Carboniferous and Ordovician systems in the Eerduosi basin and the Jurassic oil fields in the Zhungeer basin and Tuha basin are examples. On the other hand, only in the past two decades have foreign countries started to study the theory of terrigenous sequences generating oil. Some western oil companies gave the credit for their discoveries of terrigenous oil fields to the exploration experiences at the Daqing oil field in China.

In addition, we recognized that in many terrigenous basins there was widely distributed immature crude oil in the Mesozoic and Cenozoic erathem. This kind of crude oil had a lower level of thermal maturation. Even if the R_o was in the range of 0.3% – 0.5%, hydrocarbons would still be generated, which included immature–mature liquid hydrocarbon and bio-degraded gas, most of which was buried in relatively shallow traps at a buried depth of 1,000 – 2,500 m. The discovery of immature–low mature oil corrected and completed the theory of pyrolyzed kerogen generating hydrocarbons. After many years of hard work, Chinese petroleum geologists recognized the developmental patterns of terrigenous source rocks (that were discussed above), which was a contribution to the development of petroleum geology in the world.

In China, the distribution region of Paleozoic marine carbonate rocks is 300×10^4 km^2. These carbonate rocks are mainly distributed in the North China region, the Eerduosi basin, Tarim basin and Sichuan basin, with massive thickness. After many years of exploration, some large and medium size gas fields have been discovered in the marine sequences. Only a few oil fields have been discovered. Geologists are still in dispute over whether carbonate rocks can function as excellent hydrocarbon source rocks. According to Gehmen (1962), the analysis of organic carbon in carbonate rocks from 346 samples around the world indicated that the average content of organic carbon is 0.24%; the analysis of organic carbon in shale from 1,066 samples indicated that the average content of organic carbon is 1.14%. Comparing these two results, the content of organic carbon in carbonate rocks represented 1/5 of the content in shale. According to other statistical data, in Chinese carbonate rocks the content of organic carbon is usually less than 0.3%. Thus, in general, geologists agree that the calcareous argillaceous rock of basin facies could function as source rock. However, they still argued about whether shallow water carbonate rock could work as source rock. The Tarim basin is the largest craton basin in China. The lower Paleozoic carbonate rocks are widely developed in the basin. Among these carbonate rocks, the Cambrian–Ordovician system contains the commonly recognized source rocks. However, according to the analysis of more than one thousand samples from the Ordovician system, nearly 80% of total samples contained less than 0.2% of organic carbon. In addition, the main body of chloroform bitumen "A" is only 30 – 70 mg/L. These data revealed the insufficiency of hydrocarbon generation. On the other hand, the exploration and research results point out that, besides a few of the lower Paleozoic large and medium size oil and gas fields that were discovered in the Tazhong area and the Tabei uplift, geologists also discovered a huge amount of bitumen sandstone in the Silurian strata in a region of 5×10^4 km^2 in the Tarim basin. The estimated quantity of the remaining oil was nearly 400×10^8 t. By investigating the origin of the oil, geologists think that the oil came from the lower Paleozoic erathem. This is evidence that carbonate rock had generated a huge amount of oil. Therefore, in China, both petroleum exploration over a broad carbonate region and the evaluation of source rocks called for an in-depth study.

● **Both Regular and Irregular Reservoir Sequences Occurred in Sedimentary Basins, which Helped the Growth in Petroleum Reserves and Production and which also Brought Difficulties in the Exploration and Development of Oil and Gas Fields**

The reservoir sequence is the most important geological condition in petroleum exploration and the development of an oil field, which influences petroleum reserves and production in particular. Recognizing the regularity of reservoir sequences is directly related to the successful rate of petroleum exploration and the economic efficiency of oil field development. If they are conventional homogenous reservoir sequences (a porosity larger than 10% and permeability larger than 10×10^{-3} μm^2), then the reservoir sequences are easy to understand. The exploration and development will be simple. If they are non-conventional heterogeneous reservoir sequences, geologists will subsequently not only encounter enormous difficulties in recognizing the nature of the reservoir sequences, but they also will run into a complicated exploration. Because Chinese basins are of different types and have various sedimentary formations, the types of reservoir sequences are varied, including sandstone, carbonate rocks, volcanic rocks and metamorphic rocks. Moreover, because of the forceful tectonic movements that happened during the Mesozoic and Cenozoic eras, the geological processes and the gradient of the geothermal field had a great influence on the development of secondary pores. The alteration could not only transform the reservoir sequences from the excellent porous type into the dense type to become non-conventional reservoir sequences, but it also could develop the secondary pores and modify the reservoir sequences into good ones. Most of the Chinese basins have conventional reservoir sequences. For example, the oil reservoirs in the middle-shallow depths in eastern China ensured the growth of Chinese petroleum reserves and production. Conversely, there are a certain number of basins that contain non-conventional reservoir sequences, which caused many problems for petroleum exploration. For example, in the Sichuan basin, almost all the reservoir sequences are non-conventional reservoir sequences. Non-conventional reservoir sequences also occur in the Bohai bay basin, the Carboniferous–Permian systems on the southern side of the North China basin, the Triassic and Jurassic systems in the Eerduosi basin, the Permian system and the Triassic–Jurassic systems (in the deeply depressed region) in most of the Zhungeer basin. In eastern and central China, most of the deeply buried sequences in the Mesozoic and Cenozoic basins are non-conventional reservoir sequences. In recognizing and exploring these non-conventional reservoir sequences, we paid a great price. For example, at the end of the 1950s, after the discovery of the Jurassic oil reservoir in the Sichuan basin, because geologists did not recognize that Jurassic sandstone contained a fractural type of reservoir, many unsuccessful exploration wells were drilled. Later on, an exploration breakthrough was obtained during the "three years exploration campaign". During the early 1960s, on the northwestern margin of the Zhungeer basin, the nature of conglomerate reservoir sequences with volcanic pebbles of alluvial fan facies was recognized after geologists learned that the dissolution of zeolite could improve the storage ability. In the Liaohe basin, some buried hill types of oil reservoirs were made of Archean metamorphic rocks. After the distribution pattern of dissolutive pores and fractures were recognized, the exploration began to succeed. In the Bohai bay basin, the reservoir sequence was made of Tertiary basalt. Understanding the nature of this reservoir sequence was not an easy task. In the Eerduosi basin, the reservoir sequences were made of Paleozoic carbonate rocks. Geologists walked a rocky path in recognizing the distribution pattern of natural gas in the karst type of pores and fractures on the weathering surface. The Sichuan basin was the original region where geologists initially discovered the fractural type of reservoir sequences in carbonate rock. This knowledge was obtained after many years of exploration practices. It is no exaggeration to say that reservoir sequences form the central issue for Chinese petroleum exploration and the development of petroleum fields. The non-conventional reservoir sequences represent more than 45% of the total Chinese reservoir sequences. Through hard work, Chinese petroleum geologists repeatedly studied conventional and non-conventional reservoir sequences. Finally, we recognized and understood the patterns and the nature of these reservoir sequences, so as to achieve great success in increasing our oil and gas reserve and production.

- **Active Nature of Tectonic Development on Paleo Plates Influenced the Development of the Paleozoic Reef Complex Types of Oil and Gas Fields**

Over the world, many discovered bio-reef type oil and gas fields contain a huge quantity of reserves and have a high production rate. For example, the Alberta basin in Canada contains the Devonian bio-reef type of oil field. The size of the recoverable oil reserve is in the range of $0.7 \times 10^8 - 1.3 \times 10^8$ t. Another example is the Pre-Caspian basin in Kazakhstan; the Permian–Carboniferous Kalaqiaganak reef type of gas field (an anticlinal trap) contains 1.6×10^{12} m^3 of recoverable gas reserves. If the bio-reef type of oil field were to be discovered in a basin, then large size oil and gas fields would be discovered in groups. This is the lifetime dream of many explorers.

Conversely, in China the situation was different. The distribution region of Paleozoic carbonate rocks covered 300×10^4 km^2 of land with a depositional thickness of a few thousands meters. During the Paleozoic era, the three paleo plates of the Tarim, Zhongchao and Yangtze were all located in the lower latitude region of Tethys. The warm climate and healthy organisms benefited the development of the bio-reef. Accordingly, the large scale bio-reef type of oil and gas field should have been developed in these regions. However, after more than 40 years exploration, geologists only discovered 13 small size sponge bio-reef gas fields in the Upper Permian Changxing limestone in western Hubei province and in eastern Sichuan province. Most of the bio-reefs were the patch reef and tower reef. These reefs are of a small size. Usually the diameter is less than 1 km and the height is in the range of $12 - 110$ m. In the heterogeneous reservoir sequences, the thickness of a single layer is in the range of $2 - 45$ m, the average porosity is 6%, the permeability is 29×10^{-3} μm². The single reef entity usually has a high production rate and a small reserve (3×10^8 m³ in general). The bio-reefs were spread all over the place and they were hard to find. Using the seismic method to discover bio-reef gas reservoirs, the success rate of the preliminary exploration well was only 20%.

In China, the Paleozoic marine carbonate rocks not only had the material conditions for developing the bio-reef, but they also had an adequate paleo geographic environment. Why has a large scale bio-reef type of oil and gas field not been discovered? The reasons may be related to the smaller size of the paleo plate and the frequent tectonic activity throughout geological eras. Moreover, the great buried depth of Paleozoic marine sequences was maybe another objective reason. And the fact that there have been few exploration wells is perhaps another reason why a large scale reef complex type of oil and gas field has not been discovered in these three craton basins.

● **Various Types of Petroleum Traps and Diversified Oil and Gas Accumulative Belts were Formed**

The oil and gas reservoir is the basic unit for oil and gas to accumulate in. Usually it was located in a single trap. The oil and gas trap must have excellent storage space, a non-permeable upper seal and a non-permeable lateral seal. These elements made up the unified geothermal system, pressure system and interface between oil (gas) and water. Other types of trap indicated more complicated geological conditions and a more difficult level of petroleum exploration. Because of the complicated nature of Chinese plate tectonics and because of the variation in sedimentary formations, also because of the variability in the Mesozoic and Cenozoic terrigenous sequences and the power of the alterations that happened in these terrigenous sequences in the late geological period, of the 964×10^4 km² of Chinese landmass, although sedimentary rocks cover 527×10^4 km² of land in 424 basins, only 15 basins have an area larger than 10×10^4 km². 81% of the basins have an area smaller than 1×10^4 km². In eastern China, the basement of the Mesozoic and Cenozoic small basins was cut by a basement fault to form many uplifts and depressions. Some depressions have an area of less than 1×10^4 km². Each of them has its own petroleum system and several types of trap. This phenomenon indicated that the brokenness and intricacy of Chinese regional tectonics increased the difficulty of petroleum exploration.

Utilizing the geological factors that developed traps as the classification standard, designating the patterns of trap, the sealing conditions and the types of reservoir lithology as the categories of the sub-division, Chinese oil and gas reservoirs can usually be divided into six large categories, which include the anticlinal type of reservoir, the faulted type, the lithologic type, the stratigraphic type, the mixture type and the water dynamic type. Every large reservoir category can be further divided into many sub-categories. In total there are 21 sub-categories of reservoir.

● **Intricate Topography Resulted in a Difficult Exploration Environment**

In China, during the Cenozoic era, the Himalayan movement was powerful and it caused the complexity of Chinese topography. In western China, there is the Tibetan plateau with an average altitude greater than 4,000 m. In eastern China, the Quaternary system covered the large size plains of Songliao. In northern China, there is the Shangan loess plateau, in northwestern China there is the desolate Gobi desert and the great desert region of the "deadly sand ocean". The marine territories include the continental shelves of the Yellow Sea, East China Sea and South China Sea, which often have to endure typhoons. According to incomplete statistical data, a large quantity of Chinese gas resources are distributed in regions that have poor natural conditions (Kong et al., 1998). For example, the desert region contains 26% of the total gas resource, the mountain region contains 25%, the Loess Plateau contains 12% and the marine territories contain 21%. In a poor geographic environment and severe climate, Chinese petroleum exploration has encountered great difficulties and technical problems. Even in some countries of great oil wealth, it is unusual to encounter all of these complicated exploration conditions at one time.

References

Hu, J.Y., Huang, D.F., et al., 1991. The Terrigenous Petroleum Geology of China. Petroleum Industry Press, Beijing.

Jia, C.Z., 2005. The Problem of Chinese Petroleum Geology and the Strategy of Onshore Exploration for the Beginning of the 21st Century. Petroleum Industry Press, Beijing.

Jia, C.Z., 2007. The Latest Discovery of Petroleum Exploration and Related Experiences in China. World Petroleum Industry, Beijing.

Jia, C.Z., et al., 2000. Oil and Gas Exploration in Foreland Thrust Belt. Petroleum Industry Press, Beijing.

Jia, C.Z., Zhao, W.Z., Zhou, C.N., et al., 2004. Two Essential Technologies for Exploring the Lithostratigraphic Reservoir. Petroleum Exploration & Development, Beijing.

Jia, C.Z., et al., 2008. Geological Theory and Exploration Technology for Lithostratigraphic Type of Oil and Gas Reservoir. Petroleum Industry Press, Beijing.

Qiu, Z.J., Fang, H., 2009. The Great Development of Natural Gas–the Second Frontier for Chinese Petroleum Industry. Natural Gas Industry, Beijing.

Qiu, Z.J., et al. (Ed.), 1999. Oil and Gas Exploration in China. The Geological Publishing House, Beijing.

Qu, H., Zhao, W.Z., Hu, S.Y., 2006. The Capability of Chinese Petroleum Resources and Related Exploration Domain. China Petroleum Exploration, Beijing.

RIPED (The Research Institute of Petroleum Exploration and Development), 1994. The Method of Petroleum Resource Evaluation and Related Technology. The Report of the Second Conference for Oil and Gas Resource Evaluation of China (Vol. II).

Wu, S.C., 1994. The Introduction of Petroleum Resource Evaluation. Petroleum Industry Press, Beijing.

Zhao, W.Z., Zhang, G.Y., Wang, H.J., 2005. The Latest Developments in Petroleum Geology and its Significance in Expanding the Exploration Domain. Acta Petrolei Sinica, vol. 1.

Zhao, W.Z., Hu, S.Y., Dong, D.Z., et al., 2007. The Development of Chinese Petroleum Exploration during the "Fifteenth Five-Year Plan" and Important Exploration Domain for the Future. Petroleum Exploration & Development, Beijing.

Part II

Lithostratigraphic Reservoir in China

4

Main Types of Petroleum Reservoir and Characteristics of Oil and Gas Traps

Since the 1950's, along with the discoveries of the Daqing oil field and exploration in the Bohai bay area, terrestrial petroleum geology exploration has been gradually developed. The theory of oil generation in a terrestrial basin and multiple oil-gas accumulation zones is the core of terrestrial petroleum geology. This new hypothesis not only improved the theory of petroleum geology, but it also effectively directed oil and gas exploration in the Chinese terrigenous basin. In addition, numerous structural types of oil and gas reservoirs were discovered, more than a dozen oil and gas production bases were developed, which include Daqing, Bohai bay, Changqing and Xinjiang, and which effectively supported the construction and development of the Chinese national economy.

4.1 Introduction

Entering the 21st century, petroleum exploration in the Chinese terrigenous basin has entered into a new phase that emphasizes both structural reservoir and lithostratigraphic reservoir exploration. In some basins, the exploration of the lithostratigraphic reservoir is in progress. The exploration domains of the lithostratigraphic reservoir are being continually expanded into four new territories: from the terrigenous rifting basin expansion into the terrigenous depression, foreland craton basin and marine craton basin; from clastic reservoir sequence expansion into marine carbonate and volcanic reservoir sequences; from the in situ petroleum reservoir expansion into above source and below source reservoirs; from eastern China, that has a high exploration level, expansion into western-central China that has a low exploration level. According to analysis of the latent capacity of remaining resources, the exploration of the lithostratigraphic reservoir will be the most important and practical exploration domain for a considerable period of time in China. Since 2000, the additional explored oil reserve of the lithostratigraphic reservoir is $(2.5 - 3) \times 10^8$ t per year. Its contribution to the total explored reserve increased from 45% to 67% (Fig. 4.1). Currently, this percentage is still rising every year.

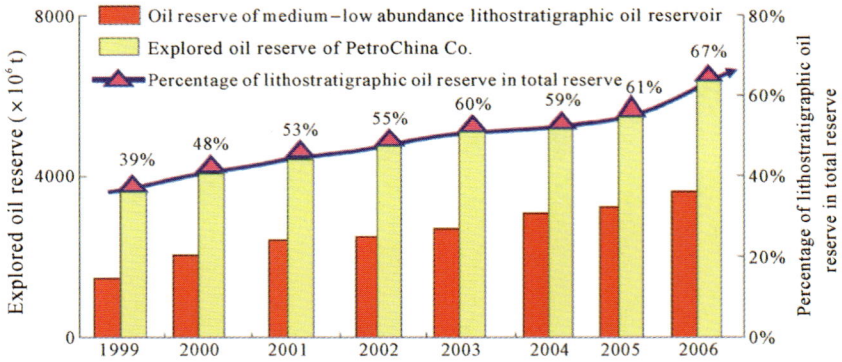

Fig. 4.1. Growth diagram of explored oil reserve in lithostratigraphic reservoir of PetroChina Co., Ltd.

The development of the Chinese sedimentary basin and its tectonic evolution process basically went through four phases, which were the form of the Precambrian basement, the drafting of small continental crusts and deposition in marine basins during the Paleozoic era, the convergence of continental crusts and deposition in terrigenous basins during the Mesozoic era, and the orogenic movements and creation of basins on the intracontinent during the Himalayan tectonic period (see Chapter 1). Because of the geological history, the Chinese terrigenous petroleum basin has two distinctive characteristics. One is the long period of multi-cycle tectonic evolution, which generally developed superimposed basins of different tectonic types. Vertically, the major petroleum basins contain several sets of source sequences and reservoir sequences. Large scale petroleum basins developed on the North China plate, the Yangtze plate and the Tarim plate, such as the Bohai bay basin, Eerduosi basin, Sichuan basin and Tarim basin. These basins have two sets of petroleum bearing systems in the Paleozoic marine sequences and in the Mesozoic–Cenozoic sequences. Secondly, we have the convergence of small terrains and strong tectonic movement on the intracontinent, which resulted in poor reservoir conditions in marine sequences. Widely distributed Mesozoic–Cenozoic terrigenous basins predominately produce oil. The terrigenous basin is the most important domain for Chinese oil and gas exploration and related research work. Currently, more than 80% of the geological reserve of crude comes from the terrigenous reservoir. According to the tectonic background of a basin and its geodynamic environment, and considering the original structure, the characteristics of deposition and genetic mechanism of a basin, four basic prototype basins developed on the Chinese landmass. These include the craton basin of marine facies, the foreland basin of terrestrial facies, the terrigenous rifting basin and the terrigenous depression. (From a sedimentology point of view, a foreland basin should be called a rejuvenated foreland basin that totally or mainly contains terrestrial deposits. However, other types of Chinese foreland basin– peripheral foreland basin and back-arc foreland basin contain marine depositions or marine-terrigenous deposits. Please see Part III.) The differentiations in depositional features are distinct. The terrigenous rifting basin contains a series of parallel or obliquely intersected grabens and half grabens, which are controlled by normal faults. If, on a large scale, regional, detached normal faults developed, an intracontinental rifting basin could be developed with extensive territory. All grabens and half grabens can be considered as a structural system that looks like an extensional chain that is located on the hanging wall of a large detachment fault. The Palaeogene basins in the Bohai bay basin are of this type. Large and medium size rifting basins contain several depressions and uplifts; every depression is a depositional unit. For the most part, the framework of a depression is a half graben that can be subdivided into a steep slope zone, gentle slope zone and depression zone. On the steep slope side, along the strike, either there is a single steep slope that is controlled by the boundary fault, or there is a steep slope zone that consists of multiple faulted steps. On the gentle slope side, either there is a single gentle slope of simple structure, or there is a complicated gentle slope that was cut by a growth fault. Due to strong activities of the boundary fault and major fault, the structural amplitude is great, the depositional rate is high and can be 2.4 – 12.5 mm/a. The growth fault systems of different depositional periods controlled the distribution of sequences and developmental position of different domains. Because the faulted blocks were positioned at different elevations, the depth of water and intensity of energy were different. Moreover, they controlled sedimentary systems and the distribution of facies. For example, on the steep slope zone that consists of multiple faulted steps, from the area near the substance source to the center of the lake, the same sedimentary sequence can develop into the following facies: alluvial fan–fan delta–underwater fan–deep water turbidite. On the gentle slope zone, from the area near the substance source to the center of the lake, the same sedimentary sequence can develop into the following facies: alluvial fan–braided river–braided delta–underwater fan. Various types of sand bodies that were developed in the terrigenous rifting basin during depositional periods were good for developing the lithologic reservoir. At the same time, during the evolution procession of the rifting basin, buried hills that were widely created at the basement were appreciated for forming the stratigraphic reservoir, such as a buried hill oil and gas reservoir.

There are two types of terrigenous depression: an intra-craton depression and post rifting depression. The Eerduosi basin during the Late Triassic is an example of the former. The Songliao basin during the Late Cretaceous is an example of the latter. A terrigenous depression contains a large composite delta and the sand body on the delta front is useful for developing the lithologic reservoir.

Inside a basin, the relief of a craton depression's basement controlled distributions of positive sedimentary units and negative sedimentary units. The inherited, long duration, elevated and subsidence movements controlled the paleo-geographic environments of a sedimentary basin. The overall sedimentation rate of secular subsidence was low with a relatively thin sediments layer and small gradient of thickness, such as in the Eerduosi basin, where the Late Triassic sedimentation rate was 0.58 mm/a. Inside a basin, because there was no elevated and subsidence movement, the topography was wide open and gentle and the slope gradient was small. These features are an indication that, inside a basin, the facies belts in secondary geomorphic units were reasonably wide. During the same tectonic evolutionary period, large scale delta systems in all stages were continually developing. Basically, the delta system served as a depositional center.

The Late Cretaceous deposits in the Songliao basin were typical rifting depression deposits; the lake basin was wide open with a small slope gradient (In the Songliao basin, the slope gradient is $0.5 \times 10^{-4} - 1.4 \times 10^{-4}$ in the K_2 sequences). The distribution of sedimentary layers was stable. The large size delta system continually developed and, inside the delta system, secondary facies belts were relatively wide. Compared with the craton depression in the Eerduosi basin, the rifting depression in the Songliao basin was relatively active with a high subsidence rate (In the Songliao basin, the depositional rate of the Upper Cretaceous series is 0.73 mm/a). The framework of the basin was influenced by the rifting activity. In the central depression area, above the rifting valley, there was a subsiding depositional center. Terraced bench or growth faults divided the central depression from the slope area or the elevated area and, inside the slope area and elevated area, growth faults of different periods controlled sedimentary deposits.

Chinese foreland basins predominately occurred in the western-central region. They were formed in a strong compressive stress field that was brought on by the collision between the Eurasian plate and the Indian plate during the Mesozoic–Cenozoic eras. Large scale subsidence happened in the transitional zone that was located between an orogeny and a stable terrain. Subsequently, the thrust movement shifted over, in a direction from the orogeny to the basin, and created a foreland depression that was parallel to the orientation of the orogeny. Similar foreland depressions include the Kuche depression in the Tarim basin (P_2-T_3; Kz), the western region of the Sichuan basin (Mz-Kz), the Jiuxi basin (N), the southern margin of the Zhungeer basin (Mz-Kz) and the Chaidamu basin.

Compared with a classic foreland basin, a Chinese foreland basin obviously has its special characteristics. Beside a lack of marine deposits, a Chinese foreland basin also has the following distinctions: (1) The orogeny does not contain Type B or Type A subduction belts of the same period because the compressive force came from the far-field effect of the collision between the Indian plate and the Eurasian plate. Therefore, the deformational gradation of the thrust belt was stronger in the southern region than in the northern region. (2) The development of a foreland basin was due to paleo-orogeny reactivation, large scale uplifting, the thrust movement toward the direction of the craton and, finally, the formation of a rapidly flexural subsided foredeep belt. (3) The evolutionary stages included multi-cycles during the various periods. The foreland basin was superimposed on top of the Mesozoic rifting basin or depression that was not on the passive continental margin or in the back-arc environment. (4) The late tectonic movement brought on strong deformation, and the Himalayan movement in particular showed strong tectonic activities.

Same as for the classic foreland basin, the Chinese terrigenous foreland basin also consists of the foreland thrust belt, the foredeep depression and the forebulge. The cross section of the Chinese foreland basin shows an outline of an asymmetrical winnow. From orogeny to craton, this cross section can be further divided into the thrust belt, the depression zone, the slope zone and the forebulge. The subsidence scale of the foreland basin was gradually reduced towards the craton. The foreland depression of Longmen mountain in western Sichuan province is a good example. This depression has a steep slope on the western side with massive deposits and a gentle slope on the eastern side with thin strata; the thickness of strata in the Upper Triassic depositional center is 3,000 – 4,000 m; on the eastern slope zone, the thickness of strata is 1,500 – 8,000 m; in the uplifting area, the thickness of strata is 500 – 800 m. Their depositional rates were 1.00 – 1.48 mm/a, 0.29 – 0.55 mm/a and 0.18 – 0.29 mm/a, respectively. Based on the experience of exploration in foreign countries, a large scale delta was developed on the slope zone that tilted from the forebulge to the depression. The delta is an advantageous area for oil and gas exploration of the lithostratigraphic reservoir.

Chinese cratons generally contain craton basins, which mainly contained marine deposits, but which might have marine–terrigenous facies during some periods of the late Paleozoic. In the early Paleozoic marine basin, the platform was a primary depositional environment. The carbonate platform generally occurred during the Cambrian–Ordovician periods. The clastic platform mostly appeared during the Silurian period. From a stable and wide open platform to a narrow and active sea trough, the basic paleo-geomorphic framework includes shallow platform, shallow–semi shallow sea slope and deep water basin. Inside a platform, the sedimentary facies include carbonate platform facies, carbonate semi-blockage platform facies, evaporite platform facies and bio-reef marginal platform facies. With a different sedimentary background, the alternative deposits of carbonate rock and argillaceous rock might occur on the slope zone. In addition, a debris flow deposit could also appear on the slope zone. Either the argillaceous deposits or turbidite deposits could dominate the basin area. Researchers recognized that sedimentary deposits in an undercompensation basin contain the highest abundance of organic materials. Then the depression inside the platform and the down slope of the basin contain the second highest abundance of organic materials. The marginal platform slope (up slope) contains a relatively low abundance of organic materials. The platform carbonate rock and marginal platform bio-reef are good reservoir sequences, such as the slope fan facies. Several high quality reservoir sequences developed during the Silurian period in the tidal flat facies, littoral facies and river delta facies. A perfect combination of hydrocarbon generation and storage is the major geological requirement for forming a lithologic reservoir. The paleo-topography that was formed during the Caledonian movement and that experienced long lasting erosion and dissolution events, was the geological foundation for developing an unconformable stratigraphic reservoir.

The Tazhong, Lunnan, Yingmaili and Yaha oil and gas fields have this kind of reservoir.

During the late Paleozoic, many Chinese craton basins transferred into marine–terrigenous basins. For example, after the Caledonian movement, during the Silurian–Devonian periods, because the Eerduosi basin was elevated to become a landmass, the sedimentary deposits of this periods were absent. During the Carboniferous–Permian periods, the Eerduosi basin slowly subsided again. The tri-provinces region of Shaanxi–Gansu–Ningxia (the eastern part of northwestern China) and the North China region were joined together as a united depositional area. The sedimentary deposits of marine–terrigenous facies and nearshore terrestrial facies contained coal beds. During the late Permian, the depositional environment was an intracontinental basin. Coal measures in the Carboniferous–Permian systems are the most important source rock for natural gas, which combined with reservoir sequences of river delta facies in the same stratigraphic unit to form the lithologic gas reservoir in situ, which also combined with dolomite reservoir sequences in the Ordovician system to form the buried hill type stratigraphic gas reservoir that generated natural gas in upper stratigraphic sequences and stored gas in lower stratigraphic sequences. The Tarim basin and the Sichuan basin have a similar evolutionary history. During the Middle–Late Ordovician epoches, the Tarim basin was partially uplifted. During the Silurian–Devonian periods, the elevated region was gradually enlarged. However, the entire basin was not uplifted. During the Carboniferous period, the entire Tarim basin subsided. During the Permian period, the Tarim basin was transformed into a terrigenous basin. During the Early–Middle Silurian epochs, most of the Sichuan basin was uplifted and during the Late Silurian epoch–Devonian period, the entire Sichuan basin was elevated. During the Carboniferous period, most of the Sichuan basin was in an elevated state. During the Permian–Triassic periods, the entire Sichuan basin subsided again.

Because distinctions between the four proto-type basins are very noticeable, in the following we will discuss petroleum reservoirs, traps and the features of zones in the Chinese lithostratigraphic reservoir according to the proto-type basin and we will discuss the distribution patterns of the lithostratigraphic reservoir in every type of basin.

4.2   Main Types of Petroleum Reservoir in Four Prototype Basins

A rifted basin, depression, foreland basin and marine craton basin contain different deposits and various types of reservoirs (Table 4.1).

Table 4.1   Depositional settings and related reservoirs in different types of basins

Basin type	Genetic mechanism of basin and its geometry	Basin's framework	Origin of sedimentary substance	Major depositional systems and distribution of facies	Major types of reservoir	Example
Terrigenous rifted basin	Large deep rifting valley, extension and strike-slip; controlled by boundary faults; most of depressions are elongated shape	Complicated structure, alternation of multi-uplifts and multi-depression with various reliefs	Most substances came widthwise with short distance; some substances came lengthwise with long distance	Sand bodies of alluvial fan, fan delta, braided delta and underwater fan, or lacustrine mudstone; facies belt was poorly diversified with narrow width. Subsidence center was very close to the growth fault on steep slope; depositional center was shifted to the center of lake; deep water area could occupy 50% of lake region during deep subsidence period	Most of reservoirs were positioned in fan delta, braided river delta and underwater fan systems; they also could occur in igneous rock and metamorphic rock	Bohai bay basin (E); Erlian basin (K_1)
Terrigenous depression	Mantle thermal subsidence, basement descended; wide open lake basin with different geometries, such as rectangular and diamond shapes	Simple structure, large scale uplift and depression, or gentle slope	Majority of substances came lengthwise with long distance; some substances came widthwise with short distance	Major depositional system was long axis river–delta system; facies belt was well diversified with wide width. Secondary system was short axis braided delta system. Subsidence center and depositional center were located at same place in the center of the basin; deep water area occupied 10%–15% of lake region	Most of reservoirs were in delta system; some reservoirs were in braided river delta, river and fan delta systems	Songliao basin (K_2); Eerduosi basin (Mz)
Terrigenous foreland basin	Thrusting compression, mountain front subsiding; basin has elongated narrow shape and is parallel to orogeny	Simple structure, generally including thrust belt, subsiding zone and slope zone	Primary substances came from folding belt and slope zone widthwise	If it came from folding belt, depositional system was fan delta with narrow facies belt; if it came from craton, depositional system was river–delta with wide facies belt. Subsidence center was located at subsidence zone on mountain front; depositional center was shifted toward craton	Most of reservoirs were in alluvial fan, fan delta and braided river delta; some reservoirs were in river–delta system	Western Sichuan depression (Mz); Kuche depression (Mz)
Marine craton basin	Shallow water continental shelf, epicontinental sea and oceanic basin on continental margin	Complicated structure, includes island arc system, continental shelf, continental slope	Majority of substances came lengthwise around oceanic basin	Marine depositional systems include littoral system, bio-reef, shallow water continental shelf, neo-deep water and deep water. In the marine–terrigenous environment, there were delta systems and estuary systems	Sand body of littoral facies and carbonate rocks (including bio-reef bank) were major types	Tarim basin (Pz); Sichan basin (Pz)

4.2.1   *Rifted Basin Containing Reservoir Types of Fan Delta and Underwater Fan Facies*

The most important rifted petroleum basins occur in eastern China (Hu et al., 1991), such as the Mesozoic–Cenozoic basins of the Bohai bay basin (E), the Erlian basin (K_1) and the Subei–South Yellow Sea basin (E). Rifted basins were formed by extending or pulling apart a rifting valley (Lu et al., 1997). Their paleo-geographic background incorporated numerous uplifts and many depressions, multiple substance sources and nearby material supply center. Because it could create a large depositional space in a short time interval, a rifted basin contained widely-distributed sand bodies of alluvial fan, fan delta, braided river delta, underwater fan and beach bar facies (Xue et al., 2002; Xue and Ying, 1991; Wu and Xue, 1992). During its early rifting phase and its declining phase, a rifted lacustrine basin primarily contained deposits of alluvial fan facies and river facies. During the peak of the subsidence phase of a lacustrine basin, because of the great depth of water and the high elevation of mountains and because the slope gradient was great, the deposits of gravity flow were favorably developed. These included a near shore underwater fan that was distributed on a steep slope zone and which also included an infra-littoral fan that had supply channels on the gentle slope zone. An axial gravity flow channel was included as well. During the contraction phase and early subsidence phase of the lacustrine basin, because the lake region was relatively small and water was shallow, a rifted lacustrine basin mainly contained all kinds of deltas. A fan delta appeared on the steep slope zone; a braided river delta occurred on the gentle slope zone. During the constant subsidence phase of the lacustrine basin, because the lacustrine basin had been filled with sediments previously, the topography tended to be gentle. The lake was wide open and water was shallow; the beach-bar sand bodies of littoral-shallow lacustrine facies and carbonate deposits were favorably developed (Cai et al., 2003). Along with the evolution procession of the rifting valley, especially during the early rifting phase, there were multiple phases of basic volcanic eruptions, which created unique volcanic reservoirs.

In rifted lacustrine basins, there are all kinds of sandy conglomerate reservoirs of river, delta, near shore underwater fan, sub-lacustrine fan, beach-bar and alluvial facies (Fig. 4.2). Because these sandy conglomerate reservoirs were near hydrocarbon source rocks, they offered excellent accumulation conditions. Using a statistical study of explored petroleum reserves in all kinds of reservoirs in the major depression of the Bohai bay basin and the Erlian basin, geologists learned that the sand bodies in a fan delta, braided river delta, near shore underwater fan and sub-lacustrine fan are major types of lithostratigraphic petroleum reservoirs. Together they contain 68% of total explored petroleum reserves. In addition, the volcanic rocks and lacustrine carbonate rocks also have accumulative petroleum potential.

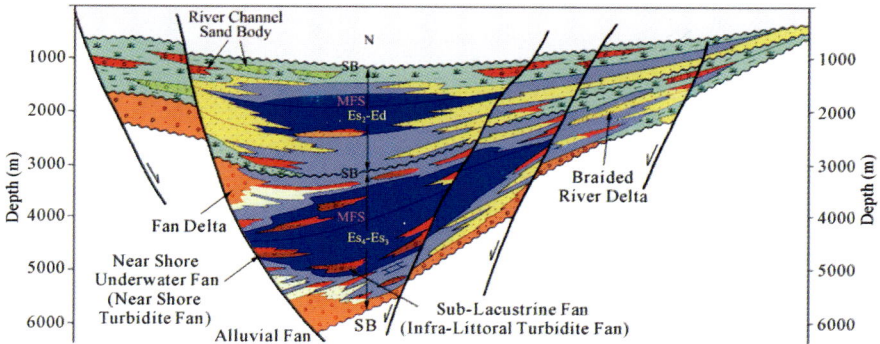

Fig. 4.2.   Major types of sand body and vertical distribution model in terrigenous rifted basin N, upper Tertiary; Ed, Dongyun formation of lower Tertiary; Es, Shahejie formation of lower Tertiary; SB, Stratigraphy boundary; MFS, Maximum Flood Surface

4.2.2   Depression Basin Containing Large Scale Sandstone Reservoirs of River–Delta Facies

In China, the Mesozoic–Cenozoic depression basins were commonly developed on shore, such as the Songliao basin, Eerduosi basin, Zhungeer basin. Because a depression basin was generally formed in the thermal subsiding region of a craton, its internal structure was relatively simple, which either tilted from the marginal slope to the intra-depression (such as the Songliao basin), or alternated between low relief bulges and depressions (such as the Eerduosi basin and Zhungeer basin). Generally speaking, a depression basin not only had a wide open, gently sloped lacustrine region, but it also had a relatively stable, long distance aquosystem source supply. Therefore, the differentiations in lithology and facies are distinct and the alteration in the lithologic thickness is small. Additionally, sedimentary systems are on a large scale and have a low depositional rate. The facies belt appears as a circle, the subsidence center and the depositional center are located in the same area and a deep water region is in the center of the lacustrine. Because some depression basins contained faulted uplifts on one side, in the circumstances the facies belts commonly developed into an asymmetrical circle. The subsidence center and the depositional center were apart. A deep water region was located near the transitional zone that was between the mountain front depression and the intra-depression of the basin.

Along the long axis of the lacustrine depression, the sand bodies belonged to alluvial plain facies–meandering river facies–delta facies (Qiu et al., 1997), such as in the northern depositional system in the Songliao basin (Fig. 4.3). In the direction of the short axis, the steep slope commonly has sand bodies of alluvial fan facies–fan delta facies. A good example is the Yingtai depositional system on the western slope of the Songliao basin. Also, in the direction of the short axis, the gentle slope has sand bodies of alluvial facies–braided river facies–delta facies. For example, in the Eerduosi basin, the Ansai depositional system contains a large scale shallow water delta that has a thickness of hundreds of thousands of meters in vertical distribution (Fig. 4.4). A characteristic of a depression basin is large size sand bodies of meandering river facies–delta facies that were developed on the gentle slopes, either along the direction of the long axis or along the direction of the short axis. In a depression basin, because the deep water region was far away from the substance source area, long distance source supply, fine grain turbidite sandstone and fluxoturbidite sandstone of delta front facies were developed here.

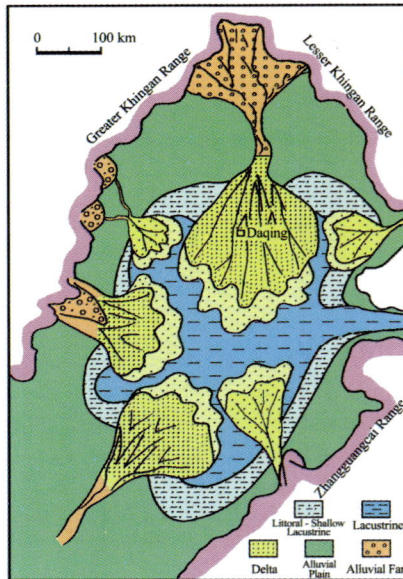

Fig. 4.3. Depositional systems of depression stage in the Songliao basin

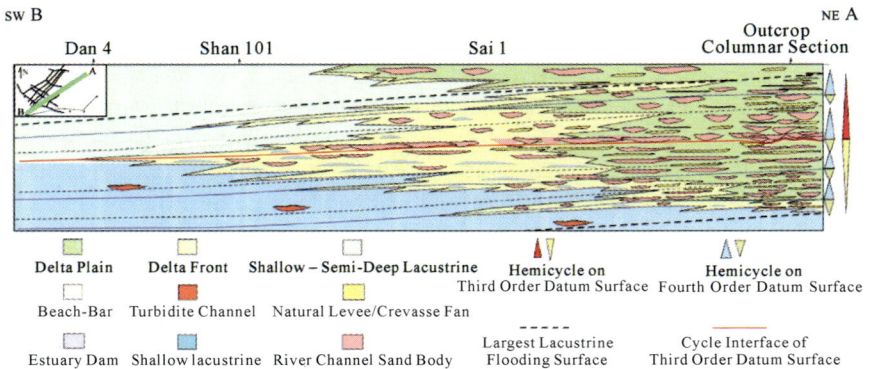

Fig. 4.4. Stratigraphic sequences model of large scale, shallow water delta in the Ansai, the Eerduosi basin

In a terrigenous depression basin, a sand body reservoir of delta facies developed next to the hydrocarbon generation area of the central depression. Thus, the major type of petroleum sand body is a delta. For example, in the Songliao basin, the Eerduosi basin and the Zhungeer basin, the reservoir of delta facies occupied 80% of the total petroleum reserve in each basin. A fan delta and braided river delta are secondary reservoir sequences. The petroleum bearing potential of a turbidite sand body needs further study.

4.2.3   Foreland Basin Containing Sandstone Reservoir of Alluvial Fan, Fan Delta and Braided River-Delta Facies

A foreland basin was distributed along an orogeny, which was located at the transitional zone between an orogeny and a stable craton. At the mountain front, subsidence was very strong. Towards the craton, the magnitude of the subsidence gradually reduced. The basal plane of sediments appeared as a slope. The cross section of the foreland basin shows an asymmetrical winnow pattern; from the orogeny to craton, the cross section can be divided into thrust belt, subsidence zone, slope zone and uplifting zone. In a foreland basin, the thickness of sediments, thickness gradient and depositional rate noticeably changed. For example, in the western Sichuan basin, the foreland basin of Longmen mountain was steep on the northwestern side and gentle on the southeastern side; the sediments were thicker on the northwestern side than on the southeastern side (Fig. 4.5).

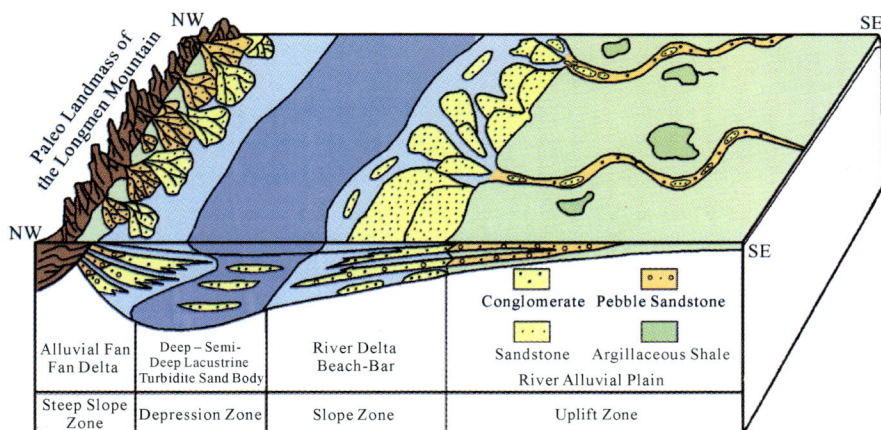

Fig. 4.5.   Depositional model of the Late Triassic foreland basin in the western Sichuan basin

On the thrust belt side of the foreland basin, the sand bodies of alluvial fan, fan delta or braided river delta were well developed in progressive sequences.

Vertically, the grains of sedimentary sequences become coarse in ascending order; from a sandy conglomerate from the substance source nearby to a medium–fine grain sandstone from the substance source distant from the mudstone. Horizontally, the shape of the sand body is a fan pattern. The thickness and lithology change quickly. Within a 10 km distance, the thickness can be reduced from 800 m to 300 m. On the slope zone side of the foreland basin, sand bodies of the transgressive delta developed. Vertically, the grains of sedimentary sequences become fine in ascending order. The depositional sequences of delta front facies consist of medium–fine grain clastic quartz sandstone distant from the substance source and lacustrine mudstone. Horizontally, this has the typical pattern of a delta and the thickness and lithology change gradually.

In a foreland basin, the major reservoir types are alluvial fan, fan delta, braided river delta. For example, on the northwestern margin of the Zhungeer basin, the major reservoir types in the Kelamayi oil field are sand bodies of alluvial fan and fan delta. Another example is in the Kela 2 gas field, in the Kuche foreland basin. The gas layers and sections are sandy conglomerate reservoirs of braided river delta facies, which is positioned in the Bashijiqike formation of the Palaeogene–Cretaceous systems.

4.2.4   *Craton Basin Containing Carbonate Reservoir and Sandstone Reservoir of Littoral Facies, or Marine–Terrigenous Delta Facies*

A marine craton basin contains several types of petroleum reservoir, which not only include sandstone reservoirs of delta, littoral and submarine fan facies, but also include a carbonate rock reservoir and bio-reef reservoir. Clearly, compared with all kinds of sandy conglomerate reservoirs in a terrigenous lacustrine basin, the reservoirs in a marine basin are very different. The Chinese marine basins mainly contain sandstone reservoirs of marine–terrigenous facies and carbonate rock reservoirs (that include bio-reef reservoirs). For example, in the platform region of the Tarim basin, the major reservoirs are the Devonian–Carboniferous beach-bar sand bodies of littoral facies (the Donghe sand body) and carbonate rocks in every oil field (Fig. 4.6). In the Sichuan basin, the major gas reservoir in the Permian system is a bio-reef. In the Eerduosi basin, the major gas reservoirs in the Carboniferous–Permian systems are delta sand bodies of marine–terrigenous facies.

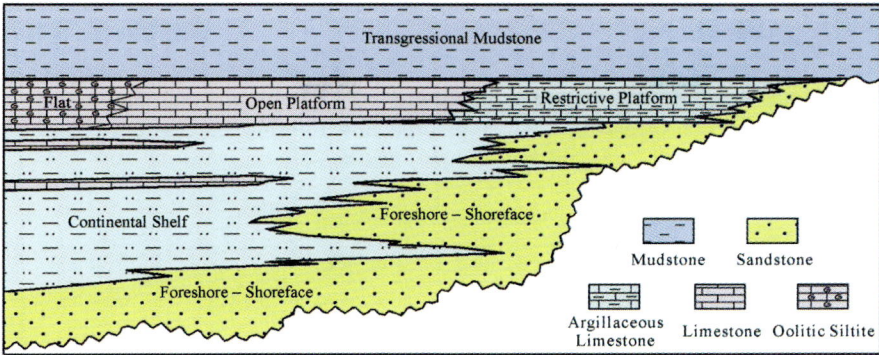

Fig. 4.6. Depositional models of the Paleozoic marine facies and marine–terrigenous facies in the Tarim basin

Littoral facies can be further divided into barrier type and non-barrier type. Coastal sand dunes, beach levees, beach sand bars and coastal sand bars were formed under the non-barrier type littoral environment. On the other hand, barrier island sand bodies, tidal channel sand bodies, tidal delta sand bodies and tidal flat sand bodies formed under the barrier type littoral environment.

Carbonate rock is the most important intra-source sedimentary rock in a marine basin. Compared with terrigenous clastic rock, carbonate rock undergoes a very different developmental process. Carbonate rocks are predominately deposited on the continental shelf within shallow water and in an open environment. Usually this formed a carbonate platform (Qi et al., 1978). For the genetic mechanism of carbonate rocks, the biochemistry was a primary function and the post diagenesis added assistance during the late stage of development. In a marine basin, carbonate rocks are the best developed petroleum reservoirs. Several large scale oil and gas fields were discovered in the carbonate sequences worldwide. Furthermore, a bio-reef and beach-bar are special types of carbonate rock that are commonly developed in a marine basin. Bio-reefs and their complex can easily form an effective trap for oil and gas accumulation.

4.3   Types of Lithostratigraphic Traps and Preconditions for Making These Traps

The establishment of a lithostratigraphic trap is the result of intricate alterations that include the alteration of the depositional conditions, the alteration of lithology and facies, the alteration of storage space either vertically or horizontally and the unconformity seals that are either above or below the reservoir sequences (Levorsen, 1936; Levorsen et al., 1941).

4.3.1   Types of Lithostratigraphic Traps

According to the methods and conditions of establishing a petroleum trap, a lithostratigraphic trap can be further divided into three types that are lithologic trap, stratigraphic trap and composite trap (Table 4.2).

Table 4.2   Classification of lithostratigraphic traps

Type	Sub-type	Mechanism of creating petroleum trap	Example
Lithologic Trap	Lithology pinch upward	Facies of reservoir sequences changed into mudstone or other dense lithology in up dip direction to form cap rock	Gaosheng in Liaohe
	Lithologic lens	Lens reservoir was enveloped by non-permeable lithology to form a trap	Niuzhuang in Dongying
	Channel deposit	Sand bodies in river channel, incised valley, and various aqua channels were sealed by non-permeable lithology of inter-channel to form a trap	Liangjialou in Dongying
	Bio-reef	Bio-reef was restricted by surrounding non-permeable lithology to form a trap	The Permian system in Eastern Sichuan
	Igneous rock	Igneous rocks (eruptive rock or intrusive rock) served as reservoir sequences that were enclosed by dense lithology, such as mudstone, to form a trap	Xujiaweizi in Songliao
	Diagenesis	Constructive diagenesis formed effective reservoir sequences; condensed diagenesis formed seal rock; together, they formed traps that include dissolution type, karst type, dolomitization type and cement type of traps	Daanzhai in Central Sichuan
	Fractures	Fractures developed in dense strata to form a trap	Central Sichuan
Stratigraphic Trap	Overlap unconformity	Reservoir sequences developed on top of an unconformity surface; non-permeable lithology overlapped reservoir sequences to form a trap	Qijia in Liaohe
	Buried hill	Buried hill was formed by weathering and leaching all kinds of basement rocks (carbonate, clastic and metamorphic rocks), non-permeable rock covered top unconformity surface of buried hill to form a trap	Renqiu in North China
	Unconformity seal	Above unconformity surface, non-permeable lithology or viscous crude oil/bitumen sealed reservoir sequences under the unconformity surface to form a trap	Qigu in Liaohe
Composite Trap	Structure–lithology	Alteration of lithology dominated this combination; structures (anticline, fault, or diapir) assisted restriction or sealing condition to form a composite trap	Haituozi, Jinlin
	Structure–stratigraphy	Mainly attached to unconformity surface, structures (anticline or fault) assisted restriction or sealing condition to form a composite trap	Ya 13-1
	Lithology–stratigraphy	Mainly attached to unconformity surface, dense lithology served as seal rock to form a composite trap	Shuguang in Liaohe
	Stratigraphy–lithology	Alteration of lithology dominated this combination; unconformity surface assisted in sealing condition to create a composite trap	Zhongshi City

4.3.2 Preconditions for Establishing Lithologic Traps

The establishment of a lithologic trap was due to the changes in either lithology or physical properties. Non-permeable sequences surrounded reservoir sequences. They either restricted or laterally covered reservoir sequences to create a trap. The basic requirements for establishing a lithologic trap are the creation of reservoir sequences under sedimentation, diagenetic or magmatic processes, the factors that determine the cap rock and the background of the sedimentary structure.

● **Sedimentation**

Sedimentation is a precondition and a dynamic foundation for alterations to lithology and facies. Sand bodies that are deposited in a delta, river channel, coastal zone, underwater fan, sub-lacustrine fan and turbidite zone have good reservoir properties, especially sand bodies that are deposited either on the delta front or in the river channel. Combined with the mudstone either above or on the side, the beneficial reservoir sequences could form an excellent petroleum system to form a lithologic petroleum trap.

In western China, the sand bodies of littoral facies were widely deposited in a marine craton basin. On the other hand, in eastern China, the sand bodies of delta facies and underwater fan facies were generally deposited in a terrigenous depression and rifted basin, which combined with non-permeable rocks in the surrounding area to form a wonderful petroleum system, and which offered excellent conditions for establishing all kinds of lithologic traps. Carbonate sedimentation also directly controls the establishment of lithologic traps (Zhou and Tao, 2007). Exploration practices indicated that the carbonate lithologic traps mostly developed in a particular paleo-depositional environment with particular lithology or facies.

● **Diagenesis**

Diagenesis preformed dual functions when establishing a lithologic trap (Zhou and Tao, 2007). The first function was constructive diagenesis, which could form beneficial storage spaces by altering non-reservoir sequences into efficient reservoir sequences, such as dissolution, dolomitization, TSR and karst. If there were efficient cap rocks in the same neighborhood, together they could form petroleum traps. The second function was condensed diagenesis. Because of cementation, silicification or re-crystallization in strata, a dense lithology was formed. Furthermore, it could serve as an impenetrable seal. A petroleum trap was formed if it was in combination with a large acreage age, effectual sand bodies or other reservoir sequences. In addition, dissolution and leaching processes could alter elevated strata to create numerous pores. If a weathered buried hill was sealed by non-permeable sequences either above or to the side, a buried hill type of trap was established (Fig. 4.7).

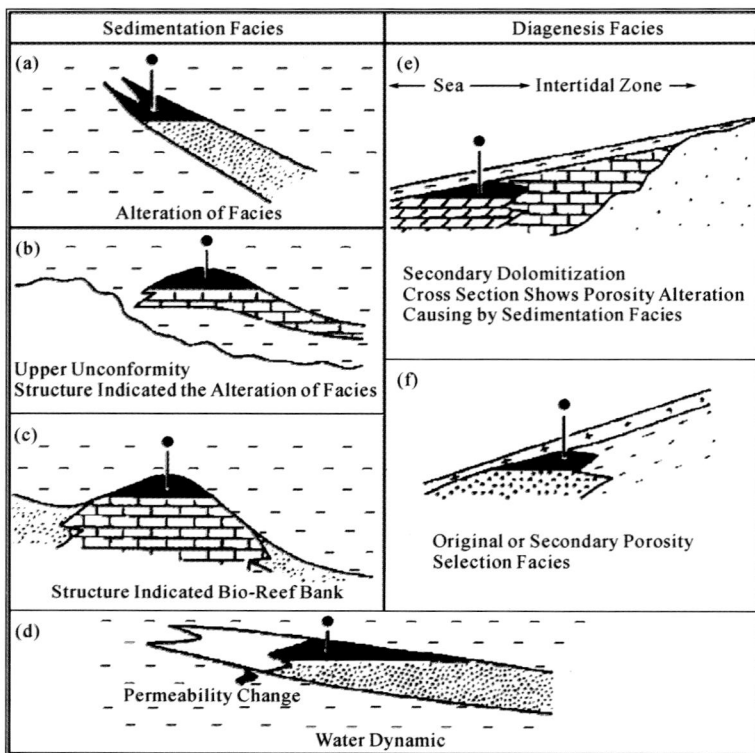

Fig. 4.7. Diagenesis and creation of lithostratigraphic traps in carbonate rocks

● **Magmatism**

During the creation of petroleum traps, the function of magmatism was that igneous rock could be used as an efficient reservoir sequence that was sealed by adjacent non-permeable lithology. Together, they formed a trap. Otherwise, because of heterogeneity or latter phase diagenesis, a petroleum trap was formed inside the igneous rocks.

● **Tectonic Background**

Generally speaking, a lithologic trap by itself does not have any significant commercial value, such as a lens trap. The most important lithologic traps were created under the structural background. For example, a nose-like lithologic trap and fault-lithologic trap were developed either on top of, or on the side of, positive and negative structures in a different order of the structural system. A statistical study of numerous lithologic traps and petroleum reservoirs indicated that most of the effective lithologic traps were developed either on the side wing of a positive structure or on the slope zone of a negative structure. Thus, certain kinds of structural background were a precondition for establishing lithologic traps.

The combination of preconditions discussed above could result in several types of lithologic traps. Traps were formed either by the lithology being pinched upward or by an alteration to the lithology on the side. These consist of lens traps, traps in water channels, bio-reef traps, igneous rock traps, diagenetic traps (the major types include the optimization of physical properties in an isolated portion of the dense layer, isolated densification either on top of, or on the side of, reservoir sequences to form an enclosed type of trap) and fractural traps (the major types include mudstone fractures that relate to the diagenesis and structure, inter-layer fractures of mudstone, fractures either in carbonate rock or in sandstone).

4.3.3 Preconditions for Establishing Stratigraphic Traps

When reservoir sequences were restricted or sealed by non-permeable sequences along the unconformity or erosion surfaces, a stratigraphic trap was formed under the geologic progression of sedimentation hiatus, erosion or overlapping deposits and weathering and leaching. These geological processes were caused by tectonic movement. The types of stratigraphic trap include stratigraphic overlapping trap, unconformity sealed trap and buried hill trap.

Sedimentation hiatus and stratigraphic unconformity are the preconditions for creating a stratigraphic trap. Regardless of when a hiatus happened or how long a hiatus lasted, despite the scale of the unconformity surface, this had the ability to rearrange previous traps and to create new stratigraphic traps. In addition, the developmental level of reservoir sequences and a certain background of sedimentary structures were the requirements for controlling the creation and scale of a stratigraphic trap.

● Stratigraphic Overlapping Traps

When marine or lake transgression reached the marginal slope of a basin or reached the lateral wing of a bulge, overlapping cyclic deposits were developed on top of the unconformity surface. At the bottom of the sedimentary cycle, the reservoir sequence overlapped the non-permeable sequence that was beneath the unconformity surface. On the other hand, the reservoir sequence was covered by another non-permeable sequence of continual deposits. Therefore, a stratigraphic trap has excellent roof and floor seals. Together they created a stratigraphic overlapping trap. However, not all overlapping zones meet the preconditions for a petroleum trap. When the reservoir sequence was deposited on the unconformity surface with roof and floor sealing sequences, and only when an overlapping line intersected with a structural depth contour line, could the petroleum trap be formed. The figure of this kind of trap was controlled by intersected lines with an irregular pattern.

● **Unconformity Sealed Traps**

Reservoir sequences endured erosion in various degrees. Both a previous trap and paleo-uplift not only received different kinds of damage, but they were also covered by non-permeable sequences of late deposits. When the unconformity line intersected with the structural depth contour line on top of reservoir sequences, the unconformity sealed trap was formed. If mudstone or another non-permeable layer covered the upper portion of unconformity, an excellent sealing condition developed. Moreover, if bitumen or viscous crude oil sealed the top layer of reservoir sequences or unconformity surface, a stratigraphic trap could be formed too.

● **Buried Hill and Topographic Traps**

This type of trap is related to the relief of the unconformity surface. Therefore, it is called a paleo-topographic trap or a "paleo-buried hill trap". This kind of trap was controlled by the unconformity surface, fault and non-permeable lithology. The unconformity surface was the foundation for establishing a petroleum trap and fault and non-permeable sequences were secondary preconditions. Paleo-topographic strata (that formed during the creation of a basin) served as a basement (which includes all the lithologies that formed before the creation of a basin). The strata that were deposited during the basin development operated as the cover layers. The reservoir sequence of the trap was made of basement strata. According to different diagenetic methods applied to the reservoir sequences, the traps can be further divided into surface leaching stratigraphic traps and underground permeated stratigraphic traps. The former is the so called "paleo-topographic buried hill trap"; the latter is the so called "internal buried hill trap" or underground permeated stratigraphic trap.

4.3.4   Creation of Composite Traps and Their Types

The creation of a composite trap was controlled by the combined mechanism of two or more factors. Here, we only consider the three factors of structure, lithology and stratigraphy.

● **Important Factors for Establishing a Composite Trap**

Structural Factor

The structural factor primarily includes the anticline (nose like bulge/uplift), fault and late stage inversion structure. These factors involved the process of the creation of a petroleum trap. If structural factors were the predominate mechanism for creating a trap, then this kind of trap should belong to the category of structural trap.

Lithologic Factor

The lithologic factor includes lithology, facies, alteration of physical properties and fractures. These factors combined with a dense lithology seal to form petroleum traps. If lithologic factors were the predominate mechanism for creating a trap, then this kind of trap should belong to the category of lithologic trap.

Stratigraphic Factor

Stratigraphic factor includes stratigraphic overlapping, pinch out and buried hill draping. If the stratigraphic factors were the predominate mechanism for creating a trap, then this kind of trap should belong to the category of stratigraphic trap.

● **Types of Composite Traps**

The following discussion includes the composite traps that are mainly related to the combined mechanism of lithologic and (or) stratigraphic factors. The structural factors were only an assistant force. This discussion leaves out composite traps that are mainly related to structural factors.

According to the variety of trap creating factors, the types of composite trap can be further divided into a two factors composite trap and a three factors composite trap. The former indicates a trap that was created by the combined mechanism of any two types of factors out of structural factors, lithologic factors and stratigraphic factors; for example, the structural–lithologic trap, the structural–stratigraphic trap, the lithologic–stratigraphic trap, the stratigraphic–lithologic trap. The latter points to a trap that was created by the combined mechanism of structural factors, lithologic factors and stratigraphic factors; such as, the structural–lithologic–stratigraphic trap, the structural–stratigraphic–lithologic trap.

4.4   Analysis of Controlling Factors when Looking for Petroleum Traps: "Six Lines and Four Surfaces"

The development of a lithostratigraphic trap is the precondition and important factor for developing a lithostratigraphic petroleum reservoir. A statistical study of oil and gas reservoirs in different types of basin indicated that the creation of a lithostratigraphic trap was mainly controlled by factors in the two categories that are called "six lines" and "four surfaces" (Fig. 4.8).

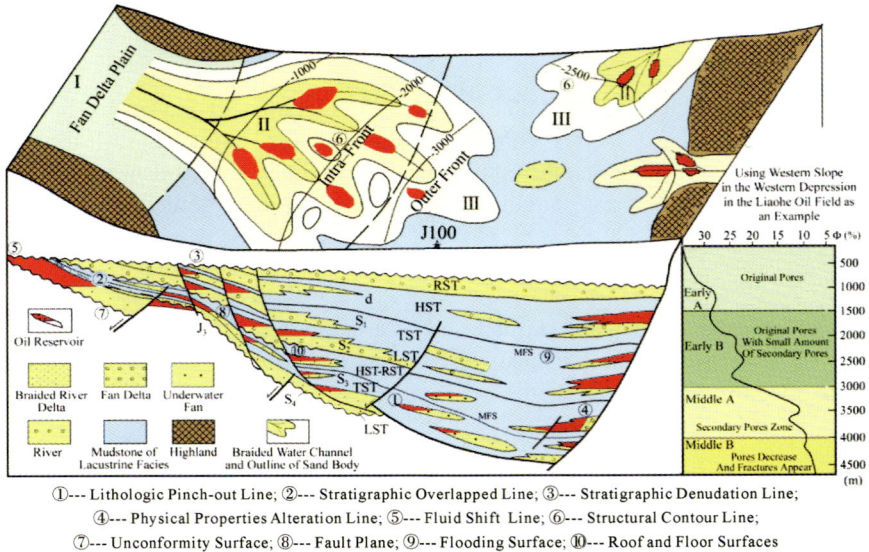

①--- Lithologic Pinch-out Line; ②--- Stratigraphic Overlapped Line; ③--- Stratigraphic Denudation Line;
④--- Physical Properties Alteration Line; ⑤--- Fluid Shift Line; ⑥--- Structural Contour Line;
⑦--- Unconformity Surface; ⑧--- Fault Plane; ⑨--- Flooding Surface; ⑩--- Roof and Floor Surfaces

Fig. 4.8.   Major controlling factors ("six lines" and "four surfaces") for the creation of lithostratigraphic trap

4.4.1   "Six Lines" Controlled Creation of Petroleum Traps

The "six lines" are lithologic pinch-out line, stratigraphic overlapped line, stratigraphic denudation line, physical properties alteration line, fluid shift line and structural contour line.

● **Lithologic Pinch-out Line**

The lithologic pinch-out line is the most important precondition for creating a lithostratigraphic trap. The pinch-out line not only occurs in a lithologic trap, but it also appears in a stratigraphic trap, such as in a stratigraphic overlapped trap. The pinch-out line controlled the creation of this kind of petroleum trap. Particularly in a lithologic trap, the lithologic pinch-out line was the required condition for creating a lithologic trap. For example, when the sand bodies extended toward a depression or basin, they made contact with non-permeable lithology to form a lithologic pinch-out line (or belt). These kinds of sand bodies appeared on the delta front, distal fan, beach bar on a littoral–shallow lake and in the river channel. In the direction of upward dipping, because of the alteration of facies, the lithology changed unexpectedly. Thus the lithologic pinch-out line was formed. If there were cap rocks on the side, it was most likely that various types of lithologic trap were formed. Furthermore, the traps that were located around the lithologic pinch-out line offered the best spaces for oil and gas to accumulate and to form a lithologic petroleum reservoir. For example, according to a statistical study of major oil and gas fields in the Songliao basin and the Eerduosi basin, when the thickness ratio of sandstone and strata is $0.2 - 0.4$, numerous thin sandstone layers, lithologic traps and petroleum reservoirs are found. However, in the structural–lithologic traps and petroleum reservoirs, the thickness ratio of sandstone and strata is $0.4 - 0.6$. In addition, most of the traps and petroleum reservoirs were discovered around the lithologic pinch-out belt (Zhou and Tao, 2006).

- **Stratigraphic Overlapped Line**

A stratigraphic overlapped line is a required condition for creating a stratigraphic overlapped trap. At the same time, this area is also a beneficial zone for developing a lithologic trap. In general, a stratigraphic overlapped line and a lithologic pinch-out line have a related creative mechanism. However, a stratigraphic overlapped line is distributed along the sea coast or lacustrine shoreline. A stratigraphic overlapped trap was formed along the stratigraphic overlapped line on the gentle slope of a depression. In addition, the traps on a gentle slope were the best ones for capturing hydrocarbons and for developing oil and gas reservoirs. For example, a statistical study of different types of basin indicated that a stratigraphic trap on a gentle slope is on a relatively large scale. On the other hand, in the Bohai bay basin, a statistical study of explored oil reserves from different structural belts in nine petroleum depressions shows that on a gentle slope, the oil reserve is 21.7×10^8 t, which is 56.7% of the total oil reserve. On a steep slope, the oil reserve is 9.8×10^8 t, which is 24.4% of the total oil reserve. In a trough, the oil reserve is 7.3×10^8 t, which is 19.1% of the total oil reserve. Evidently, a gentle slope not only contains the best developed stratigraphic overlapped traps, but it also has plenty of oil and gas.

- **Stratigraphic Denudation Line**

A stratigraphic denudation line is an important precondition for creating a stratigraphic trap. On the steep slope or gentle slope of a depression, tectonic movement created an unconformity surface, which reduced the thickness of strata or which pinched out strata to form many stratigraphic traps. Oil and gas can migrate along an unconformity surface and accumulate in a stratigraphic trap to form an oil and gas reservoir. For example, in the Zhungeer basin, Bohai bay basin and Erlian basin, the study of lithostratigraphic traps or petroleum reservoirs shows that all the stratigraphic traps or petroleum reservoirs were distributed above or below the unconformity surface. Buried hill traps or petroleum reservoirs were mostly developed in the trough bulge. Conversely, the traps or petroleum reservoirs of stratigraphic pinch-out type and of lithologic upward dipping style pinch-out type often appeared on a steep slope and on a gentle slope. Among these traps or petroleum reservoirs, most of the lithostratigraphic traps or petroleum reservoirs were distributed along the stratigraphic pinch-out line.

● **Physical Properties Alteration Line**

There are two kinds of physical properties alteration line. One is the boundary between the permeable layer and the relatively non-permeable layer, which was caused by destructive diagenesis (that formed a high density cemented zone) in a large area of reservoir sequences. Another one is the zone of high porosity and high permeability that was formed by constructive diagenesis in lithology of poor porosity and poor permeability. Mostly this is a secondary zone of high porosity. The former one formed a sealing element, the latter formed a reservoir element. For example, in Shanbei province, laumontite and feldspar were dissolved to create a secondary porosity belt in the Mesozoic feldspar sandstone during the middle phase of the diagenetic process. This means laumontite–feldspar dissolution facies are the best lithologic facies for a petroleum reservoir in this area. The dissolution reactions mainly happened near the oil generation side of a trough depression. Toward the edge of a lacustrine basin, dissolution reactions gradually decreased. Highly dissolved sandstone has good physical properties and oil bearing ability. Usually, the porosity is between 13% – 16% and the best can reach 18%. The best permeability is 50×10^{-3} μm^2. On the other hand, laumontite cemented facies and other cemented facies served as a high density sealing belt to form a diagenetic trap.

   In terrigenous basins, such as the Songliao basin and Bohai bay basin, the study of sedimentary diagenesis indicated that the alteration of physical properties that was caused by diagenesis controlled vertical distribution of the lithostratigraphic trap.

   At a middle to shallow depth (2,000 – 2,500 m), because the hydrocarbon source rock released a huge amount of organic acid, the dissolution reaction generally happened at the contact area between the sand body and hydrocarbon source rock (Fig. 4.9). Excellent physical properties appear on the edge of the sand body and thin shell shape, pinch-out upward, lithologic traps were formed. At this depth, all the sand lenses contain oil. Conversely, at a middle to deep depth

(>2,500 m), because hydrocarbon source rock released relatively less organic acid, the dissolution reaction was relatively weak. However, diagenetic cementing was strong and this formed a high density external shell. Inside the sand body, physical properties were good. As a result, sandstone formed a trap due to the alteration in physical properties. Because of cementing, the sand lenses at a middle to deep depth do not contain any oil.

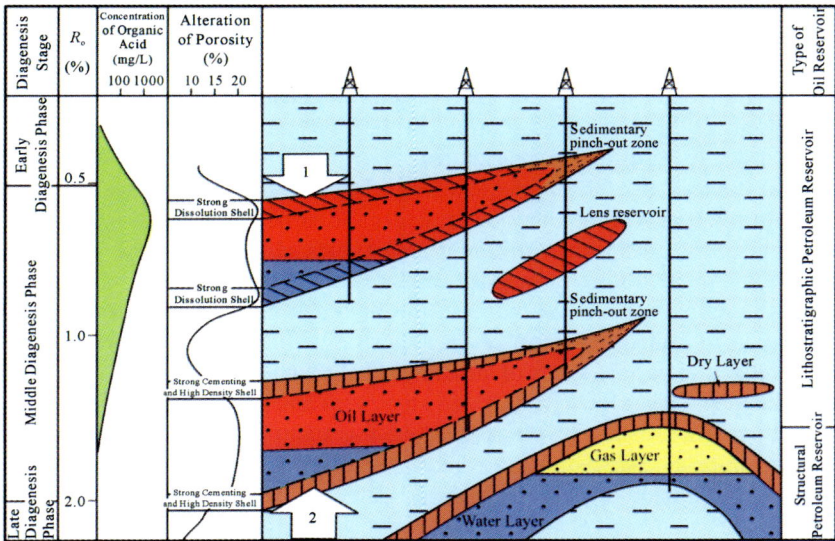

1. Large amount of acidic bedding water were released and dissolution reaction happened;
2. Ca, Mg, Fe were released with bedding water, cementing procession happened

Fig. 4.9. Diagram of underground dissolution shell, high density shell and the creational mechanism for lithostratigraphic trap

● **Fluid Shift Line**

In special circumstances, the alteration in the fluid is the important factor in the creation of a lithologic trap and petroleum reservoir (Hu et al., 1991), such as viscous crude oil and bitumen functioning as sealing elements and water dynamics working as a barricade. For example, when the Palaeogene strata were deposited on the western slope of the western depression in the Liaohe oil field, the strata gradually overlapped on top of the basement. Because a tectonic movement happened when the Dongying formation was deposited, the western slope had been unevenly elevated. The tip of the upward strata was cut off across the region. The Neogene Guantao formation was in direct contact with the Palaeogene formations. Because the unconformity surface was buried at a shallow depth, the petroleum accumulation zone was formed by utilizing viscous crude oil and bitumen as the sealing layer. On the western slope, three large scale nose-like structures (the Huanxiling, Shuguang and Gaosheng) combined with the viscous crude oil and bitumen sealing zone and stratigraphic overlapped zone to form various types of lithologic traps and petroleum reservoirs and a high abundance of stratigraphic traps and related oil and gas accumulation belts.

● **Structural Contour Line**

The structural contour line reflected the relief pattern of the strata. Usually, when it is combined with certain kinds of structural background, such as a pericline or slope, a lithologic trap would be created. In general, besides meeting the requirements of lithologic preconditions or the physical properties of the pinch-out line, in order to form a lithostratigraphic trap the strata must intersect with the structural contour line at an adequate angle. Otherwise it is not sealed off and cannot form a trap. On the other hand, under the same (or similar) sedimentation and diagenetic conditions, different types of lithostratigraphic trap were distributed around different types of structural belt. For example, the study of lithostratigraphic traps or reservoirs in the Songliao basin, Bohai bay basin and Erlian basin indicated that most of the lithostratigraphic traps or reservoirs were closely related to certain kinds of structural background. This kind of relationship mainly happened near the pericline and slope area and was distinctively presented as a belt pattern. Usually, on the steep slope that contains multiple fault steps, the lithologic traps or reservoirs most likely occurred in a group. These traps or reservoirs were made of delta front sand bodies of the fan delta or braided river delta. On the higher level of a gentle slope, a stratigraphic overlapped trap or reservoir, unconformity sealed trap or reservoir, buried hill trap or reservoir were commonly created. On the lower and middle level of a gentle slope, normally many lithologic traps or reservoirs were formed. The reservoir sequences were upward dipping and pinch-out sand bodies and sandstone lenses, which were deposited on the delta front or in the river channel. In the trough zone, many lens-like lithologic traps or reservoirs were commonly formed in sub-lacustrine fan and turbidite. Inside a depression, and near a bulge, lithologic traps or reservoirs included buried hill, beach-bar sand body and sand bar.

4.4.2   *"Four Surfaces" Managed Establishment of Petroleum Traps*

The "four surfaces" are fault plane, unconformity surface, lacustrine flooding surface and roof and floor surfaces.

● **Fault Plane**

The fault plain controlled the creation and distribution of a lithostratigraphic trap and reservoir in the following three areas. Firstly, a fault, especially a growth fault, not only controlled the development of sand bodies, but it could also make unexpected lithologic alterations and facies alterations on both sides of faulted walls, if the fault throw was relatively large. In addition, the lithologic trap or reservoir could be formed with a seal. For example, the Leijiazhuang oil field is located on a steep slope in the western depression of the Liaohe oil field. On the foot wall, the upper portions of the sand body of fan delta or turbidite deposits were sealed off by non-permeable alluvial fan deposits from the hanging wall to create lithologic traps or oil reservoirs. Secondly, at the turning point of a boundary fault in a terrigenous rifted basin or depression, there are usually two split faults with a nose-like structure at the joint end of the split faults. Many stratigraphic overlapping traps or lithologic up-dip and pinch-out traps developed in this area. Because these traps were very close to the source rocks, many lithostratigraphic reservoirs developed in this area. For example, in the Erlian basin, many lithologic traps or reservoirs were discovered in the Bayindulan depression, the Saihantala depression and the Jiergalang depression. Next, the fault plain was a major vertical passage for oil and gas migration in most of the terrigenous basins. In non-hydrocarbon generating rocks, oil and gas reservoirs utilized the hydrocarbon supply fault as a connecting channel. For example, in discovered lithologic traps or reservoirs in the Dongying formation and the Neogene system in the Bohai bay basin, in the Jurassic and the Cretaceous systems in the Zhungeer basin, these traps or reservoirs generally contained faults to supply hydrocarbon vertically. Otherwise, these traps are ineffective traps.

● **Unconformity Surface**

Stratigraphic cut off, pinch-out and overlapping and lithologic pinch-out happened along the unconformity surface. In addition, a weathering and leaching process also took place on this surface, which controlled the development of a high quality reservoir. Thus, unconformity related lithologic alteration, physical property alteration or pinch-out created the preconditions for developing petroleum traps. On the other hand, an unconformity surface was an excellent passage for oil and gas migration (Pan et al., 1998). It played an important role during the creation of a lithostratigraphic reservoir. Because of the multiple cycles of the tectonic movement and multiple orders of the tectonic cycle, in the terrestrial strata there are several unconformity surfaces or sedimentation hiatuses. When the reservoir sequences occurred on the unconformity surface with upper and lower sealing layers, only if the reservoir overlapped line intersected with the structural contour line were stratigraphic traps formed and distributed either on top or beneath the unconformity surface. If there was a petroleum migration passage (such as a fault or unconformity surface) between the traps and the hydrocarbon generating rock, a stratigraphic overlapped reservoir or unconformity blocked stratigraphic reservoir could be formed. The best oil and gas accumulations were located at the base and the top of the boundaries of major lacustrine transgressional sequences. These two boundaries were characterized by two regional unconformity surfaces that controlled vertical distributions of traps or reservoirs in a depression. For example, in the Bohai bay basin, the discovered stratigraphic traps or petroleum reservoirs in every depression were distributed either on top or beneath the regional unconformity surface of the Palaeogene. The stratigraphic overlapped traps or petroleum reservoirs predominately developed on the basal unconformity surface of the Palaeogene. If the pre-Palaeogene system had good reservoir conditions and a non-permeable sequence covered the unconformity surface, the "buried hill type" of unconformity blocked traps or petroleum reservoirs could be formed. The top unconformity surface of the Palaeogene mainly contained unconformity blocked traps or petroleum reservoirs. The stratigraphic overlapped traps or petroleum reservoirs were scarce here.

● **Lacustrine Flooding Surface**

Recently, along with broad applications of terrigenous sequence stratigraphy, the study of lithostratigraphic traps and related petroleum reservoirs produced an important achievement. The understanding of the lacustrine flooding surface that controlled the distribution of lithostratigraphic traps and related petroleum reservoirs has been enhanced (Jia et al., 2002). This new concept has been confirmed by the discovery of lithostratigraphic traps and by petroleum exploration. The largest lacustrine flooding surface reflected the broadest and deepest water body in the basin. During this time interval, dark color mudstone developed perfectly, with highly abundant organic substances. The dense lacustrine mudstone and hydrocarbon source rocks were developed during this time interval. In the terrigenous lacustrine basin, the best reservoir sand bodies were from the delta, fan delta and turbidite fan. They were distributed either on top of or beneath the largest lacustrine flooding surface. The lens shape sand bodies of the distributary river channel, estuary dam and distal sand bar had mainly developed on the front margin of these sand bodies. Horizontally, the sand bodies were distributed in a group or belt pattern. Sand bodies were enveloped by source rocks with a contact pattern in the shape of digits. This kind of contact relationship not only helped to create lithologic traps, but it also made it easier to capture oil and gas to form a lithologic oil and gas reservoir. For example, in the Eerduosi basin, the Triassic traps and related oil reservoirs were closely distributed in the region that was limited by two lacustrine flooding surfaces (Fig. 4.10). The lithostratigraphic traps or oil reservoirs were most distinctive and they were controlled by the largest lacustrine flooding surface (Qiu and Gong, 1999). As another example, in the Erlian basin 98% of all the traps or oil reservoirs were distributed in the Aershan formation and the Teng 1 member in the Lower Cretaceous between two lacustrine transgression domains. Furthermore, this characteristic was noticeable in the Songliao basin and Bohai bay basin.

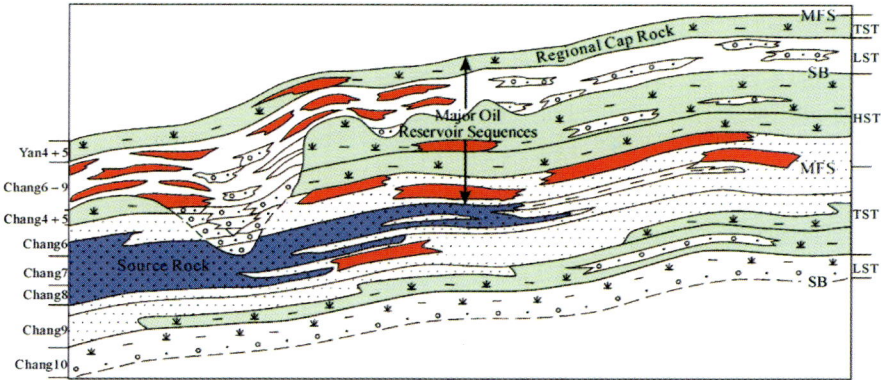

Fig. 4.10. Distribution map of the Mesozoic lacustrine transgression surface and petroleum reservoir in the central Zhungeer basin

● **Roof and Floor Surfaces**

The roof and floor surfaces mainly have a dual significance. The first is when the floor (that is located under the unconformity surface) is made of dense strata. If the floor is a permeable layer, then the trap is an inefficient trap. In particular, for a stratigraphic overlapped trap or stratigraphic unconformity trap, the requirements of floor quality demand a higher standard (Hu et al., 1991). For example, in the Jizhong depression, on the slope of the Shulu depression, the sandstone in the Palaeogene Sha 3 member overlapped on top of the Carboniferous–Permian systems. If the sandstone of the Sha 3 member overlapped on top of the sandstone of the Carboniferous–Permian systems, the drilling exploration only obtained oil shows without a good oil reservoir, such as the Nanxiaocheng stratigraphic trap. If the sandstone overlapped on top of mudstone or argillaceous limestone, then a good oil reservoir will be discovered, such as the Xicaogu stratigraphic trap and the Checheng stratigraphic trap. The second significance of the roof and floor surface is that if the roof of the sandstone consists of dense strata, this also decided the creation of a trap. If the roof is a permeable layer, then the trap is an inefficient trap.

To create a lithostratigraphic trap not only requires two or more than two factors, as discussed above (six lines and four surfaces), but they also need to be combined in a correct format. Combining these lines and surface inadequately would not create a trap. For example, when the angle between the direction of the sand body pinch-out and the direction of the strata dip is $90° − 270°$, a lithologic up-dip and pinch-out trap could be formed.

References

Cai, X.Y., Li, S.T., et al., 2003. High Resolution Sequence Stratigraphy of Continental Basins: Basic Ideas, Method and Practice for Exploring Subtle Oil Pools. The Geological Publishing House, Beijing.

Hu, J.Y., Huang, D.F., et al., 1991. Terrestrial Petroleum Geology in China. Petroleum Industry Press, Beijing.

Jia, C.Z., Zhao, W.Z., et al., 2002. New development in sequence stratigraphy research. Petroleum Exploration & Development, 29(5):1-4.

Levorsen, A. I. 1936. Stratigraphic versus structural accumulation. AAPG Bull, 20:521-530.

Levorsen, A. I., et al. 1941. Stratigraphic Type Oil Fields. AAPG, Tulsa, Oklahoma.

Li, K.Z., Qi, J.F., et al., 1997. Tectonic Model of the Cenozoic Petroleum Basin in the Bohai bay Region. The Geological Publishing House, Beijing.

Pan, Y.L., et al., 1998. Concealed Petroleum Reservoir in China. The Geological Publishing House, Beijing.

Qi, L.G., et al., 1978. Progression in Sedimentology, Carbonate Rocks. Petro-Chemical Industry Press, Beijing.

Qiu, Y.N., Xue, S.H., Ying, F.X., 1997. Terrestrial Petroleum Reservoirs in China. Petroleum Industry Press, Beijing.

Qiu, Z.J., Gong, Z.S. (Ed.), 1999. Petroleum Exploration in China: Introduction. Petroleum Industry Press, Beijing.

Xue, S.H., Ying, F.X., 1991. Depositional Facies and Reservoir Sequence in Terrigenous Basin, Petroleum Geology and Exploration Practices in Terrigenous Basin. Petroleum Industry Press, Beijing.

Xue, S.H., Liu, W.L., Xue, L.Q., et al., 2002. Lacustrine Sedimentology and Petroleum Exploration. Petroleum Industry Press, Beijing.

Wu, C.J., Xue, S.H., 1992. Sedimentology of Petroliferous Basins in China. Petroleum Industry Press, Beijing.

Zhou, C.N., Tao, S.Z., 2006. The category and distribution of litho-stratigraphic traps/reservoirs within sequence stratigraphic framework in terrigenous depressed basin: an example from the Cretaceous system in the Songliao basin. Chinese Journal of Geology (Scientia Geologica Sinica), 41(4):711-719.

Zhou, C.N., Tao, S.Z., 2007. Primary controlling factors for creating a large – medium size lithostratigraphic petroleum fields in marine carbonate rock. Chinese Science Bulletin, 52(supplement):32-39.

5

Primary Types of Lithostratigraphic Reservoir (Accumulation Region) and Their Characteristics

The development of a petroleum accumulative zone required the collaboration of a series of geological conditions, which included the source of the oil and gas, the migrating passage, conditions of the trap, conditions of the reservoir and conditions of the cap rock. According to the relationship between a lithostratigraphic petroleum system, geo-zone and traps, this chapter proposes a geo-zone concept of the structure sequences of the petroleum play and the principles of classification. The development of a lithostratigraphic reservoir was controlled by the following four factors that were the type of basin, growth structure, sequence and facies, and source rock and injection mechanism.

5.1 Reservoir-Forming Combination of Structure and Stratigraphy and its Significance

In China, oil and gas exploration has progressed into a new period that equally emphasizes both structural reservoir exploration and lithostratigraphic reservoir exploration. Therefore, the use of multiple geological factors for categorizing and evaluating the oil and gas accumulative region (belt) is inevitable. This is especially true in terrigenous basins, because these basins contain multiple types of structural framework, mixed patterns of the reservoir body, an intricate creation mechanism for petroleum traps and difficult accumulative conditions. Thus, the oil and gas accumulative region (belt) cannot be simply summarized or divided into the structural region (belt) and the lithologic region (belt). Via numerous analyses of oil and gas accumulative conditions in various regions, the author considers that the structural background, the types of sedimentary system within the stratigraphic sequence framework and the combination of hydrocarbon source rock and petroleum reservoir (or trap) are the primary factors in the process of oil and gas accumulation. Because these factors strongly emphasize the significance of structure, sequence stratigraphy and accumulation, the author initially proposes the concept of "a reservoir-forming combination of structure and stratigraphy".

A "reservoir-forming combination of structure and stratigraphy" means that sequence stratigraphy and its domain are to be found within a certain kind of structural paleo-geographic background. This combination formed a petroleum system of generation–reservoir–seal, with related formation mechanism. In addition, this combination includes reservoir bodies, traps and petroleum reservoirs, which were formed either during the structural movement in the late geological period or during the diagenetic processes, and which had similar accumulative conditions and a comparable petroleum distribution pattern. From the viewpoint of the structural background, sequence stratigraphy and the accumulative conditions, one type of basin could contain multiple types of "reservoir-forming combinations of structure and stratigraphy". For example, a sequence stratigraphy and a reservoir body could form a common type of reservoir-forming combination of structure and stratigraphy in a rifted basin, because they not only shared the same substance source and same aquatic system, but they were also within the same background slope structure. In addition, they were controlled by the same oil and gas migration system. This kind of combination is called a type of gentle slope–lacustrine transgression and high altitude braided river delta, underwater fan, which is equal to a petroleum bearing region and which belongs to the first order. Next, on a wide-ranging slope area, the alteration of the secondary structural background could happen. At the same time, the clastic materials that were brought to the slope by the aquatic system naturally diversified to form different facies belts, such as a braided river delta plain, an inner belt and outer belt of a braided river delta front. Therefore, different types of lithostratigraphic petroleum accumulative belts were formed at different locations on a slope (Zhang, 1997). On the high ground of a slope, the unconformity petroleum reservoir and viscous crude oil belt often appeared in distributary river channels. In the middle portion of a slope, where the slope gradient changed or where the growth fault belt and associated structures occurred, structural, composite and lithologic petroleum accumulative belts could be formed in the inner belt of the braided river delta front. On the lower level of a slope, the lithologic petroleum accumulative belt could be formed in the outer belt of the braided river delta front and in the turbidite sand body, which belong to the second order. The third order is the specific exploration target, such as the combination of a lithologic upward dipping and pinch-out trap or a nose-like structure on the side of a delta front, with some kinds of micro facies. The study of a reservoir forming combination of structure and stratigraphy in different types of basin has a significance in the understanding of the distribution pattern of a lithostratigraphic reservoir and for offering both sequence stratigraphy and sedimentation information for a petroleum exploration project.

A "reservoir-forming combination of structure and stratigraphy" is the development of the application of sequence stratigraphy in petroleum exploration research. It is a new improvement in Chinese terrestrial sequence stratigraphy, which emphasizes the characteristics of the structural background, sequence stratigraphy and structure in major developmental sections and the types of reservoir layer and traps. It also highlights the comprehensive evaluation of oil and gas migration, accumulation and distribution. From an oil and gas exploration point of view, one kind of reservoir-forming combination of structure and stratigraphy represented a specific environment for creating a lithostratigraphic reservoir, which is important geological evidence for separating and evaluating the favorable region (belt) of a lithostratigraphic reservoir.

On that basis, we propose that a fourth category of industrialized maps should be prepared within the third order sequence stratigraphic framework, which includes a micro facies map inside the isochronal stratigraphic frame, a distribution map of effective hydrocarbon source rock, a top surface structural map for the major targets region and an exploration level chart for different sequences. These maps are for comprehensive evaluation of favorable targets in lithostratigraphic regions. These are the major maps for separating and evaluating a lithostratigraphic petroleum accumulative region (belt). They are different qualitatively if compared with a second level structural unit map that was made during the structural investigation.

5.2 Classification of Reservoir-Forming Combination of Structure and Stratigraphy and its Characteristics

5.2.1 Principles of Classification

The three basic factors for categorizing the reservoir-forming combination of structure and stratigraphy are the framework of a basin, sequence stratigraphy and the construction of system tract and the type of sedimentary system (creation mechanism of reservoir sequence).

The framework of a basin controlled the distribution of the stratigraphic sequence, shape of 3D space and pattern of stratigraphic sequence. For a rifted basin, the steep slope zone of a depression is along the side of a boundary fault of the depression; the gentle slope zone is located on the opposite side; the deep subsidence zone is located at the median portion of the basin. Along the boundary fault of a depression, the extensional subsidence is the major factor that controlled deposits of the stratigraphic sequence. The deep subsidence zone is the depositional center of subsidence. In a simple dust pan shaped depression, from the central axial line of the deep subsidence zone to both sides of the steep slope and gentle slope, there are two different kinds of basic sequences, which also represent two different kinds of oil and gas accumulative environments. In a wide open depressed basin, the major sequence stratigraphy and oil and gas accumulative environment appear on the long axial gentle slope zone.

Different orders of stratigraphic sequence would form their own combination of generation–reservoir–seal; the petroleum source sequence, oil and gas sequence and seal sequence were located in different system tracts within the framework of the stratigraphic sequence. In different basins or the same basin, the different zones have their own features; here we simply discuss the basic developmental features of a gentle slope zone in a rifted basin.

On a gentle slope zone, a secondary stratigraphic sequence has a relatively completed system tract. The alluvial sediments and oxidized littoral–shallow water sediments constructed the base of sedimentary sequences (LST), which occurred on the middle–lower elevation of a gentle slope. The middle–lower portion of the sedimentary sequence (TST_1) contains a braided river, regressive braided river delta, shallow water lake, semi-deep water lake deposits that overlapped on top of basal sediments and that reflected a lacustrine basin extension. The middle–upper portion of the sedimentary sequence (TST_2) contains underwater fan and deep water lake deposits; the underwater fan has supplying channels. This portion of the sequence reflected the fact that the lacustrine basin continually extended and subsided. The upper portion of the sedimentary sequence (HST–RST) contains a regressive braided river delta or fan delta deposit that reflected the reduction of the lacustrine basin and uplifting events. During depositional procession of the second order of the sedimentary sequence, because of relative alterations in the subsidence rate and sediment settling rate, the next order of the lacustrine transgressional deposit and the third order of the sedimentary sequences were formed. The third order of the system tract was formed in two different combinations. The first one contains three divisions of LST, TST and HST; the LST could be the underwater fan that was located on the subsidence side of a growth fault. Another one only contains two divisions of TST and HST.

Associated with the development of a basin, and in addition to the creation of a second order of sequences, related traps were formed accordingly. Vertically, the lithologic trap and stratigraphic overlapped trap often appear in a lower portion of the sequence (LST, TST_1); the lithologic trap that relates to the underwater fan and composite trap occur at a middle portion of the sequence (TST_2); the unconformity trap that relates to uplifting and erosion events occurs in the upper portion of the sequence (HST). Horizontally, the stratigraphic trap commonly appears at a higher level of a gentle slope; the composite trap and lithologic trap occur at a middle section of a gentle slope; the lithologic trap often occurs at a lower section of a gentle slope.

When we designate a name to a tract, we should point out which system tract and what types of sedimentary system major petroleum generation and reservoir sequences were posited in. Concerning the sedimentary system and its distribution, the sedimentology provides a detailed study, which can offer a reference and example (Table 5.1).

Table 5.1   Combination types of sedimentary system and their distribution in terrigenous basin

Sedimentary System		Alluvial Fun	Alluvial Fun	Meandering River	Delat	Littoral-Shallow Lake/ Beach Bar	Near Shore Underwater Fan	Underwater Fan With Supplying Channel	Turbidite Lsoce (Deep Water)	Distribution of the Combination of Sedimentary System	
										Major Tectonic Background	Major System Tract
Combination of Sedimentary System	1	▨				▨				All Kinds of Basin	LST,RST
	2	▨	▨			▨				All Kinds of Basin	LST,RST
	3	▨							▨	Steep Slope in Rifed Basin	LST,HST
	4	▨	▨		▨				▨	Gentle Slope in Depression, Rifted Basin	LST,HST
	5	▨	▨	▨	▨				▨	Gentle Slope in Depression, Axial Rifted Basin,Gentle Slope in Foreland Depression	LST,HST
	6					▨				Gentle Slope in All Kinds of Basin,Underwater Paleo Uplifting	LST
	7	▨					▨		▨	Steep Slope in Rifted Basin, Deeply Subsided	LST,LST
	8	▨						▨		Gentle Slope in Rifted Basin, Deeply Subsided	LST,LST

Note: LST, lowstand systems tract; TST, transgressive systems tract; HST, highstand systems tract; RST, regressive systems tract

5.2.2   Primary Types of Reservoir-Forming Combination of Structure and Stratigraphy and Their Characteristics

Using an analysis of structure, stratigraphic sequences and petroleum accumulation, conditions in four prototypes of Chinese petroleum basins (rifted basin, depressed basin, foreland basin and craton basin) (Dai, 1983; Hu et al., 1986; Dai et al., 1992; Gao and Cai, 1997; Gao and Zhao, 2001; Cai et al., 2003; Jia et al., 2000), we totally identified 14 types of major reservoir-forming combinations of structure and stratigraphy (Figs. 5.1 and 5.2). Among these combinations, four of them are from a rifted basin; three of them are from a depressed basin; two of them are from a foreland basin and five of them are from a marine craton basin. The following discussions will describe the characteristics of these combinations.

Types of Basin	Structural Background	Structural – Sequence Stratigraphic Accumulative Combination	Typical Example
Combination Types in Rifted Basin	Steep Slope	Faulted steps on steep slope – The combination of lacustrine transgression, high altitude fan delta, and underwater fan	The steep slope belt in the western depression, the Liaohe oil field (E$_2$S$_4$-S$_3$) 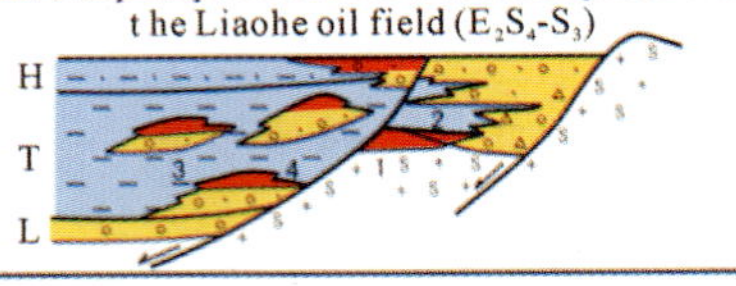
Combination Types in Rifted Basin	Gentle Slope	Multi – faults system on gentle slope – The combination of lacustrine transgression, high altitude braided river delta, and underwater fan	The gentle slope belt in the western depression, the Liaohe oil field (E$_2$S$_4$-S$_3$) 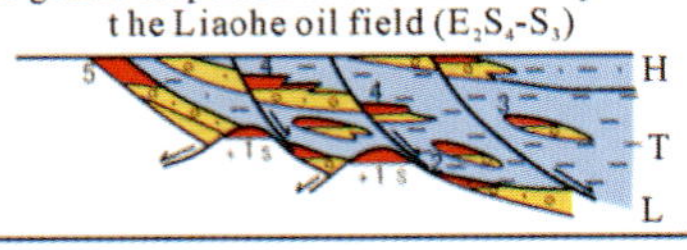
Combination Types in Rifted Basin	Central Structural Belt	Multiple central structural belts – The combination of lacustrine transgression, high altitude fan delta, and underwater fan	The central structural belts in the Huanghua depression (E$_2$S$_3$-E$_3$S$_1$) 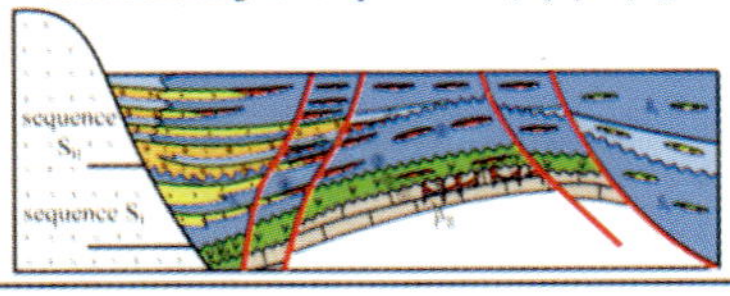
Combination Types in Rifted Basin	Deeply Faulted Structural Belt	Deeyly Faulted Structural Belt – The combination of volcanicexplosive facies and overflow facies	The rifting phase in the Songliao basin (K$_1$yc) 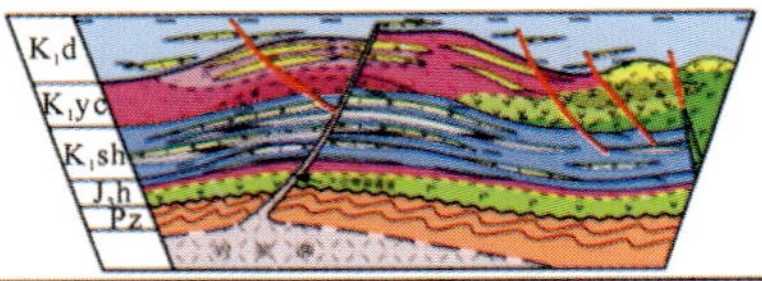
Combination Types in Depressed Basin	Long Axis	Long axis – The combination of lacustrine transgression and high latitude / low latitude river deltas	The northern axial gentle slope belt in the Songliao basin (K$_1$q-K$_2$n) 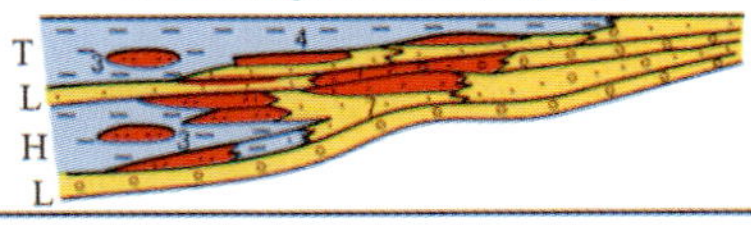
Combination Types in Depressed Basin	Short Axis — Steep Slope	Short axial steep slope – The combination of lacustrine transgression and high latitude / low latitude brained river (fan) deltas	The southwestern steep slope belt in the Eerduosi basin (T$_3$y) 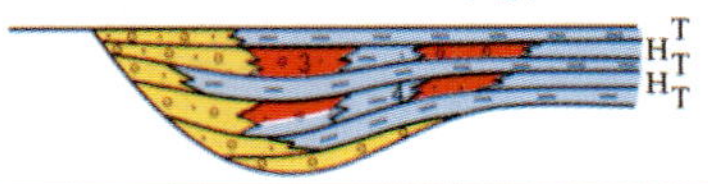
Combination Types in Depressed Basin	Short Axis — Gentle Slope	Short axial gentle slope – The combination of lacustrine transgression and high latitude / low latitude river deltas	The eastern gentle slope belt in the Eerduosi basin (T$_3$y) 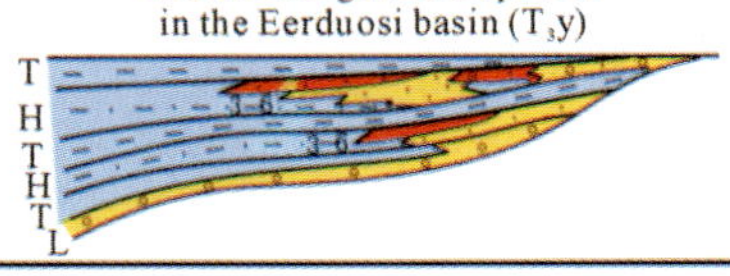

1. Buried Hill Petroleum Reservoir; 2. Stratigraphic Overlapped Petroleum Reservoir;
3. Lithologic Petroleum Reservoir; 4. Composite Petroleum Reservoir; 5. Unconformity Petroleum Reservoir;
6. Diagenesis Formed Petroleum Reservoir; 7. Structural Petroleum Reservoir

Fig. 5.1.   Types of reservoir-forming combination of structure and stratigraphy in rifted basin and depressed basin

Types of Basin	Structural Background	Structural – Sequence Stratigraphic Accumulative Combination	Typical Example
Combination Types in Foreland Basin	Steep Slope	Steep slope in foreland basion – The combination of lacustrine transgression, high altitude/low altitude alluvial fan, and fan delta	The steep slope belt on northwestern margin of the Zhungeer basin (C-T) 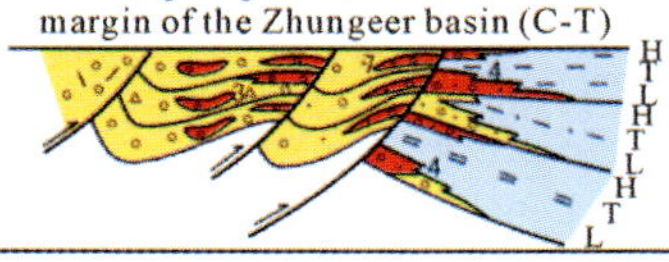
	Gentle Slope	Gentle slope in foreland basin – The combination of lacustrine transgression, river delta, and beach bar	The eastern gentle slope belt in the Sichuan basin (T₃x) 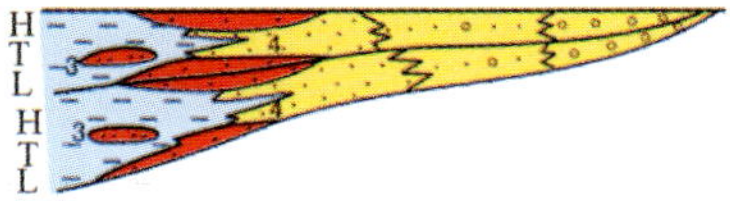
Combination Types in Craton Basin	Plateform Margin	Plateform Margin – The combination of marine transgression and reef flat	The Tazhong 1 plaeform marginal belt in the Tarim basin (O₁) 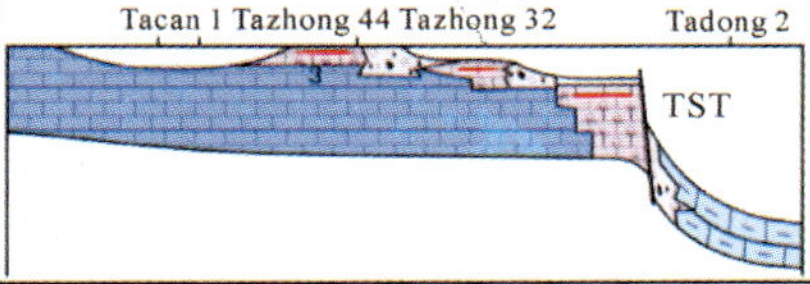
	Intra-Craton	Intra-craton – The combination of marine transgression and beach – bar	The Feixianguan oolitic beach in the northeastern Sichuan basin (T₁) 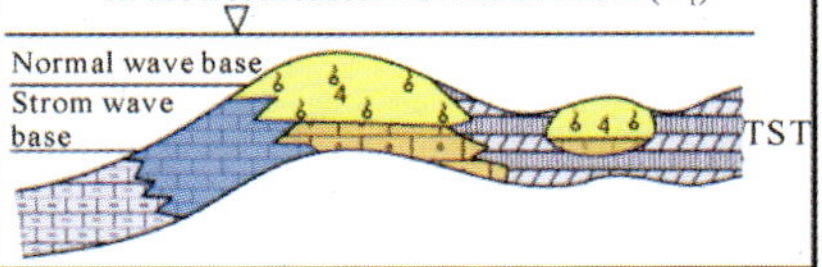
		Intra-craton – The combination of littoral beach and marine transgressional beach – bar	The Donghe sandstone in the Hadexun, Tarim basin (C) 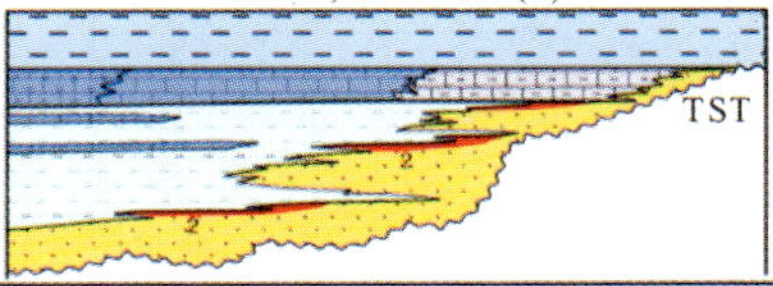
		Intra-craton – The combination of marine – terrigenous facies and high altitude delta	The delta of marine – terrigenous facies in the Eerduosi basin 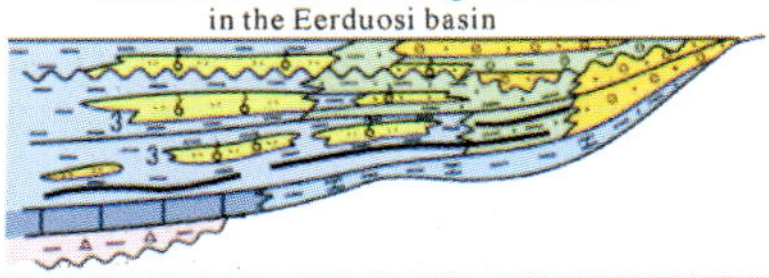
	Paleo Uplift	Paleo uplift – The combination of Karst	The Ordovician buried hill in the Lunnan, Tarim basin (O₂₋₃) 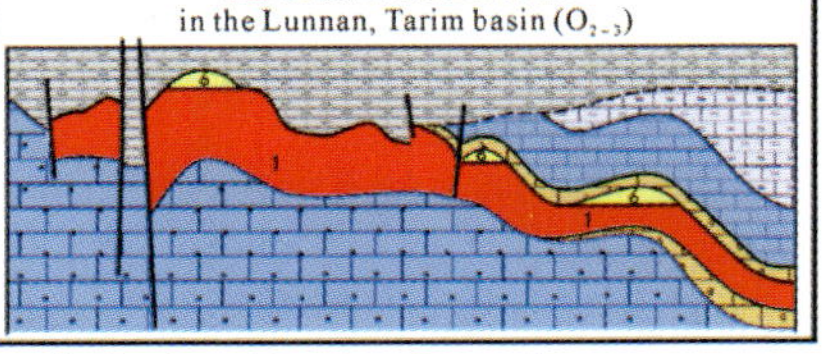

1. Buried Hill Petroleum Reservoir; 2. Stratigraphic Overlapped Petroleum Reservoir;
3. Lithlogic Petroleum Reservoir; 4. Composite Petroleum Reservoir; 5. Unconformity Petroleum Reservoir;
6. Diagenesis Formed Petroleum Reservoir; 7. Structural Petroleum Reservoir

Fig. 5.2.   Types of reservoir-forming combinations of structure and stratigraphy in foreland basin and craton basin

● **Terrestrial Rifted Basin Contains Four Types of Lithostratigraphic Combination**

Faulted Steps on Steep Slope in a Rifted Basin—The Combination of Lake Transgression and Highstand Fan Delta, Underwater Fan

Depositional Background. A steep slope was located in a rifted half-graben basin and its development was controlled by the growth fault belt (Qiao, 1986). The faulted belt could penetrate through the basement into the deep crust; it also contains several normal faults and multiple faulted steps. The dip angle of the fault is in the range of $40° - 70°$; the fault throw can be $2,000 - 5,000$ m and the fault can extend $20 - 50$ km (Fig. 5.3).

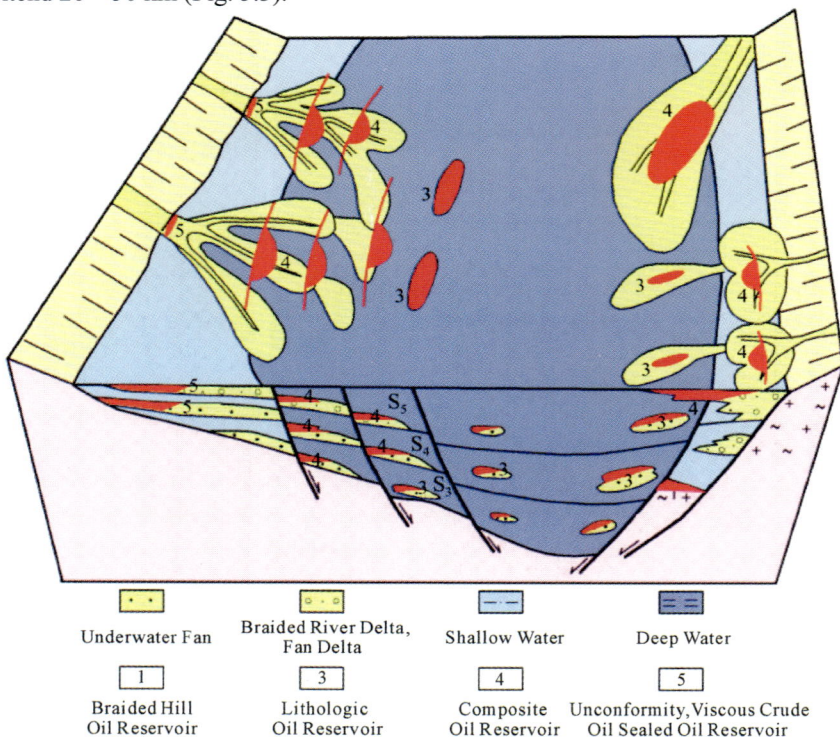

| Underwater Fan | Braided River Delta, Fan Delta | Shallow Water | Deep Water |

| 1 | 3 | 4 | 5 |

| Braided Hill Oil Reservoir | Lithologic Oil Reservoir | Composite Oil Reservoir | Unconformity, Viscous Crude Oil Sealed Oil Reservoir |

Fig. 5.3. Faulted steps on steep slope in a rifted basin—the combination of lake transgression and highstand delta, underwater fan

Characteristics of Sequences and Sedimentation. The extensional subsidence happened along border faults; the lacustrine basin gradually expanded. The magnitude of fault activity was altered between the vigorous phase and stable phase; the fault was a major geological factor that controlled the development of sequences. For example, in the western depression in the Liaohe oil field, a second order sequence ($E_2S_4 - E_2S_3$) has a complete system tract, in ascending order, which includes a lowstand systems tract (LST), an early transgressive systems tract (TST$_1$), a major transgressive systems tract (TST$_2$) and a highstand systems tract–regressive systems tract (HST – RST). The lowstand systems tract consists of red color deposits of alluvial fan, river and oxidizing shallow lacustrine deposits; A/S < 1. The early transgressive systems tract is made of delta deposits and weak reduction, semi-deep water, and argillaceous deposits; A/S≈1. The major transgressive systems tract contains a turbidite deposit and deep water argillaceous deposits; A/S > 1; this reflects the extensional period of the lacustrine basin. The highstand regressive systems tract contains the fan delta and river deposits; this reveals a decrease and uplifting of the lacustrine basin. The development of a major transgressive systems tract is related to a vigorous period of fault activity; other system tracts were developed during stable periods in fault activity. The time interval of this second order sequence is 14.6 Ma; this time interval can be further divided into five third order sequences with a sub-time interval of 2.95 Ma approximately. Fig. 5.4 shows three upper third order sequences; Fig. 5.5 shows two lower third order sequences.

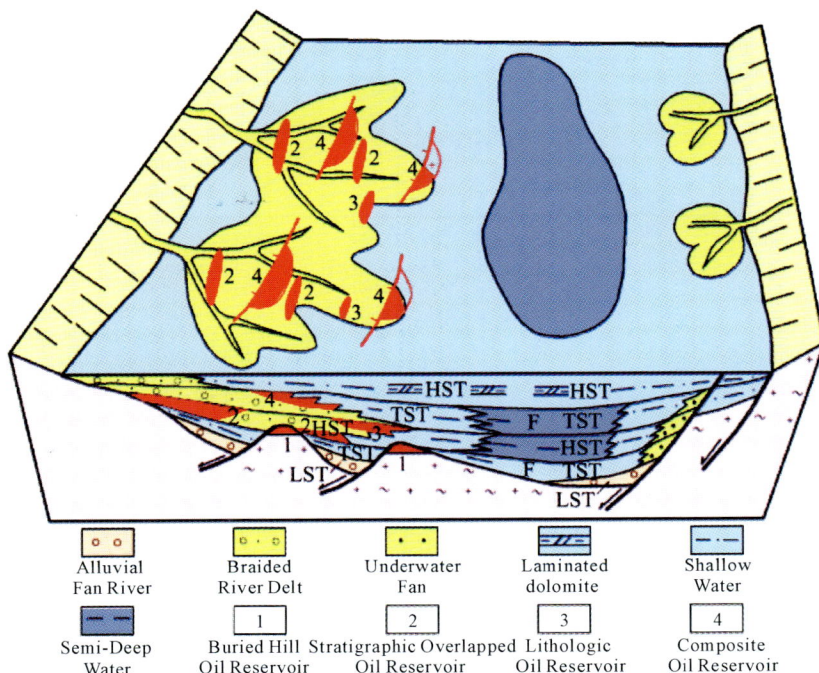

Fig. 5.4. Gentle slope with multi-faults system in a rifted basin—the combination of lake transgression and highstand braided river delta, underwater fan

Characteristics of Petroleum Accumulation. The early transgressive systems tract, major transgressive systems tract and highstand systems tract have an excellent combination of petroleum generation and storage in situ (lateral or vertical). On the foot wall of a border fault, the lithologic reservoir was developed in a sandy conglomerate of debris flow facies and turbidite flow facies in the deep water area that was created by a substantial faulted drop in the foot wall. If hydrocarbon source rock was in contact with the basement laterally at the second step, a buried hill reservoir would be created; in addition, an overlapping reservoir could form at the periclinal area of the buried hill. If a sandy conglomerate and faulted nose structure were combined together, a composite reservoir would be formed (Figs. 5.1 and 5.3).

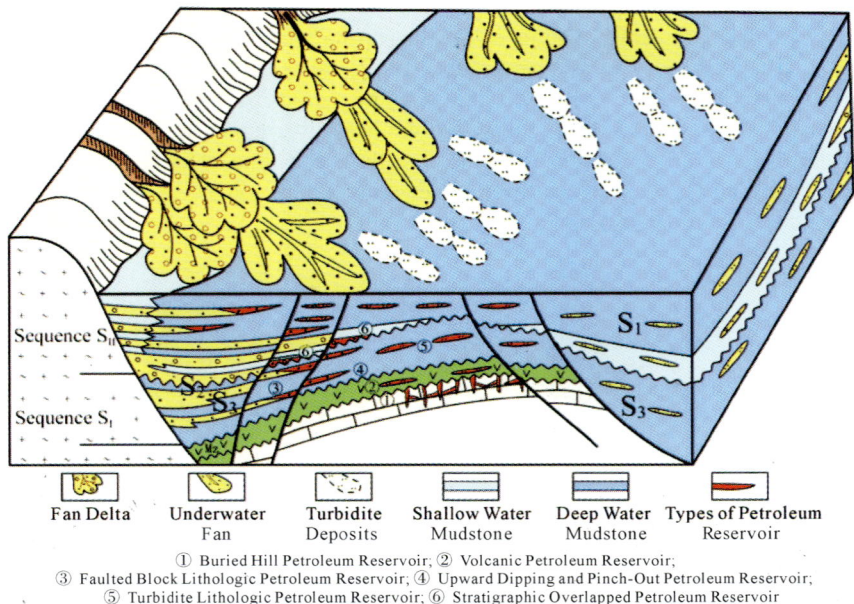

| Fan Delta | Underwater Fan | Turbidite Deposits | Shallow Water Mudstone | Deep Water Mudstone | Types of Petroleum Reservoir |

① Buried Hill Petroleum Reservoir; ② Volcanic Petroleum Reservoir;
③ Faulted Block Lithologic Petroleum Reservoir; ④ Upward Dipping and Pinch-Out Petroleum Reservoir;
⑤ Turbidite Lithologic Petroleum Reservoir; ⑥ Stratigraphic Overlapped Petroleum Reservoir

Fig. 5.5.   Central structure belt in a rifted basin—the combination of lake transgression and highstand fan delta, underwater fan

Gentle Slope with Multi-Faults System in a Rifted Basin—The Combination of Lake Transgression and Highstand Braided River Delta, Underwater Fan

Depositional Background.   The western slope in the western depression, the Liaohe oil field, is a typical example of a gentle slope belt in a multi-faults system; this gentle slope belt is oriented in an NE-SW direction with a length of 50 km and width of 18.5 – 22 km. There are three fault systems on this gentle slope, which are the NW oriented fault system, the NE oriented fault system with westwards sliding foot wall, the NE oriented faulted system with eastwards sliding foot wall. The NW oriented fault system combined with the nose-like bulge of the basement on the gentle slope to control the direction of the incoming water flow. The NE oriented fault system with a westwards sliding foot wall cut into the gentle slope to create several arrays of basement cuesta and intra-cuesta valleys; it controlled deposits of the lower sequences and alterations in the related lithology and facies. The NE oriented fault system with eastwards sliding foot wall cut the gentle slope into three belts (the higher, middle and lower belts) that controlled alterations in the lithology and facies of the upper sequences.

Characteristics of Sequences and Sedimentation.   Inside the half-graben, the development of the second order sequence was consistent; however, compared with the steep slope, the third order sequence on the gentle slope had different features that were caused by the isolation of the depositional background and substance supplies.

On the gentle slope of the western depression in the Liaohe oil field, the earliest second order (E_2S_4 – E_2S_3) includes several third sequences (four of them). In ascending order, the 1st third order sequence (E_2S_4) was made of braided river delta deposits and shallow water to semi-deep water lacustrine deposits, which represented the early deposits in the rifted basin. The 2nd and 3rd third order sequences consisted of the deposits of the supply channel underwater fan and the deposits of the deep water lacustrine basin, which represented the extensional phase of the rifted basin. The 4th third order sequence contained braided river delta deposits and shallow water lacustrine deposits, which represented the reduction phase of the lacustrine basin.

Characteristics of Petroleum Accumulation.   On the gentle slope, the 1st second order sequence was entirely overlapped layer by layer, which represented a continually subsiding slope with several sets of petroleum generation and storage combinations; this gentle slope belt was a highly abundant oil and gas accumulative area. The lower sequences contained overlapped strata and a buried hill type of oil reservoir under the overlapped strata. The viscous crude oil belt was formed on the margin of the gentle slope belt; the composite oil reservoir and the lithologic oil reservoir were formed in the middle sequence; the turbidite oil reservoir was formed at a lower level of the gentle slope; the unconformity reservoir and composite reservoir were formed on top of the sequence (Figs. 5.1 and 5.4).

Central Structure Belt—The Combination of Lake Transgression and Highstand Fan Delta, Underwater Fan

In the large rifted basin, beside the steep slope and gentle slope, there were central structure belts that were formed by underwater uplifting at the center of the depression. For example, there is the Xinglongtai structure belt in the western depression in the Liaohe oil field, the Gaoshangpu structure belt in the Nanpu depression and the Beidagang structure belt in the Banqiao–Qikuo depresion. Because they were next to the hydrocarbon generation depression, these central structure belts had plenty of oil and gas resources. The Beidagang structure belt is a good example (Fig. 5.5).

Depositional Background.   The Beidagang structure belt is located at the center of the Huanghua depression; the Banqiao depression is located on its western side and the Qikou depression is positioned on its eastern side. The exploration area covered 500 km^2; the structure belt is oriented in an NNE-SW direction and it is also a normal structure unit that has experienced multiple tectonic movements since the Mesozoic era.

Characteristics of Sequences and Sedimentation.   Under the extensional background in the lacustrine basin, during the early–middle Eocene, the Beidagang structure belt began to subside; during the end of the Eocene, because the structure belt was elevated and eroded, the 1st second order sequence (the Sha–3 section) was formed. During the middle-late Oligocene, the Beidagang structure belt subsided and was extended again; during the end of the Oligocene, because it was slightly elevated, the 2nd second order sequence (the Sha–2 and the Sha–1 sections) was formed. During the middle–late Oligocene, the Beidagang structure belt experienced large scale subsidence that formed the 3rd second order sequence (the Dongying formation). During the end of the Oligocene, the development of the rifted basin was stopped by the Himalayan tectonic movement that influenced the entire Bohai bay basin; the 4th second order sequence (the Neogene Guantao formation and the Minhuazhen formation) occurred and it represented the depressed basin stage.

The Cangxian paleo-uplift was located on the western side and northwestern side of the Beidagang structure belt; it not only functioned as a major sedimentary substance supply center, but it also controlled the development of sequences in the Beidagang structure belt. The 1st and 2nd second order sequences contained deposits of near shore underwater fan, fan delta and deep water turbidite sand bodies; the carbonate rocks were only deposited in the isolated area. The 3rd second order sequence contained deposits of river delta facies. The 4th second order sequence contained deposits of river facies.

Characteristics of Petroleum Accumulation.   The 1st second order sequence (the Sha–1 section) and the 2nd second order sequences (the Sha–2 and the Sha–1 sections) are the hydrocarbon source rocks that combined with all types of intra-sequence sand bodies to form a beneficial combination of petroleum generation and storage and develop various types of oil and gas reservoirs. In the main portion of the Beidagang structure belt (such as Gangzhong), the 1st and 2nd second order sequences primarily contain a faulted block oil reservoir; additionally, they have a faulted block–lithologic oil reservoir and lithologic oil reservoir. The 4th second order sequence contains a structure–lithologic oil reservoir (such as Gangdong). At both ends of the structure belt and on the sides, the structure–lithologic oil reservoir (the 1st and 2nd second order sequences), the faulted block oil reservoir of drape structure and the faulted–lithologic oil reservoir (3rd and 4th second order sequences) occurred; furthermore, there is the buried hill type of oil reservoir.

Deeply Faulted Belt—The Combination of Volcanic Explosive Facies and Overflow Facies

In the Mesozoic–Cenozoic rifted basins in eastern China, along with the forceful extension and subsidence events, volcanic activities occurred chronologically, forming volcanic rocks with different thicknesses. These volcanic rocks functioned as an oil and gas reservoir that has become the new oil and gas exploration domain in recent years, for example, in the Bohai bay basin, the eastern depression in the Liaohe oil field and the sedimentary layers of the rifted stage in the Songliao basin. The following descriptions reveal a large volcanic gas field by using the example of a deeply faulted belt on the northern side of the Songliao basin (Fig. 5.6).

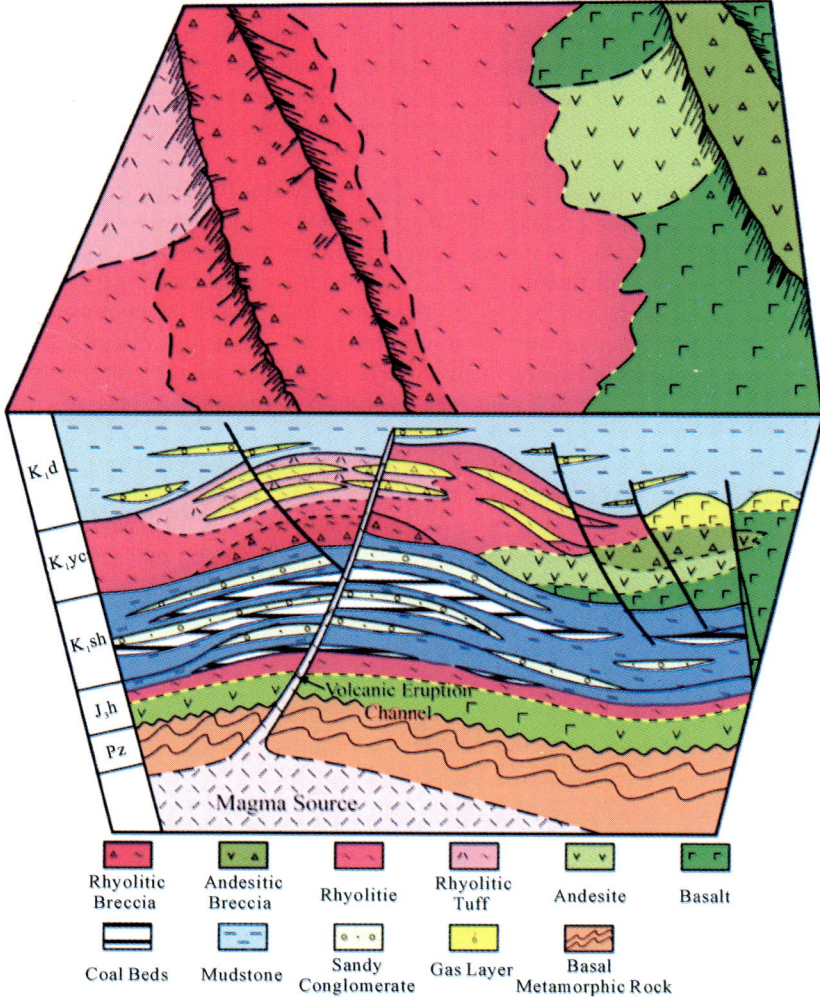

Fig. 5.6.   Deeply faulted belt—the combination of volcanic explosive facies and overflow facies

Geological Background.   On the northern side of the Songliao basin, the deeply buried strata include the Huoshiling formation (J_1h), the Shahezi formation

(K_1sh), the Yingcheng formation (K_1yc), the Denglouku formation and the 1st and 2nd sections of the Quantou formation (K_1q) in ascending order. These formations were transitional deposits between the rifted stage and the faulted depression stage. The distribution of volcanic rock was along the faulted belt; the natural gas was predominately stored in volcanic rocks in the Yingcheng formtion.

Volcanic Facies and Volcanic Reservoir.   There were two phases of volcanic activity. The early volcanic activity happened during the time interval when the Huoshiling formation was deposited; the late volcanic activity occurred during the time interval when the Yingcheng formation was deposited. The main volcanic facies include volcanic explosive facies and overflow facies. The volcanic explosive facies were distributed near the fault; the major lithologies are rhyolitic breccia and andesite.

Volcanic Breccia.   The volcanic explosive facies were distributed on the slope and on the lower ground; the major lithologies include rhyolite, andesite, basalt and rhyolitic tuff. There are three types of pores, which include original pores (such as the air holes in lava), secondary pores (such as phenocryst dissolution pores, intra-matrix dissolution pores) and fracture (such as contraction fracture, structural fracture and dissolution fracture). The acidic volcanic rocks of explosive facies contain two types of storage spaces that are the porous type and fractural type; the acidic volcanic rock is the primary volcanic reservoir in the region.

Characteristics of Natural Gas Accumulation.   The geochemical analysis demonstrates that in the Songliao basin the deeply buried natural gas has an organic origin and non-organic origin; the organic origin is the major type. The major hydrocarbon source sequence consists of lacustrine and palustrine deposits in the Shahezi formation; the content of organic carbon is 1.63% – 3.47%; the total hydrocarbon content is 220 – 1954 ppm. The organic material belongs to type II–type III and the R_o is 1.5% – 3.9%, which indicates highly mature and post mature dry gas.

Along the fault and unconformity surface, natural gas migrated into the volcanic lithologic trap and the structure–lithologic trap to form a natural gas reservoir. The major cap rock is the second section of the Denglouku formation and the mudstone of the first and second sections of the Quantou formation.

The natural gas reservoirs that were distributed along the deeply faulted belt contain a higher content of CO_2, which indicates a mixture of mantle-derived gas.

● **Terrestrial Depressed Basin Includes Three Kinds of Lithostratigraphic Combination**

Long Axial Gentle Slope in a Depressed Basin—The Combination of Lake Transgression and Highstand/Lowstand River Delta

Depositional Background.   In the Songliao basin, the orientation of the basin's long axis matches with the orientation of the uplifted Moho surface and the orientation of the Sunwu–Shuangliao crustal fault. The high point of the Moho surface corresponds to the hydrocarbon generation depressions of the Qijia–Gulong depression, the Changling depression and the Sanzhao depression. The large river delta was developed on a gentle slope in the direction of the long axis with sediments that were carried by the river over long distances.

Characteristics of Sequences and Sedimentation. In the Songliao depression, the paleo slope gradient of the long axial gentle slope was $0.5 \times 10^{-4} - 1.4 \times 10^{-4}$; the distributions of the stratigraphic sequence were stable (Xiao et al., 2005; Wang, 2001). During the peak period of the lacustrine basin, two second-order sequences were developed; one sequence was the strata sequence S_I Quantou formation–Qingshankou formation; another one was the strata sequence S_{II} Yaojia formation–Nenjiang formation. Every second-order sequence includes the system tracts of LST, TST_1, TST_2 and HST–RST. In the 1st second-order sequence, the LST contains the deposits of river facies (the first and second sections of the Quantou formation; the Nongan oil layer); the TST_1 includes the deposits of river, littoral–shallow water lacustrine and delta facies (the third and fourth sections of the Quantou formation; the Yangdachengzi and Fuyu oil layers); the TST_2 consists of the deposits of delta facies and deep water lacustrine facies with the largest lake transgressive strata (the first section of the Qingshankou formation; the lower portion of the Gaotaizi oil layer); the HST–RST are made of the deposits of delta facies and shallow–semi-deep water lacustrine facies (the second and third sections of the Qingshankou formation; the lower portion of the Gaotaizi oil layer). In the 2nd second-order sequence, the LST contains the deposits of river, delta and shallow water lacustrine facies (first section of the Yaojia formation; the Putaohua oil layer); the TST_1 includes the deposits of delta, shallow water lacustrine and semi-deep water lacustrine facies (the second and third sections of the Yaojia formation); the TST_2 consists of the deposits of deep water lacustrine and delta facies with the largest lake transgressive strata (the first and second sections of the Nenjiang formation); the HST–RST are made of the deposits of shallow water lacustrine and delta facies (the third, fourth and fifth sections of the Nenjiang formation, the Heidimiao oil layer).

Characteristics of Petroleum Accumulation. These two second-order sequences that were developed during the peak period of the lacustrine basin contained a complete petroleum system of hydrocarbon generation–storage–seal combination.

Inside the second-order sequence, the third-order sequence contains a secondary petroleum system of hydrocarbon generation–storage combination. The reservoir sequences of inherited river delta facies were widely distributed to create multiple sets of oil layers. The major hydrocarbon source rock occurred in the TST_2 in the 1st second-order sequence; the TST_2 joined with LST, TST_1, HST–RST and its own reservoir layer to form various types of petroleum accumulative combinations; the position of the source rock was beneath, inside and above the reservoir sequences.

The stratigraphic sequences of river delta facies were distributed in a broad area and they covered many elevated and dropped sedimentary units to form multi-type oil and gas reservoirs that include structural reservoirs, composite reservoirs and lithologic reservoirs (Figs. 5.1 and 5.7).

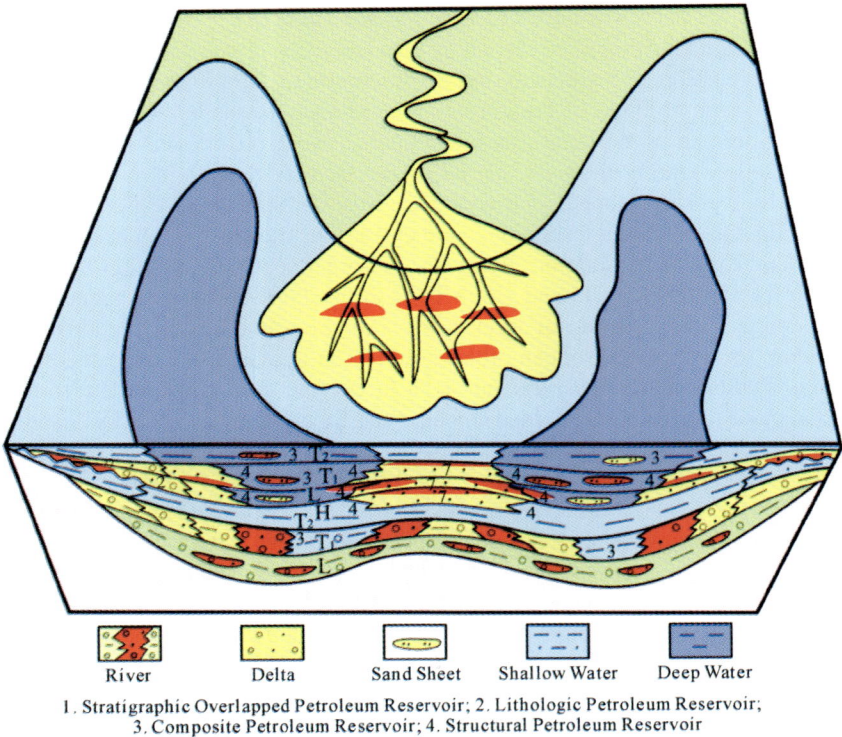

1. Stratigraphic Overlapped Petroleum Reservoir; 2. Lithologic Petroleum Reservoir; 3. Composite Petroleum Reservoir; 4. Structural Petroleum Reservoir

Fig. 5.7. Long axial gentle slope in a depressed basin—the combination of lake transgression and highstand/lowstand river delta

Short Axial Steep Slope in a Depressed Basin—The Combination of Lake Transgression and Highstand/Lowstand Braided River Delta

Depositional Background.   The western margin of the Eerduosi basin was located at the Tianhuan depression and the western margin thrust belt. Compared with the Shanbei slope, the frequency of tectonic activity was much higher on this western margin. During the Paleozoic, the Tianhuan depression was a westward dipping slope; during the Late Triassic, it was developed into a depression. In the Shigouyi area and the Pingliang area, the thickness of the Yanchang formation can be 3,000 m approximately. During the Middle and Late Triassic, the western margin thrust belt was a non-continual depression; during the middle of the Yanshanian tectonic period, because it experienced compressive and shear tectonic movement, the thrust belt was formed.

Characteristics of Sequences and Sedimentation.   During the Late Triassic, the depositional center (deep water area) of the lacustrine basin was located in Huanxian county and the Huachi area; the basin was oriented in an NW-SE direction; the delta deposits were distributed on both sides of the basin. The large lacustrine basin had relatively stable deposits; the third-order sequences and their inner framework could be correlated. Thus, the deposits of the steep slope and gentle slope can be integrated into a unified stratigraphic framework. The deposits of braided river delta facies were widely distributed on the steep slope of the western margin; from western side to the eastern side, the facies can be divided into alluvial fan, braided river, delta plain, delta front and pro-delta. In ascending order, the braided river delta front contains a progressive sequence and the grain sizes were gradually altered into coarse grains. Horizontally, the underwater distributary channel and sand bodies of the estuary dam extended into the lake in a digital shape. The sand bodies not only covered the center of the Tianhuan depression, but they also overlapped and pinched out towards the Shanbei lower slope. The turbidite sand bodies were deposited in the deep water region.

Characteristics of Petroleum Accumulation.   The structural reservoirs were prodominately formed in the western margin thrust belt and on the western side of the Tianhuan depression. Because the sand bodies on the eastern side of the Tianhuan depression and on the Shanbei slope overlapped and were pinched out ascendingly, a lithologic reservoir and stratigraphic reservoir were formed. The Chang–7 section is the primary hydrocarbon source rock; the Chang–8 and the Chang–6 sections are the major reservoir sequences. The research shows that the excessive pressure in the Chang–7 section can reach 20 MPa; on the other hand, the pressure in the Chang–8 section is 10–15 MPa. The relationship between hydrocarbon generation and storage includes two types; either lower sequence generated oil and upper sequence stored oil, or a reversed relationship. In eastern Gansu province, the sand bodies of delta facies in the Chang–8 section were widely distributed where the Xifeng large oil field was discovered. The sand bodies of distributary channel delta facies and underwater distributary channel facies were distributed in an SW-NE direction as a branched belt pattern; these sand bodies have good continuity with a length of 80 – 100 km and a width of 8 – 10 km. During early diagenesis, because a thin film of chlorite cement blocked secondary growth of quartz, the original pores were protected. Overall, the physical properties include medium porosity with low permeability, low porosity with low permeability and medium porosity with medium permeability.

The lateral and upward pinched-out sand bodies of distributary channel facies and underwater distributary channel facies and dense strata were the major sealing types for the lithologic reservoir in this region; in the deep water region, the sandstone lens of turbidite facies formed another type of trap (Figs. 5.1 and 5.8).

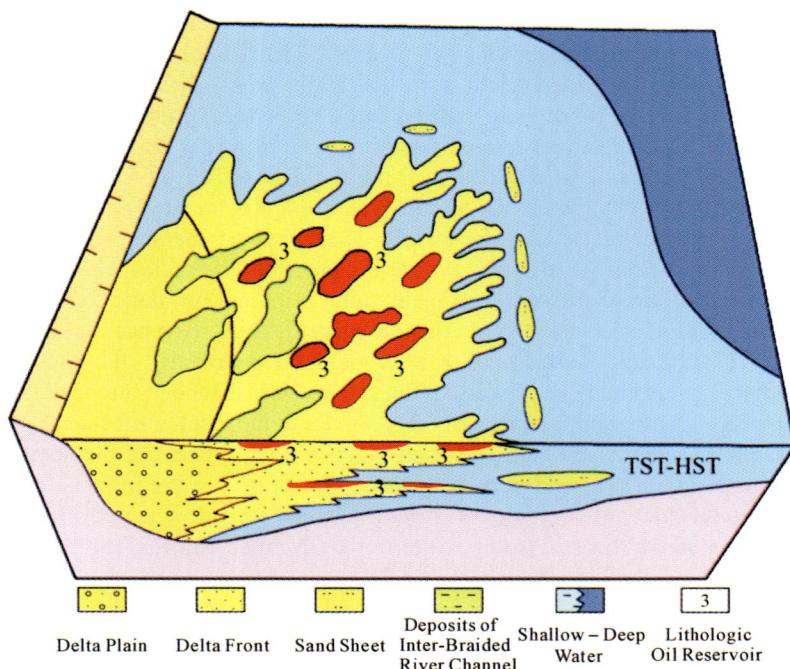

Fig. 5.8.   Short axial steep slope in a depressed basin—the combination of lake transgression and highstand/lowstand braided river delta

Short Axial Gentle Slope in a Depressed Basin—The Combination of Lake Transgression and Highstand/Lowstand River Delta

Depositional Background.   The Eerduosi basin is an intra-craton depressed basin with stable tectonic activity; the long duration, inherited elevation and subsidence activity controlled this basin. The Shanbei slope is located at the center of the Eerduosi basin and it is a gentle monocline that dips to the west. During the late Triassic, large scale river deltas were developed in the basin and they contained typical lithologic oil reservoirs.

   Characteristics of Sequences and Sedimentation.   The Upper Triassic represented the important depositional and developmental stage under the humid climate. The Lower Triassic series is a second-order sequence that contains three third-order sequences in ascending order. The 1st third-order sequence (S_1) is equal to T_3y^1 (the Chang–10 section)–T_3y^2 (the Chang–9 and Chang–8 sections). The 2nd third-order sequence (S_2) is equal to T_3y^3 (the Chang–7 – Chang–4 sections). The 3rd third-order sequence (S_3) is equal to T_3y^4 (the Chang–3 and Chang–2 sections)–T_3y^5 (the Chang–1 section). The S_1 sequence contains three system tracts that are a lowstand system tract, a lacustrine transgressive system tract and a highstand system tract. The S_2 and S_3 sequences only have a lacustrine transgressive system tract and highstand system tract. In the basin, from an NW to SE direction, the northeastern slope contained several river deltas that covered an area of $1,000 - 3,000$ km^2. Among these deltas, the Anse river delta is the largest one that contains a well developed highstand system tract in the 2nd third-order sequence (the Chang–6 section). The Anse river delta ran into the lake with several distributary channels running in an NE-SW direction. The skeleton sand body on the delta front is a sand body of underwater distributary channel facies, which overlapped on top of transgressive, semi-deep water and deep water, dark color mudstone and oil shale (the Zhangjiatan shale); horizontally, it spread out in a digital pattern (bird feet pattern) with multi-level-branches, which forms a large scale, composite river delta. Against the background of a lower sedimentation rate on the slope, the underwater distributary channels not only have inherited developmental features, but they also show the characteristics of migration. Vertically, the former one created several superimposed sand bodies in the underwater distributary channel. The latter one contains single rhythm sandstone in the underwater distributary channel–natural dam; these sandstones were distributed between the mudstones that were deposited in sags between the distributary channels.

Characteristics of Petroleum Accumulation. In the Eerduosi basin, the Upper Triassic hydrocarbon source rocks were distributed in the transgressive system tract in each sequence (the Chang–9, the Chang–7 and the Chang–4 sections); the Chang–7 section is primary source rock. The reservoir sequences (the Chang–8, the Chang–6 and the Chang 2+3 sections) in the highstand system tract combined with petroleum generation sequences in the transgressive system tract beneath to form a good petroleum system of generation–storage–combination; the delta front facies in the Chang–6 section form the important reservoir sequence with physical properties of medium porosity with lower permeability and low porosity with low permeability. The type of reservoir is a lithologic oil reservoir on a gentle slope. There are two types of reservoir sealing: the lithologic trap and the diagenesis formed high density belt trap. The former one was mudstone from the inter-distributary channel dipped upward to create a lithologic oil reservoir. The underwater overlapped siltite mudstone formed a lateral sealed oil reservoir. During the middle stage of the diagenesis process, in the delta sandstone, the laumontite and feldspar dissolved. The mineral dissolution not only created an excellent secondary porosity belt (the laumontite–feldspar dissolution facies), but it also formed the best reservoir facies. The mineral dissolution happened near the oil generating depression; however, towards edge of the lacustrine basin, the dissolution gradually decreased. The sandstone with strong dissolution has good physical properties; on the other hand, the non-dissolvable laumontite cementing facies combined with other cementing facies to form a dense lithologic belt for creating a petroleum trap (Figs. 5.1 and 5.9).

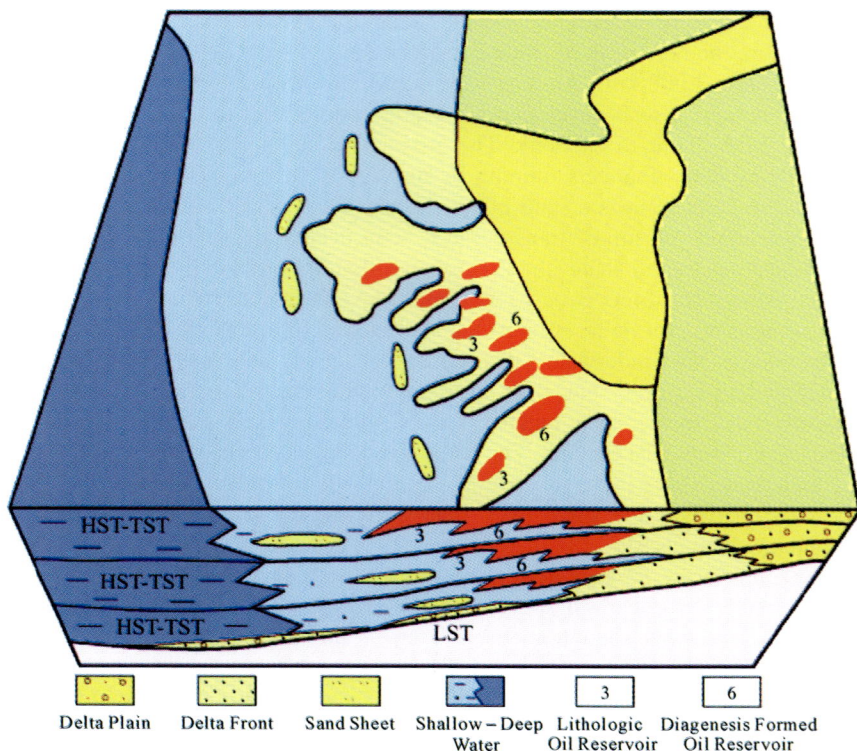

| Delta Plain | Delta Front | Sand Sheet | Shallow – Deep Water | Lithologic Oil Reservoir | Diagenesis Formed Oil Reservoir |

Fig. 5.9.   Short axial gentle slope in a depressed basin—the combination of lake transgression and highstand/lowstand river delta

● **Terrestrial Foreland Basin Contains Two Kinds of Lithostratigraphic Combination**

Steep Slope in a Foreland Basin—The Combination of Lake Transgression and Highstand/Lowstand Alluvial Fan, Fan Delta

The foreland basin is a transitional zone between the active orogeny and stable craton. In the direction from an orogeny to a craton, this transitional zone can be divided into a thrust belt, subsiding zone, slope zone and forebulge zone; in addition, the cross section reveals an asymmetrical dustpan shape. The foreland steep slope was represented by the thrust belt and the subsiding zone next to the thrust belt. The foreland gentle slope was represented by the slope zone and depositional area of the forebulge zone.

   Depositional Background.   On the northwestern margin, the fault activities were uneven; during the different tectonic periods, different structure belts demonstrated clear distinctions, which controlled development and distribution of the fans. The main characteristic shows that, during the Permian and the Triassic, the thrust activities were accompanied by the development of fans. In the Cheguai area of the Hongyiche faulted belt, the faults extended from the Laoshan area toward the inner basin (or slope), which controlled development of fans; the fan gradually progressed and migrated from the edge of the basin to the inner basin. In the Hongshanzui area, the Ke–Wu area and the Wu–Xia area, the major faults that controlled the development of fans developed from the edge of the basin toward the Laoshan area. The fan retreated from the inner basin toward the edge of the basin. The overall magnitude of overthrust activity in the Cheguai area was weaker than that in the Ke–Wu–Xia area.

Characteristics of Sequences and Sedimentation.   In the Permian system, there are three sequences that can be separated in the strata interval from the lower section of the Jiamuhe formation to the Upper Wuerhe formation. The first sequence correlates with the Jiamuhe formation–Fengcheng formation; the second sequence associates with the Xiazijie formation–Lower Wuerhe formation; the third sequence correlates with the Upper Wuerhe formation. In the first and second sequences, the broadest lake transgressive layers were the major hydrocarbon source rocks. The Triassic system is a complete second-order sequence that includes a lowstand system tract (the Baikouquan formation), a lacustrine transgressive system tract (the Kelamayi formation) and a highstand system tract (the Baijiantan formation).

In the Lower Permian Fengcheng formation, the facies of major hydrocarbon source rock formed a semi-closed lagoon bay. The major types of reservoir were alluvial fan and the fan delta. Alluvial fan was developed during the lowstand phase; fan delta and underwater fan were developed during the lake transgression and highstand phases. A single fan includes the top-fan, mid-fan and edge-fan; the top-fan contains the vertical aggradational sequence of the major channel; the mid-fan contains a lateral aggradational sequence of braided channels; the edge-fan contains fine grain sediments. Between the fans, there were inter-fan swale and flats.

Characteristics of Petroleum Accumulation.   Migrating through the faults and unconformity surface, the deeply buried oil and gas traveled into fans to form a petroleum reservoir. Thus, to form an oil and gas reservoir in a fan, the combination of fan and unconformity surface (or faults) was the crucial factor. On the faulted step zone, the fan combined with faults to form a type of vertical migration and accumulation reservoir; on the slope area, the fan combined with the unconformity surface to form a type of lateral migration and accumulation reservoir. In the fans, the scale and quality of the reservoir were decided by the size of a fan and the superimposed style of fans. The sandy conglomerates in the mid-fan and root of the fan were the beneficial areas for oil and gas accumulations. In the laterals of a fan and on the edge of a fan, the inter-fan mudstone and lacustrine mudstone functioned as the sealing layer. Where the fan body was upwards dipping (or the root of a fan), the fault combined with fine grain sediments in the lateral fan to form a sealing layer. The composite oil reservoir is the predominate type (faulted–lithologic reservoir occupied 54.8%); the next type is the lithologic oil reservoir (that occupied 17%); the last type is the stratigraphic oil reservoir (Figs. 5.1 and 5.10).

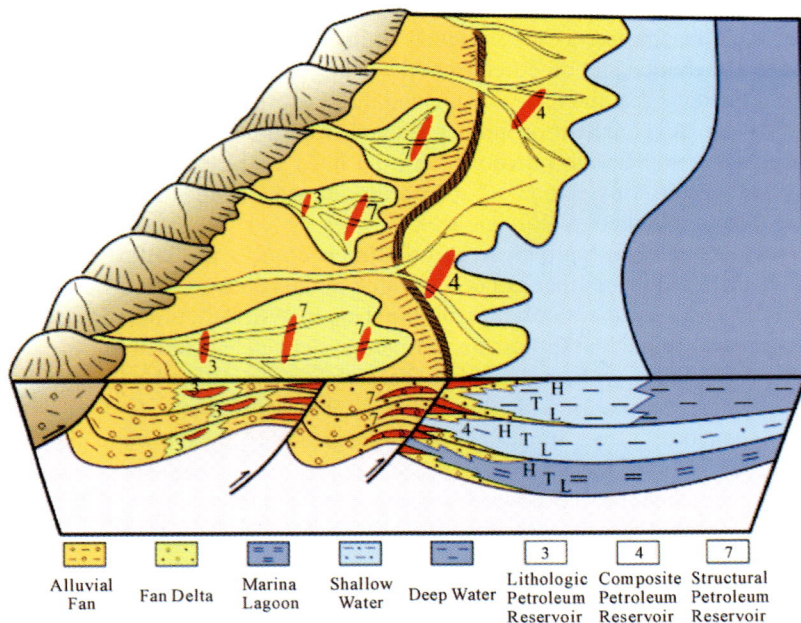

| Alluvial Fan | Fan Delta | Marina Lagoon | Shallow Water | Deep Water | 3 Lithologic Petroleum Reservoir | 4 Composite Petroleum Reservoir | 7 Structural Petroleum Reservoir |

Fig. 5.10.   Steep slope in a foreland basin—the combination of lake transgression and highstand/ lowstand alluvial fan, fan delta

Gentle Slope in a Foreland Basin—The Combination of Lake Transgression and Highstand/Lowstand River Delta, Beach Bar

The Tazhong area (lower) in the Tarim basin and the middle–eastern region of the Sichuan basin are classic foreland gentle slope zones. Here we use the Sichuan basin as an example.

Depositional Background.   During the Late Triassic, accompanied by the uplifting of Longmen mountain, the dynamic subsidence zone was formed at the mountain front of Longmen mountain. During the early phase, it linked with the ocean. However, during the time interval when the Xujiahe formation was deposited, Longmen mountain initiated an overthrusting event. The basin had an asymmetrical pattern; the steep slope was positioned on the western side and the gentle slope was located on the eastern side. In addition, water was deeper on the western side than on the eastern side. The framework of the basin included a subsidence zone and steep slope on the western side and a gentle slope on the eastern side.

Characteristics of Sequences and Sedimentation.   The Upper Triassic Xujiahe formation is an important petroleum exploration target, which contains one second-order sequence that can be further divided into four third-order sequences. Every third-order sequence consisted of a lowstand system tract, lacustrine transgressive system tract and highstand system tract. The distributions of sedimentary sequences were stable in the entire basin. The basin contained multiple substance centers and several water systems. The depositional system on the gentle slope zone was controlled by the substance centers that were located on the northeastern and southern sides of the basin. During the lowstand period, the lake region was small and the depth of water was shallow; furthermore, the deposits of river delta facies were widely distributed. During the lacustrine transgressive period, the lake was wide open and the depth of water was deep; the lacustrine transgressive periods in the second-order and third-order sequences were developmental periods for the major hydrocarbon source rocks.

Characteristics of Petroleum Accumulation.   The hydrocarbon source rocks consisted of dark color mudstone of deep lacustrine facies and deposits of limnetic facies; they combined with reservoir sequences of river delta facies from the lowstand period and highstand period to form several sets of petroleum systems in situ. The Xu–2, the Xu–4 and the Xu–6 sections are the major gas reservoirs with low porosity and low permeability. The major types of reservoir include the structural–lithologic gas reservoir and lithologic gas reservoir. In addition, because of the presentation of an unconformity surface, an unconformity surface sealed gas reservoir might be formed (Figs. 5.2 and 5.11).

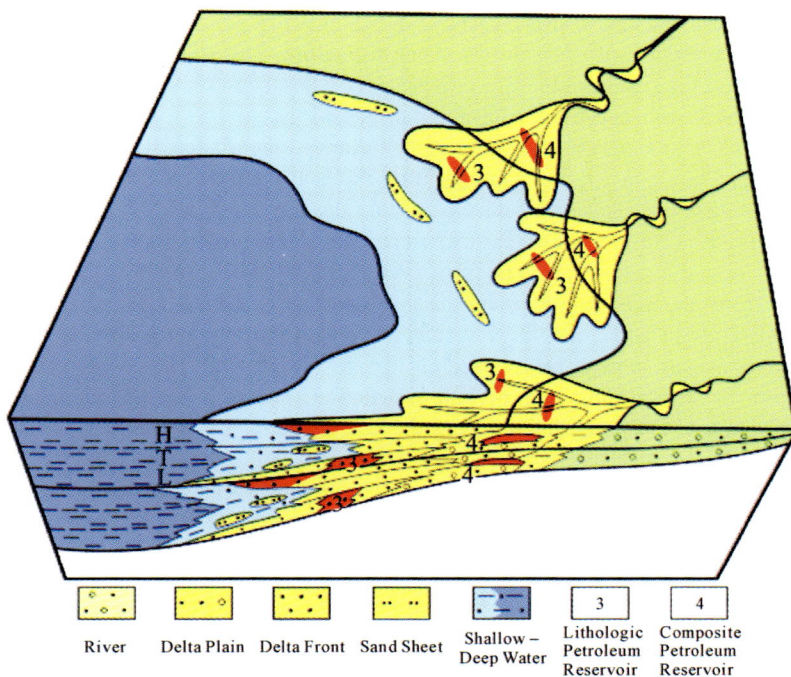

Fig. 5.11.   Gentle slope in a foreland basin—the combination of lake transgression and highstand/lowstand river delta, beach bar

● **Marine Craton Basin Contains Five Kinds of Lithostratigraphic Combination**

Platform Margin—The Combination of Marine Transgressive Reef Flat

This type of reservoir includes the Ordovician platform margin reef flat in the Tazhong area in the Tarim basin and the reef reservoir in the Changxing formation in northeastern Sichuan; the lithologic reservoir in the Ordovician reef flat in the Tarzhong area is a typical example.

Depositional Background. The Tazhong Ordovician reef–type reservoir is located on the Tazhong low bulge, on the central uplift in the Tarim basin, which is the first Ordovician bio-reef type petroleum reservoir with a hundred million tons of reserves in the Tarim basin and in the whole of China.

Characteristics of Sequences and Sedimentation. The Ordovician system includes the Upper Ordovician (the Sangtamu formation and the Lianglitage formation) and the Lower Ordovician (Yingshan formation). The granular limestone section in the Lianglitage formation is the major petroleum reservoir sequence. In the Tazhong area, the Ordovician Lianglitage formation contains one third-order sequence and eleven semi-sequences. The reservoir sequences of high energy reef flat facies were predominately developed in the highstand system tract in the third-order sequence; furthermore, the burial dissolution process in the later stage assisted in creating these superb reservoir sequences. The cap rock is the mudstone section of the Sangtamu formation. The combination of reservoir and seal is excellent.

Characteristics of Petroleum Accumulation. In the Tazhong area, the crude oil and the condensate oil predominately came from the hydrocarbon source rocks in the Cambrian–Lower Ordovician strata; the source rocks of the Middle and Upper Ordovician made a small contribution to the oil reservoirs. Most of the natural gas belongs to the sapropelinite type. Regardless of whether it is a condensate gas reservoir or oil reservoir, both the produced gas and associated gas come from the Cambrian–Lower Ordovician strata. This includes the pyrolysis gas from oil and the associated gas from crude oil. A small amount of natural gas comes from the source rocks in the Middle–Upper Ordovician strata. In the Tazhong–I slope zone, the fault system is the important factor for creating the migration system and control systems for the Ordovician reef type petroleum reservoir. In the Tazhong area, oil and gas (that originated in the Cambrian–Lower Ordovician strata) accumulated during the late Caledonian tectonic period. During the late Hercynian tectonic period, because of fault activities, the paleo oil reservoir in the underlying strata migrated upwards into the traps above; in addition, a small amount of oil and gas that was generated during the later stage also initiated migration and accumulation. The petroleum migration and accumulation during the Hercynian tectonic period was the major, effective petroleum accumulation in the Tazhong area, which formed reservoirs in the Carboniferous, the Middle and Upper Ordovician and the Silurian strata. The Indosinian–Yanshanian tectonic periods were the adjusting periods for paleo petroleum reservoirs. During the Himalayan tectonic period, near the deeply buried #1 fault belt, the paleo oil reservoir in the Cambrian–Lower Ordovician petroleum combination was altered into gas; this pyrolysis gas migrated upwards through faults that were formed during the Himalayan tectonic movement to re-distribute (Figs. 5.2 and 5.12).

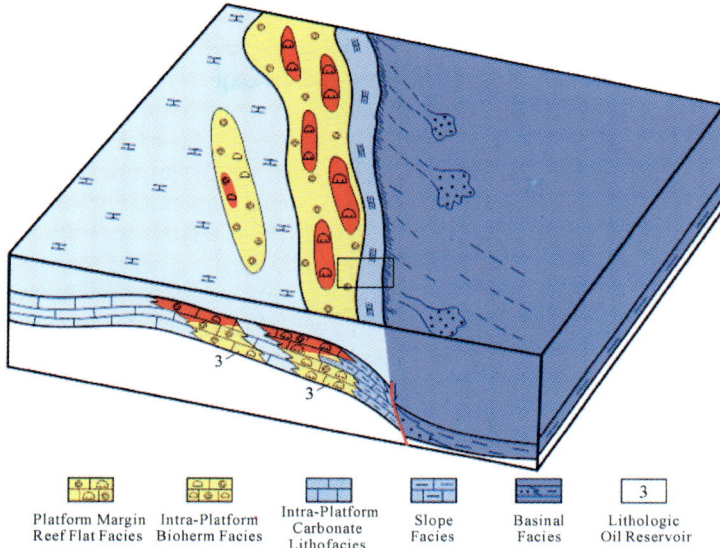

| Platform Margin Reef Flat Facies | Intra-Platform Bioherm Facies | Intra-Platform Carbonate Lithofacies | Slope Facies | Basinal Facies | Lithologic Oil Reservoir |

Fig. 5.12.   Platform margin—the combination of marine transgressive reef flat in a craton basin

Platform Margin, Intra-Platform—The Combination of Marine Transgressive Beach Bar

This type of combination includes the gas reservoir of oolite beach facies in the Feixianguan formation in the northeastern Sichuan basin and the oil reservoir in the Carboniferous bio-clasitic limestone section on the Maigaiti slope in the Tarim basin.

Depositional Background.   The lithologic–structural gas reservoir of oolote beach facies in the Feixianguan formation was distributed along the eastern side of the Kaijiang–Liangping continental shelf in the shallow sea. The Puguang gas field has been discovered on the platform margin; the Loujiazai, Dukouhe and Tieshanpo gas fields have been discovered on the intra-platform.

Characteristics of Sequences and Sedimentation.   In the northeastern Sichuan region, the Feixianguan formation includes two third-order sequences and eleven fourth-order sequences. The oolite beach was developed in the highstand system tract in the third-order sequence with a slowly alternating sea level. The highest level of the sea in the third-order sequence correlated with the peak time of the oolite beach. The paleo topography controlled the development of the oolite beach. The reservoir sequence was made of the oolitic dolomite of the Feixianguan formation, which was deposited in the oolite beach–inter-beach area on the intra-platform. During the late phase, the mixed water dolomitization and dissolution formed excellent reservoir spaces. In the northeastern Sichuan region, the cap rocks are Jurassic–Upper Triassic clay shale and a Middle–Lower Triassic gypsum–salt layer, which offered a good sealing layer for this region.

Characteristics of Petroleum Accumulation. In the northeastern Sichuan region, the natural gas in the oolite beach gas reservoir predominately came from the Upper Permian source rocks. During the Late Triassic, these source rocks progressed into hydrocarbon generation peak time, the oil and gas started to migrate and to accumulate. With oil and gas accumulation during the Indosinian–Yanshanian tectonic periods, the final petroleum reservoir adjustment happed during the Himalayan tectonic period, which formed the framework of the present petroleum system. The northeastern Sichuan region was located on the Kaijiang paleo uplift and its slope zone; this location benefited oil and gas migration and accumulation. The development of a natural gas field not only experienced a lithologic sealing event during its early phase, but it was also controlled by the structure and lithology during its later phase. After vertically migrating into the Feixianguan reservoir and influenced by paleo topography, the migrating direction of oil and gas changed from west to east. Under the high temperature, the hydrocarbons that accumulated during the early phase were gradually pyrolyzed. Finally, they formed a dry gas reservoir (Figs. 5.2 and 5.13).

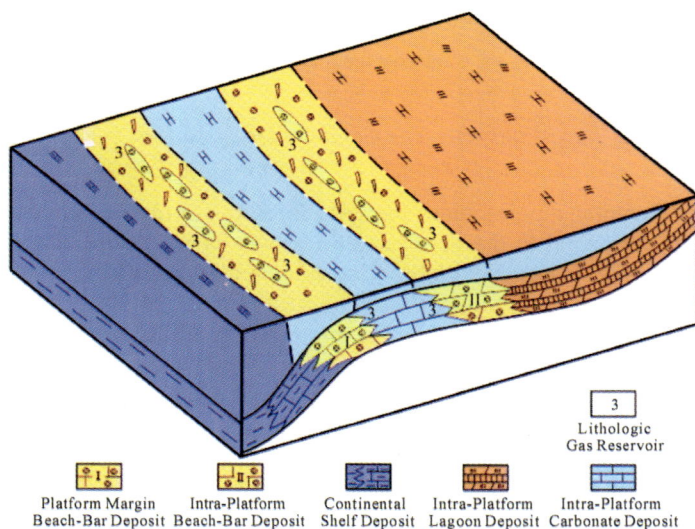

Fig. 5.13. Platform margin, intra-platform—the combination of marine transgressive beach bar in a craton basin

Intra-Platform—The Combination of Littoral Marine Transgressive Beach Bar

This type of combination includes oil and gas reservoirs of Donghe sandstones in the Hadexun area and on the eastern slope of the Lunnan area in the Tarim basin, the Silurian system in the Tazhong area and the Carboniferous oil and gas reservoir in the Eerduosi basin. The stratigraphic reservoir of Donghe sandstone in the Hadexun area is a typical example.

Depositional Background. The Hade 4 Donghe sandstone stratigraphic oil reservoir is the principal oil reservoir for the Hade 4 oil field that is the first marine sandstone oil field with more than a hundred million tons reserves in China. The Hade 4 oil field is located at the Hadexun structure belt, on the northern side of the Manjiaer depression, in the Tarim basin.

Characteristics of Sequences and Sedimentation. The Donghe sandstone was deposited during the Late Devonian–Carboniferous marine transgression. In the Donghe sandstone, the lower portion contains middle–lower shoreface facies; the middle and upper portions have foreshore to middle–upper shoreface facies. It contains a complete third-order sequence. The reservoir sequence was developed in the marine transgressive system tract. Most reservoir strata have medium porosity and medium–high permeability. The mudstone and the gypsum mudstone that were deposited during the highstand period functioned as excellent cap rocks.

Characteristics of Petroleum Accumulation. The Donghe sandstone overlapped and pinched out towards a higher elevation of the paleo uplift, which provided a precondition for creating a stratigraphic overlapped–structural composite type of petroleum trap. The oil and gas came from the Cambrian–Ordovician source rocks of marine facies. The Hade 4 Donghe sandstone stratigraphic–structural oil reservoir accumulated in the early phase and was adjusted in the late phase. The petroleum accumulated during the late Hercynian tectonic period and was adjusted during the late Himalayan tectonic period. In addition, due to the ongoing adjustment process, this oil reservoir is an active oil reservoir (Figs. 5.2 and 5.14).

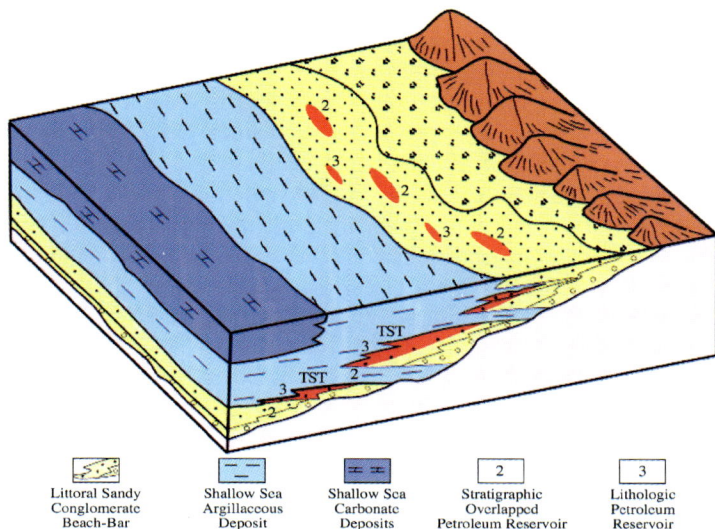

Fig. 5.14   Intra-platform—the combination of littoral marine transgressive beach bar in a craton basin

Intra-Platform—The Combination of Highstand Delta of Marine-Terrigenous Facies

The marine–terrigenous delta facies in the Carboniferous–Permian systems in the Eerduosi basin form a typical example of this type of combination, such as in the Yulin gas field, Zizhou gas field and Wushengqi gas field.

Depositional Background. The Yulin gas field is located in the central-eastern part of the Eerduosi basin; the structural position of this region is located on the northern side of the Yinshan slope and the southern side of the Yimeng uplift. The major gas reservoir is the Shan–2 section.

Characteristics of Sequences and Sedimentation. The section of Shan–2 can be further divided into three system tracts that are LST, TST and HST. The LST system tract contains medium–coarse grain sandstones that experienced multiple superimposed and washed out events in the high energy environment; these sandstones were major reservoir sequences in the region. The sandstones in the TST and HST system tracts were deposited in deep water with fine grain size and poor sorting. Because of the gentle slope and shallow water, the majority of sand bodies were deposited in the distributary channel on the delta plain, in the distributary channel on the delta front and in the estuary. The system tract and the semi-sequences controlled distribution of the sand body; the micro facies controlled the beneficial sand body; the quartz sandstone controlled the development of the superb reservoir sequence and diagenesis controlled the high porosity and high permeability.

Characteristics of Petroleum Accumulation. The hydrocarbon source rocks include coal pertrology, carbonaceous mudstone and dark color mudstone, which occur in the coal measures in the Shanxi formation and the Taiyuan formation. During the end of the Middle Jurassic, the source rocks progressed into a gas generating high peak period; in addition, the natural gas started to migrate and accumulate.

The Early Cretaceous was a rapidly buried period. The source rocks not only progressed into a high maturity–post maturity stage quickly, but also continually produced natural gas at the same time. The liquid hydrocarbons that had been generated in the early phase were pyrolyzed into natural gas. During the end of the Early Cretaceous, because of a temperature decrease and strata uplift, the entire gas reservoir was releasing natural gas. At the same time, the formation pressure gradually decreased. In the upper portion, black color mudstone in the Shan–1 section joined with carbonaceous mudstone that was interbedded with a coal layer to create a direct seal for the Shan–2 gas reservoir. In the natural gas region, the interbedded sandstone and mudstone of river flood plain facies in the lower Shihezi formation and the mudstone of river overflow lacustrine facies in the upper Shihezi formation formed regional cap rocks. On the eastern side of the Shan–2 gas reservoir, the argillaceous rock of flood plain facies formed a lithologic sealing layer in an upward dipping direction. The bauxite at the base of the Carboniferous system created a lower sealing layer. These conditions created a relatively sealed natural gas system and formed an in situ type of lithologic gas reservoir (Figs. 5.2 and 5.15).

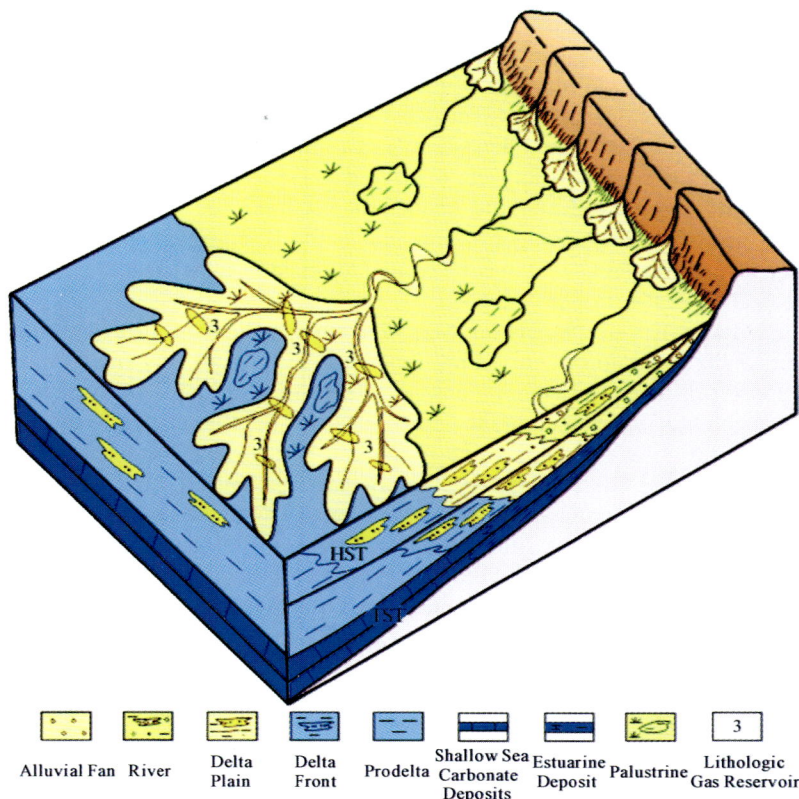

Alluvial Fan	River	Delta Plain	Delta Front	Prodelta	Shallow Sea Carbonate Deposits	Estuarine Deposit	Palustrine	Lithologic Gas Reservoir	3

Fig. 5.15.   Intra-platform—the combination of highstand delta of marine-terrigenous facies in a craton basin

Paleo Uplift—The Combination of Karst

This type of combination includes the oil and gas reservoirs of the Ordovician buried hill in the Lunnan area, the buried hill of the Zhulei belt in the Tazhong area, the buried hill in the central uplift zone in the Eerduosi basin and the Ordovician buried hill in the Bohai bay basin. The Lunnan buried hill is a typical example.

Depositional Background.   The Ordovician carbonate karst stratigraphic oil reservoir in Lunnan is located on the Lunnan low bulge, on the Tabei uplift, in the Tarim basin. The overall orientation is in an NE-SW direction. In an N-S direction, the length is 80 – 100 km; in a W-E direction the width is 30 – 60 km. The region covers an area of 4,420 km^2 with a relief of 1,550 m.

Characteristics of Sequences and Sedimentation.   The Upper Ordovician includes the Sangtamu formation, Lianglitage formation and Tumuxiuke formation. The Middle Ordovician includes the Yijianfang formation. The Middle–Lower Ordovician includes the Yingshan formation. The Lower Ordovician includes the

Penglaiba formation. The Lianglitage formation, Yijianfang formation and Yingshan formation are the important production sequences. The Ordovician system contains six third-order sequences, which experienced semi-restricted platform facies → open platform facies → platform margin facies → platform margin slope facies → mixed depositional shallow water continental shelf facies. The hydrocarbon source rocks were deposited in the transgressive system tract in the third-order sequence and the condensed section. The source rocks in the Heituao formation in the Tadong area and in the Kepingsaergan formation are high quality source rocks that may be related to the largest marine transgression of the second-level sea surface changes.

Characteristics of Petroleum Accumulation. The carbonate reservoir can be further divided into the buried hill type and the internal episodic type, which were mostly distributed in the Yijianfang formation and Yingshan formation and which were rarely distributed in the upper portion of the Lianglitage formation and the lower portion of the Tumuxiuke formation. The oil and gas occurred on the karst highland and karst slope. The type of karst includes a surface karst zone, permeable karst zone, undercurrent karst zone and bedding plane karst zone. The Ordovician carbonate karst stratigraphic reservoir is the product of multi-phase creation, destruction and adjustment. The hydrocarbons and oil reservoir of the Caledonian tectonic period have been destroyed; the hydrocarbons of the secondary generation and paleo viscous crude oil were formed during the Hercynian tectonic period; the Indosinian–Yanshanian tectonic period was the adjustment period; the Himalayan tectonic period was the oil and gas accumulation period (Figs. 5.2 and 5.16).

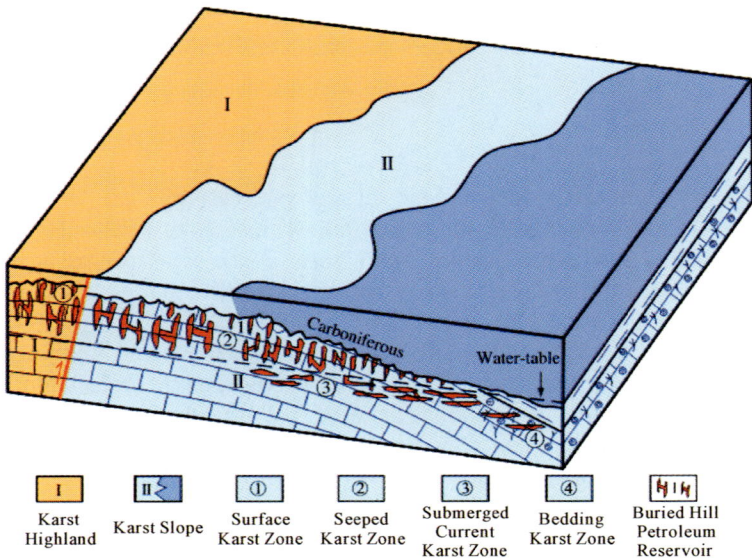

Fig. 5.16.   Paleo uplift—the combination of karst in a craton basin

References

Cai, X.Y., Li, S.T., et al., 2003. High Resolution Sequence Stratigraphy of Continental Basins: Basic Ideas, Method and Practice for Exploring Subtle Oil Pools. The Geological Publishing House, Beijing.

Dai, J.X., 1983. Oil and gas reservoirs in synclines. Acta Petrolei Sinica, 4(4):27-30.

Dai, J.X., et al., 1992. The Characteristic of a Natural Gas Belt in a Large and Medium Size Gas Field, in the Study of Natural Gas Geology. Petroleum Industry Press, Beijing.

Gao, R.Q., Cai, X.Y., 1997. Preconditions and Distribution of Oil and Gas Fields in the Songliao Basin. Petroleum Industry Press, Beijing.

Gao, R.Q., Zhao, Z.Z. (Eds.), 2001. Oil and Gas Exploration of Concealed Reservoir in the Bohai bay Basin, in the Frontier Ppetroleum Eexploration in China (vol. 3). Petroleum Industry Press, Beijing.

Hu, J.Y., Xu, S.B., Tong, X.G., 1986. The development and distribution of composite oil and gas reservoirs in the Bohai bay basin. Petroleum Exploration & Development, 13(1):1-8.

Jia, C.Z., et al., 2000. Petroleum Exploiration in Foreland Thrust Belts. Petroleum Industry Press, Beijing.

Qiao, H.S., 1986. Petroleum Exploration on Steep Slope of a Depression in the Bohai bay Basin. Petroleum Exploration & Development (vol. 01).

Wang, Y.C., 2001. The Creation and Distribution of Lithologic Reservoir in the Southern Songliao Basin. Petroleum Industry Press, Beijing.

Xiao, D.M., Chi, Y.L., Meng, Q.A., et al., 2005. The Understanding of Lithologic Reservoir and Related Exploration Practice in Synclinal Region, Northern Songliao Basin. Petroleum Industry Press, Beijing.

Zhang, W.Z., 1997. The western slope zone in the Liaohe depression - An oil and gas accumulation area in a slope zone. *In*: Major Non-marine Oilfields of China. Petroleum Industry Press, Beijing.

6

Classification of Petroleum Plays and Their Mechanism

Petroleum reservoirs and traps were carriers and storage for oil and gas migration and accumulation. However, it depended on the passage that linked the source rock and the trap as to whether hydrocarbons would accumulate. Different combinations of "source rock and trap" not only influenced the difficulty and the probability of petroleum accumulation, but also revealed clear distinctions in the petroleum injective mechanism, accumulative features and major controlling factors. Considering that the hydrocarbon source rock was the material foundation for petroleum accumulation, the traps (which were located either above source rock or beneath source rock) exhibited various dynamic features distinctively. Thus, according to the vertical relationship between source rocks and traps, this section will classify different types of petroleum play and will further discuss their mechanism and major controlling factors.

6.1 Definition of "Above-Source, In-Source and Beneath-Source" Plays

The play is a group of exploration targets that are not only connected geologically, but also have similar source rocks, reservoir sequences and trapping conditions. The play is a geological entity that consists of a migrating system, traps or petroleum reservoir, and related accumulative elements. The core of play theory is based on the theory of the petroleum system. According to the source of the oil and gas, the petroleum reservoir and process of migration, from the angle of space–time configuration of petroleum accumulative elements and geological processing, we analyze the characteristics of oil and gas accumulation and distribution in a basin so as to direct the oil and gas exploration. The play is the third level evaluation in the petroleum units estimation of "Basin–Petroleum System–Play–Prospect" that is internationally a commonly used estimation process.

Compared with a structural petroleum reservoir, the conditions for developing a trap and accumulative mechanism in a lithostratigraphic reservoir are very different. For the convenience of analysis and for describing the configuration of a source rock–reservoir sequence, control factors for accumulation, and distribution of oil and gas in a lithostratigraphic petroleum reservoir, and for better directing the petroleum exploration project, according to the vertical position of the hydrocarbon source rock and reservoir body, the play of a lithostratigraphic reservoir can be divided into three types that are above-source play, in-source play and beneath-source play (Fig. 6.1).

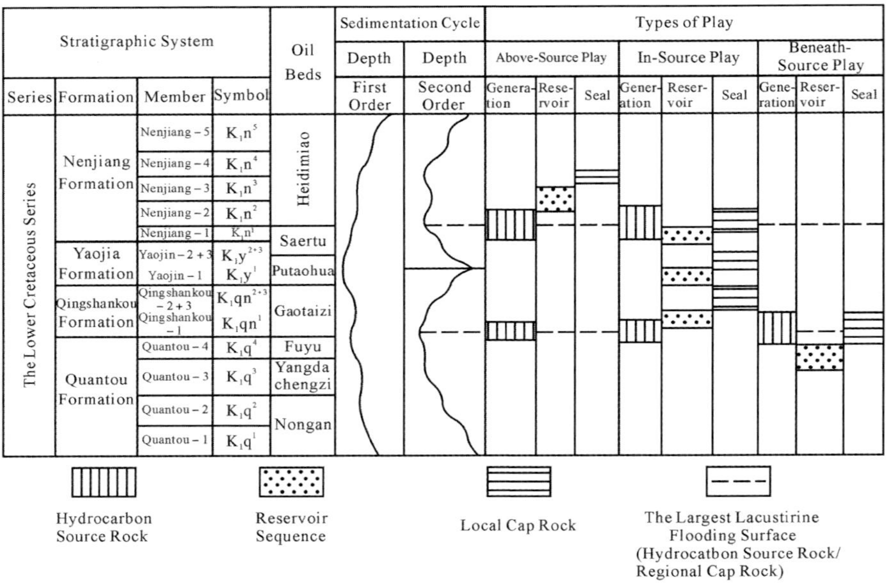

Stratigraphic System				Oil Beds	Sedimentation Cycle		Types of Play								
					Depth	Depth	Above-Source Play			In-Source Play			Beneath-Source Play		
Series	Formation	Member	Symbol		First Order	Second Order	Generation	Reservoir	Seal	Generation	Reservoir	Seal	Generation	Reservoir	Seal
The Lower Cretaceous Series	Nenjiang Formation	Nenjiang – 5	K_1n^5	Heidimiao											
		Nenjiang – 4	K_1n^4												
		Nenjiang – 3	K_1n^3												
		Nenjiang – 2	K_1n^2												
		Nenjiang – 1	K_1n^1	Saertu											
	Yaojia Formation	Yaojin – 2 + 3	K_1y^{2+3}	Putaohua											
		Yaojin – 1	K_1y^1												
	Qingshankou Formation	Qingshankou – 2 + 3	K_1qn^{2+3}	Gaotaizi											
		Qingshankou – 1	K_1qn^1												
	Quantou Formation	Quantou – 4	K_1q^4	Fuyu											
		Quantou – 3	K_1q^3	Yangda chengzi											
		Quantou – 2	K_1q^2	Nongan											
		Quantou – 1	K_1q^1												

Hydrocarbon Source Rock   Reservoir Sequence   Local Cap Rock   The Largest Lacustirine Flooding Surface (Hydrocatbon Source Rock/ Regional Cap Rock)

Fig. 6.1.   Classification of plays for the lithostratigraphic reservoir (using the Songliao basin as an example)

6.1.1   Beneath-Source Play

The beneath-source play indicates that reservoir sequences are under the hydrocarbon source rock. This type of play contains a lithologic reservoir, faulted–lithologic reservoir and stratigraphic reservoir in a depressed basin, such as in the Songliao basin, the pericline belt on the Fuxin uplift and the western slope zone. In addition, this type of play also appears in buried hill reservoirs in a rifted basin, such as all kinds of buried hill oil and gas reservoirs in the Bohai bay basin.

6.1.2   In-Source Play

The in-source play indicates that either the reservoir body is located inside of the hydrocarbon source rock, or the reservoir body and hydrocarbon source rock are positioned in the same layer. This type of play designates one set or multi-sets of major source rocks and near by reservoir sequences that were limited by regional cap rock, which had similar accumulation conditions and belonged to the same geological time interval. Usually, this type of play was controlled by the two largest flooding surfaces of second-order; for example, in the Songliao basin, the Saertu, the Putaohua and the Gaotaizi oil beds could be considered as an in-source play (Fu et al., 2002).

6.1.3 Above-Source Play

The above-source play indicates that the reservoir sequences are located above hydrocarbon source rocks. The above source rock reservoir is widely distributed in all kinds of basins and in all the exploration regions in China.

In general, the above-source play was located in the highstand system tract with the sand bodies of river, delta and fan delta facies. The common types of trap include lithologic trap, structural–lithologic trap and stratigraphic trap, such as the Cretaceous Heidimiao oil bed on the western slope, in the Songliao basin (Table 6.1). The in-source play was developed in the lacustrine transgressive system tract and the lower portion of the highstand system tract. The types of reservoir body include the fluxoturbidite fan and diluvial turbidite fan. Most traps are lithologic traps, such as the Gaotaizi oil bed in the Cretaceous Qing-1 member, in the Songliao basin. The beneath-source play appeared in the lowstand system tract and the lower portion of the lacustrine transgressive system tract. The types of reservoir body include sub-lacustrine fan, lowstand delta and fan delta. The lithologic trap and faulted–lithologic trap are principle types of trap, such as in the Songliao basin, the Fuyu oil bed in the Cretaceous Quan-4 member in the pericline belt on the Fuxin uplift. On the western slope, the stratigraphic overlapped trap could be formed.

Table 6.1 Category of accumulative combination for lithostratigraphic reservoir

Accumulative combination	Types of system tract	Major depositional facies belt	Types of common traps/oil reservoir	Example
Above-Source play	Highstand system tract	River, delta, fan delta	Lithologic trap; Structural–lithostratigraphic trap	The Cretaceous Heidimiao oil bed in the Songliao basin
In-Source play	Lacustrine transgressive system tract; Highstand system tract	Fluxoturbidite fan; Diluvial turbidite fan	Majority are lithologic traps	The Cretaceous Saertu, Putaohua, and Gaotaizi oil beds in the Songliao basin
Beneath-Source play	Lowstand system tract, lower portion of lacustrine transgressive system tract	Sub-lacustrine fan, lowstand delta, fan delta	Lithologic trap; Faulted–lithostratigraphic trap	The Cretaceous Fuyu and Yangdachengzi oil beds in the Songliao basin

6.2   Control Factors for Three Types of Play

Three types of play not only have a different relationship between source rock and reservoir sequences, but they also have different types of system tract, different types of reservoir mechanism, different types of trap and different characteristics of accumulation. The following will discuss a play's characteristics of accumulative dynamics, major control factors and the distribution.

6.2.1   *Essential Control Factors for the Beneath-Source Play*

The accumulative model of a beneath-source play is that oil and gas were generated in the upper strata and were stored in the lower strata. The vertical faults that linked with source rocks and the ultra hiqh pressure are important factors for petroleum accumulation. The pressure differentiation that drove the oil and gas migration was the dynamic mechanism for petroleum accumulation. During the petroleum accumulative process for a beneath-source reservoir, in order to drive oil and gas downward, the remaining pressure differentiation should be larger than the sum of buoyancy and capillary resistance. To carry out the generation of oil and gas in the upper strata and storage in the lower strata, the pressure differentiation between the source rock and reservoir sequence should be in an absolutely dominant position (Chi et al., 2000). Therefore, the depth and abundance of a beneath-source reservoir was decided by the magnitude of the pressure differentiation between the source rock and the reservoir sequence.

In the southern Songliao basin, the Fuyu oil bed not only is a lithostratigraphic reservoir, but it is also a typical beneath-source reservoir. The fault offered an essential migration passage that connected the sandstone and the source rock above. Plentiful oil resources and adequate fluid pressure were the necessary dynamic conditions. Oil and gas predominately migrated in a vertical direction; the horizontal migration was a secondary movement. When the dynamic of downwards migration was larger than the force of obstacles that consisted of capillary force, buoyancy and viscous force, the oil and gas migrated along the fractures from the upper strata into the lower strata; furthermore, oil and gas would conduct a lateral migration in the sandstone below to form a petroleum reservoir. In the southern Songliao basin, the oil reservoir in the Quan-4 member on the Fuxin uplift is a beneath-source oil reservoir, which was the center of attention in recent exploration activity.

● **Pressure Differentiation Between Source Rock and Reservoir Sequence**

The pressure differentiation between source rock and reservoir sequence is the crucial factor for oil and gas migrating downward. When the fluid pressure of the source rock was not only abnormally high, but when it also satisfied the requirement for the migration dynamic to be larger than the capillary force, buoyancy and viscous force in fractures, oil would migrate downward into the sandstone along the fracture. In the Fuxin uplifting area in the southern Songliao basin, because the mudstones (that were positioned from the Qing-1 member to the Nen-1 member with a thickness of around one thousand meters) capped abnormally high pressure that was produced during the hydrocarbon generation in the source rock, the oil and gas inevitably migrated downward by the pressure differentiation along the fault from the source rock into the sandstone below. After oil drove out the formation water and occupied storage spaces, the high pressure was released and a new dynamic equilibrium was attained.

The mudstone in the Qing-1 member not only was an effective hydrocarbon source rock, but it was also an excellent regional cap rock with a displacement pressure above 10 MPa. In addition, most of the overlying strata in the Qing-2 member, the Qing-3 members and the Nen-2 member consisted of mudstones with a total thickness of over thousands of meters. These overlying mudstones functioned as a regional seal to block the oil and gas that were generated by the Qing-1 member from migrating upwards to create abnormally high pressure in the Qing-1 member. At present, the stress rupture experiment in the Qing-1 member indicates that micro fractures would be formed if the ultra pressure is larger than the confining pressure (10 MPa) in the mudstone of the Qing-1 member; these fractures could help to discharge the fluid with abnormally high pressure. At present, the Qing-1 member in general contains ultra high pressure. On the slope area of the Fuxin uplift, the displacement pressure is 8 – 10 MPa (Fig. 6.2); generally, the pressure gradient is 1.2 MPa/100 m; near the depression, the displacement pressure is 10 – 12 MPa.

Horizontally, the distributions of pressure differentiation surrounded the hydrocarbon generation center and were spread out in a belt pattern; vertically, they were positioned in the argillaceous source rock sequences. In descending order, from the Nen-1 member to the Qing-1 member, the formation pressures gradually increased. However, in the base portion of the Qing-1 member, the Fuyu oil bed and the Yangdachengzi oil bed, the formation pressures suddenly dropped, which was the result of ultra pressurized fluids injected into the Fuyu oil bed and the Yangdachengzi oil bed, by utilizing the faults as a migration passage.

Because the hydrocarbon source rock and the reservoir sequence were divided by a water-contained stratigraphic layer (that has relatively high displacement pressure), oil and gas had to overcome the blockage of buoyancy in order to migrate downward from the source rock to the reservoir sequence, which required enough abnormally high pressure from the source rock. In other words, the pressure differentiation between the source rock and the reservoir sequence should be sufficient in order to overcome the blockage. The potential difference in the beneath-source reservoir functioned as another barrier; to force the oil and gas downward required sufficient pressure differentiation, which means the pressure differentiation was a determinative factor.

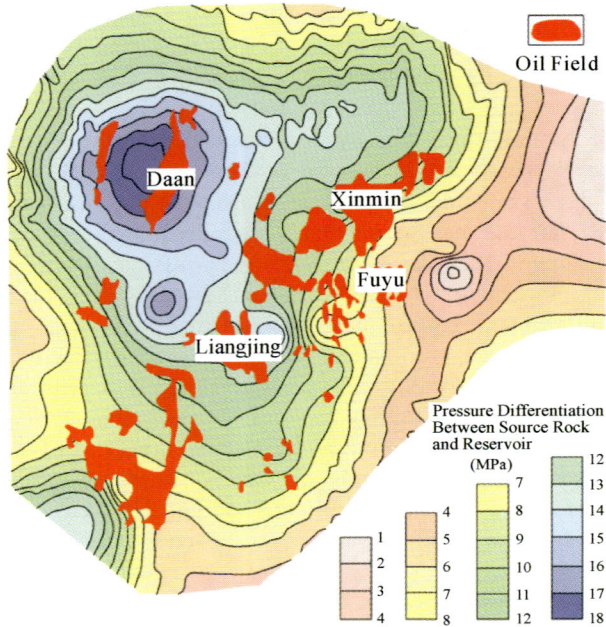

Fig. 6.2. Map of pressure differentiation between the source rock and reservoir in the K_1qn^1-K_1q^4 sequences on the Fuyu uplift and the distribution of oil reservoir beneath these source rocks

According to the calculation of the pressure sealing theory, under ultra high pressure, the formula for the fluid moving downward is

$$H = p/[(\rho_w - \rho_h)g],$$

where H, the depth of downward injection or the height of sealed oil column, m; p, pressure differentiation, Pa; ρ_w, water density, g/cm^3; ρ_h, oil density, g/cm^3; g, gravity acceleration, m/s^2.

For example, on the Fuxin uplift in the southern Songliao basin (Fig. 6.3), the pressure differentiation between the source rock and reservoir sequence is 8 – 12 MPa in the lithostratigraphic reservoir in the Fuyu oil bed. If using the formula above, the value of H is 272 – 408 m; in other words, the depth of the oil and gas downward injection is 272 – 408 m from the base of the Qing-1 member. In reality, whether the oil and gas column reached this depth or not was determined by the faulted depth and the development of reservoir sequences.

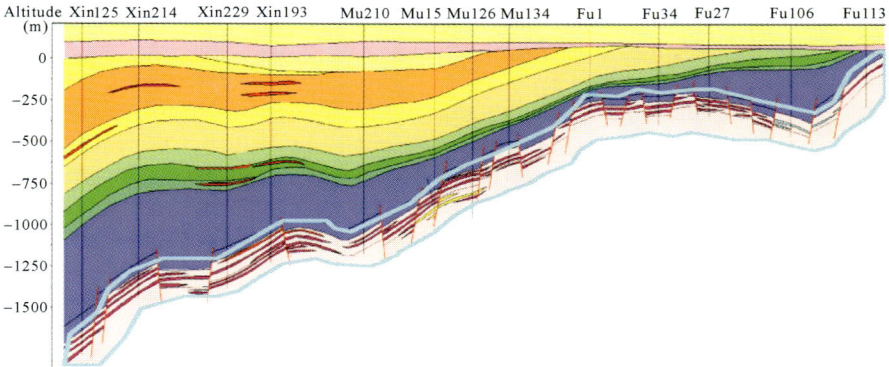

Fig. 6.3. Cross section of the Fuyu oil bed (beneath-source play) on the Fuyu uplift in the southern Songliao basin

● **Body of Sandstone**

The sand body was a required condition for oil and gas accumulation. Usually, sand bodies were well developed in the vicinity of unconformity. In a beneath-source reservoir, the sand bodies that were positioned on the unconformity surface offered a better place for petroleum accumulation. First, the sand bodies on the unconformity surface were the first ones to experience organic acid erosion and to form subsequently better porosity and permeability. Next, the oil and gas were generated on the upper strata and were stored in the lower strata; when oil and gas were injected downward by the pressure differentiation, the sand bodies on the unconformity surface were the first ones to capture oil and gas. Therefore, the intersection zone of unconformity surface and fault was a favorable place for distributing the beneath-source reservoir.

6.2.2   Fundamental Control Factors for the In-Source Play

The in-source petroleum accumulation was not only associated with well developed hydrocarbon source rock, but it was also linked with the crucial factors of the reservoir condition and the relationship of the petroleum play. In addition, the lateral migration dominated the in-source petroleum accumulation. For the gently sloped terrigenous depressed basin, the structure (such as nose-like bulge and slope) was the beneficial condition for oil and gas accumulation; the sand body, fault and nose-like bulge controlled development of the "sweet point". The accumulative mechanism of in-source sandstone reservoir or trap has been discussed in many published articles by international and Chinese researchers (Lin et al., 2004; Chen D. X. et al., 2004; Chen Z. M. et al., 1998; Hao and Dong, 2001; Li et al., 2004;

Sun and Yang, 2005; Zhang, 2001). Here we only discuss the accumulative mechanism of the lenticular sandstone reservoir.

The lenticular sandstone reservoir is a typical in-source lithologic reservoir. The accumulative dynamic of lenticular sandstone involved replacement, the fluid (formation) pressure difference, darcy flow, capillary and differential breakthrough. The creation of an oil and gas reservoir included the entire process that incorporated the oil and gas generation, expulsion, migration, accumulation and preservation. The development of a lenticular sandstone reservoir integrates three related processes. The first process is that the fluid pressure difference drove oil and gas into the transitional zone between source rock and sandstone. This is the foundation of oil and gas accumulation. The second process is that the capillary pressure difference forced the oil and gas to penetrate the transitional zone and migrate into the sand body; during the same process, formation water would be forced out of the sand body. This is the connective process for oil and gas accumulation. The third process is that as a result of the buoyancy force, oil (gas) is separated from water. This is the crucial step in oil and gas accumulation.

● Fluid Pressure Difference Drove Oil and Gas into Lenticular Sandstone

The fluid pressure difference is the remaining pressure between the overpressure in the source rock and the fluid pressure in the sand body; it is a major force for initial migration. Many terrigenous basins have overpressure (Hao et al., 2002). Because porous fluid could not be discharged on time and carried part of the overload from the formation above, overpressure was formed. Permeability was the most important controlling factor for developing overpressure. Many kinds of mechanisms can produce overpressure, which can be categorized into three types. The first type of mechanism relates to stress; the second type of mechanism relates to the alteration of fluid volume; the third type of mechanism relates to buoyancy and fluid movement. For the lenticular sandstone reservoir, predominate mechanisms that produce overpressure are uneven compaction in the mudstone and generation of hydrocarbon.

● Capillary Pressure Difference Forced Oil and Gas into Lenticular Sandstone

The capillary pressure played a determinative role during the development of the lenticular sandstone reservoir. The characteristic of the lenticular sandstone trap is that the sand body was enveloped by mudstone. There is a difference in grain size if comparing mudstone with sandstone; on the mudstone side, in the pores and throat, the capillary radius (r_{mud}) is very small; however, on the sandstone side, in the pores and throat, the capillary radius (r_{sand}) is relatively large. Therefore, oil will easily migrate into sandstone due to the size of pore and throat; at same time, the oil drop will change its shape (Fig. 6.4).

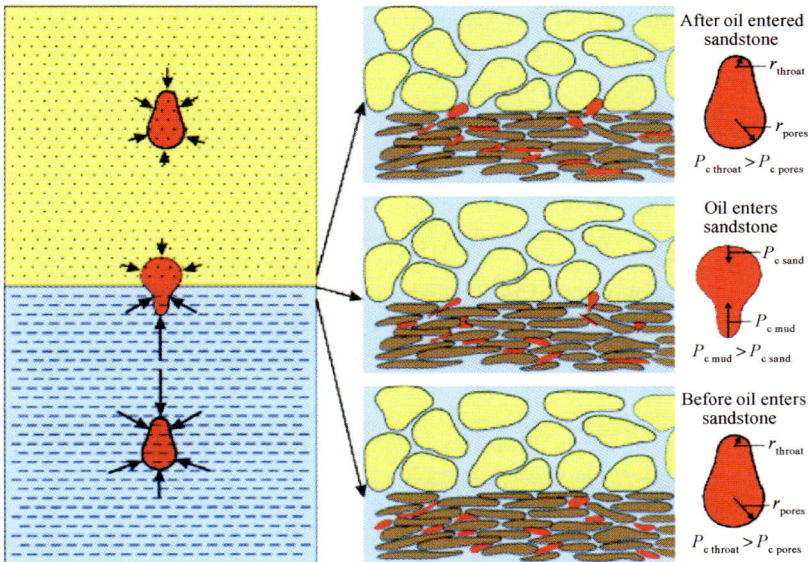

Fig. 6.4. The capillary pressure difference assists in creating the lenticular sandstone trap and oil accumulation

In the transitional zone between mudstone and sandstone, oil of the continual phase will experience a capillary pressure difference with a magnitude of several MPa, which equals a few dozen of barometric pressure. The direction of the force directly pointed to the sandstone. This magnitude of force is more than enough to overcome the viscous force and adsorption force from oil. Under the capillary pressure difference, oil would attempt to enter the pores of sandstone and occupy these spaces. Because the capacities of the pores in mudstone and sandstone are fixed, thus an equal amount of water must be discharged if a certain amount of oil enters the sandstone. In general, both mudstone and sandstone are water-wet. Therefore, on both sides of the transitional zone between mudstone and sandstone, the fluid potential is identical. As a result of the capillary pressure difference, oil drops entered the sandstone and transferred forces to water drops; utilizing the driving force from the oil drops, water drops were forced into the mudstone to occupy the spaces that were originally filled by oil drops. Discharged water did not require any additional force. Because oil and water exchanged spaces, more and more oil accumulated in the sandstone. The higher degree of oil saturation brought a lower percentage in the water content, until this exchange process achieved irreducible water saturation. At the equilibrium point of replacement, water was in a quasi-solid state and water could not be further replaced by oil. After water entered the mudstone, because the size of a water molecular is very small and because mudstone is water-wet, the water molecular passed through micro pores, throat passages, bedding surfaces and micro fractures. Water will migrate or percolate towards the outside; finally, under the weight of overlying sediments, water flows to the surface.

● **Buoyancy Made Oil and Gas Accumulate on the Top of Lenticular Sandstone**

Under the influence of the fluid pressure difference and capillary pressure difference, oil and gas not only migrated into lenticular sandstone, but were also distributed around lenticular sandstone. In lenticular sandstone, in order to accumulate a huge amount of oil (or gas), this must pass through the pores that were filled with water and move towards the middle portion and top of the lenticular sand body. This kind of migration released porous spaces to facilitate additional oil and gas entering the sand body. The migration of oil and gas was transported by the buoyancy force and was barricaded by the capillary pressure. In order to pass through the throat passage in the neighborhood, the continual phase oil must overcome the blockage of capillary pressure in order to pass through the pores.

Physical properties of the sand body are major controlling factors for the lenticular sandstone reservoir. If sand grains are of a large size with better physical properties, oil migration and accumulation in sandstone would be easy. Then, lenticular sandstone would have great oil saturation and a better reservoir. If sand grains are of a small size with poor physical properties, oil migration and accumulation in sandstone would be difficult. Then, lenticular sandstone would have less oil saturation and a poor reservoir.

● **Developmental Process of Lenticular Sandstone Reservoir**

According to different mature stages of hydrocarbon source rocks, we analyzed the developmental process of the lenticular sandstone reservoir and set up the following models (Fig. 6.5).

Characteristics of Lenticular Sandstone Reservoir with Immature Hydrocarbon Source Rock

The stage of immature hydrocarbon source rock matches $0 < R_o < 0.5\%$. This is the thermo catalytic bio-gas stage. The liquid oil had not been generated yet and only a small amount of bio-gas had formed. During this stage, the formations were buried at a shallow depth with lower temperature and lower pressure. The major accumulative factors were the fluids in the mudstone, sandstone and pores. When comparing mudstone and sandstone, because they contain different minerals and lithologic textures, their physical properties are very different. Sandstone and mudstone have different compaction strength. Sandstone has a much more compressive resistance than mudstone. Sandstone contains about $40\% - 50\%$ original porosity; after compaction, it contains approximately 20% porosity. On the other hand, mudstone contains about $70\% - 90\%$ original porosity. After compaction, it only contains approximately 5% porosity. During this stage, in mudstone, kerogen was primarily in a solid form and water was the major fluid in the pores; additionally, there was a small amount of bio-gas that was in the water–soluble phase. Sandstone contained primary pores at this stage.

The immature stage is equivalent to the early phase of diagenesis, which is the major period of compaction. The main dynamic during this stage is the remaining pressure that comes from the compaction process, which sets off a water discharge from sandstone and mudstone. Mudstone not only has a higher degree of compaction than sandstone, but it also has a higher porosity decreasing rate than sandstone. Therefore, the direction of the water discharge is from mudstone to sandstone. Discharging water from pores helps the ion exchange inside pores either in mudstone or in sandstone. In the pores of mudstone, pore water with high salinity moves towards sandstone; both the remaining interlayer water and textural water are fresh water. Therefore, after this stage, the salinity of the pore fluid in mudstone is smaller than that in sandstone. This salinity difference between mudstone and sandstone creates a precondition for fluid migration and light hydrocarbon exsolution in the subsequent process. During this stage, sandstone and mudstone have normal pressure; the fluid pressures, either inside or outside of the lenticular sandstone, are in an equilibrium state. Because primary pores developed during this stage, in sandstone and mudstone, pores, throat passage and micro bedding surface are therefore the migration channels for the fluid (water with a small amount of bio-gas).

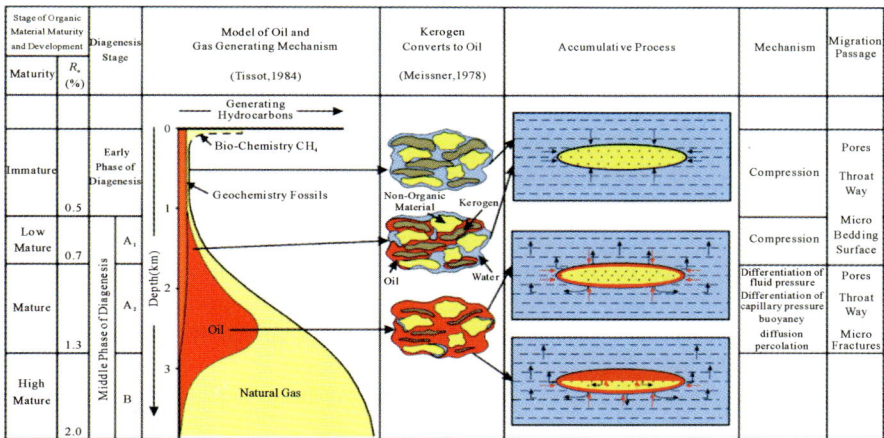

Fig. 6.5. Model for creating a lenticular sandstone reservoir

Characteristics of Lenticular Sandstone Reservoir with Low-Mature Hydrocarbon Source Rock

In this stage, the reflectance of vitrinite in hydrocarbon source rock is $0.5\% < R_0 < 0.7\%$. With an increase in buried depth, the intensity of source rock evolution also increased. A small amount of oil and natural gas started to be generated.

When the buried depth of the source rock reached the threshold depth of oil generation, solid kerogen started to transform into a liquid form and it entered pores, throat passage and micro bedding surface. However, in this stage, only a modest amount of liquid hydrocarbon was generated; mudstone still contained a large amount of pore water. Generated oil could not satisfy the absorption of mudstone to reach the saturation requirement for free phase crucial migration. Therefore, oil that was generated by kerogen still stayed in pores of mudstone; could not migrate towards lenticular sandstone. In general, when the oil saturation achieved critical migration saturation, oil could migrate as a continual oil phase. Via the experiment of a two-phase flow between oil and water in sandstone, Botset (1940) proved that if a continual oil phase migrated with water, it must occupy more than 10% of the total pore capacity. In other words, 10% is the critical migration saturation for oil migration in sandstone. Expanding on Botset's conclusions, Dickey (1975) considered that because many inner surfaces of source rock were oil-wet, then the critical migration saturation could be less than 10%; it might even be near to 1%. Although different researchers have different opinions, however, the critical migration saturation must achieve a certain level in order to initiate free phase oil migration.

For sandstone, this stage is equal to the middle phase of diagenesis (A_1). Along with an increase in the buried depth, both temperature and pressure are continually increased. When kerogen was pyrolyzed in mudstone, organic acid was formed and it entered the pores of sandstone. The organic acid dissolved the sandstone to create secondary pores.

During this stage, the remaining pressure was still a major accumulative dynamic; this remaining pressure came from the compaction process. The remaining dynamic continually discharged water from sandstone and mudstone; the direction of fluid movement is shown in Fig. 6.5. During this process, although mudstone generated a small amount of oil, overpressure could still not be reached due to the low quantity of oil that was transformed from kerogen. Thus, at this time, fluid pressure in both sandstone and mudstone was normal. The mainstream fluid was still water even though there was a very small amount of liquid hydrocarbon. In mudstone, water traveled through capillary pores, throat passage and micro bedding surfaces; however, water mainly passed through secondary pores in the sandstone.

Characteristics of Lenticular Sandstone Reservoir with Mature Hydrocarbon Source Rock

During this stage, the reflectance of vitrinite in hydrocarbon source rock is 0.7% < R_o < 1.3%. That is equal to the middle phase of diagenesis (A_2). Along with the increase in the buried depth, temperature and pressure continually increased; the porosity further decreased. During this stage, secondary pores decreased in the sandstone. Because the intensity of hydrocarbon generation escalated, a huge amount of kerogen was transformed into oil and gas; the increase in the quantity of hydrocarbons was accompanied by a decrease in water capacity (Fig. 6.5). In mudstone, because the saturation of free phase crucial migration was achieved, oil and gas started to migrate and to accumulate towards the lenticular sandstone.

During this stage, when the overpressure achieved the magnitude of a bursting pressure in the mudstone, micro fractures were created in the mudstone to form the major passage for oil and gas migration (Li, 2000). The research work of Snarsky (1962) indicated that if the magnitude of the fluid pressure in a rock or in pores was 1.4 – 2.4 times larger than the magnitude of hydrostatic pressure in surrounding rock, the mechanical strength of the lithology would be overcome by the creation of fractures. Tissot and Pelet (1971) believed that in shale stone, if the inner pressure was 1.23 times larger than the hydrostatic pressure, fractures would be formed inside the shale. Secor (1965) deliberated that if the fluid pressure (P) surpassed the product of minimum compressive stress (S_3) and tensile strength of the rock (K), fractures would be formed. Zhang (1995) considered that if pore fluid pressure in mudstone was larger than the sum of minimum compressive stress and tensile strength of the rock (S_3+K), fractures would be formed, which was $P_{\text{fluid}} > S_1 + C$.

In mudstone, because of hydrocarbon generation and an uneven compressive process, an abnormal overpressure was created; the fluid pressure difference that was caused by abnormal overpressure pushed oil and gas towards the lenticular sandstone. The oil and gas migrating passage included micro fractures, pores and a throat passage in the mudstone. In the transitional zone between mudstone and sandstone, under the influence of the capillary pressure difference, the free phase oil was transferred into the lenticular sandstone; on the other hand, the formation water of sandstone entered the mudstone simultaneously to finish the exchange process between oil and water. Initially, oil and gas only accumulated around the lenticular sandstone. However, when oil and gas accumulation reached a crucial height (Z_c), under the buoyancy force, oil and gas migrated towards the top of the lenticular sandstone. At same time, due to the displacement, the formation water in the sandstone was pushed downwards; eventually, the formation water would flow to the surface. Because hydrocarbon generation and displacement were episodic processes, the above process was also an episodic procedure. With the above process, oil and gas gradually migrated and accumulated in the lenticular sandstone where they finally filled the entire lenticular sandstone.

6.2.3 Primary Control Factors for the Above-Source Play

The accumulative model of the above-source reservoir indicates that oil and gas were generated in lower strata and they accumulated in upper strata. In the above-source play, the transportation system is the precondition for petroleum accumulation; the reservoir body offers storage space, carrier and foundation for oil accumulation; the conditions of seal and preservation are the requirements for creating a reservoir. In a depressed basin, the Jurassic and the Cretaceous oil reservoirs in the central Zhungeer basin are perfect examples, in which the creation and distribution of oil reservoirs were controlled by above-source "three surfaces" (fault plane, unconformity surface and maximum flooding surface, the same as below) and the structural-lithologic condition. In a rifted basin, the oil reservoirs in the Guantao formation and Minghuazhen formation in the Nanbao depression are good examples, in which the creation and distribution of oil reservoirs were controlled by the above-source "three surfaces" and the structural–lithologic condition. In a marine craton basin, the oil reservoirs in the Lunnan and the Donghe sandstone of the Hadexun are good examples, in which the creation and distribution of oil reservoirs were controlled by the above-source "three surfaces" and the structural–lithologic condition. Here, we use the Jurassic and the Cretaceous oil reservoirs as the example for analyzing the major controlling factors for above-source oil accumulation.

● **Transportation System**

The fault–unconformity surface/sand body constructed a transportation system that was the key element and precondition for creating the above-source reservoir. Most terrigenous depressed basins have this characteristic, whether the post rifted depressed basin (the Songliao basin) in the eastern region, or the post craton depression (the Eerduosi basin) in the middle region, or the depression on the folding basement (the Zhungeer basin) in the western region. For example, in the central Zhungeer basin, the discovered oil and gas are controlled by the transportation system of the fault–unconformity surface (Fig. 6.6).

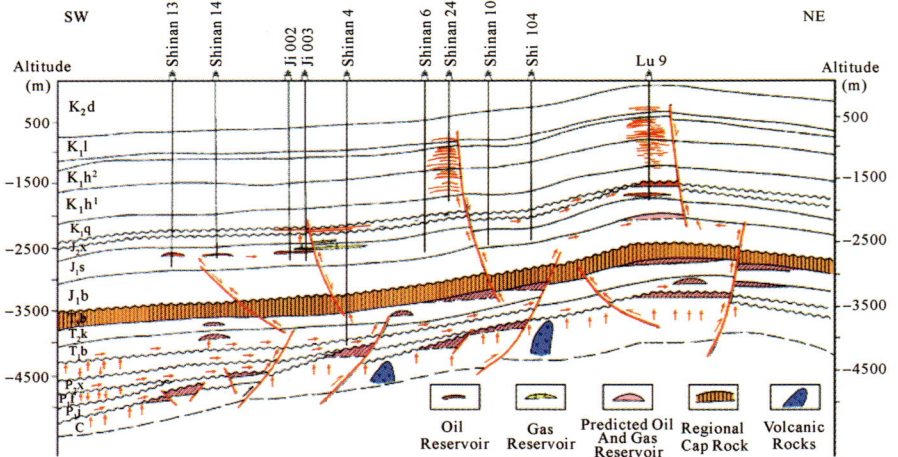

Fig. 6.6.   Petroleum migration model in the Zhungeer basin

● **Reservoir Body and Trap**

The transportation system that was made from the fault and unconformity surface is only one of the factors. The reservoir body and trap are the crucial conditions that determined the creation of oil and gas reservoirs. The beneficial sand body offered effective spaces for oil and gas accumulation. The sand bodies near the unconformity surface were the best ones for creating oil and gas reservoirs because they were located on the side of the major transportation system. Therefore, these sand bodies could capture oil and gas first. At same time, for the above-source accumulation, the existence of the trap and its effectiveness are the key factors that determined oil and gas accumulation and the success of petroleum exploration.

● **Condition of Preservation**

For above-source accumulation, the preservation condition is the necessary element. The discovered oil and gas reservoirs in the central Zhungeer basin and in the Upper Neogene in the Bohai bay basin were controlled by the lacustrine flooding surfaces, because the mudstones that correlated with the lacustrine flooding surface formed the effective regional cap rock.

6.2.4　Oil and Gas Distribution of Different Plays

Based on an understanding of the creative conditions and accumulative dynamics of all kinds of petroleum play, we conducted a comparative analysis of distributions and accumulative patterns of lithostratigraphic reservoirs with different plays.

● **Oil and Gas Distribution of Beneath-Source Play**

For a depressed basin, the main controlling factors for the beneath-source reservoir are the pressure differences between source rock and reservoir sequences, fault and sand body. The beneath-source oil reservoir primarily includes the fault–lithologic oil reservoir or lithology upward-dipping and pinch-out type of oil reservoir; the oil was generated in the upper strata and stored in the lower strata (Fig. 6.7). In the beneath-source petroleum accumulation, the pressure difference between the source rock and reservoir sequence was the critical dynamic for oil and gas migration. The fault and unconformity surface forms the migration channels; the unconformity surface was also the interface for developing a reservoir body; the dense strata that correlated with the maximum lacustrine flooding surface functioned as the regional cap rock. Therefore, the distribution of the beneath-source lithostratigraphic reservoir was controlled by the lacustrine flooding surface (hydrocarbon source rock and cap rock), fault plane and unconformity surface; this is the regularity of the "three surfaces controlled reservoir". Horizontally, the distribution of the oil reservoir was controlled by the developmental level of the unconformity surface and fault plane (the unconformity surface controlled the development of the reservoir body; the fault was the transportation channel for oil and gas downward migration under the effective pressure difference). Vertically, the distribution of the oil reservoir was controlled on two levels; the upper portion was controlled by the maximum lacustrine flooding surface; the lower portion was controlled by the magnitude of the pressure difference and unconformity surface. For instance, in the Songliao basin, lithostratigraphic oil and gas reservoirs were mainly distributed within a pressure difference in a range of between 8 – 12 MPa in the periclinal zone on the Fuxin uplift (Fig. 6.2). In this range of pressure difference, petroleum accumulation and distribution were controlled by the distribution of the reservoir body and the junction area that fault belts intersected. Therefore, the oil and gas exploration for beneath-source play should be based on the regularity of the "three surfaces", or should follow the ideal of a "discovered oil reservoir along the surfaces". Inside the space that was limited by three surfaces, analyze the correlation between the distribution of pressure difference and sand body that was connected to the source rock by a fault, in order to discover the best exploration target.

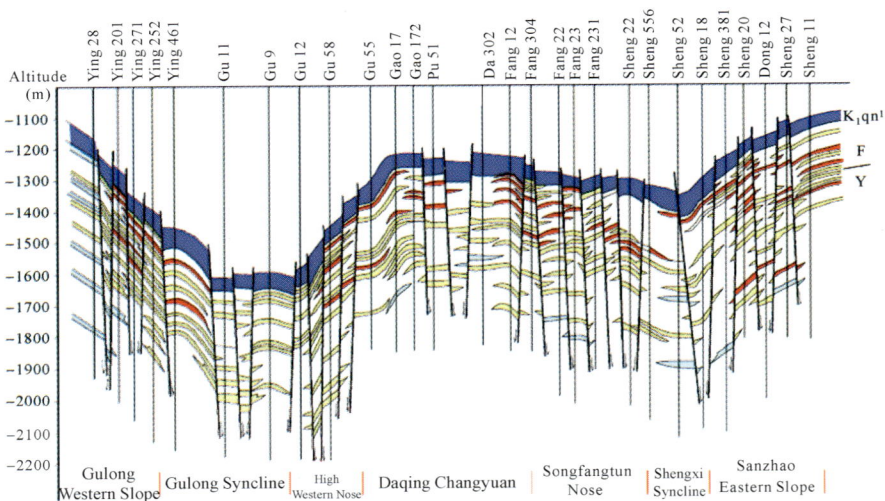

Fig. 6.7.   Cross section of the Fuyu and the Yangdachengzi oil beds from the Ying 28 well–Pu 51 well–Shang 11 well in the northern Songliao basin

For the rifted basin, the beneath-source oil and gas reservoirs were controlled by the distribution of buried hills, source rock and migration channel, and the cap rock condition. Generally, oil and gas were distributed in the buried hill region with perfect accumulative and preservative conditions. In the Damintun in the Liaohe oil field, a beneath-source buried hill with source rock of the flooding period, unconformity surface and fault plane (the "three surfaces") controlled the distribution of the buried hill type of oil reservoir. For example, in the Damintun depression, the source rocks of the flooding period in the Sha-4 member made contact with the Proterozoic strata and the Archean strata with an unconformity surface; the fault plane and buried hill were in direct contact with a plentiful source supply; the conditions of oil generation, storage, seal and trap collaborated very well to form a petroleum rich area.

- **Oil and Gas Distribution of In-Source Play**

In a terrigenous depressed basin, the in-source petroleum accumulation had a plentiful hydrocarbon source supply. The key factors for oil and gas accumulation are the perfect cooperative setting between source rock and reservoir body and the accumulative background; the delta front facies with structural background was the best area for oil and gas to accumulate. In the center of the hydrocarbon generating depression, the sand body did not develop well; far away from the hydrocarbon generation center, the formation of river facies and delta facies did not have a good transportation system. On the other hand, the best cooperative setting between source rock and reservoir body was developed on the delta front with a high probability of oil and gas accumulation. In general, here is the oil and gas rich area. Therefore, in the depressed basin, the delta front facies and lacustrine cost facies controlled the horizontal distribution of oil and gas (such as in the Songliao basin, the Lower Cretaceous Putaohua oil bed and, in the Eerduosi basin, the oil reservoir in the Triassic Yanchang formation). In the Songliao basin, the lithostratigraphic petroleum reservoirs were closely distributed on the slope zone or in the nose-like structural area with a thickness ratio of sandstone and strata of 20% – 40% in the delta front zone. Alternatively, the best cooperative setting was that of hydrocarbon source rock of the coal measure from marine–terrigenous facies and plain facies combined with superimposed river channel sand bodies, which is a beneficial area for natural gas accumulation, such as in the Eerduosi basin, the Permian gas field (Fig. 6.8) and, in the Sichuan basin, the Upper Triassic gas reservoir.

Using joint reseasch work or outcrops and in wells, the model of depositional sequences was set up. This research work discovered that the fault system and third-order sequence controlled the development of sand bodies and its characteristics in the depositional system. In the depressed basin, in-source oil and gas accumulation happened on top of the third-order unconformity surface, such as in the oil field in the Yanchang formation in the Ansai area in the Eerduosi basin. Oil and gas accumulated on top of the third-order unconformity surface; the interface of third-order sequences and overlying sand body offered an important migration channel for petroleum fluids in the basin. The characteristic of the oil and gas reservoir is that petroleum migrated along the interface of third-order sequences to accumulate in a petroleum reservoir. In the Ansai area and the Xifeng area, the oil fields display this characteristic. The natural gas accumulation also confirmed the same characteristic, such as in the Sulige gas field, the third-order unconformity surface in the Shanxi-He 8 controlled migration, accumulation and storage of natural gas.

In a depressed basin, the in-source oil reservoir includes the in situ type of lenticular oil reservoir, fault–lithologic oil reservoir and upward dipping and pinched-out oil reservoir. The structural conditions (steep slope, gentle slope, nose-like slope and fault) depositional facies (delta front zone) and storage condition not only controlled the cooperative setting of source rock and reservoir body, but they also controlled the creation and distribution of the oil reservoir. Depositional facies in the reservoir mosthy indicated that the creation of the oil reservoir was controlled by the underwater distributary channel facies and estuary dam facies on the delta front zone. The diagenesis facies controlled reservoir mainly designated some physical property traps that were developed in the area with either original pores or secondary pores and that were controlled by depth. The structural background indicated that oil and gas migration and accumulation were controlled by the nose-like bulge, slope, lower altitude uplifting and fault. Therefore, the oil and gas exploration for beneath-source play should follow the concepts of the "three facies controlled reservoir" and "discovering oil along the slope". In order to gradually approach and determine the target, we should search for the area where "three facies" collaborated well on the slope; we should also select the steep slope, gentle slope and nose-like slope inside the lacustrine coast line; in addition, we should explore the block that contained the best depositional facies and reservoir facies.

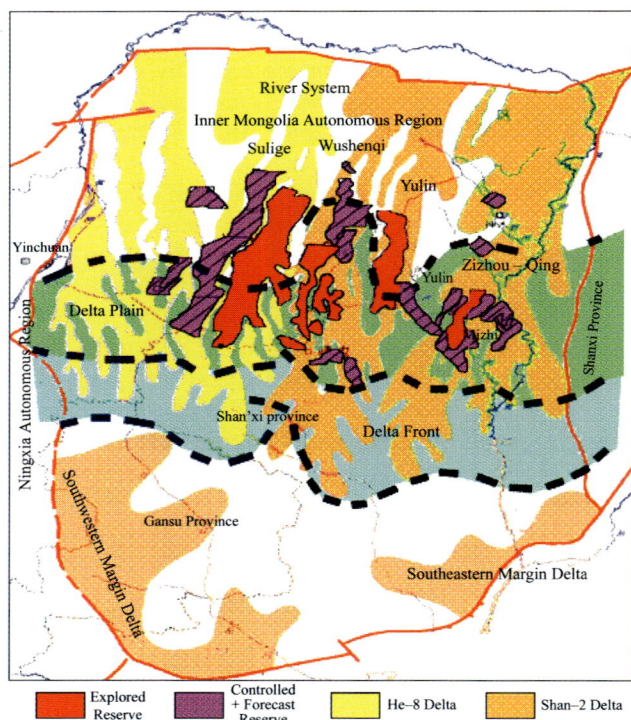

Fig. 6.8. Overlaying map of the Permian depositional system and the gas fields in the Eerduosi basin

In the rifted basin, oil and gas exploration experienced three phases, from exploring the large scale uplift → exploring the composite oil and gas accumulation zones → exploring the fully, filled, petroleum rich depression. At present, oil and gas explorations mainly target the petroleum rich depression. Comparative research and exploration practices proved that both the steep slope without faulted step or with a single faulted step and the gentle slope had poor petroleum accumulative conditions. Conversely, the framework of the faulted step obviously controlled oil and gas accumulation. For example, on the western slope in the Liaohe rifted basin, the stratigraphic oil reservoirs were concentrated on the upper portion of the faulted steps zone of an in-source circled depression; the structural–lithologic oil reservoirs occurred on the middle portion of the faulted step zone; the lithologic oil reservoirs appeared on the lower portion of the faulted step zone (Fig. 6.9). In addition, the deeply faulted zone in the circled depression was a perfect accumulative area for a volcanic gas reservoir (such as the deeply buried sequences in Xujiaweizi, in the Songliao rifted basin); the distribution of the volcanic gas reservoir was controlled by the deep fault, source rock and cap rock condition.

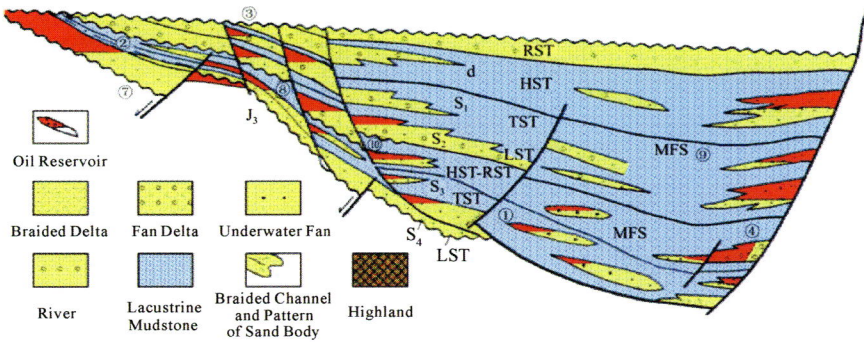

Fig. 6.9. Types of multi-faulted steps controlled oil reservoirs in the Shuguang area, on the western slope in the Liaohe oil field

Overall, in the rifted basin, the petroleum rich depression has the characteristics of a "fully filled depression containing oil", such as in the Bohai bay basin, Nanbao depression, Damintun depression, Qikou depression and Raoyang depression which demonstrate the characteristics of a "fully filled depression containing oil". Alternatively, the depression that was not rich in oil and gas has the characteristics of "major trough controlled oil", such as in the Erlian basin, where the distribution of the oil and gas reservoir was surrounded by a "major trough".

● **Oil and Gas Distribution of Above-Source Play**

The above-source oil reservoir contains a faulted–lithologic oil reservoir. The oil was generated in the lower strata and was stored in the upper strata. The fault connected with the source rock (transportation system) and the benefited trap are the major controlling factors. The depressed basin and the rifted basin have similar accumulative conditions and comparable major controlling factors. The distribution of the oil reservoir was controlled by the flooding surface, fault plain and unconformity surface. For instance, on top of the folded basement in the Zhungeer basin, the "three surfaces" of the depression controlled the distribution of the structural–lithologic oil reservoir (Fig. 6.10). In the rifted basin, such as in the Nanbao depression, the distribution of the Neogene oil reservoir demonstrated that the "three surfaces" controlled oil accumulation.

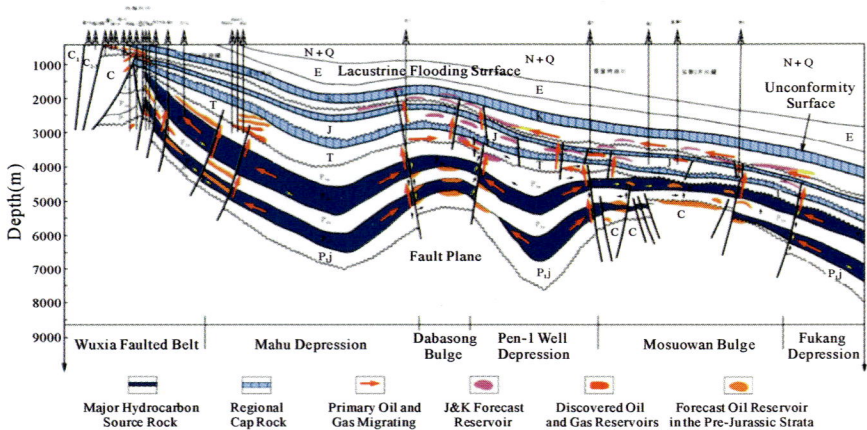

Fig. 6.10.   The "Three surfaces" controlled the above-source petroleum reservoir in the Zhungeer basin

Similar to the beneath-source oil reservoir, the distribution of the above-source reservoir also has the regularity of a "three-surface controlled reservoir". In other words, the above-source oil reservoir was controlled by the fault plain, unconformity surface and lacustrine flooding surface. The existing fault, especially the vertical existing fault, was the precondition for the above source accumulation. Therefore, the horizontal distribution of the oil reservoir was controlled by the faults that connected with the source rock and the unconformity surface. Most oil reservoirs occurred around the intersection of the fault and unconformity surface. The oil and gas accumulations were controlled by the fault plain and the distribution of sand bodies that linked with the fault. For example, in the Haituozi area, where the reservoir bodies were well developed, the lithostratigraphic traps were concentrated in the same area that contained well developed, source rock connecting faults. The vertical distribution of the above-source oil reservoir was controlled by two kinds of "surfaces". The upper lacustrine flooding surface functioned as the regional cap rock; the lower lacustrine flooding surface was the hydrocarbon source rock. These two flooding surfaces limited the vertical extension of the oil reservoir.

Thus, the exploration of the above-source lithostratigraphic reservoir should follow the regularity of "three surfaces controlling oil" in order to discover the oil reservoir along the "surface". Within the three dimensional space that was limited by three surfaces, we should analyse the position of the source rock connecting fault and its relationship with the overlying sand body to further identify a petroleum accumulative area.

References

Chen, D.X., Pang, X.Q., Qiu, N.S., et al., 2004. Accumulation and filling mechanism of lenticular sand body reservoirs. Earth Science (Journal of China University of Geosciences), 29(4):483-488.

Chen, Z.M., Zhang, Y.F., Han, Y.X., et al., 1998. A simulation and mechanism of oil accumulation in lenticular sand body. Petroleum Geology & Experiment, 20(2):166-170.

Chi, Y.L., Xiao, D.M., Yin, J.Y., 2000. The injection pattern of oil and gas migration and accumulation in the Sanzhao area of Songliao basin. Acta Geologica Sinica, 74(4):371-377.

Fu, G., Zhang, Y.F., Du, C.G., 2002. Mechanism and controlling factors of lithologic oil reservoir in the northern Songliao basin. Petroleum Exploration & Development, 29(5):22-24.

Hao, F., Dong, W.L., 2001. The development of ultra pressure system, the mechanism of fluid flow and petroleum accumulation in sedimentary basins. Advances in Earth Science, 16(1):79-85.

Hao, F., Zhou, G.Y., et al., 2002. Evolution of ultra pressure systems in sedimentary basins and preconditions for deep buried oil/gas accumulation. Earth Science (Journal of China University of Geosciences), 27(5):610-615.

Li, P.L., Pang, X.Q., Chen, D.X., et al., 2004. The mechanism and model of lenticular sandstone reservoir in the Qiyang Jiyang depression. Science in China, Series D: Earth Sciences, 34(A01):143-151.

Li, M.C., 2000. An overview of hydrocarbon migration research. Petroleum Exploration & Development, 27(4):3-10.

Lin, J.Y., Men, G.T., Huang, W., 2004. The mechanism and models of lithologic reservoir in lenticular sand body. Petroleum Geology & Oilfield Development in Daqing, 23(2):5-7.

Sun, W.G., Yang, C.S., 2005. New understanding of petroleum accumulation mechanism in lenticular sandstone. Petroleum Geophysics, 3(2):57-62.

Zhang, Y.F., 2001. Simulations of in situ lithologic reservoirs and analysis of its accumulation mechanism. Research and Exploration in Laboratory, 20(2):103-106.

7

Characteristics of Petroleum Accumulation and Distribution in the Prototype Basins

When comparing four prototype basins, they not only have different tectonic and depositional backgrounds, but they also have different petroleum accumulative (elements) conditions. Therefore, these prototype basins contain noticeable distinctions in their correlations of accumulative elements and in their accumulative processes; in addition, they also display their own accumulative features and patterns. In the rifted basin, because the petroleum rich depression contains high quality hydrocarbon source rock of large thickness and broad distribution, it thus has the characteristics of a "petroliferous rich depression". In the depressed basin, the lithostratigraphic traps were not only favorably developed on the delta front, but they also perfectly correlated with the hydrocarbon source rock that was located in the central area of the lake, which displays the characteristics of accumulating oil on the delta front. In the foreland basin, the development of a lithostratigraphic reservoir in the foreland thrust belt was controlled by faults and fans; this characteristic is represented by the northwestern margin in the Zhungeer basin. In the marine craton basin, oil and gas accumulations were controlled by the reef flat on the platform margin and other high energy facies. Understanding these characteristics and patterns will benefit oil and gas exploration in the future.

7.1 Terrigenous Rifted Basin with the Characteristics of a "Petroliferous Rich Depression"

In eastern China, there are many Mesozoic and Cenozoic rifted basins that include the Bohai bay basin, the Southern North China basin, the Erlian basin, the Hailaer basin and the Yilan–Yitong basin. Because of the extensional and rifted movements, some relatively isolated depressions were formed, which contained diversified measures for natural resources and different outcomes for petroleum exploration (Gao and Zhao, 2001; Yuan and Qiao, 2002). Some depressions contain plentiful oil and gas and they yield good exploration results. Other depressions have the preconditions for oil generation. However, only a few, small size oil and gas fields were discovered. In addition, some depressions lack the preconditions for oil generation; therefore, they are worthless for petroleum exploration. Exploration practices indicated that petroleum rich depressions are primary targets for upcoming oil and gas exploration and for increasing the petroleum reserve. Inside a petroleum rich depression, high quality hydrocarbon source rock provided sufficient oil and gas supplies. Vertically, oil and gas could be accumulated in various stratigraphic sequences and in all kinds of reservoir bodies. Horizontally, multiple sequences and different types of trap overlapped with each other to form a connected unity. Together, they demonstrate the characteristics of a "petroliferous rich depression". Here we highlight the geological setting and petroleum accumulative pattern for the characteristics of a "petroliferous rich depression" in a terrigenous rifted basin.

7.1.1   *Geological Settings for a "Petroliferous Rich Depression"*

Compared with a low abundance petroleum depression, a petroleum rich depression has perfect hydrocarbon generating conditions, excellent reservoir settings, broadly distributed traps, a highly efficient migrating passage and adequate space–time configuration.

(1) In a petroleum rich depression, a perfect hydrocarbon generating condition is the first requirement in order to achieve a "petroliferous rich depression". According to the statistical analysis of the Bohai bay basin, petroleum rich depressions have ideal conditions for hydrocarbon generation. These conditions are listed as follows: 1) The types of kerogen were generally good, types $I - II_1$ kerogen. 2) The source rock contained a high abundance of organic substances; commonly, the content of organic carbon was higher than 1%; the content of organic carbon could reach 2% – 4% in a major section of the depression. 3) The intensity of hydrocarbon generation was high; on average, the intensity of oil generation was larger than 50×10^4 t/km^2. 4) The superb hydrocarbon source rocks were broadly distributed; in general, they occupied 50% – 80% of the depressed region (Table 7.1). 5) The resource abundance was high; the average abundance of the resource was larger than $15 \times 10^4 - 20 \times 10^4$ t/km^2; in major sections of the depression, the resource abundance could be more than 40×10^4 t/km^2 (Zhao et al., 2004).

Table 7.1 Statistical table of areas between hydrocarbon source rock and depressed region in major petroleum rich depressions

Name of petroleum rich depression	Stratigraphic unit	Effective area of source rock (km^2)	Area of depression (km^2)	Ratio of area of effective source rock/depressed region (%)
Dongying depression	E	3970	5700	69
Zhanhua depression	E	1450	2800	55
Western depression of Liaohe	Es_3	1690	2560	66
Qikou depression	Es_3	2380	3835	62
Nanpi depression	E_2k	1165	1455	80
Banqiao depression	Es_3	749	971	77

(2) Widely distributed valuable sandstone reservoirs provided spaces for oil and gas accumulations. Inside the terrigenous rifted basin, the single sand body usually was of small size. However, because horizontally terrestrial clastic substances were brought into the basin by multiple water systems and because vertically lacustrine transgression and regression happened with high frequency, the sand bodies that were formed during multiple periods with multiple water/systems were superimposed on each other. In most depressions, sand bodies developed from the edge to the center of the depression. Comparing sandstones in various zones, we noted distinctions in the sandstone depending on the stratigraphic unit in which the sandstone was positioned, the types of sandstone, the thickness of the sandstone and the physical properties of the sandstone reservoir (Wu and Xue, 1992). Among all kinds of sand bodies, the sand body with excellent physical properties was usually deposited in the region with stable water. The sand bodies from the delta, fan delta, and turbidite fan have better physical properties than the sand bodies from alluvial fan. In a petroleum rich depression, the stable lacustrine not only contained a relatively large region of stable water, but it also occupied a large area within the depression, which created better hydrocarbon generating conditions and which also produced a large size, widely distributed sandstone reservoir with excellent physical properties. In the terrigenous rifted basin, the single depositional system was usually of small scale; however, because this kind of basin has a higher subsiding rate, the favorable sandstone could accumulate with a thickness of more than a thousand meters to form a large size oil field with a hundred million tons of reserves. In the western depression of Liaohe, on the western slope, the Shahejie formation contains several depositional systems from fan delta–turbidite fan in the Qijia and Xibaqian areas. The single depositional system could cover an area of $200 - 400$ km^2. These sand bodies were not only connected horizontally, but they also were in direct contact with source rock to benefit oil and gas accumulation. In this area, the discovered oil fields (with a hundred million tons of reserves) include the Gaosheng, Huanxiling and Shuguang. Up to the end of 2004, the total explored oil reserves were 11×10^8 t.

(3) The contact between source rock and reservoir body and well developed fault systems provided a highly efficient migrating passage for creating a

"petroliferous rich depression". The highly efficient network of the migrating channel, especially the large contact area between source rock and reservoir sequences and well developed faults, improved the efficiency of oil and gas migrations. Vertically, it might form multiple sets of reservoir sequences; horizontally, the overlapped reservoir sequences displayed the characteristics of a "petroliferous rich depression" (Hu et al., 1991). Usually, a petroleum rich depression experienced multiple cycles of tectonic and depositional evolutions. The alternative developments of deep water deposits and shallow water deposits created a large contact area between reservoir sandstone and source rock. In the Bohai bay basin, this contact area in a petroleum rich depression could reach 80%, even 90% of total area (Table 7.2). The large contact area between sand body and source rock produced a higher initial hydrocarbon releasing rate or higher migration constant in a petroleum rich depression; for example, in the western depression of Liaohe, the total amount of generated hydrocarbon is 235.49×10^8 t; the petroleum resource is 26.6×10^8 t; the migration constant achieved 10.06%.

Table 7.2 Statistical table of contact area between sand bodies of major water system and source rocks in a petroleum rich depression, in the Liaohe rifted basin

Name of aquosystem and stratigraphic unit		Area of sand body (km^2)	Contact area (km^2)	Ratio of contact area between source rock and sand body (%)
Qijia	Es_{1-2}	316	306	96.8
	Es_3	345	280	81.2
Xibaqian	Es_{1-2}	405	357	88.1
	Es_3	350	300	85.7
Xingleng	Es_{1-2}	303	293	96.7
	Es_3	225	215	95.6

(4) Several types of trap were developed in different stratigraphic units and in different locations; these traps provided accumulative places for a "petroliferous rich depression". Because the terrigenous rifted basin contained several depositional hiatuses and rapidly changing facies, beside structural traps (such as anticline and faulted block), many types of lithostratigraphic trap were favorably developed and widely distributed. In the terrigenous basin, because tectonic movement occurred with high frequency, in general, several unconformity surfaces and depositional hiatus surfaces were developed, which promoted the creation of a stratigraphic overlapping trap and unconformity sealed trap. The single sand body of terrestrial facies commonly was of small size and rapidly changing facies. Because lithostratigraphic traps might be created by all kinds of depositional systems, this kind of depositional environment benefited by creating numerous lithologic traps. In the Bohai bay basin, petroleum exploration focused on searching the structural trap in its early phase. In some petroleum rich depressions, several highly abundant structural reservoirs were discovered, which included buried hill–draped anticline, rolling anticline and nose-like bulge; additionally, the stratigraphic overlapping reservoir and unconformity sealed reservoir were also discovered on the slope zone in the depressions. In recent

years, numerous lithologic trapped reservoirs were discovered in the Shahejie formation that was located either on the slope zone or inside depressions.

Because the petroleum rich depression contained plentiful oil and gas resources that correlated with a highly efficient migration system, not only were lithostratigraphic traps and structural traps developed at different positions within the depression, but oil and gas might also accumulate in these positions to construct a "petroliferous rich depression". In petroleum rich depressions, because bountiful oil and gas resources are associated with faulted systems, therefore, apart from accumulating in the strata near the source rock, a large amount of oil and gas also migrated into a non-source sequence to form a petroleum reservoir. In the Bohai bay basin, petroleum rich depressions contained multiple petroleum sequences with large total thickness. Both the Raoyang depression and the Damintun depression have oil and gas in multiple reservoir sequences in the Proterozoic erathem–Tertiary system. Apart from oil and gas that accumulated in a major hydrocarbon source sequence within the Shahejie formation, some large scale buried hill type oil fields were formed in the pre-Tertiary basement. Because of strong faulted activities during the late tectonic periods, several oil fields with hundreds of million tons of reserves developed in the Neogene non-hydrocarbon generating sequences in the petroleum rich depressions of Bozhong, Zhanhua and Qikou. In some depressions that have a low abundance of petroleum reserves, such as the Beitang depression, Chezhen depression, and in the Erlian basin, the Bayindulan depression and the Wuliyasitai depression, the oil and gas beds were relatively simple; oil and gas reservoirs were distributed in the hydrocarbon generating sequences of the Shahejie formation. This is a clear distinction when compared with a petroleum rich depression.

7.1.2 Oil and Gas Accumulative Pattern in a Petroleum Rich Depression

In a terrigenous rifted basin, petroleum rich depressions have the characteristics of a "petroliferous rich depression". However, oil does not occur everywhere in these depressions. Oil and gas accumulations were not only controlled by major hydrocarbon source rocks, beneficial sand bodies and effective traps, but they were also controlled by the ideal correlations of these three geological factors. The Bohai bay is a good example.

● **Vertical Distributions of Oil and Gas were Controlled by the Palaeogene Hydrocarbon Generating Sequences**

In the Bohai bay basin, petroleum rich depressions contain several types of source rocks that include carbonate rocks of marine facies, coal measures of marine–

terrigenous facies, and source rock of lacustrine facies. Beside major source rocks of the Palaeogene, Mesozoic and Paleozoic source rocks were also found in some depressions. According to the statistical results of basin modeling for sequence analysis in a petroleum rich depression, the Palaeogene source rocks were the primary contributors to the oil and gas resources in petroleum rich depressions.

In the Bohai bay basin, the rifted valleys developed during the Palaeogene period. The lacustrine strata, from the Kongdian formation to the Dongying formation, were deposited during this period, which contained the typical characteristics of well developed hydrocarbon source rocks, multiple types of reservoir sand bodies and a perfect geological setting of source rock and reservoir sequences. Therefore, every major depression has 2 – 4 sets of petroleum plays, which were superimposed vertically. In the Tertiary system, the Sha-1 member and the Sha-2 member are the major resource sequences. The hydrocarbon generating ability of the Dongying formation was relatively weak, which was mainly distributed in a flat area and shallow sea in the Liaohe depression and the Huanghua depression. Because the buried depth was shallow, the hydrocarbons in the Dongying formation are currently still at an immature–low maturity stage. In the Bohai bay basin, because the sediments were deposited towards the ocean, the Dongying formation had the highest oil generating sequences in the Liaohe depression and the Huanghua depression. On the other hand, the Sha-1 member and the Sha-2 member were the highest oil generating sequences in the Jiyang depression, the Jizhong depression and the Linqing depression. In most depressions, the Sha-3 member contained the major oil generating sequences; in a few depressions, the Sha-4 member and the Kongdian formation were the major oil generating sequences. Therefore, in every depression, further exploration of the major oil generating sequences is definitely necessary.

In the Bohai bay basin, because the rifted faults developed before the depressions, every depression contained vertically an upper, middle and lower structural system. The middle structural system contained deposits of the Neogene rifted basin, which not only controlled distribution of major sequences of oil plays in the petroleum rich depression, but also controlled most of the oil and gas resources in these areas. In every depression, because the geological setting was different, and because it was influenced by the sediments deposited towards the ocean, the reservoir sequences were different. Relatively speaking, the oil reservoir sequences were positioned in the upper strata in the flat area and shallow sea; the Neogene system contained most of the explored oil reserves, such as in the exploration areas of the Shengli and the Dagang.

- **Multiple Sequences, Various Reservoir Bodies and Several Types of Traps Created a Distribution Frame where Oil and Gas were Superimposed Vertically and United Horizontally**

Because it contains three vertical structural systems (Hu et al., 1986), the petroleum rich depression has the accumulation conditions of multiple sequences with several types of reservoir. The lower structural system contained the pre-Tertiary strata that included the Mesozoic, Paleozoic and Mesoproterozoic–Neoproterozoic systems. Several types of reservoir developed in these pre-Tertiary strata, which include metamorphic, sandy conglomerate, carbonate rock and volcanic reservoirs. After it went through multiple tectonic movements and erosion events, the pre-Tertiary system commonly offered good storage space. The middle structural system contained the Palaeogene rifted lacustrine deposits, which included the sand bodies of river–alluvial fan–(fan) delta–underwater fan–turbidite facies, which also had bio-clastic limestone of lacustrine facies. At present, because the evolution of organic substances are at a mature stage and in the middle phase of diagenesis, the Palaeogene sandstone reservoir thus contains well developed secondary pores with better storage ability. The upper structural system contains the Neogene deposits of river alluvial facies and sand body sequences of river facies are well developed. In addition, the buried depth of Neogene system was shallow; thus, the evolution of reservoir sequences was in an early phase of diagenesis; the original pores were well developed and the storage ability was excellent. The reservoir sequences that are discussed above not only were superimposed vertically, but they also overlapped horizontally with the alterations in facies, which created a foundation of multi-sequences with various types of reservoir in the petroleum rich depression.

The developmental history of the petroleum rich depression was that a rifted event happened prior to the depressed event; thus, the asymmetrical (dustpan like) rifted framework was formed. This kind of framework not only developed different types of trap and positioned them in lower, middle and upper structural systems, but it also helped that oil and gas (that were generated in the Palaeogene system) migrated into these traps to further form a connected network between different stratigraphic sequences, vertically superimposed and horizontally overlapped. The lower structural system was the basement; it contained the buried hill type, the faulted steps type (or lower altitude buried hill type) and unconformity type of petroleum reservoirs. The middle structural system (the Palaeogene system) mostly contained rolling anticline type, faulted block type, lithologic type, stratigraphic overlapping type, or composite type of oil and gas reservoirs. The upper structural system (the Neogene system) contained draped anticline type or structural–lithologic type of reservoirs that were developed on top of the lower and middle structural system, such as on top of a bulge, buried hill, or faulted anticline zone. In the petroleum rich depression, the oil and gas accumulations were controlled by the overall structural framework to form multiple stratigraphic sequences and different types of reservoirs that were arranged in a particular pattern vertically and horizontally (Li et al., 2004). The combinations included faulted steps on a steep slope+sandy conglomerate, central uplift+draped buried hill+lithology; and faulted steps on a gentle slope+stratigraphic overlapping, pinched out; the Dongying depression is a good example (Fig. 7.1).

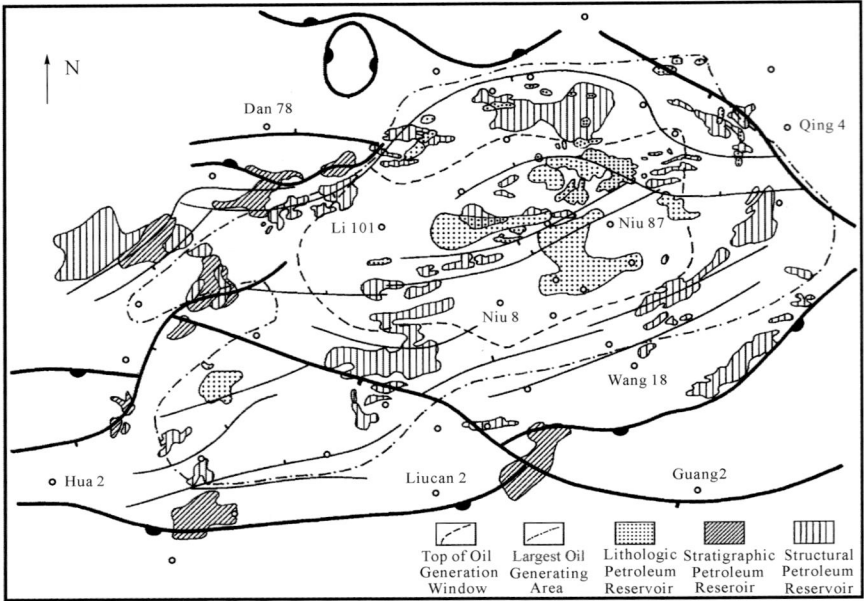

Fig. 7.1.   Types of petroleum reservoir and their distribution in the Dongying depression, in the Bohai bay basin

- **Several Types of Trap Surrounded Major Hydrocarbon Generating Sag to Form a "Petroliferous Rich Sag" and Large Scale Oil and Gas Fields**

Inside a depression, sags were the secondary downward structures, the secondary depositional centers and the subsidence centers. A single depression could contain one or several sags, such as the Dongying depression which contains five sags that are Lijin, Niuzhuang, Boxing, Yongbei and Laizhouwan. The underwater low altitude bulges or buried hills divided these sags. Most of the sags developed during the Palaeogene with mudstone and shale deposits; these hydrocarbon source rocks contained plenty of organic materials and were of great thickness. Therefore, apart from a few individual sags, the rest of the sags combined with surrounding traps of mixed types to form oil and gas generation, migration, accumulation and a balanced system. Every sag was an oil and gas generating center. Because oil and gas that were generated in the sags accumulated in the nearest traps, the distribution pattern of the reservoir was a circle or a half circle that centered on the oil generation sag. For example, in the Dongying depression, there are three oil and gas circles that surrounded the Lijin sag, the Niuzhuang sag and the Boxing sag, respectively. According to the developmental characteristics of a sag (such as the buried history of source rocks), these sags can be divided into three types, which are the inherited type, hiatus type and subsided prior to uplifting type. Among three types of source rocks, the inherited type of sag has a large area, great subsidence depth and continuity. Additionally, in this type of sag, the thickness of source rocks is large and the maturity of organic materials is higher. Thus, the inherited type of sag is the major petroleum source area, such as the Lijin sag in the Dongying depression, the Gubei sag in the Zhanhua depression, the Linnan sag in the Huimin depression and the Panshan sag in the Liaoxi depression.

In the terrigenous basin, because the aquadynamics were weak and faulted activity was strong, the lithology of lacustrine facies was altered frequently with several faults. Therefore, oil and gas that were generated in the sags only migrated a short distance into surrounding traps to simply form a large size oil field (Li et al., 2003). For example, around the Lijin sag, because of the long duration of continual development, massive source rocks and correlated reservoir sequences were deposited; the source rock in the Sha-3 member has a great thickness (1,300 m) with a good oil generation index. As a bountiful petroleum source center, the Sha-3 member could generate a large amount of oil and gas. At same time as the Palaeogene depositional center, the surrounding aquatic systems flowed into the Lijin sag and they brought a huge amount of terrestrial clastic materials. Thus, in different times, at different places, reservoir sand bodies of delta facies, nearshore underwater fan facies, turbidite fan facies and beach bar facies were formed. Furthermore, the bio-debris limestone of the Sha-4 member and the lower Paleozoic limestone reservoir sequences were developed in the Pingfangwang area, the volcanic reservoir sequences were developed in the Binnan area, the sand bodies of river facies were widely distributed in the Neogene system. Overall, the Lijin sag not only has sufficient oil resources with multiple reservoir sequences, but it also has the preconditions for accumulating large scale, multi-type petroleum reservoirs. During the long period of continual development, controlled by basement uplifting in surrounding areas and controlled by a large scale, basin-dipping, growth fault zone, many large scale structural belts developed in the Lijin sag. For example, on the northern side, under the control of the basin-dipping faults, the associated structural belts developed with the Shengtuo–Ninghai fault and the Lijin–Binnan fault. Also, on the western side, the Pingfangwang–Shangdian buried hill was associated with a drape structure belt; on the southern side, there is the Caoqiao–Chunhua paleo nose-like structural belt. The Dongying central uplift belt (that was formed by plastic extensional force) extended, from the east to the west, into the eastern side of the Lijin sag. In addition, large scale fault planes of the Shanjiasi and the Liangjialou and widely distributed sand bodies occurred in the Lijin sag. These different types of structural belts or sand bodies created different types of traps either around the sag or in the sag to offer a favorable place for oil and gas accumulations. As a result of oil and gas accumulations, two large scale oil fields, the Shengtuo and the Dongxin, were formed. Along with the in-depth exploration of concealed sand bodies, we believe that more and more lithologic petroleum reservoirs will be discovered in the structurally less-developed area in the sag. At that time, the Lijin sag will become a petroliferous rich sag.

7.2   Terrigenous Depressed Basin with the Characteristics of "Delta Front Controls Oil"

Different types of depressed basins have similar depositional models, widely distributed lacustrine mudstone and sandstone reservoirs of river–delta facies. However, because of the variations with other factors, these basins contain the distinctive characteristics of petroleum geology; these variations include the framework of the basin, environment, tectonic evolution of the late period, overlay strata and the thermal history of the basin. Therefore, in different basins, accumulative characteristics and petroleum distributions are clearly diversified. In particular, the conditions of hydrocarbon generation have noticeable variations. The comprehensive analysis of the petroleum accumulative condition in the structural layer during the depressed period in the Songliao basin, the Eerduosi basin and the Zhungeer basin indicated that, for the depressed basin that had the potential of generating sizeable hydrocarbons, the central depression was the petroliferous rich depression. Furthermore, the delta front contained vertically superimposed and horizontally united oil and gas reservoirs, such as in the Songliao basin and in the Eerduosi basin. For the depressed basins that lacked hydrocarbon generating potential, the oil and gas distributions were intricate; the oil and gas accumulated at the intersection of the paleo bulge and faulted belt, such as in the Zhungeer basin and in the Bohai bay basin.

7.2.1   In-Source Play Carried Out in Large Accumulative Area on Delta Front and Petroliferous Rich Region

The large size sand bodies of delta facies developed in a terrigenous depressed basin. On the delta front, sand bodies and lacustrine source rocks were either superimposed on each other, or they were connected horizontally within a broad region. The lithologic traps were formed in underwater distributary channels. The lower oil and gas columns and medium–low pressure system benefited the creation of sizeable lithostratigraphic petroleum reservoirs (Fig. 7.2).

- **In a Terrigenous Depressed Basin, During the Lacustrine Transgressive Period, Gentle Paleogeography Benefited the Development of Widely Distributed, High Quality Hydrocarbon Source Rocks**

A terrigenous depressed basin offered beneficial accumulative conditions over a broad area. First of all, during lacustrine transgression, source rocks were widely deposited in the terrigenous depressed basin within an extensive region. For example, the middle–shallow strata in the Songliao basin, the Mesozoic system in the Eerduosi basin and the Xujiahe formation in the Sichuan basin contain hydrocarbon source rocks of 10×10^4 km^2, 8×10^4 km^2, and 4.5×10^4 km^2 respectively; these source rocks offered material foundations for generating and accumulating hydrocarbons in a broad region. Secondly, under the gentle paleogeographic background, on the delta, source rocks and reservoir sequences broadly made contact to form a favorable petroleum system, which established the beneficial foundation for "sizeable petroleum accumulation on the delta front". Widespread, highly abundant source rocks offered hydrocarbon materials for large size petroleum accumulation. The central depression, that had great hydrocarbon generation potential, was the crucial element for the "petroliferous rich depression".

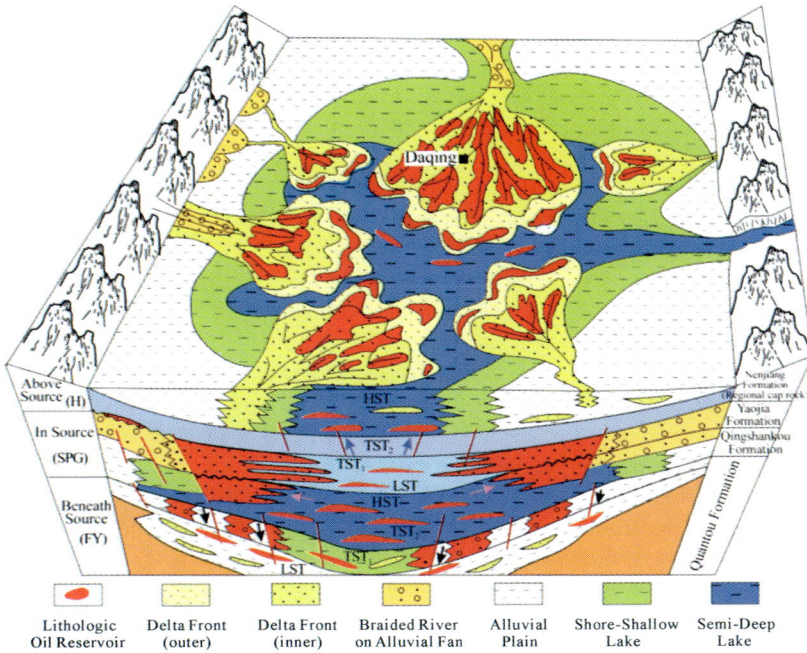

| Lithologic Oil Reservoir | Delta Front (outer) | Delta Front (inner) | Braided River on Alluvial Fan | Alluvial Plain | Shore-Shallow Lake | Semi-Deep Lake |

Fig. 7.2. Petroleum accumulation model on delta front in terrigenous depressed basin (using the Songliao basin as an example)

The large size depressed lacustrine basin had a relatively stable structure that was represented by the same lacustrine deposits. During the extension period, because the lacustrine transgressional region was broad, the center of a basin usually contained excellent hydrocarbon source rock. In the Songliao basin and the Eerduosi basin, the depressed structural system contained $1 - 2$ sets of major hydrocarbon source rocks that had great hydrocarbon generating potential; these source rocks provided huge amounts of hydrocarbons to ensure oil accumulation in all kinds of sand bodies. This was the foundation for the "petroliferous rich depression" in the central depression; the oil and gas distributions were clearly controlled by the central depressions in the Eerduosi basin and the Songliao basin (Hu, 1982).

In the Eerduosi basin, the Chang-7 member in the Yanchang formation of the

Late Triassic was deposited during the greatest lacustrine transgression; deep to semi-deep water lacustrine mudstones functioned as hydrocarbon source rock and occupied an area of 8.5×10^4 km^2, which equaled 60% of the total lacustrine region in the same period. Most of the organics consisted of type I–type II$_1$ kerogen; the average content of organic carbon is 2.17%. In the center of the basin, around the Wuqi area, the thickness of the source rock is more than 100 meters. However, the thickness of source rock decreases toward the edge of the basin. In the area of Jingbian–Zichang–Yanan, the thickness of source rock is about 30 meters. In the Songliao basin, the major source rocks (dark color mudstone) were deposited in the Qing-1 member in the Qingshankou formation; in the Binbei area that was located on the edge of the Songliao basin, the mudstone contained a higher percentage of sandstone; in the central depression, the mudstone almost covered the entire region. The thickness of the source rock was about 60 – 80 m; the average content of organic carbon is 2.207%; the organic materials consisted of type I–type II$_1$ kerogen; the effective source rocks covered an area of 6.5×10^4 km^2, which occupied 53% of the total lacustrine region.

Horizontally, the effective source rocks occupied a large percentage of the total lacustrine area. In addition, the aquatic systems that flowed into the basin from different directions produced different kinds of delta sand bodies. Therefore, these sand bodies had a higher opportunity to connect with source rocks, which benefited oil and gas migration and accumulation (Long, 1984; Wang, 2001). The statistical data (Table 7.3) indicated that, in the Songliao and the Eerduosi basins, most of the delta sand bodies in several depositional systems had more than 50% of contact area with the source rocks. In the Songliao basin, the sand bodies that were produced by three major aquatic systems (in the northern area, the Baoqian area, and the Yingtai area) included about 50% of contact area with source rocks. In the Eerduosi basin, it's over 60% of the ratio that the delta sand bodies in the northern area and south-western area sedimentary system deposited during the Yanchang period contact with resources rock.

Table 7.3   Statistic data of contact areas between the sand body of major water systems and source rock in the Songliao basin and the Eerduosi basin

Basin (depression)	Aquatic system and stratigraphic sequences		Area of sand body (km^2)	Contacted area (km^2)	Percentage of contacted area (%)
Eerduosi basin	Northern delta	T$_3$y	58,000	37,000	63.8
	Southwestern delta	T$_3$y	20,000	13,500	67.5
Songliao basin		K$_2$qn^3	12,000	5,600	46.7
	Baoqian delta	K$_2$qn^2	18,000	8,700	48.0
		K$_2$q^4	36,000	19,000	52.7
	Yingtai delta	K$_2$qn	4,500	2,000	55.6
	Northern delta	K$_2$q	44,000	23,000	52.3

Vertically, because of the oscillatory alteration of the lacustrine basin, the cycle of high frequency stratigraphic sequences meant that sandstone and mudstone interbedded with each other repeatedly to create a precondition for all kinds of lithostratigraphic traps. In the terrigenous depressed basin, because of the modification of the structure, climate and source of substances, lacustrine transgression and regression happened frequently. The sand bodies of the lowstand system tract and highstand system tract were well developed. From time to time, because water systems run into the basin from all directions, the sand bodies that are related with the water systems might cover the entire depression. On the other hand, during the maximum lacustrine transgressive period, because the basin was extended, the sand bodies were retrieved and the source rocks were widely expanded, which created the petroleum system of generation–reservoir–cap rock vertically. Along with the oscillatory alteration of the lacustrine basin, this kind of petroleum system could also be reproduced many times vertically. The sand bodies that were created by multiple water systems changed very fast horizontally; this phenomenon not only caused reservoir bodies to be in contact with source rock over an extensive area, but it also provided a precondition for creating different types of lithologic traps. During some geological intervals, because both the extensional region of source rock and the scale of sand bodies that developed towards the center of the depression were great, the source rock and reservoir sequences were interbedded with each other to create the foundation for a "petroliferous rich depression".

- **Depositional System of Inherited, Large Scale River Delta Offered Generous Reservoir Space and Trapping Conditions**

The inherited depositional system in the large scale river delta was a major controlling factor for creating a lithologic oil reservoir. The delta front was a favorable region for developing a lithologic oil reservoir. In the center of depression, a series of delta front sand bodies were deposited in the pattern of the a belt or sheet; these sand bodies were either on top of hydrocarbon source rock or inside source rock to form a connected reservoir sequence in the region (Gao and Cai, 1997). For example, in the Songliao basin, the Qingshankou formation of the Upper Cretaceous contained six major delta systems; the sand bodies were not only distributed within an area of 8×10^4 km^2, but they also were interbedded with source rocks. In the Eerduosi basin, the Yanchang formation contained two major water systems; the northern bank water systems mainly developed the meandering river delta in a northeast direction to an extent of 300 km; the southern bank water systems primarily developed the braided river delta in a northeast direction with a length of 300 km. These two water systems created the Chang-6 member of delta front facies that were in contact with the source rocks in the Chang-7 member. The contact area of 5.6×10^4 km^2 equaled 82% of the distribution area of the source rock.

    In a terrigenous depressed basin with shallow water, the sediments were often altered horizontally, which benefited the development of sizeable lithologic traps. During the course of deposition, the three dimensional alteration and effective combination of sandstone and mudstone created petroleum traps. The delta front was a favorable region for developing a thin layer, interbedded, sandstone lithologic trap over a large area. This region contained the following preconditions for creating lithologic traps. First of all, on the delta front, the sandstone and mudstone were not only interbedded with each other vertically, but they also overlapped horizontally; in addition, the front end was restricted by the lacustrine mudstone. This offered a depositional background for developing the useful lithologic trap zone. Secondly, the shallow water delta front mainly contained distributary river channel sandstones over an extensive area; the mudstones of the inter-distributary channel separated sand bodies to create lateral seals. Next, along the orientation of sand bodies in the distributary river channels, the faults or isolated structures formed seals to create petroleum traps. In the Songliao basin, in the Upper Cretaceous, the petroleum systems (oil beds of Shaertu, Putaohua and Gaotaizi) in the middle portion and the petroleum system (Fuyu and Yangdachengzi oil beds) in the lower portion were the lithologic traps on the delta front.

In the Songliao basin, the depositional background of the Fuyu and Yangdachengzi oil beds was the shallow water lacustrine–delta; therefore, all kinds of sand bodies in river channels overlapped or connected with each other horizontally, which provided extensive storage spaces for the lithologic oil reservoir. In the Gaotaizi oil bed, the large size lacustrine delta front facies controlled distributions of lithologic oil reservoirs either horizontally or vertically. In the Putaohua oil bed, lowstand delta front facies gradually expanded from the center of the basin towards the edge of the basin; thus, the sand bodies of the delta front facies were widely distributed horizontally, which caused the lithologic oil reservoirs to be connected with each other over a broad region. In the Shaertu oil bed, the development of a transgressive type of delta system meant that lithologic oil reservoirs were simply created on the side of the delta.

● **Small Oil and Gas Column and Medium–Low Pressure System Benefited Sizeable Petroleum Accumulation on Delta Front**

A large scale, shallow water delta provided favorable preconditions for creating sizeable, medium–low abundance, lithostratigraphic oil and gas reservoirs. First of all, small oil and gas columns and a medium–low pressure system lowered the requirements of the accumulative conditions, which was the foundation for accumulating a medium–low abundance, lithologic reservoir in an extensive area. Secondly, the horizontal alterations of terrestrial deposits guaranteed the creation of large scale lithologic traps; in addition, under the gentle structural background, a large size shallow water delta system created the dispersed pattern, widely distributed lithologic trap and diagenesis trap with the characteristic of low abundance. Thirdly, in a large scale shallow water delta system, the sandstone with medium–low porosity and permeability offered excellent accumulative and preservative conditions; this kind of sandstone not only could form an in-situ type of reservoir, but it also could form a large scale cluster of lithostratigraphic reservoirs to establish the medium–low abundance large scale oil and gas field (region). Oil and gas accumulations in isolated locations were controlled by the "sweet point" that was created by diagenesis dissolution facies, faulted fractures and isolated structural background. The widely spread, medium–low abundance oil regions with connected reservoir sequences occurred in the middle–shallow strata of the Songliao basin and in the Mesozoic strata of the Eerduosi basin.

- **Slope and Uplifted Background Controlled Accumulative Zone; Isolated Structure and Favorable Micro Facies Controlled "Sweet Point"**

Both the slope that was developed during a long stable duration and the uplifted structural background would benefit the creation of either widely spread structural and lithologic traps, or petroleum migration and accumulation. Therefore, together with delta front sand bodies, they controlled the cluster region of structural and lithostratigraphic reservoirs in general. For example, the central uplifted zone of the Daqing Changyuan–Fuxin and the delta sand bodies in the northern and southern depositional systems jointly controlled oil and gas distribution in the Songliao basin. The Shanbei slope and the delta sand bodies in the northern depositional system jointly controlled oil and gas distribution in the Eerduosi basin. Because of the variations in structural evolution between different basins, the primary types of petroleum reservoir varied in different basins. The Songliao basin was a post-rifting depressed basin; the structural activities of the later period were relatively strong; the types of oil reservoir included a structural reservoir, a structural–lithologic reservoir and lithologic reservoir. The Eerduosi basin was a superimposed depressed basin on top of a craton; the structural activities of the later period were comparatively weak; the lithologic reservoir was the major type of oil reservoir.

The exploration practices indicated that, in the Songliao and Eerduosi basins, the characteristics of a lithostratigraphic petroleum reservoir are low abundance with a low productive rate (Zhou et al., 2006). The following are the characteristics of this kind of petroleum reservoir. 1) Even the scale of a single reservoir was small. However, several oil reservoirs were either superimposed or overlapped with each other to form the complex of oil reservoirs over an extensive area. 2) Although the height of the oil column in a single oil reservoir was low and there were many water and oil systems, conversely the total thickness of the oil bed was great within the complex of the oil reservoirs. 3) In general, because the relief of oil beds was gentle with poor separation between oil and water, the oil production in a single well was relatively low, which influenced the effectiveness of petroleum exploration and oil field development.

In particular, there were many sets of thin layer, interbedded reservoir sequences that were superimposed or overlapped with each other; these reservoir sequences contained a low abundance of oil reserves and distinctive alterations; the relationship of oil, gas and water was intricate. These reservoir sequences were mainly distributed on the slope zones that surrounded the oil and gas generating depression; some reservoir sequences were evenly distributed in the synclinal area. The widely scattered, low abundance, lithologic petroleum reservoirs actually formed the cluster of petroleum reservoirs with this particular kind of structural background. These particular petroleum reservoirs were developed in the delta front facies and the semi-deep water lacustrine facies within a large scale, depressed lacustrine basin.

The original reservoir sequences not only had poor connective conditions and lower permeability, but they also lacked trapping conditions. Thus, oil and gas could not migrate to the sideway on a large scale. After oil and gas left the source rock, they accumulated in the nearby reservoir sequences and they resided separately. Obviously, among these widely distributed and low abundance oil and gas reservoirs, in order to improve the production rate of a single well, searching the zone with a highly abundant petroleum reserve and practicing the fractural technique in reservoir sequences will increase the economic efficiency of oil field development.

From the experiences of petroleum exploration in the Songliao basin, the distribution area of this kind of oil and gas cluster usually was larger than 100 km^2; the abundance of the petroleum reserve was $(10 - 80) \times 10^4$ t/km^2; the original daily production rate of a single well was $1 - 3$ t. After the fractural treatment, the production rate could increase $2 - 3$ times. For this kind of exploration target, the major problem during the course of exploration was the diversified abundance of petroleum reserves. In order to understand the distributions of oil and gas reserves, to provide explored reserves, and to locate a petroleum enriched area, drilling additional appraisal wells was in demand. At same time, because the petroleum reservoirs are spread about in abundance, drilling more wells will bring down the success rate of appraisal wells. Therefore, this kind of exploration target belongs to the category of progress target, which requires many years of continual search work to improve the accuracy of evaluation and to provide exploration reserves gradually. On the other hand, we can carry out experimental production before including this in the petroleum reserve. In general, this kind of petroleum reservoir needs fractural treatment in order to improve the production rate in a single well.

According to the studies of accumulation and distribution of lithostratigraphic reservoirs in recent years, the low abundance, broadly distributed, lithostratigraphic reservoir could be "spottily enriched" in the sheet form of a sand body, major sandstone belt, fractural zone and all kinds of reservoir bodies that were related to the nose-like structural background; this kind of enrichment was called "sweet point" overseas. The enriched type of structural background and the essential sealing conditions were the important controlling factors for oil and gas enrichment in isolated locations. The "sweet point" mainly occurred in the structural traps that were related to the ridge and the turning flat of the nose–like structure, high point of faulted block, high point of micro relief structure. In these structural positions, the best oil and gas bearing reservoir sequences were the sand bodies of the underwater distributary channel facies and the estuary dam facies and the sheet form sandstone. Clearly, in a particular exploration area, an understanding of the distributions of three types of lithostratigraphic reservoir will help increase the explored petroleum reserve in the lithostratigraphic reservoir.

7.2.2   *Above-Source Play Brought a Petroleum Accumulative Zone at the Intersection of the Paleo Bulge and Faulted Belt*

The exploration practices in the Songliao basin and the Eerduosi basin demonstrated that petroleum reservoirs could developed in the in-source play, beneath-source play and above-source play. However, the major controlling factors in each play were different; the in-source play offered the best conditions for creating oil and gas reservoirs. On the other hand, the above-source play was short of hydrocarbon generating potential; the accumulative mechanism and the pattern of reservoir distribution of the above-source play were clearly different if compared with the in-source play.

In the Zhungeer basin, the lithologic oil field in the Jurassic Sangonghe formation belongs to the above-source play. On the western side of the Luliang uplift, the Shixi oil field, the Shinan oil field and the Luliang oil field contained the petroleum that came from the source rocks in the Lower Wuerhe formation and the Fengcheng formation in the Permian system; the source rocks are located at the depression to the west of the Pen 1 well. The faults, unconformity surface, and horizontally connected sand bodies were significant elements in the petroleum migration and accumulation in the upper portion of the Jurassic system and the Cretaceous system. The border faults of the Chemo paleo-uplift and the unconformity surface were effective oil migrating passages. The large scale fault that extended into the depression (such as the Jidong fault) combined with the normal fault of extension–distortion in the middle–shallow strata to create the vertical migrating passage. Together, the unconformity surfaces (that were located in the Permian strata, the Triassic strata, and the top, base, and middle of the Jurassic strata) and the sand bodies created lateral migrating passages. In the Zhungeer basin, all kinds of oil migrating passages controlled accumulation and distribution of the lithostratigraphic reservoir in the above-source play. The exploration practices and research works demonstrated that, in the central Zhungeer basin, the Mesozoic lithostratigraphic reservoirs were mainly distributed in the intersection of the paleo–uplift and the fault that extended down to the source rocks (Fig. 7.3).

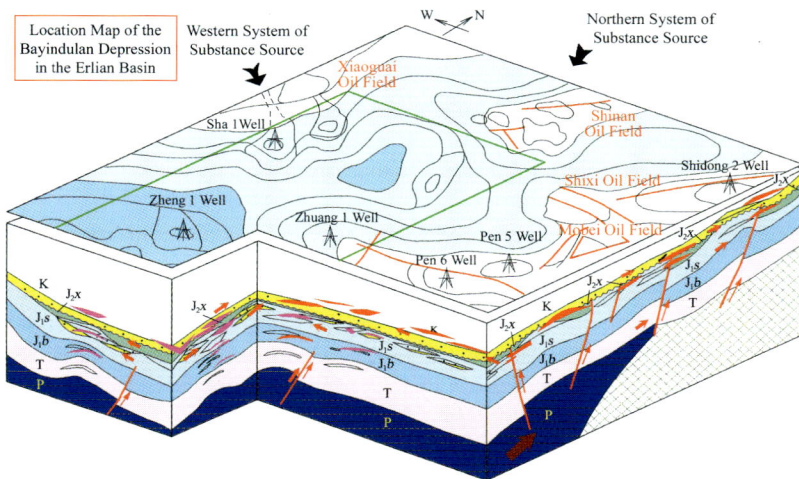

Fig. 7.3. Above-source play model in the central Zhungeer basin

7.3   Terrigenous Foreland Basin Has "Thrust Belt and Fan Body Controlled Oil"

Using the northwestern margin of the Zhungeer basin as an example, let us discuss the pattern and the characteristics of the "thrust belt controlled oil" in the terrigenous foreland basin.

7.3.1   Characteristics of Petroleum Accumulation

The northwestern margin of the Zhungeer basin contains multiple domains, various stratigraphic sequences and several types of composite petroleum reservoirs. At present, commercial grade oil and gas reservoirs have been discovered in 15 formation sequences that reside within 5 large stratigraphic sequences from the Carboniferous–Cretaceous systems. The types of reservoir are diversified; the composite types of structural–lithologic reservoir and the lithostratigraphic reservoir are major types.

On the northwestern margin, the slope zones of the hanging wall and the footwall in the faulted belt are different in the characteristics of accumulation and oil reservoirs. On the hanging wall of the faulted zone, the petroleum accumulation mainly related to the fault and unconformity surface. On the slope area of the foot wall, petroleum accumulation was predominately associated with the fault, unconformity surface and development of the fan body.

(1) The petroleum migrating pathways developed during the active period of faults. Because fault activities repeatedly adjusted or balanced the oil and gas accumulative situation, multiple oil bearing strata occurred inside the faulted belt. In general, oil and gas would migrate into the faulted strata. During an inactive period of faults, oil and gas might be accumulated and preserved in the foot wall of a fault, if the seal was presented. The oil and gas accumulative zones with multiple stratigraphic sequences developed along both sides of the faulted belt.

(2) On the hanging wall, the oil and gas distributions in the overlapped zone had a dependable relationship with the Paleozoic unconformity surface. In general, the oil beds were in direct contact with the Paleozoic unconformity surface; along with the overlapping of sedimentary layers in successive order, the geological age of oil beds gradually became younger; the type of reservoir was the bitumen sealed oil reservoir.

(3) Petroleum accumulation on the footwall slope was determined by the developmental level of the fan body. The fan body could develop the petroleum reservoir independently with its own oil, gas and water system. There are eleven fan bodies that controlled oil and gas reservoirs with a total explored reserve of 3.54×10^8 t.

1) On the northwestern margin of the Zhungeer basin, the oil and gas reservoirs in the fan body agree with the "fan body controlled oil" theory. The following are the descriptions of the characteristics. i) the development of fan body was controlled by the active growth fault; the phases of fault activity, the

magnitude of fault activity and the mode of fault activity controlled the number, scale and migrating direction of the fan body. ii) The petroleum accumulation in the fan body was determined by the combination and settings either between the fan body and unconformity surface or between the fan body and fault. On the faulted step zone, the fan body combined with faults to form a vertical migration and accumulation model; on the slope area, the fan body was incorporated with the unconformity surface to create a lateral migration and accumulation model. iii) In the fan body, the scale and quality of the petroleum reservoir were decided by the size of the fan body, type of superimposed fan body and arrangement of sub-facies. The best physical reservoir properties appear in the mid-fan and the fan root of an alluvial fan and the front margin sub-facies of a fan delta; these are the favorable places for petroleum accumulation.

2) The types of lithostratigraphic reservoir include faulted–lithologic reservoir, lithologic reservoir and anticlinal–lithologic reservoir; these types of reservoir have different accumulative characteristics and patterns. i) Faulted–lithologic oil reservoir. This type of oil reservoir was broadly distributed on the slope area; among the reservoirs, the fault–fan body oil reservoir is the principle type. Because the slope area was located on the footwall of a major fault, the diluvial fan and fan delta were perfectly developed, which combined with faults to create the fault–fan body trap and further develop the fault–fan body oil reservoir. For example, in the area around well 503, the Lower Wuerhe formation contained a typical fault–fan body oil reservoir; the reservoir sequences consisted of deposits of fan delta facies and alluvial–diluvial facies; in the north, the fault of 415 well functioned as the blockage; the P_2w^1 sequence worked as an excellent seal layer. ii) Lithologic oil reservoir. The well developed alluvial fan and fan delta on the slope area offered favorable geological conditions for creating a lithologic–fan body oil reservoir. In the eighth area, the Permian Wuerhe formation was next to the major fault and it contained a fan body–lithologic type of petroleum reservoir with an explored reserve of $8,958 \times 10^4$ t. In the fifth area, in the Ke 75 well area and the Ke 79 well area, the fan body–lithologic type of petroleum reservoir developed near the pinch-out line of the Wu-3 member, in the Permian Upper Wuerhe formation (Fig. 7.4).

7.3.2   Oil and Gas Accumulative Pattern

On the northwestern margin, the overthrusted blocks, faults and unconformity surfaces were well developed; the fault, unconformity surface, fan body and developmental level of reservoir sequences were the most important controlling factors for oil and gas accumulation.

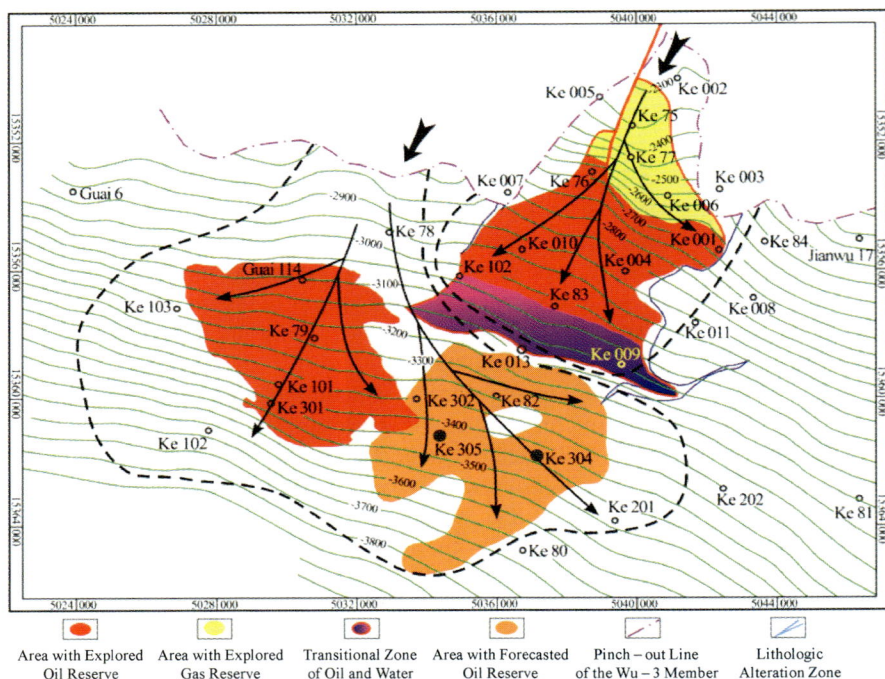

Fig. 7.4. Horizontal distribution map of lithologic petroleum reservoir on P_3w fan in the Zhongguai-5 area

● Fault and Petroleum Reservoir

On the northwestern margin, the thrusted belt contained a group of thrusts fault or overthrusted faults that were nearly parallel with each other. The oil and gas distributions were closely related with thrusted faults (Fig. 7.5). The oil and gas reservoirs appeared on both sides of the Chepaizi–Xiazijie fault; from Chepaizi in the south to Xiazijie in the north, the petroleum belt stretched more than 150 km; the petroleum rich strata of different sequences not only basically connected together, but they were also clearly controlled by the faulted belt. The major characteristics include: 1) the faults functioned as petroleum migration paths; 2) the faults also blocked oil and gas, which brought many fault obstructed types and faulted block types of oil and gas reservoirs; 3) in addition, the faults improved the storage condition of the lithology; 4) next, the faulted activity would damage the oil and gas reservoir; for example, on the northwestern margin, during the early Yanshanian tectonic period, because the faulted activities destroyed original oil and gas reservoirs, many secondary viscous oil reservoirs and surface oil seeps were widely developed. About 95% of the explored oil and gas reserves were related to faults; the fault-pump was the most important characteristic of the petroleum migration system. The favorable oil and gas accumulative region contained well developed faults and good sealing conditions in the upwards-dipping direction.

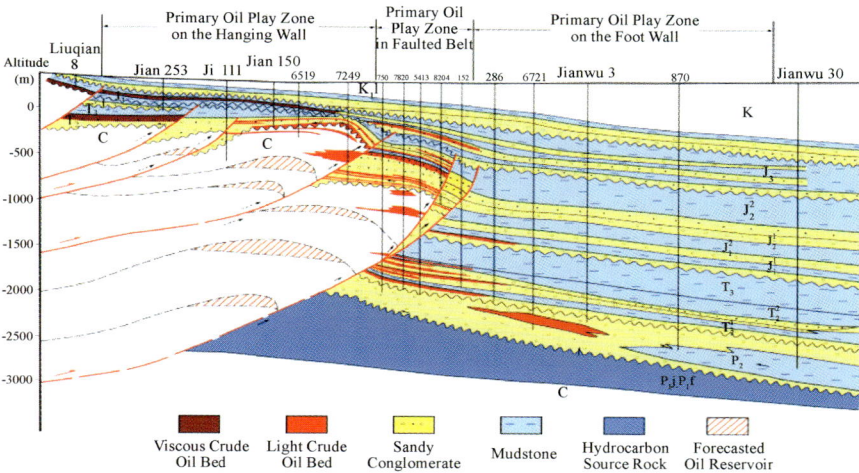

Fig. 7.5. Model of near-source play with the controls from unconformity and faults

● **Fan Body and Petroleum Distributions**

Relationship Between the Facies Belt and Reservoir Sequences

The best reservoir bodies developed in the strata of alluvial facies and fan delta facies. The oil and gas bearing potential varied, with different types of fan bodies or with different sub-facies in the same type of fan body. In the fan body controlled lithostratigraphic oil reservoir, the alteration in lithology played an important role. The sandy conglomerates, either in the mid-fan or in the top-fan, were favorable reservoir sequences for oil and gas accumulations; however, on the side of a fan, the fine grain lithology on the margin of the fan body did not provide favorable sequences for oil and gas accumulation. These sedimentary masses of the alluvial fan or fan delta could develop a petroleum reservoir independently; one fan could have one reservoir with its own system. The distributions of oil and water were not restrictively controlled by faults; most of the formation water was preserved in the inter-fan area and the margin of the fan; the basement water and lateral water of the petroleum reservoir were not active. After oil and gas accumulated, even the reservoir sequences were tilted or were folded and the oil and gas were still preserved in the formation that was an excellent place for oil and gas accumulation.

Scale of the Reservoir is Closely Related to the Thickness and Distribution of the Fan Body

In the oil and gas reservoir, the distributions of oil, gas and water were controlled by the fan body; the scale of the oil (gas) reservoir was limited by the size of the fan body (Fig. 7.4). These characteristics are as follows. In general, the distributions of petroleum reservoirs were concentrated in a particular place in the fan body; because sandstones were clustered in the middle portion of the fan body with great thickness, here was a favorable place for oil and gas accumulation. The viscous oil sealed the upward-dipping location of a fan body. The alterations in lithology happened in the inter-fan area that had poor permeability; here, the thickness of sandstone decreased and the content of clay increased, which created a lateral blockage for oil and gas migration. For example, 1) in the south of the fifth area, the oil reservoir in the Wuerhe formation can be divided into three different sized underwater fan bodies horizontally, which are the fan body of the Ke 78 well, the fan body of the Ke 75 well and the fan body of the Jianwu 13 well. In the area of the Ke 79 well, the oil reservoir of the Wuerhe formation was located in the fan body of the Ke 78 well that is of small size. The fan body of the Ke 78 well and the fan body of the Ke 75 well created the inter-fan alternating deposits near the Ke 012 well. The mainstream lines of the fan body of the Ke 78 well and the fan body of the Ke 75 well were almost parallel with each other; the direction of the substance source was in a northwest direction. The fan body of the Ke 78 well and the fan body of the Ke 75 well had similar depositional features and depositional environment; the W3 sandstone sequences were deposited near the source area of sediments, in the sub-facies of the fan root within the underwater fan facies. The lithology of the W2 sandstone sequences are basically the same as the lithology of the W3 sandstone sequences, besides the smaller grain size; the size of pebble is between 2 – 5 mm; the size of sand grain is between 0.1 – 0.25 mm, which are the deposits of mid-fan subfacies in the underwater fan. 2) In the area of the Xiaoguaiche 67 well, the oil reservoir of the Jiamuhe formation was located at the footwall of the Hongche fault. In this area, four alluvial fans developed in the Permian Jiamuhe formation, which were the fan of the Guai 5 well, the fan of the Guai 9 well, the fan of the Che 67 well and the fan of the Che 45 well. In the Che 67 well block, the reservoir sequences in the Permian Jiamuhe formation resided in the fan-root subfacies in an alluvial fan. The reservoir sequences in the Permian Jiamuhe formation contained dual mediums of pores–fractures. Compared with pores, the fractures take much less reservoir space and they were the crucial factor in the production rate of oil wells.

The thickness relationship between the oil bed and sandy conglomerate or between the oil bed and superimposed fan bodies is discussed in the following. Generally speaking, the thickest fan body would relate to the greater scale of reservoir sequences (sandy conglomerate); the sandy conglomerate layers were distributed in the fan root and mid-fan of an alluvial fan and the delta plain. The area that contained superimposed fan bodies usually has good physical properties, which not only benefited oil and gas accumulation, but which also made a transportation system for oil and gas "infiltrated migration" upwards. Additionally, in the location where the thickness of the sandy conglomerate or superimposed fan bodies was great, due to the differences in compaction, a low relief uplift would be created, which benefited oil and gas accumulation.

(1) In the area of well 530, the oil reservoir in the Upper Wuerhe formation was located on the slope zone of the northwestern margin; this reservoir sequence contained a set of course grain sediments with a thickness of more than 3,500 m. The sediments consisted of conglomerate and sandy conglomerate, which were interbedded with mudstone; on the other hand, mudstone was the primary deposit inside the basin. The principle source rock was the Permian system in the Manu depression. First, oil and gas migrated on the unconformity surface; after they reached the fault, oil and gas would migrate upward with the fault to form an oil and gas reservoir in the fan body which had good physical properties. In the area of well 530 of the eighth area, the reservoir sequences of the Lower Wuerhe formation belonged to the subfacies of fan root and mid-fan in an underwater fan body; the reservoir sequences had good physical properties; the northern side was blocked by a fault; the P_2w^1 strata functioned as perfect cap rocks. Base on this kind of geological setting, the oil reservoir of the Lower Wuerhe formation developed in the area of well 530 (Fig. 7.4). According to the research studies of structure, sedimentation, and reservoir sequences, considering the data analysis of oil tests and production tests, we believe that, in the area of well 530, the oil reservoir of the Lower Wuerhe formation is a structural–lithologic oil reservoir; it was blocked by the structure on the northern side; it was controlled by lithology and physical properties on the western and the eastern sides; and it was controlled by physical properties and the structure on the southern side.

(2) The Mabei oil field is located in the Mahu area on the northwestern margin of the Zhungeer basin; its geologic location is on the footwall of a faulted step zone on the northern slope of the Mahu depression, on the northwestern margin of the Zhungeer. The structural framework is a gentle monocline that is dipping to the south; in the isolated area, there is a nose-like structure and low relief anticline. The reservoir sequences include the Triassic Baikouquan formation and the Permian Wuerhe formation. The deposits of the Baikouquan formation belong to mid-fan subfacies within the Mabei fan body. Horizontally, sandstone was well developed in the main portion of the fan body; in addition, some sandy conglomerate blocks with high electrical resistance also formed in this location. On both sides of the fan body, due to the influence of different facies, the alterations in lithology were evident; the thickness of the sandstone layer decreased, the thickness of the interbedded layers increased and the connectivity became poor. As in the Baikouquan formation, the reservoir sequences of the Wuerhe formation belong to the mid-fan subfacies within an underwater fan body. The reservoir sequences were superimposed by a few mid-fans from several fan bodies; however, the lithology of the inter-fan flat area contains fine grain sediments with poor physical properties. From a geological point of view, the Mabei oil field is located on the superimposed position of mid-fans in the Wuerhe formation and the Baikouquan formation;

that means the reservoir sequences have good physical properties; moreover, the superimposed fan bodies also benefit vertical migration of oil and gas.

Lateral Connection of Source Rock and Reservoir Sequences, Unconformity Surface Replayed Transportation and Faulted Connection were the Crucial Factors for Petroleum Accumulation in the Fan Body

The crucial factor for oil and gas accumulation in a fan body was the connection between the source rocks and the reservoir sequences. There were two kinds of connections: one was fan body connected with source rock on the sides; another one was relayed by the unconformity surface or was connected by a fault.

According to the research studies of the Permian and the Triassic reservoirs on the northwestern margin in the Zhungeer basin, the types of petroleum reservoir include a structural oil and gas reservoir, a structural–lithologic oil and gas reservoir, a lithologic oil and gas reservoir, a stratigraphic–structural oil and gas reservoir, a faulted block type of oil and gas reservoir, and a blocky (or buried hill type) oil and gas reservoir. Among these petroleum reservoirs, most of them belong to the composite type that was controlled by multiple factors. Only a small number of reservoirs were controlled by a single factor, either a structural factor or lithologic factor; in addition, the single factor controlled reservoirs were no more than 10% of the total reservoirs. On the northwestern margin, there are eleven oil and gas reservoirs that were controlled by fan bodies and that were either lithologic reservoirs or structural–lithologic composite reservoirs. The discovered geological reserve is 3.54×10^8 t. Horizontally, these oil and gas reservoirs were distributed on the northwestern margin; vertically, they were positioned in the sequences of the T_1b, the P_3w, the P_2x, and the P_1j.

7.4   Marine Craton Basin Has "High Energy Facies Belt Controlled Oil on Platform Margin"

In recent years, petroleum explorations in the Chinese marine craton basins have achieved a series of important breakthroughs; the distribution of oil and gas was clearly controlled by high energy facies (Zhou and Tao, 2007; Zhou et al., 2005); the phenomena of "high energy facies belt controlled oil on the platform margin" displayed a petroleum distribution pattern on the number I platform margin in the Tazhong area. Besides a high energy facies belt on the platform margin, the littoral sand body, the periclinal karst zone of paleo-uplift and the high energy sand body of marine–terrigenous facies were the favorable areas for oil and gas accumulation as well.

The craton basin usually experienced tectonic evolution and depositional development over a long period. Because of peripheral plate tectonic movement and uneven subsidence inside the craton, some uplifts and depressions often developed in a craton basin; in addition, stratigraphic unconformity also occurred frequently.

The uplift structures controlled the deposition of sediments, the division of the facies zone, the division of the paleo aquadynamic zone and petroleum accumulation. The unconformity surface commonly functioned as a petroleum migration pathway. Because of inherited continual uplifting caused by tectonic movement, the paleo-uplift became the destination of oil and gas migration over a very long period; therefore, it contained plenty of oil and gas accumulation. The oil and gas accumulation was controlled by the paleo structure and the depositional factors. The common petroleum plays include the faulted horst block and buried hill type of play in the lower portion, the lithologic–structural composite type of play in the upper portion, and the periclinal type of play. Besides the factors of source rock and the preconditions of reservoir and seal (that influenced petroleum accumulation and its magnitude on the paleo-uplift in a craton basin) the time that uplift was created, the structural stability in the later period and the scale of uplift were other important factors that controlled oil distribution. Generally speaking, if a paleo-uplift was created in an earlier geological period with a longer developmental duration, if the structures were more stable during the late period, or if the paleo-uplift had a larger scale, the paleo-uplift not only could offer better accumulation and preservation conditions, but it also contained a greater quantity of oil and gas. In particular, the paleo-uplifts (that were created during the late Caledonian–early or late Hercynian tectonic periods) usually contained plenty of oil and gas if the tectonic movement of the late period was stable and if the sedimentary deposition was persistent.

7.4.1 Carbonate Platform Margin Controlled Petroleum Accumulation on Reef Flat

The reef cluster or the reef belt that was distributed along the outer zone of the platform margin was the petroleum accumulation area. At the outer zone of the platform margin, the bio-reef was not solitary; in opposition, the bio-reef appeared either in a cluster or as a belt. The bio-reef not only contained a perfect combination of reservoir sequences and cap rocks, but it was also located next to the source rock of the basin; this unique situation provided an excellent precondition for oil and gas accumulation. As the reservoir sequence, the reconstruction of the reef includes three kinds of developments, which were dissolution and karstification, dolomitization and tectonism. In some reefs, because of the cementation, the core of the reef was dense; however, the lateral position, top portion, or cap of the reef could become good reservoir sequences if they were exposed and dissolved, or were dolomitized. Thus, some researchers considered these kinds of reef as a type of diagenesis trap.

The platform margin not only offered excellent preconditions for petroleum accumulation, but it also contained abundant oil and gas reserves. The oil and gas accumulations were not controlled by the local structure. The reservoir sequence was the main factor that decided the oil and gas bearing situation; on the other hand, the regional geological framework was divided into sections, which made variations in the reservoir sequences and oil enrichment.

● **Platform Edge Belt Controlled the Distribution of Favorable Reservoir Facies**

The platform rim zone either could be developed on the platform margin, or it could consist of the sedimentation of carbonate rocks; in addition, the reservoir bodies also developed in the reef complex, bio-reef, and reef flat on the platform rim zone. For example, the Tazhong-I platform rim zone not only controlled the structural framework of the Tazhong area, but it also controlled the development of the Upper Ordovician shelf margin reef complex that was developed along the platform rim zone in the Upper Ordovician Lianglitage formation, with a high energy bio-reef and reef flat (Fig. 7.6). In the northern part of the Tazhong-I platform rim zone, the Manjiaer depression was filled by clastic rocks of trough basin subfacies; the alteration of facies consisted of sandstone and mudstone. Towards the south, the inner shelf subfacies contained the following micro facies: tide flat, inner shelf flat, inner shelf gentle slope, inner shelf sag and carbonate mud mounds; the alteration of facies was micritic limestone. The reef complex extended along the platform rim zone as a belt, the favorable reservoir sequences were broader at the western end and narrower at the eastern end. In the 62 – 64 well area of the Tazhong area, the width of reservoir sequences is 1 – 2 km; however, in the western area, the width of reservoir sequences can be 5 – 10 km. The thickness of the reef flat is around 80 – 150 m.

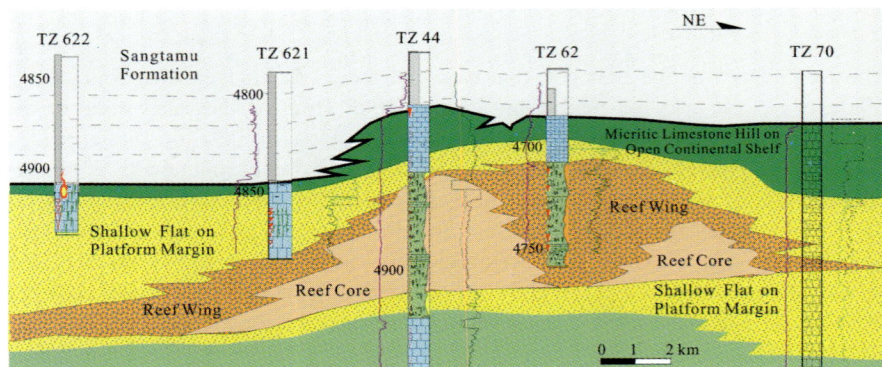

Fig. 7.6.   Cross section of the Ordovician reef flat in the Tazhong area, in the Tarim basin (paralleled with platform margin)

● **Platform Margin Contained the Best Reservoir Sequences**

For example, in the Tazhong area, the Upper Ordovician reef complex on the continental shelf margin mainly developed in the granular limestone section of the Lianglitage formation; the lithologies include skeleton reef limestone of reef flat facies, granular limestone and algal bryozoan bound stone on carbonate mud mounds. On the platform margin of Tazhong-I, the explored wells drilled into all kinds of grainstone, bio-debris limestone and reef limestone. In the Tazhong 44 well, between 4,850 – 5,050 m, there are two reef complexes with a thickness of 75 m and 11 m. The reef core micro facies include a solenopora reef and skeleton reef of sponge–solenopora. The reef-building organisms include solenopora, bryozoan, auloporidae coral, vermiporella and sponge. The micron facies of the reef base and reef cap include sparite calcarenite and oolitic limestone. In the Tazhong area, the types of reservoir spaces in the reef complex of the continental shelf are secondary pores, holes and fractures. The statistical data indicated that, in the reef flat, reef core, and grain-banks, the cements had a large porosity of 1.86%, 1.6%, and 1.29% respectively, which were larger than the porosity of limestone on the carbonate mud mounds of the inner continental shelf. If comparing permeability, the permeability of the reef complex of the continental shelf was ten times better than that in the limestone on the carbonate mud mounds of the inner continental shelf. The physical properties were excellent. The reef flat not only contained well developed cement, but it also contained dissolution pores that were formed by a buried type of karst and surface type of karst. In the Tazhong 62 well, the thickness of the Upper Ordovician reef flat is more than 60 m. The sediments were deposited in the platform margin facies. Intergranular pores, intergranular dissolution pores, intercrystalline pores and intercrystalline dissolution pores were well developed; in the lower portion, the karsts were widely distributed. The core surface porosity could reach 15%. After the well logging interpretation, the porosity is between 5% – 8%, which indicated perfect physical properties.

● **Multi-Phases of Infilling were the Characteristic of Oil Accumulation**

The Tazhong-I platform margin belt was formed during the Ordovician and its tectonic framework was basically established during the Caledonian tectonic period; in addition, this platform margin belt was next to hydrocarbon source rock of the Manxi area. The Ordovician carbonate rocks could capture oil and gas from multiple directions in several phases. All the carbonate reservoirs were covered by the Upper Ordovician, massive layer and black color mudstone. This kind of geological setting benefited oil and gas infilling and the preservation of multiple phases; the accumulative conditions were excellent. The well drilling met with a good oil and gas show; the success rate of well drilling could reach up to 67%.

● **Developmental Level of Reservoir Sequences was the Major Control Factor for Petroleum Accumulation**

The entire Tazhong-I platform margin belt is a northwest dipping slope zone that lacks a structural trap. In addition, petroleum accumulations were not influenced by local structure in this area. From the Tazhong 45 well to the Tazhong 26 well (the distance is 200 km approximately), the Upper Ordovician carbonate rocks commonly contain oil. The height differentiation between the top surface of the oil bed and the top surface of the gas bed is 1,800 m. In this region, all the reservoir sequences contained different levels of oil and gas shows. In the area of the Tazhong 45 well and the area of the Tazhong 62 well, the drilling explorations indicated that the developmental level would determine the oil and gas production rate. The characteristics of reservoir sequences controlled the distribution and abundance of oil and gas; the outstanding reservoir sequences are the main controlling factor for oil and gas accumulation and distribution in this region.

The reef complex on the Tazhong-I platform margin not only has stable thickness, but it also contained a large size oil and gas reservoir that was stretched into a belt shape horizontally and that was created almost vertically into a layer. The following are the characteristics of this type of petroleum reservoir. 1) The good oil and gas shows occurred in the upper portion of the granular limestone in the Lianglitage formation; this oil and gas interval appeared in the reef flat with a thickness of 150 m approximately. 2) The types of storage space include secondary pores and fracture–pores, both of which were controlled by the reef flat. 3) Because they contained well developed pores and fractures, the reservoir sequences have good connectivity. 4) The oil and gas reservoirs have almost the same or comparable pressure–temperature systems. 5) Lack of structural trap. In the same reservoir group, the height differentiation of the top surface of reservoir sequences could be more than 200 m. The type of reservoir is a lithologic reservoir that was controlled by the physical properties of the reservoir sequences. 6) In the reef flat reservoirs, the production rate of testing wells are stable; the basement water is inactive; the reservoir sequences contain well developed pores and fractures, which were connected over a broad area; the radius of the oil discharge is large. 7) The reef flat complexes generally contained oil; the common type of reservoir is the low abundance oil and gas reservoir.

Overall, on the Tazhong-I platform margin, the distribution of the Upper Ordovician reef flat complex of the continental shelf margin was stable, which offered excellent conditions for petroleum infilling and preservation; the characteristics of the petroleum system were that the entire reef complex contained the oil and that the reservoir sequences controlled distribution of the oil. The semi-layered petroleum reservoir was controlled by physical properties of the reservoir, which were not only semi-layered with a broad distribution, but also contained many millions of tons of petroleum reserves with low abundance.

7.4.2 Paleo-Uplift and Periclinal Karst Zone Controlled Weathering Crust Type of Petroleum Accumulation

● **Paleo-Uplift Controlled Sedimentary Facies**

The tri-provinces basin of Shaanxi–Gansu–Ningxia is a good example. The developmental history of this basin was as follows; during the Early Paleozoic, it was a continental marginal sea (on the edge of a craton) basin; during the early phase of the Late Paleozoic, it was a craton basin in the littoral–shallow sea; during the late phase of the Late Paleozoic–Triassic it was an intra-continental craton basin. At present it still retains an intra-craton basin. During the stage of the continental marginal sea, the shoulder of the Helan rifting valley was evenly uplifted to create an embryonic form of the paleo-uplift; afterwards, the uplifting event and erosion event happened alternately. Since the end of the Early Ordovician, this uplift and the North China craton endured weathering together; as the result of erosion, the karst type of erosion shell formed. During the Carboniferous, the Helan collision valley was developed; however, as the shoulder of a rifting valley, this area was uplifted; after the Late Triassic, this area was closed out. These paleo-uplifted structures (whether accompanied with a depositional event simultaneously, or whether it was an erosion type, or wherher it contained both types) always controlled sedimentations through some kinds of methods to further influence the distribution of depositional facies; thus they controlled the distribution of outstanding reservoir sequences.

● **Long Exposed Paleo-Uplift Cntrolled Development of Outstanding Secondary Reservoir Sequences**

For example, in the tri-provinces basin, the Sichuan basin and the Tarim basin, the paleo-uplifts endured erosion events of the 140 Ma, the 120 Ma and the 77 – 232 Ma; the excellent reservoir sequences developed in the erosion crust that was created by these weathering events. Here, we use the Chuanzhong paleo-uplift in the central Sichuan province as an example. Prior to the Permian period, the Chuanzhong paleo-uplift had been a peneplanation. In the weathering crust, if the basement was made of carbonate rock, 90.32% – 97.52% of the elements weathered away, which was the karst type of weathering crust. It the basement was made of either claystone or sandstone, only 25% of the elements weathered away, which was the residual weathering crust. Because it contained strong leaching, strong eluviation and a weak residual effect, the carbonate weathering crust could easily contain a karst zone (Song et al., 1995). This phenomenon also occurred on the paleo-uplift in the tri-provinces region of Shaanxi–Gansu–Ningxia. In the Tarim basin, the Cambrian–Ordovician carbonate reservoir sequences in the platform region were controlled by the weathering of the late period and the paleo karstification. In the paleo-uplift regions of Lunnan, Tazhong and the southeastern side of the Bachu area, because the Ordovician system was exposed for a long period, the reservoir conditions were therefore very good. Conversely, the Yangwu 2 well on the northern side of the Manjiaer depression and the He 3 well on the eastern side of the Bachu had relatively poor reservoir conditions because they were located on the lower section of the paleo-slopes. In addition, the inherited type of paleo-uplift that developed over a long period usually could develop multiple sets of reservoir sequences in perfect condition. That was why, in the Lunnan area, there are multiple sets of high quality reservoir sequences in the Ordovician, Carboniferous, Triassic and Jurassic systems; these reservoir sequences were closely related to the paleo-uplift that developed over a long period.

● **Cap Rock on Top of the Erosion Surface and its Evolution in the Late Period were the Control Factors for Petroleum Accumulation and Petroleum Preservation**

The sealing condition is one of the major controlling factors for petroleum accumulation and preservation in the weathering crust. For example, on the central uplift in the tri-province basin, both the bauxite at the base of the Benxi formation (C_2b) and the gypsum mudstone in the third bed, the fifth member of the Majiagou formation (O_1m^5) in the Lower Ordovician series could function as sealing layers. In addition, a series of erosion valleys (sub-channels) developed on the eastern slope; because the river deposits contained dense cement, they could function as a lateral sealing layer. During a subsidence event of the late period (K-Q) on the western side, the eastern region was elevated; when this area was transferred regionally in a westerly dipping direction, these river deposits were located on the upside of the petroleum migrating path and they served as a sealing layer to block natural gas from migrating eastward. On the Chuanzhong paleo-uplift in central Sichuan, both the Lower Cambrain Qiongzhusi formation and the Middle–Upper Cambrain clay shale could function as cap rocks to block a huge amount of natural gas during the late periods (N-Q). On the Tazhong low bulge in the Tarim basin, the mudstone section in the middle of the Carboniferous was the regional cap rock, which blocked oil and gas that accumulated in the Lower Ordovician series. Because of faulted activity during the late period (N-Q), isolated holes appeared in the cap rock; therefore, oil and gas migrated upward and re-accumulated in the delta sand body in the upper portion of the Carboniferous system.

● **On the Paleo-Uplift, the Inherited Type of Slope Zone, Stratigraphic Overlapping and Pinch-Out Zone and Facies Alteration Zone were the Best Places for Petroleum Accumulation**

In general, because the higher section of the paleo-uplift underwent the strongest tectonic alteration during the late period, the oil and gas reservoirs either were adjusted or were damaged. In this geologicalal section, secondary reservoirs commonly developed ; if the tectonic alteration was extremely strong, the oil and gas reservoirs might not even be preserved. On the lower section of the paleo-uplift or on the slope, because the tectonic activities were relatively weak, the original petroleum reservoirs were therefore preserved. Or, in this area, large scale original oil and gas reservoirs were either preserved or small size secondary oil and gas reservoirs developed. In other words, the major reservoirs were the original oil and gas reservoirs; the secondary oil and gas reservoirs usually offered small reserves. However, if comparing different paleo-uplifts, the distinctions are clear.

7.4.3 Littoral Zone that Surrounded the Paleo-Uplift Controlled Petroleum Accumulation in the Sandy Conglomerate

● **Stratigraphic Pinch-out Zone Developed on the Inherited Type of Paleo-Uplift and Peripheral Area During the Cycle of Marine Transgression and Regression**

During the late Caledonian–early Hercynian tectonic periods, the Hadexun area already had an embryonic form of uplift; in addition, prior to the Jurassic, this area belonged to the southward dipped extensional breach of the Lunnan paleo-uplift; however, an isolated high point occurred near the Hade 1 well. After the Indosinian movement, the Hadexun–Lunnan area became a nose-like uplift; one end dipped down in a northeast direction and the other end was elevated in a southwest direction; this was the Hadexun paleo-uplift. However, at the same time, in this region the Paleozoic structural sequences that included the Carboniferous system still dipped to the south. Until the Neogene, because the Kuche foreland thrust belt pushed and compressed southward, and because the northern section of the Tarim basin rapidly subsided, in the Manjiaer area the Carboniferous system modified its dipping direction to the north. Also, a large size, wide and gently sloped, nose-like uplift in the Carboniferous system was created on the western side of the Manjiaer area, which dipped down in a northeast direction and which was elevated in a southwest direction. In the northern Manjiaer–Tabei (northern Tarim) area, during the time when the Carboniferous system reversed its dipping direction from a regional dipping toward the south to a regional dipping toward the north, because it had the background of a paleo-uplift, the Manjiaer area became the "support point" of the upward dipping movement in the Carboniferous system. Therefore, during the late Himalayan tectonic period, in the Hadexun area, the event (that Carboniferous strata were upward dipped in a northerly direction) did not have a great influence on the stratigraphic traps in the Carboniferous Donghe sandstone; alternatively, north of the Hade 1 well, because of the subsidence and upward dipping, traps were created in the Carboniferous system. In other words, on the western side of the Manjiaer area, during the upward dipping of the Carboniferous system, because it carried on its uplift status, the Hadexun area was an inherited uplift of long developmental duration. On the paleo-present tectonic map of the Carboniferous system in the western region of the Manjiaer, the Hadexun area was coincidently located at the superimposed area of the paleo uplift and the present uplift. In particular, because it had the paleo-uplift or apaleo-tectonic background, and because it was relatively stable during the successive tectonic movements, the Hadexun area became the favorable area for oil and gas migration and preservation; furthermore, the paleo oil reservoirs that formed in the early period were not destroyed during the successive, stronger, repeated tectonic movement.

- **Outstanding Clastic Reservoir Sequences were Developed on the Inherited Type of Paleo-Uplift and in Peripheral Marine Facies**

Generally speaking, the paleo-uplift was near the source of sediments and it contained coarse grain deposits; either on the underwater paleo-uplift or on the upper section of its slope, because the water was shallow and the aquadynamics were strong, the sediments were rounded and sorted very well. Thus, the good reservoir sequences occurred on the paleo-uplift or around it. Away from the paleo-uplift, because the grain size became fine and the content of clay increased, the quality of reservoir sequences became poor as a result. In the northern and central Tarim basin, beside the dissolution in the late period, the main reason that widely distributed Donghe sandstone had perfect reservoir conditions was the background of the paleo-uplift while the sandstone was deposited. Therefore, in these areas, the Donghe sandstone was deposited in the environment of the littoral–shore with strong aquadynamic energy, which offered excellent original reservoir conditions. In the Hadexun area, if comparing the porosity and permeability of the Donghe sandstone from different areas, the one in the Hadexun area contained better physical properties than the one in the neighboring area of the Yuenan 1 well and the Donghetang area; again, the phenomena was related to the paleo-uplift background in the Hadexun area. For example, in the Hadexun area, the Donghe sandstone primarily contains fine grain quartz sandstone that not only is one of the best clastic reservoir sequences in the Tarim basin, but is also the major production sequence with good physical properties in the Hade 4 oil field. In the Hadexun area, many wells have drilled through this set of reservoir sequences; according to the statistical data, most of the reservoir sequences have medium–high porosity and medium–high permeability. In the Hade 4 well, the best porosity is 23.11% and 65% of reservoir sequences have a porosity greater than 15%; in addition, the best permeability is $2,040 \times 10^{-3}$ μm^2 and 50% of reservoir sequences have a permeability greater than 150×10^{-3} μm^2. The correlation between porosity and permeability is excellent, which indicates this is a great reservoir sequence.

- **Development of a Large Scale Stratigraphic Petroleum Trap was Controlled by the Peripheral Littoral Zone Around the Paleo-Uplift**

In the Tarim basin, the stratigraphic pinch-out zone of the Donghe sandstone is located at the curved lower ridge belt on the Lunnan–Hadexun–Gucheng nose-like uplift. Under the control of this particular structural setting, a series of stratigraphic traps developed to form the stratigraphic trap belt in the Donghe sandstone. In a N-S direction, the length of the trap belt is 100 km approximately; the width is about 10 – 30 km. In this trap belt, the outlined exploration area is about 2,300 km^2. At present, there are four discovered stratigraphic traps or traps that totally cover an area of 713 km^2. On the northern side of the pinch-out belt, the oil reservoir in the Hadexun Donghe sandstone is located at a lower altitude with a discovered geological oil reserve of $2,462.4 \times 10^4$ t (that included a discovered geological oil reserve of $3,068.1 \times 10^4$ t in the thin bedded sandstone within the middle section of mudstone). Thus, this pinch-out zone is an important exploration domain in the Donghe sandstone.

- **Perfect Conditions for Petroleum Migration and Accumulation**

Examples of this, in the Tarim basin, are the oil and gas that were generated by the Cambrian source rock in the Halahatang area. First, the oil and gas migrated along the faults into the Carboniferous system; then the oil and gas migrated along the unconformity surface or sandstone in a southerly upward dipping direction. The major migrating destination was a northward dipping nose-like structure that was the background of the Donghe sandstone; because the stratigraphic pinch-out zone of the Donghe sandstone was spread out along the axial area of the nose-like structure, it presented therefore good petroleum accumulative conditions.

In the Hade 4 oil field, the exploration level is relatively low; at present, there are only 9 controlling wells. Among these wells, only the Hade 4 well drilled into the interface of oil and water in the Carboniferous Donghe sandstone oil reservoir. It did not drill into the interface of oil and water either in the Carboniferous thin bedded sandstone oil reservoir or in the Donghe sandstone oil reservoir in the Hade 1 – 2 wells area. Therefore, in the Hade 4 oil field, both the Carboniferous thin bedded sandstone oil reservoir and the Donghe sandstone oil reservoir contain enormous exploration potential.

- **Sealing and Preservative Conditions of Roof and Floor in the Stratigraphic Pinch-out Zone of the Target Strata were Excellent**

The roof was the Carboniferous mudstone that was a regional cap rock; the floor was the Upper Silurian strata. The discovery of the Hadexun oil field proved that the Upper Silurian strata were non-permeable layers that provided excellent floor conditions.

According to the well data that contained adequate Silurian information (such as the Tazhong 35 well), the Silurian system could be divided into three sections. The upper section is the upper sandstone section with a thickness of 300 – 400 m; usually, this section contains the interbedded sandstone and mudstone. The middle section is the red color mudstone section with a thickness of 60 – 90 m. The lower section is the lower sandstone section that includes the bitumen sandstone portion; the total thickness of this section is 250 – 400 m. In order to understand the distributions of the Silurian strata in the Hadexun region and its peripheral area, we traced the Silurian strata. Because of the early Hercynian movement, the strata under the Carboniferous system were eroded layer by layer from the south to the north; thus, the exposed stratigraphic sequences were not only older than the Carboniferous strata, but they were also younger on the southern side and older on the northern side in geological age. On the northern side, the Yangwu area contains a lower sandstone section (the bitumen sandstone section), which is the oldest Silurian strata. On the other hand, on the southern side, the Hadexun area contains the upper sandstone section; the Carboniferous system is in contact with the younger Silurian strata in this area. The upper sandstone section was made of interbedded sandstone and mudstone. Obviously, in the Hadexun area, the underlying strata (or the floor of the Carboniferous Donghe sandstone) were the interbedded sandstone and mudstone; in addition, the sandstone consisted of argillaceous siltstone or siltstone. Via a study of the Silurian drilling cores from several wells in the Hadexun area, we learned that the sandstones commonly had unfortunate physical properties. The average porosity is 9.7%; the average permeability is 1.03×10^{-3} μm^2. However, in the area where the Donghe sandstone occurred, the Silurian sandstone had much poorer physical properties. The average porosity is 9.3%; the average permeability is 0.32×10^{-3} μm^2. The floor that controlled the stratigraphic oil reservoir (in HD 1 – 2 wells) has an average porosity of 8.87% and an average permeability of 0.21×10^{-3} μm^2. Therefore, the statistical analysis of physical properties for the Silurian system in the Hadexun area revealed that the Silurian sandstone has medium–low porosity and ultra low permeability. Because it not only has very poor physical properties, but because it was also under the Donghe sandstone, the Silurian sandstone could thus be the floor of the petroleum reservoir. The Hade 4 stratigraphic pinch-out zone has similar floor conditions to the one in the Hadexun area.

References

Gao, R.Q., Cai, X.Y., 1997. Preconditions and Distribution of Oil and Gas Fields in the Songliao Basin. Petroleum Industry Press, Beijing.

Gao, R.Q., Zhao, Z.Z. (Eds.), 2001. Oil and gas exploration of concealed reservoir in the Bohai bay basin. *In:* The Frontier Petroleum Exploration in China (vol.

3). Petroleum Industry Press, Beijing.

Hu, C.Y., 1982. Source rocks controlled distribution of oil and gas fields–the effective theology for regional petroleum exploration in terrigenous basin, eastern China. Acta Petrolei Sinica, 3(2):9-13.

Hu, J.Y., Huang, D.F., et al., 1991. Terrestrial Petroleum Geology in China. Petroleum Industry Press, Beijing.

Hu, J.Y., Xu, S.B., Tong, X.G., 1986. The development and distribution of composite oil and gas reservoirs in the Bohai bay basin. Petroleum Exploration & Development, 13(1):1-8.

Li, P.L., Jin, Z.J., Zhang, S.W., et al., 2003. The current research status and progress of petroleum exploration in the Jiyang Depression. Petroleum Exploration & Development, 30(3):1-4.

Li, P.L., Pang, X.Q., et al., 2004. The Creation of Concealed Petroleum Reservoir in Terrigenous Rifted Basin–Jiyang Depression as an Example. Petroleum Industry Press, Beijing.

Long, Y.W., 1984. Lithologic reservoir of sandstone in the northern Songliao basin. Petroleum Geology & Oilfield Development in Daqing, 3(4).

Wang, Y.C., 2001. The Creation and Distribution of Lithologic Reservoir in the Southern Songliao Basin. Petroleum Industry Press, Beijing.

Wu, C.J., Xue, S.H., 1992. Sedimentology of Petroliferous Basins in China. Petroleum Industry Press, Beijing.

Yuan, X.J., Qiao, H.S., 2002. Exploration of subtle reservoir in prolific depression of the Bohai bay basin. Oil and Gas Geology, 23(2):130-133.

Zhao, W.Z., Zhou, C.N., Wang, Z.C., et al., 2004. The signification of "Sag-wide Oil-Bearing Theory" for terrestrial hydrocarbon-rich depression. Petroleum Exploration & Development, 31(2):5-13.

Zhou, C.N., Tao, S.Z., 2007. Primary controlling factors for creating a large–medium size lithostratigraphic petroleum fields in marine carbonate rock. Chinese Science Bulletin, 52(supplement):32-39.

Zhou, C.N., Tao, S.Z., Xue, S.H., 2005. Implication of "Facies Control Theory" and its significance for petroleum exploration. Petroleum Exploration & Development, 32(6):7-12.

Zhou, C.N., Tao, S.Z., Gu, Z.D., 2006. The preconditions and distribution of large size, low abundance, lithologic oil-gas fields in China. Acta Geologica Sinica, 80(11):1739-1751.

8

Exploration Examples for Lithostratigraphic Petroleum Reservoir

In the 21st century, PetroChina has advanced the geological theory behind the lithostratigraphic petroleum reservoir and has improved exploration technology, which motivated exploration practice for lithostratigraphic petroleum reservoirs in China with positive results (Jia, 2003; Jia et al., 2005). Recently, a series of important exploration breakthroughs have been achieved in the Erlian basin, Songliao basin, Eerduosi basin, Bohai bay, Zhungeer basin and Tarim basin. Many large size lithostratigraphic petroleum reservoirs with a hundred million tons of reserves and several medium size lithostratigraphic petroleum reservoirs with $5,000 \times 10^4$ t of reserves were discovered in successive order. In the following we will discuss four exploration examples.

8.1 Exploration of Lithologic Oil Reservoir in the Bayindulan Depression, the Erlian Basin

The Bayindulan depression is located on the northeastern side of the Manite depression in the Erlian basin (Fig. 8.1); it occupies an area of 1,200 km^2. This depression not only was the origin of petroleum exploration in the Erlian basin, but was also the frontier of industrialized exploration for a terrigenous lithostratigraphic reservoir in China.

8.1.1 Exploration History

In the Bayindulan depression, the drilling exploration was initiated in 1978; however, the petroleum exploration experienced several inactive periods, which resulted in an intricate exploration history.

The first episode happened during 1977 – 1979; the geological investigation was the primary mission during this episode. In September 1977, 73.94 m of oil sand was discovered in the ZK5 hydrological investigation well during the collection of the drilling core; the maximum thickness of the oil sand layer was 25 m. From 1978 to 1979, three exploration wells were drilled around the ZK5 well, which were the Xi 1 well, the Xi 2 well and the Xi 3 well. Both the Xi 1 well and the Xi 3 well had good oil and gas shows. This discovery unveiled the curtain for a prologue for oil and gas exploration in the Bayindulan depression, the Erlian basin.

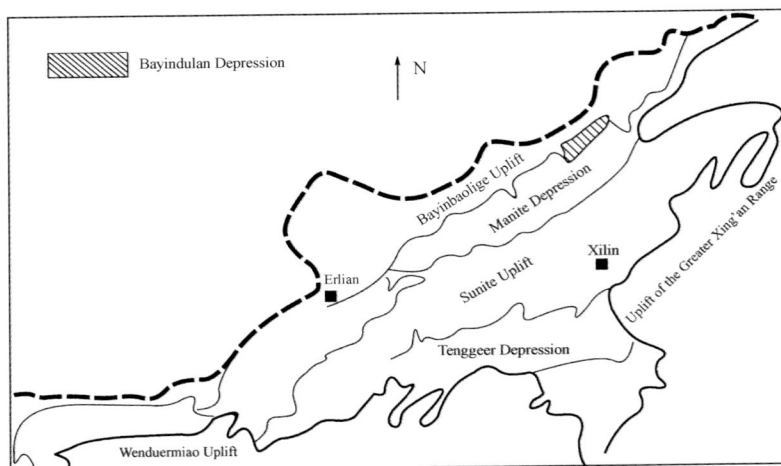

Fig. 8.1. Map of tectonic unit in the Bayindulan depression

The second episode occurred during 1980 – 1988; the main task of this episode was to determine the trough and to select the structural belt. Since 1980, drilling exploration was formerly launched in the North China oil field. Until 1988, eleven exploration wells had been drilled in the Bayindulan depression. Four wells had commercial grade oil flow, which included the Ba 1 well, the Ba 2 well, the Ba 5 well and the Ba 27 well; in addition, three wells had a low productive oil flow, which were the Ba 4 well, the Ba 9 well and the Ba 23 well. These discoveries confirm that the Bayindulan depression had the ability to generate oil and gas. The exploration results indicated that, in this depression, there are two oil generating troughs and three favorable secondary structural belts that were the Ba I structural belt, the Ba II structural belt and the Baoleng structure.

The third episode took place during 1991 – 1993; to explore the northern trough was the purpose of this episode. Around the high position of the Baoleng structure in the northern trough, eight exploration wells were drilled. Four wells had oil and gas shows, which were the Ba 32 well, the Ba 37 well, the Ba 34 well and the Badi 3 well. In the Ba 32 well, regular oil testing obtained a viscous oil flow in the interval of 175.4 – 185.4 m.

The fourth episode happened during 1993 – 1995; the exploration target was the southern trough. A 3D seismic study was designed and carried out during the exploration; the seismic network covered an area of 205 km^2. Based on the seismic information, via the selection of the target, and staying away from the procedure of pre-exploration, seven exploration wells were drilled. All of them had oil and gas shows. However, because the sandstone of major targeted sequences in the Aershan 4 member contained a thin layer, fine grains, a high content of calcium and poor permeability, the exploration did not actually achieve a breakthrough.

The fifth episode happened during 1998 – 1999; a re-exploration of the northern trough was the purpose of this episode. In 1998, because of emphasis on shallow depth exploration, on the high position of the Baoleng structure, the Ba 39 well and the Ba 201 well were drilled; in the low position of the Baoleng structure, the Ba 43 well and the Ba 45 well were drilled. A commercial grade oil flow was obtained in the Ba 43 well; however, because this was viscous crude oil with a low production rate, and because of the difficulty of mining, the discovery would bring poor economic returns.

Up until the year 2000, geologists had explored the Bayindulan depression for more than 20 years; the exploration activities had been carried out in several episodes; the targeted areas were from the northern region to the southern region and from the western region to the eastern region. The exploration project totally acquired the 2,038 km of the 2D seismic line and the 596 km^2 of the 3D seismic surveys. In addition, 37 wells were drilled; amoung these wells, 29 wells have oil and gas shows, 5 wells have commercial grade oil flow; 4 wells have oil flow with a low production rate. However, this only provided 278×10^4 t of controlled reserves. Because, in reality, the exploration did make a breakthrough, thus the Bayindulan depression became an unsolvable mystery, even it was the first place that oil and gas were discovered in the Erlian basin.

Because the exploration encountered difficulties, the research work and exploration activity have been stopped. With new knowledge, geologists agreed on the following conceptions:

Firstly, the resource evaluations from several episodes indicated that the Bayindulan depression contained an abundant petroleum resource; therefore, the expiration activities should not be terminated.

Secondly, the Bayindulan depression contained overturned structures from multiple periods; moreover, the sand bodies of different periods not only poorly coupled with current structures in space, but they also commonly occurred in a lateral position in the structures. In the past, directed by the existing ideas for exploring a structural oil reservoir, the wells were designed at the high position of the structure, which led to the failure of the exploration. With new understandings and adjusted exploration philosophy, the re-designed exploration strategy focuses on exploring the lithostratigraphic oil reservoir and emphasizes related research work.

Thirdly, a re-designed exploration strategy outlined the ideal way of exploring the lithostratigraphic oil reservoir. This highlighted macro comprehensive geologic research that was focused on the concept of "structural background, sequences stratigraphy and sand bodies". Thus, favorable exploration targets were re-selected.

Fourthly, the sand body was the study object for exploring the lithologic oil reservoir, which requires high quality seismic data. Thus, in order to identify the lithologic trap,we should improve the resolution processing of 3D seismic data and advance data quality to facilitate a good data base.

Since the year 2000, in the Bayindulan depression, according to a re-designed exploration strategy, we emphasized research into accumulation conditions for a lithostratigraphic reservoir and carried out related petroleum exploration. Finally, we obtained a breakthrough. The following discussion will demonstrate the exploration results.

8.1.2   Essential Geological Background and Favorable Exploration Scenario

● Study of Structures for Seeking the Geological Background

In the Bayindulan depression, the structures were severely overturned during the late period.The most common type of structure was the reversal structure, which included nose-like structures of the Ba I and the Ba II in the southern trough and the Baoleng structure in the northern trough; all of the structures formed a favorable structural belt for creating oil reservoirs (Fig. 8.2). These structural belts were overturned during different geological periods. From the southern side to the northern side, the ages of the structural belts become gradually younger.

● Study of Stratigraphic Sequences for Determining the Framework

According to the study results of the depositional evolution and the characteristics of lithologic conductivity, and also considering the combined factors of well data and seismic data, we set up a framework of depositional sequences for the Bayindulan depression. The lowstand system tract, the lacustrine transgressive system tract and the highstand system tract were identified in the Aershan formation and the Teng 1 member. The main exploration target was the highstand system tract in the Aershan 4 member that was positioned between two lacustrine transgressive system tracts of the middle Aershan formation and the early Teng 1 member; this target contained a good petroleum play (Fig. 8.3).

● Study of Depositional Facies in Search of Sand Body

The study results indicated that sand bodies in the Aershan 4 member were formed either in a braided channel of the fan delta or as the wedged sand body on the front margin of the fan delta; they were controlled by the flat shaped paleo-topography on the steep slope in the southern trough. In the Ba I and Ba II structural belts, two highstand system tracts of the fan delta were developed on the northern side and the southern side, which broadly extended and which contained well developed

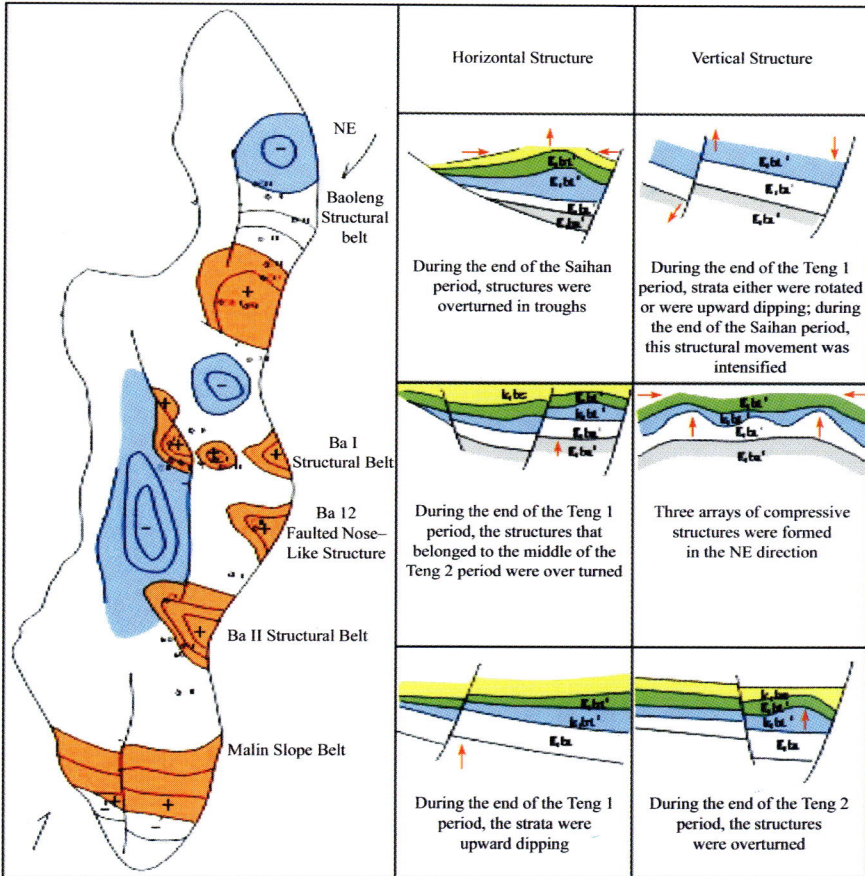

Fig. 8.2. Divisions of structural belt and their developmental model in the Bayindulan depression, the Erlian basin

braided water channels and numerous sand bodies. The distribution of the sand bodies was mainly controlled by the structural background at the time of their deposition; in addition, these sand bodies did not couple well with current structures; the sand bodies mostly consisted of fine grain sandstone with a higher content of quartz and feldspar and with a thickness of 100 m approximately.

● Study of Reservoir Body in Hunt for Reservoir Sequences

The study of diagenesis indicated that, in the Bayindulan depression, the fan delta deposits in the Aershan 4 member predominately consisted of fine grain sandstone; the sandstone was in phase A of late diagenesis procession with strong dissolution. Furthermore, the study of porosity revealed that the buried depth of the Aershan 4

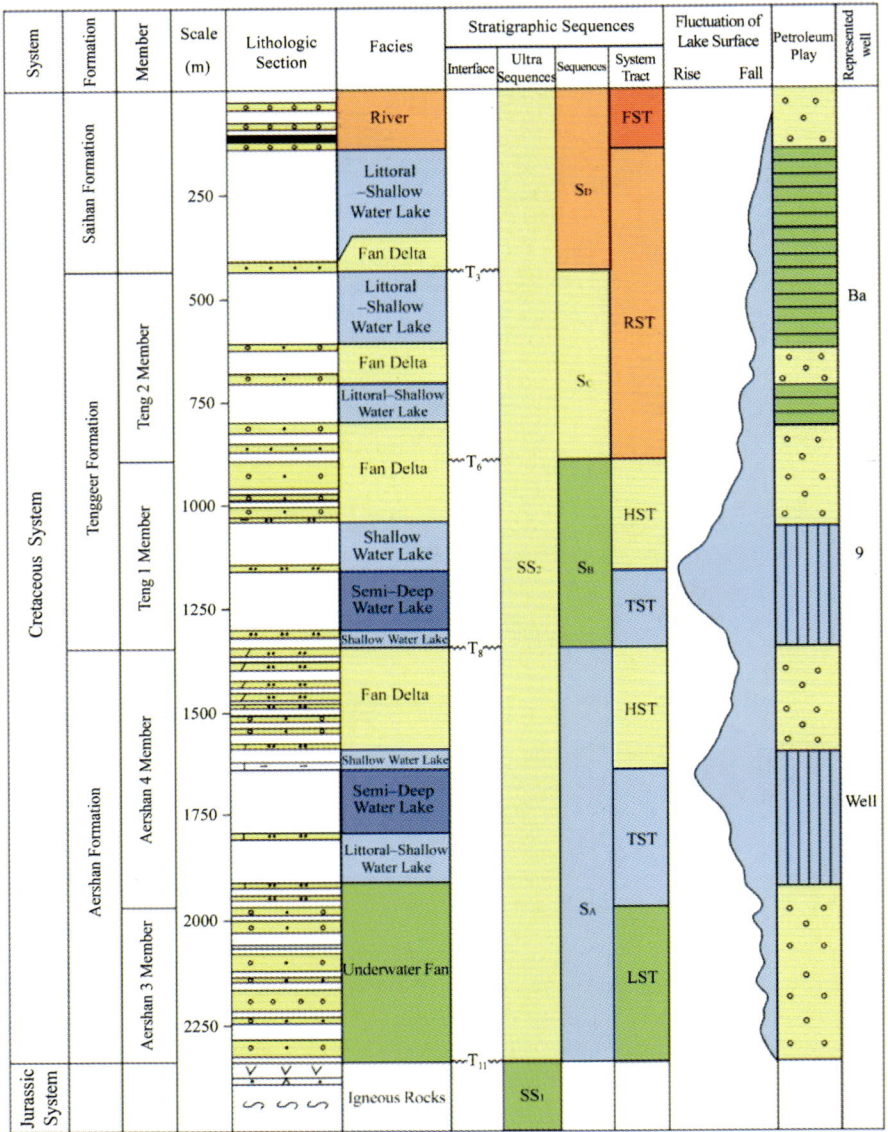

Fig. 8.3. Stratigraphic sequences analysis of the Lower Cretaceous series in the Bayindulan depression, the Erlian basin

member took place in the zone of secondary pores; the sandstones had well developed intergranular dissolution pores and intragranular pores, which offered perfect reservoir conditions. Thus, the Aershan 4 member was the primary exploration target in the depression.

● **Comprehensive Study for Selecting the Exploration Target**

The comprehensive study showed that, on the steep slope zone of the Bayindulan depression, the reversal tectonic movement formed the nose-like structures of the Ba I and the Ba II; the main portion of the fan delta sand bodies occurred at the lateral position of the structure; the sand body occurred in the highstand system tract in the Aershan 4 member and it contained the progressive sequence. The upward dipping, pinched-out type of lithologic trap would be formed in the middle and lower position of a positive structure. In addition, the sand bodies were enveloped by the dark color lacustrine mudstone that would directly supply hydrocarbons. Thus, these structures had the preconditions for the creation of a lithologic oil reservoir. According to the information, we selected the exploration targets at the lateral position of the Ba I structure and the Ba II structure on the steep slope in the southern trough and the Baoleng structure in the northern trough.

8.1.3 Emphasis on the Exploration of Lithologic Oil Reservoir in the Ba II Structure in Order to Achieve Strategic Breakthrough

The Ba II structure is located on the southern side of the southern trough, which is a large scale reversal type of nose-like structure and which is controlled by the Ba II fault. Up to the end of 2000, five exploration wells had been drilled in this structure, which were the Ba 2 well, the Ba 6 well, the Ba 16 well, the Ba 7 well and the Ba 9 well; two of the wells had an oil flow with a low production rate, which indicated this structure contained preconditions for creating oil and gas reservoirs. At same time, the discoveries also revealed two kinds of problem in this area:

The first was that the reservoir sequences at the higher position had poor physical properties; the oil test resulted in a low production rate. The second was that the relationship of oil to water was unclear; at the lateral position, the oil and water appeared in the same layer in both high and low places.

● **Reprocessed Seismic Data to Improve the Quality of Seismic Cross Section**

In 1993, on the steep slope of the southern trough in the Bayindulan depression, we collected 3D seismic data of 204 km^2; however, these were low quality seismic data because of lower resolution and poor continuity. In 2000, based on the study of exploration potential, we re-processed 3D seismic data; the data quality was significantly improved with enhanced reflective information, which constructed the data base for the in-depth research work and for discovering and identifying the lithologic object.

● **Review of Old Wells to Obtain Useful Information**

With the reviewing of old wells, we identified the problem of the oil reservoir in both the Ba 9 well and the Ba 6 well. The Ba 9 well was located at a lower position of the Ba II nose-like structure; the oil test for the Aershan 4 member pumped out both oil and water. The Ba 6 well was located at the high position of the structure; the oil test for the related strata also pumped out oil and water; there was no fault to divide them. Thus, there was a relationship problem between oil and water.

The facies analysis of a single well indicated that these two wells belong to different depositional systems. The Ba 9 well contained deposits of fan delta front facies; the Ba 6 well contained a beach bar sand body of littoral–shallow water lacustrine facies. Thus, we constructed a new depositional model.

Based on the knowledge discussed above, via the detailed sequences marking, the tracing of the sand body proved that, in the Ba 9 well, the oil bearing sand body was pinched out toward the direction of upward dipping, which was not the same sand body as in the Ba 6 well.

● **Construction of Accumulative Model of Lithologic Oil Reservoir for the Ba 9 Well**

With a comprehensive geological study, the model of the lithologic oil reservoir was re-constructed (Fig. 8.4). The exploration target was the II sandstone group in the Aershan 4 member, the Aershan formation; from the Ba 9 well to the Ba 6 well, this sandstone group gradually dipped upward and was pinched out. Therefore, the fan delta sand body in the Ba 9 well area was selected as a penetrating point for exploration of the lithologic oil reservoir in the Bayindulan depression.

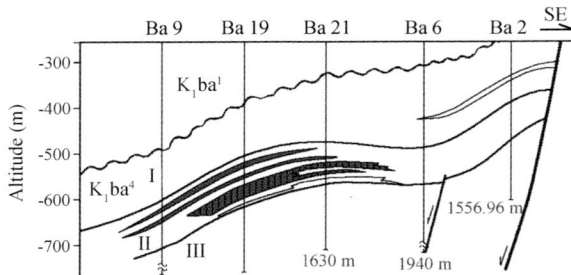

Fig. 8.4.   The accumulative model of lithologic reservoir in the Ba II structure, in the Bayindulan depression, Erlian basin

● **Forecasted Reservoir Sequences, Carefully Inspected Traps**

Using the software of Strata and ISIS, we studied this fan delta sand body; the study results clearly displayed the 3D outline of the sand body in the Aershan 4 member;

in the vertical section, the sand body is of lenticular shape; horizontally, the sand body is in a fan shape. Using seismic analysis, we forecast the distribution area of the sand body and the lithologic trap was circled within an area of 12 km^2 initially.

● **Selected Exploration Well to Carry out the Breakthrough**

With studies of the deposition, forecasting of the reservoir sequences, and the characteristics of the oil reservoir, according to the rule of "maximum similarity", in 2001 the Ba 19 well was designated at the location that had similar seismic facies as the high position of the Ba 9 well (Fig. 8.5). In the 1,430 – 1,530 m interval of the Ba 19 well, the thickness of the interpreted oil layer is 25.2 m/12 layers; the oil testing was conducted at an interval of 1,483 – 1,488 m and the thickness of the oil layer is 5.6 m. Daily oil production was 29.24 t. Another oil test was conducted at an interval of 1,527.2 – 1,530.2 m and the thickness of the oil layer was 3 m; using a 15.875 mm chock, the daily production was 21.96 t. These discoveries proved that the structural–lithologic oil reservoir of the Ba 19 contained oil, which achieved an historical breakthrough in petroleum exploration in the Bayindulan depression. After that, the beginning of the exploration of the lithostratigraphic oil reservoir in the Bayindulan depression and even in the entire Erlian basin has commenced.

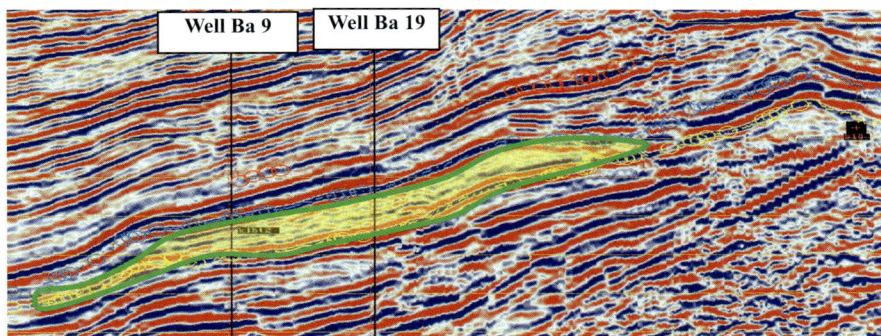

Fig. 8.5.   Seismic cross section of the Ba 9 well–the Ba 19 well in the Bayindulan depression, the Erlian basin

● **Pursued Forecasting and Drilling Exploration, Discovered Commercial Grade Oil Flow**

After the success at the Ba 19 well, in order to duplicate the exploration results at the Ba 19 well, we carried out a second round of reservoir forecasting and successfully drilled the Ba 21 well. The well logging interpretation indicates the thickness of the effective oil layer is 37 m/12 layers; the daily production rate is 20.55 t.

● **Pursued Appraisal and Drilling Exploration, Increased the Oil Reserve**

After that, the third round of reservoir forecasting was in progress. According to the exploration concept of "search the enriched accumulation in the middle position of reservoir sequences; explore the lithologic pinching out on a higher position; identify the interface of oil and water at a lower position", three wells were designed, which were the Ba 18 well, the Ba 20 well and the Ba 22 well. All of the wells successfully achieved the designed targets. In the Ba 18 well, the interpreted thickness of the oil layer was 34 m/18 layers; the oil test obtained a highly productive commercial oil flow of 58 m^3.

● **Description of Oil Reservoir, Calculation of Oil Reserve**

Using the multi-diversification identifying forecasting method (MDI) to describe the oil reservoir, the oil layer of II oil group in the Aershan 4 member in the Ba 19 well was studied in the areas of effective thickness, porosity and saturation of oil content; these studies not only provided detailed information about the reservoir, but they also determined the size of the oil reserve. Therefore, over a relatively short duration, we discovered the entire Ba 19 lithologic oil reservoir (Fig. 8.6). Within the same year, we not only identified a reservoir area of 8.2 km^2, but we also provided a geological reserve of $1,241 \times 10^4$ t; in addition, we developed the oil field with a productivity of 1.89×10^4 t. Thus, the Baolige oil field had been discovered.

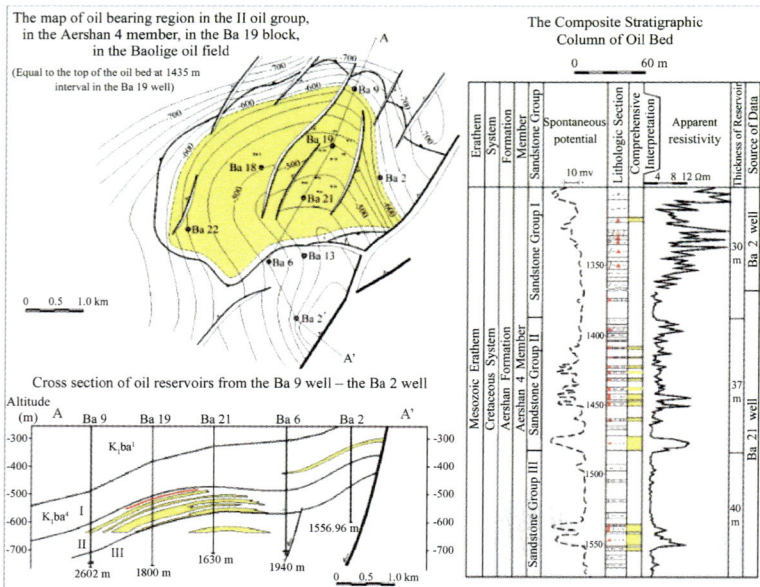

Fig. 8.6. Composite map of oil reserve of the Aershan 4 Member in the Ba 19 block, the Baolige oil field, in the Bayindulan depression, the Erlian basin

8.1.4   Enhanced Exploration Achievement Developed the Annual Production Rate of a Quarter of a Million Tons

After the breakthrough in the Ba 19 oil reservoir, a progressive study and exploration were carried out. In the east of the Ba 19 area, we discovered another lobe of the sand body that shared the same substance source as the Ba 19 fan delta sand body. In the seismic cross section, the northern side of the new sand body obviously has the potential for a pinch-out; the major portion of the sand body has middle–low frequency, poor continuality and warm like seismic facies; however, on the lateral side, it has low frequency and relatively continual seismic facies. These two sand bodies were connected on the saddle of the structure and they have similar accumulative conditions.

According to the seismic facies analysis and the result of reservoir sequences forecasting, on the southern side that contained a well developed sand body and on the northern side, we drilled the Ba 38 well and Ba 42 well with success. In the Ba 42 well, the well logging interpreted an oil layer in the Aershan formation; the total thickness of oil layers and poor oil layers were 56 m/23 layers; the swabbing test was conducted in the $1,496.4 - 1,505$ m interval; the obtained daily production rate was 55.2 m^3. In the same year, the Ba 38 oil reservoir provided a controlled oil bearing area of 6.7 km^2 and a controlled geological reserve of $1,039 \times 10^4$ t.

After the exploration breakthrough at the Ba II structure, we re-considered the Ba I structure and constructed models of the unconformity oil reservoir, stratigraphic overlapping oil reservoir and the structural–lithologic oil reservoir. We also drilled seven exploration wells with success; these wells included the Ba 48 well, Ba 36 well, Ba 40, Ba 24 well, Ba 28 well, Ba 51 well and Ba 54 well in successive order. Up to 2004, the Ba 10, Ba 48, and Ba 52 oil reservoirs together provided a total explored reserve of $1,062 \times 10^4$ t.

Up until the end of 2004, through four years of exploration practice, in the Bayindulan depression, the Baolige oil field now contains 156 developing wells with a productivity of 25.32×10^4 t. The Ba 19 oil reservoir contains 82 developing wells with a productivity of 15.21×10^4 t; the Ba 38 oil reservoir contains 48 developing wells with a productivity of 6.96×10^4 t; the Ba 48 oil reservoir contains 26 developing wells with a productivity of 3.15×10^4 t. In 2004, the total crude oil production was 21.0469×10^4 t.

In the Erlian exploration region, because of the discovery and development of the Baolige oil field, rapidly decreasing oil and gas productivity not only had been stopped, but it also increased with the stabilization.

After the exploration breakthrough in the Bayindulan depression, according to the exploration concept of "emphasis on exploring the lithostratigraphic oil reservoir; further intensification of the forecasting", we expanded the exploration success. The new discoveries were achieved in the Wuliyasitai depression, the Jiergalangtu depression and the Saihantala depression, in the Erlian basin.

8.2   Exploration of Lithologic Oil Reservoir in the Changling Depression, the Southern Songliao Basin

The Changling depression is one of four second-order structural units that are located in the central depression in the southern part of the Songliao basin; this depression covers an area of 6,712 km^2. The Changling depression is a large size, synclinal type of negative structure (Fig. 8.7), which includes an extensive exploration area. In recent years, because of emphasis on lithologic reservoir exploration, and because of the focus on two critical factors of the Changling oil rich depression and the front margin facies belt in the Baoqian depositional system, we advanced research work on petroleum accumulative conditions and highlighted studies of the accumulative model and related mechanism in the negative structure; in addition, we conducted an in-depth study of the petroleum distribution pattern. Conclusively, we reached new understandings about a "petroliferous rich depression" (Wang, 2001). At the same time, we utilized high accuracy 3D seismic technology, a fractural technique for the low permeable oil layer, and combined information about exploration forecasting and reservoir appraisal to develop a comprehensive exploration approach for "directing with knowledge, supporting with technology, and managing with efficiency". Our works confirmed that the Changling depression contains a large accumulative area of a lithologic oil reservoir, which has an exploration potential of 3.0×10^8 t. Since 2000, the total geological reserve of oil has built up to 1.5139×10^8 t; furthermore, an oil production region of a million tons has been established. The exploration practice in the Changling depression brought about the experience of exploring a large size, terrigenous, lithologic oil reservoir in China.

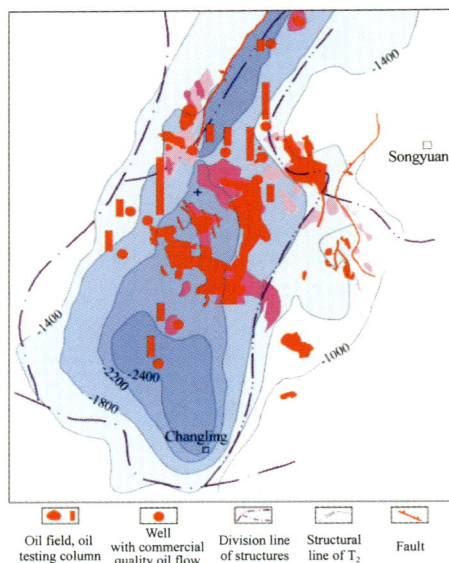

Fig. 8.7.   Map of exploration achievement in the Changling depression, Songliao basin

8.2.1   Exploration History

Petroleum exploration in the Changling depression was initiated during the 1950s. Up to date, 262 wells have been drilled in the depression, which includes all kinds of wells; the 2D high resolution seismic network has covered the entire region with a survey density of 1 km×1 km; additionally, the 3D seismic survey that covered an area of 1,555.21 km^2 has been carried out in the Qianxibei, the Daqingzi Well and the Damasu areas. Commercial grade oil (gas) flows were obtained in the Heidimiao oil layer of the upper petroleum play, in the Saertu oil layer, the Putaohua oil layer and the Gaotaizi oil layer of the middle petroleum play, and in the Fuyu oil layer of the lower petroleum play. This area is an oil bearing region with multiple reservoir sequences. At present, the discovered oil fields include the Qianan oil field, Haituozi oil field, Daqingzi Well oil field and Qianxibei oil field; the total geological reserves are 2.5032×10^8 t.

Since petroleum exploration in the Changling depression started in 1956, the exploration projects have taken a difficult course; the following will discuss this exploration history in detail.

- **1956 – 1963: Initiated Regional Geologic Survey and Discovered the Heidimiao Oil Layer**

The petroleum geological survey was initiated in 1956; by using simulated seismic technology, in 1959 both the Shuaizi Well structure (it is also called the Qianan structure) and the Heidimiao structure were discovered in the Changling depression. The study results indicated that the Changling depression was an inherited type of depression with a long developmental duration; geologists presumed that it contained good hydrocarbon generating potential. Under the guideline of a "source controlling" concept, in 1960 the Shuai 1 well was drilled in the Qianan structure; oil sand and oil bearing sandstone were discovered in the Yaojia formation. This discovery confirmed that the Yaojia formation had the ability to store oil and gas. In 1962, because the drilling activity was focused on exploring the structure, the Hei 1 well was drilled in the Heidimiao structure; On July 22nd, 1962, by using the bailing method, an oil test was conducted in the Nen 4 member; the oil test showed that the daily production rate of oil was 4.6 m^3 and the daily production rate of water was 1.8 m^3. This discovery was the first one to pass the threshold for a commercial grade oil flow in the Changling depression and it meant the discovery of the Heidimiao oil layer. After that, the Hei 2 well was drilled; however, it did not have an oil and gas show.

The acquired knowledge for this exploration episode was as follows: the oil and gas distributions were controlled by structures that were of small size; in addition, there were only small numbers of deeply buried oil layers. Therefore, the exploration risk was great. The exploration activity was stopped.

● **1975 – 1977: Expanded Exploration Region and New Discoveries Obtained in Middle and Lower Petroleum Plays**

In the 1970s, against the background of the discovery of the Fuyu oil field alongside the exploration of the Fuyu oil layer in the Fuxin uplift belt, the concept of second order structural unit controlled oil was understood. Guided by this concept, the exploration was carried out and it targeted the Fuyu oil layer in the second order structural unit of the Changling depression. Furthermore, the Qianshen 1 well–the Qianshen 10 well were drilled in the Qianan structure; good oil and gas shows appeared in the Fuyu, Gaotaizi, Putaohua and Heidimiao oil layers. The oil tests obtained a commercial grade oil flow and marginal oil flow in the Fuyu, the Gaotaizi, and the Putaohua oil layers from the Qianshen 1, 2, 3, and 4 wells. This discovery confirmed that the Qianan structure contained multiple oil layers. Because of the old-fashioned drilling and testing technologies, the oil test yielded lower productivity (less than 3.0 t). In addition, a mixture of oil and water or mixture of oil, gas and water commonly occurred. Thus, the future of petroleum exploration was questioned.

The acquired knowledge for this exploration episode was as follows: The Qianan structure contained multiple oil layers with a low production rate; the oil layers were evidently altered and deeply buried. Thus, the exploration efficiency was low. Guided by this understanding, exploration activity in the Qianan structure was held back for the second time.

● **1984 – 1989: Discovery of the Qianan Oil Field**

In the 1980s, in company with the development of exploration theory, geologists gradually understood that the delta front facies controlled oil. Because of in-depth research for petroleum accumulative conditions on the front margin of the sand body in the southwestern Baoqian area, geologists discovered the upward dipping and pinched-out type of lithologic oil reservoir, which was located on the front margin of a sand body in the Qing 3 member, and which was in the background of the Qianan reversal structure. Guided by this concept, at first geologists reviewed the old wells and utilized new technology for compressing fractures. On May 18, 1984, a compressed fractures oil test was carried out in the Qianshen 10 well; the testing target was the Qing 3 member; this oil test obtained a daily production rate of 9.3 t, which passed the threshold of high yield oil flow in the Gaotaizi oil layer of the Qing 3 member. The succeeding oil tests in the old and new wells obtained high yield oil flows of 8.841 t and 11.2 t, in the Qianshen 8 well and the Qianshen 10 well respectively. These test results further confirmed the exploration value of the Gaotaizi oil layer in the Qing 3 member, in the Qianan area. In 1985, 161.9 km^2 of an oil bearing region was confirmed and $6,902 \times 10^4$ t of geological oil reserves discovered; in addition, the Qianan oil field was established with a productivity of 20×10^4 t/a.

After discovery of the Qianan oil field, in order to increase the exploration success, the exploration activities were expanded towards the southwestern side of the Qianan area. According to normal 2D seismic data, the low relief structural trap was located and five wells were drilled; these wells included Qian 103, 104, 109, 124 and 139 wells; however, the Qian 103, 104 and 109 wells did not have a casing pipe. In the Qian 124 well, the oil test was carried out for the Gaotaizi oil layer in the Qing 1 member and the Qing 2 member, which obtained commercial grade oil flows of 3.92 t and 6.22 t respectively. In the Qian 139 well, the oil test was conducted in the Gaotaizi oil layer in the Qing 1 member; this oil test obtained a marginal oil flow of 1.1 t. These oil tests conformed that, in the Qianan area, the low relief structure contained oil.

The acquired knowledge for this exploration episode was as follows: The front margin of the Baoqian sand body controlled oil and gas distribution of the Gaotaizi oil layer in the Qing 3 member to create an upward dipping and pinched-out type of lithologic oil reservoir under the structural background. In the Heidimiao area, in the main portion of the front margin of the Baoqian sand body, the distribution of oil and gas was controlled by the low relief structure. Because the oil layer was deeply buried, confirming the low relief structure was therefore very difficult; also. the exploration wells provided a low success rate and low production rate. Therefore, the economic efficiency of the exploration was poor. Again, the exploration activity was held back for a third time.

- **1993 – 1995: Re-Exploration of the Changling Depression and Discovery of Oil Bearing Area of Qianshen 12 Well Area**

In 1993, in achieving an additional exploration success in the northern part of the Qianan oil field, under the guideline of the "delta front facies controlled oil" theory, utilizing the high resolution 2D seismic data and the results of reservoir sequences forecasting from wave impedance inversion, and aiming at the middle and lower petroleum play of the target sequences, the drilling exploration was carried out on the northern slope in the Qianan oil field. Seventeen wells were drilled in total that included the Qianshen 12 well and the Qianshen 13 well; four of the wells produced commercial grade oil flows. In the Qianshen 12 well, the oil test produced a highly productive oil flow of 13.86 t/d in the Gaotaizi oil layer of the Qing 3 member; the oil test also produced a commercial grade oil flow of 8.25 t/d in the Putaohua oil layer. In 1995, the confirmed oil bearing area was 8.6 km^2; the discovered geological reserve was 193×10^4 t.

While exploring the northern region of the Qianan oil field, expanded exploration activity in the Heidimiao area (that was located on the southern side) was also carried out. Three wells were drilled to target the Heidimiao oil layer; these wells included the Hei 40, 41 and 42 wells. Only the Hei 41 well had an oil and gas show from the Nen 4 member; however, the oil test did not produce an oil flow.

The acquired knowledge for this exploration episode: In the Qianshen 12 Well area, the Gaotaizi and the Putaohua oil layers in the Qing 3 member were controlled by lithology, which formed a lenticular sandstone oil reservoir; however, because the reservoir sequence had marginal thickness with rapid alteration, the successful rate of exploration was low. In the Heidimiao oil layer, the oil and gas distributions were controlled by lithology too. Because the alterations in reservoir sequences were evident, the successful rate of exploration was low and the risk was great. Therefore, the exploration activity was held back for the fourth time.

- **1997 to the Present: Explored Lithologic Oil Reservoir in the Negative Structure and Discovery of the Hundred Million Tons Oil Field in the Daqingzi Area, which Transformed the Changling Depression into a "Petroliferous Rich Depression"**

During 1997 – 2002, the experts summarized exploration practices in the Changling depression and further modified exploration philosophy; in addition, they also fully comprehended the synclinal type of negative structural accumulative model in the Changling depression and introduced a new theory for exploring oil in the syncline. It was considered that, if the front margin belt of the Baoqian sand body combined with the "petroliferous rich depression" of Changling, the geological setting could contain the necessary preconditions for creating a large size lithologic oil reservoir. Under the guideline of a new exploration concept, utilizing high resolution 2D seismic data for detailed interpretation, geologists carried out drilling exploration for a group of structural and lithologic exploration targets in the Daqingzi Well area. First, in the Hei 43 well that was designed to explore a structural target, the oil test obtained a highly productive oil flow of 11.38 t in the 2,242.4 – 2,239.0 m interval that was positioned in the Gaotaizi oil layer of the Qing 2 member; moreover, in the Putaohua oil layer, the oil test obtained 5.86 t of commercial grade oil flow in the 1,868.4 – 1,865.2 m interval. After the exploration success for the structural target, for swiftly confirming the exploration potential of the lithologic oil reservoir in the Daqingzi Well area, the 3D seismic survey was accomplished and high resolution 3D seismic data was utilized to describe reservoir sequences in detail, with the intention of directing the design of exploration wells. The Hei 46 well and the Hei 50 well were designed to explore the lithologic targets; the oil tests obtained a highly productive oil flow of 18.5 t/d and a commercial grade oil flow of 2.01 t/d respectively. These discoveries passed the thresholds for commercial grade oil flow and a highly productive oil flow for the lithologic oil reservoir in this area. With the evaluation of the drilling exploration and the comprehensive evaluation of the reserves, the confirmed geologic oil reserve was $12,559 \times 10^4$ t. Thus, in the synclinal area of the Changling depression, the lithologic oil exploration was initiated with a bright exploration future in the middle petroleum play of the central region.

In 2003, while exploring the central structural belt of the Changling depression, 3D seismic data were utilized and the study results of micro facies expanded exploration activity in all directions. On the northern side, the Qianan trough was explored. In the Qian 162, 163, 164 and 165 wells, the oil tests obtained a highly productive oil flow in the Gaotaizi oil layer of the Qing 3 member; the testing production rate was larger than 10 t. As a result, the Qianxibei oil field was discovered which had a discovered geological reserve of $2,580 \times 10^4$ t. On the eastern side, old wells were reviewed and exploration activity expanded. In the Qian 183, 188 and 191 wells, the oil tests obtained a good oil flow in the Gaotaizi oil layer of the Qing 3 member; the testing production rate was larger than 5 t. A forecast geological reserve of oil of $3,000 \times 10^4$ t was initially confirmed. On the southern side, the Heidimiao trough was explored. In the Hei 100, 144 and 152 wells, oil tests obtained a highly productive oil flow in the Putaohua oil layer of the Yao 1 member; the testing production rate was larger than 10 t. Initially, the magnitude of the reserve was considered to be at a level of $3,000 \times 10^4$ t. On the western side, the slope zone was explored. In the Hua 29 and 30 wells, oil tests obtained a highly productive oil flow in the Gaotaizi oil layer of the Qing 1 member; the testing production rate was larger than 10 t. The forecast geologic reserve was confirmed at $2,500 \times 10^4$ t. Thus, in the Changling depression, the exploration of the middle petroleum play was declared a success in all directions, with the potential of a "petroliferous rich depression".

During 2004 – 2005, while exploring the middle petroleum play in the Changling depression, the study was also purposely intensified of the accumulative mechanism for the lithologic oil reservoir with the low permeable river facies in the Fuyu oil layer of the Quan 4 member. This study created a new petroleum accumulative theory for a low permeable, large size, composite lithologic petroleum reservoir in the Fuyu oil layer. The theory was "the alteration of the facies controlled sand body; oil and gas migration vertically; an ultra high pressure controlled reservoir and superimposed and overlapping sand bodies connecting the petroleum reservoir". The study results indicated that the Fuyu oil layer had preconditions for the development of large size, composite, lithologic oil and gas reservoirs of river facies; the study results also led to an understanding of the "petroliferous rich depression" in the Fuyu oil layer. Guided by this concept, in the Huaaopao area, the exploration wells achieved a new breakthrough. In the Qian 182, 184 and 185 wells, oil tests achieved a highly productive oil flow in the Fuyu oil layer; the testing production rates were 13.2 t, 14.5 t and 32.2 t respectively. These discoveries indicated that the Fuyu oil layer could achieve good exploration efficiency. It was confirmed that the forecast reserve was in the area of a hundred million tons. The following exploration activities on the northern side of the Qianan–eastern side of Qianan achieved an important success that displayed the exploration potential of a hundred million tons reserve. Thus, in the Changling depression, a large size lithologic oil reservoir had been discovered, which had an oil reserve of 3.0×10^8 t and which was related to the negative structure. The "petroliferous rich depression" was discovered.

The acquired knowledge for this exploration episode was as follows: The Changling depression is a petroliferous rich depression. The reservoir sequences were superimposed vertically to form the oil bearing system with multiple reservoir sequences. The Fuyu oil layer not only had preconditions for creating a large size, composite, lithologic petroleum reservoir of river facies, but it could also be developed into a "petroliferous rich depression". The combination of the front margin belt of the Baoqian sand body and the background of the synclinal structure in the Changling depression provide conditions for the development of a lithologic oil reservoir over a large area. The superimposed region of dark color mudstone in the Qing 1 member and sandstone controlled distribution of the Gaotaizi oil layer. The distribution of the Gaotaizi oil layer had a reliable pattern; if the thickness ratio of sandstone and strata was larger than 40%, the oil reservoirs mainly belonged to the structural type; if the thickness ratio of sandstone and strata was between 10% – 40%, most of the reservoirs either belonged to the lithologic oil reservoir or belonged to the faulted–lithologic oil reservoir; if the thickness ratio of sandstone and strata was less than 10%, the oil reservoirs belonged to the lenticular sandstone oil reservoir.

During 1998 – 2004, in the Changling depression, the upper, the middle and the lower petroleum plays and two oil fields that were the Daqingzi Well oil field and the Qianxibei oil field were discovered; the total geological petroleum reserve was $15,139 \times 10^4$ t. Additionally, in the Huaaopao area and in the northern Qianan–eastern Qianan area, two reserve areas of a hundred million tons scale were also discovered. Thus, the Changling depression has an explored reserve potential of $(3–5) \times 10^8$ t.

8.2.2   *New Information on Accumulative Conditions for Lithologic Oil Reservoir over a Broad Region*

● **Large Size Petroliferous Rich Depression Provided Material Foundation for Creating Lithologic Reservoir over a Broad Region**

In the Changling depression, the hydrocarbon source rocks were mainly developed in the dark color mudstones of the Qingshankou 1 member, the Nenjiang 1 member and Nenjiang 2 member. In the Qingshankou 1 member, the thickness of dark color mudstone was around 80 – 100 m, which occupied 20% of the total thickness of the strata. This mudstone contained highly abundant organic materials with good types of kerogen (that mostly belonged to type I–type II_1); because these organic materials were at the mature stage, they therefore primarily generated oil. The results of natural resource evaluation demonstrated that the total generated oil was 45.73×10^8 t and the total generated gas was 2.004×10^{12} m^3. The plentiful oil and gas resources set up the material foundation for a large area lithologic oil and gas reservoir.

● **Large Scale Delta Front Sand Body Offered Excellent Storage Space for Petroleum Accumulation**

In the Changling depression, while the Quantou formation–the Nenjiang formation were being deposited, under the control of the Songliao basin, four different depositional systems were being developed, which were the Changchun–Huaide depositional system (I) on the southeastern side, the Tongyu–Baokang depositional system (II) on the southwestern side, the Yingtai–Honggang depositional system (III) on the western side and the Northern depositional system (IV). During different geological periods, the developmental scale of these depositional systems varied (Fig. 8.8); moreover, these depositional systems controlled the distribution of petroleum reserves.

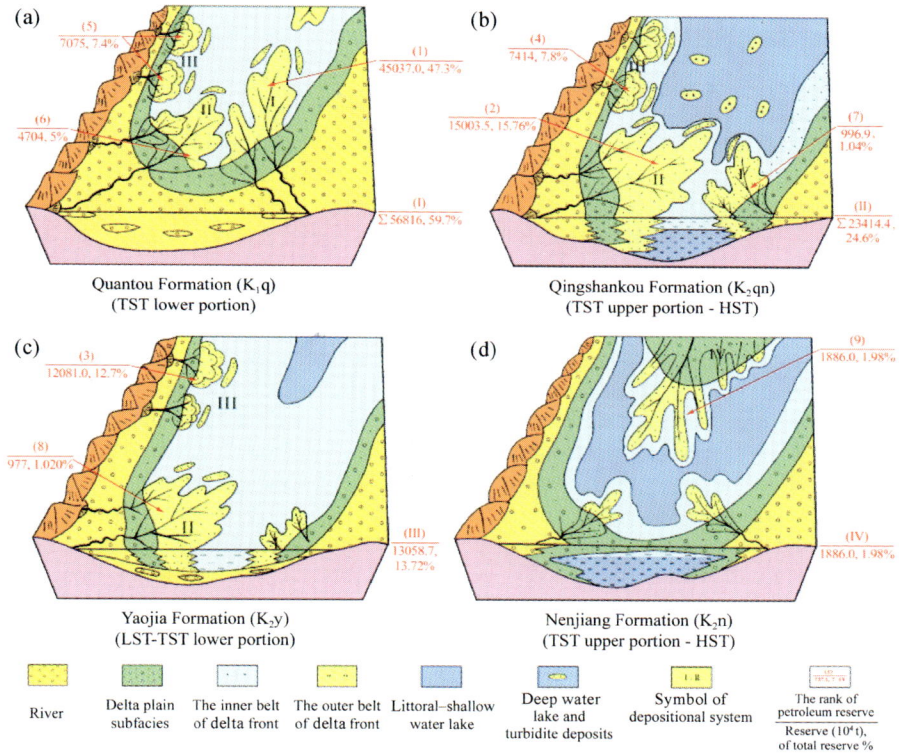

Fig. 8.8.   Illustrations of depositional model and related distribution of petroleum reserve in the southern region of the Songliao basin, from the Quantou formation to the Nenjiang formation

Changchun–Huaide Depositional Systems in the Southeastern Region

Two aquatic systems were developed on the southeastern side of the basin; one was from Huaide–Shuangtuo–Dalaoyefu–western Gudian–Qianguo; the other was from Changchun–western Nongan–Qianguo. These two aquatic systems were united at the Qianguo area; together, they run into the lake; as a result, the Fuxin delta system was developed over a broad area that was located to the east of the Fuyu–Mutou–Xinmin–Xinmiao–Xinli area. The best developmental period for the Changchun–Huaide acquatic systems was when the Quantou formation was deposited; after that, the aquatic systems became gradually latent. Therefore, the Quan 4 member (that contained shallow water delta deposits) was well developed in the southern part of the Songliao basin. The sand bodies of the Qing 1 and the Qing 2 members could extend to the area of the Qian 133 well; however, the Qing 3 member was retrieved to Shuangtuozi and the south of Nongan. The depositional system predominantly developed when the Quantou formation was deposited; after that, it gradually became latent.

Tongyu–Baokang Depositional Systems in the Southwestern Region

Along the long axis of the basin, the Tongyu–Baokang depositional system extended into the lake near the Tongliao area and the Baokang area, on the southern side of the Songliao basin. From the geological point of view, the Tongyu–Baokang depositional system passed by several first order structural units that were the Kailu depression, the southwestern uplift, the western slope and the central depression; from the geographical point of view, it was located in the areas from Tongyu–Baokang–Qianan–Changling. In the Qingshankou formation, the sand body was relatively stable in its shape, direction of the axis and the area of distribution. The orientation of the sand body was nearly parallel to the long axis of the basin. The belt of higher sand content extended towards the Qianan area along the Bao 7 well–Seven 1 well–Songnan 6 well; it was separated in the Changling area and the Daqingzi well area. During the time when the Qing 1 and the Qing 2 members were deposited, the sand body reached the Qian 110 well area. During the time when the Qing 3 member was deposited, the sand body was not only developed in the area of the Cha 22 well, but it also formed an extensive delta sand body in the Qianan area. From the south to the north, the sand body could be divided into river facies, delta facies, littoral–shallow water lacustrine facies and semi-deep water–deep water lacustrine facies. This deposition system was favorably developed during the time when the Quantou formation–Yaojia formation were deposited; it was the major depositional system in the Changling depression that contained the best developed sandstone and the largest quantity of discovered petroleum reserves.

Yingtai Depositional System

The Yingtai depositional system included deposits that were created by the Tailai–Baicheng water system. The northern boundary of the Yingtai depositional system was located to the south of the Binzhou railroad; its southern border was located near the Tongyu area; eastward, it extended into the Gulong area. Along the short axis of the basin, the Yingtai depositional system extended into the lake. In the area between Tailai and Baicheng, because between 4 and 6 parallel, inter–mountain, seasonal, braided rivers flowed into the lake, several secondary depositional systems were formed. The secondary depositional systems of Zhenlai–Yingtai and Baicheng–Honggang were of a relatively large scale. The Zhenlai–Yingtai aquatic system created a fan delta system that was perpendicular to the long axis of the basin and was located near the source area with a short traveling distance. The Baicheng–Honggang aquatic system formed a braided river delta system. The Yingtai depositional system consisted of alluvial fan facies, braided river facies, braided river delta facies, littoral–shallow water lacustrine facies and semi-deep water–deep water lacustrine facies. This depositional system was well developed during the time when the Quantou formation–Yaojia formation were deposited; however, it was on a relatively small scale.

Northern Depositional System

The Northern depositional system was the largest depositional system in the Songliao basin; along the long axis of the basin, it extended into the lake. The eastern boundary of the Northern depositional system was located in the Hailun–Qinggang area; its western boundary was positioned in the Qiqihaer area; its northern boundary was sited at the northern border of the basin and its southern boundary was placed in the Putaohua area. The paleo-topography was gentle; the Beian river and the Nahe river brought plentiful clastic materials to supply the depositional system which expanded into a broad region. In the Daqing Changyuan and both sides of this area, a high constructive delta developed. Under the differential compaction and the tectonic movement, the Daqing Changyuan structure was formed, which provided favorable storage spaces for oil and gas accumulation. In the Songliao basin, the Northern aquatic system created the best depositional system for oil and gas accumulation. In the Daqing oil field, the main productive reservoir sequences were positioned in the large size delta sand body of the Northern depositional system. From north to south, the Northern depositional system could be divided into alluvial fan facies, river facies, delta facies, littoral–shallow water lacustrine facies and semi-deep water–deep water lacustrine facies. The Northern depositional system was mainly distributed on the northern side of the Songliao basin; only during the time when the Nenjiang formation was deposited, did this depositional system influence the Changling depression.

● **Excellent Petroleum Play**

In the Songliao basin, development of the second order sequences controlled the petroleum play and its characteristics (Fig. 8.9). As reservoir sequences, different scale delta sand bodies were developed in the lowstand system tract, highstand system tract and lacustrine transgressive system tract. The source rock and the regional cap rock were developed during the late phase of the lacustrine transgressive system tract.

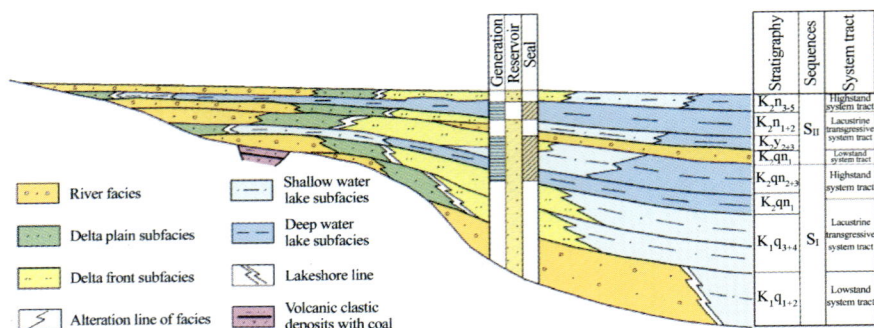

Fig. 8.9. Stratigraphic sequences (K_1q-K_2n) and related petroleum play in the southern Songliao basin

The Qing 1 member, the Nen 1 member and the Nen 2 member were the second order sequences that were the deposits of the late phase of the lacustrine transgressive system trace. The lacustrine mudstones were deposited during the two largest flooding periods and they contained abundant organic materials; these mudstones not only served as major source rock that generated oil, but they also functioned as the important regional cap rock. Vertically, these two sets of regional cap rocks separated three sets of petroleum plays in the Songliao basin. The lacustrine mudstone in the Nen 1 member and the Nen 2 member were the most important regional cap rocks in the Songliao basin, which were widely deposited during the largest flooding period. According to the statistical data, the regional cap rocks in the Nen 1 member and the Nen 2 member sealed oil bearing sequences in the Yaojia formation (that included oil layers of the Saertu and the Putaohua–SP) and in the Qingshankou formation (that included the oil layer of the Gaotaizi–G); these oil bearing sequences contained 78.1% and 14.8% of total explored oil reserves of the basin respectively; together, these oil bearing sequences constructed a middle petroleum play, which contained 92.9% of total explored oil reserves. Another regional cap rock was predominately made of the Qingshankou 1 member; which sealed the lower petroleum play (that includes the oil layers of the Fuyu and the Yangdachengzi–FY); the lower petroleum play contained 6.9% of the total explored oil reserve of the basin. The upper petroleum play was positioned above the Nen 1 member, which only contained 0.2% of the total explored oil reserve of the basin. In recent years, the exploration practices have proved this petroleum distribution pattern in general. Here, we need to point out the following: the recent exploration results indicated that, in the southern region of the Songliao basin, the regional cap rock in the Qing 1 member played a more important role and it sealed 67% of the total oil reserve in this region. On the other hand, the Nen 1 member sealed 30.6% of the total explored oil reserve in the southern region. Above the Nen 1 member, the explored oil reserve only occupied 2.42% of the total explored oil reserve. Because, in the southern region of the Songliao basin, the river–delta sand bodies were widely deposited during the time when the Quantou formation was deposited, the reservoir sequences were well developed. That was why the Quantou formation that was beneath the Qing 1 member would become the major oil bearing sequence in the southern Songliao basin.

Between the two largest flooding periods of the Qing 1 and the Nen 1, the lacustrine basin was periodically in expansion and contraction, and this formed the delta deposits in the Qingshankou formation with a highstand system tract, the delta deposits in the Yao 1 member with a lowstand system tract, and the delta sand body in the Yao 2 member and the Yao 3 member with the early phase of a lacustrine transgressive system tract. These deposits separately consisted of several first order, lacustrine transgressive–regressive, highly frequent sequences. In the Songliao basin, the recurrent alternation of highly frequent sequences controlled the secondary combination of reservoir and cap rock. The frequent alterations in the water level and the supply rate of sediments caused a frequent shifting of the shoreline of the lake and the lobe of the delta. Therefore, vertically, sandstone and mudstone were commonly interbedded with each other (especially in the front margin of a delta) to form many secondary petroleum plays; also, on the delta front, the sand bodies (that were developed in the underwater distributary river channels and estuary dam) were pinched-out at the lateral position. Thus, the lithologic traps were simply formed.

- **Perfect Match of Structural Establishing Period and Petroleum Migrating Period Assured Oil and Gas Accumulations**

The Changling depression is an NNE oriented large size syncline. At the western wing of the syncline, the east dipping slope was faulted by the NNE faults to form a framework of altered horst, graben and step alteration. The axis of the syncline is a strike in the NNE direction. The central complex consists of a series of faults that are a strike in approximately a north to south direction to create a graben zone that contains a series of small faulted grabens. At the eastern wing of the syncline, the west dipping slope with gentle gradient contains a poorly developed fault and structural trap. The petroleum accumulative characteristics of the Changling depression include the synclinal structure and an effective oil generating region near the syncline. The reservoir sequences were made of a sand body of delta front facies and a lenticular sand body of pro-delta facies; in addition, the sand body of underwater distributary channel facies also occurred. There are two kinds of situation in which petroleum traps were created. One is when, because the reservoir body was isolated inside the hydrocarbon source rocks, the lenticular lithologic trap was formed. Another situation happens at the laterals of the syncline. On the eastern wing, because the reservoir body was upward dipping and pinched-out, upward dipping and pinched-out type of sandstone traps were formed; on the western wing, because of the blockage from the faults, the faulted–lithologic oil reservoir developed. Furthermore, because the axis of the syncline was obviously controlled by the lithology and physical properties, the lithologic oil reservoir was formed.

According to the study of structural development, because of the influence from the fourth episode of the Yanshanian movement, during the time that the upper Nenjiang formation was deposited, this region had a nose-like structure, which was higher on the south western side and lower on the northeastern side, and which was on the monocline background. During the time that the upper Sifangtai formation was deposited, this region subsided on a large scale; in the Qianan area, the structures started to reverse. Up until the time that the upper Mingshui formation was deposited, the framework of the structure was established. This is the synclinal structure of the present day. During the time that the upper Nenjiang formation was deposited, the hydrocarbon source rock of the Qing 1 member had entered a hydrocarbon releasing phase; during the time that the upper Mingshui formation was deposited, the same source rock entered a hydrocarbon release peak time. The time of these two hydrocarbon releasing phases matched the time of structural developments. Thus, the high structural position was the destination of oil and gas migration at all times. The drilling exploration not only proved that the front margin of the Baoqian sand body controlled petroleum distribution, but it also indicated that the area with well developed structures was a highly productive region of oil and gas.

- **Multi-Types of Petroleum Play Provided Essential Conditions for Creating a Large Size Lithologic Oil Reservoir**

In this region, there are three types of petroleum plays: the lateral alteration type, the superimposed type and the enveloped type. The lateral type was designated to the reservoir sandstones, which were not only in the shape of a digit or in the shape of a wedge, but which also extended into the source rock (such as the oil bearing sandstone in the Qing 1 member). The superimposed type is when the source rock and the reservoir sequences contacted each other like two pieces of wood or sheets. For example, oil was generated in the lower sequences and it was stored in the upper strata, or vice versa; sometimes it was like a sandwich in which oil was generated in both upper and lower strata and was stored in the middle strata. The superimposed type widely occurred on the delta front where the thin beds of sandstone and mudstone were interbedded with each other. The enveloped type was assigned to the lenticular sandstone reservoirs that were deposited on the delta front and pro-delta and that were insulated in the middle of the source rocks to create the best framework for the petroleum play. On the delta front, these different types and different shapes of sand bodies were positioned in the source rocks; therefore, they not only have a sufficient supply of hydrocarbons and the best delivery system, but they were also a low potential area due to the combined effort of structural movement and differential compaction. Thus, this area was the major petroleum migrating destination. When oil and gas migrated from the hydrocarbon generating center to the surrounding area, the sand bodies on the front margin were the first ones that captured oil and gas to develop the large size, composite, lithologic reservoir.

8.2.3 Characteristics of the Daqingzi Well Oil Field

The Daqingzi Well oil field is located in Qianan county, Jilin province; the Qianan oil field was positioned in the northeastern part of the Daqingzi Well area. From a regional tectonic point of the view, the Daqingzi Well oil field is located in the central depressed area in the center of the Changling depression. According to the statistical data of 2005, the Daqingzi Well oil field has a confirmed oil bearing area of 331.82 km^2 and a confirmed geological oil reserve of 12,305.93×10^4 t; among the reserves, the technically obtainable reserve is 2,567.88×10^4 t.

The Daqingzi Well oil field contains the Heidimiao, Putaohua, Gaotaizi and Fuyu oil layers. The Gaotaizi oil layer possesses the major productive reservoir sequences. The oil field is located in the central depression in the center of the Changling depression, which is also the oil generation center. Oil and gas came from the hydrocarbon source rock in the Qingshankou formation; the sand bodies of underwater distributary channels on a delta, estuary dam and sand sheets were the major reservoir bodies and these sand bodies developed during multiple phases. Several sets of sand bodies were superimposed together to create multiple suites of reservoir sequences; the petroleum plays included the in situ type, the type of upper strata generated oil and low strata stored oil or vice versa. The mudstones in the Qing 1 member and in the Ren 1 member were the stable regional cap rock with good preservative conditions. In addition, combined with the structural trap and faulted block trap, the lithologic oil reservoirs formed over a broad area.

● **Characteristics of the Structure**

The regional structural background on the top surface of the Quan 4 member is a wide open, gently sloped syncline that is oriented in an NNE direction; the eastern wing and the western wing of this syncline are asymmetrical; the slope of the western wing is relatively steep and the slope of the eastern wing is rather gentle. The Daqingzi Well area is located at the center of the Changling depression which is between the Qianan secondary depression and the Heidimiao secondary depression. The depression is divided into the central faulted belt, the eastern slope zone and the western slope zone. The axis of the syncline is oriented in the direction of the Hei 43 well–Qian 139 well–Qian 110 well. In the central faulted belt, the NNE oriented fault was well developed and it brought a complication to the depression. This faulted belt consists of a series of normal faults that strike in the NNW direction; these normal faults formed graben and faulted steps in the axial area; at same time, a series of isolated structures and faulted nose structures developed on both sides of the graben. In the syncline, there are many small scale faults that extended between 2 – 4 km and that have a fault throw of 20 – 40 m. On the top surfaces of the Qing 2 member, the Qing 3 member and the Yao 1 member, the faults in the axial area were not well developed; the quantity of the fault was 60% less than that in the Quan 4 member and the Qing 1 member. The Hei 43 block is located at the southern end of the faulted belt and it contains a faulted nose structure. On the eastern side of the faulted nose structure, there is a fault that is oriented in an N-S direction with a fault throw of nearly 50 m and that extends 4.5 km. Near the Qian 124 well, there is a volcanic intrusion.

The western slope zone of the depression is located in the west of the area from the Hei 52 well–Hei 77 well; the dip angle of the slope is 2.5°. The western slope zone contains well developed faults; there are two faulted complex zones; one is along the line of the Hua 12 well–Hua 7 well–Hua 8 well; another one is along the line of the Hei 47 well–Hei 60 well–Hei 81 well; the faults extend 6 – 8 km and the fault throws are between 40 – 80 m.

On the eastern slope zone, the structure is a monocline that dips toward the west and the dipping angle is 1.5°. The upward dipping faulted blocks occur in the areas of the Hei 46 well, Hei 88 well, Hei 58 well and Hei 75 well. In the area east of the Hei 46 well, there is a large size reversed normal fault that is oriented in a N-S direction; the fault throw is about 70 m and it extends 8 km approximately.

● **Characteristics of Reservoir Sequences**

In the Daqingzi Well oil field, the oil bearing sequences include the Fuyu oil layer, the Gaotaizi oil layer and the Putaohua oil layer; the latter two oil layers occurred in the Qing 1 member and the Gaotaizi oil layer is the major oil bearing sequence. The members of the Qingshankou formation are further divided into groups and small beds; the Qing 1 member is divided into 4 sandstone groups and 16 small beds; the Qing 2 member is divided into 5 sandstone groups and 24 small beds; the Qing 3 member is divided into 12 sandstone groups and 42 small beds.

In the Daqingzi Well oil field, the reservoir sequences mainly consist of silt with a small amount of fine grain sandstone. The grain sizes are between 0.02 – 0.25 mm with medium sorting and semi-angular rounding. The content of the minerals includes quartz, feldspar and clastic. Among the grains, 30% – 42% of the grains are quartz; 32% – 45% of the grains are feldspar; 20% – 34% of the grains are clastic of feldspar sandstone. The cements of the reservoir sequences are mostly calcareous with a small amount of siliceous; the cement is of the porous and porous-regenerated type.

In the Daqingzi Well oil field, the porosity is in the range of 6% – 20% and the permeability is $(0.01 – 10) \times 10^{-3}$ μm^2, which belongs to low porosity and low permeability reservoir sequences. The types of pores include intergranular pores, dissolution pores, micro pores and remaining pores; the former three types were developed very well. Because they were filled or enveloped by the argillaceous cement, the size of a few pores became smaller or even ceased to exist, which influenced physical properties of the reservoir sequences. At same time, in this region, the leaching and dissolution of feldspar are common phenomena, which not only developed dissolution pores, but which also modified the physical properties of the reservoir sequences. According to thin section analysis, the texture of the pores is relatively intricate with a higher degree of heterogeneity; this includes medium pores and a small throat, which also have the characteristics of middle–low displacement pressure.

● **Characteristics of Oil and Gas Reservoirs**

The Daqingzi Well oil field contains a multiple set of oil bearing layers and it is a large scale oil field. The buried depth of the oil layer is between 1,800 – 2,550 m; the thickness of the oil bearing interval is between 60 – 300 m. In a single well, the effective thickness of the oil layer is between 6 – 15 m; vertically, there are 6 – 12 oil layers.

In the Daqingzi Well oil field, the relationship of oil and water is complicated. The distribution of oil and water varied between different blocks or different sequences of the same block. The distribution of oil and water was controlled by multiple factors in the reservoir sequence, which include the lithology, physical properties and structure. The lithologic oil reservoir is the primary type; the structural–lithologic oil reservoir controlled oil and gas accumulation.

In the Daqingzi Well oil field, the density of crude oil on the surface is 0.843 – 0.865 g/cm^3; the viscosity (at 50 °C) is 10 – 30 MPa·s; the condensation point is between 26 – 40 °C. In ascending order, the density and the viscosity of the crude oil are increasing. In the Quan 4 member and the Qing 1 member, the density of the crude oil is 0.851 g/cm^3 approximately; in the Qing 3 member, the density of the crude oil is 0.858 g/cm^3; in the Yaojia formation, the density of the crude oil is 0.861 g/cm^3. In the Quan 4 member and the Qing 1 member, the viscosity of the crude oil is 12 MPa·s approximately; in the Qing 2 member, the viscosity of the crude oil is 18mPa.s; in the Yaojia formation, the viscosity of the crude oil is 24 MPa·s. In general, the density of crude oil in formation is between 0.7877 – 0.8295 g/cm^3; the viscosity of crude oil in formation is between 1.82 – 9.34 MPa·s.

The salinity of formation water is in the region of 10,000 – 23,000 mg/L; the content of chloride ion is between 3,000 – 11,000 mg/L; the data indicated that the water type is of NaHCO$_3$ type. The salinity of formation water evidently changes either vertically or horizontally. In the Quan 4 member, the salinity is 13,000 mg/L; in the Qing 3 member, the salinity is 24,000 mg/L approximately; in the Yaojia formation, the salinity is 19,000 mg/L.

In the sandstone group I of the Quan 4 member, the buried depth of the middle portion of the oil layer is 2,450 m; the pressure of the oil layer is 23.2 – 26.4 MPa and the average pressure is 24.8 MPa; the temperature of the oil layer is between

97 – 101 °C and the average temperature is 99 °C. In the Qing 1 member, the buried depth of the middle portion of the oil layer is 2,350 m; the pressure of the oil layer is 20.3 – 24.4 MPa and the average pressure is 22.8 MPa; the temperature of the oil layer is between 93 – 104 °C and the average temperature is 97.3 °C. In the Qing 2 member, the buried depth of the middle portion of the oil layer is 2,250 m; the pressure of the oil layer is 20.2 – 23.6 MPa and the average pressure is 22.0 MPa; the temperature of the oil layer is between 90 – 97 °C and the average temperature is 93.8 °C. In the Qing 3 member, the buried depth of the middle portion of the oil layer is 2,120 m; the pressure of the oil layer is 20 – 22 MPa and the average pressure is 21.0 MPa; the temperature of the oil layer is between 83 – 90 °C and the average temperature is 87.9 °C. In the Yao 1 member, the buried depth of the middle portion of the oil layer is 1,890 m; the pressure of the oil layer is 18.11 MPa; the temperature of the oil layer is 81.7 °C.

In general, the pressure coefficient is between 0.96 – 1.01 and the temperature gradient is between 4.0 – 4.3 °C/100 m, which indicated a normal temperature and pressure system. The drive types of reservoir include the dissolved gas drive type, the elastic gas drive type and the water drive type.

8.3   Exploration of Lithologic Oil Reservoir in the Mesozoic Low Permeable Sandstone in the Eerduosi Basin

The Eerduosi basin is the second largest basin in China, which covers an area of 25×10^4 km². The Eerduosi basin is a large scale, multi-cycled, craton basin that has simple structures. The entire basin not only experienced tectonic events of subsidence and uplifting, but it was also depressed and shifted. The basin contains various stratigraphic systems of the Changchengian system, the Jinxian system, Cambrain, Ordovician, Carboniferous, Permian, Triassic, Jurassic, Cretaceous, Tertiary and Quaternary systems. The thickness of sediments is 6,000 m on average. Vertically, the characteristic of petroleum distribution in the Eerduosi basin is that "the Mesozoic strata contained oil and the Paleozoic strata contained natural gas". According to the characteristics, the basin can be divided into several secondary tectonic units that include the Yimeng uplift, the Weibei uplift, the thrust belt on the Western margin, the Tianhuan depression, the Shanbei slope and the Jinxi flexural folding belt (Fig. 8.10). Among these secondary tectonic units, the Shanbei slope is the primary target for petroleum exploration and development; at present, 90% of discovered oil and gas are distributed in this tectonic unit.

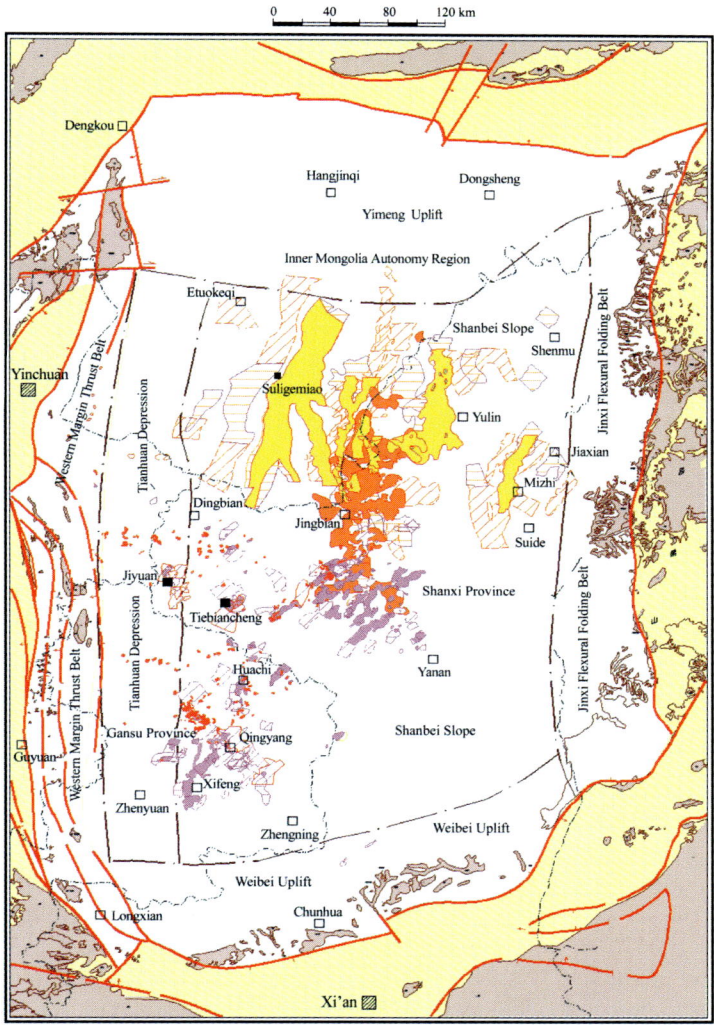

Fig. 8.10.   Map of tectonic units and related exploration achievements in the Eerduosi basin

8.3.1   *Exploration History*

Petroleum exploration in the Eerduosi basin has a rich history of over a century, which not only experienced exploration for a structure related petroleum reservoir in peripheral areas of the basin and for a paleo-topographic related oil reservoir in the Jurassic system, but which also experienced comprehensive exploration for a large size lithologic oil reservoir in the Yanchang formation in the Triassic system.

In addition, geologists discovered the Ansai oil field that has a hundred million tons oil reservoir. Since the 1990s, because exploration activities for the lithologic oil reservoir continually increased, and because geologists returned to the Shanbei area for re-exploration, in the Ansai oil field and the Jingan oil field, the oil bearing areas continually expanded and connected together to form the Shanbei oil region. Entering the 21 century, an important exploration breakthrough happened in the Longdong area, which was the discovery of the large size Xifeng oil field. Over the past twenty years, the exploration of the lithologic oil reservoir in the Eerduosi basin achieved three important successes. The first discovery was the large size, composite, oil bearing region in the Shanbei area with a geological reservoir of 7.5412×10^8 t. The second discovery was an important exploration breakthrough in the Chang 6 – 8 oil layers in the Longdong area, which was finding the Xifeng oil field and the Heshui petroleum reserve area that has third order reserves of 5.2305×10^8 t. The third discovery was a significant achievement in the Chang 4+5 oil layers in the Jiyuan area that became the new exploration target region with a controlled reserve and forecast reserve of more than 1.0×10^8 t. In addition, in the Tiebiancheng, the Baoziwan and the Xiaojianzi areas, three oil bearing regions were discovered that have a hundred million tons of reserves and that became important exploration targets.

In over 30 years, the Eerduosi basin accumulated a rich history of petroleum exploration and development. In order to suit the characteristics of a low permeable lithologic oil reservoir, continual innovations and stable development not only brought comprehensive techniques for petroleum exploration and development, but they also set up the production pattern. The continually increased oil production and the stabilized reserve–production ratio brought fast and efficient development in the Changqing oil field.

In the past five years, the supplementary explored reserve, controlled reserve and forecast reservoirs were separately over 1×10^8 t each in the Eerduosi basin. The oil reserves replacement ratio achieved 2.94 – 3.65. After oil production passed a threshold of 500×10^4 t/a in 2001, the yearly increasing production rate was over 100×10^4 t; in 2004, the oil production was 845×10^4 t. For 2005, the planned oil production was 945×10^4 t. During the "Tenth Proposal of a Five Year Plan", the oil production will increase by 100×10^4 t/a (Fig. 8.11).

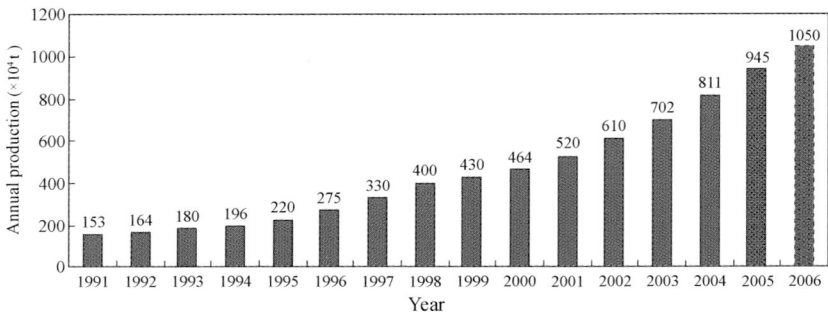

Fig. 8.11. Bar diagram of annual crude production in the Eerduosi basin, since 1991

8.3.2 Basic Geological Conditions for Developing a Lithologic Oil Reservoir in Low Permeable Sandstone

● **Abundant Hydrocarbon Supplies are the Foundation for Developing a Large Scale Lithologic Oil Reservoir**

In the Eerduosi basin, the Triassic Yanchang formation contains intra-continental lacustrine deposits on a large scale. When the Chang 7 member was deposited, because the region of the lake was the broadest and the depth of water was the deepest, it deposited a set of deep water mudstones of lacustrine facies, which had the greatest oil generating potential. The types of organic substance were saprogenic type–sapropelic type. The distribution pattern of hydrocarbon source rock is in a gourd shape and source rock dipped in a northwest–southeast direction. The thickness of effective source rock is more than 100 m; the effective source rocks occur in the areas of Yanchi, Wuqi, Fuxian, Jingchuan, Zhenyuan and Huanxian; the total distribution area is 8×10^4 km^2 approximately. The content of organic carbon is 2% – 5%; the content of chloroform bitumen "A" is 0.3% – 0.5%; the content of hydrocarbon is 1,833 – 3,505 ppm. Previously, according to the residual bitumen calculation, the hydrocarbon–generating rate was 100 kg/t TOC; people further calculated that the total amount of generated oil was 946.3×10^8 t from the Triassic Yanchang formation in the Eerduosi basin. Recently, according to a thermal simulation test, the newly calculated hydrocarbon generating rate is 400 kg/t TOC; this number increases the total amount of generated oil from the Triassic Yanchang formation to $1,996.5 \times 10^8$ t.

Two factors influenced the total amount of oil generation: one is the calculating method and another one is the improvement of the stimulation technique in laboratory analysis. In the early days, the most used calculating method was either calculating bitumen or using a hydrocarbon coefficient. This type of calculating method always used a residual hydrocarbon index in the source rock (such as organic carbon, chloroform bitumen, or total hydrocarbon content) to multiply with certain kinds of experimental coefficients; therefore, the recovered total amount of original generated oil was constantly proportionate to the amount of residual oil. In other words, if the index of residual oil was a large number, the calculation result would present a large amount of total generated oil; otherwise, the small amount of residual oil would yield a calculation result of a small amount of total generated oil. This type of calculating method ignored the fact that the residual hydrocarbon index in the present source rock was the combined action of hydrocarbon generation and hydrocarbon expulsion. The relationship of residual hydrocarbon to total generated oil is not automatically proportionate. This type of calculation method regularly used single variable equations as the mathematical equation. During the third round of natural resource evaluation, the calculation of total generated oil was based on the mechanism of the organic substance, which was adequate both in a laboratory test and hypothetically, and which brought the calculated results closer to reality. In addition, the improvement in the stimulation technique in the laboratory (especially the continual progressions in the artificial thermal simulation technique) allowed people to obtain hydrocarbon generating information directly. In the Jiefang II–674 well, the mudstone samples from the Chang 7 member were sent to the laboratory for a thermal simulation test; the following are the test results: the largest oil production rate (pyrolysis oil + bitumen "A") was 400 kg/t (residual carbons); the production rate of final gaseous hydrocarbon was 776.84 m^3/t (residual carbons). The derived curve of the production rate represented the characteristic of type I–type II$_1$ kerogen, which was mostly compatible with the results of the pyrolysis test, and which proved that, in the study area, the source rock had a good capability for generating hydrocarbons. Thus, the intensity of hydrocarbon generation was increased from $(50 - 250) \times 10^4$ t/km^2 to $(50 - 550) \times 10^4$ t/km^2. The hydrocarbon generating region was also expanded toward the margins of the lacustrine; in particular, in the Jiyuan area, the hydrocarbon generating region increased by almost 50%. Therefore, the exploration areas had been enlarged.

Based on comprehensive analysis and evaluation of the basin, and considering the exploration result in the Mesozoic strata, the quantity of the petroleum resource in the Eerduosi basin was re-calculated. The total amount of the petroleum resource was increased from 15.3×10^8 t to 85.88×10^8 t; the newly calculated number indicated that the Eerduosi basin contains good exploration potential.

The greatest lake transgression happened when the Chang 7 member was deposited, which developed a set of deep water, lacustrine mudstone. This mudstone was distributed in the areas of Dingbian–Wuqi–Qingyang–Fuxian and it covered an area of 8×10^4 km^2 approximately. At present, most of discovered oil reservoirs are located either on the slope zone on the margin of the hydrocarbon generating depression or inside the depression (Fig. 8.12).

● **Sand Body of River–Delta Facies is a Favorable Location for Accumulating a Large Scale Lithologic Oil Reservoir**

The favorable sedimentary facies zone is the important geological foundation for large scale petroleum accumulation and distribution. In the Eerduosi basin, the high point of the paleo-topography of the Late Triassic strata was located on the northeastern side and the low point was located on the southwestern side; in addition, the northeastern side has a gentle slope and the southwestern side has a steep slope. The main depression, or the subsided center, was in the shape of an asymmetrical dustpan that dipped toward the northwest. Around the Eerduosi basin, there were several paleo-uplifts that functioned as sedimentary source areas; Yinshan mountain was in the north of the basin; Alashan mountain was in the northwest of the basin; Haiyuan was in the west of the basin; Longxi was in the southwest of the basin; Qinling mountain was in the south of the basin; and Luliang

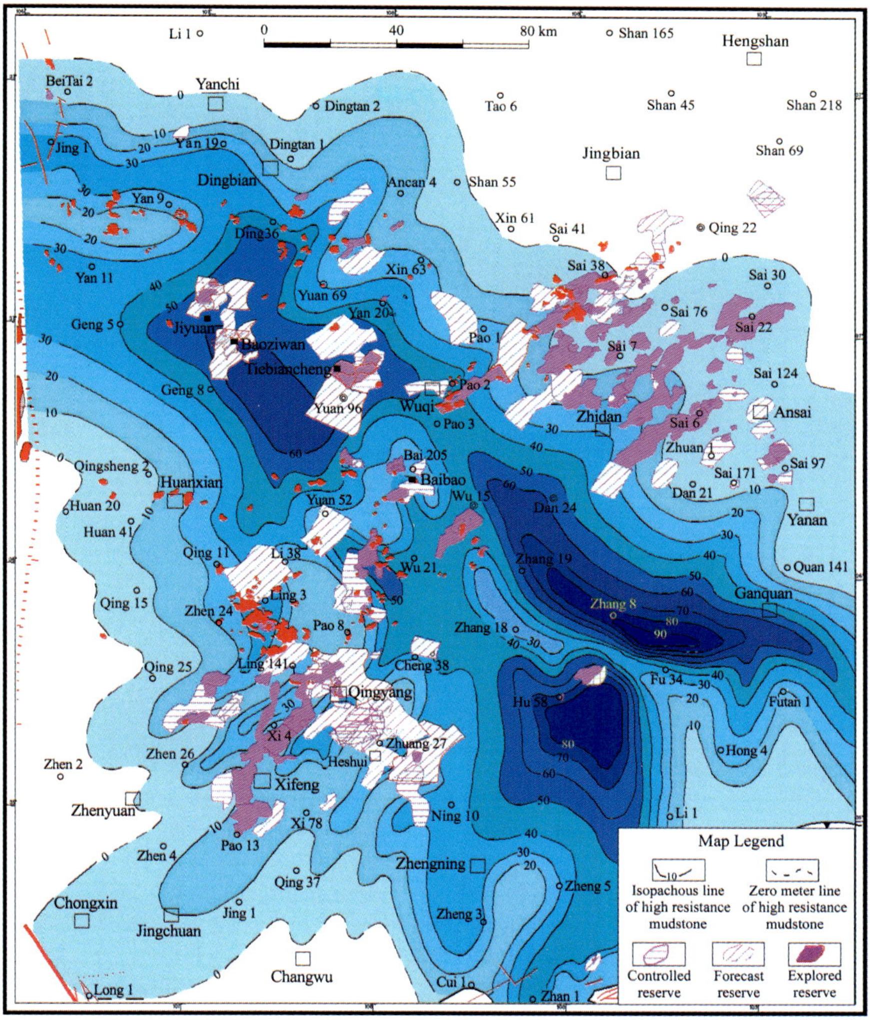

Fig. 8.12. Isopachous map of hydrocarbon source rock and the distribution of oil fields in the Chang 7 member of the Yanchang formation

was in the eastern part of the basin. During the Late Triassic epoch, the most important source areas in the basin include Yinshan mountain in the north and the Longxi paleo-uplift in the southwest. Therefore, the depositional framework consisted of river–delta depositional systems that were supplied by the source areas either in the north or in the south (Fig. 8.13). These two large scale depositional systems controlled the most important developmental characteristics of the lacustrine basin and they also offered storage places for the Mesozoic petroleum reservoirs.

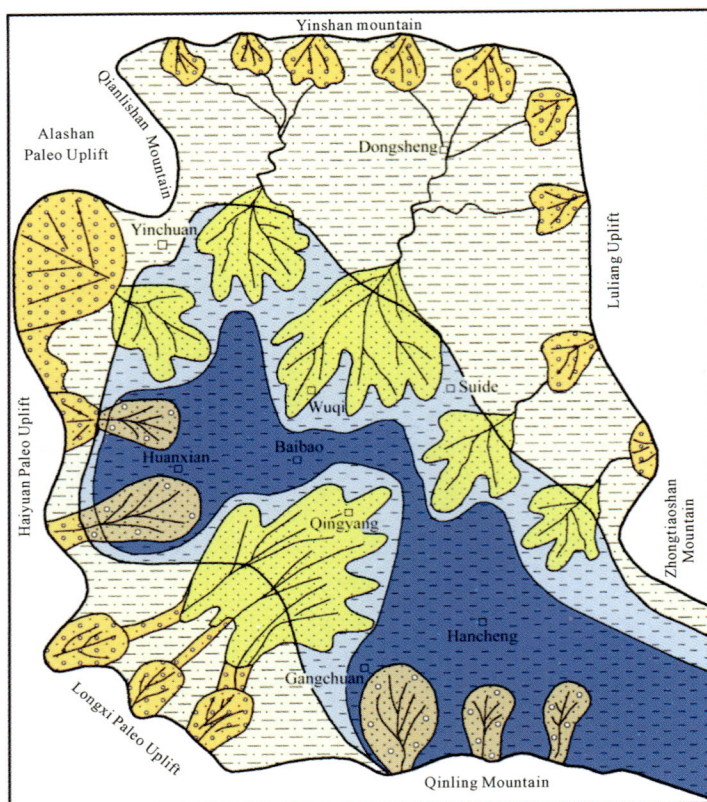

Fig. 8.13.   Illustration of Late Triassic depositional system in the Eerduosi basin

The delta sand bodies not only were on a large scale, but they were also stable in the horizontal dimension with good connectivity. In distributary river channels, the thickness of sand bodies was between 10 – 20 m with a width of 10 – 25 km. In the Longdong area, the sediments came from the Longxian–Pingliang area in a southwest direction. Because the paleo-topography was steep, the sediments were deposited either in the braided river channels or in the distributary channels on the delta; the distribution of the sand bodies was in the shape of a belt that extended a long distance; the thickness of sand bodies was in the range of 5 – 20 m with a width of 3 – 8 km.

Because the supply of sediments was sufficient and because the lacustrine basin steadily rose back, large scale, constructive deltas were formed; the delta sand bodies were distributed over a broad area and they offered the ideal place for oil and gas to migrate and to accumulate. Because there were variations in physical reservoir properties between the subfacies in the delta depositional system, excellent physical reservoir properties occurred in the delta front subfacies, which was the most favorable place for petroleum accumulation.

● **Reservoir Sequences Largely Consisted of Fine Grain Sandstone with Low Porosity and Low Permeability**

In the Eerduosi basin, the Late Triassic Yanchang formation experienced an entire depositional cycle from lacustrine progression to lacustrine regression with the deposits of lacustrine facies–delta facies; the reservoir bodies mainly consisted of delta sand bodies. The lithology includes clastic feldspar sandstone and feldspar sandstone; among the clastic materials, the average content of quartz was 45%; the average content of feldspar was 40%; and the average content of clastic was 15%. The primary filling substances include chlorite, carbonate and laumontite. The lithology had a low compositional maturity and high textural maturity; the grain size was dominated by fine grains; the types of contacting relationship between clastic grains were mostly line type, point–line type and point type. The intergranular pores were a major storage place. The next category of storage place included the dissolution pores of feldspar, dissolution pores of laumontite and all kinds of micro fractures.

In the Yanchang formation, the primary oil layers have low porosity and low permeability; the average porosity is between 10% – 14%; the permeability is $(0.1 - 2) \times 10^{-3}$ μm^2. The Chang 4+5 oil layers have an average permeability of $(0.4 - 1.0) \times 10^{-3}$ μm^2; the Chang 6 oil layer has an average permeability of $(0.5 - 1.5) \times 10^{-3}$ μm^2; the Chang 8 oil layer has an average permeability of $(0.5 - 1.2) \times 10^{-3}$ μm^2.

In the Changqing oil field, the permeability from the discovered oil reservoirs in the Yanchang formation varies. The following are the statistical data: 14.2% of the geological oil reserve has a permeability larger than 10×10^{-3} μm^2; 3.8% of the geological oil reserve has a permeability between $(3 - 10) \times 10^{-3}$ μm^2; 59.2% of the geological oil reserve has a permeability between $(1 - 3) \times 10^{-3}$ μm^2; 13.4% of the geological oil reserve has a permeability between $(0.5 - 1.0) \times 10^{-3}$ μm^2; 9.3% of the geological oil reserve has a permeability between $(0.3 - 0.5) \times 10^{-3}$ μm^2; 0.2% of the geological oil reserve has a permeability between $(0.1 - 0.3) \times 10^{-3}$ μm^2 (Fig. 8.14).

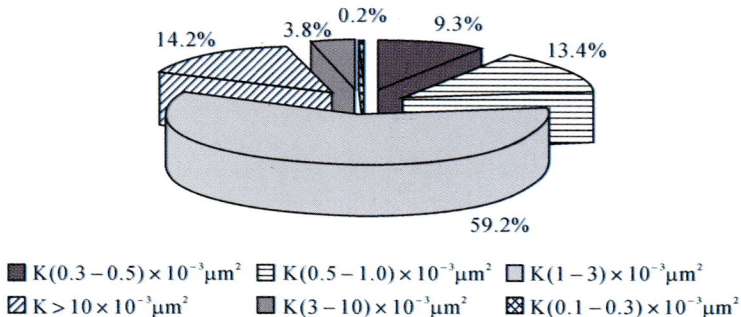

$\blacksquare$ $K(0.3 - 0.5) \times 10^{-3} \mu m^2$ ☰ $K(0.5 - 1.0) \times 10^{-3} \mu m^2$ ☐ $K(1 - 3) \times 10^{-3} \mu m^2$

▨ $K > 10 \times 10^{-3} \mu m^2$ $\blacksquare$ $K(3 - 10) \times 10^{-3} \mu m^2$ ▨ $K(0.1 - 0.3) \times 10^{-3} \mu m^2$

Fig. 8.14.   Pie chart of explored petroleum reserve in the Yanchang formation in the Eerduosi basin

According to the category of low permeability oil reservoir (either using the international standard or the domestic standard), in the Changqing oil field, 82% of the explored oil reserve is a low permeable reserve and extremely low permeable reserve. Even so, this oil reserve still is the basis for the development of the Changqing division company. With improving technology, the Changqing low permeable oil field could be efficiently developed.

The constructive diagenesis clearly influenced reservoir properties, which included dissolution, oil intrusion and authigenic clay films. The dissolution created a huge amount of secondary pores; the oil intrusion suppressed the growth of authigenic clay films; the compact resistance from authigenic clay films protected the pores.

● **Excellent Combination of Petroleum Generation–Storage–Seal**

At present, the idea that "the Eerduosi basin only has one set of source rock" has been widely accepted. The most important hydrocarbon source rocks are dark color mudstone and oil shale in the Upper Triassic Yanchang formation; these source rocks were deposited either in semi-deep water or in deep water and they belonged to the second order, lacustrine transgressive system tract. Even the Lower Jurassic, second order, lacustrine transgressive system tract did not have effective hydrocarbon source rock. However, the mudstone of lacustrine facies and the marsh peat of delta plain facies of the Yan 9 period and the marsh peat of the Yan 6+7 period created sufficient cap rocks. The large scale unconformity surface on top of the Triassic system, the sandstone of the incised river valley and the faults, jointly created an adequate petroleum system, which effectively linked the Triassic petroleum source center and reservoir sequences. In the basin, the Mesozoic Yanchang formation–Yanchang formation is the united petroleum system. The Upper Triassic, second order, lacustrine transgressive system tract provided hydrocarbon source rock; the lower Jurassic lacustrine transgressive system tract formed a regional cap rock; all kinds of delta sand bodies and the river delta (that was positioned between the strata of the Upper Triassic and the Lower Jurassic) functioned as major reservoir rocks. Together, in the Eerduosi basin, they not only formed a combination of oil generation–storage–seal in the second order sequences, but they also controlled oil and gas distribution in the Mesozoic erathem.

According to the petroleum play in the Mesozoic strata and the exploration experiences, we divided the Yanchang formation and the Yanan formation into three petroleum plays, which were the lower play, the middle play and the upper play (Fig. 8.15). In the lower play, the Chang 10 sand bodies of river and delta facies functioned as the reservoir sequences; the Chang 9 mudstone of lacustrine facies worked as source rock and the cap rock. This petroleum play generated oil in the upper strata and stored oil in the lower strata. The middle petroleum play included oil layers from the Chang 9 to the Chang 4+5. The lacustrine mudstones in the Chang 9–Chang 4+5 functioned as source rocks and the cap rocks; among these mudstones, the mudstone in the Chang 9, the Chang 7 and the Chang 4+5 also worked as regional cap rocks. All kinds of delta sand bodies and turbidite sand bodies operated as reservoir sequences to form the in situ type of petroleum play. The upper petroleum play contained oil layers from the Chang 1 to the Yan 1. In the play, the source rock was positioned in the middle petroleum play. The strata in the Chang 1 and the strata in the Yan 9–Yan 6 worked as regional cap rocks. The sand bodies of river facies in the Chang 2 and the Yan 10 and the delta sand bodies in the Chang 3 and the Yan 9 functioned as reservoir sequences. This play generated oil in the lower strata and stored oil in the upper strata.

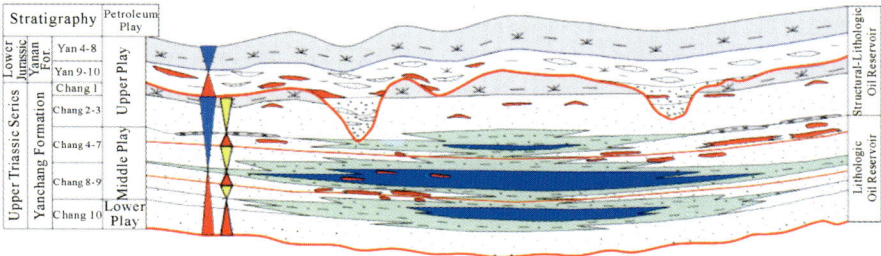

Fig. 8.15.   Model of oil reservoir distribution from the Yanchang formation–the Yanan formation in the Eerduosi basin

Among these three petroleum plays, the lower play has a lithologic oil reservoir with a lower quantity oil reserve. Up until June 2005, the lower play contained a total explored reserve of 159×10^4 t, which only equalled 1.2‰ of the total explored reserve (13.3845×10^8 t) from the Mesozoic erathem in the Eerduosi basin. The middle play contains the most abundant oil reserve in the lithologic oil reservoirs. To June 2005, the middle play contained total geological reserves of 8.0190×10^8 t, which equal 59.9% of the total explored reserve from the Mesozoic erathem in the basin. Among the oil layers in the middle play, the Chang 8 has explored oil reserves of 1.9655×10^8 t and the Chang 6 has explored oil reserves of 5.3985×10^8 t; they are the most important productive oil layers. In the upper play, the lithologic–structural type of oil reservoir and the composite (structural and lithologic) type of oil reservoir are the important types of reservoir; the lithologic oil reservoir and the structural oil reservoir belong to the next category. Up until June 2005, the upper play contained a total geological reservoir of 5.3497×10^8 t, which equals 39.9% of the total explored reserve from the Mesozoic erathem in the basin. In the upper play, the lithologic reservoir contained an oil reserve of 1.6642×10^8 t; the composite reservoir contained an oil reservoir of 3.1580×10^8 t; the Chang 3 layer had an explored reserve of $8,714 \times 10^4$ t; the Chang 2 layer had an oil reserve of $13,209 \times 10^4$ t; the Fuxian + the Yan 10 layers have oil reserves of $11,526.7 \times 10^4$ t; the Yan 9 layer had an oil reservoir of $10,470.5 \times 10^4$ t.

● **Two Accumulative Models**

During the time that the Yanchang formation was deposited, there were several sediment supply centers that surrounded the Eerduosi basin. Different depositional systems had their own depositional environment, reservoir features, the combination of reservoir and cap rock, the migration pattern and accumulative controlling factors; therefore, accumulative patterns and reservoir models also had their own characteristics. In the Eerduosi basin, the major conditions for creating an oil reservoir included the abundant hydrocarbon source rocks, excellent reservoir bodies, favorable depositional facies zone, effective traps, stable potential field, unblocked migration paths, and the geological foundation for creating a lithologic oil reservoir. Because the deposits of the Yanchang formation were controlled by two major depositional systems that were located either on the northeastern side or on the southwestern side of the basin, there were two sets of accumulative models.

Accumulative Model for the Northeastern Depositional System (Fig. 8.16)

The depositional basement was smooth with large scale delta deposits in the meandering river. The sand bodies were in sheet shape, inter-figuring shape or lobe shape and they were positioned either in distributary channels on the delta, or at the estuary dam to form favorable places for oil and gas accumulation. The laumontite dissolution pores that were formed by constructive diagenesis altered sufficient storage spaces. The hydrocarbon source rock was widely distributed in a NW-SE direction in a pear shape. Both dense lithology that was located in an upward dipping direction of the sand body and the alteration in lithologic facies offered primary sealing conditions. The delta deposits not only received lateral migrated oil and gas from the oil generating depression, but they were also related to two sets of well developed petroleum plays. One play included the Chang 7–Chang 6–Chang 4+5 strata; another play contained the Chang 3–Chang 2–Chang 1 strata. In the Shanbei area, because these two petroleum plays were superimposed together, the Chang 6 and Chang 2 strata became the important commercial oil layers. In recent years, along with in-depth research, the understandings of the petroleum play and the petroleum accumulative model in the Chang 8 and Chang 9 strata have been continually worked out in the Shanbei area.

Accumulative Model for the Southwestern Depositional System (Fig. 8.17)

The depositional basement was relatively steep; the sediments supply center was nearby; the river channel had a large ratio between width and depth. During the time that the Triassic Yanchang formation was deposited, large scale sand bodies in the Chang 6 – 8 intervals developed in the braided river on the delta plain and in the distributary channel on the delta front. Because a restriction was imposed by hydrocarbon infilling during the early phase of diagenesis and by clay film, the growth of secondary minerals was very difficult; therefore, the original pores were preserved. In addition, the micro fractions and limited dissolutions were beneficial factors for increasing permeability. The Chang 7 mudstone and oil shale generated a huge amount of oil and gas; under the driving force of abnormal pressure, through the connection of sand body and micro fractures, this oil and gas migrated into the Chang 8 reservoir sequences. After the Chang 8 sequences, the oil and gas continued to migrate eastwards, were blocked by the argillaceous or dense lithology in the inter-distributary channel and accumulated in the traps at the pinch-out side of the sand body. The type of petroleum play was that oil and gas were generated in the upper strata (the Chang 7 layer) and they were stored in the lower strata (the Chang 8 layer).

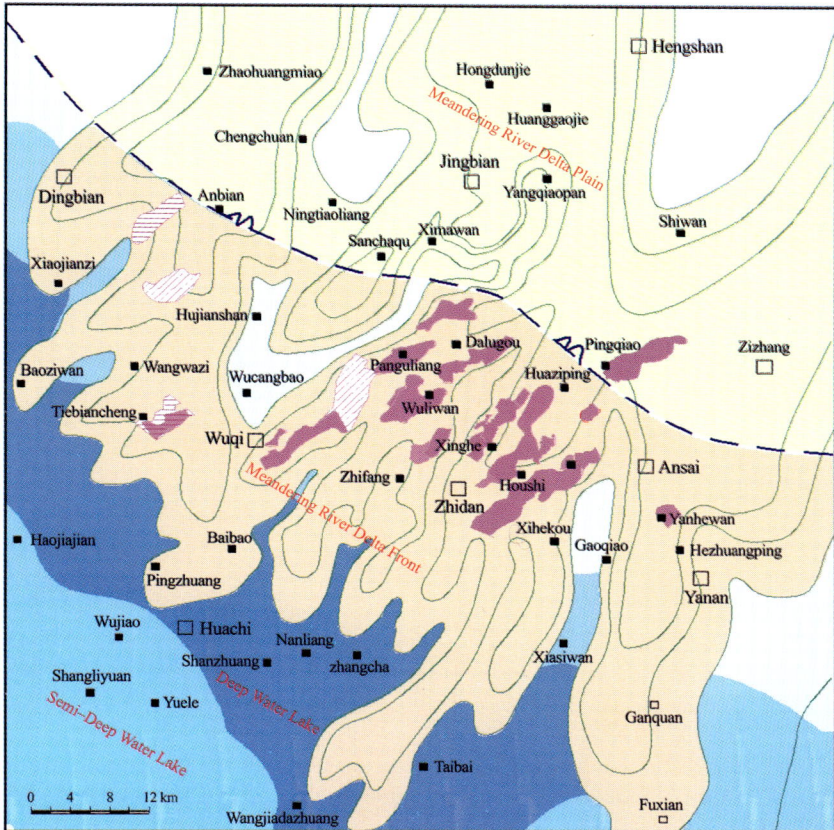

Fig. 8.16.   Accumulative model of the northeastern depositional system in the Eerduosi basin

● **Lithologic Oil Reservoir is the Primary Type Petroleum Reservoir in the Region**

The Eerduosi basin has a simple structural condition; the Yi–Shan slope is a major structure. This slope is a gentle monocline that dips westwards; the average slope gradient is 10 m/km; the dipping angle is less than 1°. On the slope, there is a series of nose-like, low relief uplifts that were created by lithologic differential compaction. Seismic exploration had not discovered a fault and complete structural trap.

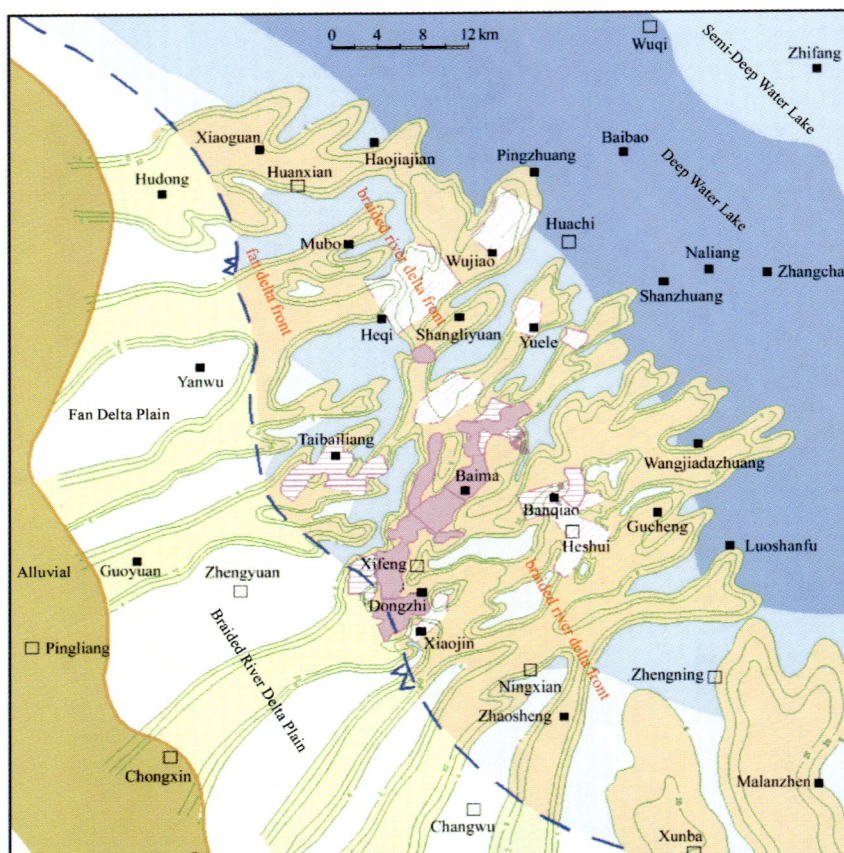

Fig. 8.17. Accumulative model of the southwestern depositional system in the Eerduosi basin

According to the condition of the petroleum traps, in the Changqing oil field most of the Mesozoic oil reservoirs were lithologic traps. The 10.6637×10^8 t of oil reservoirs were from the lithologic oil reservoir, which equals 80.8% of the total explored reservoir. In the Yanchang formation, these oil reservoirs belong to a classic lithologic reservoir, which is represented by the Chang 6 and Chang 8 oil layers and which also includes the Chang 3, the Chang 4+5 and the Chang 7 oil layers. In the middle and lower portion oil layers in the Yanchang formation (which is represented by the Chang 6 and Chang 8 oil layers), reservoir sequences have a small throat with low–extremely low porosity and permeability. The average porosity is 10.0% – 12.8%; the average permeability is 1.43×10^{-3} μm^2

approximately. The reservoir sequence contains strong heterogeneity.

In the Eerduosi basin, the oil reservoirs in the Mesozoic Yanchang formation were controlled by the lithology with an intricate relationship between water and oil. The water could come out at a high structural position and oil could be produced in a low structural position; in addition, the separation of oil and water is poor (Fig. 8.18).

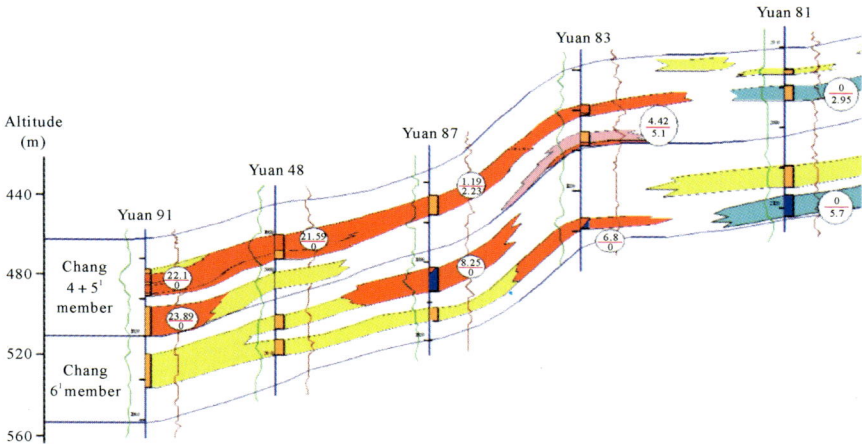

Fig. 8.18.   The cross section of oil reservoir from the Chang 4+5^2 members–Chang 6^1 member, in the Yanchang formation in the Jiyuan area, Eerduosi basin

8.4   Explored Lithostratigraphic Petroleum Reservoirs in Carbonate Reef Flat in the Central Tarim Basin

8.4.1   Exploration History

The Tazhong bulge is located in the middle section of the central uplift in the Tarim basin and it covers an area of 2.1×10^4 km^2; from a geological point of view, it is a gigantic, pre-Carboniferous, paleo uplift (Fig. 8.19). The oil reserve is 26.33×10^8 t and the natural gas reserve is 2.3×10^{12} m^3. The exploration target is the Carboniferous, Silurian, Ordovician and Cambrian systems. Up to the end of 2007, more than 160,000 km of 2D seismic survey and 9,310 km^2 of 3D seismic survey were conducted; the density of the survey network was 1 km×1 km – 2 km×2 km. A total of 258 exploration wells were drilled; in addition, 4 oil and gas fields and 12 structures that contain oil and gas were discovered and that included the Tazhong 1 structure and the Tazhong 62 structure. The Tazhong-I slope zone contains $50,424.14 \times 10^4$ t of third grade oil reserve and $2,144.8 \times 10^8$ m^3 of third grade gas reserve.

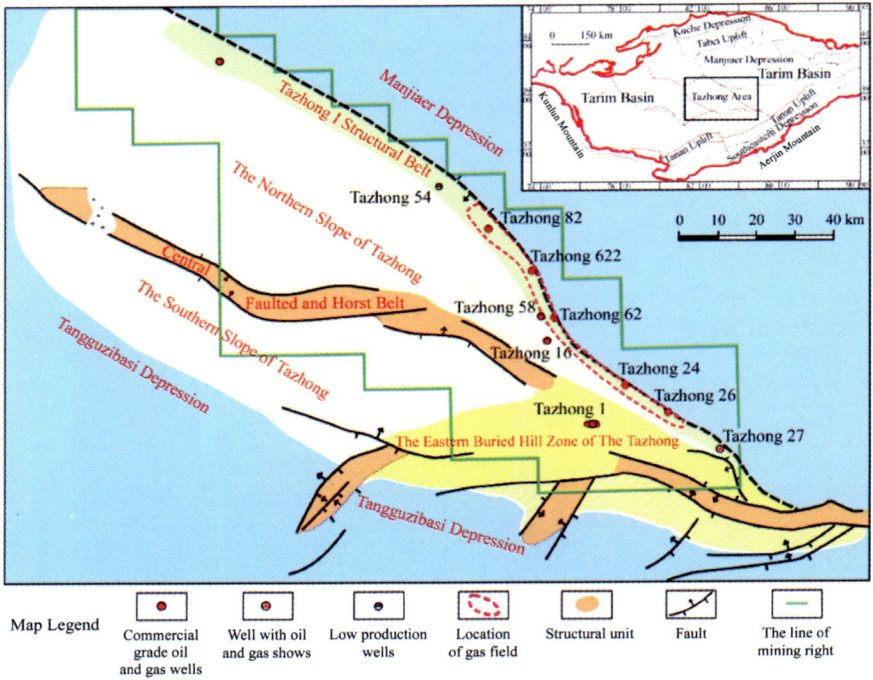

Fig. 8.19. Tectonic map of low altitude uplifts and their locations in the Tazhong area (central Tarim area) of the Tarim basin

In the Tazhong area, the initial exploration work started at 1983; on May 5, 1985, the drilling of Tazhong-1 well symbolized that petroleum exploration had progressed in all directions in the Tazhong area. The exploration of carbonate rock experienced a complicated history. In 1989, the first exploration breakthrough was obtained in the Tazhong-1 well; however, appraisal of the buried hills horst blocks in a high position was unsuccessful. During 1996 – 1997, in the Tazhong 1 faulted belt-I, a breakthrough was obtained in the Ordovician inner limestone; but again, the following exploration was fruitless. After disappointing exploration results from different domains and different types of carbonate rocks, in 2003, geologists re-explored carbonate rocks and the faulted belt-I in the Tazhong area. This time, the problems of carbonate exploration were overcome and a large size, bio-reef type of oil and gas field in the Ordovician strata on the Tazhong-I faulted slope was discovered.

● **1983 – 1991: Deadly Desert Conquered and Strategic Breakthrough Achieved**

The Tarim basin covers an area of 56×10^4 km^2; the Takelamagan desert is located at the center of the basin. Because of the harsh desert environment, this desert is also called "the sandy death sea". In 1983, PetroChina assembled seismic desert teams (two US teams and one Chinese team); the adventure of conquering "the sandy death sea" was carried out. Overwhelming all kinds of problems, the seismic desert teams surveyed 19 regional seismic sections that crossed the Takelamagan desert. In 1986, geologists not only discovered the Tazhong paleo-uplift (Fig. 8.20), but they also uncovered the structural framework of the Tarim basin, which includes three uplifts and four depressions.

With the detailed seismic survey, the Tazhong 1 anticline that is a gigantic, buried hill type of anticline was discovered. The Tg8 reflective layer (base of the Cambrain system) revealed that the trap covered 6,070 km^2 of area and the relief was 1,870 m. In 1986, the first round of natural resource evaluation was conducted; among 41 traps, the Tazhong 1 structure was on the top of the list with an oil resource of 29.8×10^8 t. At this point, the exploration philosophy began to form, which was "explore uplift in depression, drill well on buried hill type of anticline, and discover large size oil field". The Tazhong 1 structure was selected as the first exploration target in the platform and basin area.

In April 1989, the Tarim petroleum exploration headquarters was set up. Guided by the exploration philosophy of "set up two bases, operate on two targets, develop one production testing area", geologists determined to drill the Tazhong 1 well. The well was designed at the Tg5′ reflective surface (the top surface of the Ordovician buried hills); the calculated area of the trap was 6,330 km^2 at a depth of $-4,000$ m; the relief was 1,840 m. On May 5, 1989, the Tazhong 1 well was drilled initially. This well obtained good oil and gas shows in weathering dolomite in the Low Ordovician series with well developed karst and fractures. On October 18, 1989, the oil test was conducted in the interval of 3,565.98 – 3,649.77 m; by using a 22.33 mm oil choke, the daily oil production was 365 m^3 and the daily gas production was 55.7×10^4 m^3. On October 30, 1989, another oil test was carried out in the interval of 3,565.98 – 3,737.61 m; by using a 32 mm oil choke, the daily oil production was 576 m^3 and the daily gas production was 34.06×10^4 m^3 (Fig. 8.21).

During this breakthrough stage, the main exploration targets were the structure in the carbonate rocks that was positioned in the Paleozoic lower structural zone and the large size oil and gas reservoir; the Tazhong 1 condensate gas reservoir was discovered in the Ordovician buried hill. The Tazhong 1 well achieved a strategic breakthrough in the center of the desert and set up another important milestone in the exploration history in the Tarim basin. Also, this exploration breakthrough developed the exploration philosophy for the platform and basin area in the desert.

(a)

(b)

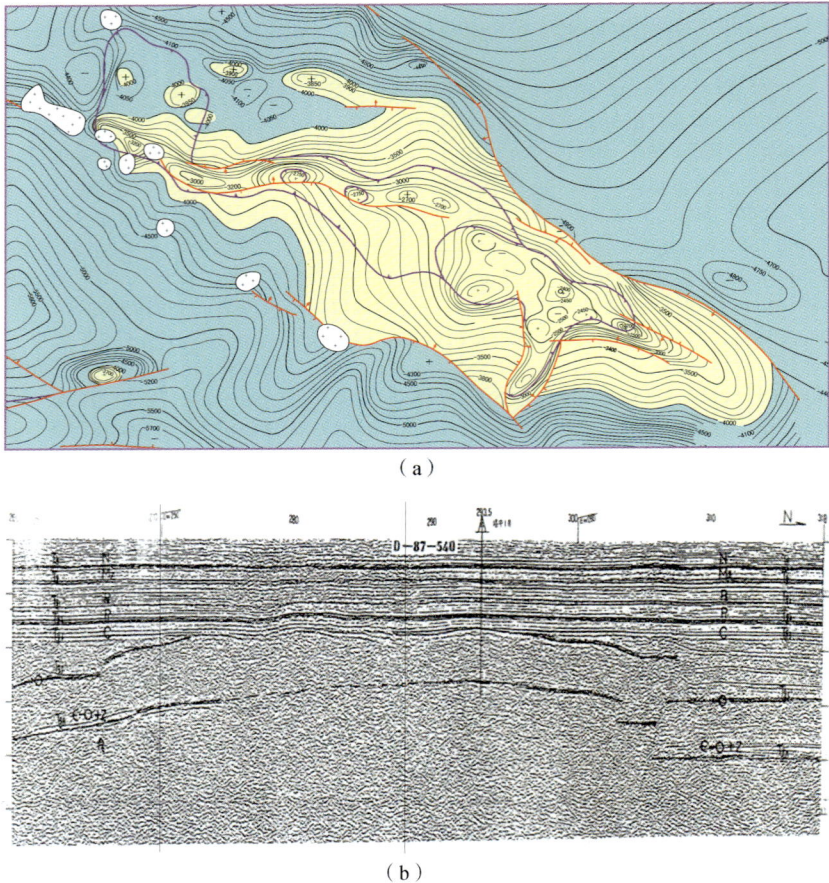

Fig. 8.20.   Designed well location of the Tazhong 1 well with the seismic cross section in the Tarim basin

● **1991 – 2001: Comprehensive Progress along with Complicated Exploration**

In order to increase the exploration achievements, in October and November 1990, the Tazhong 3 well and the Tazhong 5 well were drilled on the high positions of two buried hills that were located either 20 km east of the Tazhong 1 well or 23 km south of the Tazhong 1 well, respectively. In February 1991, these two wells drilled into the Ordovician series separately and the oil and gas shows were far less than the Tazhong 1 well; these were unsuccessful wells. The exploration failure on the buried hill in the Tazhong area indicated the heterogeneity of carbonate rock and the complicity of the oil and gas bearing situation in the large size anticline in the Tazhong area. Therefore, we should not simply apply the buried

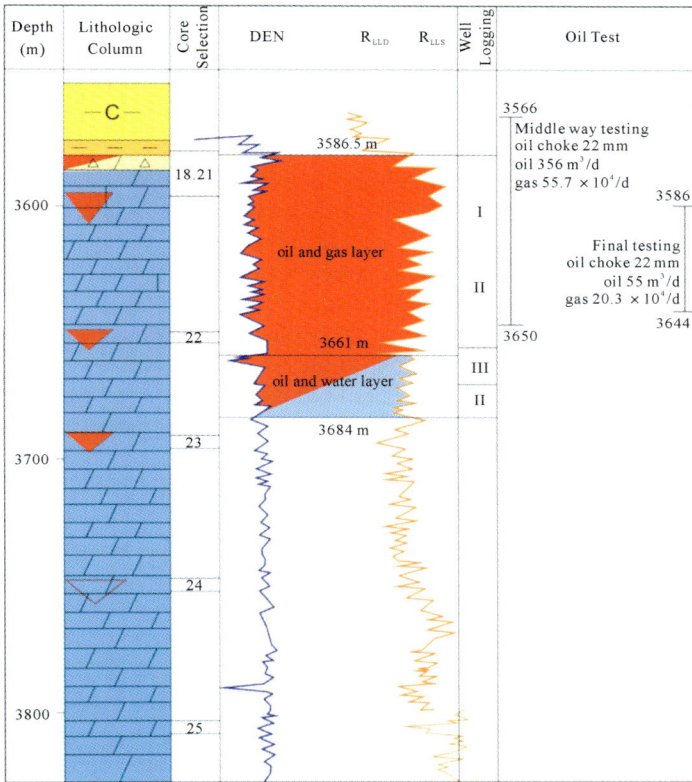

Fig. 8.21. Oil and gas shows in the Tazhong 1 well

hill type of reservoir model to direct petroleum exploration in the Tarim basin. In the Tazhong area, the exploration direction was changed and the target sequences were altered; the exploration in the carbonate rock experienced difficulties.

Persistent Exploration on the Buried Hill Yielded a Less Successful Rate

Guided by the exploration philosophy of "anticline controlled oil and buried hill controlled oil", following the exploration guideline of "mainly target the upper Paleozoic strata and also explore, explore lower Paleozoic strata", during 1992 – 1993, in the horst block zone and buried hill area, the Tazhong 2 well, Tazhong 8 well, Tazhong 401 well, Tazhong 18 well, Tazhong 9 well, Tazhong 6 well, Tazhong 17 well and Tazhong 101 well were successively drilled. The drilling explorations in the carbonate facies resulted in disappointments; the exploration of the high position of the buried hill was brought to a stop. After the disappointment, the following wells resulted in failure too, which included the Tazhong 38 well, Tazhong 19 well, Tazhong 7 well. However, in 1994, an exploration breakthrough was obtained in the Tazhong 16 structure on the northern slope.

After drilling into the Ordovician buried hill, the Tazhong 16 well had good oil and gas shows. Drilled to 4,256.74 m, there were 1.68 m of empty drilling. Drilled to 4,259.2 m, the kick happened. In the end of August, 1994, an oil test was conducted in the interval of 4,248.5 – 4,268 m; the daily oil production was 15.4 m^3. In September, after acidification treatment, by using a 4.762 mm choke, the daily production was 18.2 m^3; in 47 d, the total of produced oil was 1,416 m^3. In 1996, the Tazhong 161 well also obtained a commercial grade oil flow with a controlled oil bearing area of 37.2 km^2 and forecast geological oil reserve of 10,640,000 t. The discovery of the Tazhong 16 Ordovician oil reservoir enhanced the understanding of buried hill geology and it offered new exploration ideals.

Up to the end of 2002, 29 wells in the buried hill carbonate formation in the Tazhong area had been drilled. Among these wells, commercial grade oil flows were obtained in four wells; one well showed a low productive oil and gas flow; ten wells had oil and gas shows. The Tazhong 1 condensate gas reservoir and the Tazhong 16 oil reservoir were discovered.

The exploration in the buried hill area provided two insights: firstly that the reservoir sequences in the buried hill carbonate formation have a clear alteration and strong heterogeneity; secondly that the types of oil and gas reservoirs were complicated and the distribution of oil and gas was intricate.

Changed Exploration Target and Petroleum Production Base Built

According to oil and gas shows in the Carboniferous system from the Tazhong 1 well, and via the study of lithologic paleo-geography in the Carboniferous system, it was proposed that, because the Tazhong 1 well was located on the delta plain, the delta front zone that was located in the northwest direction should contain good physical properties and it might contain oil; at same time, it was believed that the northwestern side nears the petroleum source area and that has good preservation conditions. Therefore, in the Tazhong area, the exploration activity was transferred westward to target the Carboniferous sandstone. On November 16, 1991, the Tazhong 4 well was drilled initially. In 1992, in the Carboniferous Donghe sandstone, 191.12 m cores that contained oil were obtained. In April, in the interval of 3,597 – 3,607 m, an oil test was carried out; by using a 11.11 mm choke, the daily oil production was 285 m^3 and the daily gas production was 5.3×10^4 m^3. Thus, the first oil field (the Tazhong 4 "Donghe sandstone" oil field) was discovered in the center of the desert, which provided a commercial grade oil flow.

The exploration breakthrough of the Tazhong 4 well brought new petroleum exploration activity to the Donghe sandstone. Seven oil and gas fields were discovered, which included the Tazhong 10, Tazhong 6, Tazhong 16 and Tazhong 40. The explored oil reserve is 5,896.3×10^4 t (recalculated) and the explored gas reserve is 178.55×10^8 m^3. In addition, a petroleum production base with a productivity of 100×10^4 t was constructed; this set up a solid foundation for oil and gas exploration in the Tazhong area.

Preliminary Exploration of Faulted Belt-I Achieved New Breakthrough

When geologists explored the low relief structure in the Donghe sandstone, they also investigated the Tazhong-I faulted belt. In 1995, the drilling exploration of the Tazhong 12 well was completed; the source rock of argillaceous limestone facies was initially discovered in the Upper Ordovician strata in the Tazhong area. The lithology is dark gray, argillaceous banded limestone and brown color bio-limestone, which are widely distributed on the northern slope in the Tazhong area and which are in oil generating high peak time. The discovery of the Upper Ordovician source rocks inspired the subsequent explorations to approach the source rock and the lower Paleozoic system.

Based on the perceptions of the Tazhong area, the exploration activities altered their targets in the following three areas. The target strata were expanded from the single target of the Carboniferous system into multi-targets of the Devonian, Silurian and Ordovician systems; the exploration locations were expanded from the central faulted horst zone into the Gucheng nose-like uplift and the southern and northern slopes; the target traps were increased from just the structural trap to multiple types of traps that included the lithologic trap, stratigraphic trap and structural trap.

During 1996 – 1998, under the guidance of "move toward source rock, approach the near source petroleum play and advance the faulted belt", the Ordovician carbonate was explored in the Tazhong-I faulted belt.

In July, 1996, after obtaining a commercial grade oil flow in the Carboniferous Donghe sandstone in the Tazhong 24 well, there was continual drilling into the Ordovician carbonate rocks. Good oil and gas shows were obtained in the interval of $4,452 – 4,483.48$ m, In September, 1996, a well test was conducted in the Ordovician interior limestone in the interval of $4,461.1 – 4,522.87$ m; with acidic treatment, by using a 7.94 mm chock, the daily oil production was 15.1 m^3 and the daily gas production was 28,892 m^3. The Tazhong 24 well controls 11.4 km^2 of the oil and gas bearing area with a forecast geological oil reserve of 108×10^4 t. Following exploration clues in the Ordovician system from the Tazhong 16 well and the Tazhong 24 well, in February 1997, along the Tazhong-I faulted belt, in a direction EW along approximately 200 km, the Tazhong 26, 44 and 45 wells were drilled at the same time. Commercial grade oil and gas flows were discovered in the Upper Ordovician interior carbonate rock from these three wells successively. The total geological oil reserve (controlled reserve+forecast reserve) was $1,588.3\times10^4$ t; the gas reserve was 100.52×10^8 m^3. The exploration results confirmed that the Tazhong-I faulted belt is a petroliferous rich belt; thus, a new exploration frontier had been uncovered in the interior carbonate rock in the Tazhong area.

After that, the Tazhong 49, 54, 42, and 27 wells were drilled along the Tazhong I faulted belt and these wells were dissatisfying; only the Tazhong 451 well obtained a highly productive commercial grade oil and gas flow. At the same time, the wells that were drilled into carbonate rock not only had an unstable production rate, but also yielded oil and water simultaneously. In addition, these wells generally did not produce oil and gas (or produced a small amount of oil and gas) before acidic treatment. After acidic treatment, the oil and gas production rate and pressure were relatively high, but the pressure decreased sharply; clearly, the pressure recovery demonstrated the exhaustion of the oil and gas.

Although people recognized that the Tazhong faulted belt was an ideal place for oil and gas accumulation, however, guided by the exploration philosophy of "faulted belt controls oil and exploration of the structure", and also facing the situation that a carbonate reservoir was hard to forecast, exploration in the Tazhong-I faulted belt was stopped after 1998.

Exploration of Dolomite Produced New Discovery

Studies showed that, in the Tarim basin, the Cambrian dolomite possibly has a stratigraphic sequence and the dolomite reservoir was better than the limestone reservoir. Guided by the exploration idea of "follow the good reservoir sequences and approach the original oil and gas reservoir", during 1996 – 1997, the Tacan 1 well, the Tazhong 162 well and the Tazhong 43 well were drilled.

In the Tacan 1 well, 5 samples were collected in the interval of 5,059 – 5,113 m; this interval was positioned in the interior dolomite section under the top surface of the Ordovician limestone; 42.15 m of oil spots and fluorescent dolomite were obtained in total; the samples were immersed in light-middle weight crude oil that has a strong odor and that evaporates easily. In the Tazhong 43 well, a gas show occurred in the interior dolomite section under the top surface of the Ordovician limestone. In the Tazhong 162 well, the gas show occurred in the interval of 5,956 – 6,020 m that was positioned in the interior dolomite section under the top surface of the Ordovician limestone; the largest TG was 72%; the gas foam occupied 20% of the tank's surface; during the circulation, the height of the kick was 2.5 m and the height of the flame was 2 – 7 m; the content of H_2S was great. The well test was carried out in the interval of 5,931.12 – 6,050 m after acidic treatment; by using a 9 mm oil choke, the daily production rate was 207,000 – 164,205 m^3. The forecast natural gas reserve was 58.14×10^8 m^3.

The drilling exploration of dolomite reveals the following: the Cambrian dolomite has a certain kind of characteristic stratigraphic sequences; the interior dolomite contains a good combination of reservoir and seal and dolomite is a good reservoir body; because the interior dolomite offers good accumulative conditions, it should thus become the ideal target in the search for original oil and gas reservoirs in the future.

Investigation of Salt Dome in the Cambrian Led to Understanding of Entire Tazhong Paleo-Uplift

Guided by the exploration philosophy of "anticline controls oil and search for the original oil and gas reservoir", the Tacan 1 well was drilled in April 26, 1996 and this well was finished on February 26, 1998; the total depth of the well was 7,200 m and the floor sequence was the pre-Cambrian basement.

The Tacan 1 well is the only well that drilled through the Cambrian strata; beneath the salt layer, the samples from the interval of 7,116.5 – 7,124.8 m displayed well developed karst that was not filled (or half filled) by dolomite; when the core came out of the casing pipe, it noticeably brought with it gas, which indicated the strata has good reservoir conditions. In the Cambrian system, the lithology from the interval of 6,800 – 7,085 m included argillaceous dolomite, gypsum dolomite, and gypsum. These rocks make good regional cap rocks. During the drilling of the Tacan 1 well, even the density of slurry reached 1.6. However, in the interval of 7,108 – 7,132 m, there were four layers of gas anomaly with a head of 10 m; the largest TG was 22.25%; C1 was 15.4%; C2 was 0.34%; and C3 was 0.91%. In the Tacan 1 well, the well test was carried out in the interval of 7,015 – 7,035 m, the oil mixed with water, the production rate was 48 m^3/d. Although commercial grade oil and gas flows were not discovered in the Cambrian system beneath the salt layer in the Tacan 1 well, the gas shows actively appeared; however, in the dolomite section below the gypsum dolomite, which indicated that the Cambrian system below the salt layer was the idea domain for searching for the large scale, original oil and gas reservoir in the Tazhong area.

● 2002 to the Present: Re-Evaluation with Extended Exploration

During 1998 – 2002, geologists explored and evaluated the Donghe sandstone, the Silurian system and the Ordovician systems again; the frequent change in exploration strategy did not bring about a new breakthrough; the successful rate of exploration was low; the repeated disappointments resulted in exploration activity taking a nose-dive. In the Tazhong faulted horst belt, the only breakthrough was obtained in the Tazhong 1 well; the explorations on other buried hills ended in disappointment. In general, the Tazhong-I faulted belt has oil and gas shows and it is a beneficial area. There were oil and gas shows either in the interior dolomite or in the petroleum play beneath the salt layer; however, further exploration could not be continued. A large size oil and gas field had not been discovered in the Tazhong area. Up to this time, in the Tazhong area, even the exploration in the carbonate rock has stopped. However, geological research and technical innovation were continued.

With in-depth deduction, research and summarizing experiences, it was considered that the Tazhong-I structural belt was a large size thrust belt that was formed during the end of the Early Ordovician-early phase of the Late Ordovician; before the Upper Ordovician series was deposited, this area endured a long duration of erosion to form the complicated faulted slope zone; during the time that the Upper Ordovician Lianglitage formation was deposited, the reef complex of the platform margin facies was created along the faulted extremely steep slope zone; on the northern side, the facies were altered into the massive mudstone that was interbedded with sandstone; the reservoir sequence was the major controlling factor for oil and gas accumulation. The study shows that the Tazhong-I faulted slope zone did not have a structural trap. However, from west to east, the oil and gas were discovered over a nearly 200 km range; if comparing oil and gas layers of the eastern side and the western side, the elevation difference of oil and gas layers could reach up to 2,000 m, the oil and gas distribution was not controlled by isolated structural traps. At the low elevation area of the slope, a commercial grade oil and gas flow was obtained from the Ordovician limestone reservoir. It was also believed that the high resolution, new 3D seismic data were the effective method for exploring carbonate rock. During 2003, a 3D seismic survey was designed for the entire region; the Tazhong new 3D seismic survey was mainly concentrated on the central area; from 2002 to the present, 1,749 km^2 of 3D seismic survey have been finished.

Since 2003, a new round of exploration in the carbonate rock in the Tazhong area has been conducted. Important breakthroughs were obtained either in three domains of the Ordovician system or in the Silurian system. Thus, the new exploration wave has begun in the Tazhong area and the growth in petroleum reserves has been raised to a high peak. The petroleum exploration in the Ordovician system not only helped people reach a high production rate and stable productivity in this area, but it has outlined a large size oil field with a hundred million tons reserve.

Before 2003, the Tazhong area had a third grade reserve of 1.4×10^8 t (recalculated reserve) approximately; the carbonate rock only contained $6,082 \times 10^4$ t of controlled + forecast reserve. In 2004, in the Tazhong area, the carbonate rock contained a third grade reserve of 1.77×10^8 t; in 2005, the explored + controlled second grade reserve reached 1.39×10^8 t. After 2006, an important breakthrough was obtained in the Tazhong 83 well, which was located at the weather crust in the Ordovician Shanying formation. Until recently, via a comprehensive analysis and total evaluation of the Tazhong-I slope zone, the reef complex contained a third grade reserve of 6.36×10^8 t; among these reserves 1.02×10^8 t was an explored reserve.

Overall, through persistent research work, a short term exploration direction and target was determined in the carbonate rocks. In addition, success was achieved in three domains in the Ordovician system so as to achieve a high peak in the reserves. On the technical front, reservoir forecasting technology for Tazhong carbonate exploration was introduced. These achievements not only encouraged people to discover a major oil field in the Tazhong area, but they also help the Tazhong exploration to proceed with confidence.

8.4.2   Geologic Features of Large Scale Reef Flat Type of Oil Field in Slope Zone in the Central Tarim Basin

The Tazhong I faulted slope zone has a perfect regional structural–depositional background; it starts at the Tazhong 45 well area on the western side and it reaches the Tazhong 26 well area on the eastern side; from WE, the length is 200 km; from

SN, the width is 5 – 10 km; it covers 1,800 km^2 of land. The Tazhong I faulted slope zone is located on the northern slope and it contains the Upper Ordovician platform margin reef flat facies. The entire Tazhong faulted slope zone dips toward the west; the elevation difference between the eastern side and western side is 1,800 m; it only contains a miniature isolated structure and it offers perfect geological petroleum conditions.

- **Multi-Phases of Petroleum Infill Determined that the Central Tarim Region Contains Oil**

Multi-sets of hydrocarbon source rock and multi-phases of oil and gas infilling and adjusting created the prolific material foundation in the Tazhong area. The second round of resource evaluation showed that Tazhong has 30×10^8 t of resources; the third round of resource evaluation showed that Tazhong has 14.2×10^8 t of resources. Therefore, the Tazhong area contained adequate resources for developing a large size oil and gas field. Because the distributions of the source rock were varied, because the direction and the method of oil and gas infilling were diverse and because different blocks had dissimilar accumulative conditions, most of the Tazhong area had therefore oil and gas infilling and contained oil in general.

The study shows that there are three hydrocarbon supply centers in the Tazhong area, which supplied three phases of petroleum accumulation and two phases of adjustment (Fig. 8.22). The Middle and Lower Cambrian source rock reached an oil generation high peak during the late Caledonian movement.

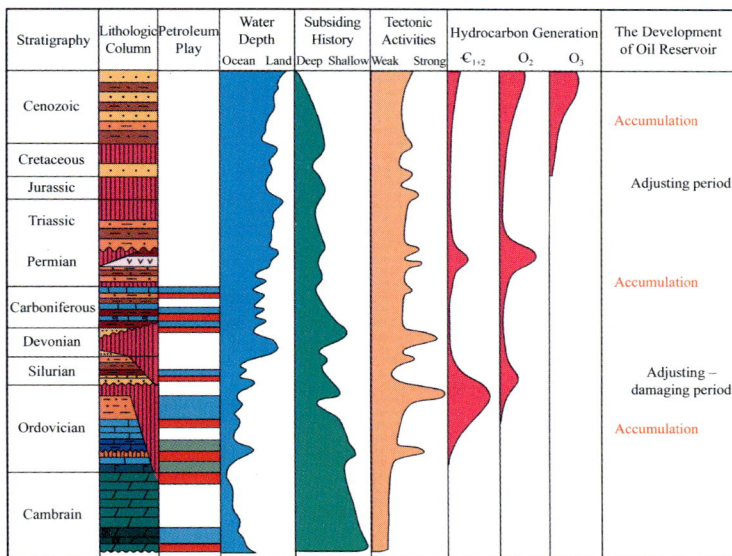

Fig. 8.22.   Composite illustration of petroleum accumulation in the Tazhong area, Tarim basin

Influenced by the uplifting and faulted movements that happened during the late Caledonian–early Hercynian tectonic movements, in the Ordovician system, the oil that was generated by the Cambrian source rock went through water washing; thus, the content of light hydrocarbon, benzene and methyl benzene was very low. As a result, in the Ordovician system, crude oil was accumulated early and mainly came from the hydrocarbon source rock in the Cambrian system.

The petroleum accumulation can be divided into three phases, which are the middle–late Caledonian phase, the late Hercynian phase and the Himalayan phase.

The first petroleum accumulation phase happened during the end of the Caledonian tectonic period. During the Late Ordovician, in the eastern Tarim basin, the Cambrian–Ordovician source rocks progressed into a mature–post mature stage and they generated a huge amount of oil and gas; this period was also the time that the Tazhong uplift was developed. The Tazhong area was the ideal place for accumulating oil and gas; oil and gas in-fillings had continued until the Late Silurian.

In the Tazhong area, the effective accumulation phase happened during the late Hercynian tectonic period. The temperature of inclusion (110 – 130 °C) also reflected the fact that oil and gas were supplemented during the late Hercynian tectonic period. During the late Hercynian tectonic period, there were newly generated oil and gas supplies other than those generated from the Cambrian system during the late Caledonian tectonic period. According to the analysis of geological history and geothermal history, the Middle and Upper Ordovician source rocks progressed into a hydrocarbon generating high peak during the late Hercynian tectonic period, which could provide a huge amount of oil and gas resources in the Tazhong area. The Tazhong I faulted slope zone was next to the Manxi depression; the oil and gas accumulations in the Tazhong I faulted slope zone were closely related to this set of hydrocarbon source rock during its late Hercynian accumulative phase.

Natural gas infilling mainly happened during the Himalayan tectonic period; the temperature of inclusion (140 – 150 °C) reflected oil and gas accumulation during this period. The buried thermal maturation history of reservoir sequences also confirmed that, during the Himalayan tectonic period, the Middle and Lower Cambrian series (that was located in the Tazhong I faulted belt) progressed into a crude oil pyrolysis stage ($R_0 > 2\%$); however, the Lower Ordovician series was still in a wet gas producing stage ($R_0 = 1.5\%$); the crude oil that either accumulated in the Lower Ordovician dolomite reservoir on the northern slope in the Tazhong I faulted belt or accumulated in the wreckage of the Upper Ordovician bio-reef of slope facies progressed into a pyrolysis stage ($R_0 > 2\%$). These sources could provide crude oil pyrolysis gas. Because fault activity was inactive during the Himalayan tectonic period, the majority of natural gas migrated laterally along the fractures that formed a limited connective net system. If natural gas that was generated in the Middle and Lower Cambrian series had a high content of hydrogen sulfide, upwards migration of natural gas was impossible. Natural gas migration mainly happened within the Ordovician strata that did not contain hydrogen sulfide; this included the migration of pyrolysis gas in the Lower and

Upper Ordovician series, which migrated upwards from the northern side of the Tazhong I faulted belt to the lithologic trap in the Ordovician reef flat complex in the Tazhong I faulted belt.

● Paleozoic Carbonate Contained Excellent Petroleum Play

The Tazhong paleo uplift was formed during the early period and it developed at a steady pace; because it was the destination for oil and gas migrating over a very long period, this uplift not only captured multiple phases of oil and gas infillings, but it also controlled a quarter of the resources in the platform and basin area; in addition, this uplift also offered perfect conditions for oil and gas to accumulate. In the Tazhong area, multiple sets of reservoir–seal combinations developed either in the lower Paleozoic carbonate rock or in the Silurian–Carboniferous systems to form various types of petroleum trap. The Paleozoic reservoir–seal combinations were next to the Cambrian–Ordovician source rocks; the Cambrian–Ordovician strata contained a well developed faulted system that offered a convenient pathway for oil and gas to migrate vertically; the mixed distributions of fractures and karsts in the carbonate rock served as a network of channels for oil and gas migrating and accumulating; the unconformity surface that was on top of the Ordovician system functioned as the major pathway for the Silurian and Carboniferous oil and gas migration. The Paleozoic reservoir–seal combinations combined with petroleum migration to form a perfect petroleum play, which had the characteristics of multiple oil layers and diversified accumulation.

● Stabilized Burying Condition of Late Geologic Period Offered Ideal Protection

Although the oil and gas reservoirs in the Tazhong paleo uplift were altered by the Caledonian tectonic movement, nevertheless, the preservative conditions for oil and gas that were generated during the Hercynian and Himalayan tectonic periods were excellent. The Tazhong uplift was finalized during the end of the Ordovician; from west to east, the Silurian and Carboniferous strata were deposited on top of the uplift; generally speaking, after the Carboniferous, there was no faulted activity; however, there were multiple phases of up-dipping events that assisted in forming stable accumulation. Therefore, the oil and gas were distributed either in the Carboniferous system or in the strata beneath the Carboniferous system; in other words, the oil and gas did not migrate into the Permian system and the Mesozoic strata.

● Reservoir Sequences were Stable

In the Tazhong faulted slope zone, the discovered oil and gas mainly accumulated in a granular limestone section in the Upper Ordovician Lianglitage formation. According to the characteristic of conductivity, geologists further divided this limestone section into three portions. 1) The clay ribbon limestone portion: gray, clay ribbon limestone; in an isolated area, limestone and mudstone were interbedded with each other with unequal thickness; the clay ribbons were discontinuous, which were either in neo-breccia shape or in spotty shape; the content of the clay ribbons decreased in descending order. Because of uneven erosion, in general the thickness of the clay ribbon limestone was 60 m approximately. Typically, this portion of limestone did not have oil and gas shows. 2) The granular limestone portion: gray or brownish gray, granular limestone (algal calcarenite and arenite calcirudite); also a bird's eye structure and the needle points were well developed. The thickness of this portion was 50 – 120 m approximately. The granular limestone portion contained most of the oil and gas shows. 3) The argillaceous limestone portion: brownish gray or grayish brown, micrite and marlaceous limestone were the primary lithology that interbedded with calcarenite and algal limestone; the limestone was tainted with a high content of clay; the clay ribbons and the horizontal sutures were well developed. Because of the overlapping deposits, the thickness of argillaceous limestone varies. This portion contained well developed source rocks.

● **Depositional Feature of Reef Flat and Characteristics of Reservoir Sequences**

In the Tazhong area, during the time that the Late Ordovician Lianglitage formation was deposited, the depositional system of the continental shelf margin was developed. Along the Tazhong-I faulted slope zone, the subfacies of the continental shelf margin formed, which include the micro facies of the high energy bio–reef, reef flat and bioherm complex. The subfacies of the inner shelf include micro facies of the tidal flat, inner shelf flat, inner shelf gentle slope, inner shelf trough and marlaceous mound. On the northern side of the Tazhong I faulted slope zone, the subfacies of the trough basin were developed that received clastic deposits.

In this area, the vertical combination of the Upper Ordovician deposits displayed multi-cycles that consisted of grain-banks subfacies, marlaceous mound subfacies and skeleton reef subfacies; overall, these cycles of subfacies were covered by massive argillaceous deposits. Usually, in every single reef flat, bioherm or reef complex, the lower portion consisted of grain-banks subfacies and the upper portion was made of marlaceous mound subfacies and (or) skeleton reef subfacies; this kind of cycle was covered by other grain-banks subfacies of the next cycle.

On the Tazhong faulted slope zone, the Tazhong 24 – 82 well areas displayed high energy facies of the continental shelf margin; the reef, bioherm, reef flat were not only superimposed vertically but they also connected together horizontally over a large area; the thickness was between 80 – 150 m; the width was in the range of 2 – 5 km; the distribution was stable.

On the Tazhong shelf margin, the reef complex contained well developed reservoir sequences. In the Tazhong 62 well, the dissolution pores were well developed in calcirudite, bio-clastic limestone and calcarenite. The thickness of reservoir sequences that had a porosity >1.8% was 41.9 m, which was 69.8% of the total thickness of interpreted strata. The thickness of reservoir sequences that had a porosity >3% was 28.5 m, which was 47.5% of the total thickness of interpreted strata. The thickness of a single layer could reach up to 11.4 m. According to statistical data, the matrix porosity in the reef flat, reef core and grain-bank is far larger than in the bioherm and trough limestone; in addition, if comparing the permeability, the former is over ten times more permeable than the latter. This evidence showed good physical properties.

Because the reef flat contained well developed matrix pores, it was beneficial for developing high quality dissolution pores in either buried karst or in surface karst. The excellent reservoir sequences in the Tazhong 44, 24 and 62 wells were all developed on the high energy reef flats and they all experienced buried dissolution.

8.4.3 Characteristics of Petroleum Reservoirs in Central Tarim Region

● **Entire Central Tarim Region Contained Oil: "Oil in the Western Region and Gas in the Eastern Region" and "Gas in the Outer Belt and Oil in the Inner Belt"**

Both drilling exploration and research work indicated that, in the Tazhong faulted belt, the Upper Lianglitage formation contained good oil and gas shows. From east to west, commercial grade oil and gas flows were discovered over a range of 200 km; the elevation difference of the top surface of the reservoir could reach up to 1,800 m. At present, from the Tazhong 26 well to the Tazhong 82 well (from east to west), the proven elevation difference of the top surface of the condensate gas reservoir is 1,200 m; the gas reservoir was not controlled by the local structure.

Overall, the Tazhong-I structural belt contained oil on the western side and gas on the eastern side; the distributions of oil and gas could also be summarized by saying that the outer circle contained gas and the inner circle contained oil. In the horizontal space, excluding the Tazhong 621 well that has a relatively low ratio of gas to oil, the distribution of the gas to oil ratio was above 1000 in the rest of the commercial grade oil and gas wells in the Tazhong 26 well–82 well area; this characteristic indicated this is a condensate gas reservoir. The Tazhong 58, 72 and 16 wells in the inner belt on the southern side have a very low ratio of gas to oil, which indicates a normal oil reservoir. The Tazhong 45 well area on the western side has a slightly evaporated oil reservoir; the distributions of oil and gas were in a belt pattern (Fig. 8.23).

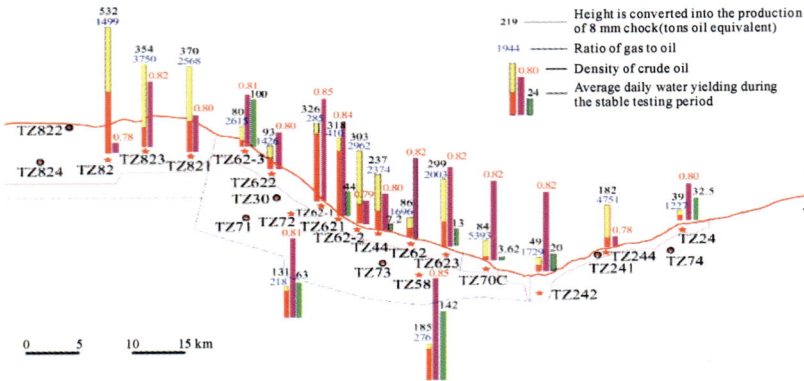

Fig. 8.23. Ratio of gas to oil, the density of crude oil and the water yielding in the eastern portion of the Tazhong I structural belt

● **Distributions of Oil and Gas were Intricate; Lateral Water and Base Water were Absent**

The crude oil of the Tazhong I structural belt has the characteristics of low density, low viscosity, low gluten+bitumen content, middle–low waxy content and middle–low sulfide content. However, if comparing different wells, there were differences in the traits of crude oil and in the ratio of gas to oil; the anomalies included high density, middle–high waxy content and a middle range of sulfide content. The Tazhong 621 well (that was located in the middle portion of the condensate gas reservoir) had the characteristics of normal crude oil in an isolated area. In the same well, if comparing different testing intervals, or if comparing different testing times, the traits of crude oil may have noticeable differences. For example, in the Tazhong 82 well, the density of crude oil in the lower oil layer was smaller than that in the upper oil layer; in addition, the ratio of gas and oil in the lower oil layer was higher than the one in the upper oil layer. Another example is the Tazhong 622 well; one year after the testing production, the crude oil density increased from 0.79 g/cm^3 to 0.85 g/cm^3.

In the Tazhong I structural belt, the constituents of natural gas have noticeable variations; the content of methane was between 80.57% – 92.5%; the content of CO_2 was between 0.1381% – 3.4782%; the content of N_2 was between 3.29% – 9.12%; the relative density of natural gas was between 0.61 – 0.68. In this area, natural gas contained hydrogen sulfide in general; the content of hydrogen sulfide had noticeable variations between the wells. The variations in natural gas constituents can be divided into belts horizontally. In the area that was located east of the Tazhong 241 well, the natural gas has a high content of N_2 and a low content of H_2S. In the area that was between the Tazhong 242 well and the Tazhong 823 well, the natural gas has a high aridity coefficient (>0.95), a middle–low content of N_2, a low content of CO_2 and a middle–high content of H_2S.

In the area of the Tazhong 82 – 54 wells, natural gas has a low content of H_2S and a middle–high content of N_2. In the area of the Tazhong 45 well, the natural gas has a low aridity coefficient, a low content of CO_2, a middle–low content of N_2, and a middle content of H_2S. The variations reflected the differences either in the creative mechanism of natural gas or in secondary alterations.

- **Variable Differences in Production Rate Reflected the Complex Nature of the Single Layer Framework in Reservoir Sequences**

The dynamic analysis shows the following: the oil pressure has the characteristics of fast decreasing, stabilizing and cyclical variation; the ratio of gas to oil not only has the traits of relative stability and slow increase, but it also shows abrupt change and a cyclical decrease; the production rate not only has the characteristics of relative stability and slow decrease, but it also shows cyclical change and fluctuation. The differences in oil and gas production rates indicate the complexity of the reservoir texture in the Tazhong carbonate rocks. If physical properties and connectivity were good, the oil and gas production rates were stable. If the reservoir sequences showed considerable alteration with intricate connectivity, the strong heterogeneity would cause variations in the oil and gas production rate.

- **Large Scale, Semi-Layered, Reef Flat Type of Lithologic Oil and Gas Reservoir**

In the Tazhong-I structural belt, although the oil and gas distributions in the Ordovician system were complicated, the entire carbonate rock of the platform margin facies contained oil, however; in addition, the developmental level of the reservoir sequence controlled oil and gas accumulations. On the eastern side, from the Tazhong 26 to the Tazhong 82 wells, the oil and gas reservoirs have unique normal temperature and pressure systems; there was no clear boundary between oil, gas and water. Vertically, the reservoir was distributed in a neo-layered pattern; horizontally, the reservoir was spread out along the platform margin facies belt. The high pressure physical properties analysis shows that this area mainly contained condensate gas with a middle–high content of condensate oil. In an isolated area, there was highly saturated and volatile crude oil. This is a large size, neo-layered, reef flat type, condensate gas reservoir (Fig. 8.24). On the other hand, the distributions of oil, gas, and water were complicated. The characteristics of oil and gas were varied, which reflected the fact that reservoir sequences were heterogeneous horizontally. Also, the creative mechanisms were intricate; the production rates of oil, gas and water were diversified and the types of oil and gas reservoirs were special.

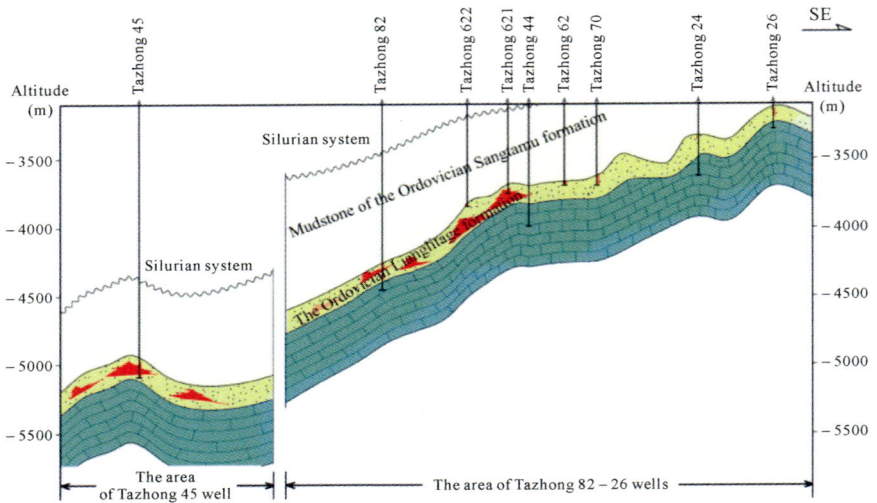

Fig. 8.24. Cross section of large scale, semi-layered, reef flat type of lithologic oil and gas reservoir in the Tazhong-I structural belt

References

Dai, J.X., et al., 1992. The characteristic of natural gas belt in large and medium size gas field. *In*: The Study of Natural Gas Geology. Petroleum Industry Press, Beijing, pp. 1-7.

Dai, J.X., Wang, T.B., Song, Y., et al., 1997. The Preconditions and Distribution of Large–Medium Size Gas Field in China. The Geological Publishing House, Beijing, pp. 184-237.

Gao, R.Q., Zhao, Z.Z. (Eds.), 2001. The Frontier Petroleum Exploration in China. Petroleum Industry Press, Beijing.

Jia, C.Z., et al., 2000. Petroleum Expiration in Foreland Thrust Belts. Petroleum Industry Press, Beijing.

Jia, C.Z., 2003. New achievements and new frontier in Chinese petroleum exploration. World Petroleum Industry, 10(3):20-25.

Jia, C.Z., Zhao, Z.Z., Zhao, W.Z., et al., 2005. The petroleum reserves and exploration potential in major onshore petroleum basin. Acta Petrolei Sinica, 26(B03):1-6.

Jia, C.Z., et al., 2008. Geologic Theory and Exploration Technique of Lithostratigraphic Reservoir. Petroleum Industry Press, Beijing.

Li, P.L., Jin, Z.J., Zhang, S.W., et al., 2003. The current research status and progress of petroleum exploration in the Jiyang Depression. Petroleum Exploration & Development, 30(3):1-4.

Wang, Y.C., 2001. The Creation and Distribution of Lithologic Reservoir in the Southern Songliao Basin. Petroleum Industry Press, Beijing.

Zhou, C.N., Tao, S.Z., Xue, S.H., 2005. Implication of "Facies Control Theory" and its significance for petroleum exploration. Petroleum Exploration & Development, 32(6):7-12.

Part III

Petroleum Exploration in Foreland Fold and Thrust Belts

Petroleum Geology of Foreland Thrust Belts in China

Since the Proterozoic, the Chinese cratons have experienced numerous geological events of divergence and convergence. Apparently, the Chinese portion of the Eurasian plate consists of core elements of the Tarim plate, the Huabei Plate and the Yangzi plate, which are attached to more than twenty micro terrains (Huang, 1954; Li, 1982; Zhang, 1991; Qiu and Gong, 1999). The tectonic setting of the Chinese continent is located at the southern margin of the Eurasia plate and northern margin of the Indian plate. It is separated from the Pacific plate by a trench–arc–basin system. The Chinese continent was united by many convergent plates during several geological events, which include three small primary tectonic plates of the Tarim plate, the Huabei plate and the Yangzi plate, in conjunction with more than twenty micro tectonic plates (terrains), such as the Junggar micro plate, the Qaidam micro plate and the Qiangtang micro plate. Compared with tectonic plates in North America or in Europe, the small plates or micro plates in China are smaller in size. For example, the area of the Tarim plate is 56×10^4 km^2; the area of the Yangzi plate is 108×10^4 km^2; and the area of the Huabei plate is 120×10^4 km^2. The largest Chinese plate is the Huabei plate. However, the area of the Huabei plate is only 6% of the area of the North American plate, or 14% of the area of the European plate. Because large scale orogenies happened on small size Chinese plates, the tectonic structures were unstable and strong tectonic movements occurred in western-central China. Small craton basins are inlaid inbetween orogenies to form complicated embedded structures. These orogenies are the Tianshan–Yinshan orogeny, the Qilian–Qinling orogeny, Kunlun orogeny and Longmen–Helan orogeny. After experiencing two stages of Paleozoic marine deposition and Mesozoic–Cenozoic terrestrial deposition, various major superimposed composite basins were formed by merging the small craton basins and which surrounded the Mesozoic–Cenozoic foreland basins. These small craton basins are enclosed by orogenies of the Hercynian–Indosinian–Himalayan tectonic phases (Jia et al., 2005) (Fig. 9.1). Among superimposed composite basins, the Ordos basin, Sichuan basin, Tarim basin, Junggar basin, and Qaidam basin have a basement of pre-Sinian acidic or intermediate-basic metamorphic rocks. Also, the interior structures of these basins are fairly gentle and the crust is about 35 – 45 km thick. Next to the orogeny, many foreland thrust belts of the Mesozoic–Cenozoic eras were developed on the margin of the basins. Therefore, compared with the large craton basins on other continents, these small craton basins of western-central China not only have similar tectonic features to large craton basins, but also demonstrate strong tectonic activity that is a unique tectonic characteristic of western-central basins in China.

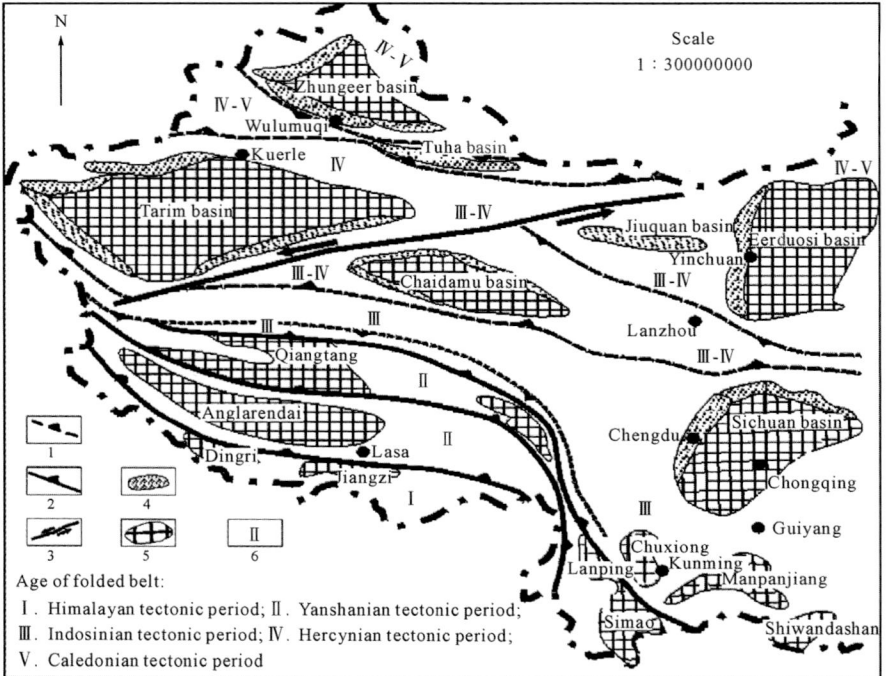

Fig. 9.1.   Miniature craton basins and foreland basin thrust belts in western-central China
1: Paleozoic suture zone; 2: Mesozoic suture zone; 3: Slip fault structure belt; 4: Foreland basin thrust belt; 5: Miniature craton or terrain; 6: Time interval for folded belt

9.1   Tectonic Background of Sinitic Foreland Basin Development

We will first introduce the basement of the paleozoic convergence between miniature craton plates.

9.1.1 Basement of the Paleozoic Convergence Between Miniature Craton Plates

● **From the Cambrian to the Silurian Periods, Small Craton Plates were Drifting Alone**

During the early Paleozoic era, the Tarim plate, Huabei plate and Yangzi plate existed as small craton plates, which were primary plates that formed the Chinese mainland. However, these three plates were separated by three connected oceans that were the Paleo-Asian Ocean, Paleo-China Ocean and Proto-Tethys Ocean (Jia, 1997; Zhang et al., 1997). Using paleomagnetic information to reconstruct the origin of ancient plates, the research shows that all three plates came from the southern hemisphere. Compared with their relative positions at the present time, the relative positions between the three plates were dislocated previously (Jia et al., 1995). During the late Paleozoic, the separated small craton plates merged into a convergent plate to form the embryonic shape of the Chinese mainland (Allen, 1991; 1992; 1993; Wei et al., 2002). From the Cambrian period to the Middle Ordovician epoch, a major portion of each of the primary plates (the Huabei plate, the Tarim plate, and the Yangzi plate) were under the sea and they developed littoral facies deposits of carbonate rocks and sandy mudstone formations. During the Late Ordovician, the Huabei plate rose above water and became a paleo-continent; in addition, the strata of the Middle Ordovician series and the Silurian system were absent.

● **From the Devonian to the Permian Periods, Convergent Events on the Southern Margin of the Eurisa Plate and a Rifted Event on the Northern Margin of the Tethys Ocean**

The late Paleozoic was an important period of tectonic evolution in China. The Junggar plate, Tarim plate and Huabei plate collided with the Siberian plate and Kazakhstan plate in consecutive order, which resulted in the disappearance of the Paleo-Asian Ocean (Xiao et al., 1991; Pan et al., 1997; Wei et al., 2002; Jin and Song, 2005; Shu et al., 2007; Jia et al., 2007). On the other hand, the Tarim plate and southern margins of the Huabei plate, which were located on the northern margin of the Paleo-Tethys Ocean, experienced a regional splitting event during the Permian period (Chen et al., 1997; 2005; Zhao et al., 2001; 2004). These two tectonic events determined basin evolution in this region.

Disappearance of the Paleo-Asian Ocean and Convergence of the Southern Margin of the Eurasia Plate

During the late Paleozoic, the Tarim plate, Huabei plate and Junggar plate drifted northward individually; they collided and converged with the Siberian plate and

Kazakhstan plate in consecutive order.

In the Aletai area of northwestern China, during the Late Carboniferous epoch, the Junggar plate was subducted northward to the Siberian plate; the Junggar plate was the middle section of the Paleo-Asian Ocean during the Devonian and Carboniferous periods. This subduction event brought about the development of the southern margin of Aertai mountain in the northern Xinjiang area. In addition, this subduction event also formed the late Paleozoic island-arc volcanic belt. From west to east, this volcanic belt can be traced at the following locations: Ashele (northern Haba River County)–Chonghuer–Taerlang–Aletai–Tiemuerte–Abagong–Mengku–northern Fuyun County. The typical lithologies were the late Paleozoic island-arc volcanic rocks that occurred at the Haba River area and other places (Chen et al., 1997; 2005; Zhao et al., 2001; 2004). Furthermore, in the northern region, this subduction event also developed a back-arc basin during the Devonian and the Carboniferous periods in the Kelan area. During the Late Carboniferous epoch, the middle section of the Paleo-Asian Ocean disappeared. The Siberian plate collided with the Junggar–Kazakhstan plate, which formed the Eerqisi–Buergen suture zone. During the Early Permian period, on the northern margin of the Junggar plate, the Aertai orogeny was uplifted on a large scale. Because the Aertai orogeny repeatedly overthrusted from the north to the south, it formed a series of southward imbricate overthrust structures. Due to compression and flexual movements, the foreland basin was formed on the south of the orogeny. The terrestrial molasse deposits settled in the foreland basin on the mountain front. At the same time, the western portion of the Junggar orogeny strongly overthrust from west to east. The crust of the foreland area flexurally subsided, which formed a large scale foreland basin on the western margin of the Junggar orogeny with more than 4000 m deposits of volcanic rocks and limestone lenses (Guan et al., 2007; Wei et al., 2004).

In the Tianshan region of the northern Tarim plate, the South Tianshan oceanic basin was subducted northward to form the middle Tianshan island-arc. The South Tianshan oceanic basin was the southern section of the Paleo-Asian Ocean and it was located between the Junggar–Kazakhstan plate and the Tarim plate. Additionally, during the Late Carboniferous epoch, the collision between island-arc and continent occured, which brought about a convergence of the Tarim plate and the Junggar–Kazakhstan plate. The South Tianshan oceanic basin extended during the Silurian period. According to the chronological study of ophiolite at the following locations: Changawuzi–Laerdundaban–Guluogou–Wuwamen–Gongbaizi–Yushugou–Kumishi–Hongliu River, the ophiolite was formed during the Middle or Late Silurian–Early Devonian epochs (Shu et al., 2007).

A characteristic of sedimentary deposits of the Upper Paleozoic erathem demonstrates that, since the Middle Devonian epoch, a large amount of thick layer limestone in the southern Tianshan region was deposited, which distinctively diverged from the sedimentary deposits of the Middle and Late Silurian–Early Devonian series. But during the Late Devonian epoch, in an isolated area, gypsum-salt formations started to occur, which indicated that the tectonic evolution of the South Tianshan oceanic basin had been transformed from an extension stage into a contraction stage. During the Early Carboniferous period, predominately muddy limestone and sandy limestone were deposited; in isolated areas, red coarse grain clastic rock and abundant gypsum mudstones were deposited. However, during the Late Carboniferous period, limestone and sandstone with frequent facies alternation were mainly deposited; in some areas, clastic rocks that were interbedded with gypsum mudstone, a thin layer coal bed and carbonaceous shale were deposited. The characteristics of lithologies show that, during the Carboniferous period, the South Tianshan Ocean had become a nearly isolated oceanic basin. In particular, during the Late Carboniferous epoch, the carbonaceous shale in the southern Tianshan area indicates that the South Tianshan Ocean had completely disappeared. During the Permian period, a molasses formation was commonly developed in the southern Tianshan region, which occurred in the following areas: Hongliu River that is located at the provincial border of Xinjiang and Gansu, the Huola mountain area of northeastern Tarim, and the cross section of the Kuqa River. The molasses formation indicates a collision between the middle Tianshan island-arc and the Tarim plate (Shu et al., 2007).

Furthermore, the paleomagnetic study shows the convergence of the Tarim plate and the Junggar–Kazakhstan plate took a course of angled collision and convergence. The paleomagnetic data analysis reveals that, during the Late Devonian epoch, the direction of the ancient magnetic pole of the Tarim plate was oriented almost perpendicularly to its present direction; also, there is about a 30° angle between the magnetic pole of the Early Triassic epoch and the current magnetic pole. The study of geomagnetic declination shows that, during the early phase of the Late Paleozoic era, the South Tianshan oceanic basin may have had an opening on its western edge. Although the South Tianshan Ocean had been closed during the Late Carboniferous period, the paleomagnetic data reveal that, during the Late Carboniferous period–Triassic period, the Tarim plate still rotated clockwise in order to adjust its position with other plates (Jia et al., 1997). This evidence demonstrates that the collision of the Tarim plate and the middle Tianshan island-arc began on the eastern side; then the scissors shape convergence gradually expanded westward to close the opening at the western end. After the convergence, the Tarim plate continually rotated clockwise to modify its position until the Mesozoic era.

In addition, during the late Early Permian epoch, the Bayingou oceanic mini-basin of the northern Tianshan region also started to disappear and it formed the southern Junggar peripheral foreland basin. The molasses formation of the Xiacangfang Canyon Group is the product of foreland compression movement.

The Rifted Event of a Passive Continental Margin on the Northern Margin of the Paleo-Tethys Ocean

On the northern margin of the Paleo-Tethys Ocean, a strong extentional event was widespread during the late Paleozoic, which can also be clearly recognized on the Tarim plate, the West Kunlun mountain, the Tuha mini-plate, the Beishan mountain area and western margin of the Yangzi plate. In the western portion of Xikun mountain, from Akesayibashi mountain to the Gaizi River, the Middle and Lower Carboniferous series was distributed in a narrow belt with an orientation of NWW. The lithologies of the lower portion are basic volcanic rocks, which primarily consist of grayish green amygdaloidal basalt and pillow basalt, massive andesite, fragmented diabase and altered olivine basalt, which are also interbedded with purple red, iron siliceous rock (chert). The lithologies of the upper portion include interbedded medium-acidic volcanic rocks and basic volcanic rocks, which consist of green, grayish green altered andesite, dacite, felsite and altered basalt, andesitic basalt and quartz diabase. The total thickness of the Middle and Lower Carboniferous series is 3,276 m. These lithologies indicate an extensional tectonic event on the southwestern margin of the basin during the Early Carboniferous period. This suite of lithology was formed by an extensional movement on a passive continental margin.

On the Tarim plate, the Carboniferous period is the important interval for marine progression and for numerous marine facies deposits in the Tarim basin. The southwestern Tarim basin is one of the locations that developed an excellent Carboniferous system that primarily includes deposits of shallow sea shelf facies and open platform facies. The thickness of sediments is about 1,000 – 2,000 min generally. The sedimentation rate is 13.9 – 27.8 m/Ma. The depositional center is located in the Kekeya–Kashi area. From northeast to southwest, the thickness of sediments gradually increases. In the Xilibili area of the Yingjisha, the Carboniferous system includes flysch deposits of deep water slope facies with a thickness of 2,057 m. The best Carboniferous system occurs in the Damusi area in Shache County, which has both the Upper and Lower Carboniferous series with a total thickness of 1,878.7 m. In this area, the Lower Carboniferous series consists of the Kelitake formation and the Heshilafu formation; the lithologies of the lower portion are gray, grayish black limestone, dolomite and shale with a total thickness of 409.4 m; the lithologies of the upper portion are gray, dark gray shale that is interbedded with sandstone with a total thickness of 594.1 m. In ascending order, the Upper Carboniferous series consists of the Kalawuyi formation, the Azigan formation and the Tahaqi formation; The lithologies of the lower portion are gray, grayish black sandstone, siltstone and muddy shale, which are interbedded with each other and which also contain black color, carbonaceous mudstone and limestone with a total thickness of 487.1 m; The lithologies of the middle portion are light gray, dark gray, grayish black dolomite, limestone and mudstone with a total thickness of 350.7 m; The lithologies of the upper portion are gray color limestone and muddy limestone (marlaceous limestone) with a total thickness of 184.8 m; In the Qipan area, the combinations of lithology are similar to those in the Damusi area. Traveling eastward to the Keziliqiman and the Piyaman areas, only the Upper Carboniferous series developed. In the Keziliqiman area, the Upper Carboniferous series includes the Kalawuyi formation, the Azigan formation and the Tahaqi formation; the lithologies are bio-clastic limestone and limestone, which are interbedded with sandstone and mudstone with a total thickness of 358.1 m. In the Piyaman area, the Upper Carboniferous series only contains the Azigan formation and Tahaqi formation; the lithology is a set of thick layer limestone with a total thickness of 475 m. Inside the basin, engineers drilled through the Carboniferous system that has a thickness of 753.5 m in the Qu 1 well; the lithologies include a set of thick layer limestone and dolomitic limestone, which are interbedded with mudstone and gypsum mudstone. At the base, there is quartz sandstone. The lithologies that are described above suggest that in the Tarim basin–the Damusi–the Xilibili (from east to west), transformations of facies in the Carboniferous system are platform–continental shelf–deep water basin. Therefore, in the southwestern Tarim basin, from northeast to southwest, there developed passive continental margin deposits of platform–shallow sea shelf–deep water slope–basin facies under the extensional tectonic background.

Inside the Tarim plate, during the Early Permian period, there are large scale magma intrusions. Because of the rifting event, a special suite of igneous rocks and a unique set of sedimentary rocks developed. In the Tarim basin, the Early Permian igneous rocks were formed in the following areas: the Await depression, the western region of the Manjiaer depression, the western region of the Tabei uplift, the Bachu uplift, the Tazhong uplift and the Taxinan depression (Figs. 9.2 and 9.3). The primary lithologies are basalt, diabase dike swarm, gabbro and alkali syenite. The study shows that igneous rocks were formed inside the plate and this represents a strong extensional tectonic movement.

Beside the strong extensional tectonic event occurring in the Tarim basin and western Kunlun mountain, there were numerous inner-plate magma intrusions during the Early–Late Permian in western China. The Early Permian inner-plate basalt occurs in the following areas: the Tuha basin and its surrounding area, the Hongliu River area, Beishan mountain area and western margin of the Yangzi plate (Xing et al., 2004; Zhao et al., 2004). Concerning the Emei mountain basalt, many researchers think it is continental overflow basalt and that it relates to the mantle plume (Xu and Chung, 2001; Zhang et al., 2002; Xiao et al., 2004; Hao et al., 2004). As for the Early Permian basalt in the Hongliu River area, Zhao et al. (2004) considered it is inner-plate basalt and its magma are a combination of depleted mantle (DM) and enriched mantle (EM). However, Xing et al. (2004) believed that, in the Tuha basin and its surrounding areas, the Early Permian volcanic rocks are the product of a rifting environment with extensional background. The rifting event happened subsequent to an orogeny movement.

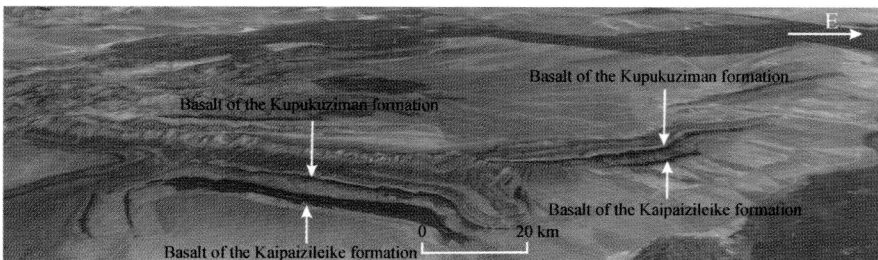

Fig. 9.2. Distribution of residual basic magma in the Tarim basin. The time interval for the magmas intrusion is the Early–Middle Permian epochs

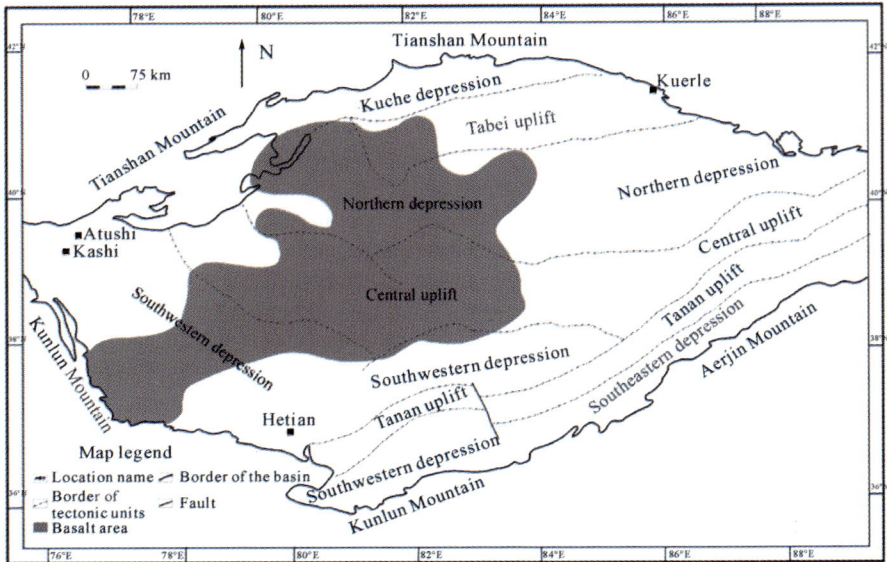

Fig. 9.3. Distribution of basalts in the Keping area. The time interval for basalt eruption is between the Early–Middle Permian epoches and they occur in the Kupukuziman formation and the Kaipaizileike formation

Therefore, on the northern margin of the Paleo-Tethys ocean, the late Paleozoic magma belt is made of the Carboniferous magma intrusion of western Kunlun mountain, the Early–Middle Permian magma intrusion in the Tarim basin, the Permian magma intrusion on the western margin of the Yangzi plate, the Permian magma intrusion in the Tuha basin and its surrounding areas, the Permian magma intrusion of the Hongliu River area and the Permian magma intrusion of the Beishan mountain area.

At the same time, in the Longmen mountain area of western Sichuan province, the characteristics of lithologies also demonstrate that, during the late Paleozoic era, most basins were rifting basins of the continental margin. During the Devonian period, alongside the Longmen mountain area, massive deep water clastic rock and carbonate rocks were deposited, which reflect a rapid subsidence event. It is assumed that, at the side of western Longmen mountain, there was a normal fault that controlled the Devonian system. The Permian carbonate rocks and fine grain clastic rocks are widely distributed on the Yangzi plate. During the Late Permian period, in southwestern Sichuan and Yunnan provinces, a continental margin rifting event occurred, which brought a large scale basalt eruption (the Emei mountain basalt). This continental margin rifting event triggered strong subsidence at western Longmen mountain and the deep water flysch deposits of the Middle and Upper Triassic.

- **Development of Paleo-Tethys Orogeny and Finalization of Tectonic Framework on the Chinese Portion of the Eurasian Plate**

During the time interval of the disappearance of the Paleo-Tethys ocean and development of the Paleo-Tethys orogeny, the tectonic framework of the Chinese portion of the Eurasian plate started to finalize its structural framework, which has an important implication.

During the late Middle Triassic and the Late Triassic, upon the disappearance of the Qinling Ocean, the Jinshajiang Ocean and the Kunlun Ocean (which are components of the Paleo-Tethys Ocean), a sequence of plates collided and converged in succession, which were the collision between the Yangzi plate and the Huabei plate, the collision between the Zhongzan mini-plate and the Yangzi plate and the collision between the Qiangtang plate and the Tarim plate. These tectonic collisions formed the eastern Paleo-Tethys orogeny (which includes the Qinling orogeny, the Kunlun mountain orogeny and the Longmen mountain orogeny), which basically established the tectonic framework of China.

Inside the Kunlun mountain area, during the Late Triassic epoch, the Paleo-Tethys Ocean (Kunlun Ocean) was continually subducted under the Tarim plate and the Qiangtang mini-terrain collided with the Tarim plate. The Qiangtang mini-terrain was located at the southern margin of the Tethys. The subduction and collision formed the Late Triassic collision suture of Tashikuergan–Kangxiwa–Muzitage–Maqin on the southern margin of the Tarim basin. Additionally, during the Late Triassic–Early Jurassic, this collision event also developed a foreland folded-thrust belt and foreland basin inside the Kunlun mountain area; it also developed a back-arc foreland basin on the southern Tarim basin and a foreland folded-thrust belt on the southern margin of the Tarim basin.

In the Qinling area, during the Late Permian epoch, the Yangzi plate and the Huabei plate initiated a collision at the eastern section; then the scissors shape collision continued its course westwards. This collision formed the Hefei peripheral foreland basin and its foreland folded-thrust belt on the southern margin; also it developed the peripheral foreland basin of northeastern Sichuan and its foreland folded-thrust belt of Micang mountain–Daba mountain.

The extinction of the Jinshajiang oceanic basin produced the northeast–southwest compression force; the Qing–Qi structure belt produced a southwards compression force. These two compression forces joined together to form the northwest–southeast compression force. East of the Songpan–Ganzi terrane, in the Longmen mountain area, this combinational force overthrusted and reversed structures inside the rifting basin that was located at the passive continental margin of the late Paleozoic era–the Early and Middle Triassic; in addition, the combinational force formed the western Sichuan peripheral foreland basin on its front margin (Fig. 9.4).

In the Helan mountain area, because the Alashan mini-plate collided with the Huabei plate, it caused the disappearance of the Helan Ocean of the late Paleozoic.

Furthermore, it formed Helan mountain and its eastern margin; additionally, it also formed the peripheral foreland basin of the Late Triassic epoch, which is located at the western margin of the Ordos.

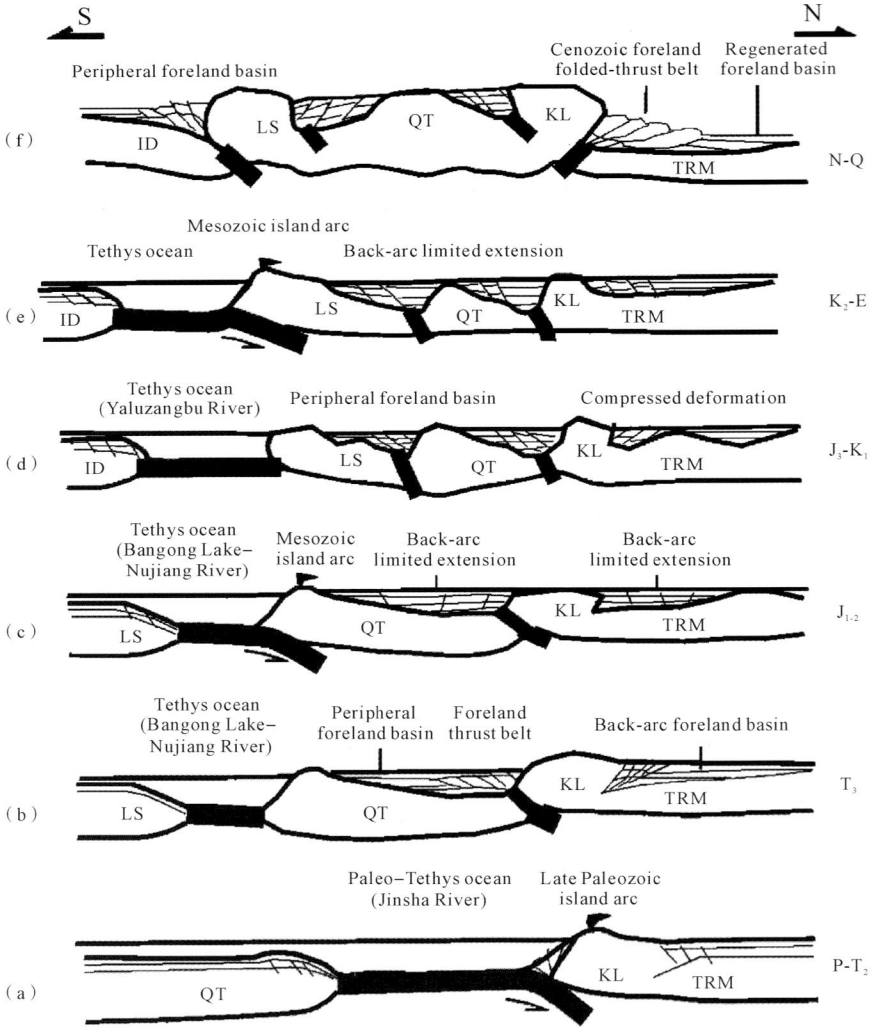

Fig. 9.4. Tectonic evolution of the Tethys plate (from (a) to (f)) and its influence on the foreland basin's development in western China. TRM: Tarim plate; KL: Kunlun terrane; QT: Qiangtang terrane; LS: Lasa terrane; ID: India plate

9.1.2 Development of the Mesozoic Tethys Ocean and Northern Margin Basin Group

● **Subduction of the New Tethys Ocean and the Jurassic Rifting (Depression) Basin in Northwestern China**

Geo-Dynamic Background

Comparing the basin structures between Jurassic basins of northwestern China (especially the Qaidam basin) and the Black Sea basin, we conclude that western-central China was in a back-arc spreading environment, which was associated with the subduction of the southern margin of the Eurasian plate at the time (Xu, 1994); thus, extension of the Jurassic basins in western-central China was related to back-arc spreading (Fig. 9.4c). The extension of the Bangong Lake–Nujiang River portion of the Tethys Ocean began noticeable tectonic activity in the Late Triassic epoch and it advanced into its major extension phase during the Early and Middle Jurassic epochs (Zhao et al., 2001). However, the ophiolite that occurs in the Bangong Lake–Nujiang River area belongs to the back-arc basin type. Possibly it was a marginal sea basin that was formed by back-arc extension of the Yaluzangbujiang River being subducted northward and it was not a fully developed portion of the Tethys Ocean. Consequently, Andes type magma activity has not been found in this area. During the Late Jurassic–Early Cretaceous, the Bangong Lake–Nujiang River portion of the Tethys Ocean was subducted northward and it disappeared. Then the Lasa mini-terrane collided and converged with the Qiangtang terrane. On its northern margin, the basins of western-central China commonly experienced a weak force compression, which caused strata deformations or erosions (Fig. 9.4d).

This regional rifting event rippled through the entire western-central China and it produced a peneplain landform. In western-central China (that includes Tianshan mountain and Qilian mountain), it deposited a suite of dark color, fine grain coal measures of limnetic facies. The sediments of Jurassics were deposited in the following basins: the Junggar basin, Tarim basin, Qaidam basin, Ordos basin and Sichuan basin. The distribution and the magnitude of the Jurassic volcanic rocks are limited in western-central China. The basic lava occurs in the Badaowan formation in the Kelamayi area with a thickness of 5.25 m. The basaltic porphyrite occurs in the Xishanyao formation around the Jiangjun Gobi desert in eastern Junggar. The basaltic-intermediate acid volcanic rocks occur in the Middle Jurassic in the Anxi–Dunhuang area. The olivine basalt occurs in the Jurassic in the Chaoshui basin. In the Kelamayi area, the Jurassic volcanic rocks are inner-plate continental basalt that is the product of a mantle partially melting; it needs a deep rifting fault to provide space for escalation and eruption of the lava. In the Chaoshui basin, the Jurassic volcanic rocks are the olivine basalt that came from the mantle; however, the intermediate-acid volcanic rocks take up a predominant portion among the volcanic rock suite. The combination of olivine basalt that came from the mantle and the intermediate-acid volcanic rocks that came from the crust are the products of eruptions in a continental rifting environment.

The average temperature gradient of the Jurassic was 3.3 °C/100 m and the calculated thickness of the crust of the Jurassic is 27 km approximately. In addition, the growth normal faults are clearly visible in the Jurassic in the Kuqa, Ejinaqi and Liupan mountain areas. In the seismic sections, there is much evidence of the Jurassic extensional graben basin. Because the Alashan terrain was slightly altered by the Cenozoic tectonic movement in the seismic section, it shows that the original profile of the Jurassic basin was controlled by a growth normal fault. The Jurassic growth normal faults controlled the distribution of sedimentary basins. These growth normal faults commonly occur in the southern margin of Junggar, Kuqa and southeastern Tarim.

Development and Distribution of Rift (Depression) Basin

According to the study of regional tectonic structures of the Early and Middle Jurassic and the characteristics of sedimentations, there are eight sedimentary geo-districts of the Early and Middle Jurassic epochs in northwestern China, which are Junggar–Tuha, Tianshan mountain, Tarim–Yanqi, Beishan mountain–Bayinhaote, Dunhuang–Huahai, Qaidam–Qilian mountain, Sichuan and Ordos. The Junggar–Tuha sedimentary geo-district is located to the north of the middle Tianshan mountain and west of the Cheerchen–Beishan fault. It includes thirteen basins or depressions (Fig. 9.5), which are the Junggar, Tuha, Chaiwobu, Balikun, Houxia, Santanghu, Kupu, Buerjin, Jimunai–Fuhai, Bukesaier, Heshituoluogai, Tacheng and Tuoli. In the Junggar–Tuha sedimentary geo-district, there are two subsidence belts. The northern subsidence belt includes Santanghu, the Wulungu depression of the Junggar basin and Heshituoluogai; the strata thickness of the Lower and Middle Jurassic is about 100 – 2,000 m; the strata are thicker in the southern area than in the northern area. The southern subsidence belt includes southern Junggar and Tuha, which is the subsidence and sedimentation center of the Jurassic, which has shore-shallow lake facies and deep water lake facies deposits. The Yili–Youerdusi sedimentary geo-district is located between North Tianshan mountain and South Tianshan mountain, which includes Yili basin, Zhaosu basin, Jinghe basin, Youerdusi basin and Yanqi basin. The Tarim sedimentary geo-district includes the southwestern Tarim depression, the southeastern Tarim depression, the Kuqa depression and Manjiaer depression. The Beishan mountain–Bayinhaote sedimentary geo-district includes the Beishan basin group (Heiying mountain, Hongtutai, Shibanjing, Gongpo spring, Zhagegaonao, Luotuo well and Zhongkouzi), Yingen–Ejinaqi basin, Yabulai basin, Chaoshui basin and Bayinhaote basin. The Dunhuang–Huahai sedimentary geo-district includes Dunhuang basin and Huahai basin.

The Qaidam–Qilian mountain sedimentary geo-district includes, from north to south, the Jiuquan–Minle subsidence belt, the Shule–Xining–Dingxi subsidence belt and the northern margin of the Qaidam–Gonghe subsidence belt. The Ordos and Sichuan basins are located on the northeastern side of the Tethys tectonic region. Compared with the western area, these two basins only experienced weak tectonic impacts; furthermore, these two basins were positioned on top of stable craton basins. They primarily contain a large area, stable depressions. In the Ordos, the depositional center of the Jurassic was located at the Helan mountain and the Liupan mountain front. However, in the Sichuan basin, the deposits of the Early and Middle Jurassic were closely related to the tectonic structures of Daba mountain in northern Sichuan province; the orientation of sediments is parallel to Daba mountain, in an east to west direction.

The Early and Middle Jurassic basins can be divided into two categories that are the faulted basin and the depressional basin. The faulted basin was the product of an extensional tectonic stage and it was controlled by a normal fault. Along the faulted-subsidence belt, the faulted basin subsided deeper and it deposited a huge suite of dark color mudstone and coal measures of semi-deep lacustrine facies, such as in Junggar, the northern margin of Qaidam, Tianshan mountain, southern Qilian mountain and Jiuquan. A depression basin usually has a stable craton basement. In the extensional environment and influenced by the harmonization adjustment of the earth, the basement of the basin continually subsided along the large fault on the mountain front. A depression basin commonly has a large depositional area, small subsidence distance and a disk shape. Because it is influenced by fault activity in its basement, the accumulation of sediments is thicker near the fault, which can be seen in the Tarim basin, Sichuan basin and Ordos basin.

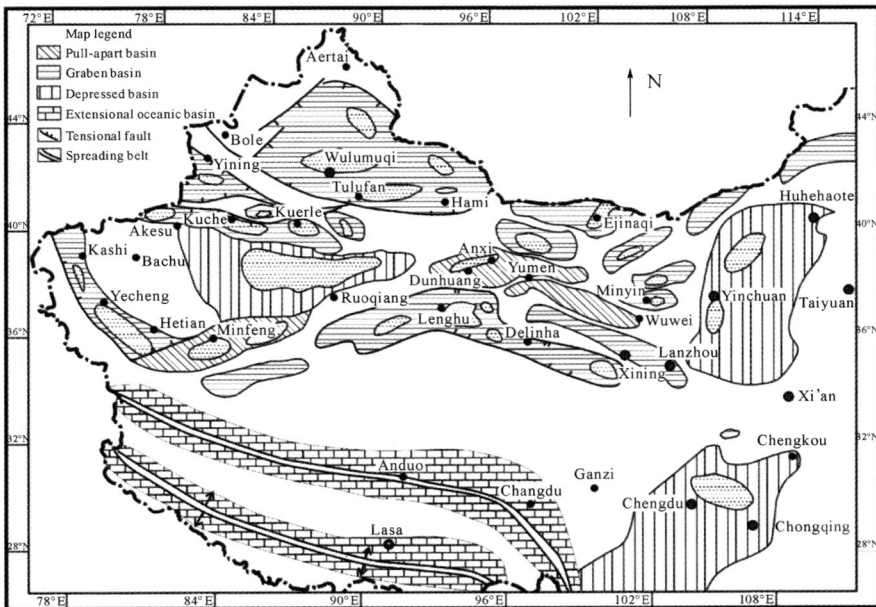

Fig. 9.5. Characteristic map of sedimentary tectonic framework during the Early–Middle Jurassic in western-central China

● Disappearance of the Bangong Lake–Nujiang River Oceanic Basin and the Cretaceous Gentle Compressive Structure

Geo-Dynamic Background

During the Late Jurassic–Early Cretaceous, the Bangong Lake–Nujiang river oceanic basin disappeared. The Lasa terrain converged northward and it collided with the Qiangtang plate (Fig. 9.4c). The following is the evidence of this tectonic event: (1) Inside the Qiangtang basin, at the Dongqiao–Anduo area, the age of the radiolarian combination zones is the Jurassic period; the radiolarias occur in chert that is a part of the ophiolite suite. In its upper portion, flysch is interbedded with limestone that contains Late Jurassic fossils. In the Ritu area, the chert also contains Middle and Late Jurassic radiolaria fossils. In the Dongqiao and Anduo area, the ophiolites have an unconformity relationship with the overlying formation of the Lower Cretaceous series. In the basal conglomerate of the Cretaceous system, most of the pebbles and cements are made of basic-ultrabasic igneous rocks of the Jurassic period, which indicates that the disappearance of the Bangong Lake–Nujiang River oceanic basin happened during the Late Jurassic–Early Cretaceous epochs. (2) During the Cretaceous period, the Bangong Lake–Nujiang River portion of the Tethys Ocean was rapidly subducted under the Gangdise island arc, which triggered extensive magma activity of the Gangdise island arc. In the Bangong Lake–Nujiang River area, in parallel on each side of the ophiolite melange, there are two strips of granite and contemporaneous calc-alkaline volcanic rocks. The age of the granite is the Late Jurassic epoch–Cretaceous period (80 – 145 Ma). (3) Along the Bangong Lake–Nujiang River tectonic suture, the Lasa terrain collided and compressed with the Qiangtang terrain. The crust was uprising. In the Qiangtang basin and on the Lasa terrain, it deposited the Late Cretaceous, red color molasse formation that was directly placed on top of deformed strata of the Jurassic, which formed an uncomformity relationship. During the Cretaceous period, in western-central China, the Tethys tectonic region (which is located at the southern margin of western-central China) still had tectonic movement; the northern portion had united into one continent without tectonic compressive stress. The tectonic compressive stress that was formed by the collision of the Lasa terrain and the Qiangtang plate traveled a long distance northward, which caused regional compressive structures in western-central China (Allegre et al., 1984; Xiao et al., 1991; Zhong and Ding, 1996; Pan et al., 1997; Jia et al., 1986; 2001a). In the western region, this tectonic movement developed an embryonic form of structure inside the following basins: Tarim basin, Junggar basin and Qaidam basin; but in the central region, Sichuan basin and Ordos basin, it caused stronger tectonic deformation, orogeny belt uprising and basins leaning westward.

Characteristic of Tectonic Deformation

The signs of tectonic movement of the Late Jurassic Period–Cretaceous period commonly exist in basins of western-central China. At first, inside the basins, the deposition areas were reduced, deposition events discontinued and the contacts of strata were in an uncomformity relationship. Second, the Cretaceous structures were considerably developed inside the basins. Third, huge suites of coarse grain clastic rock were deposited in the front margin of orogenies. Next, there were magma activities in isolated areas inside orogenies (Fig. 9.6).

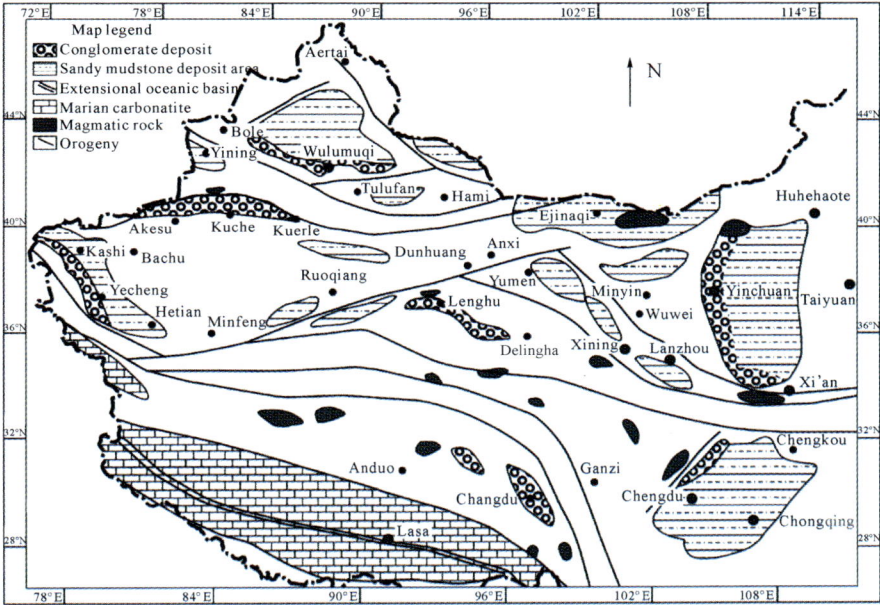

Fig. 9.6. Characteristic map of tectonic framework during Late Yanshanian tectonic period (J_3-K_1) in western-central China

9.1.3   Cenozoic Collision of the Indian Plate and the Eurasian Plate and Gigantic Basin and the Range System Around the Tibetan Plateau

● **Basin and Range System Around the Tibetan plateau, which was Controlled by the Collision of the Indian Plate and the Eurasian Plate**

Because small terrains converged northward consecutively in an earlier geological time, since the Mesozoic–Cenozoic, the region that is located in the north of the Tibetan plateau had united into one continent without noticeable compressive orogenies or thrust faults. Because the new Tethys ocean was subducted northward, the crust of the Tibetan plateau was uplifting, which set off the tectonic deformations rippling northward and eastward in successive order. Towards the north, ancient orogenies were reactivated, which triggered the deformation of the Cenozoic foreland thrust belts towards the inner basin in successive order. During the Neogene period, due to the collision of the Indian plate–Tibetan plateau, the Eurasian continent was strongly deformed and the 'wedging' of the Pamir was formed. The collision of the Indian plate–Tibetan plateau controlled development of large scale thrust belts. In the area that was far away from the collision suture zone, ancient orogenies were reactivated by the collision. In company with the Tibetan plateau uplifting and pushing northward, the thrust belts persistently spread outward (Guo et al., 1992; Chen et al., 1992; Graham et al., 1993; Hendrix et al., 1994; Jia et al., 2003a; 2003b). In western-central China, ancient orogenies were reactivated; strong deformation happened at the inner continent; from orogeny belts, the thrust belts were developed toward basins in consecutive order. For example, at the northern margin of the Tarim, the ancient Tianshan orogeny was reactivated and was uplifting again, which compressed the northern margin of the Tarim plate. The deformation of thrust faults rippled into the Tarim plate to form a foreland thrust belt of the Kuqa–Keping. Because of thrusted-overloading by orogeny, correspondingly the northern margin of the Tarim plate was subducted under Tianshan mountain. In addition, flexural structures occured within the plate, which formed a rejuvenation foreland basin (Wei et al., 2000). Therefore, in western-central China, after experiencing the Cenozoic deformation of the inner plate, the regional re-genesis foreland thrust belts were formed by reactivated orogenies during the Himalayan tectonic period. Based on the convergence of orogenies in the earlier tectonic period, the Himalayan tectonic movement formed the largest, scattering type, inner continent, tectonic deformation domain; furthermore, it also formed a gigantic basin and range system that had most energetic inner-plate deformations. Because of the heterogeneity of the orogeny's margin and heterogeneity of the tectonic stress field, while tectonic deformations were relayed throughout the Tibetan plateau as a compressive overthrust movement, in some areas they also occurred as a shear slide-thrust fault (Tapponnier et al., 2001; Wei et al., 2005).

● Foreland Thrust Belts at the Conjunction of Orogeny and Basin Around the Tibetan Plateau

There are more than ten foreland thrust belts in western-central China that are located at the northern and eastern margins of the Tibetan plateau. These thrust belts occur at the conjunctions of orogeny and basin (Fig. 9.7). At the northern margin of the Tibetan plateau, foreland thrust belts occur in the following areas: the northwestern margin of the Junggar basin in front of Zhayier mountain, the southern margin of the Junggar basin in front of Tianshan mountain, the Kuqa depression, the Kashi area, the southwestern Tarim basin in front of Kunlun mountain, the southeastern Tarim basin, the Tuha basin on the southern margin of

Bogeda mountain, the northwestern Qaidam basin in front of Aerjin mountain, the northern margin of the Qaidam basin in front of Qilian mountain and the Jiuquan basin. At the northeastern margin of the Tibetan plateau, foreland thrust belts appear in the following areas: the western margin of the Ordos basin in front of Helan mountain, the western region of the Sichuan basin in front of Longmen mountain, the northern region of the Sichuan basin in front of Qinling mountain and the Chuxiong basin in front of Ailao mountain. In western-central China, several small craton basins were surrounded by folded orogenies of the Hercynian–Indosinian tectonic periods. Influenced by the Himalayan tectonic movement, these small craton basins were modified by giant orogenies. The strongest deformations happened at the conjunction of orogeny and basin; additionally, thrust belts were developed on the margin of craton basins. The united continent, where several small craton plates of the late Paleozoic converged, experienced inner-plate deformation during the Mesozoic era. During the Himalayan tectonic period, due to the Tibetan plateau deformation, orogenies of the united continent were reactivated to form regional re-genesis foreland thrust belts by the compressive stress on the northern and eastern margins of the Tibetan plateau.

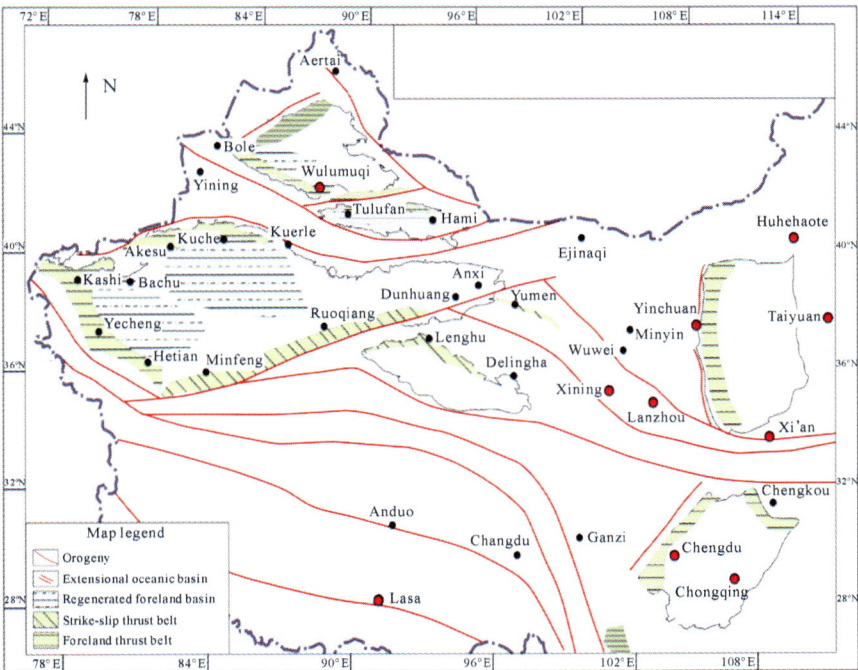

Fig. 9.7.   Distribution map of foreland thrust belts in China

Since 20 Ma approximately, the Tibetan plateau was being uplifted on a large scale, which altered the structural framework on its margin. Around 20 Ma, the left lateral motion of the Honghe River–Ailao mountain fault stopped. Comparing sedimentary environments and tectonic deformation patterns before and after 20 Ma, they were significantly different in the Kekexili area (Wang et al., 2003); in the Qaidam basin, the calcite vine in the Tertiary system shows an age of 20 – 18.4 Ma, which indicates motion of the structure and the activity of hot fluid from the magma. Around 20 Ma, the Qiangtang basin was dynamically elevated bringing planation to the area. The main duration of the Himalayan uplift started at 20 – 18 Ma (Harrison et al., 1992). From 20 to 19 Ma, the eastern part of the Tibetan plateau was uplifted dramatically with an escalated distance of 7 km. At present, strata of the Palaeogene system commonly exist on the summit of mountains and have been leveled onto a planation surface that demonstrates the geomorphology of Tianshan mountain, Bogeda mountain, Qilian mountain and Kunlun mountain. In the processing of orogeny uplifting, older strata were overthrusted on top of newer strata. For example, at Ciyaokou that is located at the northern margin of Qilian mountain, limestone of the Ordovician system was overthrusted on top of newer strata (E_1); at the southwestern Tarim basin, volcanic rocks of the Carboniferous system were overthrusted on top of the Quaternary strata. These overthrusting activities also display a dynamic tectonic movement during the late Himalayan tectonic period. According to the apatite fission track study (the samples were collected along the Du-Ku highway and the Wu-Ku highway) and according to the research on growth strata inside thrust fault belts on the northern margin and southern margin of Tianshan mountain, the uplifting of the ancient Tianshan orogeny started at 25 Ma, which set off tectonic deformation successively in nearby craton basins that were located at the northern and southern margins of the orogeny. From the orogeny to the front of the thrust belt, because tectonic activities were relayed, the age of these tectonic events was successively younger. At the same time, in company with the uplifting of orogenies and the development of foreland thrust belts, the nearby craton plates were subducted under the orogeny, causing a flexural structure. Since the Neogene period, in western-central China, major basins set off a rapid subsidence; massive coarse grain clastic rocks were deposited at the mountain front.

- **Tectonic Deformations of the Pliocene Epoch–the Quaternary, which were Controlled by the Collision of the Indian Plate with the Eurasian Plate**

Among the international geo-community, although people still disagree on when the Indian plate collided with the Tibetan plateau, most people consider the collision started no later than 40 Ma approximately. At this time, because the 'wedging' of Pamir had not been developed, the Tarim basin, the Tadjik basin and the Karakum basin were connected. At the early stage of the collision between the Indian plate and the Tibetan plateau, the Aerjin strike-slip fault did not have any activity. At this time, the Arabian plate and the Eurasian continent were still divided by the vestigial Tethys oceanic basin. Successively, the central Iranian block (that includes Yazd, Tabas, and Lut small blocks) had a large range rotation, which caused intricate tectonic alterations on the southern margin of Kopet mountain and which formed a complicated tectonic framework. However, at the same time, the Black Sea (vestigial Caspian oceanic basin) and the Mediterranean Sea (vestigial Tethys Ocean) were still connected by waterway. Therefore, in the Karakum basin of central Asia, the southwestern Tarim basin and the Kuqa depression, there were deposits of marine facies.

In western-central China, the most active period of foreland thrust deformation is the Pliocene epoch–Quaternary period. This time interval is also in the late Himalayan tectonic period. At present, the geomorphologies of Tianshan mountain, Bogeda mountain, Qilian mountain and Kunlun mountain show that the strata of the Palaeogene system regularly appear on the summit of mountains and have been leveled onto a planation surface that reveals, in western-central China, that major mountains have set off rapid uplifting events since the Neogene period. According to the apatite fission track study (the samples were collected along the Du-Ku highway and the Wu-Ku highway) and according to research into growth strata inside thrust fault belts on the northern and southern margins of Tianshan mountain, the uplifting of the ancient Tianshan orogeny started at 25 Ma, which generated tectonic deformation successively in nearby craton basins that were located at the northern and southern margins of the orogeny. From the orogeny to the front of the thrust belt, because tectonic activities were relayed, the age of these tectonic events was successively younger. Of the same duration, along with orogenies uplifting and the development of mountain front thrust belts in the neighboring area, the Tarim plate and the Junggar plate were subducted under the orogenies, which initiated the rapid subsidence of major basins in western-central China from the Neogene period. In the mountain front, massive coarse grain clastic rocks were deposited; the Kuqa and the south Junggar re-genesis basins were developed as well (Jia et al., 2000; 2003b; Lu et al., 2000; Wei et al., 2000). For example, the unconformity surface of the Kangcun tectonic period and the Kuqa tectonic period is demonstrated on a north to south oriented seismic section of the Kuqa–Baicheng depression. Also, it shows the correlation between tectonic periods. Furthermore, these unconformity surfaces and the sedimentary strata experienced another tectonic alteration and developed quaternary growth strata. The growth strata in the Kuqa area show strong tectonic deformation of the Pliocene epoch (N_{2k}); the fold cliff, which is located at the northern lateral of the eastern Qiulitage structure belt, shows that the underlying fault was developed between ±25,100 (1,900) and ±26,690 (102) years, according to thermo luminescence dating. According to thermo luminescence dating and electron spin resonance (ESR) studies (the samples were collected at the Huo–Ma–Tu fault belt on the southern margin of the Junggar), the testing results show a powerful tectonic alteration event happened during the late Himalayan tectonic movement. At Kulong mountain in the Qingxi area in Jiuquan, the strata of the Silurian system were overthrust on top of the Quaternary system; on the southern side of Saishiteng mountain at the northern margin of the Qaidam, the marbles of the

Sinian system were thrust on top of the Cenozoic strata; on the southern side of Tianshan mountain, on the eastern portion of the Bugulu fault, the limestone of the Carboniferous system was thrust on top of the strata of the Pliocene system and the Quaternary system. This is all evidence of a powerful tectonic alteration event that happened during the late Himalayan tectonic period (N_{2k}).

In China, the major deformation of foreland thrust belts has happened since 20 Ma. At this time, because of the persistent collision and subduction between the Indian plate and the Tibetan plateau, the Eurasian continent was intensively deformed. In addition, the older orogenies of the ancient Tianshan mountain, Qilian mountain and Kunlun mountain were reactivated, even though these mountains had been eroded. These stimulated ancient orogenies formed newer inland orogenies that were thrust toward basins. Thus, the basins of central Asia and western-central China were brought into the "re-genesis foreland basin" period. After 20 Ma, the ongoing collision of the Indian plate and the Tibetan plateau triggered the Pamir area to wedge northward to separate the Tarim basin and the Karakum basin. Due to the intensive foreland tectonic deformations, the Tadjik basin became gradually a vestigial tectonic basin. Since 20 Ma, the Tethys orogeny movement has had a huge impact on the basins in central Asia, the Tarim and western-central China. The uplifting of the Tibetan plateau changed the tectonic framework in its surrounding areas. For example, around 20 Ma the Qiangtang basin was dynamically elevated and that brought planation in the area; a major uplifting period in the Himalayas began around 20 – 18 Ma (Harrison et al., 1992; Alleger et al., 1984); during 20 – 19 Ma, the eastern side of the Tibetan plateau experienced a speedy uplifting of 7 km. Alongside this, comparing the sedimentary environments and the tectonic deformations before and after 20 Ma were clearly different in the Kekexili area. Tectonic events that are discussed above demonstrate the impacts of Tibetan plateau tectonic activity around 20 Ma, which formed large scale thrust faults and forceful tectonic deformations.

9.1.4   *Characteristics of Deep Structures and Geodynamics in Foreland Basin in Western-Central China*

In western-central China, there are numerous surface structures and, additionally, new structures and active structures are noticeable. At a deep level in the earth, the deformation dynamics process not only triggered substance exchange and energy exchange, but also promoted interactions between the spheres, which created surface structures with characteristics of diversity, complexity and activity. This section will discuss the latest study of deep structures and tectonic dynamics in western-central China. Using a comprehensive geophysical analysis of the study area, we will discuss the nature of the lithosphere and the mechanism of genetic dynamics in the foreland basin.

● **Characteristics of Deeply Buried Structures**

The depth distribution of Moho in the eastern Asian continent and surrounding territorial waters reveals that, in western-central China, the crustal thicknesses vary horizontally with distinct thickness variation and irregular contour lines. Overall, the crustal thickness is decreased in the basin area and is gradually increased towards surrounding orogeny belts. According to the crustal thickness, the crust can be divided into different blocks (Teng et al., 2002). The study of seismic tomography shows details of the velocity structure of the lithosphere and information concerning deep geodynamics in the Tarim basin, Tianshan mountain and Junggar basin. In the Tarim basin, the thickness of the lithosphere is about 200 – 250 km. However, in the Junggar basin, the thickness of the lithosphere is 170 km and in the Tianshan orogeny, the thickness of the lithosphere is close to 150 km. The structure of the lithosphere also has a distinctive variation horizontally (Xu et al., 2001).

The velocity of the Pn wave is influenced by the thickness of the crust, the velocity on top of the upper mantle and the anisotropic properties. Thus, it becomes the most effective medium that reveals information about the upper mantle structure and information about the regional tectonics. By utilizing numerous Pn-wave travel-time data and using seismic tomography technology, Wang et al. (2003) and Liang and Song (2004), separately, inversed the Pn-wave velocity and anisotropy of the upper mantle in mainland China (Fig. 9.8). Their study results showed basins that around the Tibetean plateau contain an abnormaly high velocity. These basins include the Junggar basin, Tarim basin, Tuha basin and Qaidam basin, Sichuan basin and Ordos basin. On the other hand, low velocity zones mainly occur in structurally active areas, such as the western region of Sichuan and Yunnan provinces, central Tibet and Shanxi graben, the North China basin, the southern region of South China and the active volcano and Quaternary volcanic province (northern Tibet). In the compressive basin, the Pn wave travels at a relatively high speed. However, in the extensional basin and graben area, the Pn wave travels at a relatively low speed. In China, the velocities of the Pn wave are distributed as a mosaic pattern that alternates between the high velocity zone and low velocity zone and which perfectly correlates with surface structures in complexity and diversity. The studies of the anisotropy of the Pn wave revealed that, in western-central China, major basins have very weak anisotropic properties. However, the orogenies that surround basins and the conjunction area between basin and range have strong anisotropic properties, which imply strong deformation or a high frequency of deformed activity in these areas.

It is well known that the characteristics of a geophysical field represent the deep buried structures in the earth. Therefore, using geophysical inversion analysis, the details of the deep structural framework and related structural information will be obtained. Due to the ambiguity of geophysical inversion analysis, joint inversion analysis combines the data of geothermal, gravity and magnetic methods, and the lithosphere structure in order to obtain information about deep structures and the lithosphere in basin areas. Detailed comprehensive geophysical

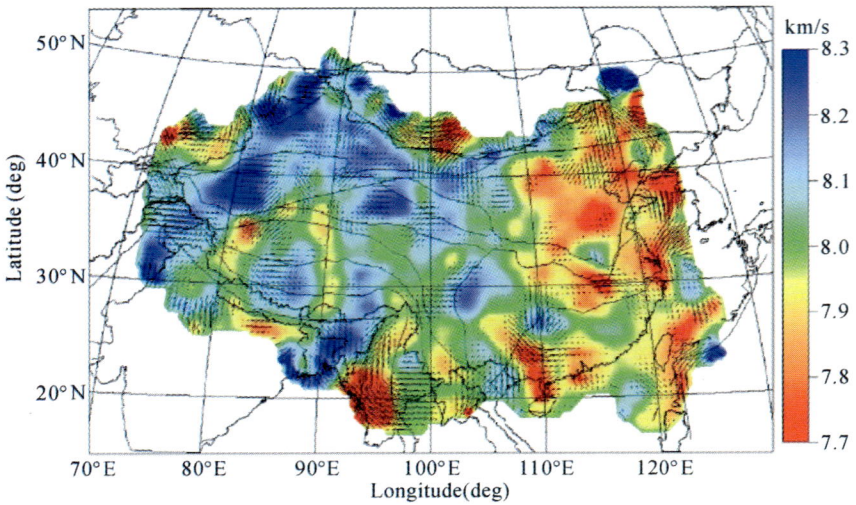

Fig. 9.8.   Diagram of Pn-wave velocity and anisotropic properties in mainland China (Liang et al., 2004) (With permission of Journal of Geophysics Research)

interpretations were carried out for the Tarim basin, Junggar basin, Qaidam basin, Ordos basin and Sichuan basin. The gravity data and the aero-magnetic data demonstrate dissimilarities in different sections of a basin. On the edge of a basin, closely packed gradient zones of gravity anomalies and high magnetic anomalies are dominant. Inside a basin, the changes are gentle, the magnetic anomaly is lower and gravity anomalies alternate between high and low volumes (Jia, 1997; Jia et al., 2004). The analysis of the geothermal field shows that the geothermal gradient, the formation temperature at a deep level and the terrestrial heat flow are relatively low in every basin. In contrast, the mountain fronts that are located on the edge of a basin have a relatively high geothermal structure (Wang et al., 1995; 2003). The distribution of the geothermal structure is heterogeneous horizontally in the basins, which reveals that the geothermal structure is controlled by the basin structures.

Wang et al. (1996) and Liu et al. (2003) systematically analyzed the geothermal structure in the lithosphere in the Tarim basin. Their study results show that temperatures of the Moho and deep mantle-derived heat flow are lower in the lithosphere of a basin area. However, the thickness of the hot lithosphere is great. The rheology study (Liu, 2004) reflects the following attributes of the lithosphere: the lithosphere of the basin has a higher intensity, the conversion depth of the brittleness-ductility is deeper and the efficient elastic thickness is great. Generally speaking, the efficient elastic thickness (T_e) is greater than the thickness of the crust. It is also greater than the conversion depth of the brittleness-ductility. The lithospheric mantle contributes more strength to the total intensity of the lithosphere than the crust does. The rheology of the upper mantle peridotite controls the rheological behavior of the whole lithosphere.

New information about seismic activity and GPS reflect the mechanical properties of the lithosphere at present. Zhang et al. (2002) conducted a statistical analysis of the seismic distribution ($M_L > 2.0$) that occurred during the period 1970 – 2000 in mainland China. The study result shows the basin area has fewer seismic activities than the conjunction area of basin and range. In mainland China, the velocity structures of Pn and Sn indicate the following phenomenon: in the intra-basin area, the velocity of the wave is high and the anisotropy is weak; however, on the marginal area of a basin, the velocity of the wave is low and the anisotropy is strong (Wang SY et al., 2001; Pei et al., 2004; Liang and Song, 2004). The GPS observation indicates that, in the displacement-field, a basin conducts translation and rotational deformation with a stable inner structure, which is the characteristic of entire deformation (Wang et al., 2001). Table 9.1 summarized the basic features of the geophysical field in basin areas in western-central China.

Table 9.1   Characteristics of geophysical field in major basins in western-central China

Name of basin	Intensity ($\times 10^{13}$ N/m)	Features of magnetic field	Features of gravity field (mgal)	Thickness of crust (km)	Thickness of lithosphere (km)	G (°C/km)	Q_0 (mW/m^2)
Qaidam	1.12	NWW	−400 ~ −240	50 – 58	160 – 180	25	52
Junggar	3.15	SE	−125 ~ −100	45 – 50	190 – 210	20	42
Tarim	3.88	NE, SW, NW	−175 ~ −125	39 – 45	170 – 190	20	45
Sichuan	2.90	SN, NE	−300 ~ −80	37 – 47	130 – 150	23	45
Ordos	1.54	NE	−170 ~ −110	43 – 46	110 – 130	23	52

G: Geothermal gradient; Q_0: Terrestrial heat flow. When intensity is larger than 1×10^{13} N/m, it has the feature of a rigid block

● **Characteristics of the Lithosphere and Geodynamics in Foreland Basin**

In a basin, the characteristics of the deep level structure reflect the nature of the lithosphere under a tectonic stress field. Combining the characteristics of a deep level structure that are discussed above, the following is a summary of the nature of the lithosphere and a discussion of the genetic mechanism of the foreland basin in western-central region:

The Lithosphere is Heterogeneous

The lithosphere of a sedimentary basin is heterogeneous. Vertically, the lithosphere can be divided into different layers. Horizontally, the lithosphere may be separated as blocks. Vertically, the upper crust of the lithosphere is a brittle layer that contains fractural deformation; the middle and lower portions of the crust are ductile layers that contain plastic deformation and elongation flow. The upper mantle of the lithosphere is a high intensity brittle layer. The characteristics of deformation in different vertical layers control the developmental process of the faults inside basins. Along the decollement horizon in the crust, a large size deep fault converges at a middle portion of the crust. Because of variations in the physical properties and litho-mechanical properties, inside sedimentary layers some decollement surfaces and ductile layers developed and numerous fault-propagated folds were created. Horizontally, the entirety of the inner basin framework is good, with minor deformation. Every layer of the lithosphere is basically parallel to each other and parallel to the sea level. However, at the conjunction area between basin and range, the transitional zone is brutally deformed, because here is the weak zone of thermal rheology in the lithosphere. The lithosphere is easily deformed by structural stress. These characteristics of the lithosphere are clearly exhibited in the gravity field, aero-magnetic field and geothermal field.

The Lithosphere is Rigid as One Unit

The studies of geothermal, seismic, lithosphere structure and rheology reveal that, in the western-central region, the lithosphere of a basin is completely rigid. This characteristic is represented by low geothermal temperature, high intensity and great thickness. Furthermore, the rheology studies point out that the lithospheric upper mantle of a basin has a very high intensity that represents most of the strength of the lithosphere; the rheological structure in a basin area is a "strong mantle–weak crust". In western-central China, most basements of a basin are paleo continental cratons. During the geological evolution, these cratons were stable with minor intra-deformations. Conversely, the accretion, convergence and welding events occurred on the margin of the cratons.

Under the Compressive Tectonics, the Lithosphere Flexurally Deformed

In western-central China, the collision between the Indian plate and the Eurasian plate was the most noticeable tectonic event during the Mesozoic– Cenozoic eras, which happened around 70 Ma – 50 Ma. Under a persistent compressive stress field, the lithosphere in the basin area was deformed and flexed because of its rigid physical properties. The rigid lithosphere was subducted under the orogenies and, as a chain reaction, the orogenies were thrust toward the basins. This phenomenon has been proved by seismic sounding data (Gao et al., 2001). In particular, deformation is represented by the center uplift in a basin and a mountain front subsidence that is located at the conjunction area between a basin and an orogeny. The subsidence was caused by structural movement and sediments overloading. For example, the Kuqa depression is located at the conjunction area between the northern margin of the Tarim basin and the southern margin of South Tianshan mountain. In addition, the southwestern Tarim depression is located at the conjunction area between the Tarim basin and West Kunlun mountain. Furthermore, the mountain front depressions in the southern Junggar basin and in the western Sichuan basin all belong to this kind of foreland basin. Because the basement of a basin was a rigid terrain previously, the types of structural deformations were therefore primarily accretion and convergence, which happened on the margin of the basement terrain. The convergent zone was an area vulnerable to structural deformation. This area was reactivated by the collision between the Indian and the Eurasian continents and a series of regenerated foreland basins were formed in this vulnerable area (Lu et al., 1994).

In a previous discussion, a comprehensive geophysics study indicated that, in western-central China, the lithosphere is a rheological heterogeneity horizontally. For an entire craton, it has strong rigidity with minor intra-deformation or no deformation. Under the tectonic stress field, the craton was deformed as a unity. On the other hand, the suture zone of a previous tectonic event and the convergent zone occur on the margin of the craton, which are the venerable areas for deformation and which would be reactivated in an adequate stress field. The western-central region is influenced by the collision of the Indian plate and the Eurasian plate and the far-field effects of persistent compression. The rigid blocks transferred the stress field and brutal deformations happened on the margin of rigid blocks. Because the lithospheric block had strong rigidity and strong intensity, consequently the lithosphere was plunged under the orogeny. Accordingly, the orogeny was thrust over to the basin (Kao et al., 2001). In the terrain, deformation was symbolized by a center flexural uplift and marginal depression that was under the dual influence of structural loading and sediments loading. The marginal depression also received sedimentary deposits (Yang et al., 2002). During the late Cenozoic era, because surrounding mountains were rapidly uplifted, the mountain front depression zone, as a result, displayed rapid subsidence, which formed a group of thrust belts that surrounded the eastern and northern sides of the Tibetan plateau (Fig. 9.9). Due to a differentiation in the physical properties of the lithology in sedimentary strata, some decollement surfaces were formed inside strata, which offer a variety of types of structural deformation. Also, a combination of fault-related folds, salt-related folds and a combination of strike-slip faults and thrust faults developed.

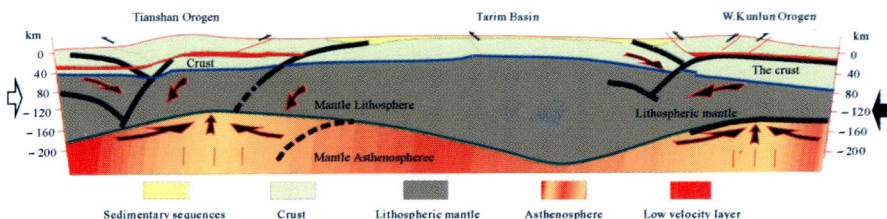

Fig. 9.9. The geodynamic model of coupling deformation in conjunction area between basin and range, western China (Xiao et al., 2004) (With permission of The Commercial Press)

Clearly, in western-central China, the development of the foreland basin was influenced by the collision between the Indian plate and the Eurasian plate and the subsequent compression. In the inner continent (far away from the edge of the plate), because of the reactivation of a paleo subduction zone and convergent zone at a structurally vulnerable area, a compressive basin developed as a consequence. Compared with a typical foreland basin, this type of foreland basin is distinct in tectonic background, structural position and evident scale. Thus, this is a new type of foreland basin. We propose to name this type of foreland basin an "Intracontinental Rejuvenated Foreland Basin", because it suitably presents the fundamental attributions of the basin. We will initially summarize the genetic mechanism of an intracontinental rejuvenated foreland basin: its structural attribution not only inherits the characteristics of the previous structure, but also has the feature of being reworked. Its structural position is located at the intracontinent, which is not only far away from the edge of the small craton plate, but also is controlled by the rheological heterogeneity of the lithosphere. Its dynamic power came from the collision between the Indian plate and the Eurasian plate and the consecutive compression.

9.2   Development of Sinitic Foreland Basin and its Four Basic Types

We will first introduce the development of a foreland basin and its major types in western China.

9.2.1   Development of Foreland Basin and its Major Types in Western China

● **Two Periods of Foreland Basin Development**

Terrestrial basins of the Mesozoic–Cenozoic eras and a typical foreland basin have some similarity in their mechanism; these terrestrial basins are formed within the compressive tectonic framework and they were widely distributed in western-central China. Compared with the definition of a peripheral foreland basin and back-arc foreland basin (Dickinson, 1976), the tectonic background of a terrestrial basin and its sedimentary features are clearly different. Thus, some international or domestic scholars assigned different names to the terrestrial basin of western-central China, for example, "Chinese type basin" (Gan, 1995), "Himalayan type basin", "re-genesis foreland basin" (Graham et al., 1993; Hendrix et al., 1994; Lu et al., 1994), "inner continental foreland basin" and "type C foreland basin" (Luo, 1984; Li et al., 2006). There is no direct connection of mechanism and timing correlation of events between the Mesozoic–Cenozoic foreland basin and subduction of plates. In western-central China, foreland basins were usually formed at the conjunction of an ancient orogeny and ancient terrain. Because ancient tectonic structures of the inner plate were reactivated, an original terrain (or a newer inner continental basin) was thrust forward along a fault on the margin of these tectonic structures (or inside structures); additionally, the thrust movement caused flexure and overload phenomenon on the front margin of these tectonic structures with massive deposits.

Since the late Paleozoic, foreland basins experienced two different periods of development, which formed foreland basins of the late Paleozoic–early Mesozoic eras and foreland basins of the Cenozoic era. The development of the late Paleozoic–early Mesozoic foreland basins were closely related to subduction and extinction events of the paleo-Asian ocean and paleo-Tethys ocean; the development of the Cenozoic foreland basins were closely related to inner plate orogenic movements. Therefore, we can divide foreland basins of western-central China into two development periods in three categories (Table 9.2). Two development periods point to the Indosinian tectonic period (late Paleozoic–early Mesozoic eras) and the Himalayan tectonic period (since the Cenozoic era); three categories are designated to the peripheral foreland basin of the Indosinian tectonic period, the back-arc foreland basin of the Indosinian tectonic period and the re-genesis foreland basin of the Himalayan tectonic period. The re-genesis foreland basin of the Himalayan tectonic period can be subdivided into three different types, which are compressive thrust type, strike-slip thrust type and thrust strike-slip type.

Table 9.2 Types of foreland basin in western-central China

Period	Category	Sub-category	Area of distribution
Indosinian tectonic period	Peripheral foreland basin		Western Sichuan, northern Sichuan, western margin of Ordos, southern Junggar, western Junggar, northeastern Junggar, Tuha
	Back-arc foreland basin		Chuxiong, southern margin of Tarim
Himalayan tectonic period	Re-genesis foreland basin	Compressive thrust type	Southern Junggar, Kuqa, western Sichuan, northern Sichuan, western margin of Ordos
		Strike-slip thrust type	Southwestern Tarim, Tuha, Jiuquan
		Thrust strike-slip type	Northern margin of Qaidam

● **Foreland Basins of the Late Paleozoic–Early Mesozoic Eras**

During the Late Paleozoic–Early Mesozoic eras, two classical types of foreland basin were developed, which are the peripheral foreland basin and back-arc foreland basin. In western-central China, the peripheral foreland basin appears in the following places: western Junggar, the eastern margin of Junggar, southern Junggar, Kuqa, the western margin of Ordos, western Sichuan, eastern Sichuan and Tuha; on the other hand, the back-arc foreland basin appears in the southwestern Tarim basin and Chuxiong basin (Fig. 9.10).

The following are discussions about foreland basins, which describe the features of the peripheral foreland basin by using the southern Junggar basin of the Late Carboniferous–Permian period as an example, and which explain the features of the back-arc foreland basin by using the southwestern Tarim basin of the Late Permian–Triassic period as another example.

In southern Junggar, the peripheral foreland basin of the Late Carboniferous–Permian period: initiated in the Late Carboniferous period, the front margin of the Tianshan orogeny that was located on the southern margin of the Junggar plate, was subducted northward into the Junggar plate. During the Late Carboniferous–Permian period, the sediments were primarily deposited in the foredeep of the orogeny, which formed mountain front depressions on the northern margin of Bogeda mountain and on the northern margin of Tianshan mountain. These depressions were parallel to the Tianshan orogeny. The cross section that is drawn along the short axis of the basin shows a wedge shape of sediments. The basins that formed during this geologic interval are mainly located on the southern margin of the Junggar plate, which expanded into the plate later. These depressions were controlled by the Tianshan orogeny and the Bogeda orogeny. The basin's framework and its characteristics of sedimentation reveal that during the Late Carboniferous–Permian period, a peripheral foreland basin was developed in southern Junggar (Fig. 9.10).

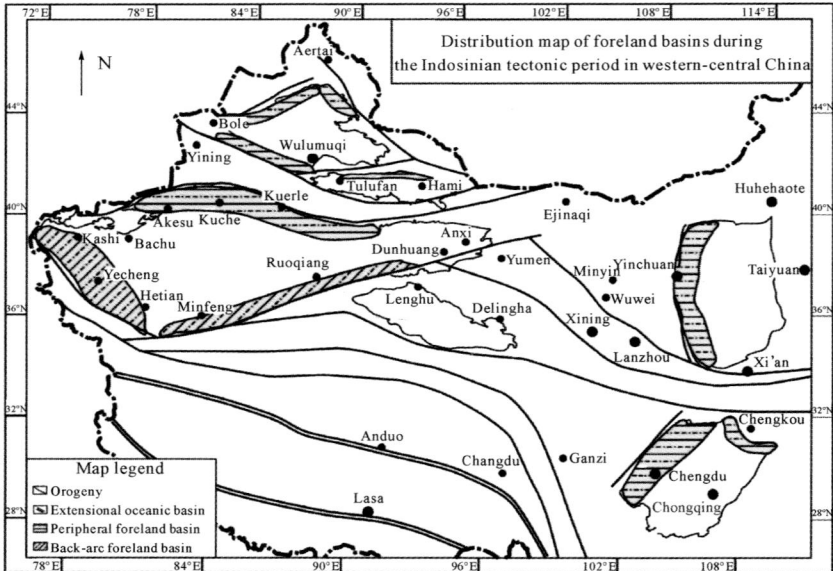

Fig. 9.10.   Distribution map of the early Mesozoic foreland basins in western-central China

In southwestern Tarim, a back-arc foreland basin of the Late Permian–Triassic period was developed. During the early Permian period, the Paleo-Tethys Ocean, which was located on the southern margin of the Tarim plate, was subducted northward under the Tarim plate in the direction of Tashikuergan–Kangxiwa. Due to the subduction of the Paleo-Tethys Ocean, magma intrusions and eruptions occurred on an extensive scale. During the Early Permian epoch, because the Paleo-Tethys oceanic crust had a relatively large subduction angle, the island-arc area and the back-arc area were under tensional stress. Within the tensional tectonic environment and beneath the basins, substances of the asthenosphere were upwelling and the lithosphere strongly protruded upward, which developed a rifting basin in the back-arc area. During the late Early Permian period, large magnitude magma activities occurred inside the Tarim basin with a distribution area of more than 100,000 km², which formed rifting type dual model volcanic rocks. That is the best example of back-arc extension. During the Late Permian period, the attributes of the inner Tarim basin reveal that the Early Permian basaltic magma activity had been stopped; furthermore, the peripheral area began uplifting. Because of the western Kunlun area constantly uplifting, the sedimentary deposits were absent in this area; however, on its front margin, it developed a delta fan deposition system that neared the source of the substance with a rapid settlement rate, which indicated it had proceeded into the developmental stage of a back-arc foreland basin in the southwestern Tarim area. The depositional center of the back-arc foreland basin was located at the mountain front on the northern side of western Kunlun mountain. Around the Yecheng area, the thickness of sedimentary deposits of the Late Permian period is more than 1,600 m (Fig. 9.11). During the Late Permian epoch, they deposited a suite of river facies–lacustrine facies sediments. It seems that the sediments of the Triassic period were absent in the southwestern Tarim area; however, according to the data from a neighboring area, the Triassic strata were possibly deposited in the southwestern Tarim area. In a cross section of Duwa in the southwestern Tarim area, Triassic pollens have been discovered in the upper portion of the Upper Permian series, which confirms the existence of Early Triassic deposits and indicates successive deposits of the Permian system and the Triassic in the area. During the Triassic period, the depositional environment of the Late Permian period was inherited, which were river facies and lacustrine facies; however, the depositional center was moved toward the northeast. During the end of the Triassic period–Early Jurassic period, these Triassic deposits were influenced by the collision between the Qiangtang terrain and the Tarim plate with the result of powerful uplifting and thrusting. This evidence indicates that, in this time interval, an island arc area and back-arc basin area were enveloped in a compressive tectonic environment. This transformation of the tectonic environment was closely related to the peripheral tectonic background. During the end of the Late Triassic–Early Jurassic, the subduction of the Paleo-Tethys ocean came to an end; the fragment of Gondwana land that drifted in the Paleo-Tethys ocean collided with the Eurasian plate and formed the Cimmerides orogeny. In the same time interval, the Qiangtang terrane, which was located on the southern side of the Tarim basin, powerfully collided with the Tarim plate in the Tashikuergan–Kangxiwa direction, which formed the Tashikuergan–Kangxiwa suture belt. Because of a strong collision between the Qiangtang terrane and the Tarim plate, during the end of the Late Triassic–Early Jurassic, on the southern side of the ancient suture belt, it formed the Tianshuihai foreland folded-thrust belt; additionally, on the northern side of the suture belt, it developed many thrust faults in the back-arc area. Together, on the southern margin of the Tarim basin, they created a thrust belt (Fig. 9.12).

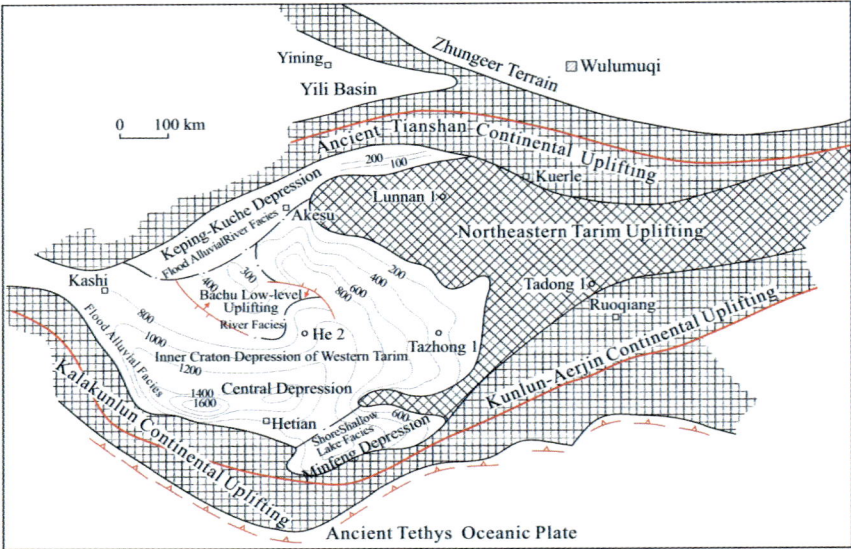

Fig. 9.11.   Map of tectonic and lithology during the Late Permain epoch in the Tarim basin (Jia, 1997) (With permission of Petroleum Industry Press)
The numbers, such as 200, show the remaining thickness of strata; the small circles, such as He 2, are the well locations

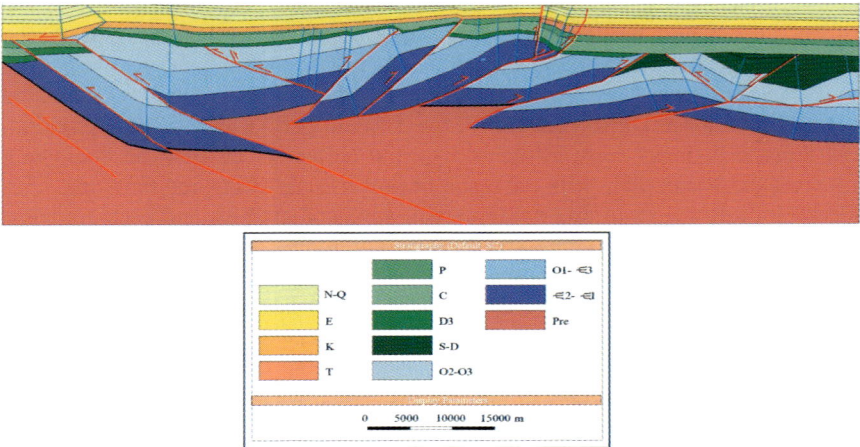

Fig. 9.12.   Thrust belts during the end of Triassic (TM1–96–07) at southern margin of the Tarim basin (Jia, 1997) (With permission of Petroleum Industry Press)

● **Cenozoic Foreland Basin**

The Cenozoic basins were formed by reactivated ancient orogenies that were influenced by the distant collision between the India plate and the Tibetan plateau. During the Cenozoic, orogenies developed inside the plate and reactivated ancient orogenies thrust toward the basin on an extensive scale, which formed foreland basins. We call it the re-genesis of foreland basins, which include southern Junggar, Kuqa, southwestern Tarim, Tuha, Jiuquan, the northern portion of western Sichuan, north eastern Sichuan, the western margin of Ordos, the northern margin of Qaidam (Fig. 9.7).

9.2.2 Developmental Characteristics of the Foreland Basin and its Combination Types in Western China

For the foreland basins of western China, petroleum generation and features of petroleum accumulation are not only hot research topics, but these issues are also closely related to oil and gas exploration. For a better understanding of petroleum generation and its pattern of accumulation, it is necessary to study the character of a foreland basin's evolution and its combinations; these basins have been formed since the late Paleozoic era. The diversity of a basin's evolution and the variety in the combination of the basins will influence the following: the combination of petroleum generation–reservoir–cap rocks, periods of oil and gas accumulation, alterations to the reservoir and oil and gas exploration. According to a basin's development periods, considering the framework and tectonic features, taking into account the spacious relationships of the basins, this study will establish the character of a basin's evolution and its combination styles.

The foreland basins of western China can be divided into two development periods of three different types, which are the peripheral foreland basin and back-arc foreland basin of the Indosinian tectonic period and the rejuvenation foreland basin of the Himalayan tectonic period. Among basins of these two development periods, some foreland basins stacked on top of another one and others did not have any spacious relationship. In some areas there was only a foreland basin of an early period. In other areas there was simply the Cenozoic foreland basin and the existence of an older foreland basin was unknown (or it never existed). Some foreland basins of an early period were altered by the thrust fault later in time and others were not. Thus, according to the realities discussed above, this study proposes four combination styles to describe the foreland basin's evolution and its combinations, which are superimposed style, modified style, presenility style and revival style (Table 9.3).

Table 9.3 Features of foreland basin's evolution and its combination types in western China

Characters/ Category	Superimposed style	Modified style	Presenility style	Revival style
Foreland basin	Back-arc foreland basin or peripheral foreland basin of the Indosinian tectonic period; Re-genesis foreland basin of the Himalayan tectonic period	Peripheral foreland basin of the Indosinian tectonic period	Peripheral foreland basin of the Indosinian tectonic period	Re-genesis foreland basin of the Himalayan tectonic period
Foreland strata sequence	$C-P_1$ passive continental margin rifting valley; J_{1-2}-E graben basin (depression)	$C-P_1$ passive continental margin rifting valley	Passive continental margin deposit before P_2	J_{1-2}-E graben basin and depression
Deformation during late period	Deformation of foreland thrust belt of the Himalayan tectonic period	Thrusted deformation of late period	Minor deformation during late period	Deformation of foreland thrust belt of the Himalayan tectonic period
Basin's alteration during late period	Foreland basins of two development periods were well preserved; Foreland basins of the Himalayan tectonic period are still in development processing	Foreland basins of early period were partially destroyed	Foreland basins of early period were well preserved	Re-genesis foreland basins of the Himalayan tectonic period are still in development process
Distribution	N. and S. sides of Tianshan Mountain, front of Kunlun Mountain	W. margin of the Ordos, W. Sichuan, NE Sichuan, Chuxiong	NW region of Junggar basin	S. and N. sides of Qilian Mountain
Typical basins	S. Junggar basin, Kuqa basin, and S. Tarim basin	Foreland basin of W. margin of the Ordos, foreland basin of W. Sichuan, foreland basin of NE Sichuan, foreland basin of Chuxiong		Foreland basin of NW margin of the Junggar, N. margin of the Qaidam, Jiuquan basin

● Superimposed Style

The main characteristic of the superimposed style is that of a foreland basin of the Himalayan tectonic period with complete framework stacked on top of another foreland basin of the Indosinian tectonic period. Vertically, the frameworks of the two basins coordinate with each other very well (Fig. 9.13). This combination style occurred on both sides of the Tianshan orogeny and the northern side of the Kunlun orogeny. For example, the Kuqa foreland basin achieved this kind of combination, which experienced the foreland basin phase of the Late Permian epoch–Triassic period, the graben basin–depressed basin phase of the Jurassic–Palaeogene periods, and the rejuvenation foreland basin phase of the Neogene–Quaternary periods. The major tectonic deformation and the alteration to the basin's framework happened after the Kuqa formation of the Neogene system was deposited (Lu et al., 2000; Li et al., 2001; Wei et al., 2002). During the Late Permian–Triassic periods, due to the disappearance of the Southern Tianshan Ocean, the Kuqa area evolved with the tectonic development of the peripheral foreland basin phase. Strata of the Triassic received sediments of great thickness at a higher settling rate. In the northern part of the basin, the thickness of strata is $1,000 - 2,300$ m, in general, and the sediments settling rate can reach $29.0 - 65.7$ m/Ma. During the Jurassic–Palaeogene periods, the basin evolved with a relaxing stress field. During the Jurassic period, the Kuqa area shifted into a graben basin phase under the extensional tectonic environment. During the Cretaceous– Palaeogene periods, the curve of the settling rate was smooth wich indicated a depression phase in a relatively stable tectonic situation. The settling rate of the Cretaceous was $12.5 - 19$ m/Ma and that of the Palaeogene was 25 m/Ma. During the Neogene–Quaternary periods, the Indian plate forcefully collided with the Eurasian plate and it kept pushing northward, which formed a rejuvenation foreland basin with asymmetrical depression, also with a fault on the northern side and an overthrust on the southern side. This rejuvenation foreland basin received $6,000$ m of sedimentary deposits. The Kuqa foreland basin reveals the character of two superimposed basins. At present, the oil and gas accumulations are controlled by the rejuvenation foreland basins of the late development period.

● Modified Style

A modified combination style is representative of the basins that were formed during the Indosinian tectonic period; also, these basins were modified by thrusted deformation of the Himalayan tectonic period. The pattern of basins that developed during the early period had been altered. However, the foreland basins of the Himalayan tectonic period did not develop, or their character was very poor. This kind of combination of basins was primarily developed in the Sichuan basin and the western margin of the Ordos basin. For example, the foreland basin of western Sichuan is a peripheral foreland basin that was formed by the extinction of the Jinshajiang ocean, which also is an early Mesozoic foreland basin that was

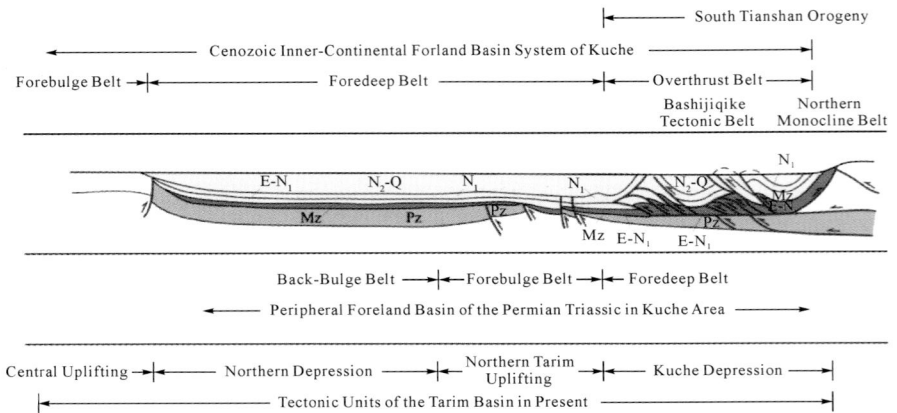

Fig. 9.13. Cross section of foreland basin in the Kuqa area (Li et al., 2001) (With permission from Xinjiang Petroleum Geology)
Q, Quaternary; N, upper Tertiary; E, lower Tertiary; Mz, Mesozoic; Pz, Paliozoic

developed on top of the late Paleozoic passive continental margin basin. During the Late Permian–Middle Triassic epochs, the western Sichuan basin experienced the development stage of a passive continental margin basin. During the Late Triassic epoch, it experienced the development stage of a peripheral foreland basin. In the foredeep depression area, it rapidly deposited the Upper Triassic coal measures as the hydrocarbon source rock. During the Early–Middle Jurassic epochs, due to the relaxing stress field, it experienced the development stage of depression. During the Late Jurassic epoch–Quaternary period, because of persistent compression, it experienced a stage of overthrusting (Jia et al., 2006). During the Late Jurassic epoch–Cenozoic era, the overthrust movement modified the western Sichuan basin with limitation, which shaped the framework of the foreland basin into its current condition. At present, from west to east, the western Sichuan basin can be divided into a foreland thrust belt (the Longmen mountain thrust belt), a foredeep depression belt and a foreland uplifting (Fig. 9.14).

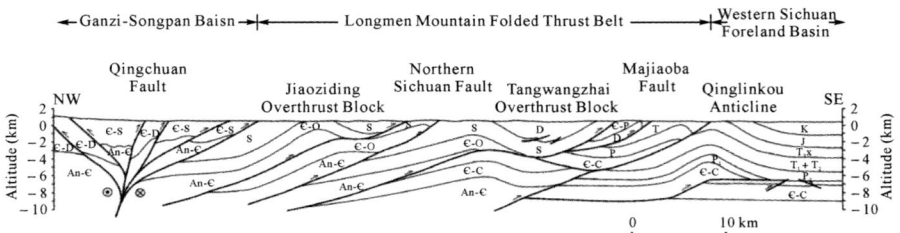

Fig. 9.14.   Cross section of Western Sichuan foreland basin
An- Є, Before Cambrian; Є, Cambrian; O, Ordovician; S, Silurian; D, Devonian; C, Carboniferous; P, Permian; T, Triassic; J, Jurassic; K, Cretaceous

● **Presenility Style**

The presenility combination style describes a foreland basin that was primarily formed during the Indosinian tectonic period. Afterwards, it experienced an extended graben basin stage and regional depression. Thrusted deformation of the Himalayan tectonic period had a very weak influence on this kind of foreland basin. In the cross section of the tectonic structure, we can clearly see that basins of the Indosinian tectonic period have the framework of a foreland basin; furthermore, the surrounding orogenies are wonderfully correlated with the structures inside the foreland basin. This kind of basin was primarily developed at the western and eastern Junggar. The foreland basin of western Junggar is an example. During the end of the Early Carboniferous epoch, the Junggar terrane collided with the Kazakhstan plate, which formed an embryonic form of Jieshan mountain on the northwestern margin. During the late Carboniferous epoch, this progressed to the development stage of a foreland basin. In the cross section, the basin can be clearly recognized; the Mahu Lake depression matches this with a foredeep depression; the Dabasong bulge corresponds with a forebulge; the depression west of the Pan 1 well coincides with a back-bulge depression (Fig. 9.15) (Xie et al., 1984; Wu et al., 2005; Guan et al., 2007; Wei et al., 2004). During the Triassic period, the basin entered a flexural subsiding stage. During the Neogene period, the foreland basin that was located at the western Junggar was not controlled by an ancient orogeny that was located on the western side of the basin. However, it was influenced by the rejuvenation foreland basin that was located on the southern margin of the Junggar. At this point, the foreland basin of western Junggar could not develop independently. Thus, we consider this combination to be of presenility style. The development of this combination of basins was decided upon by special tectonic position. In the early period of the foreland basin the orientations of the structures were almost N-S, but the direction of compressive stress of the Himalayan tectonic period was parallel to the structures of the early period foreland basin.

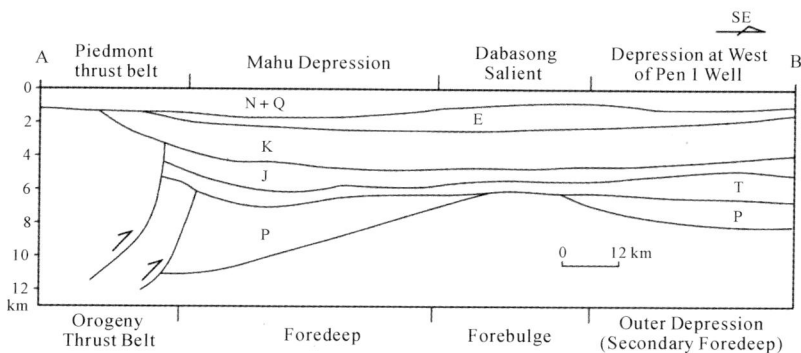

Fig. 9.15. Cross section of western margin of the Zhungeer basin (Wu et al., 2005) (With permission from Acta Geoscientica Sinica)
P, Permian; T, Triassic; J, Jurassic; K, Cretaceous; E, lower Tertiary; N, upper Tertiary; Q, Quaternary

● **Revival Style**

The revival combination style is named after the foreland basin of the Himalayan tectonic period but, underneath, the foreland basin of the Indosinian tectonic period was missing (or we do not have enough information to prove that the early period basin had existed). This kind of combination of basins was primarily developed on both sides of the Qilian mountain orogeny by the activity of the Aerjin strike-slip fault and the thrusted movement on both sides of the Qilian mountain orogeny. These tectonic activities were influenced by the distant collision between the India plate and the Tibetan plateau. For example, the Jiuquan basin had experienced a different developmental stage during various periods: the inner plate graben faulted stage of the carboniferous–Permian periods; the depression stage of the Triassic period (it is impossible to identify whether or not the foreland basin had existed); the extensional graben faulted stage of the Jurassic–Cretaceous periods; the mild extension stage of the Palaeogene period and the foreland basin development stage of the Neogene period (Huo et al., 1995; Yang, 2006). Since the Neogene period, the Jiuquan basin has demonstrated the characteristics of a typical foreland basin, which include a well developed folded-thrust belt, foredeep depression, foredeep slope, foredeep bulge and back-bulge depression (Fig. 9.16).

9.3   Features of Tectonic Deformation of Sinitic Foreland Thrust Belts

In this section, we will introduce features of tectonic deformation of foreland thrust belts and major foreland thrust belts in China.

9.3.1   *Types of Foreland Thrust Belt*

Dickenson (1976) considered a foreland basin to be of two different types, namely a back-arc foreland basin and a peripheral foreland basin. The Cenozoic foreland basins in western-central China were the products of the inner plate tectonic deformation movement. Lu et al. (1994) called them re-genesis foreland basins. As a lateral wing to a foreland basin, the foreland thrust belts also are of three different types:

Back-Arc Foreland Thrust Belt.   Typical examples are the Rocky mountain front thrust belt in North America, the Cenozoic back-arc thrust belt in South America, and the Paleo-Tethys thrust belt (northern Tethys) in Eurasia. The development of a back-arc foreland thrust belt was closely related to type A subduction. However, it was also restricted by type B subduction. During the Late Permian epoch–Triassic period, a back-arc foreland thrust belt developed on the southwestern margin of the Tarim basin.

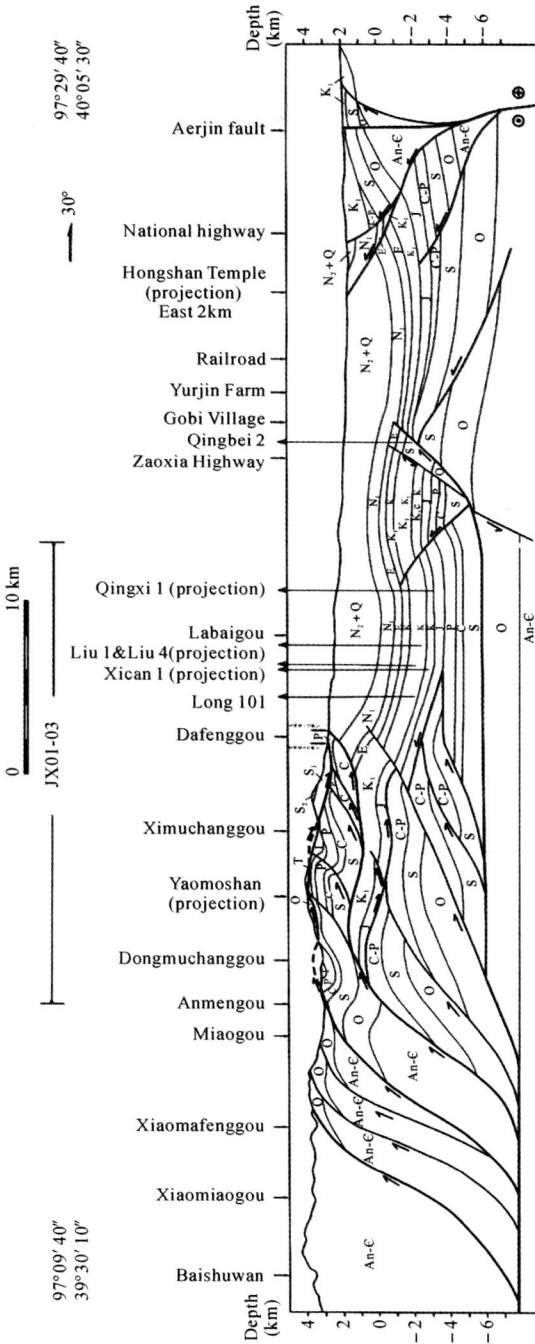

Fig. 9.16. Cross section of the Jiuquan basin

An-Є, Before Cambrian; Є, Cambrian; O, Ordovician; S, Silurian; C, Carboniferous; P, Permian; T, Triassic; J, Jurassic; K, Cretaceous; E, lower Tertiary; N, upper Tertiary; Q, Quaternary

Peripheral Foreland Thrust Belt.   Along with the oceanic crust subduction movement, two continents gradually came together. Finally, the collision happened and an orogeny developed. At the front of the orogeny, a peripheral foreland basin developed on the passive continental margin of the subducted plate. In the thrust belt that is next to the peripheral foreland basin, ophiolite and ophiolite melange commonly occur on the inner side of the thrust, which is the distinctive characteristic when compared with the back-arc thrust belt. The typical examples are the mountain front basin of the Zagros and the molasses basin.

Rejuvenation Foreland Thrust Belt.   It was developed on the active side of the re-genesis basin. The re-genesis foreland basin did not have the timing correlation with the subduction suture zone in the same vicinity; it was predominantly controlled by the distant collision of the India plate and the Eurasia plate. This tectonic collision reactivated ancient sutured orogenies and it caused powerful inner plate deformations, which formed a folded-thrust belt on the mountain front and which settled massive molasses deposits. Furthermore, the rejuvenation foreland basin does not directly attach to the collision suture zone of same period; in fact, it keeps a dissociate distance from the collision suture zone. The phenomenon that is discussed above reveals that the rejuvenation foreland basin has its own characteristics: (1) Because the basin was formed by a far away collision between tectonic plates, the direction of the stress force is from south to north. Thus, the thrust movement was stronger on the southern side. For example, the thrust deformation that occurred at the mountain front in southwestern Tarim was much stronger than the one that happened on the southern margin of the Junggar. (2) The influence on the basal structure was clearly enhanced. For example, either on the northern margin of Qaidam, or on the southern margin of Junggar, in the basement, normal faults that controlled graben basins of the Mesozoic era could be positively inversed as thrust faults during the Cenozoic era. (3) Because the size of the craton terrane is smaller than the size of the orogeny in the same neighborhood, the foreland thrust belt was therefore strongly deformed. It is possible that at the forefront of the triangular zone on the mountain front, the foreland depression or foreland slope were involved in a folded-thrusted deformation event, which created an indistinct boundary between the foreland thrust belt and foreland depression. Instead, there is a series folding zone that related to the thrust faults. As a result, a special geological framework formed that differentiated itself from a common foreland thrust belt. Good examples are the mountain front belts in the following areas: Kuqa, southern Junggar, the northern margin of Tarim and the northern margin of Qaidam.

9.3.2   Geologic Framework of Foreland Thrust Belt

● **Thin Skinned Structure and Thick Skinned Structure**

According to whether a fault cut into the basement, or according to whether the basement was involved in tectonic deformation, the structures can be divided into two types: basement-involved type and capping-sliding type (Mitra et al., 1989; 1990; 1992; 1998; Dahlstrom, 1969; Davis et al., 1983; Chester et al., 1990). The former is called a thick-skinned structure, which often occurred at the following three places: inside an orogeny (such as the back side of Longmen mountain), as an inherited activity of a basement fault in a foreland area (such as northern margin of Qaidam), or at the area of fault inversion. The latter is called a thin-skinned structure, which is a major structural type of foreland thrust belt, which was widely distributed in the Jurassic mountain, Rocky mountain, Longmen mountain, the southern margin and the northern margin of Tianshan mountain (Table 9.3).

Based on the type of structures, according to the relationship of the relative thickness between the overthrust block and underlying original strata, or according to the degree of alteration when the foreland basin was deformed by the foreland thrust belt, or according to the typical structure deformation pattern that was located at the junction area of basin and range, we further analyzed the geological framework of the foreland thrust belt (Table 9.4).

● **Displacement Transfer of Foreland Thrust Belt and its Conversion Style**

From mountain front to inner basin, the displacement (or sliding energy) of the foreland thrust belt has been gradually diminished until it fades away. In the direction of the basin, the distance of displacement in the main branch of the thrust fault also gradually decreased because the sliding energy traveled to a higher position along splitting faults, or because the sliding energy was adjusted by deformation that happened inside the thrust belt. For example, at the ramp of the main branch thrust fault, there formed a fault-bended fold that might absorb some sliding energy and the remaining sliding energy would continue to travel in the direction of the foreland area. Another possibility is that the deformation of a hanging wall gradually used up the sliding energy such as to form a splitting thrust fault, or to form a fault-propagated fold. One more possibility is that the sliding energy traveled to a higher position in the thrust fault. Thus, it formed a "front margin monocline" in the direction of the foreland area.

Table 9.4   Types of geological framework of foreland thrust belt (Jia et al., 2000) (With permission from Petroleum Industry Press)

Division	Major type
Whether fault cuts into the basement	Basement-involved type (or thick-skinned structure): it primarily occurs inside orogeny, as inherited activity of basement fault in foreland area, or as fault inversion activity
	Capping-sliding type (thin-skinned structure): it primarily takes place in foreland area, which is controlled by sliding surface, thickness of strata, and component
Relationship of relative thickness between the overthrust block and underlying original strata (such as passive continental margin, or graben basin)	Thick-thick superimposed type: thick layer of overthrust block is on top of a thick set of passive continental margin sequences. it often forms multiple sets of excellent source–reservoir–cap assemblage
	Thin-thick superimposed type: relatively thin layer of overthrust block is on top of very thick original sequences. the imbricate thrust structure is the common type of deformation. the hydrocarbons that generated from the original strata might be injected into some reservoirs that relate to thrust fault
	Thick-thin superimposed type: very thick layer of overthrust block is on top of relatively thin layer of original sequences that might directly be the basement. oil and gas are generated from overthrust block itself
	Thin-thin superimposed type: thin layer of overthrust block is on top of thin layer original sequences. fractural deformation is predominant structure type, which might form reservoir with poor storage condition
Degree of alteration when foreland basin is deformed by foreland thrust belt	Significant alteration: foreland basin is strongly cut and altered by multiple layers of overthrust blocks or numerous overthrust activities; or it might form new type of basin (such as piggyback basin)
	Medium alteration: inside basin, it was formed by multiple array of faulted-folding zones in several phases of thrust activities; however, framework of original basin might be preserved as well
	Minor alteration: subtle thrust activity was limited on the margin of a basin, which posted minor alteration to inner basin
Typical structure deformation pattern (using the junction of mountain and basin as an example)	Imbricate fault type: ancient mountain imbricately thrust toward basin and formed a series of structures at front margin of the thrust fault
	Structural wedges type (or triangle zone): from structural wedges toward foreland area, it has monocline or other type of structure
	Inversion type: boundary fault that controlled basin was inversed in the late period and it was thrust toward foreland
	Diapir type: in some foreland areas, due to compressive force, evaporite hosts diapir or imbricate fan structure

At the edge of a severe orogenic deformation belt, the intensity of strata deformation and distance of strata shrinkage were rapidly reduced. This often went together with a front margin monocline. Acute strata shrinkage primarily happened at the area of the front margin monocline. Suppe (1983a; 1983b) and Epard and Escher (1996) often used a thin-skinned structure to explain the mechanism of a monocline. In this scenario, the thin-skinned structure was developed above a low angle detachment fault and this detachment fault occurred in the basal portion of overlying strata that immediately topped the crystalline basement. One possibility is that many arrays of imbricate thrust faults developed. Together, these imbricate activities might gradually raise the elevation of the fault ramp area. In addition, a monocline may have formed on the front margin. Another possibility is the development of a "triangular zone" structure. The third possibility is a basement-involved type structure (Mitra, 1998). For example, a fault that controlled the basin in the past might be inversed, or a normal fault at the rifting valley stage during the early tectonic period might be reactivated during the late tectonic period.

Obviously, not all the sliding energy from faults would be absorbed by internal deformation. Some thrust faults might reach the erosion surface of the same tectonic period. At this erosion surface, much of the displacement of the thrust sheet might be lost because of the erosion.

● **Distributional Characteristics of Foreland Thrust Belt**

Because the displacement transfer of a foreland thrust belt and its conversion mode are different, this results in various structural deformation patterns and different styles of distribution (Jia et al., 2000). The following are eight common structural combination types:

Imbricate Thrust Belt+Fault-Related Type. Orogeny overthrust in the direction of the basin in an imbricate pattern. However, in the foreland area, a fault-related fold primarily developed that was located on top of the stepwise thrust fault. These two structures have a moderate transitional relationship.

Imbricate Thrust Sheet+Thin-Skinned Structure+Back Thrust+Front Margin Monocline Type. There was a transformation from a thick-skinned imbricate thrust fault to a thin-skinned overthrust structure. Because numerous back thrust faults developed in the direction of the backland, this evened out displacements of large scale thrust faults. There are two combinations: (a) A back thrust belt combines with a front margin triangular zone. For example, on the western margin of Ordos, a typical back thrust belt was developed. Together with thrust faults that point in the direction of the foreland, a front triangle zone (or area) was formed. (b) A lower elevation bulge combines with a fault-related anticline that was on top of a back thrust fault. A back thrust fault and a thrust fault that points in the direction of a foreland area might form a lower elevation bulge or latent anticline. For example, it appears at the mountain front of Longmen mountain in western Sichuan.

Imbricate Thrust Sheet+Foreland Basement-Involved Type Structure.
A foreland basement–involved type structure (Mitra et al., 1998; 1999) primarily
appeared as a ramp fault combination (such as a foreland thrust belt in the
northern margin of the Qaidam basin), a compression-torsion fault combination
(such as the faults in the southern margin of the Junggar basin), and a basement
fault belt (such as the front margin of the Pamir area).

Structural Wedges.   This mainly refers to a type I structural wedge that has
two sub-types. (a) Structural wedge+fault-related fold. The foreland folded–thrust
belt of Kuqa is a good example. On the frontage of the structural wedge of the
southern Tianshan mountain front, there developed three arrays of fault-related
folded anticline belts (or imbricate structural belts). In the direction of the foreland
area, along with stepwise thrust faults slowly gaining elevation, the deformed
strata gradually became younger geologically and the deformed enormity was
steadily reduced. (b) Structural wedge+(underlying) imbricate thrust sheet. In the
Keliyang–Sangzhu area of the southwestern Tarim basin, behind the triangular
structure of Guosangzhu and the triangular structure of Keliyang, in the (semi)original
sequences of the Paleozoic erathem, there developed a series of imbricate thrust
structures. Due to structural movement in the deeper position, this might have
formed a series of compressive anticlines, such as the Kekeya anticline, Heshitake
anticline and Guman anticline.

Basement-Involved Type Ramp Anticline+Imbricate Thrust Sheet.   A
basement-involved ramp anticline was formed at the mountain front, and an
imbricate thrust sheet developed at the frontier of the ramp anticline. The width of
this type of foreland thrust belt is relatively narrow and the distance of
compressive shrinkage is limited.

Multiple Combination Structure.   Actually, this is the combination of
previous types that we discussed above. In the southern Hetian area of the
southwestern Tarim basin, north of the Tiekelike overthrust block, there is a large
scale multi-zigzag growth fault and bending fold of southern Hetian. Continuing
northward, there is a superimposed structure, which consists of several overthrust
sheets that overlay collectively in piggy back format. In addition, the overthrust
block of southern Hetian tops this piggy back structure. Keeping traveling toward
the north, imbricate thrust faults have formed in the Paleozoic erathem. Overall,
this demonstrates an intricate, well balanced, deformed pattern of thick-skinned
trust fault–ramp anticline–structure wedge combination-imbricate thrust fault.

Diapir Structure.   Diapir structure is also a common structural type in a
foreland thrust belt. For example, it occurs in anticlines in the Dawanqi area in
Kuqa and in the Dushanzi area in the southern Junggar basin.

Reversed Structure.   In a foreland thrust belt, a reversed normal fault is
commonly visible. The normal fault of an early period might reverse into a thrust
fault in the following period. Also, the thrust fault could be counter-reversed in a
later period. The western margin of the Ordos basin is an example.

A foreland thrust belt not only has horizontal and vertical transformations, but
also might be altered by a strike-slip fault in general. In the area where the
strike-slip fault passed through (such as the Aerjin fault, or the Talasi–Feierganna

fault), it usually superimposed a strike-slip fault on top of the overthrust structure, which produced an even more complicated structurally deformed pattern.

9.3.3 Deformed Patterns of Foreland Thrust Belt

In a foreland thrust belt, besides developing some salt domes and mudstone diapir, the structural deformation that happens is more often brittle deformation or brittle-ductile deformation. Thus, some fault-related fold structures or superimposed structures of fault-related folds have a variety of deformed patterns (Table 9.5).

- **Fault-Related Fold**

In the foreland thrust belt, because the strata are non-uniform, it is a common phenomenon that a fault has a curved fault plane. Among the structural patterns, a stepwise thrust fault that consists of a fault flat and fault ramp is a common structural pattern. When the strata of a hanging wall move along a fault flat and fault ramp, there might correspondingly be formed a ramp fold related to the fault, which is called a fault-related fold and which has three periphery types.

　　Fault-Bend Fold.   When the strata of a hanging wall move along the turning point of an underneath fault, a fault-band fold might form. To adjust its sliding motion along the fault plane, the strata of the hanging wall formed a kink fold. The geometry of a fault-bend fold was first proposed by Rich (1934), when he studied the mechanism of a low angle thrust fault in the Appalachian mountains. Half a century later, Suppe (1983b) quantified the geometry, set up a geometrical relationship between the patterns of fault and fold, and proposed a motional model of fault sliding and fold developing.

　　Fault-Propagation Fold.   The folds that were formed at the end point of a thrust fault were called 'Leading edge folds' (Boyer and Elljol, 1982) or 'Tip anticlines' (Mitra and Namson, 1989). This kind of folding activity absorbed sliding energy. The basic characteristics of the fault-propagate fold are as follows: it has an asymmetrical pattern, the front lateral is steep and narrow and the back lateral is wide and gentle. The syncline is fixed at the end of a fault. With an increase in depth, the fold is compacted. The splitting point of the fold axial plane and the end point of the fault are in the same layer of strata. The distance between the end point of the fold axial plane on the fault plane and the turning point of the fault is the sliding distance of the fault in a dipping direction. The sliding distance of a fault decreases upwards.

　　Detachment Fold.   This is a contraction fold that formed on top of a detached litho bed or on top of several detached litho beds (Poblet et al., 1995; 1996). Because the displacement of a parallel fault was transferred into the fold that is located in the strata of a hanging wall, it formed a detachment fold. Therefore, it does not have a direct relationship with the fault ramp. However, it is

Table 9.5   Major structural type of foreland thrust belt

Structural setting	Deformed characters	Basic type	Major features	Combination type	Note
Thin-skinned foreland thrust belt — Fault-related fold / Relationship of deformation and deposit	Post-deposit deformation	Fault-bend fold (FBF)	Occurred above a fault bending area	Fold with multi-bended fault; Fold with bended imbricate fault; Dual thrust structures with structural wedge;	Three basic end points and their superimposed combination
		Fault-propagation fold (FPF)	Formed above the end of a propagated fault	The combination of bended fault, propagated fault, and detachment fault; Brittle thrust structure; Interference fold	
		Detachment fold (DF)	Developed above the end of a thrust fault inside a weak bed		Analyze tectonic history and folding mechanism. It clearly differentiates from drape fold and compaction fold
	Synchro-deposit deformation	Growth fault-bend fold; Growth fault-propagation fold; Growth fault detachment fold	Growth triangle with various patterns	Multiple growth stages	Compression-distortion environment?
	Strata have motions between the layers	Shear fault-bend fold (SFBF): Simple shear fault; pure shear fault	There is a long and gentle back tail; its dip angle is smaller than fault ramp angle		
		Salt structure	Thick salt layer float, turtle folding structure, salt welding structure	Might be overlay on top of fault-related fold?	
	Plastic deformational structure	Mudstone diapir structure			Such as the Dushanzi anticline
		Dual plastic structure (Mushwad)	Deformation propagates to food wall		Occurred at thrust belt of the Appalachian Mountains
Thick-skinned foreland thrust belt	Basement-involved structure	Basement-involved type fault-related fold	Deformed triangle zone transmitted upward	Might be superimposed with shallow layer fault-propagation fold	Such as northern margin of the Qaidam basin
	Reversed structure	Positive reversed structure	Alteration of the motion direction		

related to the distributional deformation above the detachment fault. A detachment fold has four basic features: the basal weak bed may be thickened in the core area of the fold; there is the basal detachment fault and the competent bed prior to folding does not change the thickness and length during the deformation process. Towards the top of the fold, the thickness of growth strata has been reduced and the lateral of the fold has rotated in a fan shape. Generally speaking, when the thickness of a basal weak bed is relatively thick, it is called a detachment fold, which can be a symmetrical fold or an asymmetrical fold and which is above the detachment fault. When the thickness of a basal weak bed is relatively thin, it is called a pop-up fold or a lift-off fold, which is a symmetrical fold with a steep lateral and which is above the detachment fault. The progress of a pop-up fold is due to the development of an isoclinal fold that happened in the detachment bed at the core of the fold. When the crest of the fold is flat, it is also called a box fold.

● **Foreland Basement-Involved Type Structure**

In a foreland structural environment, basement-involved type deformations often developed. Because the displacement of basal faults propagated upward, this type of structure was formed. It occurred near to overlying strata or it was near to the interface between the formation and basement (including a small amount of basal strata), which involved both overlying strata and basal strata. The folding patterns primarily appear in overlying strata. The activity of a basal fault (a major fault) triggered structural movement to form basement-involved type deformations. This activity was associated with a major fault in the basement, which extended into the overlying strata and which might vanish in the overlying strata. On the surface, it usually appears as a monocline. In the cross section, it has significant structural undulation. However, sedimentary sequences of overlying strata are uninterrupted. When the hanging wall of a major fault was uplifting, the overlying strata that covered the edge of the basement fault formed a steep lateral of a fold. The strata of a steep lateral might be overturned. The deformations in the overlying strata were concentrated at the steep lateral that might either be shortened or prolonged. On the hanging wall of the major fault, during the deformation process the overlying strata and the basement strata might either be fused together or be separated. The basement strata could be a rigidity block, or it might be folded or be sheared. In the overlying strata, the deformation primarily occurred inside the triangular zone that was broadened upwards.

● **Conjoined Structure**

If two (or more) thrust faults with related thrust sheets combined together, they formed a thrust system. In the thrust belt, geometry, kinematics and dynamics are closely related to each other. It was first proposed by Dahlstrom (1969); then it was revised by Boyer and Elliott (1982), Mitra (1986; 1992). It includes imbricate structure, structural wedges and interference structures.

Imbricate Structure.   This is the combination of two (or more) thrust sheets. The thrust sheets might propagate in a break-forward style, or they might propagate in a break-backward style. In addition, this could be the synchronized activities of deeper and shallower faults. An imbricate fault-bend fold usually has more than two fault ramps. When passing by these fault ramps, the dip angle of the strata would be changed. Thus, there are multiple clinal domains (or isoclinal areas) that occur in the front lateral and back lateral of the fold.

Structural Wedges.   This is a triangle that is encircled by two sections of different faults, or these are wedge shaped faulted blocks. The sections of faults can both be fault ramps, or one is a fault ramp, the other is a fault flat. The end point of the wedge is the crosspoint of two faults. The sliding motion along both sections of faults adjusted the deformation that was caused by propagation of the end point of the structural wedge, which formed a fold. Its basic characteristics are as follows: the existence of a synchronized fore-thrust and back-thrust; the fold occurred along the active axial plane that is fixed at the end point of the structural wedge; it would increase the structural undulation if the fold occurred at the foot wall of the back-thrust. The size of structural wedges varies over a broad range. A large size structural wedge is a common structure on the chain of the mountain front, which are also called triangle zones by Gordy or by Jones. Inside a large size structural wedge, several small structural wedges might be found. Sliding along the underlying fault in a structural wedge, a fault-related fold might form. A dual structural system often formed if multiple faults developed inside a wedge. The shape of the original fault, the direction of propagation of the end point of the structural wedge and the mechanism of folding, all influence the pattern of the structural wedge.

Interference Structures.   When two (or more) monocline style kink fold belts cross each other, they form interference structures. In the cross section, a typical anticline structure on top of a syncline often formed, which is usually called a 'rabbit ear' structure. Because the interference of the kink fold could occur in many types of fault-related folds, therefore the structural patterns are varied. Interference structures might appear in many places, for example above two turning points of a fault, above the imbricate faults or at the front lateral of fault-related folds that incline in the same direction.

9.3.4   Major Foreland Thrust Belts in China

In western-central China, because ancient craton plates had irregular boundaries and non-uniform frameworks in the basement and because the tectonic activity of the late period was very strong, the deformed foreland thrust belts on the boundary of each plate were therefore extremely complicated and the alternations of the inner plates were various. Furthermore, because ancient orogenic belts also had non-uniform boundaries and non-uniform stress fields, while structural deformations were propagated to surrounding areas of the Tibetan plateau as compressive overthrust faults, they were also propagated as a shear slide-thrust fault in some areas (Li et al., 2007).

In foreland thrust belts at the following locations of western Sichuan, northern Sichuan, Jiuquan, southwestern Tarim, the deformed patterns inside overlying strata were controlled by weak strata of a gypsum-salt layer and muddy shale. A fault-bend fold, fault propagation fold and detachment fold mainly developed. Additionally, in a combination of the three former types of fold, a variety of fault-related folds were also developed (Suppe, 1983a; 1983b; Shaw, 1993; Shaw et al., 1994a; 1994b; 1996; 1999; 2004). In the foreland thrust belt that had massive deposits of gypsum-salt layers, the weak strata were a primary factor that controlled structural deformation. In an isolated location, under compressive stress, the gypsum-salt layer might be thickened by a horizontal plastic flow, which formed the structure that related to the salt layer deformation. The salt related structure that was controlled by gypsum-salt layers and gypsum mudstone was clearly visible in the thrust belt of Kuqa (Tang et al., 2004). It possibly existed in southern Junggar, southwestern Tarim and the Tuha area. The deformed propagation type of shear slide-thrust was primarily distributed inside the thrust belts in the following areas: southern Junggar, southeastern Tarim, Qaidam and the mountain front of Kunlun mountain. Besides, it was controlled by the reactivation of an ancient orogeny, the development of a compressive-slide type of foreland thrust belt and was also related to the activities of the sliding fault in Aerjin and the eastern margin of Pamir. The echelon structures were distributed in ranges or arrays. Horizontally, the overthrust structure only spread over a narrow range. Vertically, the fault belt is very steep and, additionally, it shows a sliding tectonic movement parallel to the orogeny.

● **Foreland Thrust Belt of Kuqa**

The Kuqa thrust belt is located between the southern margin of the South Tianshan orogeny and the northern margin of the Tarim craton plate. It hosts typical salt-related structures and classic fault-related folds. Together, they formed three deformed structural layers above the salt, inside the salt, and under the salt, which have an enormous significance for the hydrocarbon accumulations and the structural deformation in this region (Fig. 9.17). In the Kuqa area, a few hundred meters of gypsum-salt layers were deposited in the Palaeogene and Neogene systems (accordingly axes T7/T6 in Fig. 9.17). Due to tectonic compression, the horizontal plastic flow might occur in some sections of the gypsum-salt layer, which not only locally enlarged the thickness of the gypsum-salt layer, but also formed a salt diapir structure. A brittle deformed fault-related fold mainly developed above the salt layers, for example, the fault-propagation fold that developed in the shallow strata of the northern Qiulitake structure north of the

Quele 1 well (Fig. 9.17). Inside the salt layer, the plastic flow caused uneven thickness of the salt bed. Above the thickened section of the salt layer, the strata had swollen. For example, in Fig. 9.17, the Dawanqi anticline and the southern Qiulitake anticline were formed by underlying the partly enlarged salt layer. Under the salt layer, thrust structures were predominantly developed, which included the structural combination of a fault-related fold in the following components: fault-propagation fold, imbricate structure, pop-up structure, dual structure, salt structural triangular zone, faulted anticline and faulted nose anticline. In the Kuqa thrust belt, the salt-related structural patterns include salt pillow, salt anticline, salt wall, salt stock, salt ridge, salt wedge, salt sheet and salt weld (Fig. 9.17).

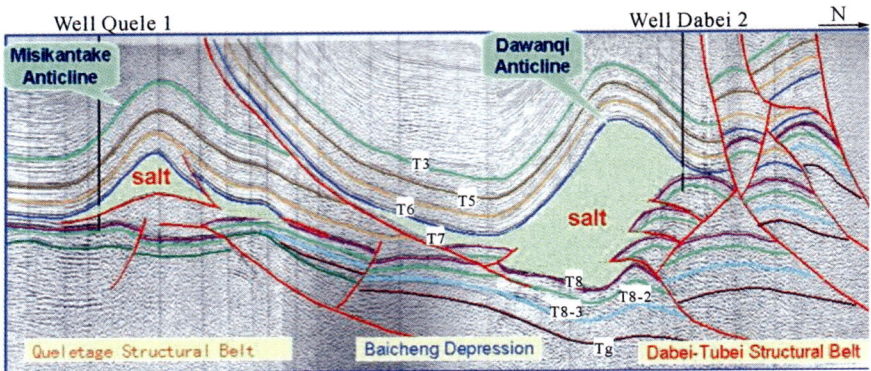

Fig. 9.17. Seismic cross section (N-S oriented) of the Kuqa thrust belt that shows salt-related structures and fault-related folds

● **Foreland Thrust Belt of Southern Junggar**

The foreland thrust belt of southern Junggar is located between the northern margin of the Southern Tianshan orogeny and the southern margin of the Junggar craton plate; the major structural deformation of this thrust belt is a fault-related fold. Horizontally, the thrust belt of southern Junggar and the boundary fault of Tianshan mountain are both oriented in an NWW direction. At the mountain front, three arrays of structural belts that are oriented in an approximately W-E direction obliquely intersect with the boundary fault of the basin and range. Under the influence of the collision between the India plate and the Tibetan plateau, the pressure direction of the compressive stress field is N-S. Thus, the N-S compressed structure predominantly developed, but it is also W-E oriented, a right sliding fault. Presumably, the boundary fault has the feature of an oblique thrust. However, in the thrust belt the structural deformations are mostly positive compressive thrust faults. In the foreland belt of southern Junggar, three arrays of anticline were clearly distributed in an echelon pattern. On the N-S oriented seismic cross section, there is not only a distinctive overthrust structure, but there also developed some high angle sliding structures that appear as a flower pattern in the cross section (Fig. 9.18). For example, Fig. 9.18 shows the feature of structural deformation. The thrust direction is from the northern margin of Tianshan mountain to the inner Junggar basin. The second and third structural belts of the inner basin can be divided into upper and lower structural systems. The upper structure system pushed overlying strata out to the surface from the south to the north, which formed a fault-propagation fold. The thrust fault plane is very steep at shallow depth, but it became a gentle slope at a lower depth. Furthermore, the detachment also occurred along the strata bed at a lower depth. The lower structure system is a structural wedge that consists of a fault-bend fold (from the south to the north) and a back thrust fault that vertically cut into the strata.

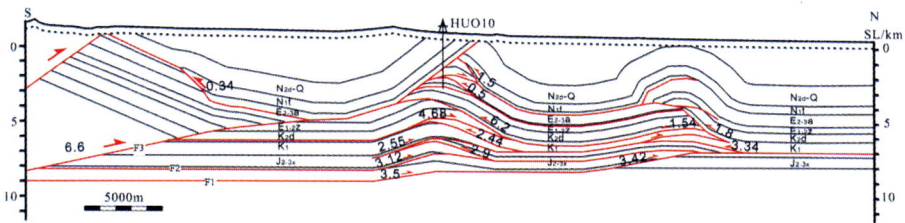

Fig. 9.18.   Geological cross sections (N-S oriented) of the thrust belt in southern Junggar basin, which show the deformed characters of thrusted-slided structure (By interpretation of Seismic NY200201)

J_2x, formation Xishanyao in Jurassic; K_1tg, Group Tugulu in lower Cretaceous; K_2d, formation Donggouzu in upper Cretaceous; E_{1+2}, formation Ziniquanzi in lower Tertiary; E_{2+3}, formation Anjihai in upper Tertiary; N_1s, formation Shawanzu in upper Tertiary; N_1t, formation Taxihe in upper Tertiary; N_2d, formation Dushanzi in upper Tertiary; Q, Quaternary

● Foreland Thrust Belt of the Northern Margin of Qaidam

The foreland thrust belt of the northern margin of Qaidam is located at the junction of the southern margin of the Qilian mountain orogeny and the northern margin of the Qaidam basin, which is a typical basement-involved type sliding-thrust structure (Wei et al., 2005) (Fig. 9.19). The basement deformation influenced overlying strata to join the tectonic processing. The structural deformation of overlying strata was controlled by the basement structural deformation; thus, both shallow and deep structures are coordinated very well. The main characteristics of the basement-involved structure are: (1) it was influenced by the Himalayan tectonic movement; (2) vertically, the thrust belt was formed by basement deformation with a poorly developed detachment plane and flower shape structural pattern; (3) horizontally, the deformation affected a relatively large area toward the inner basin; (4) thrust faults or folds were distributed in an echelon pattern and they obliquely intersected with the ancient orogeny.

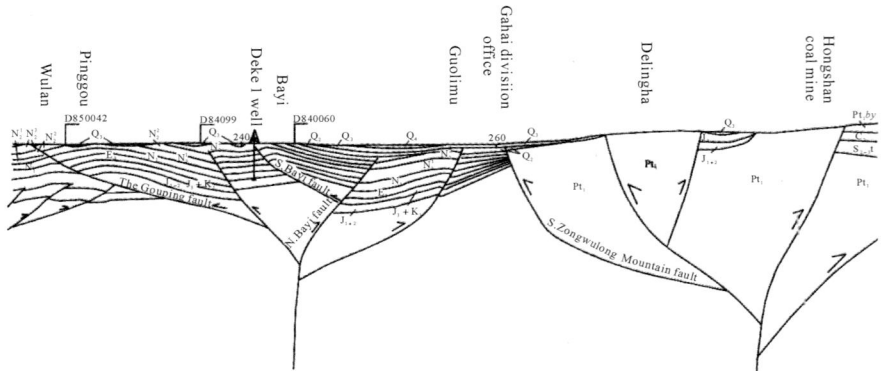

Fig. 9.19. Structural cross section (N-S oriented) in the northern margin of Qaidam basin, which shows the characteristics of basement-involved type thrusted-slided structure
Pt, Protozoic; J, Jurassic; K, Cretaceous; E, lower Tertiary; N, upper Tertiary; Q, Quaternary

9.4   Characteristics of Petroleum Geology in the Sinitic Foreland Thrust Belts and Patterns of Hydrocarbon Accumulation

We will introduce settings and features of petroleum geology in a foreland basin, geological attributes of categorized Chinese foreland basin, and oil and gas accumulation pattern of foreland thrust belts in China.

9.4.1   Settings and Features of Petroleum Geology in Foreland Basins

● **Massive Terrestrial Deposits and Foreland Sequences Interbedded with Non-Foreland Sequences**

In western-central China, the foreland basins of the Mesozoic–Cenozoic eras were developed at the conjunction of converged ancient orogenies and ancient plates (or terranes). Because of reactivation of the stress field inside plates, the ancient plates were thrust toward original terranes (or a new inner-continental basin) along the fault belts on the margins of ancient plates (or inside the ancient plates). The flexural and loading events happened on the front margin of the thrust belt, which formed massive sedimentary deposits. Therefore, this kind of basin is lacking an underlying foreland marine deposit. In the mountain front of Longmen mountain in western Sichuan, although the foreland basin has deposits of marine facies (Maantang formation) and deposits of marine-terrigenous facies (Xiaotang formation), its tectonic background should belong to a back-arc basin. The actual foreland basin deposit is the Xujiahe formation (T_3^3). After the folding-orogenic tectonic event, the back-arc basin of Songpan-Ganzi (T_3^{1-2}) thrust toward the Yangzi craton; the Xujiahe formation was the result of this thrust event. The foreland basin has multiple developmental intervals. Between these intervals, it developed a faulted basin or depressed basin and received different kinds of sedimentary deposits. Therefore, the foreland sequences and non-foreland sequences were interbedded with each other with different superimposed styles. Some sequences were smoothly correlated, and others were obliquely intersected with a high angle. The contact features of these sequences decided different combinations of petroleum generation-reservoir-seal systems in western-central China.

The following are the characteristics of sedimentary deposits of the inner continental foreland basin in western-central China:

(1) The distribution of the basins is usually along the front margin of the (reactivated) orogeny in the form of an elongated belt (the ratios of length to width are 5:1 – 3:1). Horizontally, some of foreland basins were connected with an inner-continental depressed basin at the far end of the orogeny. For example, at the boundary of the basins, there are transitional facies between the foreland basin of northwestern Sichuan (T_3^3-K) and the inner-continental depressed basin of central-eastern Sichuan. A similar phenomenon also occurs between the foreland basin of the Late Triassic epoch on the western margin of the Ordos and the inner-continental depressed basin of the Huabei in the same developmental interval.

(2) Vertically, the basins have an asymmetrical pattern. Near the orogeny side, the thickness of strata is clearly enlarged, which appears as a wedge in the cross section. Because the orogenies were persistently subducted in the direction of the basins, the boundary of the foreland basin and the foreland basin itself also continually migrated in the direction of the basin. The migration of sedimentary facies is the noticeable evidence, especially the migration of alluvial fan facies and fluvial delta facies. For example, the development of the foreland basin of western Sichuan was closely related to the southeastward migration of the Longmen mountain orogeny. The change of conglomerate components and the shifting of the depositional center demonstrate that it is a front-propagated basin that developed from the backland into the foreland. In this basin, there are three major thrust belts: 1) The Qingchuan–Maowen thrust belt started its activity at the beginning of the Jurassic period, which preserved the Early Jurassic conglomerate. The pebbles predominantly consist of Triassic and late Paleozoic rocks. Furthermore, it deposited a sandy conglomerate of braided river facies in front of the fault. 2) The structural activity of the Beichuan–Yingxiu thrust belt widely produced deposits of alluvial fan facies during the Late Jurassic–Early Cretaceous epochs. It continued thrust activity until the interval of the Late Cretaceous–Palaeogene period. The depositional center migrated from the northern section to the south-central section of Longmen mountain. The conglomerate consists of lightly metamorphic rocks of the early and late Paleozoic era, which revealed

thrust activity of the Yanshanian tectonic movement. 3) The Guanxian–Anxian thrust belt was formed during the Neogene period. The Dayi conglomerate reveals the tectonic activity of this thrust belt. Beside the Paleozoic pebbles, it also contains granite pebbles and highly metamorphic rocks of the per-Sinian system. These conglomerates demonstrated a forceful orogenic movement of the Himalayan tectonic period. Therefore, this lithological evidence demonstrates that the geologic age of the Longmen mountain thrust belt gradually became younger in the direction northwest-southeast. It not only progressed in a front-propagated style, but also formed a piggy-back basin and front margin basin (Liu and Liang, 1994). Other examples are the Jiuxi basin and Jiudong basin that are located at the Gansu Corridor, which do not have the strata of the Eocene-Upper Cretaceous series. During the Oligocene epoch, Qilian mountain thrust along the Northern Qilian fault in the direction southwest-northeast and this formed the Cenozoic flexural basin with fluviolacustrine facies deposits.

During the end of the Pliocene-beginning of the Pleistocene, the Tertiary system was deformed. Additionally, a fault was formed on the northern margin of North Qilian mountain with the strike in a northwest direction. During the early Pleistocene epoch, the deposits of the Yumen formation occurred on the northern side of a new fault. During the middle Pleistocene epoch, the Yumen group was deformed; another new fault was formed on the northern side of the Yumen group and it controlled a depositional center (Jiuquan group) of this time interval. During the late Pleistocene and Holocene, a tectonic movement tilted the Jiuquan group with mild folding. The alluvial fan also migrated toward the center of the basin (Yang et al., 2007).

(3) Sedimentary deposits predominately belong to terrestrial facies. Overall, we see an inverse cycle of sedimentary facies sequences in ascending order with gradually decreasing water depth and increasing grain sizes. Because they were influenced by intermittent orogenic activity, the coarse grain sedimentary deposits of the alluvial fan and fan delta (molasse deposits) irregularly appeared among the sedimentary sequences of fluviolacustrine facies. Vertically, in the foreland basin, the components of pebbles in the conglomerate of alluvial fan facies were inverted sequences of the source area, which were also called inverse sequences eroding sedimentation (Liu and Liang, 1994). In the foreland basin of western Sichuan, the geologic age of pebbles in the conglomerate gradually becomes older in ascending order, which is the opposite of the age order when compared with the sequences in the thrust belt.

(4) The sedimentary sources of the inner-continental foreland basin included wide-ranging of materials, which not only came from the orogeny of the island-arc, but also from the reactivated orogeny that was located on an active continental margin (such as foreland basins of the Kuqa and southwestern Tarim). The materials could also come from the folding orogeny of the latter period on the passive continental margin (for example, the foreland basin on the western margin of Ordos). Compared with a typical foreland basin, the characteristics of source

material that came from the inner-continental foreland basin are much more complicated.

(5) Some foreland basins are superimposed basins. For example, in the Kuqa area on the northern margin of the Tarim basin, a foreland basin developed during the Late Permian epoch–Triassic period and a graben-depressed basin was superimposed during the Jurassic–Cretaceous periods. However, it progressed back to a foreland basin after the Tertiary period. The foreland sequences and non-foreland sequences were deposited alternately. This feature is clearly identifiable in the following areas: southwestern Tarim basin, southern Junggar basin and northern margin of Qilian mountain.

- **Hydrocarbon Sequences of Extensional Tectonic Environment in Foreland Basin**

Passive Continental Margin Basin of the Late Paleozoic Era

A foreland basin is the major field for petroleum exploration. Usually, geologists consider it is the most beneficial type of basin for oil and gas generation. If the foreland basin only has a compressive stress field, it is not actually favorable for hydrocarbon source rock development. The major hydrocarbon source rocks were deposited in the passive continental margin, rifting valley or graben basin prior to the development of the foreland basin. For example, in the foreland basin of Zagros, hydrocarbon source rocks were 100% from the underlying marine strata, which contains an oil reserve of 238×10^8 t and a natural gas reserve of $17\times10^{12} - 18\times10^{12}$ m^3. According to the sequences of a foreland basin, in the overseas foreland basins the conditions of petroleum generation and accumulation were based on a flysch formation that was developed on the passive continental margin in the earlier period and which offered high quality hydrocarbon source rock sequences. The massive overlying molasses formation offered sand bodies to form a petroleum reservoir. Furthermore, deeply buried source rocks and the heating process accelerated the progression of hydrocarbon maturity. During the same geological period, oil and gas that were generated by hydrocarbon source rock and the traps of compressive structure together formed an excellent timing-spacious geological setting. Therefore, the foreland basin became the ideal area for oil and gas exploration.

The foreland basins of western-central China also have the beneficial features that are discussed above. This region was predominantly influenced by tectonic activities of the southern margin of the Tethys oceanic basin, which were the rifting event, subduction event, convergence event and collision event. Two developmental phrases of the Tethys ocean controlled the alternation between two different types of basins in western-central China. During the late Paleozoic, the regional extension structure formed a rifting valley and miniature oceanic basin. It also deposited sediments on the passive continental margin. The re-genesis basin of western China was controlled by the subduction event of the Tethys ocean during its early developmental period during the Jurassic. In the back-arc extension environment, fine grain coal measures were deposited that formed a suite of beneficial saprogenic source rock for natural gas. Because most extensional basins had been subducted under the orogeny, an understanding of the passive continental margin that is under the orogeny will enhance the quantity of the natural gas reserve. On the passive continental margin, because the slope area has well developed reservoir sequences, petroleum exploration may be carried out at a reasonable depth. However, the significance of hydrocarbon source rock in foreland sequences of the Indosinian tectonic period should be considered seriously. For example, in the Kuqa depression, beside hydrocarbon source rocks of the Middle and Upper Triassic, geologists have become aware of the existence of Paleozoic hydrocarbon source rock under the Mesozoic system. The natural gas source rocks of the southern Junggar basin predominately come from pre-foreland sequences of the Carboniferous–Permian systems and the Jurassic. The distribution of the natural gas reservoir was determined according to where the natural gas was generated. In the western Sichuan basin, natural gas in the Jialingjiang formation comes from underlying marine sequences; the Jialingjiang formation is located in the lower portion of the Zhongba natural gas field. Beside source rock of the Xujiahe formation in the Upper Triassic, the source rock of the upper Paleozoic erathem (C-P) might be another important natural gas generating source. Therefore, in the Sinitic foreland basin, beside abundant hydrocarbon source rock of coal measures in the Triassic–Jurassic, late Paleozoic source rock may possibly exist, which will increase the quantity of the natural gas reserve.

In the western Sichuan area, the strata of the Devonian system are nearly 6,000 m thick. The strata were controlled by a series of N-E oriented growth extensional normal faults that were developed between central Sichuan (Leshan–Longnusi) and Longmen mountain. These strata formed the passive continental margin (the western portion was thicker than the eastern portion), which developed from the paleo-continent of Yangzi to the slope of the paleo-Tethys ocean. Up to now, the following are the hydrocarbon source rocks that have been recognized: in the Lower Permian series, the mudstone contains 1.18% of organic carbon and carbonate rock contains 0.55% of organic carbon; on the other hand, the Upper Permian series contains 2.3% of organic carbon, but the carbonate rock contains 0.7% of organic carbon. In the Lower Permian series, the major reservoirs strata are the Shengxietan dolomite of the Qixia formation (P_1x) that occurs from Guangyuan–Mianzhu in northwestern Sichuan and the dolomite of the Maokou formation (P_1m) that appears at Jiangwuchang and Luzhou in southwestern Sichuan. In total it contains natural gas reserves of 400×10^9 m^3. In contrast, the hydrocarbon source rocks and the reservoirs in the Devonian system need to be further studied. Recently, in western Sichuan, drilling activity has discovered commercially acceptable natural gas shows that were supported by the Paleozoic source rock. In the Hewanchang structure, the He 1 well discovered natural gas in the Feixianguan formation (T_1f) of the Lower Triassic and the natural gas production rate is 2.73×10^4 m^3/d. Also in the same well, the Wujiaping formation in the Lower Permian series has a natural gas production rate of 49.95×10^4 m^3/d. In the He 3 well, the Maokou formation (P_1m) of the Lower Permian series has a natural gas production rate of 37.0×10^4 m^3/d. In the Heshen 1 well, the Ordovician system has a natural gas production rate of 1.88×10^4 m^3/d. In the Shejianhe structure, during the Jian 1 well drilling, natural gas flows have been discovered in the Maokou formation (P_1m) of the Lower Permian series and the Feixianguan formation (T_1f) of the Lower Triassic. Furthermore, in western Sichuan, a total of 81 oil and bitumen seeps and 4 natural gas seeps have appeared in the northern section of Longmen mountain.

In the southwestern Tarim basin, during the late Paleozoic (C-P), extensional faults were located between the Kunlun ocean and the Tarim craton plate. These faults controlled the sedimentary deposits on the passive continental margin, on the continental shelf and the continental slope. The following are the details of major hydrocarbon source rocks in the Carboniferous and the Permian strata. In the Lower Carboniferous (C_1), the strata of the Heshilafu formation II is 400 m thick, which produces oil on the western side and produces gas on the eastern side, TOC: 1% – 5.98%, R_o: 1.9% – 2.7%; in the Kalashayi formation, the mudstone is 100 m thick, TOC: 1% – 3%, R_o: 1% – 1.5%. In the Permian strata, in the Yang 1 well of the Kekeya oil and gas field, the mudstone is 100 m thick, TOC: 1% – 2.04%, R_o: 0.6% – 1.0%. The hydrocarbon–rich strata belt of the Carboniferous and the Permian (C-P) occurs on the upper slope of the Maigaiti mountain with abruptly changed lithology and a well developed reservoir sequence. Both ancient uplifts of the Qunke–Qiake and the Hetian–Minfeng have hydrocarbon-rich shoals and sand bars. The area of lithologic trap is about 6,500 km^2; the petroleum reserve is about 13×10^8 t. Up to date, the discoveries include the Qun 5 well (the natural gas source rock is Cambrian strata) and the Kekeya oil and gas field (the hydrocarbon source rock is Permian strata).

On the western margin of Ordos, during the early Paleozoic, the Helan mountain–Liupan mountain region went through a series of different developmental events. From the Cambrian–Ordovician (Є-O), it experienced a depressed rifting trough development and this time interval also equaled the Qinling–Qilian ocean tectonic period. During the Carboniferous–Permian (C-P), it experienced rifting basin development. In the Helan mountain area, the thickness of Carboniferous strata is 1,719 m. However, inside the basin, the thickness of the same period strata is only 140 m. On the western margin of the Ordos, the thickness of Permian strata is 1,179 m. Conversely, inside the basin, the thickness of the same period strata is only 700 m. In the Middle Ordovician series (O_2), black shale is the primary hydrocarbon source rock, TOC: 0.72% – 1.2%. However, the late Paleozoic hydrocarbon source rock study is limited. The reservoir sequences (which are from the early Paleozoic strata, Pz_1) mainly consist of granular limestone of shoal facies, secondary dolomite and cavity dolomite. The petroleum exploration area covered 50,000 km^2; the discovered natural gas reserves are

$3,500 \times 10^8$ m^3 approximately. In the Tianchi structure, a commercial quality natural gas field has been discovered and the natural gas comes from the lower Paleozoic strata.

Settings of Hydrocarbon Source Rock of the Early and Middle Jurassic in Graben Basin

During its early developmental phase in the Jurassic period, in western China the re-genesis basin were controlled by the Tethys ocean subduction event in the back-arc extensional environment. Additionally, it received a suite of fine grain coal measures that was beneficial saprogenic source rocks for natural gas. During the Early and Middle Jurassic epochs, because the new Tethys oceanic crust was subducted northward, it triggered a back-arc extensional event that carried out the extension of the remaining oceanic basin of the Black Sea–Caspian Sea (back-arc oceanic basin). The broad range of marine progression occurred in the Tarim basin and the Qaidam basin. The high salinity sea water and the lagoon water preserved the organics of the early period (avoiding biodegradation and oxidation). These organics have a higher hydrocarbon production rate. The regional extension event of the Early and Middle Jurassic epochs developed a graben basin, pull-apart basin and depressed basin. After the Neogene period, the re-genesis basin system was formed on the foundations of the previous regional extension event.

The organic materials predominately consist of type III kerogen that came from coal measures and the natural gas supply is plentiful. In the Cenozoic re-genesis foreland basin, during the Early and Middle Jurassic epochs, the hydrocarbon source rocks of pre-foreland sequences were controlled by the paleo-climate, paleo-vegetation, paleo-geography and sedimentary facies. The strata in the Lower–Middle Jurassic are more than a hundred meters thick. The source rocks consist of coal measures that contain type III kerogen and dark color mudstone of lacustrine facies. The organic maceral of the source rocks is mostly made of inertinite with a less amount of exinite, which indicates a good quality natural gas source rock. The thickness of hydrocarbon source rock is about 200 – 800 m. The organic abundance is 1.5% – 3.5%; the maturity of the organic materials is from high maturity to post-maturity; the kerogen is of type III category. These characteristics indicate that the source rock is in the stage of natural gas generation, hence the natural gas supply is plentiful (Table 9.6). For example, in the Kuqa depression, the hydrocarbon source rocks in the Upper Triassic–Middle Jurassic mainly consist of vitrinite, type III kerogen. After it went through the heating process, the hydrocarbon source rock has been altered to produce natural gas and the highest productive intensity can reach 120×10^8 m^3/km^2. The total natural gas reserve is 2.8×10^{14} m^3, which is the material base for the Kela 2 and the Tuziluoke natural gas fields. In the western Sichuan basin, the hydrocarbon source rock is in coal measures in the Xujiahe formation in the Upper Triassic, which mainly consists of vitrinite, type III kerogen. Due to the heating process, the hydrocarbon source rock of coal measures has been altered to produce natural gas. The highest productive intensity can reach 150×10^8 m^3/km^2. The natural gas reserve is 1.6×10^{12} m^3, which is the basis for the Zhongba, the Baimamiao and the Pingluoba natural gas fields.

Table 9.6.   Features of hydrocarbon source rock in the Lower–Middle Jurassic in the Sinitic foreland basins

Region	Lithology	TOC (%)	"A" (%)	HC (ppm)	S_1+S_2 (mg/g)	Thickness of strata (m)
Northern margin of Qaidam	Coal	55.80	0.65	2,072	70.55	500 – 900
	Carbonaceous mudstone	12.06	1.474	5,542	45.20	
	Mudstone	2.02	0.0672	510	4.84	
	Oil shale	13.40	0.16	532	17.90	
Tuha	Coal	65.00	1	2,500	120.00	300 – 500
	Carbonaceous mudstone	17.50	0.15	550	42.20	
	Mudstone	1.65	0.05	240	2.10	
Kuqa	Coal	44.60	0.748	1,302	107.99	300 – 500
	Mudstone	2.53	0.108	739	3.92	
Southern Junggar	Coal	65.40	1.86	305		300 – 800
	Mudstone	5.70	0.05	0.2		

● **Natural Gas Accumulating Domain of Northern Margin Basin Group of Tethys in the Mesozoic Era**

During the early Paleozoic, the Chinese continent and its surrounding areas were individual miniature craton terrains that were divided by the Paleo-Asian ocean and the Paleo-Tethys ocean. At the time, Junggar, Tarim, Qaidam, Huabei and Yangzi were small terrains and they were drifting in the ancient oceans individually. Until the time interval of the late Paleozoic era–Triassic period, these individual small craton terrains converged to the southern margin of the Eurasian continent and the northern margin of the new Tethys ocean in successive order to form a united continent. In a later period, influenced by tectonic development on the southern margin of Tethys, this area became the eastern section of the northern margin basin group of the Tethys.

During the late Paleozoic era (P_2z-T_2), the Paleo-Tethys ocean was subducted northward along the Jinshajiang River. This tectonic event formed a back-arc basin of the southwestern Tarim and other basins. During the Late Triassic epoch (T_3), due to the Paleo-Tethys orogenic movement, the peripheral foreland basin and the back-arc foreland basin were developed. During the Early–Middle Jurassic epochs (J_{1-2}), the subduction events of the Bangong–Nujiang ocean and the Yaluzangbu ocean happened. A graben basin (or depressed basin) was formed by the back-arc extension movement and the basins received deposits of coal measures. During the Late Jurassic–Early Cretaceous epochs (J$_3$-K$_1$), the Bangong–Nujiang miniature oceanic basin was extinct and a gentle structural deformation occurred. During the Late Cretaceous epoch–the Palaeogene period (K$_2$-E), because of the western margin of the Tethys subduction and the influence of the remaining Yaluzangbu ocean, a marine progression happened in an isolated area. Moreover, the extensional event occurred with limitation and red color gypsum mudstone was deposited. Since the Neogene period, the re-genesis basin has been formed by the long distance influence of the new Tethys orogenic movement. Overall, in western-central China, due to the influence of the tectonic development of the Tethys, the following geologic events commonly occurred: foreland basin development of the Triassic period, graben (depressed) basin development with a deposit of coal measures of the Jurassic period, widespread gentle tectonic deformation of the Cretaceous period and the re-genesis basin development with molasses deposits of the Neogene period.

The northern margin basin group of the Tethys consists of the Mesozoic sedimentary basins that are located on the northern margin of the Tethys. These basins have a similar tectonic background, comparable sedimentation history and related structural deformation events (Jia et al., 2001b; Yang et al., 2004). During the Jurassic–Palaeogene periods, because the new Tethys ocean was subducted northward, the northern margin of the Tethys progressed into the back-arc environment on an active continental margin. The back-arc extensional movement formed a graben basin and depressed basin. During the same geologic interval, multiple marine progressions happened from the west to the east. Together, a depositional system of marine-terrigenous facies formed, which predominately includes coal measures of terrestrial facies, reef limestone of shallow marine facies, a gypsum-salt combination of lagoon facies, lacustrine delta facies and sandstone or mudstone of river facies. The sediments of different facies were deposited alternately. Furthermore, in a W-E direction, these alternating facies were deposited on a more extensive scale. The marine progressions were more immense on the western side than on the eastern side. The northern margin basin group of the Tethys is a united continental basin, which has the characteristics of gentle topography, short levitated distance between the basin and range and an indistinguishable boundary between the basins. After the Miocene epoch, because of persistent collision and compression between the Indian plate and the Eurasian plate and also because of the subduction that happened inside the continent, the Pamir region wedged northward and the northern margin basin group of the Tethys progressed into a re-genesis foreland basin stage. Because of the dynamic uplifting of the inner continent orogenies and structural compression, the foreland thrust belts were formed on the margins of these basins on an extensive scale. This kind of compressive structural deformation energetically happened in the Pamir arc and gradually occurred in deformed magnitude in both the eastern and the western laterals.

In the northern margin basins of the Tethys, the unified structural system formed a regional petroleum system. The foreland basin of the early period contains hydrocarbon source rocks of marine-terrigenous facies from the late Paleozoic era (C-P) and the reservoir-seal combination of the Triassic. The foreland basin of the late period contains the hydrocarbon source rock of coal measures in the Early and Middle Jurassic, the reservoir in the Cretaceous system and the cap rock of gypsum and mudstone in the Tertiary system. The northern margin basin group of the Tethys has plentiful petroleum resources, especially natural gas reserves. The Karakum basin has an explored natural gas reserve of $66,955 \times 10^8$ m^3; the basin of the adjacent Caspian Sea has an explored natural gas reserve of $46,140 \times 10^8$ m^3 and explored oil reserve of 30×10^8 t. The northern margin basin group of the Tethys is well-known worldwide as an oil and gas accumulating region. It is usually called the coal bed methane accumulating domain of central Asia (Dai and Li, 1995). For example, large scale oil and gas fields have been discovered in the Karakum basin, the Afghanistan–Tadjik basin and the Fergana basin. In western-central China, because the foreland thrust belts were part of the northern margin basin group of the Tethys during the Mesozoic era, they should have a similar petroleum exploration potential as foreign basins have. Up to date, the Kela 2 natural gas field has been discovered in the Kuqa foreland thrust belt, which has an explored natural gas reserve of $2,840 \times 10^8$ m^3; furthermore, natural gas fields are continually being discovered in the foreland thrust belts in the following areas: the western Sichuan basin, southern Junggar basin, southwestern Tarim basin and the northern margin of Qaidam basin.

- **Gigantic Structural Trap of the Himalayan Tectonic Period Formed Useful Timing-Specious Setting for Natural Gas Accumulation**

Reservoir-Cap Rock Combination of Foreland Basin from the Mesozoic to the Cenozoic

The classic rock of terrestrial facies contains well developed reservoir sequences. At the time the Mesozoic foreland basin was developed, the vertical distance between the basin and range was significantly great. In addition, the source of sedimentary material was near the basin and the transportation stream (or distance) was short. Thus, it primarily deposited massive sandstone reservoirs of fan dealt facies, braided delta facies, river facies and lake shore facies. Because these sandstones are relatively young in geological age, the strata had lower temperatures with poor diagenesis and mostly primary pores. For example, in the Kuqa depression, the sandstones of the Bashijiqike formation in the Cretaceous system are about 200 – 600 m thick, which are broadly distributed and which have a porosity of 12% – 18% and permeability of 30 – 200 mD. The widely distributed, high quality reservoirs in the Cretaceous system are the key component in the formation of a large scale natural gas field. In order to find the next large natural gas field, we need to discover a reservoir that has similar conditions to the one in the Kuqa depression.

   During the Palaeogene, the new Tethys ocean progressed into the Tarim and the Qaidam areas; under the hot and dry climate, red color gypsum mudstone was deposited, which was an excellent regional cap rock for petroleum reservoirs of terrestrial clastic rock. Equally, the preservation conditions for regional cap rock are another crucial factor in natural gas exploration in the Chinese foreland basin of the latter phase.

Hydrocarbon Accumulations of Late Period Were Better Preserved in the Mesozoic Foreland Basin

In western-central China, the hydrocarbon source rocks of coal measures in the Upper Triassic–Middle Jurassic were covered by massive deposits of the foreland basin of the Himalayan tectonic period. During the foreland compressive tectonic period, because of rapid sedimentary settlement, both the buried depth of the hydrocarbon source rock and strata temperature increased, which brought organic materials into the hydrocarbon generating peak stage (or high-temperature pyrolysis gas stage). Under the structural compression, natural gas migrated along strata beds or faults planes into the structural traps. The structural traps of the Himalayan tectonic period were arranged in arrays or ranges. Combining with the hydrocarbon generating depressions in same neighborhood, proficient spacious geologic settings were created for hydrocarbon accumulation. The foreland basins in western-central China were formed by the combination of the inner plate craton basin and marginal foreland basin that was on the edge of a plate. From an orogeny to a craton basin, in successive order, a foreland thrust belt, foredeep depression, front slope belt and front uplift developed. Because of plentiful supplies of natural gas and cooperative geologic settings, in the foreland basin various types of petroleum reservoir might develop in different structural units. An anticlinal gas reservoir was formed in the foreland thrust belt, the deep basin gas reservoir was formed in the foredeep depression, the stratigraphic trapped gas reservoir was formed on the front slope and the faulted anticline gas reservoir or faulted nose gas reservoir was formed in the front uplift. For example, the Cenozoic strata are about 4,000 – 7,000 m thick, which were rapidly deposited in southern Junggar, Kuqa, southwestern Tarim, western Qaidam, the northern margin of Qaidam and western Sichuan. The speedy depositional event not only increased the buried depth of hydrocarbon source rock, but it also increased the stratigraphic temperature gradient (the original strata temperature gradient was about 2.2 – 2.5 °C/100 m). Therefore, it accelerated the maturing process of organic materials into the thermal maturation stage (high maturity–post maturity stage) that predominantly produces natural gas. Except for Ordos and western Sichuan areas, where hydrocarbon generation high peaks from the coal measures were in the Cretaceous period, in other foreland basins (such as in Kuqa, southwestern Tarim, southern Junggar, Tuha and the northern margin of Qaidam), the hydrocarbon generation high peaks from the coal measures happened during the Himalayan tectonic period. Because hydrocarbons were generated during the late geological period, natural gas was therefore well preserved. During the Himalayan tectonic period, because gigantic over thrust structures developed, the structural traps were well developed in the mountain front, which benefited natural gas accumulation.

Gigantic Thrust Structural Traps Benefited Petroleum Accumulation

The foreland basins in foreign countries only experienced solo tectonic movement. Based on the continental margin that was developed in an earlier period, first the subduction of the oceanic crust was preceded, then the disappearance of the oceanic basin occurred. At the ending of the tectonic cycle, an orogeny was formed. Because of the compressive stress from the tectonic movement and overloading flexure from the sediments, the foreland basin was developed between the front margin of a plate and an orogeny. Conversely, the foreland basins of China experienced dual tectonic events during the Mesozoic and the Cenozoic eras, which were the Indosinian and the Himalayan tectonic movements and which formed foreland basins of two different geological ages. The foreland basins of the Indosinian tectonic period are located in the following areas: western Sichuan, northern Sichuan, the western margin of Ordos, southern Junggar, the northern margin of Bogeda, Tuha, Chuxiong, southern Tarim and northern Tarim. The foreland basins of the Himalayan tectonic period are located in the following areas: southern Junggar, Tuha, Kuqa and southwestern Tarim. The foreland thrust zones of the Himalayan tectonic period are located at northwestern Junggar, the northern margin of Bogeda, Kashi, southeastern Tarim, the northern margin of Qaidam, Jiuquan, Chaoshui, western Sichuan, northern Sichuan, Chuxiong and the western margin of Ordos. In the foreland thrust zone usually 2 – 3 arrays (or ranges) of structural belts developed that contain well formed structural traps. In contrast, the Chinese craton plates had a small dimension but the Chinese orogenies were of large magnitude. Because of powerful tectonic compression, some foreland basins of the Indosinian tectonic period were totally involved in the foreland thrust zone of the Himalayan tectonic period. For example, the foreland basins of the early developmental period were converted into foreland thrust belts during the Himalayan tectonic period in the following areas: the southern section of western Sichuan, the Kuqa depression, the western margin of Ordos, the Kashi depression and the northwestern margin of the Junggar. The structural traps in the foreland basins are beneficial to natural gas accumulations. Numerous faults are the best migration passages for the natural gas. Abnormally high pressures are not only beneficial to the oil and gas accumulations, but also indicate well preserved natural gas. The structural traps offer unlimited oil and gas storage spaces.

Recently, foreland thrust belts, especially the forefront of foreland thrust belts, are the most important regions for natural gas exploration. Considering the storage conditions, most oil reservoirs at the back margin of the thrust belts have been damaged, which only contain oil sweeps and bitumen. The traps under the overthrust block were well preserved, the structural fractures changed the condition of petroleum migration and accumulation and, because it was near to hydrocarbon generating sources, this kind of trap was beneficial to petroleum accumulations. In particular, because of dynamic thrust-compression movement in the Chinese foreland basin, the coal measures that were the major hydrocarbon source rock and muddy shale sequences have become the detachment layers. Hydrocarbon source rock has been superimposed several times inside the detachment layers, which increased the material source of hydrocarbon accumulations. Along the detachment layers, the foreland thrust belt pushed over in the direction of the craton. This positioned thrust belts directly on top of hydrocarbon source rock. Through the vertical faults inside the thrust belt, hydrocarbons might migrate and accumulate in the traps in the fault-related folds. Because it contains well developed structural traps with relatively shallow buried depth, a foreland thrust belt area is an advantageous region for drilling exploration. According to the conclusions of natural gas exploration, six large–medium size natural gas fields have been discovered in the foreland thrust belts. The total explored reserves of natural gas are $3,706 \times 10^8$ m^3, which is on top of the reserve list of different structural types. After favorable natural gas exploration in the Kuqa foreland thrust belt, the next natural gas exploration region in the Chinese foreland basin will be in the foreland thrust belt located at the forefront areas in particular. These areas include western Sichuan, southern Junggar, and the northern margin of Qaidam. However, in these areas the geological settings are more complicated and this adds to the difficulties of recognizing the structural traps.

As with foreign foreland basins, the foreland basins in China have the same beneficial conditions for petroleum accumulation. Under the extensional tectonic background, massive hydrocarbon source rocks were deposited that established a material supply center of natural gas. Jointly, the Jurassic coal measures and the Palaeogene gypsum mudstone preserved the petroleum generation system and reservoirs. During the compressive tectonic period, because of the rapid settlement of sediments, both the buried depth of hydrocarbon source rocks and the stratigraphic temperature had increased. Thus, hydrocarbon source rocks progressed into the hydrocarbon generation peak phase (or the phase of high-temperature pyrolysis gas). Under structural compression, and along the strata bedding and faulted plane, natural gas migrated into the structural traps. Structural traps of the Himalayan tectonic period combined with hydrocarbon generating depression in the vicinity to form proficient timing-spacious geologic settings to form a petroleum system. In western-central China, foreland basins were formed by the combination (or were superimposed) of an inner plate craton basin and the foreland basin on the margin of a plate. From the orogeny to the craton, in successive order, the foreland thrust belt, the foredeep depression, the foreslope and the front uplift developed.

- **Utilizing the Regional Detachment Surfaces to Establish Three Sets of the Reservoir-Cap Rock Combination Vertically**

Throughout western-central China, the coal measures in the Jurassic, the gypsum-salt combination in the Palaeogene system, or the mudstone in the Palaeogene system might be utilized as the boundaries for establishing three major sets of the reservoir-cap rock combinations, which are found in the Carboniferous–Triassics, the Jurassic–Cretaceous system and the Cenozoic erathem respectively.

Reservoir-Cap Rock Combination in the Carboniferous–Triassic. This combination is the oldest one in the regional geological timetable in the foreland thrust belt of the western-central region. For example, the thrust belt on the northwestern margin of the Junggar is the current oil reach area in the Junggar basin. In this area, nine oil fields have been discovered, which are the Kelamayi, Baikouquan, Hongshanzui, Wuerhe, Xiazijie, Chepaizi, Fengcheng, Mabei and Xiaoguai. The explored oil reserves were discovered in the Carboniferous system, the Permian system and the Triassic. In the Permian system, the explored oil reserves came from the Jiamuhe formation, Fengcheng formation, Xiazijie formation, lower Wuerhe formation and upper Wuerhe formation. In the Triassic, the explored oil reserves came from the Baikouquan formation, Kexia formation, the Keshang formation and Baijiantan formation. The oil reservoirs in the Carboniferous system mainly consist of sequences of volcanic facies. In the Permian system, beside the Jiamuhe formation that has an oil reservoir in volcanic–clastic volcanic rocks, the remaining formations have oil reservoirs in the sequences of alluvial fan facies and fan delta facies. In the Triassic, the sedimentary facies for the reservoir sequences vary within the individual formation, which are the sequences of alluvial fan facies in the Baikouquan formation, the sequences of fan delta facies and river facies in the upper and lower Kelamayi formation and the sequences of fan delta facies in the Baijiantan formation. As a result, on the northwestern margin of the Junggar basin, the reservoir sequences of volcanic facies were only developed in the Jiamuhe formation in the Carboniferous system, which contains 12.8% of total oil reserves in the region. Conversely, the reservoir sequences in the Permian and the Triassics contain 71% of total oil reserves in the strata of alluvial fan facies and fan delta facies. Thus, on the northwestern margin of the Junggar basin, the petroleum reservoirs are closely related to fan facies. The development of fan deposits were controlled by thrust fault activity. These alluvial fans or delta fans joined with overlying mudstone to form a useful combination of reservoir-cap rock.

Reservoir-Cap Rock Combination in the Jurassic–Cretaceous Systems. This combination commonly occurred in the following areas: Kuqa, southwestern Tarim, southern Junggar, the northern margin of the Tuha basin and the northern margin of Qaidam. For example, in the Kuqa depression, reservoir sequences in the Cretaceous system primarily consist of sandstone of river facies and delta front–lake shore facies; these sandstones occurred in the Bashijiqike formation (K_2b) and the Baxigai formation (K_1b). In the Lower Cretaceous series, the thickness of sandstone in the Baxigai formation is about 200 – 400 m; in the Upper Cretaceous series, the thickness of sandstone in the Bashijiqike formation is around 200 – 400 m, too. These two layers of sandstones form excellent reservoir units. Above the sandstone, the gypsum-salt combination in the Palaeogene system is the cap rock for this combination.

Reservoir-Cap Rock Combination in the Cenozoic Erathem. This combination frequently occurred in southwestern Tarim, southern Junggar, Kuqa, the southern margin of Qaidam. For example, in the thrust belt in southern Tarim, the sandstone of river-delta facies combined with overlying mudstone to form a valuable reservoir-cap rock combination. These sandstones occur in the Ziniquanzi formation ($E_{1-2}z$), the Shawan formation (N_1s) and the Taxihe formation (N_1t).

Among these three sets of reservoir-cap rock combinations, because the lower and middle combinations were near oil generating strata (mudstone of the Permian system, hydrocarbon source rock of coal measures or mudstone of the Triassic or the Jurassics), they formed a valuable spacious combination for a petroleum system. Although the upper combination was far away from oil generation strata, but had a better timing correlation with the oil generation peak, the trap development and fault activity came together in the same geologic time interval to benefit oil generation, migration and storage. Thus, it is the best timing-matched combination, which generated oil in the lower portion of the stratigraphic column and stored the oil in the upper portion of the stratigraphic column.

9.4.2   Geological Attributes of Categorized Chinese Foreland Basin

According to their developmental character, the foreland basins in western-central China might be classified into four different combination styles, which are superimposed style, modified style, presenility style and revival style. Because every combination style had its own developmental passage, the combinations of petroleum generation-reservoir-seal are very different.

● Petroleum System in the Superimposed Style Basin

The superimposed style basin was developed on both sides of the Tianshan orogeny and on the mountain front of Kunlun mountain; the re-genesis foreland basin of the Himalayan tectonic period was superimposed on top of the foreland basin of the Indosinian tectonic period; both foreland basins have a complete framework (Fig. 9.20). The major hydrocarbon source rocks in the superimposed style basin consist of two different stratigraphic sequences: one is the Paleozoic marine strata that are beneath the foreland basin of the Indosinian tectonic period; another one is the Early–Middle Jurassic coal measures under the re-genesis foreland basin of the Himalayan tectonic period (Fig. 9.20). During the Yanshanian tectonic period, the organic materials in the upper Paleozoic erathem changed into hydrocarbons. Because the source rocks were deeply buried by the thrust–subsiding events of the Himalayan tectonic movement, during the Neogene period the coal measures in the Lower–Middle Jurassic generated a large quantity of natural gas. For example, in the superimposed style foreland basin in the Kuqa area, three sets of hydrocarbon source rocks developed. Among these hydrocarbon source rocks, the ones in the Lower Permian series contributed a lower amount of hydrocarbons; the major hydrocarbon suppliers were located in the Middle and Upper Triassic (normally they were 200 – 600 m thick) and the Lower and Middle Jurassic (generally they were 300 – 700 m thick). These two sets of hydrocarbon source rocks in the Mesozoic erathem have a broader distribution range and great thickness. The muddy hydrocarbon source rocks in the Jurassic mostly belong to III_1 and III_2 types and contain 0.4% – 37.36% of organic carbons; on the other hand, the muddy hydrocarbon source rocks in the Triassic mostly belong to III_1–II types and contain 0.4% – 10.1% of organic carbons.

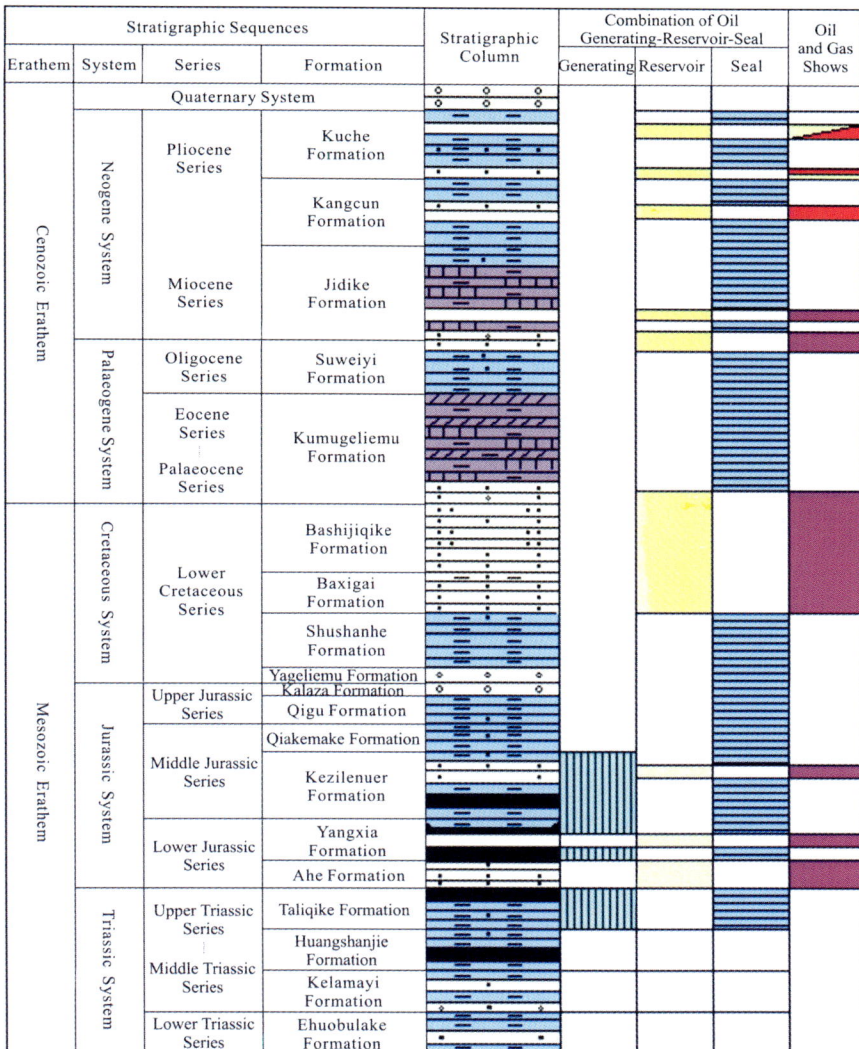

Stratigraphic Sequences				Stratigraphic Column	Combination of Oil Generating-Reservoir-Seal			Oil and Gas Shows
Erathem	System	Series	Formation		Generating	Reservoir	Seal	
		Quaternary System						
Cenozoic Erathem	Neogene System	Pliocene Series	Kuche Formation					
			Kangcun Formation					
		Miocene Series	Jidike Formation					
	Palaeogene System	Oligocene Series	Suweiyi Formation					
		Eocene Series	Kumugeliemu Formation					
		Palaeocene Series						
Mesozoic Erathem	Cretaceous System	Lower Cretaceous Series	Bashijiqike Formation					
			Baxigai Formation					
			Shushanhe Formation					
			Yageliemu Formation					
	Jurassic System	Upper Jurassic Series	Kalaza Formation					
			Qigu Formation					
		Middle Jurassic Series	Qiakemake Formation					
			Kezilenuer Formation					
		Lower Jurassic Series	Yangxia Formation					
			Ahe Formation					
	Triassic System	Upper Triassic Series	Taliqike Formation					
			Huangshanjie Formation					
		Middle Triassic Series	Kelamayi Formation					
		Lower Triassic Series	Ehuobulake Formation					

Fig. 9.20. Combination of petroleum generating-reservoir-cap rock in the Kuqa depression

Hypothetically speaking, in a superimposed style foreland basin, there are upper Paleozoic reservoir sequences of marine facies, which were not only covered by massive deposits from the re-genesis foreland basin of the Himalayan tectonic period, but were also damaged by diagenesis processing. Therefore, the primary reservoir sequences were the sandstones of the Mesozoic era. The cap rocks were coal measures in the Lower–Middle Jurassic and the red color gypsum mudstone in the Palaeogene system. In the thrust belt, the valuable source rock and trap and petroleum exploration activity, in combination, were the coal measures in the Lower–Middle Jurassic and structural traps beneath the detachment layer of the Palaeogene red color gypsum mudstone. Due to the influence of the late tectonic movement, the oil and gas might have been redistributed. Thus, the shallow petroleum reservoir might have been formed above the detachment layer of the Palaeogene red color gypsum mudstone, such as the Dawanqi and the Hutubi.

In the superimposed style basin, there are four kinds of reservoir-seal combinations, which are the combinations in the lower section (combinations I and II) and the combinations in the upper section (combinations III and IV). Combination III is the most important one in large size or super size petroleum exploration.

Combination I. In southwestern Tarim, there were passive continental margin deposits in the Carboniferous–early Permian, which not only formed the oldest hydrocarbon source rocks of marine facies or marine–terrigenous facies in the foreland basin, but also developed a self-efficient combination of petroleum generation-reservoir-seal. Both a gypsum-salt combination of marine facies and mudstones of marine-terrigenous facies were the regional cap rocks.

Combination II. The widely distributed Mesozoic hydrocarbon source rock of coal measures and regional cap rock of mudstone joined together and were overlaid one on the other to form a combination of petroleum generation-reservoir-seal. The self efficiency of oil generating and storage is the main characteristic of this type of petroleum system. For example, in the Kuqa area, the combination of petroleum generation-reservoir-seal was developed in the Triassic and the Jurassics. In southern Junggar, hydrocarbon source rocks of coal measures were developed in the Lower–Middle Jurassic and the interbedded sandstones were utilized as a petroleum reservoir. The Badaowan formation in the Jurassic operated as cap rock for this petroleum system, such as in the Qigu oil and gas field, which developed the Jurassic petroleum reservoir. Commonly, the reservoir strata in this type of combination have poor physical properties; for example, in the Kuqa and southern Junggar areas, the average porosity of reservoir strata is only 6% – 11%.

Combination III. This type of combination primarily developed in the Cretaceous–Palaeogene systems. For example, in the Kuqa foreland basin, the hydrocarbon source rocks occurred in the Triassic–Jurassics; the reservoir strata appeared in the Cretaceous and the Palaeogene systems with good physical properties (the average porosity was 12% – 16%); the cap rocks were developed in the Kumugeliemu formation, which consisted of a massive layer of gypsum (mudstone) that was deposited under the dry saline lake environment. The Kela 2, Dawanqi and Yaha are natural gas reservoirs in the Cretaceous system; the Yingmai 7 natural gas reservoir in the Palaeogene system belongs to this type of petroleum system too. In the foreland basin of southern Junggar, reservoir strata, which were developed in the Cretaceous system and the Palaeogene system, had good physical properties with average porosities of 10% – 17%, and the regional cap rocks were developed in the Anjihaihe formation in the Palaeogene system. The natural gas reservoir in the Ziniquanzi formation in the Hutubi area and the natural gas reservoir in the Palaeogene system in the Tugulu area belong to this type of petroleum system.

Combination IV. This primarily occurred in the Neogene system. In the Kuqa foreland basin, the reservoir sequences are sandstones that occurred in the Jidike formation, the Kangcun formation and the Kuqa formation. The reservoir sandstones have good physical properties with a porosity of 13% – 20%. A set of high-quality cap rocks were developed in the Jidike formation. The natural gas reservoir in the Jidike formation in the Dina natural gas field and the natural gas reservoir in the Kangcun formation in the Dawanqi natural gas field belong to this type of petroleum system. In southwestern Tarim, the Miocene series were the major reservoir sequences with an average porosity of 13% – 18%. Additionally, the cap rock was mudstone in the Miocene series as well. The Miocene natural gas reservoir in the Kekeya gas field belongs to this type of petroleum system. In the foreland basin of southern Junggar, combination IV occurred on top of the regional cap rock of the Anjihaihe formation; the Neogene reservoir strata consisted of the Shawan formation and the Taxihe formation, which have good physical properties with an average porosity of 10% – 22%; the cap rocks were the mudstone of the Taxihe formation. In the Dushanzi oil field, the reservoir in the Shawan formation is this type of petroleum system.

In the superimposed style basin, the main characteristics include the fact that reservoirs were developed in multiple geologic periods and natural gas accumulated in the late period. In the Kuqa foreland basin, the natural gas predominately came from the hydrocarbon source rocks of coal measures in the Lower and Middle Jurassic. The oil mainly came from mudstone of lacustrine facies in the Triassic and there is also a small amount of oil that was generated from coal measures. The developmental feature of superimposing two foreland basins from different geologic periods determined multiple petroleum accumulation phases in the Kuqa foreland basin. The dynamic development of the re-genesis basin determined that natural gas accumulation happened during the late period, which was also confirmed by a geochemical study of petroleum reservoirs. During the Late Cretaceous epoch–end of the Palaeogene period, the structural traps on the mountain front were formed in the compressive tectonic environment, which trapped oil and gas from the Triassic. Since the Miocene epoch, the strongest structural deformation has developed. During this time interval, the major structural framework in the Kuqa foreland basin had also been set up. From north to south, in successive order, forward propagated thrust structures were developed and many types of fault-related folds. This tectonic movement not only damaged and adjusted the petroleum reservoirs that formed in the early period, but also matched the major natural gas generation peak time and natural gas migration time in the late period. Thus, the traps that were formed during this time interval mostly accumulated natural gas.

● **Petroleum System in the Modified Style Basin**

The modified style basin referred to the foreland basin that was formed during the Indosinian tectonic period and it had been deformed by the thrust movement during the Himalayan period. The framework of the early period foreland basin had been modified to some degree, but the foreland basin of the Himalayan tectonic period did not develop or was not fully formed. The modified style basins were predominately developed in the Sichuan basin and the western Ordos basin.

In the modified style basins, the major hydrocarbon source rocks include pre-foreland passive continental margin marine deposits of the Paleozoic erathem and coal measures of the foreland period. In the foreland basin of western Sichuan, the Paleozoic hydrocarbon source rocks belong to marine facies or marine-terrigenous facies with good organic materials. However, this group of source rocks did not favorably contribute to petroleum accumulation. The oil and gas discoveries are limited to the natural gas reservoirs in the thrust belt in western Sichuan and in the uplifting belt in central Sichuan, such as the natural gas reservoir in the Leikoupo formation in the Zhongba area. The coal measures are primary hydrocarbon source rocks for the foreland basin of western Sichuan. The hydrocarbon source rocks predominately developed in the Xujiahe formation in the Upper Triassic. They include dark color muddy shale stone (that has an average thickness of $300 - 1,000$ m; in the depositional center, the thickest section can reach up to 1,400 m) and carbonaceous shale and coal petrology (that have an average thickness of $2 - 10$ m with the thickest section up to 28 m; the average content of organic material is about $1.95\% - 65\%$). Most organic materials are in the high maturity stage or post maturity stage ($R_o > 1.7\%$), which benefits natural gas generation. The gas generation strength can be 200×10^8 m^3/km^2.

There are two kinds of reservoir sequences. One is the pre-foreland period passive continental margin marine carbonate rocks and clastic rocks of the late Paleozoic to Lower and Middle Triassic (such as in the western Sichuan area with carbonate rocks of the Lower and Middle Triassic). Another sequence is the foreland period sandstone of lacustrine facies in the Mesozoic erathem. Because this was deposited in wide open, shallow lakes, the Mesozoic sandstone could be deposited as a broader sand sheet. The sandstone of the foreland period is one of the excellent reservoir sequences. In the foreland basin of western Sichuan, the reservoir sequences of the passive continental margin primarily consisted of carbonate rock of platform facies, which were distributed at the thrust belt and uplifting zone with poor physical properties, such as the Leikoupo formation in the Middle Triassic which has an average porosity of 5%. The reservoir sequences were made of sandstones of river-lacustrine facies in the Upper Triassic and the

Jurassic. Because the sediments were rapidly deposited and buried, the set of reservoir sequences were commonly tight. The performance improvement for the reservoir properties mainly depends on micro cracks. For example, in the foreland basin of Sichuan, in the Upper Triassic, the average porosity of reservoir sequences is about 5% – 13% and the average permeability is less than 10 MD. On the other hand, because this is at a relatively shallower level compared with the Upper Triassic, the Jurassic has better reservoir properties than the Upper Triassic, which has an average porosity of 5% – 13%.

The cap rocks are pre-foreland period deposits of the Middle Permian series and foreland period deposits of mudstones of the Upper Triassic and of the Jurassic. In the foreland basin of western Sichuan, there are two important regional cap rocks that have a strong ability to seal the reservoir. One is mudstone of the Xujiahe formation in the Upper Triassic, another is mudstone of the Suining formation in the Jurassic. The former not only contributes as the hydrocarbon source rock, but also functions as a cap rock. It has a stable distribution with a thickness usually large than 100 m and also has low porosity and low permeability. In addition, its bursting pressure is in the higher range, and the testing result for bursting pressure of the mudstone is 10.3 – 12.5 MPa. The latter mudstone was predominately distributed in the depression zone; the thickness of the northern section was relatively thin, but the thickness of the southern section is more than 200 m and the bursting pressure is 10.1 – 22.6 MPa.

During the Yanshanian tectonic period, the hydrocarbons were generated. Due to the tectonic movements (of thrusting, uplifting and erosion) of the Himalayan tectonic period, the oil and gas reservoirs formed during the early period were redistributed to form newer oil and gas reservoirs.

The modified style basins have three sets of reservoir-seal combinations, which are the combinations in the lower section (combinations I and II) and the combination in the upper section (combination III). These three sets of reservoir-seal combinations are important to oil and gas exploration.

Combination I.   This combination was mainly developed in the Lower and Middle Triassic or its underside. For example, in the foreland basin of western Sichuan, a set of hydrocarbon source rock developed in the Permian system, which consisted of a carbonate rock of platform facies and coal measures of marine-terrestrial facies. The reservoir sequences resulted from the carbonate rock of the Lower and Middle Triassic, which had low porosity and poor permeability. For instance, in the Middle Triassic, the Leikoupo formation had an average porosity of 5% approximately. The cap rock was a coal measure of the Upper Triassic. This kind of petroleum system combination generated oil and gas in the older strata and stored petroleum in the younger strata. For example, in the foreslope zone, the Maxi natural gas field consists of this kind of petroleum system.

Combination II.   This mainly developed in the Upper Triassic in the western Sichuan area. The reservoirs belong to a tight and improved fracture type with an average porosity of 3% – 8%. The cap rock was mudstone from the Upper Triassic. The natural gas reservoirs were made of combination II in the second section of the Xujiahe formation in the Zhongba area and in the third and fourth section of the Xiangxi group in the Pingluoba area.

Combination III.   This developed in the Jurassic in the western Sichuan area. The reservoirs were made of Jurassic sandstone of wide opening shallow lacustrine facies, which had a good porosity of 5% – 13% on average. The cap rocks were Jurassic mudstone that was overlaid with Jurassic sandstone. Combination III occurred in the Jurassic natural gas reservoirs in the Baimamiao, Pingluoba and Xinchang areas.

The petroleum system in the modified style basin has the following characteristics: the hydrocarbon source rocks matured in the early period and petroleum reservoirs were formed in a relatively earlier period too. However, the final arrangement of the reservoir system happened in the late period. In the foreland basin of western Sichuan, a multiple set of hydrocarbon source rocks entered their oil and gas generation peak stage at a different geological period. Clearly, there were two petroleum accumulation periods. The petroleum reservoirs that formed during the end of the Indosinian tectonic period were supplied by Permian hydrocarbon source rocks. The petroleum reservoirs that formed during the Yanshanian tectonic period were supplied by Upper Triassic hydrocarbon source rocks. During the Himalayan tectonic period, faults connecting deeply buried natural gas and shallower traps transferred the natural gas up to a shallow reservoir. The natural gas accumulated in the early period had been adjusted and transformed to form secondary reservoirs. Thus the final arrangement of the reservoir system was concluded.

● **Petroleum System in the Presenility Style Basin**

The presenility style basin combination is designated as the foreland basin that was formed during the Indosinian period. In the late period, an extensional graben basin and regional depression basin developed. The thrust deformations of the Himalayan tectonic period were very weak. On the structural cross section, the framework of the foreland basin of the Indosinian tectonic period can be clearly recognized. In addition, the surrounding orogenies are correlated with the elements of the inner basin as well. In the presenility style basins, the hydrocarbon source rocks are made of pre-foreland sequences of marine facies and marine-terrestrial facies. For example, in the foreland basin on the western margin of the Junggar, the hydrocarbon source rocks include the Fengcheng formation and the lower Wuerhe formation in the Lower–Middle Permian series, which are pre-foreland sequences (Fig. 9.21). This set of sequences belongs to the sapropelic type or humosapropelic type, which predominately generated oil with a small amount of natural gas all through geological history. These particular hydrocarbon rocks occur in the southwestern region that includes Lunan, Shixi, Dishuiquan and Wucaiwan areas. The reservoir sequences are predominately made of sandstones that were deposited during the following tectonic periods: the pre-foreland period, the foreland period and the depression phase of the post foreland period respectively. For example, in the foreland basin on the western margin of Junggar, the reservoir sequences mostly consist of pre-foreland period sandstone in the Lower–Middle Permian series, foreland period sandstone in the Triassic and depression period sandstone in the Jurassic. It also includes pre-foreland period reservoir strata in the Carboniferous system (Fig. 9.21). The cap rocks include mudstones that came from three tectonic periods, which are the pre-foreland period, the foreland period and the post foreland period. For example, in the foreland basin on the western margin of Junggar, the cap rocks include mudstones that came from different tectonic periods: pre-foreland period mudstone in the Upper Permian series, foreland period mudstone in the Upper Triassic and the mudstones in the Lower and Upper Jurassic (Fig. 9.21).

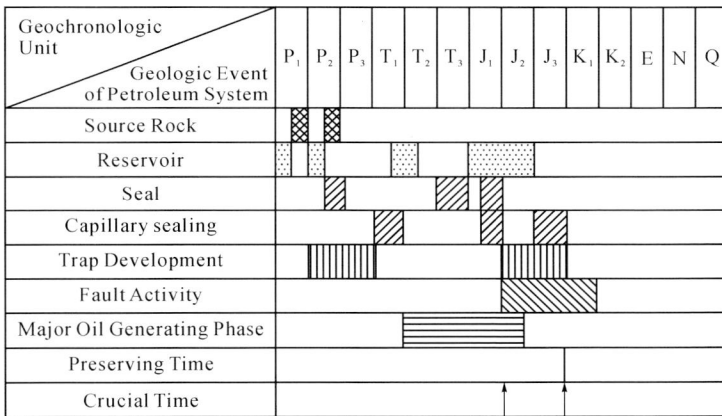

Geochronologic Unit / Geologic Event of Petroleum System	P_1	P_2	P_3	T_1	T_2	T_3	J_1	J_2	J_3	K_1	K_2	E	N	Q
Source Rock	▨	▨												
Reservoir	▨	▨			▨		▨	▨						
Seal			▨			▨	▨							
Capillary sealing				▨			▨		▨					
Trap Development				▥	▥				▥	▥				
Fault Activity								▨	▨					
Major Oil Generating Phase						▤	▤							
Preserving Time														
Crucial Time							↑		↑					

Fig. 9.21. The geologic events of petroleum system in the western Zhungeer basin (Song et al., 2002) (With permission of Chinese Science Bulletin)
P, Permian; T, Triassic; J, Jurassic; K, Cretaceous; E, lower Tertiary; N, upper Tertiary; Q, Quaternary

The presenility style basin only has the reservoir–seal combination in the lower section (Combination I). In addition, Combination I can be subdivided into two sub–combinations: petroleum was generated and stored in the same sequences (Combination I_1), or petroleum was generated in older strata and it was stored in younger strata (Combination I_2).

In the presenility style basin, there were two critical phases for oil and gas accumulations: one is the phase of original petroleum accumulation; another is the phase of secondary petroleum accumulation. For example, in the foreland basin on the western margin of Junggar, the critical phase for original petroleum accumulation was in the Permian system and the second critical phase for secondary petroleum accumulation was in the Jurassic. In the Permian system, the critical phase for original oil and gas to accumulate was during the Middle Jurassic epoch. During this time interval, the hydrocarbon source rocks in the Lower Permian series progressed into an oil generation peak period (wet gas thermal maturation). Moreover, the hydrocarbon source rocks in the Middle Permian series also entered an oil generation peak period too. The traps that were formed during the foreland tectonic period offered spaces to facilitate oil reservoir development. In the same time interval, the strata of the Upper Triassic had obtained the ability of capillary sealing, and behaved as the regional cap rocks for the original oil reservoirs in the Permian system–Lower and Middle Triassic. Multiple factors combined together to form the petroleum system beneath the Upper Triassic with accumulated original oil and gas. For the secondary oil and gas reservoirs, the hydrocarbon delivery period was the time that faults connected hydrocarbon source rocks in the Permian system and reservoir sequences in the Jurassic. The critical phase of petroleum accumulation was decided by timing related correspondence from the fault, the Jurassic traps and the capillary seal in the regional cap rock. For this particular region, the second critical phase of petroleum accumulation was during the Early Cretaceous epoch.

● **Petroleum System in the Palingenetic Basin**

The Palingenetic basin designates foreland basins that were predominately developed during the Himalayan tectonic period and the foreland basins of the Indosinian tectonic period were not fully developed, or at least current information cannot prove that the foreland basins of the Indosinian tectonic period were developed. The Palingenetic basins mainly appeared on both sides of the Qilian orogeny, such as the Jiuquan basin and the northern margin of the Qaidam basin. The main hydrocarbon source rocks include coal measures in the Lower–Middle Jurassic of the pre-foreland tectonic period or dark color mudstone in the Cretaceous system.

   Both major hydrocarbon generation and important petroleum accumulation events happened during the Himalayan tectonic period. For example, on the northern margin of Qaidam, the hydrocarbon source rocks were mainly developed in the Lower–Middle Jurassic coal measures and mudstone of lacustrine facies during the extensional tectonic stage. In the Lower Jurassic, good quality coal beds and the carbonaceous mudstone were deposited in the following areas: (1) In the Lenghu structural belt and Kunteyi depression area, the thickness of hydrocarbon source rock is about 500 – 1,200 m, with the thickest section of to 2,200 m. (2) In the area south of Nanbaxian–Lenghu 7, the thickness of the hydrocarbon source rock is about 600 – 1,000 m, with the thickest section of 1,200 m. The strata of the Middle Jurassic appeared in the Lenghu structure belt and the area north of Nanbaxian–Mahai; the thickness of hydrocarbon source rock is about 250 – 500 m, the thickest section can reach up to 650 m. The mudstone of lacustrine facies occured in the Lenghu area and the oil shale appeared in the Yuka area. As another example, in the Jiuquan basin, the Early Cretaceous hydrocarbon source rock was developed in the Chijinpu formation, Xiagou formation and the Zhonggou formation during the extensional tectonic stage. The Chijinpu formation and the Xiagou formation contain terrestrial mudstone, dolomitic mudstone and muddy dolomite, which were deposited in the alkali environment of a semi-deep to deep lake with slightly saline to semi-saline water.

The reservoir strata are sandstones in the Cretaceous and the Palaeogene systems. The cap rock is red color gypsum mudstone in the Palaeogene system. The best reservoirs are found in the Cretaceous–Palaeogene systems beneath the Palaeogene red color gypsum mudstone.

The Palingenetic basin has three sets of reservoir-seal combinations, which are combinations III and IV in the upper section and combination II in the lower section. Combination III is the most important target for petroleum exploration. At the same time, combination II in the lower section should be considered as well.

Combination II.   On the northern margin of Qaidam, the petroleum system developed within the Lower and Middle Jurassic. Coal measures function as the hydrocarbon source rocks, the strata in the Lower and Middle Jurassic function as the reservoir sequences and cap rocks. Examples are the Lenghu 3 and Lenghu 4 oil reservoirs in the Jurassic. This kind of petroleum system was located in the thrust belt near the mountain front. Because of the uplifting and deformation of the reservoir sequences, it has good storage conditions. For example, in the Lenghu 3 oil reservoir, the porosity of the Early Jurassic reservoir sequences is about 10% – 20%.

Combination III.   This material predominately developed within the Palaeogene system on the northern margin of Qaidam. The reservoir sequences are located in the lower Xiaganchaigou formation, which has good physical properties with an average porosity of 9% – 14%. The upper Xiaganchaigou formation functions as the regional cap rock. Examples are the Nanbaxian oil and gas reservoir in the Palaeogene system, the Mahai natural gas field and the Mabei 1 oil and gas reservoir in the Palaeogene system.

Combination IV.   This material includes the Shangganchaigou formation and overlying sequences. The reservoir sequences include the Shangganchaigou formation and the Youshashan formation, which have good physical properties; the average porosity is about 16% – 23% and the best porosity can be 28%. The upper Xiaganchaigou formation functions as the regional cap rocks. Examples are the Nanbaxian oil and gas reservoir in the Neogene system and the Lenghu 4 and 5 oil reservoirs in the Neogene system (N_1).

In the Palingenetic basins, the reservoirs that were formed during the late geological period are characteristic of petroleum accumulation. For example, on the northern margin of the Qaidam basin, during the late Tertiary period, in the Kunteyi–Yibeici depression, hydrocarbon source rocks in the Lower Jurassic were at low maturity–maturity stages and the early phase of oil and gas accumulation took place in the Palaeogene traps. Examples are the Lenghu 3, 4 and 5 oil fields and the Yuka oil field; in addition, the Nanbaxian structure had an early phase of oil and gas accumulation as well. With the thickness of the Neogene sediments increasing, the Jurassic source rocks entered high maturity and post maturity stages. A large amount of oil and gas migrated vertically along the faults into the proficient reservoir sequences, mixing with the oil and gas that accumulated in the early phase and most mixtures exist as condensate oil and gas. During the end of the Himalayan tectonic period, powerful tectonic movement caused oil and gas to migrate vertically along the faults to form secondary petroleum reservoirs in the Neogene system.

9.4.3   Oil and Gas Accumulation Pattern in Foreland Thrust Belt in China

● **Coal Measure Generated Hydrocarbons and Anticlinal Traps Stored Petroleum**

In western–central China, foreland basins were predominately developed during the Mesozoic and the Cenozoic eras. Currently some foreland thrust belts have been intensively studied and this has resulted in multiple discoveries of large petroleum reserves. These foreland thrust belts are located in the Kuqa area at Tarim, southwest Tarim, western Sichuan, the southern margin of Junggar and the northern margin of Qaidam. Coal measures of limnetic facies in the Triassic and Jurassics are the most important hydrocarbon source rocks in the foreland basins in western-central China. This set of hydrocarbon source rocks has a large thickness of 200 – 1,000 m in general, with higher organic abundance. The average content of organic carbon is in the range 1.5% – 3.5%. The organic materials primarily belong to type III, or else to type II. At present, the thermal maturation has usually reached the high maturity–post maturity stage. The latest research information shows that the natural gas that has been discovered in thrust belts contains a heavy weighted carbon isotope. In particular, the ethane carbon isotope is less than –28‰ in general, which indicates natural gas was generated by coal measures. The ethane carbon isotope is used to indicate the origin of kerogen. For example, in the Kuqa depression, the values of $\delta^{13}C_2$ are between –16.8‰ ~ –25.7‰ for natural gas. In the Hutubi natural gas field in southern Junggar, the value of the carbon isotope is between –27.7‰ ~ –28.2‰ for condensate oil and the value of the $\delta^{13}C_2$ is between –22.27‰ ~ –22.96‰ for natural gas. In the Xu 2 natural gas field in the Zhongba area, the value of the $\delta^{13}C_2$ is between –25.23‰ ~ –25.72‰ for natural gas and the value of the carbon isotope is –24.4‰ for condensate oil. The value of $\delta^{13}C_2$ is larger than –25‰ for natural gas in the Pingluoba natural gas field in Sichuan, and in both the Mahai natural gas field and the Nanbaxian natural gas field on the northern margin of the Qaidam basin. Comparing the value of the carbon isotope from benzene, methyl benzene and light hydrocarbons, confirmed these natural gases were related to coal measures (Song et al., 2008).

In Chinese foreland basins, the foreland thrust belts of western-central China were the most important geo-provinces for oil and gas accumulations. Petroleum predominately accumulated in compressive anticlines and in a variety of folds that were related to large scale thrust belts, such as a fault-propagated fold, fault-bended fold, detachment fold, dual thrust fault and pop-up structure. These structures were formed by powerful compressive tectonic movement during the Himalayan tectonic period (Wei et al., 2000; Lu et al., 2000). On the other hand, petroleum accumulations were closely related to the structures of early periods as well. That means the structures of early periods controlled petroleum accumulation in the foreland thrust belt.

For example, the Kela 2 natural gas field and Dabei 1 natural gas field correlated positionally with structures that formed during the end of the Cretaceous period. In the Zhongba area of western Sichuan, during the Indosinian tectonic period, the petroleum accumulated in traps that were developed on top of the ancient uplift. On the northern margin of the Qaidam basin, ancient uplifts of the Middle Jurassic epoch–the Cretaceous period existed in the Nanbaxian, Mahai and Lenghu structure belts. In the Hutubi natural gas field of southern Junggar, the traps were formed during the Himalayan tectonic period, but the oil and gas had previously accumulated in the ancient structures of the Yanshanian tectonic period. Due to fault movement, the original oil and gas reservoir in the Jurassic had been destroyed and petroleum migrated upwards into the new traps that were formed during the Tertiary period.

In the foreland thrust belts of western-central China, the discovered sizeable natural gas fields were mostly in anticline traps. Because the foreland thrust belts were located at the junction area of basins and ranges, and they went through strong tectonic movement during the late geologic period, the structural deformations were significant and arrays of structural traps developed on the mountain front. Currently, discovered sizeable natural gas reservoirs in the foreland thrust belts are located in huge structure traps. Furthermore, in the thrust belts of western-central China, oil and gas exploration is still at an early phase. Both seismic data and geological analysis encourage exploration activity to be concentrated on the anticlinal traps. In the Chinese foreland basins, the ratio of gas to oil is normally larger than 0.1 for explored petroleum reserves, but in the foreland basins of foreign countries, the ratio of gas to oil is commonly less then 0.01 for explored petroleum reserves. In the major Chinese foreland basins, the ratio of gas to oil is mostly larger than 0.5 among the natural reserves and the explored reserves. The characteristic of natural gas abundance in the Chinese foreland thrust belts was determined by Kerogen type III in source rock of coal measures and the higher degree of thermal maturation of hydrocarbon source rocks since the Miocene epoch. Kerogen type III is the agent that generates natural gas. The foreland thrust belts have become the most important region for natural gas exploration in China. Because the structure traps were well developed in the Chinese foreland thrust belts with great promise of oil and gas discovery, it therefore became the major exploration target for sizeable natural gas fields. Recently, the development of petroleum exploration in foreland thrust belts has been accelerated. Innovative exploration technology and advanced geological theories have been established to support exploration activity of sizeable oil and gas fields. In the Chinese foreland thrust belts, the estimated productive reserves of natural gas may be as much as 8.7×10^{14} m^3, which is equivalent to 30% of the total natural gas reserves in the basins of the western-central region. Up to now, the explored natural gas reserve is 8.3×10^{12} m^3, which is equivalent to 28% of the total natural gas reserves in the basins of the western-central region. During the 1990 s, the Kela 2 natural gas field was discovered in the Kuqa thrust belt. This natural gas field covers an area of 47 km^2, the height of the gas column can reach up to 500 m, the explored reserve is $2,840.29 \times 10^8$ m^3 and the productive reserve is $2,290 \times 10^8$ m^3. The designed annual production rate is 107×10^8 m^3. With the progression of a natural gas pipe line project, the foreland thrust belts in western-central China have become the most important natural gas producing region. In the western-central region, because most areas contain abundant natural gas, these areas have thus become the major targets for upcoming natural gas exploration. With the development of geological theories and the innovation of exploration technologies, the Chinese foreland thrust belts will develop into one of the principal natural gas producing areas within the northern margin basin group of the Tethys tectonic domain.

- **Petroleum Migration and Accumulation were Closely Related to Abnormal Pressures**

Abnormally high pressures commonly appear in the foreland thrust belts in the western-central region. The distributions of the petroleum reservoir were closely related to the distributions of abnormal pressures. There are two kinds of distribution relationship between abnormal pressure belts and petroleum reservoirs. One is represented by the foreland thrust belt of Kuqa, which produces natural gas from high pressure strata. Another one is represented by the natural gas fields in southern Junggar and western Sichuan, which produce natural gas from the normal pressure strata. But the development of the natural gas reservoir was closely related to the high pressure strata directly above and beneath the natural gas reservoir.

In the Kuqa foreland thrust belt, the high pressure strata exist below the massive gypsum-salt combination of the Tertiary period. These high pressure sequences include a gypsum-salt combination and the underlying strata, which are reservoirs in the Cretaceous system and hydrocarbon source rocks in the Triassic and Jurassics. Above the gypsum-salt combination, there is the normal hydrostatic pressure system. In the Kuqa area, discovered natural gas fields have high pressure reservoirs in general, such as the Kela 2 natural gas reservoir in the Kelasu structure belt where the average pressure coefficient of the Palaeogene system is $1.96 - 1.89$ and where the average pressure coefficient of the natural gas reservoir in the Bashijiqike formation in the Cretaceous system is $1.87 - 1.56$. The pressure coefficient of the reservoir in the Dabei natural gas field is 1.61 and the pressure coefficient of the reservoir in the Palaeogene system is $1.75 - 1.65$. Under the gypsum-salt combination that occurs in the Jidike formation in the Neogene system, the pressure coefficient of the natural gas reservoirs in Dina 1 and Dina 2 is about 1.8 as well.

In the Hutubi natural gas field in southern Junggar, the reservoir strata were made of normal pressure sequences that occurred in the Ziniquanzi formation in the Palaeogene system. However, both the overlay strata of the Anjihaihe formation and the underlay hydrocarbon source rocks in the Jurassic were high pressure sequences. Oil and gas that were generated by the hydrocarbon source rocks of coal measures in the Jurassic migrated upward into the reservoir sequences that were directly beneath the high pressure strata of seals. In the natural gas field in western Sichuan, natural gas was predominately produced in the high pressure hydrocarbon source rocks in the Xujiahe formation and low pressure sequences in the Penglaizhen formation that overlaid the hydrocarbon source rocks (Li et al., 2002).

Due to structural compression, rapid hydrocarbon generation and sealing off by the gypsum mudstone and coal measures, in the foreland thrust belts the discovered oil and gas reservoirs commonly have an abnormally high pressure (Table 9.7). For example, in the Kela 2 natural gas field in the Kuqa area, under the gypsum-salt combination or under the plastic mudstone layer, at the point where the structural stress is concentrated, the reservoir strata close to the hydrocarbon source rocks have a pressure coefficient above 2. In the Nanbaxian area on the northern margin of Qaidam, in the deeply buried reservoir under the plastic gypsum mudstone, the salinities are 20,445 – 44,509 ppm and the pressure coefficient is between 1.58 – 1.68.

Table 9.7 Stratigraphic pressure of petroleum reservoirs in foreland thrust belts of China

Basin	Natural gas field	Stratigraphy	Pressure coefficient
Kuqa	Kela 2	Cretaceous	2.02
Kuqa	Dina 2	Palaeogene	2.01
Kuqa	Dabei 1	Cretaceous	1.62
Western Sichuan	Weicheng	U. Triassic	2.21
Western Sichuan	Jiulong mountain	U. Triassic	1.96
Western Sichuan	Xinchang	M. Jurassic	2.03
Southern Junggar	Huoerguosi	L.M. Palaeogene	2.40

● **Petroleum Accumulated During Late Geologic Period**

In the western-central region, except for the foreland thrust belt on the northwestern margin of Junggar basin, which contained petroleum reservoirs that accumulated during the early period (T_3-J), most foreland thrust belts presented the characteristic of petroleum accumulating during the late geological period. These oil and gas reservoirs have accumulated since the Neogene period and sizeable natural gas reservoirs have accumulated since the Quaternary period (Jia et al., 2005; 2007; Jia, 2005).

The maturity study of natural gas and crude oil indicated that there were three phases of petroleum accumulation in the Kuqa foreland basin. Considering the geological timing interval during which local structures were formed and the correlation of hydrocarbon generation history, the three documented phases are as follows. The first phase was when petroleum started to accumulate during the Neogene period when the Jidike formation was deposited, which predominately accumulated oil and gas of low maturity. The second phase was when petroleum began to accumulate during the depositional period from the Jidike formation to the Kuqa formation, which accumulated mature–highly mature oil and gas. The third phase represents the dry gas accumulation period, which established reservoirs during the depositional period from the Kuqa formation to the Xiyu formation. The organic inclusion indicated that a large amount of natural gas migrated and accumulated during this depositional period of 5 – 1 Ma approximately.

The following are characteristics of petroleum accumulation during the late geological period:

(1) The structures were formed during the late geological period. Since the Pliocene epoch, the movements of foreland thrust belts were very strong and this formed fault-related folds and other superimposed structures spreading out in arrays and regions. Therefore, the traps were established in the late geological period.

(2) Hydrocarbon generation high peaks emerged in the late geological period. Because of the overlay of massive strata of the Neogene–Quaternary period, the hydrocarbon source rocks in the Mesozoic or Palaeogene systems were buried very deep and with rapid speed, which thus accelerated thermal maturation processing. As the end result, the hydrocarbon source rocks were brought rapidly to a hydrocarbon generation high peak. Due to the strong compressive force and the very well developed fault system, the hydrocarbons migrated efficiently.

(3) The regional cap rocks contained abnormally high pressure. The Palaeogene gypsum mudstone and mudstone formed regional seals of the petroleum system. With the strong compressive stress field, an abnormally high pressure area also developed in the petroleum system. Most discovered petroleum reservoirs contained high pressure. In particular, during the natural gas accumulation period, high pressure was present as well. Therefore, the high pressure anticlinal reservoir is the common type of reservoir that appeared in the foreland thrust belts in western-central China.

In the Kuqa area, the strata above 3,000 m have normal pressure; on the mountain front, the structure belt of the Dawanqi–Kelasu–Yinan contains abnormally high pressure below 3,000 m with a pressure coefficient larger than 2.0. Both the Yangtake structure belt and the Kalayuergun structure belt contain abnormal pressure below 4,500 m with a pressure coefficient around 1.5. The Yaha structure belt contains abnormal pressure in the 3,600 – 5,000 m interval. From the piedmont belt to the front uplift belt, the strata pressure gradually decreases to a normal state (Jia et al., 2000). These instances reveal that there is some kind of mechanism relationship between abnormally high pressure and the strong structural compressive force on the mountain front.

In the Kelasu area in the western region, the abnormally high pressure strata are primarily located under the Neogene gypsum–salt combination, which include the gypsum–salt combination, the reservoir in the Cretaceous system and hydrocarbon source rocks in the Mesozoic erathem. In the Yiqikelike area of the eastern region, the abnormally high pressure strata include the Neogene massive gypsum mudstone in the Jidike formation and the sequences that are located below the gypsum mudstone. It even includes the hydrocarbon source rocks in the Triassic–Jurassics. Generally, the overlay sequences have hydrostatic pressure. These examples reflect that the regional seal (it also is the major detachment layer) influences the creation and distribution of the high pressure field in the strata.

(4) The critical conditions for petroleum accumulation are well-matched. During the late geologic period, petroleum accumulation required compatible accumulative conditions. In the Kuqa area, the sandstone reservoirs were well developed with overlaid massive gypsum mudstone as the cap rocks. The crucial elements for petroleum accumulation are the entirety of structures under the salt layer and the migration channels for oil and gas to rise. On the southern margin of Junggar, the crucial conditions for petroleum accumulation might be the compatibility between the large scale structure and the sandstone reservoirs. In the western Sichuan area, the crucial conditions for petroleum accumulation are the compatibility between the ancient traps above the primitive uplift and the reservoir sequences that have excellent porosity and permeability.

● Petroleum Distribution has a Predictable Pattern

There is a coupled system between the development of an orogeny on the outside and the developments of a foreland basin and structural units within the basins. The creation and development of an orogeny belt might influence the following five areas inside the basin, which are structural units, depositional conditions, the distribution of hydrocarbon source rocks, regional thermal maturation and the development of the petroleum reservoir. Therefore, the foreland basin and surrounding areas became one system that has distributed petroleum reservoirs in a predictable pattern.

On the mountain front, the distribution of oil and gas has reliable patterns. As the depth of the reservoir increases, the reservoir distribution pattern is from oil → oil and gas → condensate gas. In a depression of great depth, as the depth of the reservoir decreases, the reservoir distribution pattern gradually changes from condensate gas → oil and gas → oil.

During the process of petroleum generation, migration and accumulation, there is a timing differentiation between the oil process and the gas process. Therefore, in the folded thrust belt that has well developed structures, the orogeny contains oil reservoirs alongside the side close to the basin, but on the opposite side natural gas has accumulated. For example, in the Kuqa area, the Kelasu–Yiqikelike structure belt predominately accumulated natural gas of high maturity during the late geological period. However, the Yangtake–Yaha–Tiergen structure belt mainly accumulated condensate oil of low maturity during the relatively early geologic period. This phenomenon commonly occurs in the foreland thrust belt of North America. For example, in the Rocky mountain thrust belt of the western USA, natural gas fields were developed near the Crawford thrust, which were the Anschutz Ranch gas field, Cave Creek gas field, Yellow Creek gas field, Red Valley gas field and Whitney Valley gas field. However, on the western side of the Rocky mountain thrust belt, oil fields were developed, which were the Lakeman Creek oil field, Clear Creek oil field, Painter Reservoir oil field, Anschutz Ranch East oil field and Pineview oil field. Also, the same phenomenon occurs in the Appalachian–Ouachita–Marathon structure belt of the eastern USA. Natural gas accumulated in the deformed zone on the mountain front and the oil accumulated in the non-deformed strata that distended from the orogeny.

In western-central China, some foreland thrust belts have a relatively long development period (such as the northwestern margin of Junggar, or the foreland thrust belt in southwestern Tarim). Generally, either in the ancient thrust belt or in shallow strata, oil is mainly produced. However, both in the newer thrust belt and in deeply buried strata, natural gas is primarily produced. This phenomenon possibly derives from the following two reasons. Firstly, in the newer thrust belt, hydrocarbon source rocks were deeply buried or had a relatively longer development period. Thus, there is a high degree of thermal maturation and this mainly produced natural gas. Secondly, this relates to the kerogen type in the hydrocarbon source rocks (coal measures), for example in the Kekeya condensate gas field and the Hutubi gas field. On the other hand, the petroleum reservoirs in the ancient thrust belt were formed during the early geological period and this was in a state of uplifting for a long period. Therefore, oil was mostly produced, such as in the Kelamayi oil field and the Xiazijie oil and gas field.

Because of an abundant natural gas supply and the beneficial setting of time and space, in the foreland basin different structures might lead to the development of different types of petroleum reservoir. The anticline trap with natural gas accumulation was developed in the foreland thrust belt, the deep basin natural gas reservoir was developed in the foredeep depression, the stratigraphic trap with natural gas accumulation was developed on the front slope and the faulted anticline trap and faulted nose trap were developed in the front uplift (Fig. 9.22). The reservoir distribution patterns have a significant meaning when it comes to petroleum exploration planning.

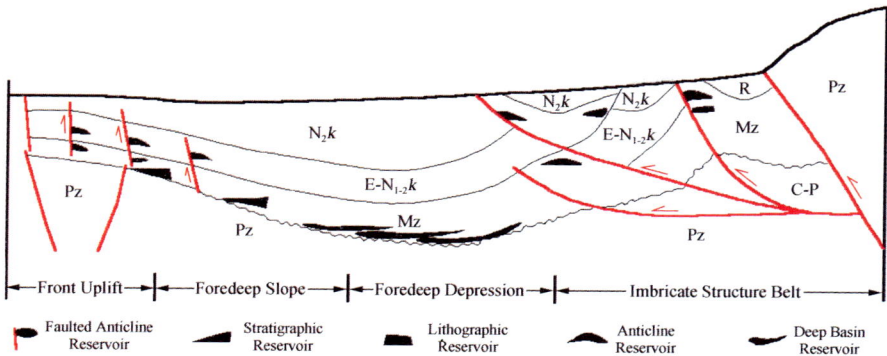

Fig. 9.22.   Structure models and the reservoir types in the Chinese foreland basins (Jia, 1997)
(With permission of Petroleum Industry Press)
Pz, Paliozioc; C, Carbonniferous; P, Permian; Mz, Mesozoic; K, Cretaceous; R, Tertiary; E, lower
Tertiary; N, upper Tertiary; Q, Quaternary

● **Diversified Reservoir Distribution Pattern Associated with Various Types
of Foreland Thrust Belt**

The superimposed relationship between a foreland thrust belt and an effective
hydrocarbon source region decided the distribution and accumulation of oil and
gas. Hydrocarbon source rocks were commonly developed in non-foreland
sequences as the underlay strata. Deformed foreland thrust belts were
superimposed on top of hydrocarbon source rocks. Therefore, the type of
association between foreland thrust belts and depressions that generated
hydrocarbons is parallel or nearly parallel one to the other, obliquely correlated or
at a high angle one to the other, and perpendicularly connected one to the other.
Various superimposed relationships brought a differentiation of oil and gas
distribution, such as oil and gas concentrated sequences, or an oil rich section.
   Parallel Foreland Thrust Belt.   The strike direction of structures in foreland
thrust belts is parallel to the distribution direction of the depression that generated
hydrocarbons or effective hydrocarbon source rocks. Because of the thrust
structure, the critical conditions for accumulation and distribution of the petroleum
reservoir were determined by the following elements in the foreland thrust belt,
which are the facility of the oil and gas supply (petroleum migration channels) of
the structures, the storage capacity in the structures and the overall preservation
condition. Oil and gas fields were distributed in zones. Several thrust belts have
this feature, which are the mountain front thrust belt at Longman mountain, the
thrust belt on the western margin of the Ordos basin, the Wuerhe–Xiazijie faulted
belt on the northwestern margin of the Junggar basin. For example, in the
mountain front thrust belt at Longman mountain, there are currently six natural
gas fields that have been discovered in the following areas: Zhongba, Jiulong
mountain, Hewanchang, Pingluoba, Daxingxi and Baimamiao. In addition, there

are thirteen structures that contain natural gas that have been documented in the following areas: Shejian River, Siyichang, Bailongchang, Zhebachang, Wenxingchang, Laoguanmiao, Weicheng, Fengguzhen, Daxingchang, Hanwangchang, Zhougongshan, Guanyinsi and Songhuazhen. In the western Sichuan area, natural gas accumulations were controlled by the ancient traps and the reservoir sequences that have good porosity and permeability (which include the area with well developed fractures). These large scale ancient traps were formed by the compression-distortion stress field during the early geological period. The traps and reservoir sequences were well matched in the western Sichuan area. The secondary natural gas reservoirs in the Jurassic were supplied by underlay sequences of the Xujiahe formation and natural gas was stored in the stratigraphic traps on top of the structural background.

Obliquely Correlated Foreland Thrust Belts. The Kelamayi–Baikouquan faulted belt on the northwestern margin of Junggar basin was obliquely intersected by the underlay Mahu depression that generated hydrocarbons. Furthermore, the faulted folding structures on the southern margin of Junggar basin also had an oblique contact with underlay hydrocarbon generating depressions of the Permian, Jurassic and Neogene periods.

The accumulation and distribution of oil and gas reservoirs were controlled by the petroleum migration channels and the geological setting between hydrocarbon source rocks and reservoir strata. There was high accumulative efficiency and abundant resources.

Vertically Superimposed Foreland Thrust Belt. The Jiuxi basin in the Gansu Corridor area represents this type of thrust belt. The Jiuquan basin is located in the middle section of the Gansu Corridor. It is one of the basins where oil and gas exploration in the foreland thrust belt were carried out during the earliest exploration phase. Under the stress field of the thrusted force oriented in a N-S direction and a strike-slip force on the western side, the Jiuxi depression experienced a medium strength deformation event that brought about structures of imbricate thrust belts and fault-related folds, which belonged to the basement-involved type structural system. These structures can be divided into three structural belts, which are the mountain front thrust folding belt, the Laojunmiao thrusted-folding belt and the Baiyanghe thrust belt. In the Laojunmiao thrust belt, three oil fields of Laojunmiao, Yaerxia, and Shiyougou were discovered. In the Baiyanghe thrust belt in the northern part of the Jiuxi depression, the Baiyanghe oil field and the Shanbei oil field were discovered. In the thrust belt on the northern margin of Qilian mountain, a fractural type reservoir at Liugouzhuang–Kulong mountain and the Kulong mountain oil field were discovered.

On the Laojunmiao anticlinal belt, a series of anticlines were not connected with their axis and a chain of local structures occurred as an echelon structure, which formed several oil and gas fields in the tertiary system. In the western Jiuxi depression, under the tertiary structure of Yaerxia, a lithologic reservoir in the Cretaceous system and a buried hill reservoir in the Silurian bedrock were discovered. As a thrusted block in the southern part of the basin, it was superimposed on top of the Cretaceous graben basin, the western section faced the Qingnan sub-depression, the eastern section made contact with the Shibei sub-depression, the Shibei sub-protrusion and the Dahongquan sub-depression. The oil and gas fields were developed in the superimposed position of a normal fault with an NE strike and a structure belt with an NW strike.

References

Allegre, C.J., Courtillat, V., Tapponnier, P., et al., 1984. Structure & evolution of Himalaya-Tibet orogenic belt. Nature, 307:17-22.

Allen, M.B., Windly, B.F., et al., 1991. Basin evolution within and adjacent to the Tien Shan range, NW China. Journal of the Geological Society of London, 148:369-378.

Allen, M.B., Windly, B.F., Zhang, C., 1992. Paleozoic collisional tectonics and magmatism of the Chinese Tien Shan, central Asia. Tectonophysics, 220:89-115.

Allen, M.B., Windly, B.F., 1993. Evolution of the Turfan basin. Chinese Central Asia Tectonics, 12:889-896.

Boyer, S.E., Elljol, D., 1982. Thrust systems. AAPG Bulletin, 66:1196.

Chen, F.J., et al., 1992. Structure and geodynamic setting of the Mesozoic–Cenozoic petroleum basins in China. Geoscience, 6(3):317-327.

Chen, H.L., Yang, S.F., Dong, C.W., et al., 1997. Distinction of basic igneous rocks belt of the Permian period in the Tarim basin and its tectonic significance. Geochimica, 26(6):77-87.

Chen, H.L., Yang, S.F., Li, Z.L., et al., 2005. Geochemistry and tectonic setting of the early Late Paleozoic felsic volcanic rocks in the Altai orogenic belt. Acta Geologica Sinica.

Chester, J., Chester, F., 1990. Fault-propagation folds above thrusts with constant-dip. Journal of Structural Geology, 13:903-910.

Dahlstrom, C.D.A., 1969. The upper detachment in concentric folding. Bulletin of Canadian Petroleum Geology, 17:336-344.

Dai, J.X., Li, X.Q., 1995. Attribute of coal bed gas accumulation domain in eastern Central-Asia. Petroleum Exploration & Development, 22(5):1-7.

Davis, D., Suppe, J., Dahlen, F.A., 1983. Mechanics of fold-and-thrust belts and accretionary wedges. Journal of Geophysical Res., 88:1153-1172.

Dickinson, W.R., 1976. Plate tectonic evolution of sedimentary basin. *In*: Plate Tectonics and Hydrocarbon Accumulation. AAPG Educational Series.

Epard, J.L, Escher, A., 1996. Transition from basement to cover: a geometric model. Journal of Structural Geology, 18:533-548.

Gan, K.W., 1995. The discussions of foreland basin. *In*: Ding, G.M. et al. (eds), Development of Oil and Gas Exploration (vol. 1). Petroleum Industry Press, Beijing, pp. 72-76.

Gao, R., Li, P.W., Li, Q.S., et al., 2001. Deep level collisional deformation on northern margin of the Qinghai-Tibet Plateau–An inspiration of deep seismic exploration. Science in China, Series D: Earth Sciences, (S1):66-71.

Graham, S.A., Hendrix, M.S., Wang, L.B., 1993. Collision successor basins of western China: impact of tectonic inheritance on sand composition. Geo. Soc. of Amer. Bull., 105:323-344.

Gu, J.Y., Zhang, G.Y., Li, X.D., et al., 2002. The innovation of geologic theory and related technology and development of petroleum exploration. Acta Petrolei Sinica, 23(1):6-10.

Guan, S.W., Li, B.L., Hou, L.H., He, D.F., et al., 2007. New petroleum exploration domain in footwall concealed structures on northwestern margin of the Junggar Basin. Petroleum Exploration & Development, 35(1):17-22.

Guo, L.Z., Shi, Y.S., Lu, H.F., et al., 1992. Two types of distant tectonic effects of the Indo-Tibetan collision. Symposium of the Researches on Modern Geology (vol. 1). Nanjing University Press, Nanjing, pp. 1-8.

Hao, Y.L., Zhang, Z.C., Wang, F.S., et al., 2004. Petrogenesis of high-Ti and low-Ti basalts from the Emeishan large igneous province. Geological Review, 50(6):587-592.

Harrison, T.M., et al., 1992. Raising Tibet Science, 255:1663-1670.

Hendrix, M.S., Dumitru, T.A., Graham, S.A., 1994. Late Oligocone-early Miocene Unroofing in the Chinese Tian Shan: An early effect of the India-Asia collision. Geology, 22(6):487-490.

Huang, T.K. (Trans.), 1954. On Major Tectonic Forms of China. Men. No A 20 Geol. Sutv. China. The Geological Publishing House, Beijing.

Huo, Y.L., Tan, S.D., 1995. Petroleum Geology and Exploration Practice in the Jiuquan Terrigenous Basin. Petroleum Industry Press, Beijing, pp. 5-60.

Jia, C.Z., 1997. Structural Characteristic of the Tarim Basin and Related Petroleum System. Petroleum Industry Press, Beijing, pp. 1-110.

Jia, C.Z., 2005. Attributes of foreland thrust belt and related natural gas accumulation in western-central China. Petroleum Exploration & Development, 32(4):9-15.

Jia, C.Z., 2007. Characteristics of intra-continental deformation and hydrocarbon distribution under control by the Himalayan tectonic movements in China. Earth Science Frontiers, 14(4):96-104.

Jia, C.Z., Shi, Y.S., 1986. The study of type A subduction zone of Yanshanian tectonic period in the Eastern Qinling mountains. Journal of Nanjing University (Natural Sciences), 22(1):120-128.

Jia, C.Z., Wei, G.Q., Yao, H.J., et al., 1995. Structural Evolution and Regional Tectonics in the Tarim basin. Petroleum Industry Press, Beijing.

Jia, C.Z., He, D.F., Lei, Z.Y., et al., 2000. Petroelum Exporation in Foreland Thrust Belts. Petroleum Industry Press, Beijing, pp. 1-116.

Jia, C.Z., Hu, Y.Y., Tian, Z.J., et al., 2001a. The exploration of large scale natural gas field in the Kuqa depression, Tarim basin. *In*: Gao, R.Q., Zhao, Z.Z. (eds.), Frontier Petroleum Exploration in China (vol. 1). Petroleum Industry Press, Beijing, pp. 1-254.

Jia, C.Z., Yang, S.F., Chen, H.L., Wei, G.Q., et al., 2001b. Structural Geology and Natural Gas in Northern Margin Basin Group of the Tethys. Petroleum

Industry Press, Beijing, pp. 1-161.

Jia, C.Z., Chen, H.L., Yang, S.F., Lu, H.F., et al., 2003a. The Late Cretaceous uplifting and related geologic effects in Kuqa Depression. Acta Petrolei Sinica, 24(3):1-5.

Jia, C.Z., Wei, G.Q., Li, B.L., et al., 2003b. Two developmental phases of foreland basin in western-central China and related natural gas. Acta Petrolei Sinica, 24(2):13-17.

Jia, C.Z., Wei, G.Q., Li, B.L., 2005. Characteristics of superimposed-composite micro-craton basins and their petroleum bearing systems in western-central China. Geological Journal of China Universities, 11(4):479-492.

Jia, C.Z., He, D.F., Shi, X., et al., 2006. Characteristics of China's oil and gas pool formation in latest geological history. Science in China (Series D, Earth Sciences).

Jia, C.Z., Li, B.L., Zhang, X.Y., et al., 2007. Formation and evolution of the Chinese marine basins. Chinese Science Bulletin, 52(Supp):1-11.

Jia, D., Wei, G.Q., Chen, Z.X., et al., 2006. Longmen Shan fold-thrust belt and its relation to the western Sichuan basin in central China: New insights from hydrocarbon exploration. AAPG, 90(9):1425-1447.

Jin, J.Q., Song, J.G., 2005. Plate tectonics, development of petroleum basin and petroleum distribution in China. Oil and Gas Geology, 26(1):2-8.

Kao, H., Gao, R., Rau, J., et al., 2001. Seismic image of Tarim basin and its collision with Tibet. Geology, 29:575-578.

Li, B.L., Jia, C.Z., Pang, X.Q., et al., 2007. The deformation pattern of foreland thrust belt in the basin and range system around the Tibetan plateau. Acta Geologica Sinica, 81(9):1-8.

Li, C.Y., 1982. Instruction of Tectonic Map of Asia. The Geological Publishing House, Beijing.

Li, J.M., Liu, S.G., Li, B.L., et al., 2006. The Creation and Development of Type–C Foreland Basin and Related Petroleum Accumulation in Western China. Petroleum Industry Press, Beijing, pp. 1-278.

Li, Y.J., et al., 2001. The coupling relationship between the South Tianshan orogeny and foreland basin in northern Tarim and the Kuqa depression. Xinjiang Petroleum Geology, 22(5):376-382.

Liang, C.T., Song, X.D., 2004. Tomographic inversion of Pn travel times in China. J. Geophys. Res., 109(B11).

Liu, H.P., Liang, H.S., 1994. Structural patterns of the Longmen mountain thrust system and foreland basin development in western Sichuan province. Acta Geologica Sinica, 68(2):101-118.

Liu, S.W., Wang, L.S., Li, C., et al., 2003. Thermo-rheological structure of the lithosphere and its geodynamic significance on northern margin of the Tarim basin. Science in China, Series D: Earth Sciences, 33(9):852-863.

Liu, S.W., Wang, L.S., Li, C., et al., 2005. Thermo-rheological structure of the lithosphere and the Cenozoic thermal regime in the Tarim basin: An inspiration for basin evolution. Acta Geologica Sinica.

Lu, H., Howell, D.G., Jia, D., et al., 1994. Rejuvenation of Kuqa foreland basin, northern flank of the Tarim basin, Northern China. International Geology Review, 36:1151-1158.

Lu, H.F., Chen, C.M., Liu, Z.H., et al., 2000. The origin and structural attributes of the Kuqa rejuvenation foreland thrust belt. Acta Petrolei Sinica, 21(3):18-24.

Luo, Z.L., 1984. Discussion of Chinese type (Type-C) thrust belt and related petroleum exploration. Oil and Gas Geology, 5(4).

McClay, K.R., 1992. Glossary of thrust tectonics terms. *In*: McClay, K.R. (ed.), Thrust Tectonics. Chapman & Hall, New York, pp. 419-433.

Mitra, S., 1986. Duplex structures and imbricate thrust systems: Geometry, structural position and hydrocarbon potential. American Association of Petroleum Geologists Bulletin, 70:1087-1112.

Mitra, S., Namson, J.S., 1989. Equal-area balancing. American Journal of Science, 289:563-599.

Mitra, S., 1990. Fault-propagation folds: geometry, kinematic evolution and hydrocarbon traps. American Association of Petroleum Geologists Bulletin, 74:921-945.

Mitra, S., 1992. Balanced structural interpretations in fold and thrust belts. *In*: Mitra, S., Fisher, G.W. (eds.), Structural Geology of Fold and Thrust Belts. John Hopkins University Press, Baltimore, pp. 53-77.

Mitra, S., Mount, V.S., 1998. Foreland basement-involved structures. AAPG Bulletin, 82(1):70-109.

Pan, G.T., Chen, Z.L., Li, X.Z., et al., 1997. Establishment and Development of the Eastern Tethys Structure. The Geological Publishing House, Beijing, pp. 1-218.

Poblet, J., Hardy, S., 1995. Reverse modeling of detachment folds: application to the Pico de Aguila anticline in the south Central Pyrenees (Spain). Journal of Structural Geology, 17:1707-1724.

Poblet, J., McClay, K., 1996. Geometry and kinematics of single-layer detachment folds. American Association of Petroleum Geologists Bulletin, 80(7): 1085-1109.

Qiu, Z.J., Gong, Z.S. (eds.), 1999. Petroleum Exploration in China (vol. 1). Petroleum Industry Press, Beijing, pp. 1-19.

Rich, J.L., 1934. Mechanics of low-angle overthrust faulting as illustrated by Cumberland thrust block, Virginia, Kentucky and Tennessee. American Association of Petroleum Geologists Bulletin, 18:1584-1596.

Shaw, J.H., 1993. Active Blind-Thrust Faulting and Strike-Slip Fault-Bend Folding in California. Doctoral Dissertation, Princeton University, Princeton, NJ, p. 216.

Shaw, J.H., Bischke, R.E., Suppe, J., 1994a. Relations between folding and faulting in the Loma Prieta epicentral zone: Strike-slip fault-bend folding in Loma Prieta, California, Earthquake of October 17, 1989. U.S. Geological Survey Professional Paper 1550-F, p.131.

Shaw, J.H., Hook, S.C., Suppe, J., 1994b. Structural trend analysis by axial surface mapping. American Association of Petroleum Geologists Bulletin,

78(5):700-721.

Shaw, J.H., Hook, S.C., Suppe, J., 1996. Structural trend analysis by axial surface mapping-Reply. American Association of Petroleum Geologists Bulletin, 80(5):780-787.

Shaw, J.H., Suppe, J., 1996. Earthquake hazards of active blind-thrust faults under the central Los Angeles basin, California. Journal of Geophysical Research, 101(B4):8623-8642.

Shaw, J.H., Bilotti, F., Brennan, P., 1999. Patterns of imbricate thrusting. Geological Society of America Bulletin, 111(7):1140-1154.

Shaw, J.H., Connors, C., Suppe, J., 2004. Seismic interpretation of contractional fault-related folds: An AAPG seismic atlas. American Association of Petroleum Geologists Special Publication, pp. 1-270.

Shu, L.S., Wang, B., Zhu, W.B., 2007. Geologic age of radiolarian fossils from ophiolitic mélange, South Tianshan mountain and its tectonic significance. Acta Geologica Sinica, 81(9):1161-1168.

Song, Y., Jia, C.Z., Zhao, M.J., et al., 2002. Controlling factors for large gas field formation in thrust belt of Kuqa coal derived hydrocarbon foreland basin. Chinese Science Bulletin, 47 (Supp.):55-61.

Song, Y., Wei, G.Q., Zhao, M.J., et al., 2008. Petroleum Geology of the Foreland Basins in Western-Central China. Science Press, Beijing, pp. 1-236.

Suppe, J., 1983a. Geometry and kinematics of fault-propagation folding. Ecologic Geological Helvetiae, 83(3):409-454.

Suppe, J., 1983b. Geometry and kinematics of fault-bend folding. American Journal of Science, 283(3):684-721.

Tapponnier, P., Xu, Z., Roger, F., et al., 2001. Oblique stepwise rise and growth of the Tibet Plateau. Science, 294(23):1671-1677.

Teng, J.W., Zeng, R.S., Yan, Y.F., et al., 2002. The depth distribution of Moho in eastern Asian continent and surrounding territorial waters and structural framework. Science in China, Series D: Earth Sciences, 32(2):89-100.

Tang, L.J., Jin, Z.J., Jia, C.Z., et al., 2004. A large complex of salt nappe in foreland fold-thrust belt in the Kuqa. Acta Geologica Sinica, 78(1):17-23.

Wang, L.S., Li, C., Shi, Y.S., 1995. Geothermal distribution in the Tarim basin. Chinese Journal of Geophysics, 38(6):855-856.

Wang, L.S., Li, C., Yang, C., 1996. Geothermal structure in formations in the Tarim basin. Chinese Journal of Geophysics, 39(6):795-803.

Wang, L.S., Li, C., Liu, S.W., et al., 2003. Geothermal gradient distribution in the Kuqa regenerated foreland basin on northern margin of the Tarim. Chinese Journal of Geophysics, 46(3):403-407.

Wang, Q., Zhang, P.Z., Freymueller, J.T., et al., 2001. Present-day crustal deformation in China constrained by Global Position System measurements. Science, 294:574-577.

Wang, S.Y., Hearn, T.M., Xu, Z.H., et al., 2001. The Pn wave velocity structure on top of the upper mantle beneath the Chinese continent. Science in China, Series D: Earth Sciences, 31(6):449-454.

Wei, G.Q., Jia, C.Z., Shi, Y.S., et al., 2000. Tectonic characteristics and petroleum prospects of the Cenozoic compound rejuvenated foreland basins in the Tarim. Acta Geologica Sinica, 74(2):123-133.

Wei, G.Q., Jia, C.Z., Li, B.L., et al., 2002. Silurian to Devonian foreland basin on the south edge of Tarim basin. Science Bulletin of China, 47(Supp.):42-46.

Wei, G.Q., Li, B.L., Xiao, A.C., 2005. Characteristics of strike-thrust structures and petroleum exploration concept in northern Qaidam Basin. Earth Science Frontiers, 12(4):397-402.

Wei, Y.J., He, D.F., Lei., Z.Y., 2004. The Permian thrust fault and related sedimentary deposits in foreland thrust belt on northwestern margin of the Junggar basin. Acta Geologica Sinica, 78(5):612-625.

Wu, K.Y., Zha, M., Wang, X.L., et al., 2005. Further researches on tectonic evolution and dynamic setting of the Junggar basin. Acta Geoscientica Sinica, 26(3):217-222.

Xiao, L., Xu, Y.G., He, B., 2003. Interaction between the Emei mantle plume and sub-continental lithosphere: Sr-Nd and O isotopic evidences from low-Ti and high-Ti basalts. Geological Journal of China Universities, 9(2):208-217.

Xiao, X.C., Tang, Y.Q., Li, J.T., et al., 1991. Tectonic Evolution on Southern Margin of the Paleo-Asian Composite Mega Suture Belt. Beijing Science & Technology Press, Beijing, pp. 1-29.

Xiao, X.C., Liu, X., Gao, R., 2004. The Crustal Structure and Tectonic Evolution in Southern Xinjiang Province. The Commercial Press, Beijing, pp. 4-9.

Xie, H., Zhao, B., et al., 1984. Oil-bearing features of overthrust belt on northwestern margin of the Junggar basin. Xinjiang Petroleum Geology, 5(3):1-15.

Xing, X.J., Zhou, D.W., Liu, Y.Q., et al., 2004. Geochemistry of the Early Permian volcanic rocks and their tectonic setting in the Turpan–Hami basin. Xinjiang Geology, 22(1):51-55.

Xu, J.H., 1994. Back-arc collisional orogeny and related tectonic facies. Journal of Nanjing University, Earth Sciences, 6(1):1-11.

Xu, Y.G., Chung, S.L., 2001. Petrologic and geochemical constraints on the petrogenesis of Permian-Triassic Emeishan flood basalts in southwestern China. Lithos, 58:145-168.

Xu, Y., Liu, F.T., Liu, J.H., et al., 2001. Attributes of deep level continental collision in northwestern China and their dynamic significance. Chinese Journal of Geophysics, 44(1):40-47.

Yang, S.F., Jia, C.Z., Chen, H.L., et al., 2002. The development of the Tethys structure belt and the creation of the northern margin basin group, as well as the implication of natural gas exploration in the Tarim basin. Chinese Science Bulletin, S1:36-43.

Yang, S.F., 2006. The Structural Attribute of Thrust Belt and the Implication of Petroleum Exploration on Northern Margin of Qilian mountain. Science Press, Beijing, pp. 1-159.

Yang, Y.Q., Liu, M., 2002. Cenozoic deformation of the Tarim basin and the implications for mountain building in the Tibetan Plateau and Tianshan. Tectonics, 21(6).

Zhai, G.M., Wang, J.J., 2000. The regularity of oil and gas distribution. Acta Petrolei Sinica, 2l(1):1-9.

Zhao, W.Z., Xu, D.F., Zhao, C.J., et al., 1998. Category of structural deformed sequences in Kuche depression and their significance for oil and gas exploration. Acta Petrolei Sinica, 19(3):1-5.

Zhao, Z.H., Guo, Z.J., Zhang, Z.C., et al., 2004. The geochemical characteristics and tectonic setting of the Lower Permian basalts in the Hongliuhe area, between Xinjiang and Gansu provinces. Geological Journal of China Universities, 10(4):545-552.

Zhao, Z.Z., Li, Y.T., Ye, H.F., et al. (eds.), 2001. Tectonic Characteristics and Related Basin Evolution on the Qinghai-Tibet Plateau. Science Press, Beijing, pp. 102-132.

Zhang, G.M., Wang, S.Y., Li, L., et al., 2002. The depth of earthquake center in Chinese continent and its tectonic significance. Chinese Science Bulletin, 47(5):663-669.

Zhang, G.Y., Xue, L.Q., 2002. Petroleum distribution and exploration implication in foreland basin, western-central China. Petroleum Exploration & Development, 29(1):1-5.

Zhong, D.L., Ding, L., 1996. Uplift of the Qinghai-Tibet plateau and its mechanism. Science in China, Series D: Earth Sciences, 26(4):289-295.

Zhang, G.Y., Chen, F.J., Wang, X.W., 1994. Deformation style and its distribution in northern Tarim basin. Earth Science (Journal of China University of Geosciences), 19(6):755-768.

Zhang, K., 1991. Discussion of divergence, drifting, collision and convergence of Chinese plates and development of petroleum basin. Xinjiang Petroleum Geology, 12(2).

Zhang, Y.C., Zhang, H., Sun, Z.C., et al., 1997. Analysis of Prototype Petroleum Basins in China. Nanjing University Press, Nanjing, pp. 1-449.

Zhang, Z.C., Wang, F.S., 2002. Geochemistry of two types of basalts in the Emeishan basaltic province: evidence for mantle plume-lithosphere interaction. Acta Geologica Sinica, 76(2):229-238.

10

Oil and Gas Exploration in Chinese Foreland Fold and Thrust Belts

In western-central China, oil and gas exploration in the foreland fold and thrust belts experienced five exploration episodes. Before 1949, the preliminary surface investigation was the first episode of Chinese petroleum exploration. From the 1950s – 1960s, the exploration stepped forward to the initial reorganization episode and the basic advances in techniques episode. From the middle of the 1960s to the middle of the 1980s, the exploration was in recess. Since the middle of the 1980s, the exploration has progressed into a fully operational mode. Because of the complexity of mountains and ranges, seismic data were hard to obtain and seismic images were poor. Also, the complication of structural deformation inserted additional difficulties to seismic data processing and data interpretation. Furthermore, abnormally high pressure underground brought a higher drilling risk. Therefore, the high risk of exploration, the complexity of the topography and the complication of the geological conditions determined that, in the foreland fold and thrust belts, oil and gas exploration is an intricate process. Recent exploration results reveal that foreland basins contain large scale oil and gas reservoirs that have sufficient reserves and a higher production rate per single well. For example, the Kela 2 gas field in the Kuche depression contains an explored natural gas field of 48.1 km^2, an explored natural gas reserve of $2,840.29 \times 10^8$ m^3, and a reserve abundance of 59×10^8 m^3/km^2. Recently, petroleum exploration achieved an important break through in the Yilake structure in the Kuche depression and in the Kulong mountain structure in the Jiuxi basin, which revealed a great potential for oil and gas exploration in foreland fold and thrust belts of western-central China.

10.1 Oil and Gas Exploration in the Kuche Foreland Fold and Thrust Belt and the Kela 2 Natural Gas Field

The Kuche depression in the Tarim basin is a foreland basin of the South Tianshan orogeny. It associates with the South Tianshan orogeny in the north and borders with the Tabei uplift in the south. From east to west, the Kuche depression includes the Yangxia depression, the Baicheng depression and the Wushi depression. In the direction east to west, the length of the Kuche depression is about 550 km; in the direction north to south, the width of the Kuche depression is in the range of 30 – 80 km. The Kuche depression covers 28,515 km^2 in total. The Kuche depression can be further divided into four structural belts and three depressions (Fig. 10.1); the Kela 2 natural gas field is located in the central part of the Kelasu structural belt. Under strong tectonic compression of the South Tianshan orogeny, the folds and fold and thrust belts were formed in arrays or in ranges in the Kuche depression (Fig. 10.1) (Jia et al., 2001; Zhang et al., 1996; Zhao et al., 1999). Inside these structural belts, all types of fault-related folds developed, such as fault-bend folds, pop-up folds, fault-propagation folds and duplex structure. The Kela 2 structure is a pop-up fold under a duplex structural background. The characteristics of structure distributions not only demonstrate belt patterns from the north to the south, but these can also be divided into eastern and western sections. In addition, these can be separated into upper level and lower level structures as well. In an W-E direction, the sectional division is relatively distinctive. By using the Kuche River and the Kalayuergun fault as boundaries, the fold and thrust belts can be divided into three sections that are the eastern section, middle section and western section. Because compressive stress altered its strength and shifted its direction along the strike direction of the structures, this kind of sectional division was formed. The upper level structure and the lower level structure are related to decollement layers in the Neogene Jidike formation, in the Palaeogene gypsum, gypsum mudstone and mudstone Jurassic mudstone, shale and coal beds. The variations in structural distribution decided distributions of large and middle size natural gas fields.

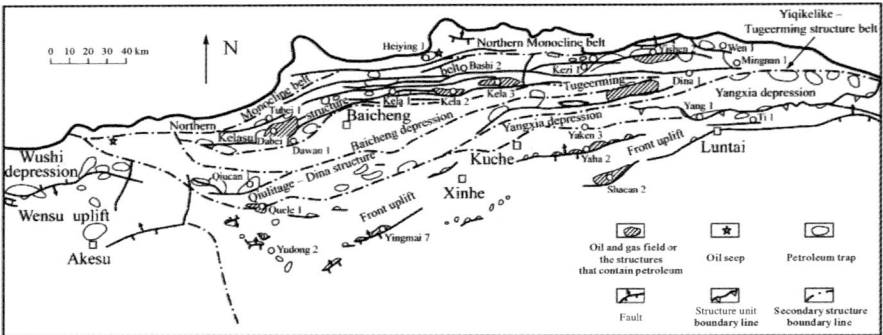

Fig. 10.1.   The structural units in the Kuche foreland basin

10.1.1   Oil and Gas Exploration History of the Kuche Foreland Fold and Thrust Belt

In the Kuche depression, the initial exploration in the mountain front fold and thrust belt began in the 1950s, which mainly targeted surface or shallow structures.

Until the 1990s, the exploration progressed into a new era. Guided by the theory of fault-related folds and targeting the structures under the salt layer, we solved the problems of seismic data collection, processing and interpretation and we created ten structural interpretation models that improved structural interpretation accuracy. In addition, for effectively drilling and testing gas wells in an abnormally high pressure area, we worked out the techniques for drilling in extremely steep structures, high pressure well logging and other related well testing. These technical advancements offered accessibility to oil and gas exploration in the Kuche area. In 1995, drilling at the Kecan 1 well discovered an excellent reservoir-seal combination that is made of the Palaeogene gypsum-salt layer and Cretaceous sandstone. On March 25, 1997, the drilling of the Kela 2 well was initiated. On January 20, 1998, well testing was carried out in the Palaeogene dolomite section. By using a 6.35 mm nozzle, the daily production rate was 27.71×10^4 m^3. After the discovery of natural gas in the Kela 2 well, highly productive gas flows were obtained in the Kela 201 well, the Kela 203 well and the Kela 204 well. Thus, the Kela 2 natural gas field was discovered. Up to now, the Kela 2 gas field is the largest sandstone natural gas field in China, which contains the highest abundance and has the best entirety of a petroleum system. The drilling of the Kela 2 well stopped on May 24, 1998. The total depth of the well is 4,130 m. At the base of the well, the strata belong to the Shushanhe formation in the Lower Cretaceous; the drilling encountered a reservoir thickness of 341 m. The discovery of the Kela 2 natural gas field was an important milestone for oil and gas exploration in the mountain front structural belts in China. Successively, several natural gas structures were discovered, which included Tuziluoke, Dina, the Dabei 1 gas field, Kela 3, Yinan 2 and the Yeyun; additionally, the Yilake structure was discovered in the Wushi depression, which was a large scale structure and contained both oil and gas (predicted geological reserve of natural gas was $3,447.12\times10^8$ m^3). With these natural gas discoveries, the Kuche depression not only became one of the four largest natural gas regions in China, but is also established the material foundations for the initiation of the Chinese "West-East Natural Gas Transmission Project".

10.1.2   Geological Settings and Petroleum Systems in the Kuche Foreland Fold and Thrust Belt

● **Geological Features of the Kuche Foreland Fold and Thrust Belt**

Evolution and Development of the Basin

The Kuche foreland basin was formed during the Mesozoic–Cenozoic with mostly terrestrial deposits. The basin went through three developmental stages, which were the foreland basin stage during the Late Permian–Triassic, the intra-continental depression stage during the Jurassic, and the re-generated foreland basin stage during the Cretaceous–Quaternary (Jia, 1997; 2005; Jia et al., 2001). The Triassic foreland basin had a large subsidence rate, huge settling volume and massive sedimentary deposits. However, crossing the basin from north to south, the settling volumes were gradually reduced. The basal conglomerate of the Ehuobulake formation in the Lower Triassic series is a set of molasse deposits of alluvial fan and fan delta facies, which reveals a flexural subsidence event that happened rapidly during the basin's early development stage. During the Middle and Upper Triassic, due to the under compensation of sediments supply, the Kelamayi formation and the Huangshanjie formation were formed by find grain sediments in a relatively deep water environment. During the Upper Triassic, the tectonic subsidence rate was gradually reduced. The supply of source materials and the subsidence rate of the basin reached an equilibrium point; in addition, the depth of water became shallower. The Upper Triassic Taliqike formation was formed by coarse grain sediments of meandering river facies and flood plain facies. During the Early and Late Jurassic, the basin was in a state of rapid subsidence. However, during the Middle Jurassic, the basin was relatively stable. From the Cretaceous–Palaeogene, the re-generated foreland basin experienced a mild extension during its early development stage. During the Early Cretaceous, the subsidence rate was slow and steady, the depositional environment developed into a widely open, shallow, oxidizing lake. Throughout the Late Cretaceous, the basin was upraised and it experienced various erosion events. During the Palaeogene, the Kuche depression experienced a marine transgression event that brought a set of sediment deposits from brackish facies to river facies. During the Neogene, influenced by South Tianshan orogeny reactivation, a strong flexural subsidence event happened and it brought a re-generated foreland basin. From the late Neogene–Quaternary, the basin was deformed with folds. At same time, on the foreslope, faulted-blocks were formed by normal faults.

Characteristics of Structural Geology

The Kuche foreland fold and thrust belt is a northern monocline belt that is very close to the South Tianshan orogeny. The basement-involved type thrust fault system and structure wedge that are controlled by back thrust faults are formed at a deep level. In addition, the deep level structure is presented as a monocline belt. Towards the foreland, the second array of the structure belt includes the Kelasu structure belt and the Yiqikelike–Tugeerming structural belt. On the surface and at a shallow depth, the Kelasu structure belt has two ranges of anticlines, southern array and northern array anticlines, which mostly are fault-propagated folds on top of the salt layer. They also could develop an overthrust structure along the decollement layer of the Palaeogene gypsum-salt layer. The salt layer can form salt domes. Under the salt layer, the major structures are an imbricate fold and thrust belt and duplex structures. The fault-bend anticlines and overthrust blocks are common structures. Deeply buried structures are not superimposed on the shallow structures. The Yiqikelike–Tugeerming structure belt is a relatively narrow one, which characteristically contains an imbricate thrust system and the basement-involved type structure. Towards the foreland at a further distance, the last array of the structure belt is the Qiulitage–Dina structure belt. In the eastern section, the Dina–eastern Qiulitage structure belt primarily includes fault-propagated folds above salt layers and overthrust structures above the Palaeogene salt layer, but it predominately contains a passive-roof duplex structure under the salt layer. In the western section, above the salt layer, the Quele–western Qiulitage structure belt contains fault-propagated folds, which are controlled by ramp faults and which are recognized as the northern and southern Qiulitage anticline belts. However, under the salt layer, it has a basement-involved type faulted block structure that is related to the reactivation of previous basement structures. Inside the salt layer, the salt structures include salt pillow and salt anticline. Towards the front bulge, there is a series of faulted blocks that are controlled by normal faults.

Characteristics of Sediments and the Combination of Reservoir and Seal

The Kuche depression predominately contains Mesozoic and Cenozoic erathem. Total stratigraphic thickness can reach up to 10,000 m. In general, the thickness of the Mesozoic strata is about 2,000 – 3,000 m and, in isolated locations, the thickness has reached as much as 4,000 m. The depositional environment diverged from lacustrine facies–palustrine facies–river facies. The Mesozoic lithology is gray, interbedded sandstone and mudstone. In addition, in the middle and lower portions of the Mesozoic stratigraphic column, there are interbedded coal beds, carbonaceous mudstone and oil shale. Commonly, the thickness of the Cenozoic strata is about 3,000 – 5,000 m. However, in isolated locations, the thickness can reach up to 8,000 m. The depositional facies are lacustrine facies and river facies. The Cenozoic lithology is brown, interbedded sandstone and mudstone. The middle and lower Cenozoic strata are also interbedded with two sets of gypsum mudstone and salt layers. Near the base, the strata are interbedded with thin layered dolomite.

In ascending order, three major reservoir-cap rock combinations developed in the Kuche depression. (i) In the Jurassic, the combination is located in the upper Yangxia formation–lower Kezilenuer formation, the upper Yangxia formation and fine grain sandstone and pebble sandstone in the middle Yangxia formation. This combination primarily occurs in the northeastern part of the depression with the discovery of the Yinan 2 natural gas reservoir. (ii) The reservoir-seal combination includes the Palaeogene gypsum-salt layer and gypsum mudstone, the Palaeogene sandstone and the sandstone of the Bashijiqike formation in the Cretaceous system. This combination mostly appears in the western-central part of the Kuche depression, which is the best reservoir-seal combination in the Kuche foreland basin. Documented natural gas reservoirs are the Kela 2 and Kela 3, the Dabei 1 and the Dina 1 and Dina 2; in addition, the Quele 1 oil reservoir. (iii) In the Neogene system, the reservoir-seal combination incorporates gypsum mudstone, siltstone, fine grain sandstone and sandy pebble stone in the Jidike formation. This combination appears in the eastern and southern parts of the Kuche depression.

The discovered gas reservoirs are the Tuziluoke, the Dina 1 and Dina 2 gas reservoirs.

● **Settings of Petroleum Geology and the Features of Petroleum System**

Hydrocarbon Source Rock

In the Kuche depression, the hydrocarbon source rocks of coal measures are widely distributed in the Triassic–Jurassic with massive thickness and higher organic abundance. Most organic substances are of humus type within the high end of post-maturity. These source rocks primarily generate the natural gas and they are the material foundation for regional natural gas fields in the Kuche area (Zhao et al., 1999; Zhao and Zhang, 2001). In addition, these source rocks also generate a certain amount of liquid hydrocarbons. In ascending order, the Upper Triassic hydrocarbon source rocks include the Huangshanjie formation and the Taliqike formation and the Jurassic hydrocarbon source rocks include the Yangxia formation, the Kezilenuer formation and the Qiakemake formation. Among the formations, the Huangshanjie formation, the Yangxia formation and the Kezilenuer formation are the most valuable source rocks. The lithology of the Huangshanjie formation is dark color mudstone of lacustrine facies. The lithology of the Yangxia formation and the Kezilenuer formation primarily contains the coal measures that include carbonaceous mudstone, mudstone and coal petrology. The hydrocarbon source rocks were distributed in a range of $12,000 - 14,000 \text{ km}^2$. The cross section, from the Kuche River–Kapushaliang River, reveals the maximum thickness of the source rock is about 1,000 m. In the Huangshanjie formation, the hydrocarbon source rock is dark color mudstone that contains 0.4% – 5.58% of total organic carbon. In the Yangxia formation, the hydrocarbon source rock of limnetic facies contains 0.81% – 5.98% of total organic carbon. In the Kezilenuer formation, the primary hydrocarbon source rock is dark color mudstone that contains 0.75% – 5.96% of total organic carbon. The additional source rocks are carbonaceous mudstone with 6.1% – 39.9% of the total organic carbon and coal petrology with 41.5% – 93.2% of total organic carbon. Among the organic maceral of hydrocarbon source rocks, the content of exinite+sapropelinite is between 20% – 40%, the content of vitrinite+inertinite is larger than 60%, most of H/C are less than 0.8 and the kerogen carbon isotope is larger than –25‰. The organic maceral data are favorable to the type III organic substance. According to thermal pressure simulation of hydrocarbon generation analysis for the Triassic and Jurassic source rocks, mudstone, carbonaceous mudstone and coal petrology predominately produce natural gas. However, the mudstone of lacustrine facies in the Huangshanjie formation also generates a significant amount of oil. At present, the maturity of hydrocarbon source rocks is higher on the western side than on the eastern side of the basin. In the western part of the basin, the Triassic source rocks reached an oil generation peak during the end of the Palaeogene and reached a natural gas generation peak during the Neogene. The Jurassic source rocks reached an oil generation peak during the middle Miocene and reached a natural gas generation peak during the end of the Neogene. Currently, the source rocks of the western part predominantly produce dry gas. On the eastern side of the basin, the Triassic source rocks started to generate oil during the late Palaeogene. They started to generate natural gas during the end of the Miocene and reached a natural gas generation peak from the Pliocene to the present. The Jurassic source rocks reached an oil generating peak during the end of the Miocene and reached a natural gas generation peak during the Pliocene. The source rocks on the eastern side generate condensate gas–dry gas at present. In conclusion, the source rocks in the Triassic and Jurassic mainly generated oil in the early stage, have entered the natural gas generation peak phase since the Pliocene, and have formed the natural gas reservoir at the late stage. In the Kuche depression, the Triassic and Jurassic source rocks of coal measures were superimposed together to form a hydrocarbon supply center. This not only accumulated oil and gas reservoirs in the Kuche depression, but also built up reservoirs on the foreslope through lateral migration. Together, they constituted the Kuche oil and gas system (Li et al., 2000). This petroleum system has the characteristic of a high abundance of natural gas. The highest gas generation intensity of the Triassic and Jurassic hydrocarbon source rock may reach 280×10^8 m^3/km^2 and there are more than 10,000 km^2 of gas fields with a gas generation intensity larger than 100×10^8 m^3/km^2. Therefore, this is a super injected petroleum system. After the new-round of petroleum resource appraisal, the Kuche oil and gas system is said to contain an oil reserve of 4.1×10^8 t and a natural gas reserve of 2.23×10^{12} m^3.

Reservoir Sequences

In the Kuche depression, the primary targets for exploration are the Jurassic, the Bashijiqike formation in the Cretaceous system, the Palaeogene system and the sandstone in the Jidike formation in the Neogene system. Among these targets, the Bashijiqike formation in the Cretaceous system and the Palaeogene sandstone are the most valuable reservoir sequences. In the Jurassic, the Ahe formation, the Yangxia formation and the Kezilenuer formation are primary reservoir sequences. Predominantly, the reservoir is made of feldspar, clastic sandstone with a total thickness of 407 – 1,336 m. Comparing the thickness of sandstones, the northern portion is thicker than the southern portion. The reservoirs are largely made of sand bodies that were deposited in a braided channel and underwater braided channel on a braided delta. These sand bodies have considerable thickness, few interbedded mudstones, and a stable distribution range. Comparing the physical properties, the eastern portion has better porosity and permeability than the western portion. On the eastern side, from the Tugeerming anticline to the Tuziluoke canyon, the porosity is 10.7% – 17.25% and the permeability is $(0.71 – 418) \times 10^{-3}$ μm^2. The types of pore include the remaining primary intergranular pores, granular dissolution pores and intergranular dissolution–enlarged pores. On the western side, the types of reservoir include tight reservoir sequences and fractured tight reservoir sequences, which commonly have a porosity of 6% – 8% and a permeability of less than 1×10^{-3} μm^2. Because primary intergranular pores have nearly vanished, the granular dissolution pores are the major type of pores. The Cretaceous system contains braided delta facies and shore–shallow lacustrine facies deposits that are feldspar and clastic sandstone. The reservoir sequence mainly consists of the sandstone in the Bashijiqike formation with a thickness of $100 - 300$ m. This sandstone has a lower compositional maturity and intermediate texture maturity. In descending order, the sandstone can be divided into three litho sections. The first and the second litho sections have good physical properties, and generally have intermediate porosity with intermediate permeability, or intermediate–low porosity with low permeability. The average porosity is $9.13\% - 13.74\%$ and the maximum porosity is 22.4%. The average permeability is 49.42×10^{-3} μm^2 and the maximum permeability is $2,340 \times 10^{-3}$ μm^2. The primary type of pore is the remaining primary intergranular pore. When the sandstone was deposited, the paleo-topography was relatively flat and various material sources joined together to form a braided fluvial plain with widely distributed sand bodies. The reservoir model is one of "superimposed sand bodies", which have good connectivity between the sand bodies and less of an interlayer, and which is the primary gas reservoir for the Kela 2 gas field. The Tertiary system includes deposits of river facies, shore–shallow lacustrine facies and saline lake facies within an arid environment. The lithology consists of pebble stones with various amounts of gypsum, sandstone, mudstone, micritic limestone and a gypsum layer. Generally, the Palaeogene reservoir sequences are thicker with coarser grain sediments on the eastern side. However, they are thinner with fine grain sediments on the western side. The basal sandstone is an important natural gas reservoir for the Kela 2 gas field, which has a porosity of $3.997\% - 18.14\%$ and an average porosity of 12.3%. Also, it has a permeability of $(0.03-202) \times 10^{-3}$ μm^2 and an average permeability of 52.6×10^{-3} μm^2. The major type of pore is a secondary intergranular dissolution pore. In the Jidike formation in the Neogene system, the thickness of the reservoir sequence is about 120 m. The formation is featured by coarse grain, a clastic sandy conglomerate of alluvial fan facies–lacustrine facies. The formation can be divided into five litho sections with a stable distribution range horizontally. The reservoir sequences commonly appear in the sandy mudstone section and basal conglomerate section. Field investigation in the eastern Kuche depression indicates that, in the sandy mudstone section, mudstone of shore–shallow lacustrine facies and gypsum layers are well developed. The thickness of the single sand body is moderately thin, the vertical connectivity is poor and the framework of the sand body belongs to the multi-layer type. The sandstone reservoir has a porosity of $8\% - 16\%$, a permeability of $(0.1 - 10) \times 10^{-3}$ μm^2 and a maximum permeability of 209×10^{-3} μm^2. The primary physical properties are intermediate porosity with intermediate permeability and intermediate porosity with lower permeability. The next category is lower porosity with lower permeability.

Cap Rocks

In the Kuche depression, there are three sets of primary regional cap rocks. In the

Neogene system and the Palaeogene system, the cap rocks include gypsum-salt layers and gypsum mudstone. In the Jurassic, the cap rock exclusively consists of coal measures. Among these cap rocks, the one in the Palaeogene system and the one in the Neogene system are two important sets of regional cap rocks, which sealed off a large quantity of oil and gas. In the Palaeogene system, the cap rocks are located in the Kelasu area, west of the Kuche River. The cap rocks include gypsum-salt layers and gypsum mudstone with a thickness of more than 300 m. The diffusion coefficient of gypsum mudstone is 9.98×10^{-13} m^2/s. However, the diffusion coefficient of mudstone is 1.89×10^{-11} m^2/s in general. The test data shows that gypsum mudstone has a very good sealing ability and this is the cap rock for the Kela 2 natural gas field. In the Neogene system, the cap rock in the Jidike formation consists of gypsum-salt layers and gypsum mudstone, which are located at the eastern part of the depression and which also joined with the Palaeogene cap rock to cover the entire Kuche depression. In the Jurassic, the cap rock of coal measures mainly includes a thick layer, carbonaceous mudstone and mudstone interbedded with coal layers, which appears in the lower Kezilenuer formation-the upper Yangxia formation with a thickness of 330 m.

Structural Traps

In the Kuche depression, the fold-fold and thrust belt was formed during the Neogene-Quaternary periods (Wei and Jia, 1998). The tectonic deformations gradually progressed from north to south, which produced successive structures that were older on the northern side and younger on the southern side (Lu et al., 2000). This tectonic movement developed the front propagated thrust faults and a variety of fault-related folds. In the fold-fold and thrust belt, there are compressed anticline, faulted anticline and fault nose trap (Liang and Jia, 1999). Due to the great possibility of discovering oil and gas, the most beneficial structures are anticline and faulted anticline. The faulted nose trap has relatively poor conditions for preserving oil and gas. On the foreslope, faulted anticline traps that relate to normal faults developed. In addition, due to stratigraphic overlapping, there are unconformity traps. In the fold-fold and thrust belt, the type of trap above the salt layer is different from the one under the salt layer. Above the salt layer, the structures include an asymmetrical or overturned, fault-propagated anticline, a salt dome anticline and faulted nose structure, which are formed by a large scale slipping-thrust force. The traps above the salt layer have relatively poor conditions for oil and gas accumulation. Under the salt layer, within the Mesozoic erathem, the structures include a fault-bent anticline, a fault-propagated and fault-bent anticline and a fault-propagated anticline, which are related to the passive-roof duplex structure and the imbricate thrust fault. The best structure trap is the fault-bent anticline that relates to the passive-roof duplex structure, such as the Kela 2 structure.

Characteristics of Oil and Gas Pools

In the Kuche depression, the hydrocarbon accumulation happened during the late geological periods and it included three phases, which were the Palaeogene tectonic phase, the Jidike–Kangcun tectonic phase and the Kuche–Xiyu tectonic phase. These phases were equivalent to the early, middle and late Himalayan tectonic episodes respectively. During the early Himalayan episode, the Triassic source rocks were at the oil generating stage. In the same period, on the northern margin of the Kuche depression, the fold-thrust fault belt, the foreslope structures and the stratigraphic trap initiated their developmental process. Inside the fold-thrust fault belt, most of the accumulated oil and gas was destroyed by later tectonic movement and some of the accumulated oil was turned into secondary residual oil reservoirs, such as the Yiqikelike oil field. During the middle and late Himalayan episodes, hydrocarbon source rocks in the Triassic and Jurassics reached a peak of natural gas generation, which primarily accumulated natural gas with small amounts of light crude. On the slope of the forebulge, during the middle and late Himalayan episodes, oil and gas continually accumulated from the source rocks of the Triassic and the Jurassics.

The regional geologic background determines the Kuche depression as a gas rich region. The Tarim basin not only accommodated the Kuche depression, but it also joined with other central Asian basins, such as the Karakum basin and the Afghanistan–Tajikistan basin, to form a basin group on the northern margin of the Paleo-Tethys Sea. These basins are of similar basin type, comparable deformational features and related major hydrocarbon source rocks (Graham et al., 1993; Hendrix et al., 1994; Ritts et al., 1999). Together, they formed a natural gas region on the northern margin of the Tethys Sea. In the natural gas region on the northern margin of the Tethys Sea, the Middle and Lower Jurassic coal measures of terrestrial facies and littoral facies were primary source rocks with high organic contents and great natural gas generating potential. The hydrocarbon source rocks in the Middle and the Lower Jurassic series commonly entered a high maturity stage, or even a post-maturity stage. Various gypsum-salt layers and mudstones functioned as cap rocks with a broad distribution range and high quality. In the Kuche depression, coal measures of the Middle and Lower Jurassic series were the essential substance for generating a large scale natural gas field. The sandstones of the Jurassic and Cretaceous systems joined with the Palaeogene gypsum-salt layers to form a perfect reservoir-cap rock combination. In the fold-thrust fault belt, the anticlines developed into a format of arrays or ranges. These anticlines were formed at an ideal period to match the major gas generating peak. These geological settings determined that the Kuche depression had the geological environment to accumulate a large scale gas field and to become an important part of a natural gas region on the northern margin of the Tethys Sea.

The best oil and gas accumulations are under the gypsum-salt layers and the gypsum mudstone in the Tertiary system. In the Kuche depression, cap rocks in the Jidike formation in the Neogene system and two sets of massive gypsum-salt layers and gypsum mudstone in the Palaeogene system are beneficial factors and crucial elements for developing large and medium size natural gas fields of ultra-higher pressure (Zhou, 2001). In addition, these cap rocks combined with the thin layer sandstone of the Jidike formation in the Neogene system, dolomite of the Kumugeliemu formation in the Palaeogene system, the basal conglomerate section in the Palaeogene system and the sandstone of the Bashijiqike formation in the Cretaceous system to form the best reservoir-cap rock combinations. Most of the discovered oil and gas fields belong to this kind of reservoir-cap rock combination.

In the fold-thrust fault belt, the second and third structural arrays produced the best anticline traps that were located in the passive-roof duplex structure, under the salt layer. A series of thrust faults developed in the fold-thrust fault belt in the Kuche depression. The strike of thrust faults was nearly east-west and the thrust direction was to the south. Among thrust fault controlled duplex structures and passive-roof duplex structures, the widely extended, low angle, fault-bent anticline provided the best oil and gas traps in the Tertiary system, under the salt layer. The fault that forms duplex structures connects the Triassic–Jurassic source rock downward and the Cretaceous–Tertiary reservoir upwards However, the fault stops at the gypsum-salt cap rock. This kind of trap offers the best oil and gas preservation environment. The Kela 2 anticline is a good example (Fig. 10.3). The thrust faults that were developed above the salt layer extend into the shallow strata or to the surface. The most common structures are fault-propagated anticlines. In the area where the salt volume was increased by plastic flow, arch extension was formed, which might create anticlines and faulted nose structures in the overlaying strata. Furthermore, above the salt layer, the conditions of hydrocarbon accumulation and preservation were poor, and only secondary residual oil and gas were preserved. Therefore, within the Dabei–Kelasu–Yinan structure belt and the Qiulitage–Dona structure belt, under the salt layer, the passive-roof duplex structure and fault-bent anticline offer the best hydrocarbon traps. However, the imbricate thrust faults developed on the northern margin of these two structure belts and a shallow decollement fold appeared on their southern side. Therefore, the conditions for oil and gas accumulation were poor.

10.1.3   Characteristics of the Kela 2 Gas Field

The Kela 2 natural gas field is located in the middle section of the Kelasu structure belt in the Kuche depression. It is next to the Baicheng depression to the south and it is adjacent to the Northern Monocline belt to the north (Fig. 10.1). The creation and evolution of the Kelasu structure belt is mainly controlled by the North Dayuanqi–Kelasu fault belt. Inside the Kelasu structure belt, there are different types of fault-related folds, such as fault-bend fold, pop-up fold, fault-propagate fold, and duplex structure. The Kela 2 structure is a pop-up fold under the duplex structure background.

● **Attribute of Reservoir Sequences**

In the Kela 2 gas field, primary reservoir sequences include sandy conglomerate in the Palaeogene system and the Bashijiqike formation in the Cretaceous system. The porosity is in the range of 8% – 20%; the maximum porosity is 22.4%; the peak value is 15%; the average is 12.56%. The permeability is in the range of (0.1 – 1,000)×10^{-3} μm^2 in general; the maximum permeability is 1,770×10^{-3} μm^2; the average permeability is 49.4×10^{-3} μm^2. Porosity and permeability have a good linear relationship that shows the evenly distributed pore and throat volume and excellent physical properties. Compared with the Bashijiqike formation, the physical properties of the Baxigai formation are poor. The porosity is in the range of 2.67% – 13.7%; average porosity is 9.9%. The permeability is in the range of (0.04 – 6.62)×10^{-3} μm^2; average permeability is 0.78×10^{-3} μm^2. In the Kela 2 natural gas field, the distribution and thickness of all micro facies are stable inside the Cretaceous Bashijiqike formation, the continuity of sand bodies is very good, and the distribution of mudstone is scattered or partial. When the Bashijiqike formation was deposited, there were multiple material sources of plentiful supply, and accommodation spaces were reduced to a minimum. The fluvial deposition was extensive and the erosion of the river channel was strong. The primary facies were braided fluvial plain facies and delta front facies (Jia, 2000; Jia et al., 2001). This depositional environment produced a set of tabular sand bodies, which had good continuity horizontally and great thickness vertically and which was distributed in a broad range.

In the Palaeogene Kumugeliemu group, the lithology in the dolomite section contains sparite doloarenite, micrite doloarenite and arenite dolomicrite, micrite bio-dolomite, micrite dolosiltite, politic dolomitic gypsum and gypseous dolomite. In the Palaeogene dolomite, the average porosity is 11.4% and the average permeability is 3.6×10^{-3} μm^2. The primary type of reservoir space is of the porous type, and the fractures offer secondary reservoir space. The reservoir type belongs to the porous type.

The pore types are bio-mold pores, intergranular dissolution pores, intercrystalline pores, intercrystalline dissolution pores and non-fabric selective dissolution pores. The bio-mold pores occur in micrite doloarenite and bio-dolomite, intergranular dissolution pores appear in sparite doloarenite and intercrystalline dissolution pores and intercrystalline pores occur in dolosiltite.

● Features of Structural Trap

The Kela 2 structure is controlled by two back thrust faults on the northern side and on the southern side (Fig. 10.2). On its northern boundary, the strike of controlling fault is oriented in a WE direction and it turns into an NE direction. The fault throw is about 1,000 m and the fault plane dips to the south or the southeast. This is a large scale, regional fault that extends 23 km. On its southern boundary, the controlling faults consist of a series of thrust faults. The strikes of these faults are oriented in a WE direction and in an NE direction. The fault planes dip to the north and the northwest; and the faults extend 5 – 15 km. Compressed by the compression-distortion stress field that was oriented in an NS direction, three sets of minor faults were formed, which strike in an NWW-SEE direction, an

NNW-SSE direction and an NE-SW direction. These minor faults were either conjugate shear faults or tensional normal faults that occurred along the axis of the structure. In total, there are 34 minor faults with a fault throw of 10 – 90 m and an extension of 1 – 4 km. Compared with the thickness of reservoir strata and height of the gas column, these minor faults are relatively small in scale and they are covered by a massive gypsum-salt layer. Therefore, minor faults do not influence the reservoir pattern and the design of the production well.

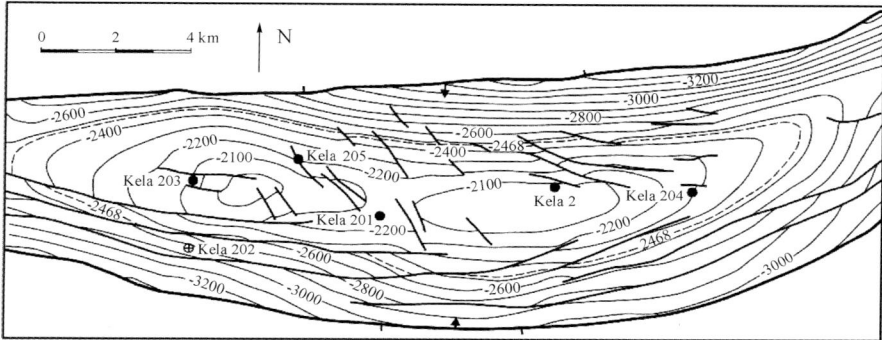

Fig. 10.2.   Upper surface structure of Dolomite in the Palaeogene system in the Kela 2 gas field

The Kela 2 structure is a symmetrical anticline and its long axis is oriented in a W-E direction. The length of the long axis is 18 km and the extent of the short axis is 3 – 4 km. The ratio of the short axis to long axis is 1:4.5. On the northern wing, the dip angle of strata is between 16° – 20° and on the southern wing, the dip angle of strata is between 19° – 22°. The enclosure loop for the trap is defined at a contour line of –2,480 m. This trap covers 49.56 km² with a structural relief of 510 m. The seep point is located at the west end of the anticline. If using the contour line of –2,468 m as the enclosure loop line, the size of the gas trap is 48.06 km² (Table 10.1).

Table 10.1   Structural trap attributes of the Kela 2 gas field

Name of reservoir sequences	Elevation of peak point (m)	Structure relief (m)	Elevation of enclosure loop (m)	Acreage of enclosure (km²)	Acreage of reservoir (km²)
Top surface of Dolomite section in the Palaeogene system	–1,970	510	–2,480	49.56	48.06
Top surface of sandstone and conglomerate section in the Palaeogene system	–2,000	510	–2,510	49.35	45.00
Bashijiqike formation in the Cretaceous system Top surface of the 3rd litho-section	–2,230	480	–2,710	37.07	14.96
Top surface of the Baxigai formation in the Cretaceous system	–2,330	440	–2,770	30.50	4.36

● **Characteristics of Gas Reservoir and Fluid Properties**

In middle section of the Kela 2 gas reservoir, the stratigraphic temperature is 100.58 °C; the geothermal gradient is 2.188 °C/100 m; the temperature system is normal. The dolomite section and sandy conglomerate section in the Palaeogene system and three litho-sections in the Cretaceous system belong to the same pressure system. In the middle section of the gas reservoir, the stratigraphic pressure is 74.41 MPa; the average pressure coefficient is about 1.95 – 2.20. Therefore, this is a super high pressure natural gas reservoir. There are several reasons that might explain the high pressure in the gas reservoir. Considering the characteristic of strong tectonic compression that the Kuche foreland basin obtained during the Xiyu period of the Himalayan movement and also considering the reconstruction of the paleo stress field in this region, from the Cretaceous to the present, the compressive force is gradually increasing in an NS direction. The structural compression brought minerals close together, which reduced porous space and increased fluid pressure. In addition, the seal of the massive gypsum-salt layer blocked the leakage of pressure. Thus, abnormally high pressure in the reservoir formed (Magara, 1981; Zhang et al., 1996; Ma et al., 2000). In the Kela 2 gas field, the sandy conglomerate section in the Palaeogene system and the sandstone in the Cretaceous system belong to the same pressure system. The hydrostatic pressure system is above the regional cap rock. Most people consider that, in the Kelasu structural belt, the abnormally high pressure was primarily caused by the sealing system of the massive gypsum-salt layer and the strong tectonic compression that brought uplifting. In the Kela 2 gas field, the thickness of the regional cap rock of gypsum mudstone and the salt layer is more than 400 m, which provides an excellent sealing condition. In general, when the pressure coefficient reaches 1.96 in the mudstone, the cap rock will naturally burst. Hence, this pressure coefficient value is called the "death line" of petroleum accumulation. Because it has strong plastic rheological properties under the compressive stress field, gypsum mudstone has a higher breakthrough pressure point. Therefore, oil and gas can accumulate when the pressure coefficient reaches 1.96. During the Miocene–Pliocene, the Kela 2 structure was continually developed and at the same time natural gas regularly accumulated in the structure. Eventually, the natural gas reservoir was formed during the late Himalayan tectonic period. During the late Himalayan movement, because of powerful tectonic compression, abnormally high pressure developed in the reservoir sequences that were positioned under the salt layer in a totally sealed–partially sealed condition. During the compressive process, the strata were uplifted with erosion. For example, the thickness of eroded strata in the Kela 2 well is in the range of 1,500 – 2,750 m. This uplifting event not only preserved high pressure under the regional cap rock, but it also set out an obviously visible stratigraphic pressure.

The molecular weight of natural gas is in the range of 16.4 – 16.7, which is the lowest number in the Kuche foreland basin. The density of natural gas is 0.569 g/cm^3, which is also at the relatively lower end. In the natural gas, the content of methane is 97.265% on average. The content of heavy hydrocarbon (C$^+$)

is small, the content of nitrogen (N_2) is 1.58% on average, the content of acidic gas is very small and the content of carbon dioxide (CO_2) is 0.686% on average. The aridity coefficient (C_1/C_1^+) is high, which indicates this is dry gas. Because of the high content of methane and low content of non-hydrocarbon gas, the Kela 2 gas field produces high quality natural gas. Under the original accumulating conditions, the deviation coefficient of natural gas is relatively high, in the range of 1.4612 – 1.4900; the volumetric factor is in the range of $(2.516 – 2.587) \times 10^{-3}$ m^3/m^3; viscosity is in the range of $(3.403 – 3.591) \times 10^{-2}$ mPa/s. The physical properties of high pressure fluid (PVT) indicate abnormally high pressure. In the Kela 2 gas field, carbon isotopes are relatively heavy and the content of $\delta^{13}C_1$ is in the range of –27.07‰ ~ –27.8‰ and the content of $\delta^{13}C_2$ is in the range of –17.87‰ ~ –19.4‰. These numbers indicate the natural gas is generated by highly mature coal beds. According to the relationship of carbon isotopes to maturity, we presume the maturity of the source rock (R_o) can achieve up to 2.2%. The natural gas is primarily generated by the Kezilenuer formation and the Yangxia formation in the Lower–Middle Jurassic series. The type of formation water is $CaCl_2$ with a density of 1.082 – 1.111 g/cm^3. The density of chloride is $(7 – 10) \times 10^4$ mg/L; the total salinity is in the range of $(12 – 16.5) \times 10^4$ mg/L. These data indicate an excellent sealing condition.

In the Kela 2 gas field, the reservoir and cap rock combination is made of a stable dolomite reservoir in the Palaeogene system, a sandstone reservoir in the Palaeogene–Cretaceous system and a gypsum-salt layer and mudstone in the Palaeogene system. In addition, they form an integrated anticline with small faults in the upper portion. These small faults dislocated the mudstone section that was about 20 m thick and that was located between the dolomite section and the sandy conglomerate section in the Palaeogene system. The shattered mudstone section not only allowed natural gas to migrate freely between reservoir strata, but it also formed the uniform pressure system, single gas and water interface and an integrated natural gas reservoir (Fig. 10.3). In the Kela 2 gas field, the gas reservoir has a large amplitude. On the map of the top surface of the Palaeogene dolomite section, the elevation of the highest point is –1,970 m and the elevation of the gas and water interface is –2,468 m. The amplitude of the gas reservoir is 498 m, the trap closure is 510 m and the trap is fully filled with natural gas. The thickness of reservoir strata is 561 m (the Kela 2 well), which is larger than the amplitude of the trap. The characteristics indicate a chunk reservoir section with water at its base.

The base water in the Kela 2 gas reservoir is inactive, because the third litho-section of the Bashijiqike formation in the Cretaceous is a conglomerate with poor physical properties and several layers of mudstone. The test result shows the third litho-section is a dry bed with poor productivity and it also limits the base water flow upwards. On the other hand, the lateral water is possibly active with enough energy. However, considering all the conditions, Kela 2 is a middle scale water-driven gas reservoir.

In the Kela 2 gas field, all five wells (Kela 2 well, 201 well, 203 well, 204 well and 205 well) showed a high production rate, a commercial quality natural gas flow with excellent production rates per single well and high formation pressures. In the Kela 2 natural gas field, the geological reserve of dry gas is calculated by using a volumetric method applying 3D seismic data and testing data of six controlling wells. The explored reserve is $2,840.29 \times 10^8$ m^3 and the recoverable reserve is $2,130.22 \times 10^8$ m^3. According to the Chinese natural gas evaluation standard, Kela 2 is a large size natural gas field that is deeply buried with high abundance and an excellent production rate.

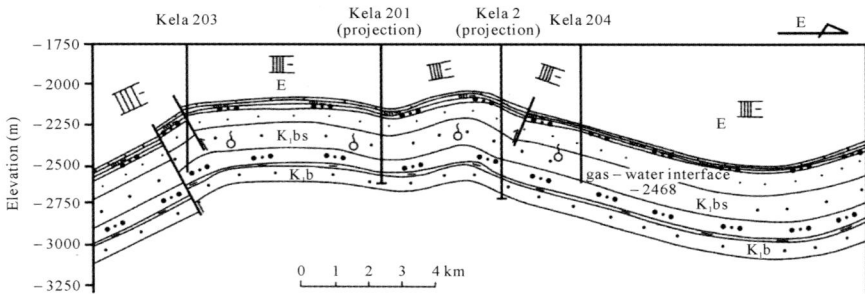

Fig. 10.3.   The cross section (in WE direction) of a reservoir in the Kela 2 gas field
K$_1$b, Formation Baxigai of lower Cretaceous; K$_1$bs, Formation Basijiqik of lower Cretaceous; E, Lower Teriary (Gypsum salt)

10.2   Oil and Gas Exploration in the Jiuquan Foreland Fold and Thrust Belt and the Qingxi Oil Field

10.2.1   Oil and Gas Exploration History of the Jiuquan Foreland Fold and Thrust Belt

The oil and gas exploration in the Jiuquan Foreland Fold and Thrust Belt experienced the following four phases.

● **During the 1980s, Discovery of Oil Shows in the Lower Cretaceous in the Xican 1 Well in the Jiuxi Depression**

Before 1970, Yumen Oil Co., Ltd. discovered several oil fields in the Jiuxi depression. These oil fields include Laojunmiao, Yaxi, Baiyanghe, Shiyougou and Shanbei. After that, the exploration activity had been stopped, which directly slowed down the growth of oil reserves. Until the early 1980s, after examination of the exploration level and exploration history of the Jiuxi depression, geologists considered the Qingxi depression was the most important oil generation depression in the Jiuquan basin. On the southern side of the Jiuquan basin, all of the discovered oil and gas was generated from the Qingxi depression. In addition, there were plenty of remaining resources in the depression. Therefore, a new approach to exploration was aimed at a deep level in the depression. In 1983, the Xican 1 well was drilled, which was the first parameter well in the Qingxi depression. The oil and gas shows occurred in mudstone of lacustrine facies in the Xiagou formation in the Lower Cretaceous and the test results showed a low productive oil flow of 7 m^3/d. From 1983 to 1989, in the Liugouzhuang area that is located on the front margin of the foreland fold and thrust belt, seven oil wells were drilled. Among these wells, three of them yielded a commercial quality oil flow and the Liugouzhuang oil reservoir was discovered. The Liugouzhuang oil reservoir is a deep buried, complex, fractural oil and gas reservoir. The major target formation is the Lower Cretaceous that was buried 4,000 m under the surface. Because people did not fully understand reservoir controlling factors, and because of a lack of efficient exploration technology, the exploration project confronted a low success rate and poor economic benefit. Therefore, in 1989, the exploration of the Liugouzhuang reservoir was stopped.

- **Using New Technology, Discovery of Highly Productive Oil and Gas Flows in the Liu 102 Well and a Commercial Quality Reservoir in the Argillaceous Dolomite**

In the middle and late 1990s, Yumen Oil Co., Ltd. continued petroleum exploration in the Qingxi depression. At the same time, they implemented reservoir condition analysis. Yumen Oil Co., Ltd. considered that the Liugouzhuang oil reservoir was a fractural dolomite reservoir with structural background and the level of fractural development was a crucial factor in petroleum exploration. Thus, the geological research emphasized a forecast of lithology and fractural development. On August 8, 1998, the well test of Liu 102 showed commercial quality oil and gas flows of 54 m^3/d and, after acidification, the production rate increased to 126 m^3/d. After that, Yumen Oil Co. progressed into a new era of petroleum exploration in the Jiuquan basin. The achievement of the Liu 102 well proves that there are high abundance, original oil and gas reservoirs in the Qingxi depression. Subsequently, several development wells also achieved success. In 2000, Yumen Oil Co. submitted an explored reserve of 6.54×10^7 t for the Liugouzhuang area, which marked a new era in petroleum exploration history.

- **Practicing Foreland Fold and Thrust Belt Theory, Expanding the Exploration Domain to the Overthrust Belt and Discovery of the Qingxi Oil Field**

After exploration success in the Liugouzhuang area, Yumen Oil Co. devoted more attention to 3D seismic data and they obtained supplementary detailed knowledge of the comprehensive, litho-fractural oil reservoir in the Qingxi depression. After research work, they believed that the southern Kulong mountain structure was a large size anticline that had been complicated by several thrust faults and the Liugouzhuang nose-like structure was the northern wing of this large size anticline. If they explored a higher position on the structure in the south, a better outcome might be obtained. In 2000, the Long 4 well was drilled on the higher section of the Kulong mountain structure. Highly productive oil and gas flows were obtained in a sandy conglomerate section in the Xiagou formation. In the same year, highly productive oil and gas flows were also discovered in the Long 5 well. Both wells confirmed that the Kulong mountain structure was a high abundance petroleum structure. The sandy conglomerate in the Xiagou formation contains well developed fractures, a thicker reservoir sequence and higher productivity per single well. Compared with the Liugouzhuang oil reservoir (in fractural dolomite), the sandy conglomerate is a new reservoir sequence that is located in a different section of the Xiagou formation. In 2001, supported by a joint stock company, researchers utilized 2D seismic data and practiced foreland fold and thrust belt theory to explore the overthrust. The Long 8 well obtained excellent oil and gas flows in the Xiagou formation (K_1g_0) on the footwall of the overthrust. The oil production rate is 201 m^3/d and the gas production rate is 32,000 m^3/d. This discovery reveals the characteristics of reservoir geology in the Kulong mountain structure: the whole structure contains oil and the high abundance reservoir is located in an isolated area. Up to the end of 2004, the accumulated geological reserve in the Qingxi oil field is $5,750 \times 10^5$ t. A new oil field with hundreds of millions of tons of reserves was discovered.

- **Exploration Extended into the Deeply Buried Formation, the Chijinpu Formation in the Lower Cretaceous is the Imminent Exploration Target**

In the Kulong mountain structure belt, drilling activities proved that the K_1g_0 section contains oil, The deepened drilling in old wells led to the discovery of an oil bed in the K_1g_0 section. These wells included the Long 104J well and the Qin 2-14J well. In the Long 104J well, the initial oil production rate was 64 m^3/d and initial gas production rate was 2.6×10^4 m^3. Using comparative analysis for sources rock, geologists think the oil came from a deep buried source formation and they presume the Chijinpu formation contains a larger oil and gas reservoir.

Yumen Oil Co. Ltd. targeted the entire Jiuquan basin that has the longest exploration history with a relatively low investigation level among Chinese petroleum basins. Previously, the Laojunmiao oil field was the largest one, that has an oil reserve of $6,000 \times 10^4$ t. At present, considering the capacity of natural resources in the basin, it is possible to discover a large size oil and gas field with an oil reserve of $(1-2) \times 10^8$ t or larger. Every depression contains a major oil field, which is the normal pattern for oil and gas distribution in this area. In the Jiuquan basin, after the discoveries of Laojunmiao, Yaerxia, Shiyougou, Baiyanghe, and the Shanbei oil fields (the total geologic reserve is $9,006 \times 10^4$ t), geologists only discovered a fractural type oil reservoir in the Cretaceous dolomite in the Liugouzhuang structure during the middle of the 1980s. After that, there was no important oil and gas discovery for almost 20 years. From 1998, and especially since 1999, inspired by the discovery of the Kela 2 gas field in the Kelasu structure belt in the Kuche foreland basin, the exploration focus in the Jiuquan basin turned to the fold and thrust belt on its southern margin (or on the northern margin of Qilian mountain). Due to the influence of the Himalayan tectonic event, the structure is intricate in this region. The changes of topography are distinct, which include high elevation, huge landscape relief and severe terrain cuttings. The geological conditions are complex, such as overthrust block, high angle formation and high angle thrust fault. Distressed by multiple unbeneficial factors, the quality of seismic data was very poor, which was unable to support the verification of oil and gas traps (Fig. 10.4). After the discovery of the Kela 2 gas field, in the Jiuquan basin, on southern side of the Jiuxi depression, guided by the theory of a fault-related fold based on the characteristics of the structure in the fold and thrust belt and the reconstruction model of the deformed structure and resolving the seismic problems, using the technologies of image well logging, under-balance drilling and deep level acidification, we uncovered the Kulong mountain thrust structure on the southern margin of the Jiuquan basin (or on the northern margin of Qilian mountain). In succession, several highly productive (100 t) oil wells were discovered, which include the Liu 102 well and the Long 101 well. These discoveries turned a new page in oil and gas exploration in the Jiuxi depression.

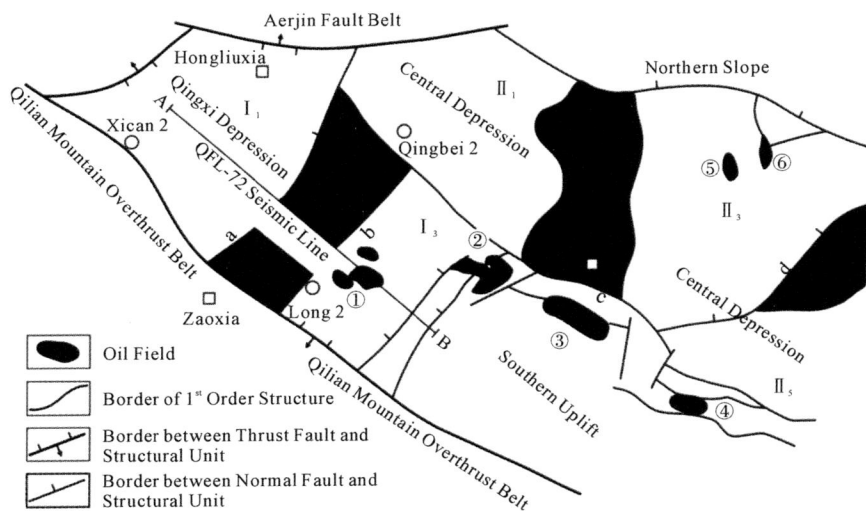

I_1 Hongnan Depression; I_2 Qingxi Low Bulge; I_3 Qingnan Depression;
II_1 Chijin Depression; II_2 Yabei Bulge; II_3 Shibei Depression; II_4 Shibei Bulge; II_5 Dabei Depression
Fault: a. Qingxi II; b. Qingxi I; c. Miaobei; d. Shibei II
Oil Field: ① - Liugouzhuang; ② - Yaerxia; ③ - Laojunmiao; ④ - Shiyougou; ⑤ - Baiyanghe; ⑥ - Shanbei

Fig. 10.4. Structural units and oil and gas fields in the Jiuquan basin

10.2.2   *Geologic Settings and Petroleum Geology in the Jiuquan Foreland Fold and Thrust Belt*

● **Characteristics of the Geology**

The Jiuquan basin covers an area of 2,700 km^2 and is located at the junction area of the Tarim terrain, the Alashan terrain and the Caledonian folding belt of Northern Qilian mountain. It is bordered and controlled by large scale faults (Fig. 10.4). The basement of the basin consists of the Silurian system, the Devonian system and the metamorphic rocks of the Carboniferous system. The strata of the Cretaceous–Quaternary systems were developed on the basement of a pre-Cretaceous faulted block. The Jiuquan basin experienced two tectonic events, which were the extensional faulted depression event during the Mesozoic and the compressed deformation and foreland basin development event during the Himalayan tectonic period. The Jiuquan basin went through two phases of basin development, which were the rifting basin development phase during the Yanshanian tectonic period (the Cretaceous period) and the depression development phase during the Himalayan tectonic period (the Tertiary–Quaternary periods). The inherited and long-term tectonic activities of the overfold and thrust belt of Qilian mountain and the Aerjin strike-slip fault have controlled the developmental history of the Jiuxi basin since the Cretaceous period. The derivative secondary faults in the basement controlled structural frameworks, depositional environments and deformations in the basin during various geological periods.

In the Jiuquan basin, the Yanshanian movement was mainly represented by a faulted block movement that formed a half-graben rifting basin The eastern side of the basin was deeper than the western side. Faults developed on the eastern side of the basin and overlaps occurred on the western side. During the Early Cretaceous, a series of NEE oriented, extensional or extension-distortional growth faults developed in the basement. These faults controlled sedimentary deposits during the Early Cretaceous. The Himalayan movement brought the ramp structures. The structures were stronger on the southern side than on the northern one. The NW and NNW oriented thrust faults of the Himalayan tectonic period were perpendicular to the NE and NNE oriented fault of the Yanshanian tectonic period. These two group faults formed a new structural framework in the basin: structural blocks occurred from the west to the east, structural belts were distributed from the north to the south and the depressions and bulges were spaced in-between. The major structural units included the southern uplift, the central depression, the northern slope and the Qingxi depression (Fig. 10.4).

In the Jiuquan basin, the sedimentary deposits include the Middle–Lower Jurassic series, the Lower Cretaceous, the Tertiary system and the Quaternary system. The Jurassic outcrop in the Hanxia area with stable thickness and distribution of the strata is perhaps limited. The Jurassic lithology mainly includes clastic rocks of river facies interbedded with a small number of coal beds. The maximum thickness of the Cretaceous system can reach 5,000 m. The Cretaceous system not only contains the major hydrocarbon source sequences of the basin, but it also has the reservoir sequence and the cap rocks. The Tertiary system has the red clastic rocks that served as the reservoir and the cap rocks with an unstable thickness of 1,000 – 3,000 m. The Quaternary system mostly contains a conglomerate of piedmont facies with a maximum thickness of 1,000 m.

The Jiuquan basin contains three major tectonic units that are the Jiuxi basin, the Jiayuguan uplift and the Jiudong basin. The Jiudong basin and the Jiuxi basin are different. The Jiuxi basin is distinctively controlled by two large overfold and thrust belts. One is the Laojunmiao fold and thrust belt and another one is the northern margin fold and thrust belt of Northern Qilian mountain. Recently, the Qingtoushan structure was discovered in the Laojunmiao fold and thrust belt which extends from Kulong mountain to the east and to the west. It even extended to the eastern side of Jiuquan basin (Fig. 10.4).

● **Conditions of Petroleum Geology**

Condition of Hydrocarbon Source Rock

In the Jiuquan basin, lithology in the Cretaceous depression is a set of deep water mudstone with abundant organic materials and large thickness. The contents of organic carbon are 1% – 2% and the maturity is in an adequate middle range. The Cretaceous depressions appear to be the remaining basins. They contain about 1,000 km^2 of effective source rocks. In other types of basins, the effective source rocks only occupy 200 – 300 km^2 among the 1,000 km^2 total basin area and they are surrounded by marginal facies. Therefore, the Cretaceous depressions can form small, oil rich basins with good exploration potential. In the Jiuquan basin, the hydrocarbon source rocks were only formed in the Chijinpu formation, the Xiagou formation and the Zhonggou formation in the Lower Cretaceous. Other formations do not contain hydrocarbon source rocks. The Liugouzhuang is located in the central region of the Qingnan depression. Many exploration wells in the Liugouzhuang area are drilled into the Zhongtao formation (with a thickness of 300 – 500 m) and the Xiagou formation (with a drilling exposed thickness of 1,000 m). The acquired data demonstrate that, in the central region, the Xiagou formation predominately contains mudstone, dolomitic mudstone and argillaceous dolomite. It only contains a small amount of thick bedded siltstone. The thickness of hydrocarbon source rocks occupies more than 90% of the total thickness of the formation. At present, the Chijinpu formation has not been discovered in the central region. However, on the northern side of the basin, much drilling data and the outcrops confirm that the Chijinpu formation has great thickness and favorably developed source rocks. The drilling data come from the Shibei depression, the Dabei depression and edge of the Qingnan depression. Thus, in the Qingnan depression, the hydrocarbon source rocks should be highly developed in the

Chijinpu formation. In the Hongnan depression, the exploration drilling activities have not encountered the Lower Cretaceous. However, in the Hanxia cross section that is located on the southern margin of the Hongnan depression, the thickness of the Lower Cretaceous source rocks is about 300 m. In addition, on the Hongliuxia cross section that is located on the northern margin of the Hongnan depression, the thickness of the Lower Cretaceous is about 2,250 m. Among the revealed Lower Cretaceous, the thickness of the Chijinpu formation is about 1,420 m (including 600 – 700 m of source rocks in the upper section that occupy approximately 50% of the total formation thickness). The thickness of the Xiagou formation is about 720 m (including 500 m of source rocks that occupy approximately 80% of the total formation thickness). The thickness of the Zhonggou formation is about 100 m (including 50 m of source rocks). The depositional system analysis and the sequences analysis reveal that the original basin in the Hongnan depression has depositional synchronicity and sedimentary similarity with the Qingnan depression. Thus, the Hongnan depression should have deep water mudstone of lacustrine facies with great thickness and broad distribution. According to sedimentary facies, seismic data, drilling information, outcrop, cross section and drilling data from other depressions, the deduced maximum thickness of the Lower Cretaceous source rocks is more than 2,500 m in the Qingnan depression and the Hongnan depression. In the Jiuquan basin, the contents of organic matter in the Lower Cretaceous are 0.6% – 2.5%; the distribution of kerogen H/C is broad. Overall, the Xiagou formation contains the best type of organic matter, the Chijinpu formation contains the second best type of organic matter, the Zhonggou formation contains Konrogen type II of organic matter.

Condition of Reservoir Sequences

In the Jiuquan basin, there are two sets of exploration targets. The first one is an upper target that includes the Baiyanghe formation and the Huoshaogou formation in the Tertiary system. These formations have good physical properties and they have a secondary oil and gas reservoir. The structural traps are distributed along the fault with high abundance. The second one is a lower target that includes the Xiagou formation and the Chijinpu formation, which contain original oil and gas reservoirs with a broad distribution area and multi-type reservoirs to form a large size oil and gas field. The entire Lower Cretaceous is a normal depositional cycle. The regressive sandy conglomerates of lacustrine delta facies were deposited near the Qingxi low bulge, the 509 fault and the southern margin of the basin. These conglomerates were near the substance sources and they were distributed in a fringe pattern. On the southern side of the Qingxi oil field, the best reservoir sequences are gray color sandstone, pebble sandstone and conglomerate, which belong to the distributary channel micro-facies, delta front subfacies. The dark color mudstone, dolomite and argillaceous dolomite were deposited in a semi-deep lake basin. From the southern side of the Qingxi oil field to the center of the Qingnan secondary depression, the sedimentary systems are delta plain–delta front–semi deep lacustrine. There are two types of reservoir. One is the fractural reservoir, another one is medium–low porosity, low permeability sandstone in the Chijinpu formation.

Condition of Structural Traps

During the Himalayan tectonic period, under the influence of the fold and thrust belt on the northern margin of Qilian mountain, a northwest oriented structural belt developed on the southern side of Jiuquan basin. This latest tectonic movement controlled the development of oil and gas traps. On the footwall of large fold and thrust belts, most of the oil and gas traps are structural traps. Oil and gas are mainly distributed along large fold and thrust belts of the Himalayan tectonic period. In other words, the major fold and thrust belts are a primary controlling factor for oil and gas distribution. A large fold and thrust belt and small graben basin are primary characteristics of a foreland belt. The displacement of the overthrust can be more than 15 km; the orientation of the structural lineament is in the direction NWW. The structural blocks were divided from east to west. The structural belts were formed from north to south (Fig. 10.4). The types of structure include a fault-related fold, overthrust block and inter-layers.

10.2.3   Discovery of the Qingxi Oil Field and Characteristics of Oil Reservoir

● Discovery of the Qingxi Oil Field

In the Jiuquan basin, after the discoveries of the Laojunmiao, Yaerxia, Shiyougou, Baiyanghe and Shanbei oil fields (the total geologic reserve is $9,006 \times 10^4$ t), geologists only discovered a fractural type oil reservoir in the Cretaceous dolomite in the Liugouzhuang structure during the middle of the 1980s. After that, there was no important oil and gas discovery for nearly 20 years. In 1997, when oil and gas exploration in the Jiuxi basin restarted, instead of making a geophysical exploration plan, the first project was geological evaluation, After systematically evaluating and studying geological information from the Jiuquan basin, the exploration target was selected. The reason for the disappointment with the explorations in the Liutaozhuang area in the Qingxi depression during 1983 – 1997 became clear. An improved exploration proposal for the Jiuquan basin was presented. After geological evaluation and exploration preparation, the Liu 102 well was drilled, which was the first exploration well for the Jiuquan basin re-exploration project. On August 8, 1998, the well test at the Liu 102 well showed oil and gas flows of 54 m^3/d. After acidification, the production rate increased to 126 m^3/d. The Liu 102 well was not only the first high production well (100 t/d) in the Qingxi depression, but it was also the first high production well that the Yumen Oil Co. discovered in more than a dozen years. After the success at the Liu 102 well, based on in-depth research work, a strategy for exploring the Kulong mountain structure was proposed, because it was located at a higher structural position with complete structural framework and a large area of trap. From 1983 – 1997, three wells were drilled in the Kulong mountain structure; only one well showed an oil and gas flow. In 1984, the Long 2 well was drilled at the high point of the structure and it was a dry well. In 1997, the Long 3 well was drilled and, again, it was a dry well too. In the circumstances, the study found out the reason for failure and proposed a re-exploration of the Kulong mountain structure. Until 1998, the exploration focused on the fold and thrust belt on the southern margin of the Jiuquan basin (that is also located at the northern margin of Qilian mountain). Based on the characteristics of the structure in the fold and thrust belt and the reconstruction model of a deformed structure, resolving the seismic problems using the technologies of image well logging, under-balance drilling and deep level acidification, petroleum geologists explored the Kulong mountain structure on the southern margin of the Jiuquan basin (Fig. 10.5) and succeeded. At the end of 1998, the Long 101 well was set up on the Kulong mountain structure. During drilling of the Long 101 well, they discovered in total 207 m of reservoir sequences. After a productivity test for the second reservoir layer and acidification, the oil production rate was 104 m^3/d and the daily gas production rate was 58×10^4 m^3. Therefore, oil and gas exploration in the Kulong mountain structure progressed into a new era. The good oil and gas shows in the Long 101 well confirmed the possibility of discovering a large oil field. Immediately, the exploration was intensified towards the west. The Long 4 well was drilled in the middle of the structure. A conglomerate reservoir was discovered in the Lower Cretaceous Xiagou formation with a good oil and gas layer. The well test showed that the oil and gas flow is 108 m^3/d. After acidification, the production rate increased to 253 m^3/d. Thus, another type of highly productive oil and gas reservoir was found. Because the Long 4 well was drilled at quite a high position on the Kulong mountain structure, researchers recognize that the high productivity of the Long 4 well is due to structural control. The structure is a major factor that controls oil and gas accumulation. The Kulong mountain oil reservoir is a structural reservoir, not a lithology reservoir nor a synclinal reservoir. Therefore, they proposed that the Long 5 well should be drilled at a high point in the Kulong mountain structure. However, in 1985, the Long 2 well was drilled at a high point in the Kulong mountain structure and it was a disappointment. The distance between the Long 5 well and the Long 2 well is 253 m. After reviewing the data of the Long 2 well, researchers recognize the failure of the Long 2 well was due to technology. If they used today's exploration technology, the Long 2 well should have high productivity. Thus, they decided to drill the Long 5 well on the high point of the Kulong mountain structure to test the result of a comprehensive study. On December 19, 2000, an oil and gas flow of 500 m^3/d was discovered in the Long 5 well. This discovery proved that the Qingxi oil field contains a structural reservoir that was controlled by the Kulong mountain structure. The oil and gas were accumulated at the high position of the structure. After that, oil and gas exploration in the Jiuxi depression progressed into a new phase.

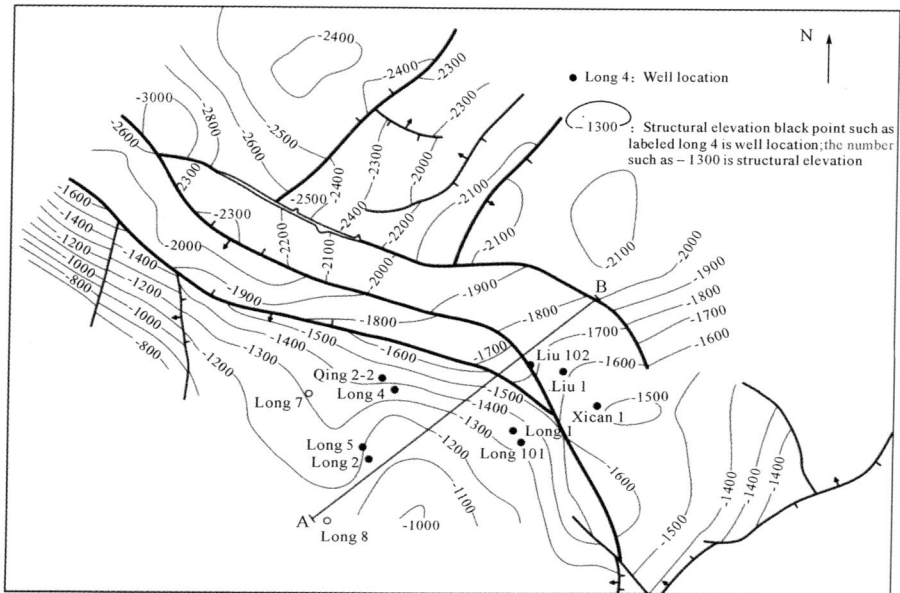

Fig. 10.5.   Structures and wells in the Xiagou formation (TK$_1$g) in the Qingxi oil field

The Kulong mountain structure is located in the middle section of the Kulong mountain overfold and thrust belt on the southern margin of the Jiuquan basin (or on the northern margin of Qilian mountain). The entire structure is an anticline; the top surface of the anticline trap is 103 km^2, which was positioned in the Xiagou formation (K$_1$g$_1$) in the Lower Cretaceous (Fig. 10.5). The distance between the structure and Yumen city is about 70 km. The geomorphologic trend demonstrates the southern side is higher than the northern side. Yaomo mountain is the highest mountain on the southern side with an elevation of 4,586 m. The reservoir types include fractured argillaceous dolomite and sandy conglomerate. There are three sets of discovered reservoir formations, which are the Baiyanghe formation in the Tertiary system, the Zhonggou formation and the Xiagou formation in the Cretaceous system. In 2004, at the eastern block of the Kulong mountain structure, good oil and gas shows and 64 m^3 of natural gas were obtained in the Long 104J well, the Q2-14J well and the Long 10 well. The reservoir sequence belongs to the Xiagou formation of the Lower Cretaceous. The Long 104J well and the Q2-14J well are deeper drilled, based on the Long 104 and Q2-14 wells. This discovery not only added a new petroleum accumulation block to the map, but it also enlarged the oil and gas reserve in the Kulong mountain structure. At same time, it revealed a bright future for oil and gas exploration in the deeply buried Chijinpu formation. Up to the end of 2004, in the Kulong mountain structure, the geological reserve of oil was 5,750×10^4 t, the controlled geological reserve of oil was 1,719×10^4 t, the forecast oil reserve was 2,892×10^4 t, and the third-order reserve was 1,036×10^4 t. The exploration breakthrough in the

Kulong mountain structure may modify the productivity of the Yumen oil field to 100×10^4 t.

● **Characteristics of Qingxi Oil Reservoir**

The Qingxi oil field is located at the footwall of the Kulong mountain fold and thrust belt that is the western section of the overfold and thrust belt on the southern margin of the Qingxi depression in the Jiuquan basin and on the northern margin of Qilian mountain. It has a deeply buried, fracture type lithology oil reservoir and petroleum was generated and stored in situ (Fig. 10.6). In the Qingxi oil reservoir, the major control factors are lithology, structure and fracture. The oil and gas accumulated in the isolated traps that were formed by the best combination of the structure, the facies and the fractures. The isolated traps controlled oil and gas accumulation. The Kulong mountain anticline is the major petroleum structure in the Qingxi oil field; the third-order and fourth-order faults not only controlled the development of fractures but they also controlled oil and gas accumulation in an isolated area and the productivity level. The Qingxi oil field includes the Kulong mountain anticline and the Liugouzhuang faulted nose structure with a total estate of 146 km^2, which includes about 60 – 80 km^2 of area that was buried under mountains due to the overthrust movement. The structural high point is located at the Long 8 well. The structural axis is oriented in an NWW direction from the Long 8 well to the Long 107 well (Fig. 10.5). On the Kulong mountain structure, formations on the northern wing have a relatively high dip angle with concentrated fractures; formations on the southern wing have a moderate dip angle and they are buried under the Paleozoic base rock of Qilian mountain due to overthrust movement (Fig. 10.6). The major structural system of the Kulong mountain structure contains an NW oriented thrust fault and a nearly NS oriented adjusting fault. These faults belong to the Himalayan tectonic period and they play important roles in creating traps and developing fractures. The nearly NS oriented adjusting fault is a strike-slip shear fault. During the Tertiary period, the adjusting fault was very active with powerful energy. It caused the complexity of the Kulong mountain structure and, in addition, it controlled oil and gas distribution and accumulation. The strike orientation of the third-order and the fourth-order faults are in an NE-SW direction and in an NW-SE direction respectively. For the inclined fractures, the primary strike orientation is in an NE-SW direction and the secondary strike orientation is in an NW-SE direction. The two groups of fractures have a good correlated relationship. The developments of the inclined fractures are controlled by the third-order fault, the fourth-order fault and local structures. Due to the intersection of two sets of faults, the two groups of fractures are also inter-sliced. The intersection of fractures modified the conditions of the reservoir and the ability of migration in isolated locations. They further controlled the accumulation of oil and gas in isolated locations. That is the main reason for a difference in production between different wells.

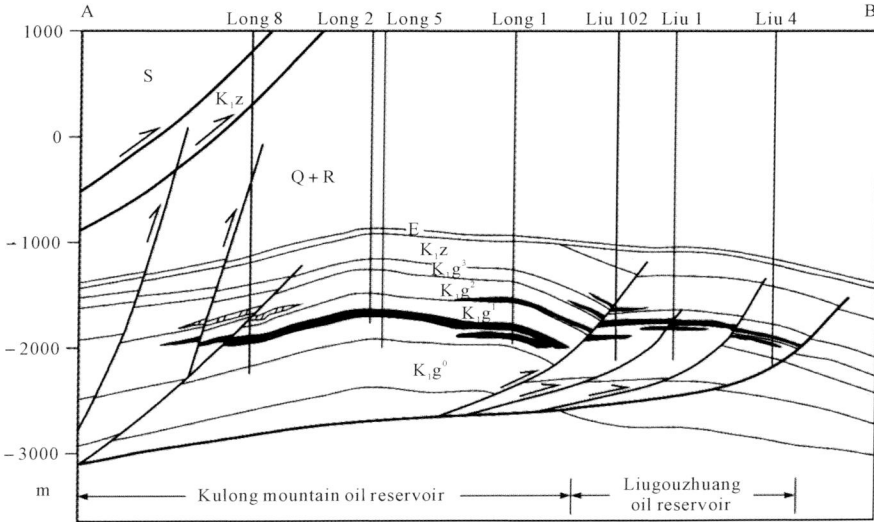

Fig. 10.6. The cross section of oil reservoir from the Long 8 well to the Liu 4 well in the Qingxi oil field (For the location see the line A-B in Fig. 10.5)
S, Silurian; K_1z, Formation Zhonggou of lower Cretaceous; Q, Quaternary; R, Tertiary; E, Lower Teriary; K_1g, Formation Gangou of Lower Cretaceous

The sandy conglomerate is the best reservoir sequence in the southern Qingxi oil field. The sandy conglomerate belongs to micro facies of a distributary channel on the fan delta front. In addition, argillaceous dolomite of semi-depth lacustrine facies produces beneficial reservoir sequences in the Qingxi oil field. The Qingxi oil field contains three types of reservoir sequence, which are argillaceous dolomite, conglomerate and pebble sandstone. Via an analysis of petrology, porous type, porous texture, reservoir physical properties and reservoir oil-bearing properties, the test results demonstrate that sandstone and a conglomerate of fan delta front subfacies contain adequate ankerite, 8% – 15% in general. The porous types include intergranular pores, intergranular dissolution pores, intercrystalline pores, micro pores and fractures. Primarily, fractures have a high dipping angle and they interconnect with pores. An analysis of the porous texture shows a relatively low displacement pressure. A saturation analysis and fluorescence thin section analysis show good oil-bearing properties. The sandstone and conglomerate of fan delta front subfacies are the best reservoir sequences in the Qingxi oil field. The reservoirs of the Long 6 well and other high production wells are located in the gray color conglomerate within the micro facies of distributary channels on the fan delta front. The multi-colored conglomerates of fan delta plain subfacies have high clay content; the highest clay content can reach 44%. The porous types include gravel seams and dissolution pores and the connectivity between the two types of pores is poor. Analysis of the porous texture shows high displacement pressure; saturation analysis and fluorescence thin section analysis show poor oil-bearing properties. The multi-colored conglomerate of fan delta plain subfacies is a relatively poor reservoir sequence in the Qingxi oil field. The argillaceous dolomite of semi-depth lacustrine subfacies contains multi-types of pores, which include dissolution pores, intercrystalline pores and fractures. The principal fractures are interlayer fractures and inclined fractures. The fractures and pores are interconnected with each other. The analysis of the porous texture shows a relatively high displacement pressure. The saturation analysis and fluorescence thin section analysis show moderate oil-bearing properties. The argillaceous dolomite of semi-depth lacustrine subfacies is the beneficial reservoir sequence in the Qingxi oil field.

An effective combination of matrix pores and fractures formed a highly productive petroleum reservoir with a stable production rate. In the Qingxi oil field, the conglomerate reservoir sequence has strong heterogeneity. The reservoir forecast depends on comprehensive analysis. The examined factors include distribution of beneficial facies, distribution of effective pores, a fault concentrated area, local beneficial structure and distribution of effective fracture. The block of the Long 4–Long 8 is located at the high point of the Kulong mountain anticline and it is sandwiched by nearly N-S oriented faults. This area is also the distribution area of gray color conglomerate of fan delta front facies. The study of micro facies showed that the primary reservoir sequence was deposited in distributary channels on the fan delta front. Most pebbles had grain support. The porous types include intergranular pores, intergranular dissolution pores and intragranular micro pores. The dissolution pores are very well developed with a high porosity rate. These pores connect with nearly N-E oriented fractures to form an efficient fracture-pore system. These fractures not only extend a long distance, but they also have a high dip angle. This block is the stable, highly productive area in the Qingxi oil field.

The factors that controlled the reservoir include structure, lithology and fracture. Structure influences petroleum accumulation; lithology dominates petroleum distribution; fracture manipulates percolation capacity. Structure controls oil and gas accumulation. An isolated, high abundance reservoir is located at the high point of the structure within a fault concentrated area. Lithology controls distribution of oil and gas; primary reservoir sequences include gray color sandy conglomerate of fan delta front facies and argillaceous dolomite of semi-depth lacustrine facies. Fractures control the production rate of a single well and the fractures are the major percolation passage for oil and gas. Inclined fractures control single well productivity and the major productive formations contain well developed inclined fractures. For example, the Long 4–Long 8 block has a high production rate, and reservoir sequences are located in the Lower Cretaceous (the $K_1g_1^{1}$ and the K_1g_0 sections) with stable distribution and good correlation.

At the end of 2004, in the Qingxi oil field, the explored resource area for oil was 24.4 km^2 and the geological reserve of oil was 5,750×10^4 t. The recoverable reserve of oil was 945.2×10^4 t. The geological reserve of solution gas was 114.34×10^8 m^3 and the recoverable reserve of solution gas was 22.96×10^8 m^3. The controlled geological reserve for oil was 1.719×10^4 t, the controlled recoverable reserve for oil was 275.2×10^4 t and the remaining forecasted geological reserve was $2,770 \times 10^4$ t. The third order geological reserve for oil was $10,239 \times 10^4$ t. In 2004, crude oil production was 44×10^4 t and developed productivity was 58×10^4 t. Petroleum exploration progressed into the era the discovery of a hundred million tons oil field.

References

Graham, S.A., Hendrix, M.S., Wang, L.B., 1993. Collision successor basins of western China: impact of tectonic inheritance on sand composition. Geo. Soc. of Amer. Bull., 105:323-344.

Hendrix, M.S., Dumitru, T.A., Graham, S.A., 1994. Late Oligocone-early Miocene Unroofing in the Chinese Tian Shan: An early effect of the India-Asia collision. Geology, 22(6):487-490.

Jia, C.Z., 1997. Structural Characteristic of the Tarim Basin and Related Petroleum System. Petroleum Industry Press, Beijing.

Jia, C.Z., 2005. Attributes of foreland thrust belt and related natural gas accumulation in western-central China. Petroleum Exploration & Development, 32(4):9-15.

Jia, C.Z., Hu, Y.Y., Tian, Z.J., et al., 2001. The exploration of large scale natural gas field in the Kuqa Depression, Tarim Basin. *In*: Gao, R.Q., Zhao, Z.Z. (Eds.), Frontier Petroleum Exploration in China (vol. 1). Petroleum Industry Press, Beijing.

Jia, J.H., 2000. The study of depositional sequence and reservoir of the Cretaceous Bashijiqike formation in the Kuqa foreland basin. Earth Science Frontiers, 7(3):133-143.

Li, X.D., Zhang, G.Y., Tian, Z.J., et al., 2000. Petroleum System and Petroleum Distribution in the Tarim Basin. Petroleum Industry Press, Beijing.

Liang, D.G., Jia, C.Z., 1999. The achievements of natural gas exploration and its prospects evaluation in the Talimu basin. Natural Gas Industry, 19(2):3-12.

Lu, H.F., Chen, C.M., Liu, Z.H., et al., 2000. The origin and structural attributes of the Kuqa rejuvenation foreland thrust belt. Acta Petrolei Sinica, 21(3):18-24.

Ma, Q.F., Chen, S.Z., Zhang, Q.M., et al., 2000. Ultra Pressure Basin and Petroleum Distribution. The Geological Publishing House, Beijing.

Magara, K., 1981. Compaction and Fluid Migration, Chen, H.L., et al., translated. Petroleum Industry Press, Beijing.

Ritts, B.D., Hanson, A.D., Zinniker, D., et al., 1999. Lower-middle Brassic non-marine source rocks and petroleum systems of the Northern Qaidam basin. AAPG Bullein, 83(12):1980-2005.

Wei, G.Q., Jia, C.Z., 1998. Structural attributes of thrust belts and oil & gas in the Tarim basin. Acta Petrolei Sinica, 19(1):11-17.

Zhang, Q.M., Liu, F.N., Yang, J.H., et al., 1996. Ultra pressure system and petroleum migration in the Yingge sea basin. China Offshore Oil and Gas (Geology), 10(2):65-75.

Zhao, M.J., Zhang, S.C., 2001. Genetic type of natural gas generation and environment of natural gas accumulation in the Tarim basin. China Petroleum Exploration, 6(2):27-31.

Zhao, M.J., Zhou, X.X., Lu, S.F., 1999. Tarim basin rich in natural gas. Natural Gas Industry, 19(2):13-18.

Zhou, X.X., 2001. Petroleum accumulation and reservoir model in the Kuqa petroleum system. Petroleum Exploration & Development, 28(2):8-l0.

Part IV

Deeply Buried Petroleum Reservoirs in China

Characteristics of a Superimposed Basin and the Promise of a Buried Petroleum Play

The term deeply buried oil and gas reservoir was designated for particular oil and gas reservoirs that have a buried depth deeper than 4,000 m. From a geological point of view, this depth is equivalent to the special depth where liquid hydrocarbon was transformed into gaseous hydrocarbon. In general, geologists consider that, in a basin, the deeply buried oil and gas reservoirs identify a petroleum reservoir in the new strata that is positioned beneath the existing productive oil and gas layers. More than 30 years of worldwide in-depth exploration practices confirm that there are plentiful of oil and gas reservoirs in the deep positions of basins. In-depth petroleum exploration not only is an important method for increasing the production rate for an old oil field, but it is also a valuable exploration domain for increasing the production rate in a new oil field (Gan et al., 1988; Guo et al., 1996). Preliminary estimations show that the reserves of deeply buried oil and gas are equal to 28% of the total oil and gas reserves in China. In China, large scale petroleum basins have the preconditions in a deep position to form oil and gas reservoirs. Several deeply buried oil and gas reservoirs have been discovered. Emphasizing oil and gas exploration in deeply buried strata has extraordinary significance for Chinese petroleum exploration.

Usually, a superimposed basin is made of several monotype basins that experienced several phases of tectonic reformation and were superimposed together in multiple directions. Superimposed basins commonly developed in China. Because monotype basins of different geological periods had various development mechanisms, a superimposed basin has the following characteristics: (1) The framework is complicated and variable with multiple sets of petroleum play. (2) In different sections of the formations, several phases of hydrocarbon generation and migration occurred and multiple phases of hydrocarbon accumulation and deformation also occurred. (3) The progression of hydrocarbon generation and accumulation was intricate and of extensive duration; in addition, the sources of hydrocarbon were diversified and hydrocarbons accumulated in multiple phases. (4) In some formations or regions, a composite oil and gas system developed. Alteration and adjustment happened in various regions with multi-cycles; the target formations were deeply buried.

11.1   Characteristics of Superimposing and Types of Superimposed Basin

11.1.1   *Concept of Superimposed Basin*

Oil and gas exploration has been rapidly expanded into deeply buried formations in the Chinese basins. In addition, a series of large scale oil and gas fields have been discovered in the following basins: the Sinian–Ordovician system in the Bohai bay basin, the Ordovician system–Upper Paleozoic erathem in the Eerduosi basin and the Tarim basin, the Paleozoic erathem and the Middle–Lower Triassic system in the Sichuan basin, the Mesozoic volcanic rocks in the Songliao basin (He et al., 1996; Hu et al., 1997; Luo et al., 1996). Since the 1980s, using exploration of deeply buried formations and the discoveries that are listed above, discussions about superimposed basins gradually increased and our insight became progressively clearer. Although some parts of the definition of a superimposed basin still vary or are even contradictory, there is agreement about the major aspects of a superimposed basin, with only minor differences. For example, first of all, Zhang et al. consider that a superimposed basin has an intricate framework, which not only experienced multiple phases of tectonic alteration, but which was also made of several monotype basins by vertically overlaying basins in various directions. As a second idea (not the same as the idea of a remaining basin or an altered basin), Liu presented a definition of a superimposed basin, which was that of "relatively independent basins of different geological periods depositionally superimposed together to form a collective basin". The third idea is "the superimposed basin is a type of altered basin". Pang and Jin defined a superimposed basin as "in the same tectonic unit as the depositional basins that were of different types, different formations and different geological periods and were superimposed together; these basins were vertically associated by depositional hiatus or unconformities". For the next idea, Zhao et al. described a superimposed basin as being "in a negative structural unit, experiencing multiple phases of structural alteration, with several monotype basins superimposed together in various directions to form a basin with a complicated framework". Finally, in 2004, according to the characteristics of multi-phase tectonic movement and the activity theory of tectonic history, He et al. considered that a "superimposed basin" has an overlaid geological framework and it is made of prototype basins that are at different basin stages and that experienced multiple alterations with different movement systems (that included a tectonic system and thermal system). This concept presents the most comprehensive summary of a superimposed basin in the 21st century and it was used in this book.

To the comprehensive summary that was described above, we may add the following deductions. In a specified developmental stage and relatively stabilized tectonic environment, under a dominant depositional mechanism, the deposited sediments and their related geological boundary was called a "prototype" or "prototype basin" at this stage. The prototype basin of a former developmental stage worked as a foundation for basin development in the succeeding stage. Under the new tectonic system, some parts of the basin were inherited and others were altered. Obviously, the prototype basin of a former stage was replaced by the one from a latter stage; the superimposing event not only overlaid the sediments, but it also placed the tectonic movements on top of each other.

The prototype basins were formed during different tectonic stages, which had continuality in time duration and which were superimposed together in 3D spaces. The combined new basin not only had a superimposed framework and varied types of combination, but it also experienced multi-phases of alteration in the dynamic system, which included the following characteristics: (1) It experienced various stages of geo-dynamic evolution; every geo-dynamic stage had its own thermal–structural system. (2) At the same geo-dynamic stage, it might create one prototype basin. Also, it might develop an assemblage of either the same kind of prototype basins or different kinds of prototype basins. (3) The prototype basins of different geo-dynamic stages can be superimposed together, not only inheriting some features from a previous prototype basin, but also creating a new prototype basin. The differences were exhibited in the type of basins, in the framework of the basins and in the boundary of the basins. (4) Because the geo-dynamic evolution was in cycles, for example the diversion–conversion cycle or the open–closed cycle, a faulted basin or a continental marginal basin that was created during a particular extensional period could be reformed into a trough–arch–basin system or into a foreland basin system during the subsequent conversion movement. When the evolution progressed into a new diversion–conversion cycle, the basement of the basin, the magnitude of the basin, the filling sediments and the subsidence mechanism were obviously changed, although it was possible to form a similar type of basin again. (5) Because geo-dynamic conditions changed, when the prototype basins were superimposed, the orientation, the region, the dynamic mode and the pattern of geometry could have many variations. Between the prototype basins, either a constructive event (such as buried or inherited events) or destructive event (such as folding, faulting, uplifting and eroding) could happen. (6) If they only experienced two stages of basin evolution, the overlaid basins usually had a distinctive geological framework, which could be called a simple "superimposed basin"; if they experienced more than two stages of geo-dynamic movement, the combined basins commonly had a complicated framework which could be called a complicated "superimposed basin". However, in a relatively stabilized environment, although it experienced multiple stages of basin evolution, the superimposed basin could have a simple framework. For example, Zhang (2002) considered that the Eerduosi basin is a superimposed basin that was made of several prototype basins that experienced various tectonic movements. Zhang's proposal is identical to the common theory that "the Eerduosi basin is a combined intra-continental basin that is located at the margin of a craton; furthermore, it contains multi-cycles of petroleum plays. (7) A superimposed basin was not simply the placing of prototype basins on top of each other. Under a new dynamic system, a new prototype basin was created and the previous basin was reformed; therefore, the framework of the new basin was not the same as the one either beneath it or above it. From a petroleum geology point of view, the petroleum system was not only created on the superimposed interface, but it was also developed in assorted accumulative modes that would not appear inside the prototype basin. Thus, oil was generated in old strata and it was accumulated in younger strata, or vice versa. Or oil was generated in the lower portion of strata and it was stored in the upper portion of strata.

A superimposed basin required that the prototype basins (that were either controlled by different dynamic systems or created at different tectonic stages) should not only have continuality in time duration, but also be stacked up on top of each other. The Chaidamu basin is an example. Although the prototype basins that were either controlled by different dynamic systems or created at different tectonic stages had continuality of time duration, these prototype basins (the Jurassic faulted basin on the northern margin of the Chaidamu basin, the Tertiary strike-slip and pull-apart basin on the western Chaidamu basin and the Quaternary depressed basin) were not superimposed vertically. These prototype basins were next to each other horizontally. Therefore, the Chaidamu basin is not a superimposed basin.

11.1.2   Tectonic Background

Since the concept of the classification of a basin (Kingston et al., 1983) and the idea of a prototype basin (Zhang et al., 1997) were proposed, Chinese petroleum geologists have found it very difficult to define onshore basins in China by using the single type of basin or solo kind of prototype basin, because both the traditional classification of a basin and the taxonomy of a prototype basin were founded on the single tectonic cycle of Wilson. From the viewpoint of the active theory of tectonic history, Chinese tectonic movement had several phases. In most Chinese basins, the multi-cycled developmental history determined how the framework of a basin was superimposed (Jia et al., 1995).

The Chinese landmass was located in a transitional tectonic domain that was between Siberia (it became Laurasia later) and Gondwana. The Chinese landmass was made up of several small landmasses and many micro landmasses; these small landmasses that include North China, the Yangzi, South China and Tarim were surrounded by a few gigantic orogenies. From the early geological period to the late geological period, the Chinese landmass was controlled by three major geo-dynamic systems that included the Paleo-Asia Ocean (Laurasia Ocean), the

Tethys–Paleo-Pacific Ocean and the Indian Ocean–Pacific Ocean; the northern part belongs to the tectonic domain of the Paleo-Asia Ocean, the southern part belongs to the Tethys tectonic domain, and the eastern part belongs to the tectonic domain of the Pacific Rim. These three geo-dynamic systems were not only superimposed over time and space, but they also combined together or interfered with each other. In northern and northeastern China, the Pacific Rim domain was superimposed on top of the Paleo-Asia ocean domain. In the region of Tianshan–Baikal and the region of Kunlun–Qilian–North Qinling, the Tethys domain was superimposed on top of the Paleo-Asia ocean domain. In southern China and the South China Sea region, the Tethys domain was superimposed on top of the Pacific Rim domain. Chinese sedimentary basins were developed either on the landmasses that were described above or on the margin of these landmasses. During different tectonic periods, different prototype basins were superimposed (Jia et al., 1997; 2000). The multi-cycled tectonic evolutions resulted in a complicated "multi-cycled superimposed basin". These three dynamic systems were superimposed and combined together; the framework of the basins was overlaid. For example, typical superimposed geological frameworks were uncovered in the Zhungeer basin, Tuha basin, Tarim basin, Chaidamu basin, Eerduosi basin, Sichuan basin, Bohai bay basin, Songliao basin, Subei–South Yellow Sea basin (Fig. 11.1).

In the past, people conducted numerous research work to study the characteristics of multiple tectonic cycles and to investigate the superimposed framework of basins. These studies established a solid foundation of geological theory for subsequently proposing the concept of a "superimposed basin". Li (1941), Chen (1959) and Zhang (1959) believed that the Chinese tectonic system experienced a complicated developmental process. From a different point of view, Huang (1945), Zhang (1958), and others, proposed the idea of multi-cycled tectonic movements. The understanding of Chinese tectonic evolution (that had various phases and multiple cycles) made many geologists become aware of a superimposed framework in Chinese basins. Zhu (1984; 1986) and other petroleum geologists considered that in China, the Paleozoic basins and the Mesozoic–Cenozoic basins belonged to two separate depositional phases throughout geological history. These two sets of sedimentary basins were related to two kinds of worldwide thermal–tectonic systems. The Paleozoic basins and the Mesozoic–Cenozoic basins were superimposed in most regions. Especially over the last decade, along with intensified petroleum exploration (that was represented by the Tarim basin) in west-central China, geologists obtained numerous first hand seismic data and drilling data. Jia et al. (1995; 1997; 2000) studied the tectonic characteristics of the Tarim basin and the feature of oil and gas accumulations. Incorporating study results from other basins, gradually they fully discovered the small scale superimposed craton basin, the structural characteristics of a composite petroleum system and petroleum distribution patterns (Tian et al., 1990).

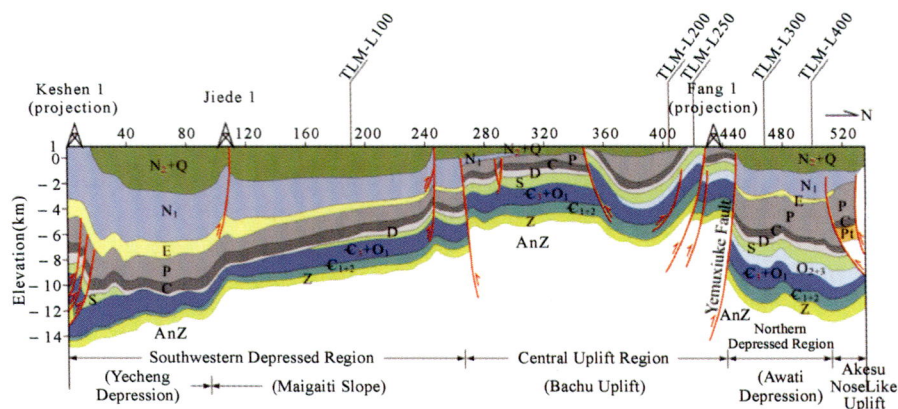

Fig. 11.1. Cross section from the Yecheng depression–Bachu uplift–Awati depression (along the TLM–Z10 seismic line) in the Tarim basin

11.1.3   Characteristics of Superimposed Basin

Many scholars have discussed the characteristics of an onshore superimposed basin (Jin, 2005; Jia, 2006; Han et al., 2006; Liu, 2007; Zhang et al., 2008). These characteristics can be summarized in six aspects and can be further classified into two categories.

● Complicated Framework and Intricate Structures

Because the components of a basin were superimposed in multiple dimensions, the framework and structures of a basin were extremely complicated. 1) Because Chinese geological history contains multiple developmental cycles, geological entities that are from different periods with various intricate natures were superimposed repeatedly, such as superimposed systems of the depositional system, the structural system and the oil and gas system. 2) Many unconformity surfaces of different periods were superimposed together; the underlying unconformity surface was affected by a strong erosion event in the late period, which was a noticeable feature of a superimposed basin. In the Tarim basin, there are nine unconformity surfaces, in ascending order, which included the Sinian basement, Ordovician basement, Carboniferous basement, Permian basement, Triassic basement, Jurassic basement, Triassic basement (Ma et al., 1990). These unconformity surfaces were stacked in an uplifting area or bulged regions. 3) The Chinese landmass was located between the Paleo-Asia domain, Tethys domain and Pacific Rim domain. Influenced by joint structural actions from three domains, the

Chinese superimposed basin had various kinds of merged faulted system that was created by different stresses in multi-phases. In addition, these intricate faulted systems cut into unconformity surfaces to form complicated faulted unconformity systems.

● **Unique Geological Conditions**

Hydrocarbon Generation, Migration, Accumulation and Evaporation Happened in Various Strata within Multiple Phases

Multi-phases of tectonic alteration were the main reason for multiple phases of hydrocarbon generation, migration, accumulation and evaporation. Several sets of hydrocarbon source rocks not only arrived at the threshold of generating and migrating at different times, but they also reached the high peaks in different periods, which benefited multiple strata generation and the release of hydrocarbons. This also assisted the multiple phases of oil and gas accumulation. In a superimposed basin, after experiencing various structural alterations that included distorted compression, pulled extension, folding and erosion, the oil and gas distributions were very complicated. Because of the appearance of a fault or poor sealing conditions, a large scale anticline or uplifting does not have exploration significance. On the other hand, a syncline or a slope zone can become the exploration target because of the alteration in the lithology and because of the seal of a fault.

Mixed Oil and Gas Accumulation Came from Various Sources and Multiple Phases

Controlled by the framework of the prototype basin, the oil and gas system that was based on a single hydrocarbon source rock experienced a relatively independent migration and accumulation event. However, when it was superimposed, an exchange of fluids often happened; in other words, the phenomena of "mixing resources" occurred. Therefore, the basic characteristics of a superimposed basin are "diversified hydrocarbon sources, various thermal centers", "multiple accumulative phases with various reservoirs" and "oil and gas from mixed resources".

Adjustment Happened in Numerous Areas and Various Cycles

In a superimposed basin, the diversified hydrocarbon source, various thermal centers and multiple accumulative phases occurred horizontally, which must lead to multiple accumulation zones. In addition, multi-cycled tectonic evolution, especially the multi-phased tectonic movements that happened after the creation of the original oil and gas reservoirs, would modify, alter, or even damage the original oil and gas reservoirs, resulting in the development of secondary oil and gas reservoirs. The outcome of multi-cycled development in a superimposed basin was that oil and gas reservoirs experienced a series of complicated historical events that included alteration, adjustment and redistribution, damage and restoration.

In a superimposed basin, the multi-cycled tectonic evolution created a complicated and inconsistent superimposed framework, a merged unconformity surface, a complicated fault system, various structural deformation patterns and multi-types of depositional systems. Therefore, several oil and gas systems (that had diversified sources, various thermal centers, multi-phases of accumulation, and several adjustments) were connected, mixed, superimposed, merged or crossed with each other, which determined the complicity and diversity of the creation mechanism for the oil and gas reservoir and its distribution. Furthermore, oil and gas exploration in a superimposed basin, particularly in deep formations of a basin, is more troublesome than in a monotype basin. It requires an intense study of geological theory in a superimposed basin in order to effectively direct oil and gas exploration, especially for exploration in deeply buried strata.

11.1.4   *Types of Onshore Superimposed Basin*

Overall, onshore Chinese basins contained three tectonic sequences, in ascending order, which included the lower craton sequences (the early Paleozoic marine faces strata), the upper craton sequences (the late Paleozoic marine and terrigenous faces), and the Mesozoic terrestrial faces (the Triassic–Quaternary terrestrial clastic strata). In different areas, these three sequences had different combinations. In the same tectonic sequences, the prototype basins and their combinational types were also varied. Some basins contained all three sets of tectonic sequences, such as the Eerduosi basin and the Tarim basin. Although the Sichuan basin contained all three sets of sequences, the upper craton sequence (the marine and terrestrial facies) was not well developed in the Sichuan basin (Liu et al., 1989; 2000). Other basins only contained one or two sets of tectonic sequences; for example, the Zhungeer basin only had the upper craton sequences (the late Paleozoic marine and terrestrial facies) and the Mesozoic–Cenozoic terrestrial sequences (the Triassic–Quaternary terrestrial clastic strata). The Songliao basin and the Bohai bay basin mainly contained the Mesozoic–Cenozoic terrestrial sequences that were made of superimposed strata from different tectonic phases (Xie et al., 1984; Zhao et al., 1989; Zhang et al., 1997).

Huaizeng Huang and others considered that the major types of superimposed basins were made of the following combinations: foreland basin–craton basin, rifted basin on the continental margin–rifted basin on back arch, rifted basin–strike-slip basin and foreland basin–strike-slip basin. According to the nature of the basins, framework and superimposed style, Wenzhi Zhao and others classified Chinese superimposed basins into three basic categories. (1) The superimposed basin was on top of the Precambrian cratons, which included the inherited–modified type of superimposed basin that was made of a craton basin with Paleozoic marine facies–marine and terrestrial facies and the Mesozoic–Cenozoic terrestrial foreland–intracontinental depressed basin (such as the Tarim basin, Eerduosi basin and Sichuan basin), which also included an altered type of superimposed basin that was made of the Paleozoic craton basin and the Mesozoic–Cenozoic rifted basin (such as the Bohai bay basin, Subei–South Yellow Sea basin and Jianghan basin). (2) The superimposed basin was on top of the Hercynian–Indosinian folding basement, which included the foreland (intracontinental depression) inherited type of superimposed basin (such as the Zhungeer basin and Tuha basin), which also included the modified type of superimposed basin that was made of a pulled apart basin (faulted basin) of the early period and the rejuvenated foreland basin of the late period (such as the Chaidamu basin and Jiuquan basin). (3) The modified type of superimposed basin was in the rifted valley of the continental margin–passive continental margin (such as the East China Sea, Pearl River Estuary, Beibu Gulf and Ying–Qiong basin).

From the viewpoint of oil and gas exploration, according to the dynamic nature of a basin and the superimposed method of filled sediments in the basin, here we simply classify the Chinese superimposed basins into two categories. (1) One type of superimposed basin was on top of carbonate strata in the Paleozoic small carton basin; commonly, the combination of basins included either craton basin–foreland basin or craton basin–rifted basin. The former superimposed combination occurred in the Tarim basin, Eerduosi basin and Sichuan basin; the latter superimposed combination appeared in the Bohai bay basin. In this book we simply called it the "carton superimposed basin". (2) Another type of superimposed basin was developed in eastern China, which was beneath the Cenozoic rift basins. This type of superimposed basin is called the eastern China rifted basin in this book, which was commonly made of the Cenozoic depressed basin and the Mesozoic faulted basin; for example, the Cretaceous–Cenozoic depressed basin was superimposed on top of the Jurassic volcanic faulted basin to create the Songliao superimposed basin.

11.2   Geologic Features of Deeply Buried Petroleum Systems and Their Probable Resources

Here, the deeply buried petroleum system is also equivalent to the middle and lower petroleum systems in the superimposed basin. These exploration targeted sequences were developed in a sedimentary basin of the early period and they were positioned in the deep position of a superimposed basin; in addition, they were not only covered by the present prototype basin, but they were also superimposed and altered by later tectonic movement.

11.2.1   Geologic Features

● **Independent Petroleum System**

In an onshore superimposed basin, the middle and lower petroleum systems were influenced to different degrees by tectonic events; the original oil and gas reservoirs were re-modified or damaged; moreover, petroleum reservoirs with mixed sources emerged. In a superimposed basin, compared with the petroleum system in the upper portion, the petroleum systems in the middle and lower portions were relatively independent, although they experienced tectonic alteration or a superimposed event.

In a superimposed basin, the middle and lower petroleum systems contained one or several sets of regional cap rocks. Superimposing the Paleozoic prototype basins might create multiple sets of regional cap rocks: gypsum mudstone, mudstone and coal measures. Most of the oil and gas was still preserved in the traps that were sealed by these regional cap rocks. The well preserved Paleozoic marine basin and the Mesozoic–Cenozoic foreland basin were superimposed together in the Tarim basin, Eerduosi basin, Sichuan basin and Zhungeer basin. Vertically, the Paleozoic strata were commonly covered by sandy mudstone; horizontally, the Mesozoic and the Cenozoic foreland basins (that were developed on the periphery of the superimposed basin) effectively blocked the oxidation of the oil and gas system by air and fresh water. In general, the reservoir sequences that were deposited on the weathering crust of the lower Paleozoic erathem were sealed by regional clay shale or a gypsum–salt layer. The upper Paleozoic sandstone reservoirs were covered by the regional coal measures. For example, in eastern Sichuan, they discovered $2,777.5 \times 10^9$ m^3 of natural gas in the Carboniferous strata that was covered by a very thick and stabilized gypsum–salt layer in the Lower Triassic Jialingjiang formation. In the eastern Sichuan area, where the Triassic gypsum–salt layer in the Jialingjiang formation appeared, abnormally high fluid pressure would usually occur and gas reservoirs might be discovered in the strata beneath. In the areas where the gypsum–salt layer was missing, the anomaly of fluid pressure and gas reservoir was absent; alternatively, there was a transitional zone of air and fresh water. These occurrences indicated that the gypsum–salt layer in the Triassic Jialingjiang formation was a set of excellent regional cap rocks. In the Tarim basin, four sets of regional cap rocks appeared in the Middle and Lower Cambrian series, the Middle and Upper Ordovician series, the Silurian system and the Carboniferous system, separately. These four sets of cap rocks were related to the development of the early Paleozoic restricted–evaporation platform in carbonate rock, the submergence of the carbonate platform, the transition of the humid siliciclastic shelf–arid shelf, and the wide-ranging transgression that was based on the peneplanation of the basin. In addition, these four sets of cap rocks determined that in the Tarim basin the oil and gas distributions were dispersed more vertically. These cap rocks decreased oil and gas evaporation from the Paleozoic strata; therefore, the oil and gas would accumulate and be preserved to form large scale oil and gas fields. In the area that was covered by the Carboniferous gypsum mudstone, recently discovered commercial grade oil and gas flows or oil and gas shows were all beneath this set of cap rock. However, in the area where the Carboniferous gypsum mudstone was absent, commercial grade oil and gas flows or oil and gas shows were discovered in various sequences that were above this set of cap rock. This indicates that the Carboniferous gypsum mudstone was a set of excellent regional cap rocks. The natural gas field in the central Eerduosi basin contained two sets of regional cap rocks. The first set of cap rocks was lacustrine mudstone from the Shiqianfeng formation and the Shihezi formation in the Permian system; the thickness of cap rock was between 240 – 350 m and it sealed the Ordovician weathering crust type of reservoir sequences. The second set of cap rocks was at the base of the Benxi formation in the Carboniferous system, which included bauxite mudstone, calcareous mudstone and sandy mudstone, which also represented more than 70% of the total thickness in the Benxi formation; this set of cap rocks sealed the upper Paleozoic sandstone reservoirs. In the Songliao basin, the volcanic rocks in the Ying 1 member was the principal reservoir sequence in the middle and lower petroleum systems, which was covered by mudstone from the Quan 1 member, the Quan 2 member and the Deng 2 member. The cap rocks covered the entire faulted basin and they secured natural gas in the strata beneath to form natural gas reservoirs.

In the middle and lower petroleum systems, the independent hydrocarbon source rock and reservoir sequences were developed under the regional cap rocks. Hydrocarbon source rock that was made of clay shale and marlite of slope facies usually developed in the lower petroleum system of the craton basin. Among the oil and gas resources in the Sichuan basin, 78.3% of resources came from the Cambrian and Silurian argillaceous rocks of slope facies; the discovered natural gas reservoirs in the central and eastern Sichuan basin came from these hydrocarbon source rocks. In the Tarim basin, the Middle–Upper Ordovician clay shale and the argillaceous limestone of marlaceous mound facies on the platform margin slope functioned as primary hydrocarbon source rocks. These source rocks were distributed in the northern Tarim basin and on the northern slope in the central Tarim basin. On the western margin of the Eerduosi basin, the Ordovician mudstone and argillaceous limestone of slope facies had a high content of organic carbon that might function as valuable source rocks. The weathering crust type of carbonate reservoir sequences developed in the lower petroleum system in the craton basin. In the Sichuan basin, 73.8% of the discovered natural gas reserves were distributed in the layered porous type of reservoir sequences; moreover, the carbonate reservoir sequences represented 90% of the layered porous type of reservoirs. For example, the Leshan–Longnusi uplifting (the Weiyuan–Ziyang) contained the neo-layered weathering crust type of reservoir sequences. In the Tarim basin, the Lunnan–Tahe oil field in the Paleozoic erathem mainly contained Ordovician karst–fractural type of reservoir sequences. In the central Eerduosi basin, a large scale gas field was discovered in the neo-layered weathering crust type of reservoir sequences. In the Renqiu oil field in northern China, the buried hill type of oil reservoir was discovered in the Proterozoic and the lower Paleozoic karst strata.

In a craton basin, the Carboniferous–Permian hydrocarbon source rock of coal measures developed in the upper petroleum system. For example, in the Eerduosi basin, even in the entire northern China region, upper Paleozoic massive layered coal measures were deposited in the Shanxi formation and the Taiyuan formation; these coal measures provided hydrocarbon substances for the creation of natural gas fields in the Sulige–Yulin, Wenliu and Suqiao regions. Several sets of terrestrial clastic reservoir sequences or marine reef flat reservoir sequences developed in the upper petroleum system of a craton basin; for example, Carboniferous Donghe sandstone in the Tarim basin, the sandstone of river channel facies on a delta in the Shanxi formation and the Taiyuan formation in the Eerduosi basin, Carboniferous algal dolomite in eastern Sichuan, Triassic oolite beach limestone in the Feixianguan formation in eastern Sichuan and Permian bio-reef reservoir sequences in southern Sichuan. Most of the clastic rocks were sand bodies in a littoral zone; the creating mechanism of the sand body related to the early phase of the rise in sea level. These were typical transgressive sand bodies that had broad, stable distribution and excellent connectivity.

In the eastern Songliao basin, the Mesozoic petroleum system was made of deeply buried sequences that contained two sets of regional primary hydrocarbon source rocks; one was dark color mudstone and coal measures in the Shahezi formation that developed during the rifting period; the other was dark color mudstone in the Deng 2 member that developed during the depressed period. Furthermore, in some isolated areas, insignificant hydrocarbon source rocks developed in the Carboniferous–Permian strata, the Yingcheng formation and the Huoshiling formation. The primary reservoir sequence was the volcanic rock in the Ying 1 member; additionally, in some shallow buried places on the southern side of the basin, the sandy conglomerate in the Quan 1 and Quan 2 members with good physical properties and in the Denglouku formation might function as reservoir sequences.

• Well Developed Hydrocarbon Source Rocks and Abundant Natural Gas Resources

In the superimposed basin, the middle and lower petroleum systems mainly contained Paleozoic hydrocarbon source rocks. In some basins, the Cenozoic strata functioned as major source rocks for the middle and lower petroleum systems. According to the lithology, the hydrocarbon source rocks of the middle and lower petroleum systems included mudstone of marine facies and terrestrial facies, carbonate rocks and coal measures. The Tarim basin and the Sichuan basin contained well developed lower Paleozoic marine carbonate rocks and mudstone to work as the hydrocarbon source rocks for the middle and lower petroleum systems. In western-central China, the basins commonly contained upper

Paleozoic hydrocarbon source rocks that were represented by the Carboniferous and Permian coal measures. The Songliao basin and Bohai bay basin contained well developed argillaceous hydrocarbon source rocks in the Jurassic–Lower Cretaceous strata.

The following are noteworthy facts about the hydrocarbon source rocks discussed above. Because they were formed during the early period and they experienced an extended period of thermal evolution, at present these hydrocarbon source rocks are deeply buried and they are at a gas generating high peak. Accordingly, the primary exploration target for the middle and lower petroleum systems is the natural gas reservoir because these petroleum systems contain the natural gas resources to supply the creation of large or extra large gas fields (indeed, some basins have oil resources, such as the Zhungeer basin and Tarim basin that contain paleo-oil reservoirs that survived tectonic adjustment and, in northern China, the buried hill type of oil reservoir has oil that was generated in younger strata and stored in older strata).

Hydrocarbon source rocks could generate either oil or gas, in which the thermal maturation was an important controlling factor. In western China, the Paleozoic craton basins contained two kinds of hydrocarbon source rocks that generated natural gas, which were the highly mature–post-mature source rocks of marine facies and the saprogenic type of source rocks of mudstone and coal measures of marine and terrestrial facies. Because they were deeply buried and they experienced an extended period of thermal maturation, the early Paleozoic hydrocarbon source rocks of marine facies progressed into highly mature and post-mature stages and they generated abundant thermal pyrolysis natural gas in deep layers. For example, in the Tarim basin and Sichuan basin, R_o is usually > 2.0%. However, the highest R_o has exceeded 4%. In the circumstances, the oil that accumulated during the early periods has been pyrolyzed into natural gas. In the Eerduosi basin, the Paleozoic hydrocarbon source rocks were at a highly post-mature stage and they mainly generated natural gas. Even though the Ordovician hydrocarbon source rocks mainly contained type I organic materials, because they were at a high thermal maturation stage and because the R_o of hydrocarbon source rocks inside the craton basin could reach 1.2% − 5.0%, the oil had been pyrolyzed into natural gas. Therefore, the Paleozoic strata mainly contained natural gas (Fig. 11.2).

Another factor that triggered the hydrocarbon source rock to generate natural gas was the type of kerogen. During the late Paleozoic era, because the oceanic basins were closed and the foreland basins developed, hydrocarbon source rocks of marine and terrestrial facies and of terrigenous facies were deposited. These two types of hydrocarbon source rocks basically contained type III kerogen that indicated high quality gas generating source rocks. The characteristics of these source rocks included multi-types, various sources and several gas generating stages that also determined multi-phases of gas accumulation.

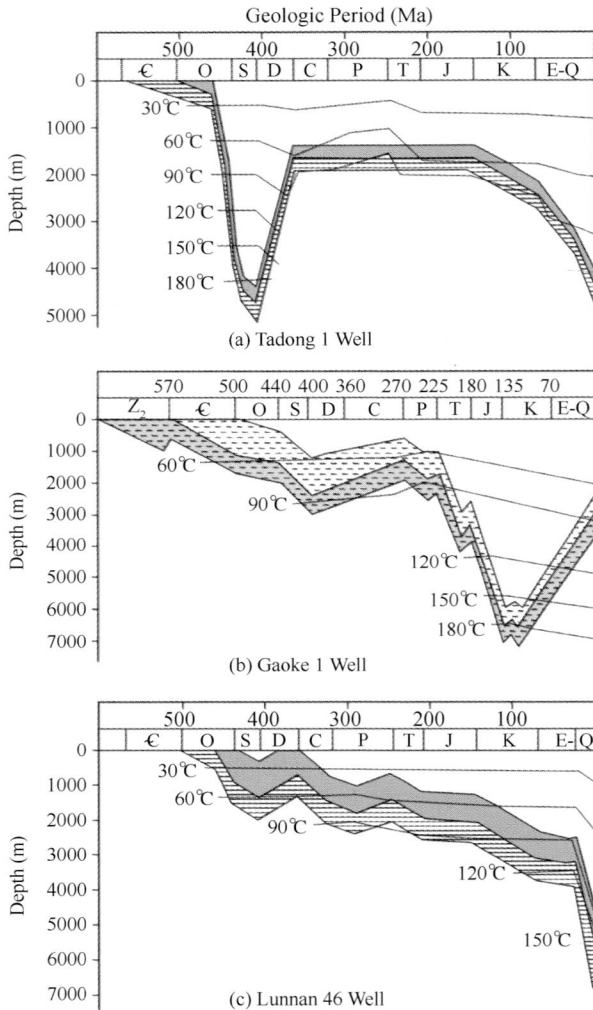

Fig. 11.2. Standard buried history and thermal history of hydrocarbon source rocks in the Paleozoic marine facies

Large scale natural gas fields may also be present in the rifted basin (deep layers in the Songliao basin), in eastern China. There are 19 large size faulted basins in the deep levels of the Songliao basin; on the southern and northern sides of the basin, the total prospective area for petroleum exploration is about 60,000 km^2. The forecast resource is about 2×10^{12} m^3; the forecast geological natural gas resource is more than 1.1×10^{12} m^3. The Songliao basin had the preconditions for developing large and medium size natural gas fields, which have the following characteristics. First, the thickness of the gas layer is great; from the drilling data of the Xushen 1 well and the Xushen 2 well, the effective reservoir thickness is greater than 200 m; in addition, the thickness of the interpreted gas layer is larger than 200 m too. Second, the volcanic rocks were distributed over a broad range; the 3D seismic survey covered about 200 km^2 and the seismic interpretation identified eight volcanic craters that covered a total of about 118 km^2. The size of a single anomaly was between 10 – 36 km^2; these anomalies were maybe connected together because they accrued in clusters. According to the analysis of reservoir abundance in the Xushen 1 well, every square kilometer contained about 2×10^9 m^3 of natural gas. Therefore, this region should have the preconditions for the development of a large size natural gas field with a magnitude of 10×10^{10} m^3.

In the lower petroleum system, the characteristics of hydrocarbon generation obviously influenced the accumulative phases of oil and gas. In the Manjiaer depression in the Tarim basin, as the earliest phase, the Cambrian–Lower Ordovician source rocks progressed into a hydrocarbon generating high peak during the Devonian period. However, as the latest phase, the Middle and Upper Ordovician hydrocarbon source rocks moved into an oil generating high peak during the Cenozoic era. In the eastern Sichuan basin, the Cambrian source rocks entered a hydrocarbon generating high peak during the Devonian; the Silurian source rocks entered a hydrocarbon generating high peak during the Jurassic; the Permian source rocks entered a hydrocarbon generating high peak during the end of the Jurassic and the Cretaceous. In the middle and lower petroleum systems in every superimposed basin, the primary hydrocarbon source rocks entered the major gas generating phase during the Cenozoic era, which commonly brought late phase accumulation. For example, in the platform region and the foreland basin in the Tarim basin, many discovered oil and gas reservoirs developed during the late period. Among these petroleum reservoirs, the developmental period for the reservoirs in the foreland basin was about 5 – 2 Ma. The statistical data shows that in a Chinese superimposed basin, large size oil fields developed during the Cenozoic era and large and medium size natural gas fields developed during the Neogene period.

- **Reservoir Sequences Have Dual Natures: Strong Heterogeneity and High Quality Large Scale Reservoir Sequences**

Exploration practices and studies showed that many types of high quality reservoir sequences developed in the deep levels of the middle and lower petroleum systems in the superimposed basins. The physical properties of some pressure resistant rocks (such as carbonate rock, igneous rock and metamorphic rock) were slightly influenced by the buried depth; however, these rocks had strong heterogeneity. Four types of reservoir sequences developed in the middle and lower petroleum systems, which were carbonate rock, volcanic rock, sandy conglomerate and metamorphic rock. Among these sequences, the widest distributed sequences were marine carbonate reservoir sequences that developed in the low Paleozoic, upper Paleozoic and Mesozoic carbonate rocks in the Sichuan basin and some areas in the Tarim basin. The carbonate rocks contained four types of beneficial reservoir sequences that included reef flat, weathering crust, karst and dolomite and mainly occurred in the Tarim basin, Eerduosi basin, Sichuan basin and Bohai bay basin. Volcanic reservoir sequences were often related to the faulted depression and they were mainly distributed in northern Xinjiang, Tarim, Northeastern China, North China and the lower Yangtze regions. The distribution of metamorphic rocks was limited. Because clastic rocks had poor pressure resistance, the deeply buried clastic reservoir sequences usually had poor physical properties; however, they were distributed over a broad range, such as the upper Paleozoic sandstone of river facies and delta facies in the Eerduosi basin. On the other hand, because of low formation temperature, excessively high pressure, burial in the late phase and fractures, the clastic rocks contained acceptable porosity and could develop into high quality reservoir sequences, such as the sandy conglomerate on the margin of the deeply buried faulted depression in the Songliao basin and the Cretaceous conglomerate on the Kulong mountain front in the Jiuquan basin. Thus, high quality, large scale reservoir sequences could still be developed at deep levels to form multi-types of reservoir sequences.

In western China, craton basins contained abundant natural gas resources; the natural gas could accumulate easily when high porosity and high permeability were present. The Cambrian–Ordovician carbonate reservoir sequences were the most important type. In particular, the Caledonian movement transformed the Ordovician weathering crust type reservoir sequences into a major oil and gas productive layer, such as in the central gas field in the Eerduosi basin, the Lunnan–Tahe oil and gas field in the Tarim basin. In addition, there were various types of reservoir sequences at deep levels of the Tarim basin, which included the karst type, the fracture type that was controlled by faults, and the bio-reef type. In the Sichuan basin, the reservoir sequences of the limestone of shallow flat facies, dolomite, algal dolomite of tidal flat facies, dolomite of bio-reef facies, brecciaed dolomite from collapsed karst, and fractured limestone were widely distributed in great volumes. These carbonate rocks are an important exploration domain too. For example, in the Sichuan basin, natural gas exploration practices in the upper layer showed that the quality of reservoir sequences obviously controlled the magnitude of the gas reservoir and the results of exploration. In the Sichuan basin, among 14 discovered gas layers, the proven gas reserves are $5,503.8 \times 10^8$ m^3. Seven of them were the fractured–porosity type of reservoir sequences that contained 38 natural gas reservoirs with gas reserves of $4,170.7 \times 10^8$ m^3; these gas reserves represented 76% of total proven reserves. These natural gas reserves were clustered in the Carboniferous porosity type of dolomite reservoir sequences in the eastern Sichuan basin and in the porosity type of dolomite of oolite flat facies in the Triassic Feixianguan formation in the northeastern Sichuan basin.

In the Paleozoic craton superimposed basins, the paleo weathering crusts, sandstones and oolite beach were the primary reservoir sequences for the large and medium size oil and gas fields. (1) On the paleo uplift, the weathering crust that had high porosity and high permeability, and the secondary porosity type of reservoir sequences, were the ideal places for oil and gas to accumulate. The long duration and stabilized tectonic evolution of a craton basin created preconditions for developing the large scale inherited type of weathering crust on the paleo uplift. In the central Eerduosi basin, the karst type of reservoir sequences on the paleo uplift functioned as major reservoir sequences for the Changqing Ordovician gas field. In the Sichuan basin, the dissolute porosity type reservoir sequences on the Leshan–Longnusi paleo uplift controlled the creation and development of the Weiyuan Sinian natural gas field. In eastern Sichuan, a Carboniferous gas reservoir was formed in dissolute reservoir sequences in the paleo weathering crust that was on top of the Luzhou–Kaijiang paleo uplift. In the Tarim basin, the oil and gas fields in Tahe–Lunnan, Hetian River, Yakela and the central Tarim areas were related to unconformity surfaces and isolated reservoir sequences that had high porosity and high permeability on the Bachu paleo uplift, the Tabei paleo uplift and the Tazhong paleo uplift, respectively. (2) The late Paleozoic river channel sandstone of marine and terrestrial facies formed natural gas contained sand bodies that were distributed over a broad range. In the Eerduosi basin, the excellent combination of reservoir and seal was made of the widely occurring Carboniferous–Permian reservoir sequences (that included beach bar sandstone, delta sandstone and river channel sandstone) and mudstone in the upper Shihezi formation above. In the Tarim basin, Carboniferous Donghe sandstone was a perfect place for oil and gas to accumulate; the Donghe sandstone was deposited in the littoral–shallow sea environment. (3) The oolite beach on the platform margin formed a gas reservoir for a natural gas accumulation belt in northeastern Sichuan. During the early–middle Triassic, in the northeastern Sichuan basin, a large area of oolite beaches developed in the Feixianguan formation on the margin of a carbonate platform in the Kaijiang–Liangping sea trough, which not only had good physical properties but were also widely distributed on the margin of the Kaijiang–Liangping sea trough. In addition, these oolite beaches worked as excellent natural gas reservoirs in the northeastern Sichuan basin. In general, the porosity of these dolomitic oolite reservoir sequences was between 5% – 12%. The discovered oolite beach type of natural gas reservoirs were located in the following areas: Tieshan, Dukou River, Tieshanpo, Shuangjiaba, the Tiandong 5 well area, Luojiazhai, Shiyougou, Fuchengzhai and the Bandong 5 well area.

In eastern China, the reservoir sequences in the Mesozoic petroleum system were made of volcanic rocks and conglomerates. The volcanic rocks in the Hou 2 member, the Ying 1 member and the Ying 3 member were three sets of principle reservoir sequences in the Xujiaweizi faulted depression in the Songliao basin. The volcanic reservoir sequences were of various types. The natural gas layers could occur from andesite to rhyolite; the types of volcanic reservoir sequences not only contained a lava group, but they also included volcanic clastic rocks. The types of storage included all kinds of pores and fractures of various creative mechanisms. In addition, the types of combination were varied. The volcanic facies controlled the original distribution of volcanic rocks. The lava in the upper subfacies of extrusive facies was the best facies belt for storage under original circumstances. The next category was the subfacies from the pyroclastic flow of explosive facies; the primary pores were matrix contractive fractures and dissolved phenocryst; the secondary dissolute pores were created by weathering and eluviation either during the intermission of a volcanic eruption or during an uplifting period. Besides the fine grain clastic tuff, the other subfacies of extrusive facies and explosive facies also had some storage ability. Especially when assisted by fractures, these subfacies could function as good reservoir sequences. The storage ability of volcanic rocks was noticeably improved by tectonic movement, weathering and eluviation. The diagenesis brought dual influences to the reservoir capability of volcanic rocks.

• Oil and Gas Reservoirs Have Various Fluid Pressures

In a superimposed basin, the middle and lower petroleum systems were covered by regional cap rocks that separated fluids of the upper system and lower system. Under the regional cap rocks, there was an independent pressure system. Inside oil and gas reservoirs, there was abnormally high fluid pressure, nonstandard low fluid pressure and normal fluid pressure.

In the Sichuan superimposed basin, under the blockage of regional cap rocks, abnormally high pressure or normal pressure formed in the middle and lower petroleum systems. In the eastern Sichuan basin, there are two regional abnormal pressure zones in the natural gas region in a vertical direction. Under the cap rocks of the Triassic gypsum–salt layer, the upper abnormally high pressure zone appeared in the Permian reservoir sequences that were made of bio-clastic limestone, silt-crystalline limestone and gypsum dolomite; the pressure coefficient was between 1.39 – 1.92. For example, in the Wolong River gas field, the pressure coefficient in Upper Permian strata is 1.92 in the Wo 117 well; the pressure coefficient in Lower Permian strata is 1.52 in the Chi 12 well in the Dachi dry well gas field. Under the mudstone blockage, a lower abnormally high pressure zone occurred in the Cambrian–Silurian strata and it was formed by depositional compaction and by matured organic material generating hydrocarbons. Currently, in the Zuodongya structure, the pressure coefficient in the Cambrian strata is 1.89 in the Zuo 3 well. In the Weiyuan gas field in central Sichuan, Silurian clay shale is the regional cap rock and Lower Cambrian shale is the direct cap rock. However, in the Sinian reservoir sequences, the pressure coefficient is around 1.00, which almost equals the hydrostatic pressure.

In the Eerduosi basin, under the cover of coal measures, reservoir sequences not only had uniquely low formation pressure, but they also contained abnormally low pressure oil and gas reservoirs. These sequences occurred in the lower Shihezi formation–the Shanxi formation in the upper Paleozoic erathem. On the northern side of the basin, the pressure coefficient is 0.746 – 0.981; in the central region of the basin, the pressure coefficient is 0.787 – 0.998; on western side of the basin, the formation pressure is normal and the pressure coefficient is 0.938 – 1.01.

In the Lunnan area in the Tarim basin, the pressure coefficient is 1.36 in the Carboniferous gas reservoir in the Jilake gas field. In the Lunnan–Tahe area, the

Ordovician buried hill type of oil and gas reservoir has a normal pressure system and its fluid distribution system is complicated. Different kinds of crude oil and natural gas were distributed in separated sections horizontally; the oil reservoir has normal pressure. In the central Tarim basin, the Carboniferous and Permian systems had ultra high pressure that was caused by under compaction; in the Carboniferous system, the ultra high pressure was between 9.46 – 16.55 MPa. In the Hetian River gas reservoir, the Carboniferous bio-clastic limestone functioned as reservoir sequences. Above the interface of gas and water, the pressure coefficient is 1.16 – 1.00; below the interface of gas and water, the pressure coefficient is less than 1.00.

11.2.2   Probable Resources

In the superimposed basin, the capacities of the resources at deep levels were mainly determined by two factors. One was that deeply buried formations had preconditions for generating and preserving the oil and gas. Another was the difficulty of exploration, the lower level of exploration and the large amount of remaining resources.

● **Solid Substance Foundation**

The deeply buried strata in a superimposed basin had the preconditions for establishing oil and gas reservoirs, which include source rocks, reservoir sequences and traps. The general characteristics of these conditions have been discussed in section 1 of this chapter. In conclusion, in a superimposed basin, the middle and lower petroleum systems commonly contained valuable hydrocarbon source rocks. Because most of them were of high quality and high maturity, they mainly generated natural gas. The oil and gas were accumulated in various phases; the late accumulative phases were the major events. In the middle and lower petroleum systems, there were four types of reservoir sequences that included carbonate rocks, volcanic rocks, sandy conglomerate and metamorphic rocks. The physical properties of these rocks were slightly influenced by the buried depth. In other words, excellent reservoir sequences could still be developed at great buried depth. In the middle and lower petroleum systems, the major types of oil and gas reservoirs included the lithologic type, stratigraphic type and composite type; the complicated substance facies were the result of multi-phases of accumulations. In the middle and lower petroleum systems, the distributions of oil and gas were closely related to developments of paleo uplifting, sedimentary or volcanic facies belts, faulted belts and the unconformity surface. Therefore, in the future, the middle and lower petroleum system in a superimposed basin will be one of the important domains for onshore petroleum exploration in China.

● **Lower Exploration Level with Large Quantity of Remaining Resources**

In Part I, Chapter 3, the author points out "the characteristics of oil and gas distribution in geological periods." The Mesozoic and the Cenozoic strata mainly produced oil with reserves of 760×10^8 t, which represent more than 80% of the total reserves. The Paleozoic marine strata mostly produced natural gas (if we do not consider the accompanying gas from the tertiary oil field) with gas reserves of 18×10^{12} m^3, which represent 47% of total reserves. The proven oil reserves equal 32%; the remaining oil reserves are 326.87×10^8 t. The confirmed natural gas reserves only equal 12.5%. Compared to the Cenozoic erathem, the exploration level was very low; the total remaining gas reserves are 21.05×10^{12} m^3. As previously discussed, in China there are two types of superimposed basins. One is a craton basin superimposed either by a foreland basin or by a rifting valley; the other one is a faulted basin superimposed by a depressed basin. In these two types of superimposed basins, there are three kinds of petroleum systems that include the lower system in a craton basin, the upper system in a craton basin and the deeply buried system in the Mesozoic erathem. The Paleozoic marine strata are part of the middle and lower petroleum systems only. Thus, we can imagine the potential of deeply buried petroleum resources in a superimposed basin.

11.3   Deeply Buried Petroleum Systems

The deeply buried petroleum system of a superimposed basin is also called the middle and lower petroleum system of a superimposed basin. The name was designated to the exploration targeted strata that were deposited during the early developmental phases of a basin, not only located at the deep level of a superimposed basin, but were also covered by the present prototype basin and, moreover, were altered by tectonic movement during the late period. According to the developmental features of a superimposed basin and the knowledge that was acquired from exploration practices, an onshore superimposed basin can be further categorized into three types: (1) the lower craton petroleum system that was mainly made of the Proterozoic–lower Paleozoic strata; (2) the upper craton petroleum system that was made of the upper Paleozoic strata; (3) the Mesozoic petroleum system in eastern China.

11.3.1   Lower Craton Petroleum System

The lower craton petroleum system was positioned in the Proterozoic–lower Paleozoic strata of marine facies. During the early Paleozoic, the craton plates of Tarim, Sichuan, Eerduosi and North China were not only drifting in the ocean, but they were also subsiding at a stable rate. The carbonate deposits of basin facies mainly developed in the intra-craton plate. The aulacogen, the passive continental margin and the peripheral depressed basin appeared on the peripheral area of the craton plate. The lithology was a set of carbonate rocks that was interbedded with sandy shale, which indicated the depositional environment of deep water in the continental marginal sea. Furthermore, high quality hydrocarbon source rocks of marine facies also developed here. In this type of petroleum system, the discovered oil and gas fields currently include the Lunnan–Tahe oil field in the Ordovician system in the Tarim basin, the Central Region gas field in the Ordovician system in the Eerduosi basin, the Renqiu oil field that is a buried hill type of oil field in North China, and the Weiyuan gas field in the Sinian system in the Sichuan basin.

● Tectonic Background

The development of lower craton sequences was controlled by sedimentary deposits in small craton basins that were drifting in the ocean during the late Proterozoic–Ordovician. During the late Neoproterozoic era, because the paleo continent of Rodinia was divided, several small landmasses broke away from Rodinia, which included North China, the Yangtze, South China and Tarim. During the beginning of the early Paleozoic, these small landmasses were located in a low altitude area of the southern hemisphere; then they drifted a long distance northwards. The Tarim basin, Sichuan basin and Eerduosi basin were in the stage of aulacogen deposits during the Sinian period–the beginning of the early Paleozoic era. For example, the Mandong area in the Tarim basin and the eastern side of Qingyang in the Eerduosi basin were the well developed aulacogen regions. During the middle and late early Paleozoic, these regions received platform carbonate deposits; horizontally, the facies belts were from lagoon facies of the intra-platform-platform dolomite facies-platform margin facies-shallow sea facies. The carbonate deposits from a shallow depression or platform developed in a craton landmass; in the peripheral areas of the craton, the carbonate deposits of the continental shelf or slope developed (Song, 1989).

Compared to the craton basin either in Siberia or on the North American continent, the middle and lower craton petroleum systems had similar tectonic features. The strata were leveled with a few magma activities and poorly developed faults, without a noticeable folding event or missing metamorphic activity. The uplifting and subsidence events were the major tectonic movements, which not only formed multi-phases of unconformity surfaces, but which also developed the paleo karst reservoir sequences. Among the craton plates, the North China plate was above water and it experienced a long period of erosion during the Late Ordovician–Early and Middle Silurian; in addition, the multi-phases of unconformity surfaces developed in the Tarim and the Yangtze craton basins. However, because they were of small size and poor tectonic stability, these small or micro size Chinese craton basins were rather active internally, and this was represented by well developed paleo uplifting inside the basins.

● Characteristics of Petroleum Geology

The tectonic and sedimentary features that were discussed above determined the characteristics of petroleum geology in the lower stratigraphic sequences of a craton basin. For the hydrocarbon source rocks during the early Paleozoic in western China, the craton basin received a set of deep water carbonate rocks that were interbedded with sandy shale in the continental marginal sea. In addition, high quality source rocks of marine facies also developed here and they were thermally matured to become high-post matured natural gas source rocks. For the reservoir sequences, the karst reservoir sequences were the predominant type.

Regarding preservation conditions, the three large size craton basins of Tarim, Sichuan and Eerduosi had relatively stable internal structures; moreover, multiple sets of regional cap rocks developed in these craton basins. For the accumulative pattern, the paleo uplifting in the lower stratigraphic sequences offered an ideal place for natural gas to accumulate. Here, we mainly discuss the developmental characteristics of the reservoir sequence that was the crucial factor for oil and gas exploration.

In the lower craton petroleum system, the reservoir sequences were predominately made of non-sandstone type of karst sequences. Firstly, the carbonate strata and dissolved unconformity surfaces were widely developed; secondly, only carbonate karst reservoir sequences were found in the discovered large and medium size oil and gas fields; thirdly, because of a longer diagenesis period and great buried depth, the sandstone had lost a considerable amount of pores. According to their creative mechanisms, the lower Paleozoic karst reservoir sequences can be further categorized into four different types, which are the syngenesis interlayer karst, denuded weathering crust karst, buried compacted karst and deeply buried hot water karst. The most important reservoir sequence is the denuded weathering crust karst. Therefore, in the lower craton petroleum system, the reservoir sequences were mainly made of marine carbonate rocks that contained karst and secondary pores.

● Oil and Gas Accumulative Pattern

In the craton basin, the discovered large size oil and gas fields had a structural reservoir on the paleo-uplift. The oil and gas distributions were clearly controlled by paleo uplifting and its slope; paleo-uplifting experienced a long developmental period and it was the designated area for oil and gas accumulation.

In the lower craton petroleum systems, paleo-uplifting controlled oil and gas accumulations. The structural belts on the paleo-uplift were mainly distributed in the Sichuan basin, Eerduosi basin and the Paleozoic craton basin in Tarim. Because they experienced a long period of stabilized development, carton basins had the preconditions for creating and developing the large size inherited type of paleo-uplift and sloped structure. In China, all the large carbonate gas fields were closely related to paleo-uplifting. The worldwide statistical data (Zhang, 1990) showed that, in extra large size and large size natural gas fields, 30.6% of total gas reserves were accumulated in a paleo-uplift. If considering gas fields that were based on paleo-uplifting and that were modified lately, then in extra large size and large size natural gas fields, 75% of the total natural gas reserves in the world are related to paleo-uplifting. There are four reasons why paleo-uplifting came to control oil and gas accumulations. (1) Paleo-uplifting helped the development of the karst type of reservoir sequences; the fractures in the high position of a structure would improve the storage ability (Fig. 11.3). Because paleo-uplifting was above water, the strata not only experienced eluviation and dissolution from air and fresh water, but they also developed the karst type of reservoir sequences. At same time, paleo-uplifting was in a high position on the structure where the structural stresses were concentrated and fractures were well developed. Therefore, the storage ability of the brittle carbonate rock was improved. In the northern Tarim basin, although oil and gas distribution in the Ordovician buried hill was complicated, the whole geological entity contained oil and gas. The reservoir sequences in the weathering crust karst controlled oil and gas distribution and accumulation, which formed the super large, buried hill type of oil and gas field in the Lunnan–Tahe area. The pores that developed on the salt dissolving paleo highland controlled natural gas accumulation in the lower Paleozoic in the Eerduosi basin. (2) A paleo-uplift was the designated place for natural gas migration and accumulation. Whether in the Permian basin of North America or in the Triassic basin in North Africa, the central uplifts are the main places for oil and gas accumulation at present. In China, commercial grade oil and gas reservoirs have been discovered in the uplifts in the Tarim basin, Sichuan basin and Eerduosi basin. (3) The unconformity surface and faults that were related with paleo-uplifts worked as major channels for oil and gas migration. The top surface of the Carboniferous system on the Tabei uplift in the Tarim basin and the top surface of the Sinian system on the paleo-uplift in the Leshan–Longnusi area in the Sichuan basin worked as major migration passages for oil and gas migration and accumulation. In the Eerduosi basin, the upper Paleozoic natural gas directly passed through the unconformity surface to accumulate in the Ordovician karst type of reservoir sequences. (4) Paleo-uplifting assisted petroleum accumulated in the early period and it helped reservoir adjustment during the late period. The Weiyuan gas field that was located on the paleo-uplift of Leshan–Longnusi (Caledonian tectonic period) in the Sichuan basin and the group of natural gas fields on the Kaijiang paleo-uplift (Indosinian tectonic period) in eastern Sichuan are examples. These gas fields accumulated on paleo structures during the early period and they were adjusted during the late period.

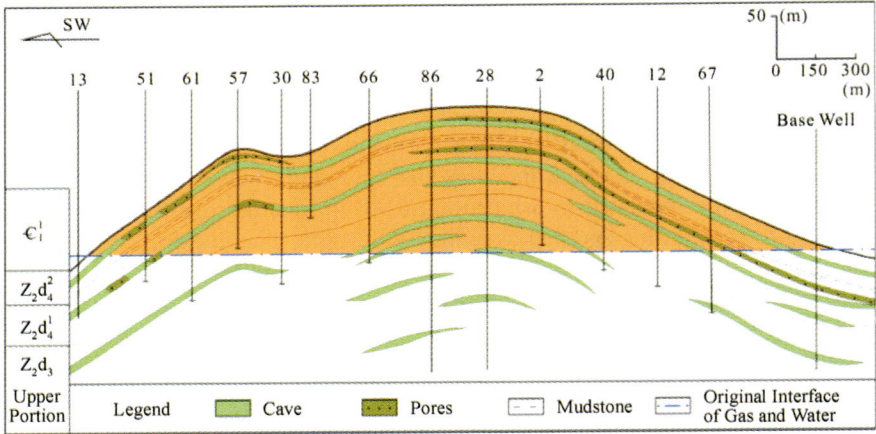

Fig. 11.3. Cross section of the Sinian gas reservoir in the Weiyuan gas field

11.3.2 Upper Craton Petroleum System

This type of petroleum system was positioned in stratigraphic sequences of marine and terrestrial facies, which were mainly made of the upper Paleozoic strata. During the late Paleozoic era, the original oceans were gradually closed and craton plates progressively converged together. Limited separation events happened on the margin of a craton and a restrictive oceanic basin or rifting valley developed in Tianshan, Qilian mountain, south Kunlun–south Qinling. The depressed basin at the intra-craton and the rifted basin on the margin of a craton received sediments of marine and terrestrial facies during the Carboniferous–Permian periods. Marine transgression happened during the Early Carboniferous. Although the North China and Yangtze plates still behaved as a paleo landmass, the Tarim paleo plate had received carbonate and clastic deposits of a shallow sea. This depositional event continued through the Middle and Late Carboniferous epochs to create Carboniferous hydrocarbon source rocks. During the Middle and Late Carboniferous epochs, the main portion of the Yangtze paleo plate still behaved as a paleo landmass. However, during the Permian period, the coal measures of marine and terrestrial facies (that were dominated by marine facies) were deposited on the Yangtze plate. During the Permian period, the North China paleo plate received massive marine and terrestrial formations that contained coal. At present, many oil and gas fields have been discovered in the upper craton petroleum system, which include the Sulige–Yulin gas field in the Eerduosi basin, the Hetian River gas field and the Tazhong oil and gas field in the Tarim basin, the Shixi oil field in the Zhungeer basin, the Carboniferous gas fields in the eastern Sichuan basin and the Feixianguan oolite beach gas field in the northeastern Sichuan basin, as well as the Wenliu gas field in the North China region.

● Tectonic Background

The development of stratigraphic sequences in the upper portion of a craton was controlled by tectonic events that included the closing of the late Paleozoic oceanic basins and the converging of small craton plates. The Paleo-Asia ocean was completely closed prior to the Late Permian epoch. The small landmasses of North China, Zhungeer–Tuha and Tarim converged on the southern margin of the Siberia plate. Because of the closing of the original oceanic basin, several small oceanic basins or orogenies (that included Aertai Mountain, Tianshan Mountain, Qinling Mountain, Kunlun Mountain and Helan Mountain) were formed between the small landmasses and they eventually converged together to form the united Paleo-Asia continent. During this time interval, the craton basins of North China, Zhungeer–Tuha and Tarim mainly received clastic sediments of marine and terrestrial facies; the thickness of sedimentary strata was stable; reservoir sequences of layered sandstone and cap rocks of a gypsum–salt layer were deposited in a broad region inside every craton basin. For example, the Tarim basin mainly received deposits of littoral sand bodies, gypsum mudstone of littoral lake facies, and a mudstone combination. On the other hand, after it received sediments of marine and terrestrial facies during the Benxi tectonic period, the Eerduosi basin received a set of coal measures of limnetic facies near the ocean and the deposits of the delta system.

● Characteristics of Petroleum Geology

Because, during the late Paleozoic era, the tectonic background controlled the development of stratigraphic sequences that were primarily made of the sediments of marine and terrestrial facies and coal measures, the associated preconditions for petroleum geology were controlled by the structural–depositional characteristics. For hydrocarbon source rocks, the upper Paleozoic mudstone of marine and terrestrial facies and the saprogenic type of coal measures functioned as source rocks for gas fields. For reservoir sequences, coarse grain sandy conglomerate and oolitic rocks that had intergranular pores worked as the reservoir sequences. For petroleum preservation conditions, multiple sets of regional coal measures that operated as cap rocks developed in a stabilized craton basin. For the petroleum accumulative pattern, the widely distributed sandstone (reef flat facies) offered an ideal place for natural gas to accumulate. Here, we mainly discuss the reservoir sequences that were a crucial factor in petroleum exploration.

In the upper craton petroleum system, the reservoir sequences were not only made of clastic rocks and carbonate rocks, but they also contained intergranular pores. The late Paleozoic tectonic–paleogeographic evolutions created preconditions not only for developing the porous type of carbonate rocks of beach bar facies and bio-reef facies, but also for developing the sand body of delta facies. Furthermore, because the harmful diagenesis events were weak, the original pores were preserved. The late Paleozoic upper craton sequences primarily contained sediments of marine and terrestrial facies, which included the sandstone of river channel facies, beach bar facies and littoral facies, and which contained well developed original pores. Because they experienced a relatively short evolution procedure, and because the diagenesis compactions were weak, the intergranular pores were well preserved in these sequences. For example, in the Eerduosi basin, the sandstone of beach bar facies, delta facies and river channel facies developed in the Carboniferous–Permian systems; in the Tarim basin, the Donghe sandstone of littoral–shallow sea facies developed in the Carboniferous system; in the northeastern Sichuan basin, the Feixianguan oolite reservoir sequences developed on the carbonate platform margin in the Kaijiang–Liangping sea trough.

- **Oil and Gas Accumulative Pattern**

In the upper craton petroleum system, the widely distributed lithologic reservoir sequence characteristically contained oil and gas; for example, the upper Paleozoic low permeable oil and gas reservoir in the Eerduosi basin, the Lower Triassic oolitic beach reservoir in the Feixianguan formation in the eastern Sichuan basin, and the Donghe sandstone in the Tarim basin. The coarse grain sand bodies or beach bar directly controlled oil and gas accumulation. Compared with the Tarim and the Yangtze landmasses, the Eerduosi landmass was relatively stable. During various geological periods, the entire basin was either uplifted or subsided with a few faults; the reservoir sequences were distributed over a broad region with relatively poor physical properties (average porosity was 4% – 12%, average permeability was $0.1 \times 10^{-3} – 2 \times 10^{-3}$ μm^2). Hydrocarbon source rocks were widespread, oil and gas only needed to migrate a short distance vertically. Therefore, in the Eerduosi basin, reservoir sequences contained oil and gas over a broad region (more than 90% of the exploration wells contained gas). Similarly, in the Sichuan basin, under the control of the Kaijiang–Liangping sea trough, the Lower Triassic oolitic beaches in the Feixianguan formation were distributed over a broad area. The Upper Permian source rocks were also scattered over an extensive area with a short migration distance; thus, oil and gas were widely distributed in the Feixianguan formation. At present, the discovered gas fields include the Sulige gas field in the Eerduosi basin, the oolitic beach gas field in the Triassic Feixianguan formation in the Sichuan basin, and the Carboniferous gas field in the Hetian River area, Tarim basin.

11.3.3 *Mesozoic Petroleum System in Eastern China*

This type of petroleum system was positioned at deep levels in the Songliao basin, which included structural sequences in the Upper Jurassic–Lower Cretaceous faulted basin and structural sequences in the Lower Cretaceous depressed basin. The structural sequences in the faulted basin were made of the Upper Jurassic Huoshiling formation, the Lower Cretaceous Shahezi formation and the Yingcheng formation. These sequences appeared in more than 30 faulted basins and they were controlled by the faults that controlled the faulted basin and that mostly extended in an NNE direction and in an NW direction. The faulted basins developed along three faulted belts that were the Nenjiang faulted belt, the Sunwu–Shuangliao faulted belt and the Haerbin–Siping faulted belt. These three faulted basin belts were oriented in an NNE direction and they were separated by the uplifted basement rocks. Relatively large size faulted basins were distributed in the following areas: Xujiaweizi, Yingshan–Shuangcheng, Changjiaweizi–Gulong and Lindian–Heiyupao. The structural sequences in the depressed basin included the Denglouku formation, the Quan 1 member and the Quan 2 member. The Qijia–Gulong area had the most complete stratigraphic sequences of great thickness.

● Tectonic Background

The Mesozoic petroleum system of eastern China was controlled by the deposits of the back arch extensional basin when the Pacific plate was subducted under the Eurasian continent. During the late Triassic–the beginning of Jurassic in China, a series of terrigenous rifted basins developed in the southeastern costal region that was located inside the tectonic domain of the Pacific Rim. During the late stage of the Middle Jurassic, once again the Paleo-Pacific ocean was subducted under the Paleo Asia landmass. An NE oriented, gigantic tectonic belt developed in eastern China and, subsequently, the J_3-K_1 terrestrial lacustrine basins developed over a broad region. After the Early Cretaceous, the extensional tectonic system was the main mechanism in eastern China. During the Late Cretaceous–Palaeogene, NNE oriented faulted basins developed. During the Neogene–Quaternary, these faulted basins were transformed into depressed basins.

The deeply buried stratigraphic sequences in the Songliao basin experienced double tectonic movements that were characterized by unconformity surfaces in the upper and lower Yingcheng formation. Horizontally, if the Yingcheng formation (especially the volcanic rocks in the Ying 1 member) was designated an exploration target, then the primary control factor was the rifted basins that were controlled by two groups of faults that were oriented in an NE and NW direction. When the final layers of strata in the Yingcheng formation were deposited, a tectonic movement happened and most of the strata were eroded; the remaining faulted basins were the residual product of the tectonic movement. The Ying 1 member in these remaining basins is the major exploration target at present. In the southern Songliao basin, if utilizing the Quan 1 member, the Quan 2 member and the Denglouku formation as exploration targets for determining a favorable exploration region, the superimposed relationship between the upper and lower structures is very complicated. Thus, we should not only analyze the Early Cretaceous faulted basin, but should also consider the structures of the late period. The Songliao basin experienced the development of a rifted basin and three compressive alterations in the late period; the NNE oriented structural features were superimposed on top of a tectonic framework of the "twice uplifted and twice depressed" type. This superimposed tectonic movement created the ideal place of "triple vertical zones and double horizontal belts" for oil and gas to accumulate.

● Characteristics of Petroleum Geology

There is agreement about the theory behind the deeply buried petroleum system in the Songliao basin. The coal measures in the Shahezi formation worked as primary source rocks that included coal and dark color terrestrial mudstone. Subsequently, the Carboniferous–Permian systems, the Yingcheng formation and the Huoshiling formation also contained hydrocarbon source rocks. The volcanic rocks in the Ying 1 member served as the most important reservoir sequences; furthermore, in isolated areas with a shallower buried depth, the reservoir sequences may include sandy conglomerates in the Quan 1 member, the Quan 2 member and the Denglouku formation. The mudstones in the Quan 1 member, the Quan 2 member and the Deng 2 member functioned as cap rocks.

The Songliao basin contained two sets of regional hydrocarbon source rocks; one is the dark color mudstone and coal measures of the Shahezi formation that developed during the rifting period; the other is the dark color mudstone of the Deng 2 member that developed during the depressed period. From the viewpoint of abundance, type and the maturity of the organic materials, the Denglouku formation had worse source rocks than the Shahezi formation. However, these source rocks were still of medium–good grade. The dark color mudstone of the Shahezi formation was mainly distributed in the relatively large sized faulted basins that included Xujiaweizi, Yingshan–Shuangcheng, and Changjiaweizi–Gulong; in general, the thickness of the mudstone is between 200 – 500 m and the estimated thickness can be more than 1,000 m. The lacustrine mudstones of the Deng 2 member were mainly distributed in the Qijia–Gulong area with a forecast distribution range of 660 km^2; the greatest thickness of the mudstone can reach 900 m. Together, three sets of volcanic rocks and four layers of clastic rocks made a complicated petroleum distribution pattern in the geological space.

In the northern Songliao basin, the Xujiaweizi faulted basin mainly contained three sets of deeply buried volcanic reservoir sequences in the Hou 2 member, the Ying 1 member and the Ying 3 member. The volcanic rocks of the Hou 2 member were distributed in layers at the bottom of the faulted basin. In isolated areas, these volcanic rocks extended beyond the border of the faulted basin. The thickness of volcanic rocks was relatively great in the Xujiaweizi–Zhaozhou area with a greatest thickness of 1,000 m; the high thickness value sectors were distributed in a belt pattern with great variation horizontally. The shield-like combination of volcanic rocks and layered volcanic rocks were formed together. The combination of facies was complicated, the thickness varied. Overall, the strata were deformed in a west dipping monocline. The drilling exploration demonstrated that the intermediate volcanic rocks of the Huo 2 member were relatively dense with poor physical properties. However, in the higher position of the structure, because of erosion, good reservoir sequences could be developed in these volcanic sequences. The strata of the Ying 1 member almost covered the entire Xujiaweizi faulted basin. The massive strata (thickness was greater than 600 m) were mainly distributed to the west of the Yushulin–Xujiaweizi area; the greatest thickness could reach 900 m; the strata were smoothly leveled to cover the Shahezi formation, the Huoshiling formation and the bed rocks. On the western side of Xujiaweizi, near the controlling fault area, commercial grade gas flows were discovered in several wells, where the main portion of reservoir sequences in the basin were positioned. The layered volcanic rocks were of the primary type, which developed into a belt pattern, shield-like combination of volcanic rocks. The extrusive facies were the major type of facies. In isolated areas, explosive facies also occurred. In addition, the reservoir sequences of the Ying 1 member commonly had good physical properties. The volcanic sequences of the Ying 3 member only occurred inside a syncline that was located to the west of the Yushulin and Songzhan areas. On both the eastern wing and the western wing, the volcanic sequences overlapped on the top surface of the Ying 1 member and the Songxi controlling fault. The thickness of these reservoir sequences was between 300 – 500 m and there were relatively minor alterations regarding thickness. The combination of volcanic rocks was nearly horizontal. However, in the Songzhan area, a shield–like combination and a cone- shaped combination appeared. In the area from the Wang 903 well–Shengshen 2 well–Shengshen 201 well, the acidic volcanic rocks of explosive facies and extrusive facies had good physical properties. Three sets of hydrocarbon source rocks (in the Huo 1 member, the Shahezi formation and the Ying 2 member) and three sets of volcanic reservoir sequences (in the Huo 2 member, the Ying 1 member and the Ying 3 member) were not only placed on top of each other vertically, but they were also intertwined with each other horizontally to create a complicated combination of reservoir and cap rocks in the Xujiaweizi faulted basin. The Songzhan–Yuxi area contained well developed source rocks and reservoir sequences. On the western, central and southern sides of the Xujiaweizi faulted basin, the reservoir sequences of the Huo 2 member and the Ying 1 member united with the Shahezi formation and the Ying 4–Ying 2 members to form a combination of reservoir and cap rocks (Fig. 11.4). On the other hand, on the eastern margin of the paleo central uplift and on the northern side of the Shengping–Wangjiatun structure, the volcanic rocks of the Huo 2 member were directly covered by the Deng 2 member, which created a simple combination of reservoir and cap rock.

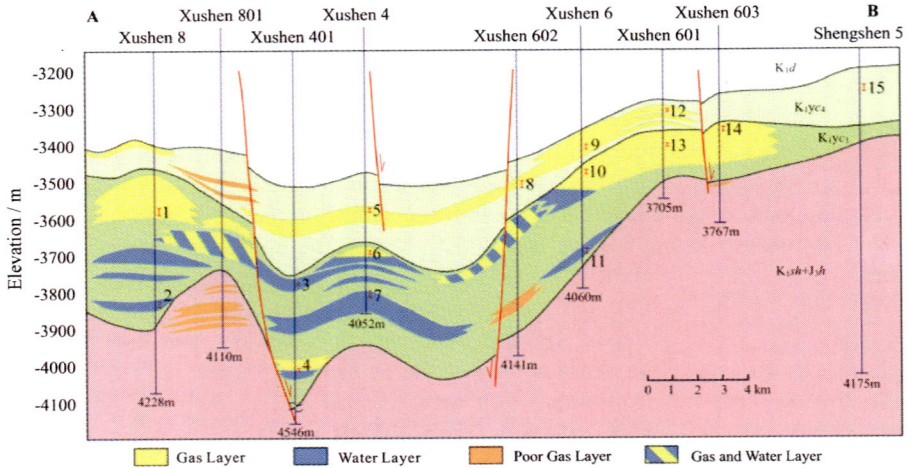

Fig. 11.4.   Cross section of gas reservoir, the Xushen gas field

Legend: Gas Layer | Water Layer | Poor Gas Layer | Gas and Water Layer

1 -- MFE-II Artesian Well, Daily Gas Production: 226234 m³, Commercial Grade Gas Layer;

2 -- MFE-II Testing, Daily Water Production:1.5 m³, Water Bed;

3 -- 3950-3958 m, Daily Gas Production: 244 m³, Water: 36.48 m³;

4 -- Daily Gas Production: 62861 m³, Water: 110.13 m³;

5 -- 3774-3783 m, Artesian Gas Flow after Pressure Treatment,Daily Gas Production: 33427 m³,

6 -- 3873-3881 m, Daily Gas Production: 77315 m³, Daily Water Production: 115.2 m³,

7 -- Daily Gas Production: 450 m³, Daily Water Production: 120 m³,

8 -- Artesian Gas Flow after Pressure Treatment, Daily Gas Production: 9838 m³, Low Productive Gas Layer;

9 -- Producing Gas after Fractural Treatment, Daily Gas Production: 522676 m³,

10 -- Artesian Gas Flow after Pressure Treatment, Daily Gas Production: 105689 m³, Daily Water Production: 124.8 m³;

11 -- MFE-I+T Testing, Daily Gas Production: 280 m³, Daily Water Production: 18.2 m³;

12 -- Artesian Gas Flow after Pressure Treatment, Daily Gas Production: 262641 m³, Commercial Grade Gas Layer;

13 -- Artesian Gas Flow after Pressure Treatment, Daily Gas Production: 232098 m³, Commercial Grade Gas Layer;

14 -- Artesian Gas Flow after Pressure Treatment, Daily Gas Production: 156780-249970 m³, Commercial Grade Gas Layer;

15 -- Dry Layer

● **Oil and Gas Accumulative Pattern**

Gradually, we acquired knowledge about the lithologic oil and gas reservoirs in the rifting valley, eastern China. The following are the conclusions regarding the volcanic gas reservoirs in the Ying 1 member. (1) The discovered reservoir sequences had dual medium strata that contained both pores and fractures; in addition, some matrixes had fairly good porosity and permeability. (2) In general, the gas reservoirs not only have normal pressure, but they also have massive layers with water at the bottom. The data from several wells indicate a massive gas reservoir of great thickness and water at the bottom; however, these reservoirs have strong heterogeneous features. (3) The traps were special lithologic traps, most of which were volcanic traps or even eruptive volcanic traps; in addition, a structural trap or litho-structural trap also appeared here. (4) Most of the natural gas was dry gas mixed with a large amount of non-organic gas. The non-organic gas also occurred in isolation. Most of the natural gas was generated by organic substances. Coal generated, highly mature–post mature natural gas was the major type. (5) The buried depth of the natural gas reservoir was between 2,000 – 4,300 m.

11.4 Exploration Direction for Deeply Buried Oil and Gas in the Superimposed Basins

The exploration domains in the superimposed basins included lower craton stratigraphic sequences, upper craton stratigraphic sequences and the Mesozoic erathem in eastern China. The oil and gas exploration in the lower craton stratigraphic sequences mainly concentrated on the paleo uplifts and slopes; the major exploration targets included the Tabei uplift, the Bachu uplift, the Tazhong uplift, the Tadong low uplift, the Yimeng uplift, the central paleo uplift and slope in the Eerduosi basin, and the Leshan–Longnusi paleo uplift. In addition, the buried hill type oil and gas reservoirs in the North China area were as important as other exploration targets. On the other hand, the oil and gas explorations in the upper craton stratigraphic sequences mostly targeted the stratigraphic trap, the lithologic trap or structural–lithologic trap, because these places contained coarse grain sedimentary rocks. The major exploration targets included the Upper Paleozoic sand bodies of delta river facies in the Eerduosi basin, the Triassic oolite beach dolomite in the northeastern Sichuan basin, the Carboniferous Donghe sandstone in the Tarim basin, the Carboniferous volcanic rocks in the Zhungeer basin, the Permian limestone of bio-reef facies or bio-clastic beach facies in the Sichuan basin, the upper Paleozoic sandstone of delta river facies in the North China area and the deeply buried volcanic rocks and conglomerate in the central region of the Zhungeer basin. Recently, oil and gas explorations in the Mesozoic erathem in eastern China focused on the deeply buried reservoir sequences of volcanic rocks–conglomerate in the Songliao basin and the Bohai bay basin.

11.4.1 Tarim Basin

In the Tarim basin, the exploration of the middle and lower petroleum system in the superimposed basin focuses on the karst type of reservoir sequences that were developed on the paleo uplift and related slopes in the lower craton stratigraphic sequences. At the same time, the exploration also paid attention to the lithostratigraphic traps that were positioned in the upper craton stratigraphic sequences.

On the northern side of the Tabei uplift, the oil and gas came from the Kuche area; the roof-like faulted blocks controlled petroleum accumulation. On the southern side of the Tabei uplift, the oil and gas came from the Manjiaer area. The oil and gas accumulated in the stratigraphic and lithologic traps that were based on the anticlinal foundation. In the Tabei area, detailed petroleum exploration recently aimed at two domains; one domain is Carboniferous lithostratigraphic traps; the other is the deeply buried interior carbonate rocks–structural type of petroleum reservoir in the Cambrian system. These two domains were distributed north of the central line in the Awati–Manjiaer depression and south of the Yingmaili–Yaha belt. The whole exploration area includes 6.21×10^5 km^2 of land. The exploration targets included the Cambrian, Ordovician and Carboniferous systems. There are 15.8×10^8 t of oil resources and 7.7×10^8 t of equivalent gas resources. The Carboniferous lithostratigraphic trap zones were distributed on both sides of the Carboniferous erosion belt that was located between the Halahatang depression and the Caohu depression. There were two large size stratigraphic overlapping and pinch out belts that were made of Carboniferous Donghe sandstone. At the end of 2000, 1.85×10^8 t of proven reserves in the Carboniferous system were discovered. It is a porous and layered type of petroleum reservoir that could form large scale oil and gas reservoirs. There are $1 \times 10^8 - 2 \times 10^8$ t of forecast reserves in the Carboniferous lithostratigraphic belt. The deeply buried interior carbonate rocks–buried hill type of petroleum reservoir contains 1.2×10^4 km^2 of effective exploration area. This type of exploration domain has abundant oil and gas resources; there are $13 \times 10^8 - 15 \times 10^8$ t of preliminary forecasted resources still underground.

The Tazhong uplift contained an exploration area of 2,500 km^2. In the Carboniferous, Silurian and Ordovician systems, there are $10,783 \times 10^4$ t of proven reserves and $2,737 \times 10^4$ t of controlled reserves. Exclusively, there are 7.7×10^8 t of total resources. The targets for future exploration include the low relief structure in the Donghe sandstone on the northern slope in the Tazhong area, the deeply buried internal episodic type of large size anticline in the Tazhong area, the lithostratigraphic traps on the eastern side of the Tazhong area, the Ordovician$_{2-3}$ #1 fault, the Carboniferous system in the Tazhong–Manxi area, the low relief structure in the Silurian system and the stratigraphic traps. At present, petroleum explorations in the Tazhong area are in the #1 faulted belt on the northern slope and in the Silurian sandstone, which are located at the Tazhong lower uplift and which have an exploration surface area of 7.25×10^4 km^2; the primary targets are the Ordovician and Silurian strata. The petroleum resources include 15.36×10^8 t of oil resources and $7,200 \times 10^8$ m^3 of equivalent gas resources. According to the calculations of the petroleum reservoir scale sequence method, there are 9.40×10^8 t of recoverable oil reserves and 6.12×10^8 t of recoverable equivalent gas reserves. In the Tazhong area, we have forecast that the faulted belt #1 contains $3 \times 10^8 - 5 \times 10^8$ t of proven reserves on the northern slope. The sandstone contains $2 \times 10^8 - 3 \times 10^8$ t of proven reserves in the Silurian system.

The Bachu paleo uplift is located on the western edge of the Tazhong uplift and it is next to the Keping thrust belt. The Bachu uplift contains 4.16×10^4 km^2 of exploration land and more than 30 exploration wells. The discovered Hetian River gas field contains 617×10^8 m^3 of proven gas reserves. In the San 1 well, the controlled gas reserve is up to 433×10^8 m^3. The targets for the future exploration phase include the Haimiluosi structure and the southern part of Mazhatake; in addition, the Tumuxiuke structure is also worth exploration attention. In this area, we estimate there are $2,500 \times 10^8$ m^3 of proven gas reserves. Furthermore, the Carboniferous lithologic trap is another attractive exploration domain for the near future, which is located on the western section of the Maigaiti slope, on the southern margin of the Bachu uplift.

11.4.2 Eerduosi Basin

In the Eerduosi superimposed basin, petroleum exploration in the middle and lower petroleum systems is mainly designed to discover natural gas. Because there are 10.7×10^8 m^3 of total resources in this region, and because in total $11,142.69 \times 10^8$ m^3 of proven gas reserves have been discovered, the exploration potential in this region is excellent. The lower craton petroleum system was represented by a large size gas field in the Ordovician system in the central region of the basin; the upper craton petroleum system was represented by the Sulige gas field in the Carboniferous–Permian systems. Currently, petroleum exploration is aimed at discovering highly efficient reservoir sequences in the Shan 2 member and the Ma 5 member, in the central and eastern parts of the basin. The expanded exploration regions for the future include: (1) Zizhou–Qingjian, (2) the eastern side of the Central Region gas field, (3) the Yongning district on the southern side of the Central Region gas field, (4) Huangling–Huanglong, (5) Gannan on the southern side of the basin, (6) Wushenzhao, (7) the northern section of Tianhuan and (8) the southern section of the western margin thrust belt.

The Zizhou–Qingjian region is located in the eastern part of the basin; it is 50 km to the Yulin gas field in a southeast direction. From the geological point of view, it is the 1st order tectonic unit in the basin and it is located on the eastern section of the Yishan slope, with 7,000 km^2 of explorable land, approximately. In the northeastern part of the Zizhou–Qingjian region, the Mizhi area and the Zhenchuangpu area mainly have the He 6–He 8 members and the Taiyuan formation. We have discovered 358.48×10^8 m^3 of proven gas reserves and 142.58×10^8 m^3 of controlled gas reserves. In addition, we estimate there are 803.14×10^8 m^3 of gas reserves underground. Estimating the natural resources in this region, there are $4,500 \times 10^8$ m^3 of total gas resources underground. Because this region contains well developed sandstone reservoir sequences with good physical properties, it was possible to develop $1,000 \times 10^8$ m^3 of gas reserves.

The eastern portion of the Jingbian buried platform is located on the eastern side of the Jingbian gas field, which was the eastward extension of the Ordovician buried paleo platform of Jingbian; the favorable exploration area covered 3,500 km^2 of land approximately; on the eastern side of the buried platform, 58 exploration wells were drilled. In general, the eastern side of the buried platform contained natural gas. Even with this great exploration result, this area still has reasonable exploration potential. In the isolated area that contained well developed fractures, a highly productive and highly accumulative reservoir might be created. According to the preliminary estimation, the 1+2 sections in the Ma 5 member contain $1,100 \times 10^8$ m^3 of the controlled gas reserves+the forecast gas reserve. The exploration blocks of Butu, Huangcaomao and Balasuna have better control in the wells. After picking up the pace of drilling, the proven gas reserves will be increased.

The Yongning district is located on the southern side of the Central Region gas field. In the eastern section, the Ma 5 weathering crust type of gas reservoir contains 20,000 km^2 of favorable exploration land. In the central section, the better explored area covers 9,000 km^2 of land with $2,900 \times 10^8$ m^3 of proven gas reserves. In the southern section, the preliminarily explored area covers 5,000 km^2 of land with $2,000 \times 10^8$ m^3 of forecast gas resources; the Yongning exploration target area was selected from here, which includes 1,800 km^2 of land.

Similar to the Etuokeqi–Dingbian region, the Huangling–Huanglong region was located at the conjunction of the North China Sea and the Qinling Sea, which was the ideal place for developing dolomite reservoir sequences (that covered 3,500 km^2 of land). The dolomite was below the upper Paleozoic Ganquan sandstone on the southern side of the basin.

The Ganquan region is located on the southeastern side of the Eerduosi basin, which covers 1.1×10^4 km^2 of land. In this region there are $1,500 \times 10^8$ m^3 of potential gas resources; the objective of the exploration is to discover the upper Paleozoic gas reservoir in the Shan 2 member and also to investigate the He 8 member and the Taiyuan formation.

The neighborhood of the northern section of the Tianhuan is from the Yi 8 well on the northern side to the Yi 25 well on the southern side, and from the E 6 well on the eastern side to the E 3 well on the western side, which includes 3,500 km^2 of favorable exploration land. In this region there are $2,800 \times 10^8$ m^3 of potential gas resources; the objective of the exploration is to discover the upper Paleozoic gas reservoir in the Shan 2 member and also to investigate the He 8 member and the Taiyuan formation.

11.4.3   Sichuan Basin

Both geological theory and exploration practice confirmed that the Sichuan superimposed basin is a petroliferous rich basin. The organic substances in the hydrocarbon source rock were thermally matured into high–post maturational stages and they mainly produced gaseous hydrocarbons. Vertically, in the Sinian–Jurassic systems, 19 natural gas layers were discovered that contained commercial grade gas flows. Horizontally, 275 gaseous structures were distributed all over the basin and 100 gas fields were discovered. The superimposed basin contained multiple sets of petroleum systems in a vertical direction. Also, gas fields have several gas producing layers vertically. Usually, there are three natural gas layers. However, in the Wolong River gas field, nine natural gas layers were stacked up vertically. At present, the petroleum exploration mainly focuses on the Feixianguan oolite beach reservoir sequences in the northeastern Sichuan basin. In addition, on the paleo uplift in central Sichuan, the petroleum exploration also actively investigates the reservoir sequences of bio-reef facies and bio-clastic flat facies in the lower Paleozoic erathem and the Permian system. Moreover, in the eastern Sichuan basin, the petroleum exploration examines the Carboniferous dolomite at depth.

In the eastern Sichuan basin, the Feixianguan oolite beach contained Paleozoic marine strata which has a promising future for exploration purposes. Since the discovery of natural gas from the bio-reef in the Shibao 1 well in 1984 (which had a gas production rate of 37×10^4 m^3/d during the well testing), large bio-reef type gas reservoirs have been discovered in Tieshan, Wubaiti and Huanglongchang areas in recent years. We have discovered 707×10^8 m^3 of geological gas reserves and $1,892.1 \times 10^8$ m^3 of third grade gas reserves. Around the peripheral area of the Kaijiang–Liangping sea trough, an intra-platform oolite beach developed on flats with a thickness of 20 – 140 m, which covered 820 km^2 of land and which contained $6,970 \times 10^8$ m^3 of gas resources. At present, exploration activities are being carried out on 7,000 km^2 of land; in addition, there are 13,000 km^2 of land waiting to be explored. Therefore, the exploration domain is vast. By the year 2015 we should have discovered $3,000 \times 10^8$ m^3 of the geological gas reserves. Furthermore, $4,000 \times 10^8$ m^3 of the proven gas reserves will be discovered at the end of the exploration process. The exploration targets for the near future include: (1) the Wenquanjing–Tieshanpo area positioned on the northeastern side of the Kaijiang–Liangping sea trough with 3,750 km^2 of exploration land, (2) The Tieshan–Huangnitang area, positioned on the southwestern side of the Kaijiang–Liangping sea trough, which contains a platform margin oolite bar of the Feixianguan formation, and which covers 3,500 km^2 of land and (3) the Tongjiang–Wanyuan area.

The Leshan–Longnusi paleo uplift is located in the central Sichuan basin. According to the 2nd round of natural resource evaluation, the lower Paleozoic erathem (Z–∈) contains $15,567 \times 10^8$ m^3 of gas resources in the Sichuan basin. At present, there are 400×10^8 m^3 of proven gas reserves, 102×10^8 m^3 of controlled gas reserves, 338×10^8 m^3 of the forecast gas reserves and $15,567 \times 10^8$ m^3 of gas resources. In this region petroleum exploration produced positive results. The natural resources offer great potential and the magnitude of petroleum accumulation has been estimated. At present, searching the fractural reservoir sequences is the crucial factor for advancing the exploration process. The Daxingchang structure and the Hanwangchang structure were at high positions on the original paleo uplift. In the western Sichuan basin, because it was compressed by a stronger stress, the largest anticlinal trap in the Sinian system contained well developed fractures. Here is the ideal place for petroleum exploration.

The Permian bio-reef complex in the eastern Sichuan basin was a byproduct of the late Permian extensional tectonic movement. During the late phase of the Permian, in the eastern Sichuan area, because the magnitude of extensional tectonic movement at the intra-platform was intensified, the Kaijiang–Liangping sea trough was created, which was oriented in an NWW direction and in an NE direction. As a consequence, excellent reservoir entities developed around the sea trough, reef complexes developed on the platform margin and patch reefs developed on the gentle slope at the intra-platform. The marginal reef complexes developed in the high energy environment on the margin of the sea trough and the patch reefs developed on the gentle slope at the intra-platform. This facies belt covered 3,000 km^2 of land. According to the reconstruction of the taphrogenic belt, the horizontal spreading pattern of the Permian bio-reef complexes were related to three paleo extensional faults. (1) On both sides of the Kaijiang–Liangping sea trough, four gas reservoirs have been discovered on the platform margin reef complexes. (2) On the surface of the Huayingshan faulted belt, the reef complexes in the Laolongdong and Wenxingchang areas were related to the Huayingshan paleo extensional fault. This was deduced by examining the evidence from the Tieshan anticline to the Huayingshan basalt. (3) In the concealed fault zone that was orientated in an NW direction from the Linshui–Fuling, the limestone in the Changxing formation contained 5% – 6% of argillaceous substances, which indicated the possible occurrence of a deep water syneclise. Above this concealed fault zone, gas reservoirs in the reef complex were discovered in Bandong, the Wolong River and Shuanglong areas.

The Permian bio-clastic flats are located in the western Sichuan basin. According to the current exploration status in the Sichuan basin, the Lower Permian dolomite of bio-clastic flat facies in the Qixia formation has a promising future for petroleum exploration in the western Sichuan basin. Among the distribution range of bio-clastic flats, dolomite covers 6,920 km^2 of the land and weak dolomitic limestone covers 3,280 km^2 of the land; together they totally cover 10,000 km^2 of land. The Qixia dolomite has perfect geological conditions for the accumulation of natural gas reservoirs. (1) It is a good hydrocarbon source: the Lower Permian strata were the major source rocks for the Sichuan basin; the bio-clastic flats in the Qixia formation had plenteous hydrocarbon resources and this condition could benefit the creation of an in situ type of gas reservoir. (2) There is a widely distributed fracture–porous type of reservoir sequences: the dolomite reservoir sequences contain dissolution pores; they not only have a constant thickness of between 20 – 60 m, but they also cover 6,929 km^2 of land. (3) There are various types of trap: the lithologic trap could develop in the dolomite that had a lenticular shape and dissolution pores. The anticlinal trap or structural–lithologic traps could develop in the concealed structure. (4) In general, the gas shows commonly occurred in the dissolution pores dolomite, which indicated that dolomite has the precondition for natural gas accumulation. (5) The buried depth of target sequences is between 3,500 – 5,000 m, which is a suitable depth for drilling.

11.4.4 Deep Level in the Zhungeer Basin

Petroleum exploration has been carried out in the Zhungeer basin for more than 50 years. We have discovered 38 volcanic reservoirs in the deeply buried Carboniferous–Permian strata that contain $15,802 \times 10^4$ t of geological oil reserves and 139.96×10^8 m^3 of geological gas reserves. These oil and gas reservoirs are mainly distributed on the northwestern margin of the basin; a few reservoirs occur in the central region and on the eastern side of the basin. Most of the reservoirs are distributed in the Carboniferous strata, then in the Permian strata. A volcanic reservoir has not been found in other stratigraphic sequences. The types of reservoir include: the faulted block type, the lithologic–structural type and the stratigraphic–structural type. The reservoir lithology includes: andesite, basalt, tuff and volcanic breccias. All the discovered oil and gas reservoirs were hidden under the unconformity surface formed by the Mesozoic strata overlapping. In the past, because exploration wells that targeted the Carboniferous and the Permian volcanic rocks were fewer, we did not make an important discovery in this exploration domain. Recently, in the increasing number of exploration wells that were designed for discovering volcanic reservoirs, oil and gas exploration also achieved some breakthroughs.

The Carboniferous and Early Permian stratigraphic sequences were the places where volcanic rocks were largely deposited. Also, in these two sets of stratigraphic sequences, the reservoir lithology was mainly made up of volcanic rocks that were the important exploration targets. There are two types of volcanic reservoir sequences: one is eruptive rock and the other is clastic rock. In the eruptive rocks, the types of pores and their combination were complicated, which included dissolved pores, intracrystalline dissolution pores, intragranular dissolution pores, intergranular dissolution pores, metasomatosis pores, air holes, micron fractures and fractures. The combination of pores consisted of intracrystalline dissolution pores–fractures. The clastic rocks include the basaltic type and the andesitic type. They contain more than 50% of clastic materials without chemical cement. These clastic materials were compacted together or were cemented by lava (brecciated lava). These reservoir sequences had small size dissolution pores with low porosity and low permeability. At deep levels, primary exploration domains include the faulted step zone on the northwestern margin, the Mosuowan uplift, Baijiahai uplift, Zhongguai uplift, Sangequan uplift, Xiayan nose-like uplift, Jidong nose-like uplift, Shinan faulted uplift, Mobei uplift, Dixinan faulted uplift, Zhangbei faulted folding belt and the mountain front structural belt of Northern Tianshan Mountain. Furthermore, in the Zhungeer basin, the Mosuowan anticline, Mahu anticline and southern Mahu anticline are large scale, deeply buried structures and they have not been investigated. Therefore, these deeply buried structures are worth exploration attention.

11.4.5   Deep Level in the Bohai Bay Basin

In the Bohai bay basin, the major reservoir sequences in the middle and lower petroleum systems include the Mesozoic erathem, Paleozoic erathem, Neoarchean erathem and pre-Sinian system. The hydrocarbon source rocks for deeply buried petroleum reservoirs include the Palaeogene systems, the Carboniferous–Permian systems and the Ordovician systems that generated the secondary hydrocarbons. In the Palaeogene system, the Kong 2 member, Sha 4 member, Sha 3 member, She 1 member and the dark color mudstone in the Dongying formation functioned as source rocks. There were two types of accumulative method. (1) Under the high pressure in the hydrocarbon source rocks, the oil and gas were generated in the younger strata and stored in the older strata, such as the oil and gas reservoirs in the following areas: Shuguang, Qianmiqiao, Balizhuang, Zhuangxi and the lower buried hill type of reservoir in Futai. (2) Another type was the secondary oil and gas reservoirs that "generated oil and gas in old strata and stored them in old strata too". For example, in the Bohai bay basin, the natural gas reservoir in the Wushen 1 well was supplied by the upper Paleozoic source rocks and the oil reservoir in the Konggu 2 well was supplied by the lower Paleozoic source rocks.

In the Bohai bay basin, the future exploration targets include: (1) Searching for the buried hill type of oil and gas reservoirs in the lower stratigraphic sequences that were covered or contacted by deeply buried hydrocarbon source rocks. In this kind of reservoir, oil and gas were generated in the younger strata, but they were stored in the older strata. For example, in the Shengli oil field, the Futai buried hill type of oil and gas reservoir was discovered in the Chezhen depression with more than $3,000 \times 10^4$ t of geological oil reserves. (2) In order to discover the deeply buried oil and gas reservoirs, we should intensify the comprehensive study and evolution of the Paleozoic sequences. The Ordovician original type of oil reservoir with commercial grade oil flow was discovered in the Konggu 3 well; the oil layers occurred in limestone that was positioned in the Ordovician Fengfeng formation. At different intervals the oil production rates vary; in the interval between 3,485 – 3,483 m, the daily oil production rate is 7.5 t; in the interval between 3,501 – 3,547 m, the daily oil production rate is 2.55 t.

11.4.6   Deep Level in the Songliao Basin

In the Songliao basin, the evaluations of large size belts in the faulted basin and for targeted sequences are comprehensive. In the Songliao basin, the imminent exploration areas for deeply buried natural gas include four large faulted basins and two large upliftings. The four large faulted basins are the Xujiaweizi faulted basin, the Changjiaweizi faulted basin, the Yingshan–Shuangcheng faulted basin and the Changling faulted basin. The two large upliftings are the paleo central uplifting belt and the southeastern uplifting belt in the southern part of the basin.

Because the research work was preliminary, there was little discussion about the southeastern uplifting belt. However, there are many gas shows in the southeastern uplifting belt. The first type is the uplifting in the basin; in other words, it is the uplifting structure in the four large size faulted basins such as the central faulted uplifting belt in the Xujiaweizi. The second type is the lithologic belt that was created by the irregularity of the lithology inside the faulted basin. For example, in the faulted basin, the gas rich belt might be created in volcanic rocks. The third type is the faulted blocks and higher positions on the paleo uplifts. In the future, many detailed belts or zones may possibly be identified in the four large faulted basins and in the two large uplifts. However, these three types of belt are the most important types. We have an amenable understanding of the target sequences: the first sequence is the volcanic rock in the Ying 1 member. Then there are the Quan 1 member, Quan 2 member and sandstone in the Denglouku formation, which were shallowly buried.

Discussions about upcoming explorations are ongoing and need to be detailed in depth. Along with an increase in exploration activity, in order to perfectly direct the exploration practices, we should categorize, summarize and refine the acquired data over time.

References

Gan, Z.G., et al., 1988. Formation time of folds and its control to oil and gas accumulation in the Sichuan basin. Natural Gas Industry, 8(4):32-40.

Guo, Z.W., Deng, K.L., Han, Y.H., et al., 1996. The Creation of the Sichuan Basin and its Development. The Geological Publishing House, Beijing.

He, D.F., Li, D.S., Lu, X.X., 1996. Tectonic types of petroliferous basins in Northwestern China. Acta Petrolei Sinica, 17(4):8-17.

Hu, G.C., Xie, Z.X., et al., 1997. Carboniferous Gas Fields in High Steep Structures of Eastern Sichuan. Petroleum Industry Press, Beijing.

Jia, C.Z., Wei, G.Q., Yao, H.J., Li, L.C., 1995. Tectonic Evolution and Regional Structural Geology in the Tarim Basin. Petroleum Industry Press, Beijing.

Jia, C.Z., et al., 1997. Structural Characteristics and Oil and Gas in the Tarim Basin. Petroleum Industry Press, Beijing.

Jia, C.Z., et al., 2000. Petroleum Exploration in the Foreland Thrust Belt. Petroleum Industry Press, Beijing.

Liu, D.L., Song, Y., Xu, A.M., et al., 2000. Comprehensive Study of Structures and Natural Gas Accumulative Belts in the Sichuan Basin. Petroleum Industry Press, Beijing.

Lu, H.F., et al., 1989. Classification and mechanism of overthrust structure in foreland basin, on the Longman Mountain front. Journal of Nanjing University, 4.

Luo, Q.H., et al., 1996. Research on natural gas enrichment conditions of major coal-bearing state of the Triassic system in the centre-west part of Sichuan basin. Natural Gas Industry, 16(suppl.):9-17.

Ma, B.L., et al., 1990. Creation and Development of Sediment Rocks in the Tarim Basin and Related Oil and Gas. Science Press, Beijing.

Song, W.H., 1989. Discussion of Nappe structure at northern sector of Longmen mountains and its oil and gas prospects. Natural Gas Industry, 9(3):2-9.

Tian, Z.Y., Chai, G.L., Li, L., 1990. Creation and development of the Tarim basin. Xinjiang Petroleum Geology, 1(4):259-274.

Xie, H., Zhao, B., Lin, L.D., You, Y.M., 1984. Characteristics of oil bearing in the overthrust belt, on the western margin of the Zhungeer basin. Xinjiang Petroleum Geology, 5(3):1-5.

Zhang, Y.C., et al., 1997. Prototype Analysis of Chinese Petroleum Basin. Nanjing University Press, Nanjing.

Zhao, B., 1979. Structural nature and characteristics in the Zhungeer Basin. Petroleum Exploration & Development, (2):18-26.

12

Petroleum Geology of Carbonate Reservoir

Carbonate rocks are important for oil and gas reservoir and petroleum production globally; the oil and gas production from carbonate rocks represented 60% of total production worldwide (Roeh and Choquette, 1985). In China, carbonate rocks occur in an area of 250×10^4 km^2 that represent 55% of the total area covered by sedimentary rocks. Carbonate rocks mainly occur in the Sichuan basin, the Tarim basin and the Eerduosi basin (Fu et al., 2002; Zhao et al., 2002; 2004). In China, the total thickness of stratigraphic sequences from the late Proterozoic eonothem to the Cenozoic erathem is approximately 27,000 m; the thickness of the carbonate rocks is approximately 11,000 m, which represents 41% of the total stratigraphic thickness in China (Zou et al., 2007; Lu et al., 2000; Dai et al., 1997; 1999; 2003; Zhang et al., 2003; Lou et al., 2001; Dou et al., 2003). Most of these are marine carbonate rocks and the lacustrine carbonate rocks only appear in the Mesozoic–Cenozoic strata.

12.1 Depositional Features of Reservoir Sequences

12.1.1 Classification and Characteristics of Sedimentary Facies

Although carbonate rocks that were deposited in all kinds of environments would develop into reservoir sequences by depositional or diagenetic processes, different types of carbonate rocks not only have different storage potential, but they also have different porosity characteristics. Fig. 12.1 is a depositional model of carbonate rocks, which demonstrates the carbonate characteristics that were influenced by the water depth of the continental shelf, from shallow water to a basin of deep water. The depositional environments include the nearshore, coastal plain, intra-platform, platform margin, platform slope and semi-deep water–deep water basins. The local or regional depositional environments controlled the nature and pattern of reservoir sequences; furthermore, they also controlled the relationship between the reservoir sequences and other co-existing geological entities. The relationship between the depositional environment and reservoir sequences can also be seen in other areas to include the sediments and types of carbonate rocks, the grain sizes and relative contents and the primary type of pores and the relative quantity (Wei et al., 2004; Zhang et al., 2004).

Fig. 12.1.   Depositional model of carbonate rock

The carbonate rocks that were deposited in the high energy facies belt (reef flat on the platform margin) are valuable exploration targets. In this facies belt, the development of reservoir sequences was closely related to the palaeogeography; the reservoir sequences have excellent quality, wide distribution range and high predictability. The carbonate rocks in the high energy facies belt were distributed in the Eerduosi basin, the Upper Permian–Lower Triassic series in the Sichuan basin, the Cambrian–Ordovician systems in the Tarim basin.

● **Reservoir Sequences of Beach Facies**

The grain beach (shoal) was deposited in an environment of shallow water and high energy (or relatively high energy). During the diagenetic process, some original pores in the grain beach (shoal) facies reservoir were filled up; however, other grain beach (shoal) entities experienced dolomitic and dissolved processing to become the reservoir sequences that contained the intragranular pores or dissolution pores. In China, the reservoir sequences of grain beach (shoal) facies mainly consisted of dolomite, then limestone. The intragranular pores basically disappeared in the limestone; as an alternative, there were syngenetic intergranular dissolution pores.

There are two kinds of beach facies, the intra-platform beach and the platform margin beach. The platform margin beach has great thickness and a narrow facies belt. On the other hand, the intra-platform beach has small thickness and a broad distribution range. In China, most discovered gas reservoirs of beach facies were in the intra-platform shallow beach facies.

In the eastern Sichuan basin, the carboniferous gas reservoirs mainly contained the strata of grain beach (shoal) facies, then the strata of warm clastic beach facies (Luo et al., 2000). The main lithology is sandy dolosparite that contained dissolution pores. The depositional facies formed the transgressive shallow beach; the dolomite developed in the brackish lagoon. The syngenetic exposure type of dissolution pore favorably developed. Because of the lack of gypsum cover, these dissolution pores were well preserved. Average porosity is 5.27% – 17.0%; average Klinkenberg (gas) permeability for the matrix is $5.67 \times 10^{-5} - 2.99 \times 10^{-2}$ μm^2; average effective thickness is 13.8 m. These gas reservoir sequences are widely distributed and they form the most important gas reservoir in the Sichuan basin.

In the northern Sichuan basin, the Feixianguan oolitic beaches were widely distributed along the Kaijiang–Liangping paleo sea trough and the Exi–Chengkou paleo sea trough (Fig. 12.2) (Han et al., 1995). Recently, large size gas fields were discovered in the Luojiazhai, Dukouhe, Longgang and Puguang areas. The reservoir sequences of oolitic beach facies were created by dolomitization and dissolution.

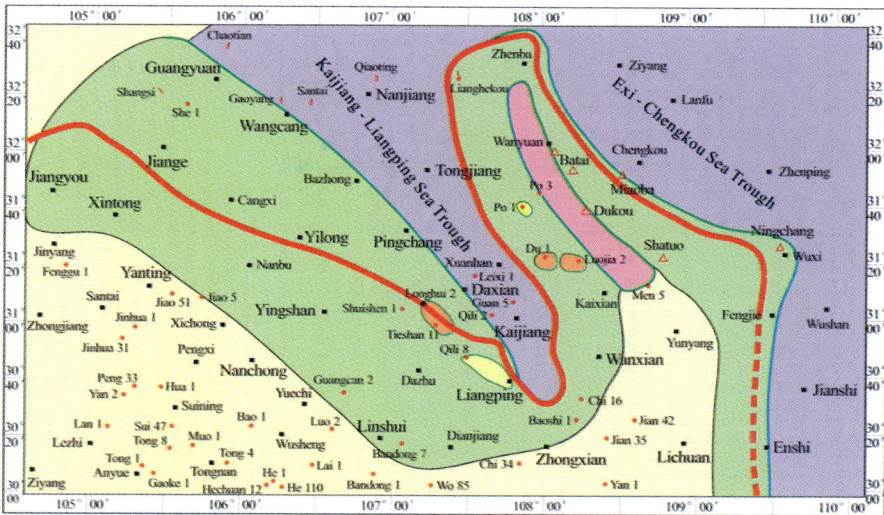

Fig. 12.2. Distribution map of oolite flat in the Feixianguan formation, northern Sichuan

In the Tarim basin, the Carboniferous reservoir sequences in the Hetian River gas field consisted of sediments of bio-beach facies, which were widely distributed in the Bachu area in the central Tarim basin with a stable thickness of 30 – 60 m. The lithology included gray–light gray color bio-clastic sparite, silty limestone, dolomitic limestone, calcareous dolomite and dolomite. The limestone contained well developed gypsum chunks. The lithology indicated the depositional environment was an intra-platform beach on a semi-restricted platform. The porosity is in the range of 0.01% – 19.11%; average porosity is 3.55%. The permeability is in the range of $0.005 \times 10^{-3} - 128 \times 10^{-3}$ μm^2; average permeability is 2.23×10^{-3} μm^2. The reservoir belongs to the low porosity and low permeability type. The types of pores include intracrystalline pores, intracrystalline dissolution pores, intragranular dissolution pores, extra large pores and fractures.

● Reservoir Sequences of Reef Facies

The independent carbonated entity is called a reef, which contained a certain amount of in situ type reef framework and which had the ability to resist storm waves (Fan et al., 2005). In addition, it often had a protruding topography. In the Yudong and Exi regions, the gas layers in the Changxing bio-reef sequences are not limestone of reef core facies; they are clastic carbonate rocks of flat facies that include a reef base flat, lateral reef flat, post reef flat and reef roof flat. During diagenetic processing, dolomitization and dissolution created pores and holes in the clastic dolomite that became the major reservoir sequences for natural gas. In order to become a natural gas reservoir, the reef flats had to undergo dolomitization and dissolution. It is very important to understand this process. In the past, people learned hard lessons from exploration practices. Initially, we considered that a bio-reef contained a petroleum reservoir. Then, according to the seismic data, we identified the Changxing formation and drilled wells into the protruding structure. As a result, the bio-reefs that were drilled not only contained low porosity and low permeability, but they were also dry wells when tested. These dense carbonate rocks did not contain petroleum reservoirs. In fact, we discovered natural gas when we drilled into dolomite that contained only dissolution pores.

In the Tarim basin, the Tazhong I faulted belt consisted of the Ordovician bio-reef that contained the sponge and receptaculites. The facies belts of the bio-reef include reef core facies, reef flat facies and reef roof facies. These reefs developed on the gentle slope of the Changxing carbonate rocks. According to the developmental situation of reef facies belts, these reefs were further classified into patch reef and marginal reef. Patch reefs are scattered, random and small; they have a symmetrical pattern with less sparite filling in the skeleton, which indicates a depositional environment of low energy. However, when the sea level slightly decreased, the water depth became shallower and the energy level increased; as a consequence, shallow beach facies would develop. Marginal reefs were distributed in a relatively concentrated region with a large reef body and an asymmetrical pattern; they could be further divided into a front reef facies belt and a post reef facies belt (Fig. 12.3). The skeleton of the marginal reef contained many sparite cements that indicated a high energy environment during the growth of the bio-reef.

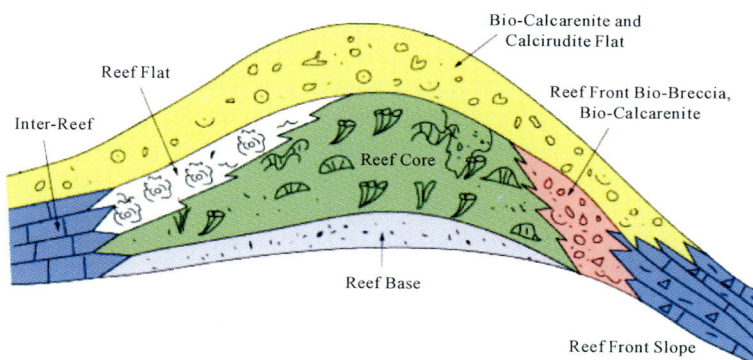

Fig. 12.3. Reef facies model of the slope zone I in the central Tarim basin

In the marginal reef, the porosity of porous dolomite is in the range of 2% – 24.34%; average porosity is between 5% – 10%. Klinkenberg (gas) permeability in the matrix is between $4 \times 10^{-5} - 6.3 \times 10^{-2}$ μm^2; average permeability is in the range of $5 \times 10^{-4} - 4 \times 10^{-3}$ μm^2. The effective thickness is between 4.45 – 69.27 m. Because it has a lenticular shape, the continuity of the marginal reef is poor.

● **Reservoir Sequences of Algal Flat Facies**

Because it was deposited in a very shallow water environment (such as coastal tidal flat, brackish epeiric sea and brackish lagoon), and because it was created by a biochemical mechanism and contained pores, cryptomonas dolomite is called algal flat facies reservoir sequences.

The Weiyuan Sinian gas layers in the Luzhou area, Sichuan basin, consisted of algal dolomite that was deposited in the brackish epeiric sea. The Zhongba gas field in the western Sichuan basin, the Lower Triassic Jialingjiang formation and some reservoir sequences in the Middle Triassic Leikoupo formation (the 3rd section) are also algal dolomites that were deposited in the brackish epeiric sea. In the Eerduosi basin, some reservoir sequences in the Ordovician Majiagou formation are algal dolomites that were deposited in the brackish lagoon.

The blue-green algae and bacteria are the main organisms that formed algal dolomite in brackish water. There two types of dolomite: epigenetic dolomite and primary dolomite. According to the study (Pu et al., 1987), the Sinian algal dolomite was the primary dolomite that was formed by the bacterial algae related bio-chemical mechanism. The following is the explanation. Firstly, the dolomite has stable stratigraphic sequences and great thickness; in general, the thickness is between 450 – 1,000 m; in the upper Yangtze platform and the Ezhong platform, the distribution area of dolomite is 70×10^4 km^2 approximately. Such a huge dolomite entity cannot be explained by the epigenetic theory of dolomite. Secondly, dolomite coexisted with evaporate rocks; the pseudocrystal chunks of anhydrite were distributed along the strata layers in some algae rich layers of stromatolite algal dolomite in an isolated area. When the husk-like structure was formed, the salinity of the sea water was above 230 g/L and monoclinic gypsum crystals formed; these gypsum crystals were covered by organic mud that consisted of bacterial algae and bio-chemical dolomite. Also, there was much pseudocrystal gypsum and gypsum crystal of mold porosity. Because dolomite is an evaporate mineral, if it coexisted with gypsum this dolomite must be a primary dolomite. Thirdly, the analysis results of the scanning electron microscope (SEM) show that the grain size of the dolomicrite is 0.8 – 2.5 μm and the size of bacterial colonies is 0.01 – 0.5 μm inside the crystal. These test results indicate that when the sediments were deposited, there was abundant bacterial algae in the water; the decay of the bacterial algae increased the concentration of magnesian ion in the water and formed the bacterial algae related lithological combination of bio-chemical dolomite and other evaporate minerals. Fourthly, according to the diffraction analysis of the X-ray, the degree of order in the dolomite was relatively low, $(7\pm)d(10.4)$; $2,883\times10^{-4} - 2,880\times10^{-4}$ μm. This test result can correlate with the primary dolomite in the Coulomb brackish lagoon, Australia. On the other hand, bacterial algae related primary dolomite was discovered on the beach of the Lesser Chaidan Lake in Qinghai province by Wenjie Xia and others; this discovery was direct evidence of the existence of primary dolomite. Thus, because they also co-existed with gypsum, the Carboniferous and Triassic algal dolomite are primary dolomites too.

In the ocean, different types of algal dolomite formed, due to the variation in energy levels. Algal dolomicrite, wavy stromatolite algal dolomites, stratifera algal dolomite and spongiostromate algal dolomite formed in shallow water with a low energy level. Algal pellet and algal spherules dolomite, oncolite algal dolomite and thrombolite algal dolomite formed in the agitating environment. Columnar stromatolite algal dolomite and algal sand dolorudite formed in the high energy environment.

Algal dolomite contains well developed pores. It not only has the primary type of pores (that include intragranular pores of intraclast dolomite, algae bounded window-like pores, decayed pores inside the algal layer and inter-grains, bird eye holes and intracrystalline pores in addition), but it also has the secondary type of pores (that include intergranular dissolution pores, crystal mold of gypsum, dissolution pores and dissolution holes). Among various factors that influenced the nature of reservoir sequences, the geological factors are the most fundamental and the most important ones. The geological factors include the depositional feature and the diagenesis attribute. The depositional feature controlled distribution of lithologic facies and the diagenesis alteration controlled the nature of reservoir sequences (physical properties).

Throughout geological history, the sea level rose and fell repeatedly. The changes in sea level caused variations in sedimentation, which generally controlled the distribution of reservoir sequences and their nature. Therefore, it is very important to understand the history of sedimentation and the relationship of facies belts when we study reservoir sequences or forecast reservoir sequences.

12.1.2 Classification and Characteristics of Reservoir Sequences and Controlling Factors of Sequence Development

● **Types of Reservoir Sequences**

With years of working experience, we think that the reservoir classification of *E. Cmexob* is a better one. In this classification, the single name represents not only a space that has worked as a storage space, but that has also functioned as a fluid channel, such as a porous type and fractural type reservoir. For the combined name, the name in front of the dash line means that this space has larger permeability (Wei et al., 2000; 2003; 2004). For example, in the fractural–porous type or fractural–cave type reservoir, the fractures worked as a major fluid channel. However, porous or cave types not only functioned as a major storage space, but also have smaller permeability in the matrix than the ones in fractures. According to this statute, in China the carbonate rocks can be categorized into four types.

    Porous Type. Reservoir sequences contain well developed pores with a wide throat passage. Also, the matrix has high permeability. Without press-opened fractures, a commercial grade gas flow will be produced by pores and the throat passage.

    Fractural–Porous Type. The narrower throat passage brought low permeability in the matrix; moreover, a commercial grade gas flow could not be produced just by permeability in the matrix. Therefore, when opening fractures became the major fluid channel, a commercial grade gas flow could be produced. This type of reservoir is called a fractural–porous type.

    Fractural–Cave Type. In this type of reservoir, the cave is the main storage space and the fracture is the major fluid channel; the reservoir sequences have strong heterogeneity.

    Fractural Type. Because of the dense matrix, the lithology is categorized into non-reservoir sequences. However, the opening of the fractural network not only worked as storage space, but also functioned as the fluid channel. This type of reservoir is called the fractural type.

● **Basic Features of Reservoir Sequences**

In China, large or medium size gas fields were distributed in the Sichuan basin, Eerduosi basin, Bohai bay basin, Tarim basin, Pearl River estuary and Chaidamu basin. In these basins, carbonate rocks have been (or will) become the important oil and gas producing sequences.

    (1) In a relatively stable tectonic environment, reservoir sequences were widely distributed

    The stable depositional environment and stable diagenetic environment indicate the widely spreading epicontinental sea, the entire vertical movement and diagenetic environment of fresh water dissolution. In this kind of environment, the reservoir sequences developed over a broad range.

(2) Multi-types of storage space

The types of storage space mainly included the fractural–porous type and the fractural–cave type. The reservoir sequences have strong heterogeneity. In the carbonate reservoir sequences, beside pores and throat passages, the spaces also include open fractures and caves. The open fracture and cave changed the nature of carbonate reservoir sequences that originally contained low porosity, low permeability and small throat passages. This alteration made carbonate reservoirs either produce a commercial grade gas flow or become oil and gas reservoirs that yielded high production rates.

(3) Carbonate rocks have the ability to generate natural gas and to create in situ type or multi-source types of gas reservoirs

When the content of residue organic carbons was larger than 0.08%, the carbonate rocks would become effective source rocks for natural gas. The carbonate rocks in the following stratigraphic sequences and locations contained 0.09% – 0.58% of organic carbons and they could become source rocks for natural gas, which offered strong evidence of the theory that marine carbonate rocks had the ability to generate gas. These strata and locations are the Ordovician strata in the Tarim basin, Triassic strata, Permian strata, Carboniferous strata and Sinian strata in the Sichuan basin and Lower Ordovician strata in the Eerduosi basin.

(4) The reservoir sequences have noticeable dynamic change

During the development of a gas field, the alteration in storage ability and permeability in carbonate reservoir sequences is called a dynamic change in reservoir sequences. In the original state, in a natural gas reservoir there was a balanced relationship between the fluid pressure in spaces and the pressure from overlaid strata. After a natural gas reservoir developed, the fluid pressure decreased, the net stress (effective pressure from overlaid strata) increased; the net stress is the difference between the pressure of overlaid strata and the fluid press. Therefore, the spaces will be deformed in shape and shrink in size; in addition, the relationship between base water and natural gas will become complicated. In carbonate rocks, the dynamic change is more obvious than the change occurring in clastic rocks.

● Major Controlling Factors in Reservoir Development

The controlling factors for creating the carbonate reservoir include structure, sedimentation and diagenesis.

Structural Factor

The tectonic activity controlled sedimentation of reservoir sequences. When the Upper Sinian–Middle Triassic carbonate rocks were deposited in the Sichuan basin, and when the lower Paleozoic carbonate rocks were deposited in the

Eerduosi basin, the crust of the earth was at a stable stage; the depositional base plane was relatively smooth with gentle waves and the depositional rate was small. Therefore, the reservoir sequences that were deposited in this kind of environment had small thickness and broad distribution range.

The paleo depositional structures included a paleo growth fault, syngenetic paleo-uplifting and a syngenetic paleo depression, all of which controlled paleo geography during the depositional periods. On both sides of paleo growth fault, the position of stratigraphic sequences, thickness of strata, lithology and reservoir sequences were different. In general, the reservoir sequences on the hanging wall of the paleo growth fault were better than the ones on the foot wall. The syngenetic paleo uplifting had relatively higher topography with a shallow depth of water. Here, the reservoir sequences of beach facies were well developed, such as the paleo-uplift in the Jiayi area, Luzhou. Conversely, if the paleo topography had a lower relief, the depth of water was great and the reservoir sequences were poorly developed.

When tectonic subsidence occurred, a depositional event would occur. If the subsidence was slower, the thickness of reservoir sequences would be small; if the subsidence was faster, the thickness of reservoir sequences would be large. If the crust of the earth was elevated, the depositional event would be stopped. Shortly after the strata were deposited, and if they were elevated above the water surface, syngenetic dissolution reservoir sequences formed. If the strata were exposed after the sediments were compacted, dissolution reservoir sequences were created. The compressed movement in the earth would create traps and fractures, which benefited petroleum accumulation. However, excessive compression would destroy the gas reservoirs.

About 70% of the anticlines and nose-like upliftings contained better reservoir sequences at the higher positions and poor reservoir sequences in the lower positions. Also, on the anticlines and nose-like upliftings that developed during the Himalayan tectonic period, the better reservoir sequences occurred at the high positions and the central axial area; the poor reservoir sequences appeared in the lateral regions. This kind of situation occurred in the Sinian system in Weiyuan, the Ordovician system in the Tarim basin and Eerduosi basin and the Kela gas field in France. This was the result of inherited control from the tectonic movement. In other words, a high position on the paleo geography during the depositional period became a high position on a present anticline. However, some anticlines did not possess inheritability.

Depositional Factor

Carbonate rocks can be formed in various environments that included a lake environment, shallow sea environment with clear and warm water, and a deep ocean environment.

The oceanic environment can be further divided into littoral sea, shallow sea and deep ocean. In the littoral sea, backshore and foreshore belong to a supratidal zone and intertidal zone (in 1989, Feng pointed out that both supratidal zones and intertidal zones belong to tidal flats; however, a subtidal zone does not belong to a tidal flat). Carbonate reservoir sequences were formed in some special environments. For example, flats and reefs developed at a higher position with high energy and shallow water in the ocean and lakes; the algal sheets developed in an intertidal zone, or very shallow water, either in the ocean or in the lake; gypsum, salt nodules, marl, or dolomitic marl developed in restricted shallow saline water. These special environments were one of the control factors that controlled development of carbonate reservoir sequences (Fig. 12.4).

There are two kinds of mechanisms that account for the change in sea level; one is caused by the tide, which happens near the shore or around an island when the sea level changes with a rise and fall in the tide. Another mechanism is a global or regional event, caused by an increase in the water volume or by tectonic movement. The former mechanism only created a small amount of reservoir sequences that were distributed near the shore. The latter one brought about large scale changes in sea level, which created widely distributed carbonate reservoir sequences which had stable stratigraphic positions.

The salinity of water was an important control factor in the development of carbonate reservoir sequences. In the ocean, limestone would be formed if the salinity was normal. In China, the limestone only contained an insignificant gas reservoir. Dolomite would be formed when the sea water was salinized; in China, dolomite contained the most important gas reservoirs. If sea water was excessively concentrated or was over salinized, gypsum and salt were formed. In other words, reservoir sequences would not be formed. When setting up the depositional model of reservoir sequences, one must consider the salinity of the water.

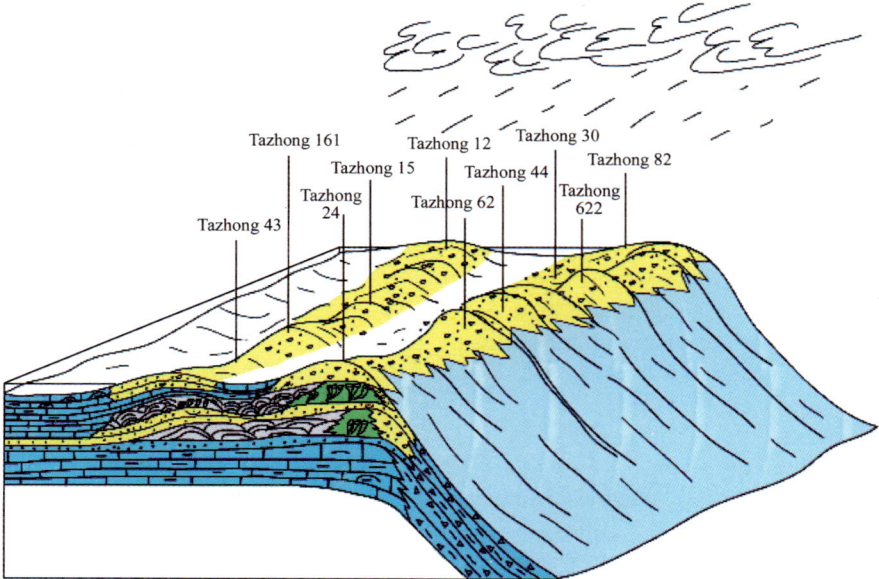

Fig. 12.4.   Developmental model of the Ordovician bio-reef in the Tazhong I faulted belt

The warm water in the tropical and subtropical oceans was the crucial factor for sedimentation of carbonate rocks. However, when carbonate sediments were exposed during the syngenesis period or solidified carbonate rocks were exposed during the period, the dissolution processing required a huge amount of rain water or fresh water in order to create secondary pores in the reservoir sequences.

Because the tides have a relatively low energy level, the marls are the primary sediments here. Even grainy sediments would develop in the tidal channel or tidal canal; however, they were distributed in very narrow ribbons. After diagenetic processing, these grainy sediments could not form the widely distributed reservoir sequences. The algal sheet in the intertidal zone could form reservoir sequences with a narrow facies belt.

In the shallow epicontinental sea, the storm wave energy is much larger than the tidal energy. The storm waves were important for widely distributed grain stone reservoir sequences of beach facies and reservoir sequences of reef facies. Because the lake was not influenced by the tide, the reservoir sequences of grainy beach facies and reef facies could be explained by storm waves only. With adequate water depth, a shallow beach or reef would be formed on higher positions of micron topography, where the storm waves would agitate.

When the crust of the earth was stable, the subsidence rate was small; supplied sediments were few. Thus, the thickness of porous reservoir sequences was small. This was one of the reasons why Chinese carbonate reservoir sequences had a narrow thickness.

Diagenetic Factor

The diagenetic period was the interval between the end of sediments deposition and before strata metamorphism. In the diagenetic period, diagenetic changes occurred with non-organic materials and organic materials. Overall, the sediments solidified into carbonate rocks; porosity and permeability decreased gradually. However, in some diagenetic stages, secondary pores could be created, for example, the dissolution in the syngenetic and supergene periods. When organic materials were transformed into hydrocarbons, the decarboxylation produced organic acid that could create dissolution pores and dissolution holes. In addition, the fractures that were produced by tectonic movements also benefited the development of reservoir sequences.

The porosity of modern carbonate rocks is between 40% – 70%. On Hainan Island, the beach rock (carbonate rock with uncemented grains) from the syngenetic period was of 35% – 37% tested porosity; the littoral reef was of 55% porosity. Therefore, it is reasonable that some researchers consider original sediments were of 40% porosity.

During diagenesis, the following diagenetic processes will decrease porosity in the rocks, which includes cementation, filling, compaction, pressure solution, dedolomitization and recrystallization. The argument about dedolomitization is endless. In the Yudong area, the drilling practices in the Carboniferous strata proved that, because dedolomitization created secondary limestone, the porosity in the dolomite was decreased by 2%; in addition, dry wells commonly showed up in this kind of dolomite. Furthermore, the recrystallization did not create reservoir sequences.

Under some diagenetic processes that decrease the original porosity in the rocks, the porosity in carbonate rocks was 1% – 2%, if there was no special kind of diagenesis that would increase the porosity.

(1) Dolomitization

Dolomitization means that the calcite, aragonite and high magnesium calcite in the marls or limestone were replaced by dolomite.

The debate about the origin of dolomite is rather intense. Because the original dolomite has not been produced by experiment, in certain periods people not only completely denied the existence of original dolomite, but they also rejected the idea that original dolomite contained reservoir sequences. This theory considered that the original sediments were marl only; dolomitization created dolomite that was a secondary mineral that developed during epigenesis in the diagenetic period. Recently, some researchers, such as Xu and Wang (1995), considered that dolomite has two formats, original dolomite and epigenetic dolomite. Through observations, we think the latter opinion is reasonable.

Either in the Ordovician carbonate rocks in the Eerduosi basin, or in the Sinian–Middle Triassic carbonate rocks in the Sichuan basin, the pores mainly developed in the dolomites. Dolomite even became the major reservoir sequences. However, not all dolomite could work as reservoir sequences; the dense dolomite could not function as reservoir sequences. Why did dolomite that contained well developed pores or small holes become reservoir sequences? The author thinks that because dolomite crystal had a slow growth rate, or because there were not enough dolomitic supplies, the original or secondary pores and holes could be preserved in original dolomite or epigenetic dolomite. Conversely, in the limestone, the original intragranular pores were mostly filled up; the residual pores were the intergranular dissolution pores of the syngenetic period. Other than that, under the depositional hiatus surface, the karst caves and holes developed by supergene weathering either in the Ordovician limestone in the Lunnan area or in the bio-reef in the Changxing formation in the Sichuan basin (He et al., 2007). Commercial grade oil and gas flows were produced in the limestone sequences that contained karst caves and holes. Compared with dolomite, the quantities of gas reserves in the limestone were much less. Therefore, in the strata below the Middle Triassic series, limestone only worked as insignificant reservoir sequences.

In the Eerduosi basin, the period when the Ma 5 member of the Ordovician was deposited was the developmental period of the gypsic saline lake in the restricted sea. Along the gypsic saline lake, the semi-syngenetic dolomite of beach facies that contained gypsum nodules was widely deposited inside the basin; this dolomite developed into a semi-syngenetic, pores type of dolomitic facies belt that functioned as reservoir sequences after the tectonic uplifting (Fig. 12.5).

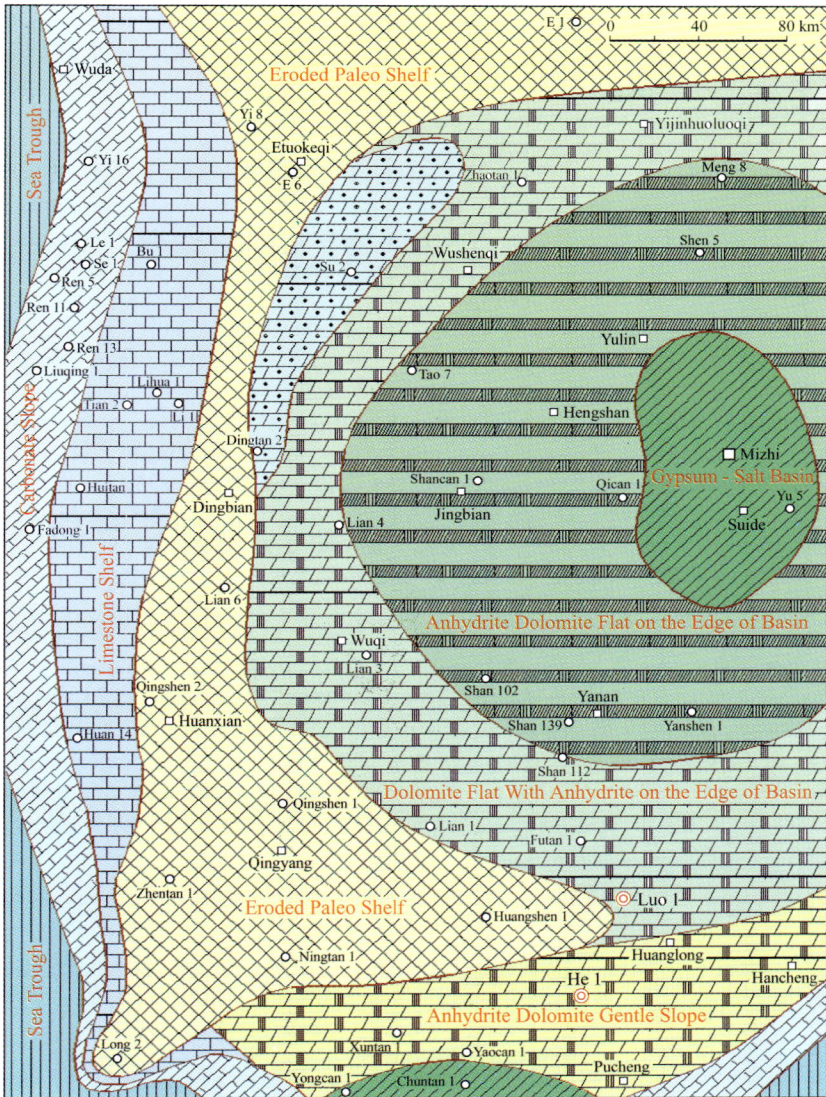

Fig. 12.5.   The map of lithofacies and paleo-geography of the Ma–5 member in the Eerduosi basin

(2) Dissolution

In the syngenetic period, the sediments contained excellent porosity and permeability; in an environment of meteoric water or mixed water, the dissolution would happen completely. During the entire diagenetic processing, the secondary pores of carbonate rocks mainly developed during the syngenesis.

During the supergene period, the carbonate sediments had been solidified into rocks. The dissolutions happened along fractures or in the original pores, which created strong heterogeneity spaces, such as caves, underground rivers and channels, which also formed weathering fractures and dissolved fractures. However, the dissolution spaces of the supergene period had much smaller capacity than the ones that were created during the syngenetic period.

During the hydrocarbon–generating period, dissolution from organic acid would form some pores. The organic acid was formed by decarboxylation when organic substances were transformed into oil. Some people considered that in the strata that did not experience the supergene weathering process, the intragranular dissolution pores were the evidence of organic dissolution. We think this opinion is right.

The deeply buried dissolution events happened during the Himalayan tectonic movement. After the Himalayan tectonic movement, in the Sichuan basin, the carbonate strata had not only been folded and faulted, but their buried depth had also become shallower. The formation water (that was in the geochemical equilibrium environment) migrated along the opened fractures into a new environment of imbalanced geochemistry. As a consequence, dissolution occurred. In the thin section, we can see that within 1 cm distance on both sides of the Permian opened fracture, there are dissolved pores; additionally, on the drilling cores, around the fault of the Himalayan tectonic period, there are dissolved pores and dissolved holes with a diameter of 0.6 m. In Weiyuan, from the Sinian interface of gas and water to the water bed that was extended 9 m below this interface, there are 1 – 3 parallel layers of karst caves, dissolved fractures and dissolved pores, which were distributed horizontally. The diameters of the caves are between 0.5 – 3 cm; the thickness of a single layer is between 0.2 – 0.5 m. Because the gas reservoir developed during the Himalayan tectonic period, these layers of karst, dissolved fracture and dissolved pores (which were distributed in parallel with the interface of gas and water) developed in the base water dissolutions after the gas reservoir developed.

12.2 Mechanism, Distribution and Exploration Domain of a Carbonate Reservoir

12.2.1 Classification of a Carbonate Reservoir and its Features

According to primary controlling factors, carbonate reservoirs can be categorized into two subdivisions that are lithostratigraphic reservoirs and structural anticline reservoirs. The anticlinal reservoirs included the structural anticline reservoir, the faulted block reservoir and the faulted anticline reservoir; according to the developmental mechanism, the lithostratigraphic type of reservoir can be further divided into the buried hill type of reservoir and the internal episodic type of reservoir. The primary mechanism for the buried hill type of reservoir was the weathering crust dissolution; the developmental mechanism for the internal episodic type of reservoir included sedimentation, dolomitization and buried dissolution. According to the developmental features of the carbonate reservoir sequence and the geometric patterns of a petroleum reservoir in cross section, the two types and related subtypes that were described above could also be categorized as the layered type, the semi-layered type and the massive type (Zhou and Tao, 2007).

● Layered Oil and Gas Reservoir Sequences

This type of reservoir often appeared in layers. The thickness and physical properties of the reservoir sequences were controlled by depositional facies belts and by original pores. The reservoir sequences consisted of layered limestone or dolomite and the roof and floor were controlled by mudstone. In the cross section, the oil and gas reservoirs commonly appeared in layers with water on the side. These types of oil and gas reservoirs were mainly distributed in the Carboniferous bio-clastic limestone section and in the Neogene limestone strata; for example, the Hetian River gas field and the Tazhong 4 oil field were in the Carboniferous bio-clastic limestone section; the Kekeya oil and gas field and the Kela 2 gas field appeared in the Neogene dolomite.

● Semi-Layered Oil and Gas Reservoir

The semi-layered oil and gas reservoirs also belonged to the lithostratigraphic type of reservoir. The oil and gas reservoirs were distributed in massive carbonate rocks; in addition, the reservoir sequences had a certain degree of heterogeneity. The oil and gas layers contacted with underlying dense carbonate rocks; the fractures and faults connected with base water in the deep level. In the cross section, the oil and gas reservoirs demonstrated an irregular layered pattern. According to the diversity of reservoir sequences and controlling factors, there were two kinds of situations. One was where the weathering dissolution controlled the physical properties of reservoir sequences. Furthermore, the distribution of the oil and gas reservoir was controlled by the weathering crust. In other words, this was the weathering crust type of reservoir; such as the Lungu oil and gas fields. Another situation was where the elements of sedimentation and deeply buried dissolution controlled the physical properties of reservoir sequences; the distribution of oil and gas was controlled by the reef complex, such as the reservoir in the Tazhong I slope belt (Fig. 12.6). In general, because this type of reservoir could be developed into a large or extra large oil and gas reservoir, it was very important for petroleum exploration.

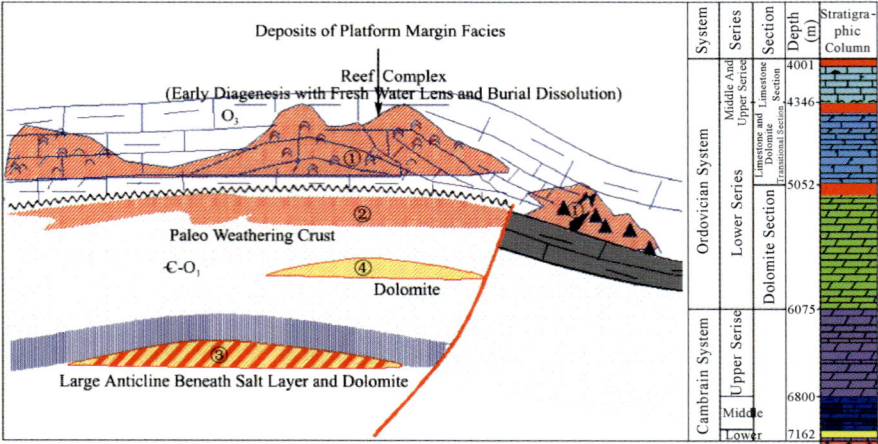

Fig. 12.6. The cross section of exploration model from the Cambrian–Ordovician systems, the central Tarim basin

● **Massive Oil and Gas Reservoir**

This type of carbonate reservoir has massive thickness; in the cross section, the oil and gas reservoir is blockish in structure; such as the Cambrian oil reservoir in the Yingmai 32 area and the Ordovician oil reservoir in the Yingmai 2 area.

12.2.2   Primary Control Factors for Development and Distribution of Carbonate Reservoir

According to the primary controlling factors for carbonate reservoir development, Zhou and Tao (2007) discussed this in detail and they suggested there were six major controlling factors.

● **Excellent Hydrocarbon Source Rocks were the Material Foundation that Controlled Creation of the Carbonate Reservoir**

Widely distributed highly productive hydrocarbon source rocks of exceptional quality indicated an intensified gas generating area of extensive range, which was the comprehensive expression of the thickness of the source rocks, the abundance of organic materials and the type of organic material and its maturity.

This kind of source rock not only continually supplied highly abundant natural gas, but also compensated for the dissipation of natural gas during the migration.

If there was a trap, the natural gas would be easily accumulated to form a large size gas field. In China, a large size gas field usually occurred in the area where natural gas generating intensity was larger than 20×10^8 m^3/km^2. In the discovered large size Chinese gas fields, the gas generating intensity is above $(50 - 60) \times 10^8$ m^3/km^2; (according to the new classification of PetroChina, the largest size gas field indicated that recoverable reserves are larger than 250×10^8 m^3). Therefore, widely distributed highly productive hydrocarbon source rocks of exceptional quality were the foundation and precondition for hydrocarbon generation and accumulation, in order to develop large or medium size lithostratigraphic types of oil and gas fields.

In China, most of the marine carbonate rocks were positioned in the middle–lower petroleum systems in the superimposed basin. The hydrocarbon source rocks had a prolonged developmental history of hydrocarbon generation and accumulation. The accumulated oil and gas not only revealed the complex problem of various sources with multiple accumulative phases, but they also came from several types of hydrocarbon source rocks. In particular, the natural gas that accumulated during the late period not only contained pyrolysis gas from kerogen, but it also contained pyrolysis gas that came from crude oil of the paleo oil reservoir. For example, in the Wubaiti, Wolong River and Shapingchang gas fields in the eastern Sichuan basin, the carbonate strata contained pyrolysis natural gas that came from large size paleo oil reservoirs of the Indosinian tectonic period that accumulated in the traps of the Himalayan tectonic period; in addition, the Feixianguan oolitic gas reservoir in northeastern Sichuan also contained pyrolysis natural gas that came from crude oil. Thus, the size and distribution of the paleo oil reservoir determined the characteristics of natural gas accumulation and distribution.

- **Paleo Uplift and its Periclinal Belt not only Controlled the Developments of Ideal Reservoir Sequences and Lithostratigraphic Traps, but also Controlled Oil and Gas Accumulation**

In the marine craton basin, the paleo uplift commonly controlled natural gas accumulation. In the paleo uplift and surrounding preclinical area, the structural trap and lithostratigraphic trap not only developed in the early period, but they also worked as the destination area for hydrocarbon migration and accumulation. They were ready to accept the hydrocarbon supplies from different hydrocarbon generating high peaks. This was the best time–space setting for hydrocarbon generation and accumulation. At same time, because of its special topography, the paleo uplift also controlled the development of a high-energy facies belt in the shallow water, the development of a stratigraphic pinch out belt, and the eroded leaching event during the late geological period; in other words, through controlling the sedimentation, the paleo uplift could control the development and distribution of high quality reservoir sequences. To be more specific, the following three areas reveal how a large size inherited paleo uplift controlled natural gas migration and accumulation.

(1) The periclinal belt of the paleo uplift controlled the development of carbonate rocks in the high-energy facies belt and lithostratigraphic pinch out belt. During the long period of uplifting, the erosion clearly altered carbonate reservoir sequences. Furthermore, these alterations defiantly controlled development of the lithostratigraphic trap and petroleum accumulation.

Inside the craton and on its peripheral areas, the tectonic framework of uplift and depression not only controlled distribution of the depositional facies belt, but it also determined the thickness of every depositional sequence in the direction of the uplift, the alteration in lithology and its pinch out line and the relationship of overlapping. In addition, it also controlled distribution of the reservoir entity and the lithostratigraphic trap. On the paleo uplift, because the water depth was shallow, the carbonate rocks that developed in the high-energy environment (such as sparite, calcarenite, sparite calcirudite and oolitic limestone) not only contained well developed original pores, but they were also altered by the late phase of diagenesis to create secondary dissolution pores. Because the paleo uplift was frequently above the water surface, mixed dolomitization and evaporation often happened to develop the dolomite that was a good reservoir sequence. The Hetian river gas field and the Shan 1 well gas reservoir were distributed on the margin of the Hetian paleo uplift. Because the higher positions on this uplift experienced a prolonged erosion period, the reservoir sequences contained well developed karst and good physical properties; therefore, these were ideal places for oil and gas to accumulate.

(2) The uplift was the place for a large amount of oil and gas to accumulate because it was the destination of petroleum migration over a long period.

In China, the Paleozoic uplifts usually inherited the developmental process. They not only functioned as the developmental region to create a petroleum trap and reservoir sequences, but they also worked as the final destination for oil and gas migration. Thus, the uplift area was the ideal place for creating an oil and gas reservoir. The time that the uplift was elevated, and its magnitude, controlled the petroleum migration and distribution. Because the higher positions on the paleo uplift were easily influenced by the tectonic movement, oil and gas commonly adjusted migration and accumulation to develop secondary oil and gas reservoirs in the traps of the late period. On the other hand, the original lithostratigraphic type of oil and gas reservoirs developed in relatively lower positions and the slope areas on the uplift.

(3) The large size stable sedimentary basin contained the inherited type of paleo uplifts that had an extended developmental history, well developed traps and time–space settings of source rocks and a major hydrocarbon generating period. Therefore, this not only benefited natural gas accumulation and preservation, but it also offered ideal conditions for secondary petroleum accumulation.

In the craton basin, the paleo uplift was the crucial controlling factor for initial oil and gas accumulation and secondary accumulation. The paleo uplifts that related to large and medium size gas fields were located in the continually developing uplift belt, in the oil and gas accumulative region. The pattern of the paleo uplift could either be placanticline or a bulging upwards shape. In addition, the uplift was usually located in the gas generating center or on its periphery. The Kaijiang paleo uplift in the Sichuan basin, the Urengoy placanticline and the Nizhnevartovsk uplift in the Siberian basin were paleo structures that were located inside the petroleum region and developed over an extended period. As a result, large and medium size oil and gas fields were discovered in the Wubaiti area and other places in Sichuan. Also, the Urengoy large gas field and the Samotlor gigantic oil field were discovered in the Siberian basin. For example, the Kaijiang paleo uplift in the Sichuan basin was formed during the Indosinian tectonic period; it contained the oil and gas that matured in the early geological period and that were generated by the Silurian hydrocarbon source rocks. The area that contained natural gas covered 24,541 km^2 of land and the total gas reserve is $15,000 \times 10^8$ m^3. The gas reservoir was distributed in a layered and broad range. This paleo gas reservoir became the major source of secondary gas accumulations for eight large and medium size gas fields that were distributed in the structural belts of Datianchi–Mingyuexia, Tongluoxia–Qilixia and Dachi Dry Well–Gaofengchang. These structural belts were created during the Himalayan tectonic period.

In the Eerduosi basin, the lower Paleozoic weathering crust type of reservoir sequences were distributed along the paleo uplift; from the west to the east, they were displayed horizontally as three arched belts; from an exploration point of view, they developed two sets of natural gas systems in the Ordovician strata. (1) The arched belt of weathering crust (in the Jingbian area): The Ma 5$_1$ gypsum dolomite flat was distributed along the Ma 5 gypsum lake in the Mizhi area and it was the ideal facies belt for the reservoir sequences. (2) The dolomite arched belt (in the Dingbian–Huanglong area): the Ma 4 dolomite entity made ideal reservoir sequences that were distributed in an arched shape across the Eketuoqi–Dingbian–Huachi–Luochuan–Huanglong areas; this location was positioned on the inner side of the "three uplifts and two saddles". (3) The intricate arched belt of weathering crust and dolomite (in the Tianchi–Zhenyuan area): in the Zhuozishan formation and the Kelimoli formation, the dolomite made beneficial reservoir sequences that were distributed in an arched shape along the Bulage–Tianchi–Lucan 1–Huan 14–Zhenyuan–Xunyi–Yijun; this location was positioned on the outer side of the "three uplifts and two saddles".

- **Large Unconformity Surface Controlled Modification of Reservoir Sequences, Development of Migration Network, or Creation and Distribution of Large Size Stratigraphic Reservoir**

The unconformity surface fundamentally controlled the distribution and alteration of the reservoir entity, oil and gas migration and the development of the large size stratigraphic type of oil and gas reservoirs. In the Tarim basin, Eerduosi basin and Sichuan basin, the development of oil and gas reservoirs in the marine facies were related to the unconformity surfaces of carbonate rocks (Wu et al., 1998; Zhang et al., 1999).

Unconformity Surface Modified the Physical Properties in Reservoir Sequences

The secondary dissolution and karst developed along the large unconformity surfaces. Firstly, during the developmental process of an unconformity surface, folds and faults could create many structural fractures that improved connectivity and permeability between the pores. Secondly, after the unconformity surface was created, the meteoric water seeped into the ground through fractures of an earlier creation to dissolve the underlying strata. Many weathering fractures, dissolutive pores, or dissolutive holes formed under the unconformity surface; the dissolutive breccia developed above the unconformity surface. Therefore, the storage capacity was largely improved. Next, the unconformity surface was the precondition for developing the paleo weathering crust. When the unconformity surface was developed, the underlying strata endured erosion. In the carbonate rocks, dissolution was a major erosion process that created the weathering crust with high porosity and high permeability. Although it altered all kinds of reservoir sequences, the unconformity surface significantly modified carbonate strata. These kinds of examples widely occurred in the world, such as the Fahud oil field in Oman, where the distributions of reservoir sequences and high-quality oil and gas layers clearly related to the unconformity surface. In the Sichuan basin, the multiple sets of the karst type of reservoir sequences in the Sinian–lower Paleozoic strata were related to the unconformity surface (Fig. 12.7).

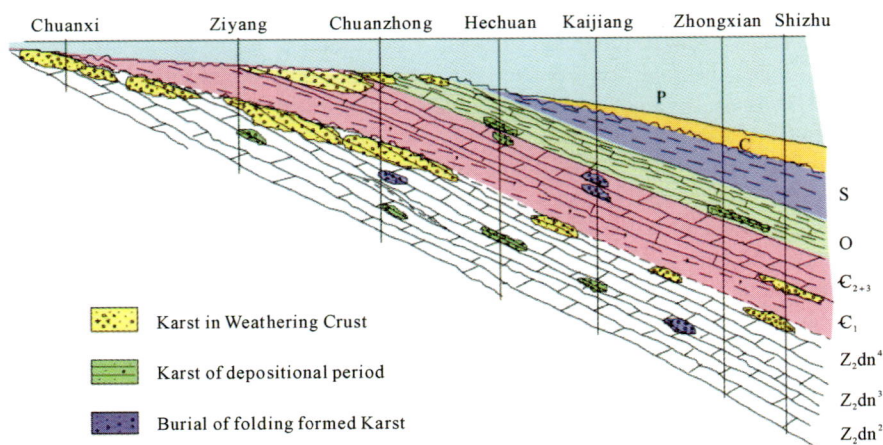

Fig. 12.7. The distribution model of the Sinian–lower Paleozoic karst in the Sichuan basin

Unconformity Surface Controlled Oil and Gas Migration

The unconformity surface not only recorded tectonic movements or sea (lake) level changes, but it also signified different degrees of modifications when the older strata were altered by the later geological processes. The heterogenic stratigraphic modification (and the overlay strata that were deposited during subsidence and transgression events in the later period) occurred in the layered structures in the unconformity surface or in the strata above and below. Along the unconformity surface, the weathering fractures and folds favorably developed; the eluviation from the surface water was very strong; the secondary fractures, dissolutive pores and dissolutive holes also developed well. Carbonate rocks and volcanic rocks formed more easily a thick layered belt that contained secondary pores and played an important role in controlling oil and gas migrations.

Unconformity Surface Controlled Oil and Gas Accumulation

In the carbonate rocks, the unconformity surface played an important role that controlled the oil and gas accumulations. In the Tarim basin, the Paleozoic oil and gas reservoirs were commonly distributed around the unconformity surface, because the unconformity surface provided a network of passages for oil and gas migration, and because the good reservoir sequences usually occurred either above or below the unconformity surface. The Tazhong 4 oil and gas field and the Donghetang oil field occurred on the unconformity surfaces between the carboniferous strata and underlying strata. In the Yingmai 32 oil and gas reservoir, because it was elevated by the late tectonic movement, the dolomite reservoir sequences were exposed on the ground several times and they went through eluviation and erosion caused by meteoric water; therefore the weathering crust type of oil and gas reservoir was created and was sealed by the Cretaceous mudstone. Near the surface of the buried hill, dissolved pores were well developed; the main types of storage space include dissolved pores, intracrystalline pores, intracrystalline dissolution pores and fractures; the porosity of reservoir sequences was between 2% – 10.4%. At same time, the network of fractures connected dissolutive holes and caves to form a migration system that provided a perfect passage to transport oil and gas.

● **High-Energy Facies Belt Controlled Development and Distribution of Beneficial Reservoir Entity and Lithologic Trap**

In carbonate rocks, the development and distribution of reservoir sequences on the reef flat were controlled by the high-energy facies belt on the platform margin. For example, in the Tarim basin, the Middle and Late Ordovician carbonate rocks in the bio-reef complex on the platform margin developed in three areas that included the platform margin, topographic turning zone at the intra-platform and isolated bio-reefs that appeared on the platform. In the Bachu area, the best Ordovician bio-reefs in the Tarim outcropped area were perfectly developed on both sides of the Dawuzitage mountain. In the Lunnan area, the Middle Ordovician barrier reef in the Yijianfang formation developed on the platform margin.

In the high energy-facies belt, ideal lithologic traps developed in carbonate rocks of reef flat facies and in grain limestone of flat facies; in addition, these traps could develop into oil and gas reservoirs. For example, in the northern Tazhong area, during the Ordovician period, the high-energy beach flat facies developed on the carbonate platform–platform margin. In the Lianglitage formation, the thickness of strata in beach flat facies is between 100 – 300 m. The thickness of effective reservoir sequences is between 100 – 300 m. A typical heterogenic carbonate karst type of petroleum trap developed in this set of reservoir sequences. This trap covered 1,000 – 2,000 km^2 of land and it contained an extra large primary lithologic type of integrated oil and gas field. This oil and gas reservoir offered excellent storage conditions. In the Manjiaer depression, it was the first choice for petroleum accumulations during various phases, especially during the late Hercynian tectonic period. In recent years, several oil and gas wells were discovered in the Middle and Upper Ordovician platform margin facies belt. These wells included the Yangwu 2, Tazhong 44, Tazhong 45, and other wells. These wells demonstrate a bright future for oil and gas exploration in the bio-reef complex on the platform margin. In the Tazhong area, the high-energy facies belt on the northern slope not only controlled development of original pores and fractures, but it also influenced the developmental level of secondary pores and fractures. For the bio-reef type and reef flat type of petroleum reservoirs, the most important types of pores were dissolutive crystal pores and the casting holes of crystal. In the bio-reef flat, the original intragranular pores often developed in the grain supported carbonate rocks; in addition, the micro pores commonly developed in the argillaceous reef flat. In the area along the Tazhong 26–Tazhong 161–Tazhong 44–Tazhong 35 wells and in the area around the Tazhong 45 well, many wells that had good oil and gas shows were located in the Middle and the Upper Ordovician carbonate rocks that were deposited on the continental shelf margin, especially in the outer zone on the shelf margin. This was closely related to the marginal deposits of the continental shelf, which were placed in the peripheral area of carbonate rocks, typically along the continental shelf margin. These included various kinds of grain beaches and bio-reefs. Inside this facies belt, beside the best original pores and fractures, the carbonate reservoir sequences also contained secondary dissolutive pores and dissolutive fractures that developed along the original or the remaining pores and fractures. Moreover, they could create ideal secondary dissolutive holes. The Tazhong 44 well was located at the outer belt of the laced depositional entity of the reef flat on the margin of the continental shelf. Buried dissolution was the most influential diagenetic process that affected carbonate reservoir sequences. The platform margin belt on the gentle slope had the nature of a superimposed migration; the reservoir sequences of reef flat facies were distributed over a broad region, which benefited from the development of the lithostratigraphic type of oil and gas field that was widespread with low abundance. For example, in the Sichuan basin, the Tieshan gas field developed in the oolite beach zone in the Feixianguan formation on the western side of the Kaijiang–Liangping sea trough (Yang et al., 2007). The platform margin belt on the steep slope has a superimposed structure, with massive thickness and good physical properties, and which could develop into a large scale lithostratigraphic type of highly abundant oil and gas field. For example, the Puguang gas field and the Qilibei gas field were developed on the northeastern side of the Kaijiang–Liangping sea trough; the Longgang gas reserve was developed on the western side of the sea trough (Fig. 12.8).

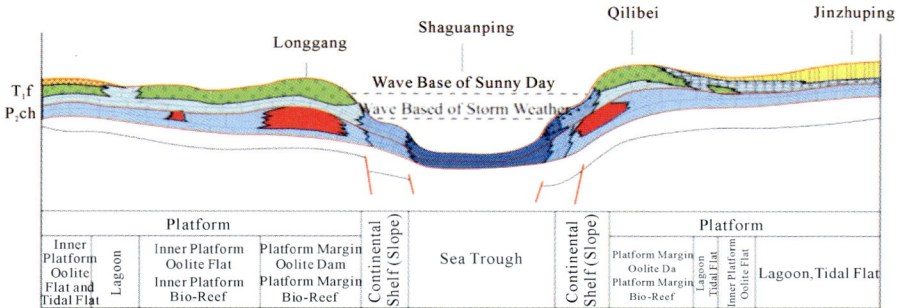

Fig. 12.8.   Depositional model of the Changxing–Feixianguan formations in the Kaijiang–Liangping sea trough

- **Secondary Dissolved Eluviation Zone Controlled Development and Distribution of Lithologic Trap that Contained High Porosity and High Permeability**

The secondary dissolution and eluviation zone directly controlled development and distribution of high quality reservoir sequences. No matter if the reservoir sequences were in the weathering crust with buried hill type of karst, or were in deeply buried internal episodic dolomite, the secondary dissolution and eluviation processes were the crucial factor for creating the secondary pores, secondary holes and secondary fractures in the reservoir sequences. To modify reservoir sequences, the main diagenetic processes included the surface karst, the buried dissolution and underground eluviation. When the internal episodic carbonate rock related to the porous type of reservoir sequences or the fractural–porous type of reservoir sequences, the storage spaces contained matrix pores, pores, holes and fractures, for example the carboniferous strata and Triassic strata in the Sichuan basin, the lower Paleozoic strata in the Shengli oil field. These internal episodic reservoir sequences were controlled by the secondary dissolution or secondary eluviation processes.

Although the reservoir sequences in the Tahe oil field were mainly controlled by epidiagenesis, a buried dissolution process occurred following the surface karst process (Lin, 2002). The buried dissolution happened in a relatively sealed system, which also coupled with a buried filling (cementing) process, which was the process of redistributing and adjusting the sediments and spaces during the burying stage. Therefore, in the area that contained a well developed buried dissolution process, the primary reservoir sequences were altered to improve their storage capacity. Because the buried dissolution created well developed pores, the possible storage spaces were in the fractural–porous type and porous type reservoir sequences. On the other hand, in the peripheral area of the buried dissolutive region, the buried filling (cementing) process also blocked the original reservoir sequences to form a layer of non-permeable sealant that created the precondition for developing the lithologic trap.

Another stage of secondary dissolution and eluviation processes included the fact that either the original reservoir sequences were eluviated by fresh water to develop secondary pores during the earlier diagenetic period, or sea water mixed with fresh water to form the dolomite reservoir sequences that were the favorable precondition for developing large or medium size gas fields. These types of reservoir sequences occurred in the following locations: the eastern Sichuan basin, the Jiyang depression in Shandong province, the basin of the Pearl River estuary, the Eerduosi basin and the Tarim basin. In the Scurry bio-reef oil field in Texas, USA, the majority of pores were secondary eluvial holes that consisted of connected smaller holes. The eluvial zone and non-eluvial zone alternated in the reef complex, which indicated this reef complex was periodically exposed to rain and was eluviated by fresh water. Across the world, most of the carbonate reservoir sequences were modified by rain water during the diagenesis process at the surface or near the surface, to become better reservoir sequences. For the mechanism of buried dissolution, common opinion considered that corrosive acidic fluid was the essential element for buried dissolution; moreover, buried dissolution was closely related to organic diagenesis. The mature processing of organic material produced organic acids, the CO_2 and the H_2S, which were the main sources of acidic fluid. During the surface dissolution period, the crude oil that accumulated during the early period was bio-degraded to produce organic acid. Also, the hydrocarbon thermochemical sulfate reduction was considered as another reason why reservoir sequences contained high concentrations of organic acids (strong acidic fluid) at higher temperature at great buried depth. In the reservoir sequences, when they experienced thermal catalytic pyrolysis and sulfate reduction, hydrocarbons were modified to produced bitumen and hydrogen sulfide. Hydrogen sulfide can be easily dissolved in water to form hydrosulphuric acid; the following is the reaction equation:

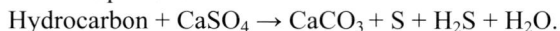

$$Hydrocarbon + CaSO_4 \rightarrow CaCO_3 + S + H_2S + H_2O.$$

- **Faulted Fracture Belt Controlled Distribution and Accumulation of Carbonate Lithostratigraphic Petroleum Fields**

The faults and fractures not only controlled modification of reservoir sequences, but they also noticeably controlled petroleum migration and accumulation. During their active period, faults could function as petroleum migrating passages. At the same time, the fault movements would break the strata, which obviously altered reservoir sequences. Fractures can be categorized into structural fractures and diagenetic fractures. Both types of fracture clearly modified reservoir sequences, especially the reservoir sequences that had low porosity and low permeability.

Fault Controlled Distribution and Accumulation of the Lithostratigraphic Type of Oil and Gas Fields in Carbonate Rocks

In many large and medium oil and gas fields around the world, the positional relationship between the distribution of the oil and gas field and the fault system demonstrated that they obviously depended on each other. The fault utilized its controlling nature to influence petroleum migration and accumulation in the following four areas: firstly, it offered a vertical passage for oil and gas migration; secondly, the fault activity developed the faulted breccia belt that clearly altered physical properties of the reservoir sequences; thirdly, the hot fluid that was brought up along the fault positively reconstructed reservoir sequences around the fault; fourthly, fault-related high relief traps were created (such as a faulted horst block). In China, the most typical examples are in the Sichuan basin (gas fields in the Feixianguan formation and the Leikoupo formation) and in the Paleozoic platform and basin region, in the Tarim basin. Fault activity clearly advanced the process of karst, which mainly related to the structural stress of the faulted belt and the nature of fault activity. The multiple episodes of fault activity helped pull open the faults, which made water circulation much easier and which benefited the development of karsts along the faulted belt. Moreover, faults and the unconformity surface often interacted with each other, which intensified the controlling power and process of karst formation. Because faults mainly worked as vertical passages for oil and gas migration, and because the unconformity surface offered a lateral migration path, an ideal karst system was constructed that not only contained zoning belts in a vertical direction, but which also connected to form a sizeable area in a lateral direction. On the other hand, the deep fault also controlled the alteration that was caused by fluid from a deep level; this kind of alteration developed a fault related, deeply buried, dissolutive type of reservoir sequences. The composite carbonate area of the high-energy facies belt and the faulted belts was the favorable place to discover the deeply buried, dissolutive type of reservoir sequences. For example, in the Tazhong 44 well and the Tazhong 45 well, the Ordovician reservoir sequences were related to deeply buried dissolution.

Fractures Controlled the Distribution and Accumulation of Carbonate Lithostratigraphic Type of Oil and Gas Fields

In certain kinds of situation, fractures or fractural reservoir sequences provided the crucial aspect in oil and gas migration and accumulation. Especially for reservoir sequences that contained low porosity and low permeability, the fracture was the crucial element in oil and gas accumulation. In the Middle East, this kind of reservoir sequence was especially appreciated. For example, micro fractures occurred in the Thamama group of the Early Cretaceous, in the United Arab Emirates. In the Fahud oil field and the Natih oil field in Oman, the Natih formation (the Albin–Cenomanian phases of the Cretaceous period) contained many micro fractures. Furthermore, after the Natih formation was deposited, there was a regional unconformity surface that clearly increased dissolution and porosity. In Iran, the fractured reservoir sequences consisted of Asmari limestone; the thickness of the limestone is between 100 – 800 m; the porosity is between 5% – 15%; the permeability is $(1 - 10) \times 10^{-3}$ μm^2. According to the study results from geochemical research, the source rock is the mudstone in the Kazhdumi formation of the Cretaceous. During the Pliocene–Pleistocene epochs, the tectonic movement fractured the marlite and shale stone (that were deposited during the Campanian in the Late Cretaceous–Oligocene) and transferred oil into the Asmari limestone. The Gachsaran evaporite worked as a good seal above the Asmari limestone. Therefore, the Asmari limestone became the primary reservoir sequences for oil and gas accumulation in Iran. Classic examples of carbonate oil and gas fields that were distributed along the fractural belt included the Ordovician internal episodic type of carbonate oil field in Lunnan, the Tarim basin and the Mondak oil field in the Williston basin, USA. Under the tension of regional stresses (extension, compression and distortion), inside carbonate rocks the fractural belt commonly formed at the folding and distortion positions, such as the axial area on the anticlinal structure. In the Tazhong 45 well, the Ordovician carbonate reservoir sequences occurred in the Lianglitage formation in the Middle and Upper Ordovician strata. In the upper portion, the limestone section (in the interval of 6,034 – 6,080 m) contained a micro fractural–matrix porous type of reservoir sequences. Most of the holes and fractures were filled by calcite. However, some micro fractures on the horizontal level not only favorably developed, but they also connected with matrix pores in the grainstone. In the lower portion, the fluorite limestone section (in the interval of 6,080 – 6,150 m) contained a fractural–porous type of reservoir sequences; the structural fractures were interfaced with fluorite cleavage fractures; the dissolutive holes were well developed; thus, it has excellent storage conditions (Fig. 12.9).

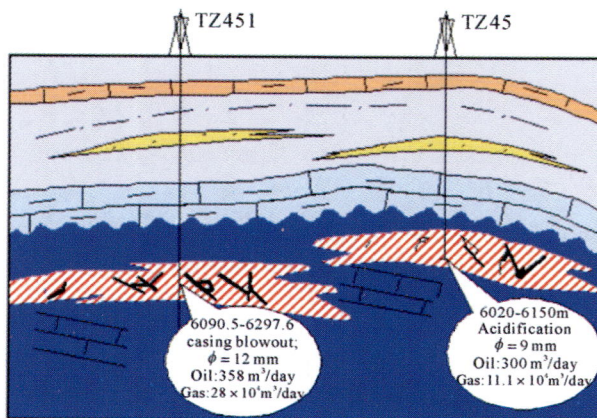

Fig. 12.9. The cross section of oil reservoirs in carbonate rock, from TZ451–TZ45 wells

12.2.3 Major Exploration Domain in Carbonate Rocks

● **Carbonate Domain in the Sichuan Basin**

Sinian–Lower Paleozoic Weathering Crust Type of Dissolutive Reservoir Sequences

Caledonian Paleo Uplift in the Leshan–Longnusi Area. The paleo uplift contained the Lower Cambrian hydrocarbon source rocks. In the paleo uplift, there were two important units of paleo karst that occurred in the Sinian–Cambrian strata and in the Ordovician–Permian strata, and that offered a precondition for developing the reservoir sequences. This paleo uplift also contained well developed structural traps from past to present. At present, the discovered geological reserves of natural gas are 400.5×10^8 m^3; the remaining resource is $4,119.85 \times 10^8$ m^3 Because of the low discovery ratio, the exploration potential is great. Although there are six sets of petroleum systems, the best one occurred in the Sinian Dengying formation. A comprehensive study revealed that the best geological precondition for petroleum accumulation occurred in the eastern section of the Leshan–Longnusi uplift, which included the Gaoshiti–Moxi structural belt and the Ziyang–Anyue area.

    Structural Belt in the Gaoshiti–Moxi Area. The traps developed in the early period and they were finalized during the Caledonian tectonic period. During the late geological period, this structural belt experienced an inherited continual development that benefited oil and gas accumulation. The source rocks contained good conditions and the late hydrocarbon generating peak time of the Middle Jurassic–Early Cretaceous. The reservoir sequences consisted of algal cemented dolomite from the Dengying formation. The paleo weathering type of karsts were well developed; fractures developed in an isolated area, which formed a network for reservoir space. In addition, the Cambrian weathering crust type of reservoir sequences were also favorably developed. The cap rock was in the Lower Cambrian clay shale in the Qiongzhusi formation; the thickness of the clay shale is more than 100 m.

    Lithologic Belt in the Ziyang–Anyue Area. Because it was located on a karst slope, this belt was a favorable place for developing reservoir sequences that contained the remaining fractures–porous type of storage space. During the hydrocarbon generating period, this belt was in the axial area of a structure which offered the preconditions for oil and gas accumulation. In the Ziyang area, we have discovered a highly productive gas well in the lithologic type of gas reservoir.

Reservoir Sequences of Reef Flat Facies in the Changxing–Feixianguan Formation

The reef flat type of natural gas reservoirs were discovered in the Changxing formation–Feixianguan formation in the following locations: the Dukou river,

Tieshanpo, Luojiazhai, Jinzhuping, Longhuichang, Gunziping, Qilibei. The characteristics of these gas reservoirs included a large capacity of reserves and high productivity per single well. The exploration results were remarkable; the proven reserve is $1,940.86 \times 10^8$ m³; the controlled reserve is 892.8×10^8 m³; the forecast reserve is 296.23×10^8 m³. From the third round natural resource evaluation, the total resource is $10,102 \times 10^8$ m³, the remaining resource is $9,161.14 \times 10^8$ m³. But the rate of discovery is only 9.3% of the total amount Therefore, this area still has enormous exploration potential.

In the Changxing–Feixianguan formation, the reef flat type of reservoir sequences occurred on the platform margin reef flats that were distributed around the Kaijiang–Liangping sea trough and along the Exi–Chengkou sea trough (Yang et al., 2001; 2002; 2003; 2007).

The great promise of a lithologic–structural type of gas trap was revealed in recently discovered natural gas reservoirs, which included the Longgang reef flat gas reservoir, the Qilibei gas reservoir, the Huanglongchang gas reservoir and the Puguang gas reservoir (that was discovered by Sinopec). These discoveries increased the exploration value of the reef flat gas reservoir in the Changxing–Feixianguan formation. At present, petroleum exploration is carried out in the Poxi area in northeastern Sichuan and in the west of the Longgang–Jiulong mountain region that is located at the western side of the Kaijiang–Liangping sea trough. The reef flat reservoir sequences from the Changxing–Feixianguan formation were widely distributed in the Exi–Chengkou sea trough. Because they contained plentiful natural resources, this region should be an exploration target in the near future.

Carboniferous Dolomite Reservoir Sequences in the Eastern Sichuan Basin

In eastern Sichuan, the exploration region of carboniferous strata covered 5.5×10^4 km² of land; 375 wells have been drilled. Among these wells, 160 were gas wells. There are 33 discovered structures; 12 of them contained natural gas. We have carried out 48,311.167 km (insert squared symbol) of 2D seismic survey and 1,851.76 km² of 3D seismic survey. Since the exploration obtained a breakthrough in 1997, the proven reserve is $2,802.6 \times 10^8$ m³, the controlled reserve is 380.4×10^8 m³, the forecast reserve is 154.82×10^8 m³, the total natural resource is $7,953 \times 10^8$ m³, the remaining resource is $5,150.81 \times 10^8$ m³, and the rate of discovery is 35.2%.

Based on the favorable depositional facies belt, the carboniferous gas reservoirs were controlled by the Indosinian paleo uplift, which resulted in the creation of a very steep structural framework that controlled gas. In the Sichuan basin, these kinds of stratigraphic sequences not only contained most of the discovered gas reservoirs of large–medium size, but they also contained most of the discovered gas reserves. Recently, the areas with a lower exploration level include the area east of the Yangtze River (Fangdoushan structural belt), the Nanmenchang–Huangnitang structural belt, Yun'anchang structural belt, Hot Spring Well structural belt and Heimenlou structural belt. Using the preliminary exploration, we may obtain a new breakthrough. Although the Datianchi–Mingyuexia structural belt and the Qilixia–Liangshuijing structural belt have a relatively higher exploration level, these structural belts contain plentiful natural gas. Through comprehensive exploration, we can still make new discoveries. Additionally, in the eastern Sichuan area, the discovered gaseous structures can be the next exploration targets for increasing the gas reserve. Overall, the carboniferous system contained abundant natural resources. Because the discovered natural resources have continually increased through recent exploration activities, comprehensive exploration has great potential.

Reservoir Sequences in the Intra-Platform Beach and in the Intra-Platform Dolomite in the Jialingjiang–Leikoupo Formation

The exploration area covered 10×10^4 km^2 of land approximately. Multiple sets of the proven fractural–porous type of natural gas layers were positioned in $T_1j_2^1$–T_1j_1, $T_1j_2^2$, $T_1j_2^3$, $T_1j_4^1$–T_1j_3, and $T_1j_4^3$ strata. There are many traps. The reservoir sequences and petroleum migration were controlled by the Indosinian Luzhou paleo uplift. The exploration practices in recent years indicated that this area still has great exploration potential. On the other hand, the gas layers were shallowly buried (a few hundreds meters – 2,500 m) and they were located near the pipe line and market. The proven reserve is 609.49×10^8 m^3, the controlled reserve is 78.31×10^8 m^3, the forecast reserve is 98.36×10^8 m^3, the total natural resource is $4,017 \times 10^8$ m^3, the remaining resource is $3,407.21 \times 10^8$ m^3, and the rate of discovery is only 15.2%.

The favorable exploration areas include the Dawoding–Datachang area in southwestern Sichuan and the Longnusi–Anyue area in the transitional region of central Sichuan–southern Sichuan. The Dawoding–Datachang area covered 3,000 km^2 of land approximately. According to the calculation method of reserve abundance from the Wolong River (that is 1.66×10^8 m^3/km^2), this area contains 500×10^8 m^3 of resources. In the near future we expect to discover 250×10^8 m^3 of reserves. The Longnusi–Anyue area covers 5,000 km^2 of land approximately. According to the calculation method of reserve abundance from the Wolong River (that is 1.66×10^8 m^3/km^2), this area contains 850×10^8 m^3 of resources. In the near future we expect to discover 400×10^8 m^3 of reserves.

● **Carbonate Domain in the Eerduosi Basin**

Bio-reef Type of Oil and Gas Reservoir

The bio-reefs mainly occurred in the Middle and Upper Ordovician strata that were distributed on the southwestern margin of the paleo central uplift in the Eerduosi basin. In addition, algal reefs and other bio-formations developed in the Lower Ordovician strata on the western side of the Luliang uplift and on the southern slope of the Yimeng uplift.

These kinds of reservoir sequences occur in the Sandaogou formation, Pingliang formation, Beiguoshan formation (3) and the upper portions of the Jiagou formation and Fengfeng formation. These formations developed on the southwestern margin of the paleo central uplift. The bio-reefs were distributed along the beddings discontinuously in dome shape, lenticular shape and irregular shape. The height of the bio-reef is between 5 – 20 m; the height of largest one is 70 m. The bio-reefs have a distinctive boundary, and they made contact with local rocks in abrupt fashion. The reef core usually developed on top of the grain beach limestone and it consisted of coral, the skeletons of stromatoporoids and algal cemented rocks, and buried reef deposits. The lateral wing consisted of collapsed breccia from the reef body and grainstone. Inside the reef body, instead of beddings, there is a blockish structure and taxitic structure. In general, the bio-reef complexes are of large size. Their thickness is more than 300 m and their expansion is more than 10 km. The bio-reefs have a distinct seismic reflection image. Chaotic reflection and semi-parallel reflection are the major types that obviously distinguish themselves from parallel reflection–continual reflection and divergent reflection. The latter two kinds of reflection represent the continental shelf and slope facies. On the southwestern margin of the basin, the Middle Ordovician bio-reefs developed from the northern section of Tianhuan to the northern section of Weibei.

Inside the Eerduosi basin, the Lower Ordovician bio-reefs were mainly distributed in the middle portion of the Ma 5 member of the Majiagou formation that was located either on the western side of the Luliang uplift or on the southern slope of the Yimeng uplift. In the basin, the Ordovician bio-reefs were concurrent with all kinds of grain beach and bio-clastic beach. Although the size of a single reef body was limited, the magnitude of the reef complex was enormous. The petrological study indicated that, after the alteration of dolomitization and secondary dissolution, the bio-reef related reservoir sequences contained well developed pores. In general, the porosity of reservoir sequences is between 5% – 12%, the permeability is $(1 - 10) \times 10^{-3}$ μm^2, and the greatest permeability is above 100×10^{-3} μm^2. This kind of reservoir sequence has excellent permeability. Because the bio-reefs were close to the hydrocarbon source rocks, the natural gas supplies were adequate. Close to the reservoir entity, the dense rocks could form sufficient lithologic traps. On the southern slope of the Yimeng uplift, in the area of the Shan 196 well, the first discovered bio-reef type of gas reservoir in the Eerduosi basin was the Ma 5 gas reservoir.

Comparative analysis shows that, on the southwestern margin of the Eerduosi basin, the Weibei uplift and the Tianhuan depression contained massive reef complexes. Moreover, because the reservoir sequences have good physical properties, a large size gas reservoir would easily develop. Therefore, these regions should be considered as valuable exploration domains in the near future.

Weathering Crust Type of Dissolutive Reservoir Sequences

In the central region and the eastern sides of the Eerduosi basin, the regional paleo erosion surface developed on top of the Majiagou formation in the Lower Ordovician series by a tectonic uplifting movement during the Caledonian tectonic period. Under the paleo erosion surface, the carbonate rocks not only experienced inter-bedding dissolution and weathering crust dissolution, but they also endured dissolution from the hot water and from the compressively discharged acidic water of the burring period. Therefore, all kinds of paleo karst type of reservoir sequences were created, which provide crucial preconditions for natural gas accumulation. Because the weathering crust type of karst intensively erodes strata, and because gypsum dissolution happened on a large scale, in the central region and eastern sides of the basin, various types of second order and third order topographic units developed. The second order topographic unit included karst platform, saddle, bench and basin. After it was buried by the sediments, the weathering crust system was altered by the acidic pressure–bearing fluid and the deep level thermal fluid. Finally, several types of storage spaces developed, which included dissolutive holes, dissolutive fractures, dissolved gypsum pores, intragranular dissolutive pores, intracrystalline dissolutive pores and intracrystalline pores. Furthermore, these pores and fractures formed a complicated combination of pores and fractures, which created five different types of karst reservoir sequences: dissolutive holes–dissolutive pores type, dissolutive pores–intracrystalline pores type, fractural–pores or holes type, intracrystalline pores type and micro pores–fractural type. In general, the thickness of the weathering crust type of reservoir sequences is 25 – 40 m, the average porosity is 5%, the greatest porosity can reach 20%, the average permeability is 5×10^{-3} μm^2, and the greatest permeability can reach 100×10^{-3} μm^2. In the central region and eastern sides of the basin, the weathering crust type of reservoir sequences occurred on top of the Majiagou formation. The overlying strata were the coal measures in the carboniferous and Permian systems. The underlying strata were the hydrocarbon source rocks of algal dolomite and argillaceous carbonate rocks of the Majiagou formation. Since the Mesozoic, because these regions were in an area of low potential for oil and gas migrations, they functioned as the destinations for natural gas migration over a long duration. In other words, they offered good preconditions for petroleum accumulation. In the central region, besides the extra large gas field called "The Changqing Gas Field", there are several favorable exploration targets.

Using preliminary analysis, we selected several areas that possess similar geological conditions to the Changqing gas field, which include the Shenmu–Yulin area, the area north of Wushenqi, the Zhengning–Huangling area, Yanchang–Ansai area and the area east of Etuokeqi. These areas should be the main exploration targets in the future.

Dolomite Reservoir Sequences Beneath the Evaporate Rocks

In the eastern Eerduosi basin, gypsum and salt were deposited during the Early Ordovician epoch. When the Ordovician Majiagou formation was deposited, the Ma 1, Ma 3 and Ma 5 members represented the high peak of the evaporite deposit.

The gypsum–salt layers covered 5×10^4 km^2 of land. Influenced by climate change and sea level fluctuation, the sediments of the salt lake and the sediments of the tidal flat were deposited alternately. Therefore, in the cross section, the depositional sequence showed that dolomite and the gypsum–salt layer alternated. The semi-syngenetic dolomite, recrystallized dolomite and bio-clastic debris, the oolite beach and the residual grain dolomite of storm beach facies commonly developed in the middle and lower portions of the Ma 5 member, the upper portion of the Ma 4 member, the Ma 3 member and the middle and upper portions of the Ma 2 member. These rocks worked as reservoir sequences. The dolomite reservoir sequences mainly contained intracrystalline pores, intracrystalline dissolutive pores, casting holes of gypsum and salt crystal, needle shape dissolutive pores and porphyritic dissolutive pores. In addition, they also contained diagenesis fractures, structural fractures and karst fractures. While the greater percentage of porphyritic dissolutive pores had been filled up, the rest of the pores were not filled or were only partially filled. The surface porosity usually is 1% – 6%; the highest surface porosity can be 8% – 10%. According to the quantity of pores and fractures, the state of occurrence and the relationship beneath the gypsum and salt layer, the dolomite can be further classified into three types: porous type, fractural–porous type and fractural type. In general, the porosity of reservoir sequences is 1% – 8% and the greatest porosity can be 12.8%. The common permeability is $(0.005 – 1) \times 10^{-3}$ μm^2, the greatest permeability can be 24.1×10^{-3} μm^2. In the eastern Eerduosi basin, under the evaporite rocks, the thickness of carbonate hydrocarbon source rocks is 500 m, most of the kerogen is of the sapropelic type, the abundance of organic carbon is 0.2% – 0.25%, and the level of thermal maturation is high ($R_o > 2\%$). This information indicated that the hydrocarbons have entered the matured dry gas stage with great hydrocarbon generating capacity. By this calculation, the total amount of generated hydrocarbons is 54.1×10^{12} m^3; the amount of released hydrocarbons is 35.8×10^{12} m^3.

The result of seismic exploration indicated that multiple types of traps developed under three sets of gypsum–salt layers. The types of traps included the nose-like salt dome, the faulted–lithologic type of trap, the stratigraphic pinch out type of trap and the lenticular type of trap. The gypsum–salt layers occurred in the Ma 1 member, Ma 3 member and Ma 5 member. Among these traps, the salt dome and the stratigraphic type of trap have better exploration potential.

Fractural Carbonate Reservoir Sequences

In the Yusui area, in the eastern Eerduosi basin, the fractural carbonate reservoir sequences developed in the Taiyuan formation in the Permian system. These reservoir sequences consisted of four sets of bio-clastic limestone. The thickness of a single layer of limestone is 5 – 15 m and the total thickness is 40 m. The debris of bio-clastic limestone includes fragments of echinoderm, bryozoan, foraminifera, calcareous sponge spicules and brachiopods. Because the content of the bio-debris was low, after intensive diagenesis the original pores were basically filled up and therefore the limestone did not have storage capacity. Throughout geological history, the eastern Eerduosi basin was a tectonically active region, especially during the late Yanshanian tectonic period. Influenced by the uplifting of Luliang mountain on the eastern margin of the basin, a series of uplifts and faults formed. Because the limestone was dense and brittle, in the direction of the obliquely crossed compressive stresses, a huge number of shear fractures were created. Via the obviations at the outcrops and in the drilling cores, the Taiyuan limestone mostly had "X" patterned fractures with a higher level of openings and fewer fillings. The large scale fractures could cut through overlying sandstone and mudstone to a depth of tens of meters. Under the microscope, the limestone primarily contained micro fractures that were either perpendicular or at an angle to the bedding surfaces. Some of the fractures were filled by calcite and argillaceous materials. Even the surface porosity is 0.2% – 0.5% and the permeability is good. However, the storage capacity was limited. Because the fractures would better assist the discharge of acidic water from the coal measures above, during the late diagenesis stage, a huge network of dissolutive pores developed in company with the fractures in the Taiyuan limestone. The porous types included needle shape dissolutive pores, intragranular dissolutive pores and bio-casting holes. This kind of combination of pores and fractures effectively altered storage capacity and permeability in the limestone. Since the Taiyuan limestone was enveloped by the hydrocarbon source rock of coal measures, the supply of natural gas is sufficient. Also, because the traps were controlled by the fractures, the region that contained well developed fractures was the accumulative zone for natural gas. At present, Taiyuan limestone has been discovered in a few dozen exploration wells. Among these wells, 21 of them produced natural gas, 4 of them yielded a commercial grade gas flow with a medium–high production rate.

- **Carbonate Domain in the Tarim Basin**

In the Tarim basin, there are abundant petroleum reserves in carbonate rocks with various types of petroleum reservoirs and a distinct boundary with the accumulative zone. The important accumulative domains include the unconformity and weathering crust type, the bio-reef and reef flat type, the internal episodic type of dolomite and the deep level fluid modified type.

Unconformity and Weathering Crust Type of Petroleum Accumulative Zone

In northern Tarim, central Tarim and Hetian areas, the Middle and Lower Ordovician carbonate rocks contained three phases of weathering crust type of karst caves on the paleo uplifts, all of which were the products of the karst process that happened near the surface during the early Hercynian tectonic period.

In the central Tarim basin, the unconformity and weathering crust type of carbonate reservoir sequences were distributed in the central faulted horst block zone, the buried hill type structural belt on the southern side and in the area of the

Tazhong 1 well–Tazhong 5 well–Tazhong 48 well. In the areas of the Tazhong 1 well and Tazhong 16 well, the oil and gas reservoirs belonged to this type. In the northern Tarim basin, these types of oil and gas reservoirs were mainly distributed in the faulted horst block zone in the Lunan area, the higher position on the uplift in the Yingmaili area and the northern slope.

In the weathering crust type of reservoir sequences, dissolution is commonly displayed in vertical zones. In descending order, these zones are the surface karst belt, the permeable fluid karst belt, the underground fluid karst belt and the deep level slow fluid karst belt. The major fractures and caves usually developed under the weathering surface within a depth of 200 m, where the surface karst belt–the permeable fluid karst belt–the underground fluid karst belt (Guo, 1999; Yang, 2005) developed. In addition, the study of the Ellenburger formation in the Lower Ordovician series, central Texas (Loucks et al., 2004) indicated that when the karst system was buried, influenced by the mechanical pressure and subsequent fault activities, a collapsed paleo karst system was formed and many fractures connected karst caves. Therefore, this became the exploration target.

In the Tarim basin, the unconformity and weathering crust type of reservoir sequences contained a well developed system of fractures and caves. The best reservoir sequences occurred under the unconformity surface within 200 m of depth expansion. On the slope of the uplift, the oil and gas accumulative belts usually appeared connected on a horizontal level. In the northern Tarim basin, the central Tarim basin and the Hetian area, the slopes of the uplift are the favorable regions for oil and gas exploration. The discoveries of oil and gas in the Yingmai 32 well and the Yingmai 33 well indicated that the northern slope of the Tabei uplift contains huge potential for exploring the paleo weathering crust type of petroleum reservoir.

Reef Flat Type of Reservoir Sequences

In the Tarim basin, the Middle and Late Ordovician carbonate platform bio-reef and the reef complex developed in three positions that included the platform margin, the topographic turning zone at the intra-platform, and the isolated areas at the intra-platform (Gu et al., 2005). In the Bachu area, on both sides of Dawuzitage mountain, the Ordovician bio-reefs were well developed, which were also the best ones in the outcrop area of the Tarim basin. Moreover, in the Lunnan area, the buried reef complex of platform margin facies occurred in the Yijianfang formation in the Middle Ordovician series. On the northern slope in the central Tarim basin, the skeleton reef facies and reef flat facies from the middle and late phases, the marlaceous mound of platform margin facies from the early phase, and argillaceous beach facies, developed in the Lianglitage formation in the Upper Ordovician series. In recent years, oil and gas wells (that included the Yangwu 2 well, Tazhong 44 well and Tazhong 45 well) were discovered in the Middle and Upper Ordovician strata of platform margin facies. These discoveries revealed a bright future for petroleum exploration in the bio-reef and reef complex, which developed on the platform margin.

The bio-constructive reef and reef flat types of petroleum reservoir mainly developed in the central Tarim basin. The depositional environment of the carbonate platform margin offers favorable preconditions for developing the carbonate reef complex facies and reef flat facies. These facies could develop into ideal accumulative facies belts. On the northern slope in the central Tarim basin, the skeleton reef facies and reef flat facies from the middle and late phases, the marlaceous mound of platform margin facies from the early phase, and argillaceous beach facies, developed in the Lianglitage formation in the Upper Ordovician series. In the bio-constructive reef and reef flat types of oil and gas reservoirs, the secondary pores that were created during the late period were more valuable. The physical properties of reservoir sequences were influenced by secondary dissolutive pores, casting holes and fractures that were created by faults.

Internal Episodic Type of Dolomite Reservoir Sequences

In the Tarim basin, dolomite reservoir sequences were discovered in the central Tarim basin; massive dolomite strata developed beneath the Lower Ordovician series. The internal episodic type of dolomite reservoir sequences in the Tazhong 162 well were the important type of reservoir sequences for petroleum accumulation, which consists of buried dolornite that was constructively altered by the deep level dissolution. This kind of dolomite developed on the northern slope in the central Tarim basin, especially on the western side of the Tazhong-I fault. In addition, this kind of dolomite also occurred in the Tazhong 43 well and the Tacan 1 well. In the central Tarim basin, the internal episodic type of dolomite reservoir sequences in the Lower Ordovician series were mainly controlled by the diagenetic environment. The dolomite reservoir sequences are of the fractural–porous type, which not only contains intracrystalline pores, dissolutive holes and pores and fractures, but which also has better storage ability and permeability when compared with the limestone in the same stratigraphic sequences. In the Tacan 1 well, from 5,044.4 – 5,131.0 m, the porosity of the dolomite reservoir sequences is 2.22% and the permeability is 30.1×10^{-3} μm^2. Firstly, the dolomite has better strength to resist the compacted pressure and compressed dissolution than limestone. Thus, with an increase in buried depth, the rate of porosity reduction in the dolomite was smaller than that in limestone. Secondly, under the same natural conditions, dolomite was much more easily fractured than limestone. Next, inside the dolomite, during the dedolomitization process and the selective dissolution, intracrystalline dissolutive pores were easily created along the gaps between the crystals. All the factors that were discussed above contributed to the better storage capacity of dolomite reservoir sequences.

In the Tarim basin, the internal episodic type of dolomite reservoir (that was represented by the Tazhong 162 well) is the important type of exploration target. This kind of dolomite developed on the northern slope in the central Tarim basin, especially on the western side of the Tazhong #É fault. Reservoir sequences were mainly controlled by the diagenetic environment. The dolomite reservoir sequences are of the fractural–porous type, which contain intracrystalline pores, dissolutive holes and pores, and fractures.

Deep Level Fluid Modified Type of Carbonate Reservoir Sequences

The Tarim basin contains the secondary reservoir sequences that were related to the deep level fluid and geothermal fluid mechanism. Firstly, fluorite with well developed holes was discovered in the Tazhong 45 well; moreover, Permian volcanic rocks were discovered in the Tazhong 21 well and the Tazhong 18 well (the Tazhong 21 well was located near the Tazhong 45 well in an eastern direction and the Tazhong 18 well was positioned in a southwestern direction from the Tazhong 45 well). Superimposed fluorite caves were discovered in the Ordovician carbonate rocks in the south of the Yijianfang area, on the western side of the Tarim basin. Also, Permian volcanic rocks were exposed on the surface at a distance of 7 – 8 km from the fluorite caves. Next, in the Yingmaili area, a large amount of dolomite with a geothermal fluid mechanism was discovered in the wells on the Tabei uplift. This is another important kind of reservoir sequence subsequent to the weathering crust type of reservoir sequences, which had been discovered in the Yingmai 7 well, Yingmai 16 well and Yingmai 32 well. In the Yingmaili area, volcanic activity was frequent. On account of this special kind of reservoir sequence and petroleum reservoirs that were first discovered in the Tazhong 45 well and that were created in the carbonate rocks by deep level fluid modification, Jin et al. (2005) named them the deep level fluid modified type of carbonate reservoir sequences and the deep level fluid modified type of petroleum reservoir.

In the Tazhong 45 well, the fluorite related reservoir sequences contained dissolutive fractures and holes, which were a typical example of deep level fluid modification. Because of the deep level fluid alteration, in these types of reservoir sequences, the storage capacity would be excellent. On the other hand, these types of carbonate reservoir sequences had very strong heterogeneity. An understanding of the development and distribution of these types of reservoir sequences is solely dependent on the reorganization of deep level fluid activity. The study of the deep level fluid modification type of reservoir sequences forecast a bright petroleum exploration future in the central and northern Tarim basin, especially in the areas that were influenced by the Permian volcanic activities, in the western section of the Tazhong-I faulted belt and in other faulted belts.

References

Dai, J.X., Wang, T.B., Song, Y., et al., 1997. The Preconditions for Developing the Large-Medium Sized Gas Fields and Related Distribution Pattern in China. Geological Publishing House, Beijing.

Dai, J.X., Xia, X.Y., Hong, F., et al., 1999. The primary aspects that controlled development of large size Chinese gas field in cold measure. Chinese Science Bulletin, 44(22):2455-2464.

Dai, J.X., Chen, J.F., Zhong, N.N., et al., 2003. The Large Size Chinese Gas Fields and their Source Rocks. Science Press, Beijing.

Dou, L.R., Wang, Y.G., 2003. The development and distribution of the Paleozoic marine carbonate reservoirs in China. Petroleum Geology & Experiment, 25(5):419-425.

Fan, J.S., 2005. The characteristics of carbonate reservoir sequences in the world and the primary aspects that controlled petroleum accumulation in carbonate petroleum fields. Earth Science Frontiers, 12(3):21-30.

Fu, S.T., Huang, J.S., Yan, X.X., 2002. New exploration domain in the Paleozoic marine carbonate rocks, in the Eerduosi basin. Natural Gas Industry, 22(6):17-21.

Han, K.Q., 1995. The development of large–medium size gas fields on the Kaijiang paleo uplift in the Eastern Sichuan Basin and Related Petroleum Exploration Target. Natural Gas Industry, 15(4):1-5.

He, J., Han, J.F., Pan, W.Q., 2007. The accumulative mechanism of the Ordovician buried hill type of oil and gas reservoirs on the Lunnan Paleo Uplift. Acta Petrolei Sinica, 28(2):44-48.

Lin, Z.M., 2002. The characteristics of the ordovician carbonate reservoir sequences and their accumulative conditions in the Tahe Oil Field. Acta Petrolei Sinica, 23(3):23-26.

Lu, X.X., Jin, Z.J., 2000. Distribution patterns of oil and gas fields in the carbonate rock. Acta Petrolei Sinica, 21(3):8-12.

Luo, Z.L., Liu, S.G., Liu, S., 2000. The discussion of favorable gas exploration target and new domain in the Sichuan Basin. Natural Gas Industry, 20(4):10-13.

Luo, Z.L., Lou, P., Liu, S.G., Zhao, X.K., Li, G.R., 2001. The new concept of oil and gas exploration in the Paleozoic Erathem, in the Tarim Basin. Xinjiang Petroleum Geology, 22(5):365-370.

Wei, G.Q., Jia, C.Z., Song, H.Z., Shi, Y.S., Lu, H.F., Li, Y.H., 2000. The Ordovician structural–depositional model and the forecast for the fractural type of carbonate reservoir sequences in the Central Tarim Basin. Acta Sedimentologica Sinica, 18(3):408-413.

Wei, G.Q., Chen, G.S., Yang, W., et al., 2003. The identification and application of the interface in carbonate sequences. Petroleum Exploration & Development, 30(6).

Wei, G.Q., Chen, G.S., Yang, W., et al., 2004a. The platform-trough type of sedimentary system in the Feixianguan formation in the lower Triassic series, in northern Sichuan Basin and its evolution. Acta Sedimentologica Sinica, 22(2):254-260.

Wei, G.Q., Yang, W., Wu, S.X., 2004b. The distribution forecasting for Oolite beach type of reservoir sequences in the Feixianguan formation and the selection of favorable exploration region in the Northern Sichuan Basin. China Petroleum Exploration, 9(2):38-43.

Wu, Y.J., Zhang, S.A., Ai, H.G., 1998. The types of unconformity and their relationship with petroleum reservoir in the Tarim Basin. Xinjiang Petroleum Geology, 19(2):101-105.

Yang, W., Wang, Q.H., Zhao, R.D., et al., 2001. The diagenesis of carbonate rocks and the quantity evaluation of carbonate control factor on reservoir sequences. Acta Geoscientica Sinica, 20(5):441-446.

Yang, W., Wang, Q.H., Wang, Y., et al., 2002. The Karst process in the Hetian River Gas Field. Xinjiang Petroleum Geology, 23(2):124-126.

Yang, W., Wei, G.Q., Wang, Q.H., et al., 2003. The characteristics of the ordovician carbonate reservoir sequences and constructive diagenesis in the Hetian River Gas Field. Natural Gas Geoscience, 14(3):191-195.

Yang, W., Wei, G.Q., Wang, Q.H., et al., 2004. Two types of premium hydrocarbon source rocks in the cambrian system and related petroleum system, in the Tarim basin. Oil and Gas Geology, 25(3):263-267.

Yang, W., Wei, G.Q., et al., 2007a. The primary aspect that controlled the Oolite beach type of reservoir sequences in the Feixianguan formation and related model of creating mechanism, in The Northeastern Sichuan Basin. Natural Gas Geoscience, 18(2).

Yang, W., Wei, G.Q., et al., 2007b. The diagenesis and porous evolution in the Oolite beach type of reservoir sequences in the Feixianguan formation, in the Northeastern Sichuan Basin. Geology in China, 34(5).

Zhang, L., Wei, G.Q., Wang, Z.C., et al., 2004. The accumulative model of the Dengying formation in the Sinian system, in the Gaoshiti–Moxi Structural Belt, in the Sichuan Basin. Natural Gas Geoscience, 16(6):584-589.

Zhang, S.A., Wu, Y.J., 1999. Preconditions for accumulating the unconformity type of gas field in the Tarim basin and related distribution pattern. Xinjiang Petroleum Geology, 20(1):5-17.

Zhang, K., Wang, D.R., 2003. The inspiration of Chinese petroleum exploration in marine strata. Petroleum Exploration & Development, 30(2):9-16.

Zhao, J.Z., Li, Q.M., 2002. The stage and history of petroleum accumulation in the craton marine strata, in the Tarim basin. Chinese Science Bulletin, 47(S):116-121.

Zhao, J.Z., Li, Q.M., Wang, Q.H., et al., 2004. Development and distribution of large and medium size petroleum fields in the Tarim basin. Journal of Northwest University (Natural Science), 34(2):212-217.

Zou, C.N., Tao, S.Z., 2007. The primary aspects that controlled development of large and medium size lithostratigraphic type of petroleum fields in marine carbonate rocks. Chinese Science Bulletin, 52(S1):32-39.

13

Geological Aspects of a Volcanic Petroleum Reservoir

13.1 Characteristics and Distribution of Volcanic Sequences in the Chinese Petroleum Basins

In petroleum basins around the world, strong volcanism not only controlled geo-dynamic and tectonic evolution at a deep level, but it also influenced deposition and accumulation in the basin during the late period. Using research works and statistical study of volcanic rocks in rifted basins that are related to the continental margin around the world, Plank (2003) believed that all of the rift basins experienced strong volcanism if they are related to the continental margin. Among these rifted basins, many of them have become important targets for petroleum exploration. At present, the study results and exploration practices indicate that volcanic rocks could function as excellent reservoir sequences. Large size volcanic petroleum reservoirs have been discovered in many countries around world, for example the Jatibarang basaltic oil and gas field in Indonesia (which has oil reserves of 1.65×10^8 t and natural gas reserves of 764×10^8 m^3); the Scott Reef basaltic oil and gas field in Australia (which has oil reserves of $1,795 \times 10^4$ t and natural gas reserves of $3,877 \times 10^8$ m^3); the Kudu basaltic natural gas field in Namibia (which has natural gas reserves of 849×10^8 m^3); the Cenozoic volcanic oil and gas reservoir in Japan; the Muradkhanli oil and gas reservoir in trachybasalt and andesite in Azerbaijan. Because these volcanic oil and gas fields have thick production layers, a high production rate and huge petroleum reserves, they have become important exploration targets (Petford and McCaffrey, 2003; Zhao, 2008; Zhou, 2008).

In China, petroleum exploration in volcanic regions has been carried out for more than 50 years. Many volcanic oil and gas fields have been discovered on the northwestern margin of the Zhungeer basin, in the Bohai bay basin and in nine other basins. Since the year 2000, important exploration breakthroughs have been achieved in the Bohai bay basin, Songliao basin, Erlian basin, Zhungeer basin, Santanghu basin, Tarim basin and Sichuan basin. At present, in-depth investigations have been carried out in the volcanic regions. In eastern China and northern Xinjiang, the volcanic oil and gas fields have started to take their initial shape.

13.1.1 Distribution Pattern of Volcanic Rocks

In the Chinese petroleum basins, the volcanic rocks were broadly distributed either inside the basins or in the peripheral regions (Fig. 13.1). From the Precambrian to the Cenozoic era, four sets of volcanic rocks developed in the petroleum basins. (1) The Precambrian volcanic rocks covered 4.6×10^4 km^2 of land approximately, which mainly occurred in peripheral regions in the Tarim basin, the Sichuan basin, the North China basin and the Jianghan basin; in addition, they also appeared in the Jiangxi–Fujian area. (2) The Paleozoic volcanic rocks were distributed over 84.5×10^4 km^2 of land approximately, mainly in western and southwestern China, such as in the sedimentary basins of Tianshan mountain and Qilian mountain, the sedimentary basins on the Tibetan plateau and the intra-basin areas from the Sichuan basin–Nanpanjiang basin and related peripheral regions. Important exploration breakthroughs were obtained in the Paleozoic volcanic rocks in the Zhungeer basin, the Tarim Basin, the Santanghu basin and the Sichuan basin. In addition, Paleozoic volcanic rocks also occurred in the isolated peripheral areas around the Erlian basin and the Songliao basin, in northeastern China. (3) The Mesozoic volcanic rocks were distributed over 81.2×10^4 km^2 of land approximately, mainly appearing in the sedimentary basins either in northeastern China or on the Tibetan plateau. Petroleum exploration breakthroughs were achieved in the Mesozoic volcanic rocks in the Songliao basin. Furthermore, Mesozoic volcanic rocks also frequently occurred in southeastern China, such as in the area from Zhejiang–Fujian provinces (Lin and Xu, 1995; Lu et al., 2003). (4) The distribution of the Cenozoic volcanic rocks was relatively scattered and only covered 81.2×10^4 km^2 of land approximately, mainly appearing in the sedimentary basins on the Tibetan plateau and in the sedimentary basins in northeastern China, which also were scattered at the peripheral area of the Songliao basin. Recently, the Cenozoic volcanic regions have become the new exploration domains.

13.1.2 Characteristics of Volcanic Reservoir Sequences

The overall characteristics of volcanic reservoir sequences are noticeable in many aspects: firstly, the reservoir sequences contained strong heterogeneity; secondly, the swift alteration in lithology and lithofacies happened horizontally and vertically; next, the reservoir sequences not only have poor permeability, but they also have a large dart coefficient and large variation coefficient (Shu et al., 2007; Wan et al., 2003; Wang et al., 1987; 2003; 2008; Zhao et al., 1996). According to the developmental feature of the volcanic rocks (Deng et al., 2000), the volcanic lithofacies were determined by the type of eruption. The type of volcanic lithology was decided by the volcanic lithofacies; the development level of storage spaces and their type were controlled by the type of volcanic lithology. During the late phase of volcanism, the late alteration not only influenced the evolution of storage spaces, but it also modified the physical properties of the reservoir sequences. Therefore, the volcanic reservoir sequences were closely related to the type of eruption, the lithofacies, the lithology and the type of storage space and its evolution during the late phase.

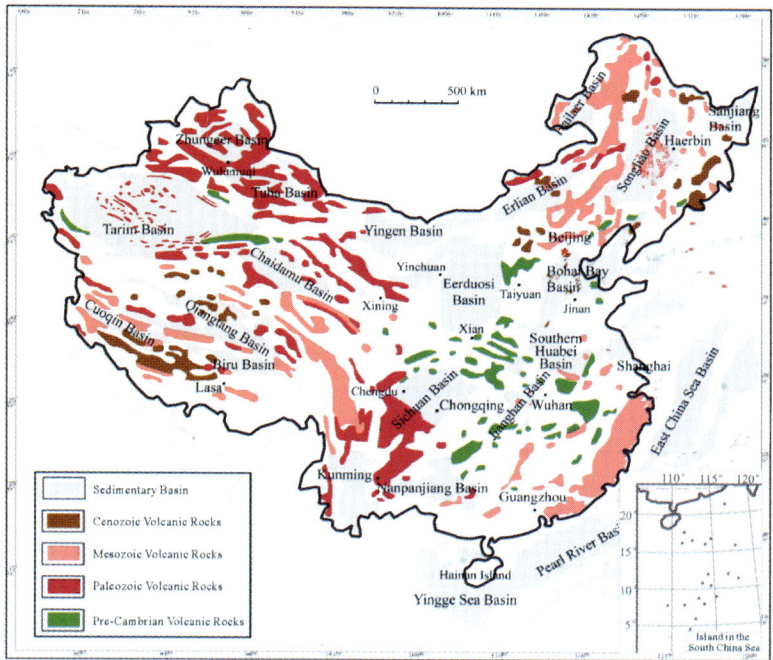

Fig. 13.1.   Distribution map of volcanic petroleum basins in China

● **Types of Volcanic Rocks and Related Lithologic Facies**

Classification of Volcanic Lithology

According to the drilling cores from the Chinese petroleum basins and the observation of the outcrops, and in relation to textural petrogenesis, the volcanic rocks were divided into four major types: volcanic lava, pyroclastic lava, pyroclastics and layered pyroclastics. Considering the contents of SiO_2 and other petrochemical components, the second order of classification includes the basaltic type, basaltic andesitic type, andesitic type, trachyandesitic type, dacitic type and rhyolitic type. Finally, according to the composition of substances, textural features, grain size of volcanic clasts and their contents, the third order of classification of specified types of volcanic rock is established as shown in Table 13.1 (Wang et al., 2003; 2008).

Type of Volcanic Lava. It has the texture of volcanic lava; after it extruded to the surface, the lava condensated and consolidated to form volcanic rocks. This type of rock usually contained fine grains with porphyritic texture and vitroporphyritic texture. The single phenocryst mineral can be identified by the naked eye (or by using a hand lens); however, most of the crystals in the matrix could not be identified by the naked eye. In general, the volcanic lava contained a vitric substance, cryptocrystalline, a felsitic texture, spherulite texture, micrographic texture, aplitic texture, pilotaxitic texture, trachyte texture and intergranular texture. The common structures included a rhyolitic structure, vesicular structure, amygdaloidal structure and a massive structure. In the matrix, the content of volcanic clasts is less than 10%. If considering the content of SiO_2, the lavas could be subdivided into basalt, basaltic andesite, andesite, dacite, trachyte and rhyolite; these volcanic rocks commonly occurred in the Chinese petroleum basins.

Type of Pyroclastic Lava. It is the fused volcanic clastic lava that contained a pyroclastic lava texture and a welded texture. The former indicates that volcanic clasts were fused by lava to form the rock in a transitional form between the lava and the pyroclastics. In fact, this type of volcanic rock belongs to volcanic lava. No matter if it is rigid rock, or it is lava that welded the rigid rocks together, their petrogenesis still belonged to the "condensate and consolidate" type of process. However, in the matrix, the content of volcanic clasts is larger than 10%. According to the grain size of volcanic clasts, the rocks can be subdivided into agglomeratic lava (the grain size > 64 mm), brecciated lava (grain size is 2 – 64 mm) and tuffaceous lava (grain size < 2 mm). The latter one mainly indicates that after the lavas were broken by the volcanism, they were turned into different kinds of fragments that included hot plastic lava fragments (magmatic clasts), vitric fragments, volcanic dust and crystal fragments. In general, this type of lava also contained hot rigid litho-clasts. Under the combined influence of the gravity flow and the loading pressure, the hot pyroclastic flow (that consisted of rigid substances and plastic materials) swiftly moved across the ground and accumulated. During this process, the volcanic clasts (especially the plastic components) were changed in shape, were flatted, welded and condensated; finally, these volcanic clasts were welded to each other and were "condensated and consolidated" to form the volcanic rocks. In the matrix of pyroclastic lava, the content of volcanic clasts is larger than 10%. According to the grain size of volcanic clasts, the rocks can be subdivided into welded agglomeratic lava (the grain size > 64 mm), welded brecciated lava (grain size is 2 – 64 mm) and welded tuffaceous lava (grain size < 2 mm). Again, considering the content of SiO_2, this type of lava can be subdivided into basaltic pyroclastic lava, andesitic pyroclastic lava, dacitic pyroclastic lava and rhyolitic pyroclastic lava.

Table 13.1 Classification of volcanic facies (Wang et al., 2008) (with permission from Science Press)

Facies	Subfacies	Lithologic Features	Textural Features	Structural Features	Phase Sequence & Phase Rule
Facies of volcanic vent (Located at lower position of volcanic edifice)	Subfacies of Cryptoexplosive Breccia	Cryptoexplosive Breccia	Cryptoexplosive Brecciated Texture, Auto-Porphyrocxxlastic Texture, Cataclastic Texture	Tube, Layered, Veined, Multi-Branched, Filled Fractures	Around the crater, top of the subvolcanic rock, or penetrated into country rock
	Subfacies of Subvolcanic Rock	Subvolcanic Rock, Porphyrite and Porphyry	Porphyritic Texture, Seriate Holocrystalline Texture	Condensation Rim Structure, Planar Flow Structure, Streamline Structure, Columnar Joint, Tabular Joint	It was located at lower position of volcanic edifice, from a few hundreds meters to 1,500 m; it has cross-cut-like contact relationship with other litho-facies and country rock
	Subfacies of Volcanic Plug	Lava, Fused Breccia, Tuffaceous Lava, Tuff, Breccia	Porphyritic Texture, Welded Texture, Brecciated Texture, Tuffaceous Texture	Piled Structure, Ring Joints or Radial Joints, Lithological Zoning	Diameter could be a few hundreds meters. occurrence is almost vertical, cut through other lithology
Explosive Facies (Formed during the early phase of volcanic cycle)	Subfacies of Pyroclastic Flow	Fused Tuff (Lava): Magma Fused, Polymictic Conglomerate	Welded Tuffaceous Texture, Pyroclastic Texture	Lump, Normal Graded Bed Sequence, Inversed Graded Bed Sequence, Vesicular Structure, Volcanic Glass, Elongated Orientation	It commonly appeared during the early phase of volcanic cycle, the upper portion of the explosive facies transited into extrusive facies
	Subfacies of Thermal Base Surge	Crystal Fragment, Vitric Fragment, Magmatic Fragment, Crystal Tuff, Vitric Tuff, Magmatic Tuff	Pyroclastic Texture (crystal tuffaceous texture is major type)	Parallel Bedding, Cross Bedding, Regressive Sand Wave Bedding	It was located at the middle–lower portions of the explosive facies, or it was interbedded with subfacies of airfall deposits. The thick bed occurred in a low-cavity place. In ascending order, grain size became small and bed became thinner. It was draped over the paleo topography.
	Subfacies of Airfall	Agglomerate, Breccia, Crystal Tuff	Agglomeratic Texture, Brecciated Texture, Tuffaceous Texture	Grain Support, Normal Graded Bed Sequence, Ballistic Drop Stone	It was located at the lower portion of explosive facies. In ascending order, grain size became small and bed became thinner. Also, it can interbed with others.

(To be continued)

(Table 13.1)

Facies	Subfacies	Lithologic Features	Textural Features	Structural Features	Phase Sequence & Phase Rule
Extrusive Facies (Formed during the middle phase of volcanic cycle)	Upper Subfacies	Stomatal Rhyolite	Spherulite Texture, Aplitic Texture	Vesicular Structure, Amygdaloidal Structure, Lithophysal Structure	Upper Portion of the Flowing Unit
	Middle Subfacies	Rhyolite Structure/Rhyolite	Aplitic Texture, Porphyritic Texture	Rhyolite Structure, Vesicular Structure, and Amygdaloidal Structure	Middle Portion of the Flowing Unit
	Lower Subfacies	Fine-Grained Rhyolite, Autobreccia Rhyolite	Vitreous, Aplitic Texture, Porphyritic Texture, Brecciated Texture	Lump, Intermittent, Deformed Rhyolite Structure	Lower Portion of the Flowing Unit
Facies of Intrusive Vein (Formed during the late phase of volcanic cycle)	Subfacies of Outer Zone	Brecciated Lava with Deformed Rhyolite Texture	Welded Brecciated Texture, Welded Tuffaceous Texture	Deformed Rhyolite Structure	It was located on outside of the dome of intrusive vein facies; it could also be intergraded into the lithology of extrusive facies
	Subfacies of Intermediate Zone	Massive Perlite, Fine-Grained Rhyolite	Vitreous Texture, Perlitic Texture, Less Porphyritic Texture, Mortar Texture	Massive Structure, Layered Structure, Lenticular Structure, Draped Structure	It was located in middle zone of the dome of intrusive vein facies
	Subfacies of Inner Zone	Pillow Perlite and Spherical Perlite	Perlitic Texture, Less Porphyritic Texture, Mortar Texture	Nodule Structure, Pillow Structure, Domal Structure	It was located at the center of the dome of intrusive vein facies
Volcanic Sedimentary Facies	Tuff Interbedded with Coal Layer	Tuff Interbedded with Coal Layer	Volcanic/Terrigenous Clastic Texture	Rhythmic Bedding, Horizontal Bedding	It was located in swamp zone near the volcanic dome
	Re-transported pyroclastic deposits	Layered Pyroclastics, Tuff	Pebble Was Rounded Without External Debris, Pyroclastic Texture	Cross Bedding, Trough Cross Bedding, Graded Bedding, Massive Structure	It commonly appeared in low lying areas between the volcanic edifice domes. It also appeared in the combination of near-source in the volcanic edifice
	Pyroclastic Deposits with External Debris	Tuff and Tuffaceous Sandy Conglomerate with External Debris	Pebble Was Rounded with External Debris, Volcanic/Terrigenous Clastic Texture		It appeared in low lying areas between the volcanic edifice domes

Type of Pyroclastics. All kinds of volcanic clasts (that were made by volcanism) were compacted and consolidated to form this type of volcanic rock; in other words, after the volcanic clasts were deposited (usually they were not transported), via multiple petrogenesis processes, the volcanic clasts were compacted and consolidated together to form the pyroclastics. Apart from containing more than 90% of volcanic clasts, the pyroclastics also contained the components of lava and exogenous clasts to create the transitional types of volcanic rocks that were either close to typical volcanic lava or to normal sedimentary rock. The volcanic clasts could be crystal fragments, vitric fragments or litho fragments. According to the grain size of the clast, this type of volcanic rock can be subdivided into three categories: the agglomeratic type (grain size > 64 mm), brecciated type (grain size is 2 – 64 mm) and tuffaceous type (grain size < 2 mm). According to the content of SiO_2, the pyroclastics can be subdivided into basaltic pyroclastics, andesitic pyroclastics, dacitic pyroclastics and rhyolitic pyroclastics.

Type of Layered Pyroclastics. The layered volcanic rock is a transitional type of rock between the pyroclastics and the sedimentary rocks, which was formed by dual processes of volcanism and sedimentation. The content of volcanic clasts is 50% – 90%; the main petrogenesis is "compaction and consolidation"; this type of rock has the clastic texture from both sedimentary rock and volcanic rock. The volcanic clasts mainly consisted of crystal fragments and vitric fragments; sometimes, it also contained a few litho fragments. The depositional tuff is the main type of lithology and the depositional agglomerate and depositional breccia were less common.

Volcanic Lithofacies

Volcanic lithofacies possess the overall characteristics of the different types of lithology and different kinds of lithosome, which reveal the distribution pattern in space and the mechanism relationship between different lithologic combinations. Previously, much research was conducted for classification of volcanic lithofacies and several kinds of classification proposals were offered. Recently, Wang (2003) and Zhou (2008) offered the best classification proposal that contained five facies and fifteen subfacies (Fig. 13.2; Table 13.1).

● **Developmental Characteristic of Volcanic Reservoir Sequences**

Lithologic Features of Volcanic Reservoir Sequences

Generally speaking, in petroleum basins in eastern China, the Mesozoic volcanic rocks are of the intermediate–acid type and the Cenozoic volcanic rocks are of the intermediate–basic type; on the other hand, the petroleum basins in western China mostly contain the intermediate–basic type of volcanic rocks (Zhou et al., 2008). In the Chinese petroleum basins, volcanic reservoir sequences contained many kinds of lithology; the major lithology of the lava includes basalt, andesite, dacite, rhyolite and trachyte; the major lithology of pyroclastics includes agglomerate, volcanic breccia, tuff and welded pyroclastics.

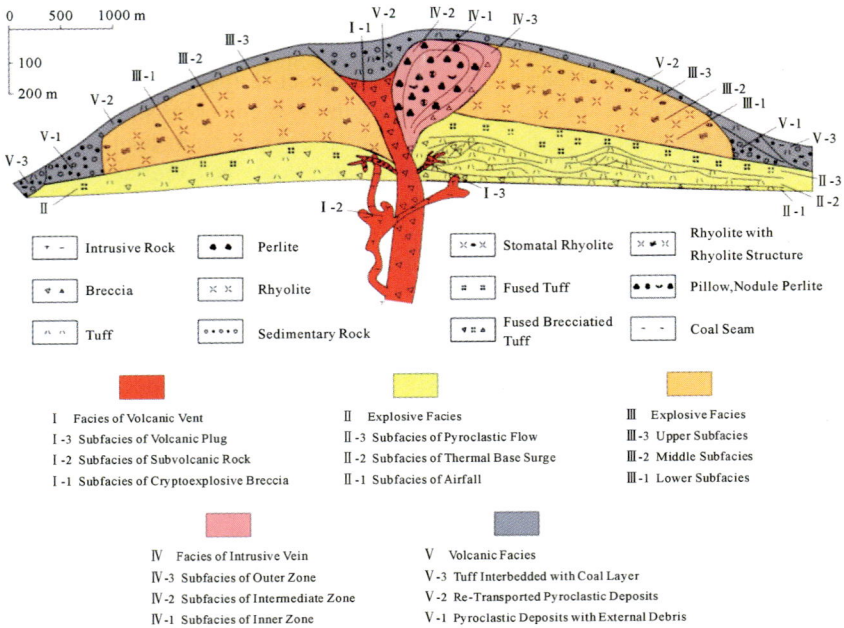

Fig. 13.2.   Model of volcanic lithofacies (Wang et al., 2008) (With permission of Science Press)

In the Songliao basin, the major lithologic types of volcanic rocks include rhyolite, andesite, dacite, basalt, basaltic andesite, trachyte dacite, rhyolitic brecciate tuff, rhyolitic volcanic breccia, dacitic volcanic breccia, basaltic andesitic volcanic breccia, andesitic crystal-fragment tuff and depositional volcanic breccia (Guo et al., 1996; 1997; Hu et al., 1999). Among the samples of volcanic rocks, 86% of them are of the intermediate–acid type, and 14% of them are the basic type of volcanic rocks. Overall, these rocks belong to the alkaline series and calc-alkaline series of volcanic rocks (Fig. 13.3).

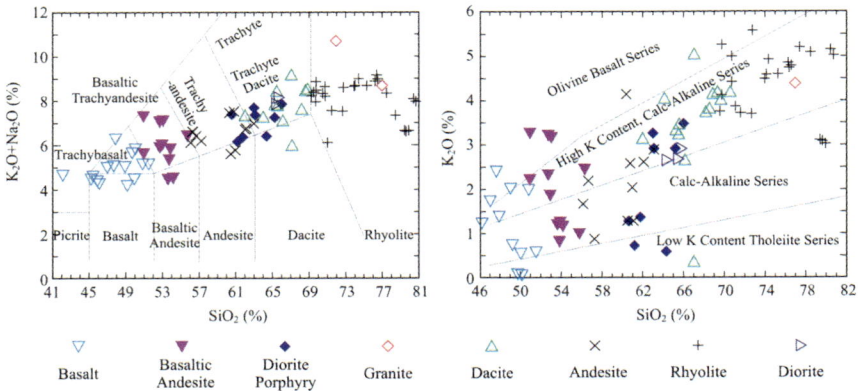

Fig. 13.3.   The diagrams of total alkali and SiO_2 in the Songliao basin

In the Bohai bay basin, the primary volcanic rocks are basalt, andesite and trachyte; for example, in the Liaohe basin, the Mesozoic volcanic rocks mainly consist of andesite and the Palaeogene volcanic rocks consist of basalt and trachyte. In the Jizhong depression, the Jurassic system mainly contains dark purple and gray colored andesite that interbedded with tuff. Basalt, andesitic breccia and pyroclastics are on the top of the andesite. In the Cretaceous system, the lower portion contains multicolored volcanic breccia and the upper portion contains gray color tuffaceous sandy conglomerate, sandstone and andesitic breccia. In the Dongying depression, basic volcanic rocks, submerged volcanic rocks and pyroclastics are general types; the principal lithologies include olivine basalt, basalt, basaltic porphyrite, tuff and volcanic breccia. In the Fenghuadian area in the Huanghua depression, the major types of volcanic rocks include rhyolite of alkaline series, dacitic rhyolite, rhyolite and rhyolitic dacite. In the Nanbao depression, the volcanic rocks include basic pyroclastics, intermediate pyroclastics and basalt. In the Gaoyou depression, there is grayish black, grayish green and grayish purple colored basalt. In the Jianghan basin, in the Cretaceous–Palaeogene systems, the main types of volcanic rocks include quartz tholeiite, olivine tholeiite, basaltic porphyrite; the next types on the list are diabase and pyroclastics (Zhou et al., 2008).

In the Erlian basin, the primary volcanic rocks include autoclastic brecciated andesite, vesicular–amygdaloidal lava, massive lava, tuff, breccia and agglomerate.

In the Chagan depression in the Yingen basin, major volcanic rocks include the intermediate–basic type of basalt, trachyte andesite and andesite; also, there are small amounts of tuff, welded breccia and diabase. In the Sichuan basin, the Permian volcanic rocks include plagioclase basalt, tuff and tuffaceous breccia. In the Tarim basin, the Permian volcanic lavas mainly include basalt and dacite that form the major type of volcanic lava and that represent 80.3% of the total thickness in volcanic rocks. Secondly, there are brecciated dacite and small amounts of brecciated basalt, brecciated tuffaceous dacite, brecciated tuffaceous basalt, tuffaceous breccia and volcanic clastic breccia, crystal fragments and vitric fragment tuff, crystal fragments and litho fragment tuff and crystal fragment tuff, depositional tuff, depositional breccia, tuffaceous argillaceous siltstone, a small amount of pebble tuffaceous mudstone and pebble tuffaceous siltstone. In the Zhungeer basin, basalt, andesite, dacite, rhyolite, volcanic breccia and tuff appear in the Ludong–Wucaiwan area; moreover, on the northwestern margin, the major lithologies in the Carboniferous system include andesite, basalt, andesitic basalt, volcanic breccia, tuffaceous breccia, welded breccia, tuff and agglomerate. In the Santanghu basin, the Permian volcanic rocks include basalt, andesite, dacite, rhyolite, tuff and volcanic breccia (Zhou et al., 2008).

In the Hailaer basin, the Xinganling group can be subdivided into three sections in ascending order. The lower section mainly contains the intermediate–acidic type of lava, pyroclastics, grayish yellow rhyolitic porphyry, trachyte and grayish green tuff. The middle section has the intermediate–acidic type of volcanic rocks that are interbedded with coal layers; the lithologies include grayish purple colored andesite and andesitic basalt that is interbedded with coal layers. The upper section has the intermediate–basic type of volcanic rocks; the lithologies include black–grayish black colored basalt that is interbedded with thin layered, black color mudstone (Zhou et al., 2008).

Porous Features of Volcanic Reservoir Sequences

In general, because they contained original pores, secondary pores and fractures, volcanic rocks have a certain amount of storage space. Compared with sedimentary rocks, the volcanic rocks usually have poor storage ability. However, because the major petrogenesis of volcanic rocks includes condensation and consolidation, the porosity was less influenced by the buried compaction as compared with sedimentary rocks. When the buried depth was deeper than a certain level, the storage ability of volcanic rocks was better than that in sedimentary rocks and they became the major reservoir sequences. The volcanic reservoir sequences have intricate porous structures, such as the dual porosity reservoir that has various physical porosities and strong heterogeneity. In the volcanic reservoir sequences, according to the creating mechanism, the storage spaces can be divided into three types: original pores, secondary pores and fractures.

The original pores include the original air holes, lithophysa cavum pores, inter-amygdaloid pores, intragranular/intracrystalline pores, fractures from the contractive matrix, and cracks caused by minerals. The secondary pores mainly indicate the dissolutive pores and fractures of the late period, which include intercrystalline dissolution pores, dissolutive pores in the matrix and intra-breccia pores from the faulted breccia. The secondary pores developed during the erosion and dissolution processes in the volcanic entity (Fig. 13.4).

In the volcanic rocks, the fractures were the primary channels for all kinds of fluids. Without the connection of fractures there would have been insufficient volcanic reservoir sequences. The factors that influenced fracture included primary and secondary factors. The primary factors indicated all kinds of conditions and elements during the course of fractural development, including lithology, composition, texture and structure, temperature and stress. The secondary factors indicated the factors that altered the former fractures during the late geological process after the fractures were created. During the course of fractural development, the primary factors were in the dominant position, which was the foundation for the creation of fractures. On the other hand, the secondary factors were the force behind the alteration that either remade or modified former fractures. According to the nature of the stress that created the fractures, the fractures in volcanic reservoir sequences can be categorized into tensional fracture, sheared fracture, compressive fracture, tensional sheared fracture and compressive sheared fracture.

The fractures could also be sorted into different kinds of categories. The volcanism and petrogenesis created eruptive cracks and contractive fractures. The structural stress either altered the shapes of the volcanic rocks or dislocated volcanic rocks, which created structural fractures. The processes of weathering and dissolution either assisted or extended the erosion of volcanic rocks that was initiated by the structural stress. Even if the volcanic rocks were covered by overlying strata, a huge amount of water or a solution of organic acid could seep into the rocks via the faults and fractures. Thus, dissolution could happen at a deep level, which created the dissolutive pores and the dissolutive fractures. The fault activity could control the development of fractures to some degree; the fractures were well developed near the fault and they were poorly developed if far away from the fault.

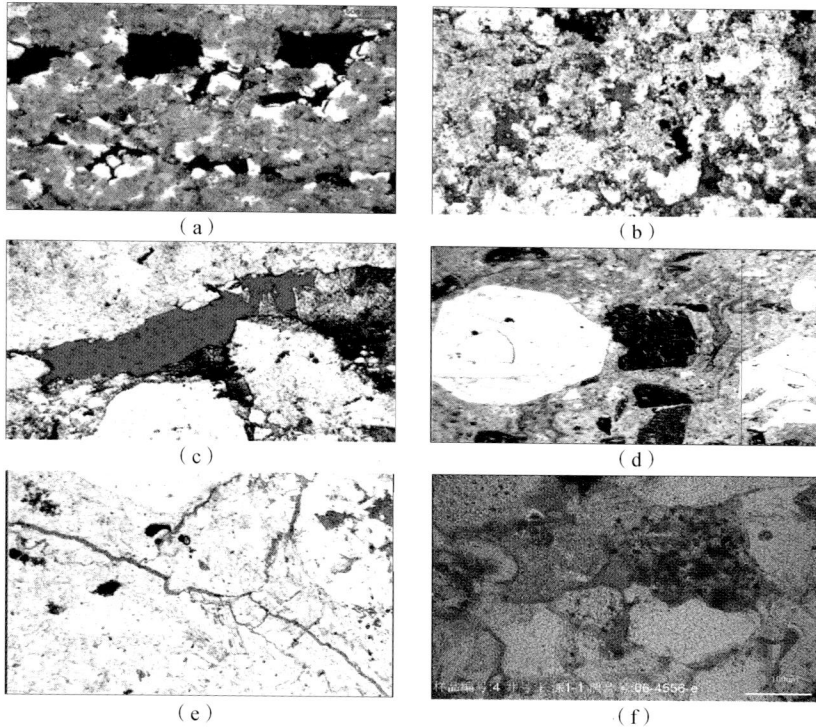

Fig. 13.4.   Photos of porous development in volcanic rocks. (a) Pores in rhyolite, upper portion of extrusive facies; (b) Intercrystalline pores in the rock of explosive facies; (c) Inter-pebbles pores in agglomerate of explosive facies; (d) Intra-matrix micro pores; (e) Micro fractures; (f) Dissolution pores

In the volcanic reservoir sequences, the air holes and the dissolutive pores usually contained a moderate amount of oil. On the other hand, the structural fractures and the weathering fractures worked as connective channels that linked the air holes, the dissolutive pores and other types of storage spaces. Moreover, they also functioned as the major transportation channel during the oil and gas migrations. Although the structural fractures and the weathering fractures could operate as storage spaces, they only contained a small amount of oil. The storage spaces did not appear individually, they always occurred in some kind of storage combination. The combination of pores, fractures and holes could form ideal spaces for oil and gas accumulations. Various reservoir sequences contained diverse storage combinations. The type of geological process, the varied storage combinations and the interaction of these two factors determined the development and evolution of volcanic reservoir sequences. The volcanic reservoir sequences experienced various and intricate forms of progress, which included development, blockage, alteration and recreation. Among these developmental processes, the major geological processes that either influenced or controlled the development of volcanic storage spaces are the tectonic movement during the late period, the weathering and eluviation processes and the alterations caused by fluids.

In the Chinese petroleum basins, the volcanic rocks commonly developed in the Paleozoic, Mesozoic and Cenozoic erathems with the characteristics of broad distribution and longer duration throughout geological periods. Regardless of whether it was the basic type, the intermediated type, or the acidic type of volcanic rock, no matter if it was volcanic rock or intrusive rock, whether it was lava or pyroclastics, the good reservoir sequences occurred in the Cenozoic erathem– Archean eonothem. For example, the following strata contained volcanic reservoir sequences: the Yingcheng formation in the Songliao basin, the Suhongtu formation in the Yingen basin, the Xinganling group in the Abei oil field in the Erlian basin, the Mesozoic and the Cenozoic erathems in the Bohai bay basin, in the Jianghan basin and in the Subei basin, the Carboniferous system in the Kelamayi oil field in Xinjiang and in Wucaiwan in Ludong and the Permian system in the Tarim basin, the Santanghu basin and the Sichuan basin.

● **Primary Factors that Controlled the Development of Volcanic Reservoir Sequences**

The evident feature of volcanic reservoir sequences was that the porosity was less influenced by the buried depth. Because the internal skeleton framework of volcanic rocks was strong if compared with the clastic sedimentary rocks, the volcanic reservoir sequences had stronger resistance to compaction and they were less influenced by the mechanically compacted force during the burying process. Most of the Chinese petroleum basins contained volcanic reservoir sequences that were not only distributed widely, but also had great thickness. According to the developmental process of volcanic reservoir sequences, Caineng Zhou and others (2008) proposed three categories that were the volcanism process, the petrogenesis process and the tectonic process. Next, according to the characteristics of the creative mechanism, the secondary categories were the lava type of reservoir sequences, the pyroclastic type of reservoir sequences, the dissolutive type of reservoir sequences and the fractural type of reservoir sequences. All types of reservoir sequences varied in the positions in which they occurred, in the distributive patterns and in the type of pores and physical properties.

The volcanic storage spaces had an intricate developmental process that included the creation, development, blockage and recreation. The original pores and fractures were controlled by the original eruptive conditions; in other words, they were controlled by the volcanic lithofacies (or subfacies). After the volcano erupted, although the volcanic rocks (that were formed either by condensation and fusion or by compaction and consolidation) contained original air holes, the volcanic rocks still did not have permeability if these air holes were not linked together. Only after they went through all kinds of alteration in different geological processes during varied developmental stages during the late period, did these volcanic rocks subsequently obtain storage capacity. Overall, volcanism, tectonic movement and alteration during the late period were the most important geologic processes that either influenced or controlled the development of reservoir sequences.

Volcanism Controlled Reservoir Sequences

Volcanism constrained the pattern and size of the reservoir body and the interaction between the reservoir bodies; in addition, it also controlled the types of storage space and the composition of minerals in the reservoir sequences. In the same area, the storage capacity of volcanic reservoir sequences was mainly controlled by the types of lithology and lithofacies. Among the volcanic reservoir sequences, different types of reservoir systems developed in different kinds of lithology.

Volcanic lithofacies were an important factor that influenced reservoir sequences; different lithofacies or subfacies had different types of pores. Even in the same facies, different subfacies might also have various physical properties, because the litho-textures and structures varied in each subfacies. The litho-texture and structure controlled the combination and distribution of original pores, secondary pores and fractures. The volcanic lithofacies (or subfacies) controlled the physical properties of reservoir sequences, the types of storage space and their characteristics and alteration. In general, the storage spaces in volcanic vent facies and explosive facies were better than the ones in the extrusive facies and intrusive vein facies (Fig. 13.5).

The storage spaces in volcanic vent facies mainly consisted of isolated air holes and intra-pyroclast pores; for example, in the Jiyang depression, the lava of volcanic vent facies occurred in the interval between $2,541.71 - 2,548.27$ m in the lower portion of the Shan 7426 well. The measured porosity of the lava is $13.1\% - 16.4\%$ and the permeability is 106.95×10^{-3} μm^2; in addition, the lava contained well developed columnar joints. Therefore, ideal storage spaces developed in the lava. The pyroclastics served as the characteristic of the explosive facies. When the volcano erupted, the explosive energy broke the roof and surrounding rocks to create many fractures and micro cracks. At same time, it also formed volcanic breccia with well developed intra-breccia pores and air holes; moreover, because the explosive facies usually appeared at the paleo eroded highland, the rocks often experienced a weathering process; therefore, the dissolutive pores (or holes) and the dissolutive fractures were well developed, which benefited the creation of the ideal accumulative region. For example, in the Jiyang depression, the volcanic breccia occurred in the interval between $1,975 - 1,979.6$ m in the Shang 74212 well. The porosity of the volcanic breccia is $0.7\% - 33.1\%$ and the permeability is

$11.2 \times 10^{-3} - 140 \times 10^{-3}$ μm². In the Shang 7426 well, the volcanic breccia occurred in the interval between 1,830 – 1,838 m, the porosity of the volcanic breccia is 6.7% – 37.4% and the permeability is $0.988 \times 10^{-3} - 3,170 \times 10^{-3}$ μm². This facies belt is also an ideal accumulative zone.

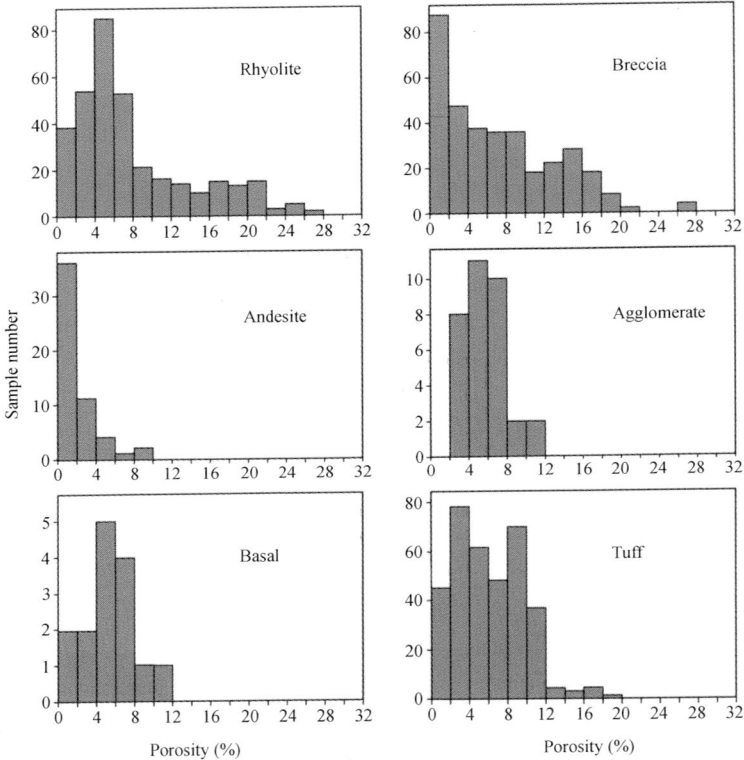

Fig. 13.5. The diagrams of porosity development in volcanic rocks, the Songliao basin (The explosive facies are better than the extrusive facies. Among the rocks of extrusive facies, rhyolite is the best one)

Extrusive facies developed during every stage of a volcanic eruption; the lava contained well developed original pores, the secondary pores included dissolutive feldspar and contractive pores and fractures caused by devitrification (when the feldspar and quartz grew from vitreosity). According to statistical data, in the Anda and Shengping areas in the Songliao basin, the reservoir sequences in the upper subfacies of extrusive facies contained the best physical properties (Fig. 13.6). In the intermediate subfacies of intrusive vine facies, because the storage spaces included fractures, dissolutive pores, intracrystalline pores and micro pores, the reservoir sequences had good physical properties and would function as the ideal accumulative facies belt (Wang et al., 2008).

In the volcanic reservoir sequences, the volcanic lithology also controlled the development of pores. Generally speaking, reservoir sequences would contain favorably developed pores if the rocks consisted of more acidic substances. Because the increase in the viscosity in volcanic rock was proportional to the increase in the content of SiO_2 in the magma, the greater viscosity of the magma, the less ventilation materials would escape from the magma. As a result, the volcanic rocks did not only have well developed pores, but they also commonly contained large size pores. In the Songliao basin, the characteristic of the volcanic reservoir sequences in the Yingcheng formation showed that the rhyolite of extrusive facies contained the best developed pores. Secondly, the intermediate-acid contained a reasonably good amount of pores too. Compared with intermediate–basic volcanic rocks, rhyolite not only contained the original pores, but it also had well developed secondary dissolutive pores. The more acidic the magma became, the more the alkali feldspar grew in the volcanic rocks. Thus, the dissolutive pores of the late period were better developed.

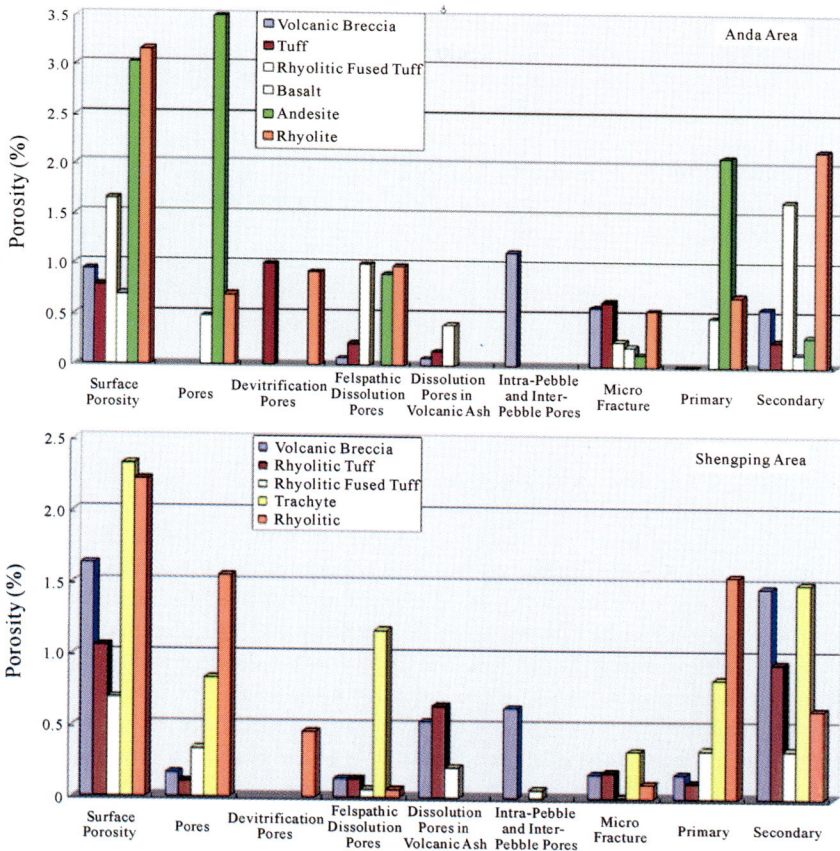

Fig. 13.6. The diagrams of porosity development in volcanic rocks in the Anda and Shengping areas, Songliao basin

Usually, the air holes had the tendency to come together on the top of volcanic rocks. In general, the thinner the positioned layer of magma, the broader the distribution of the pores. When the magma had a greater thickness (> 3 m), the air holes basically came to the top of the volcanic rocks. Thus, the top structure of the volcanic rock was the concentration belt of air holes. In the Songliao basin, the classic examples were found in the thick layered rhyolite near the Yingcheng coal mine and at a depth of 2,960 m in the Wan 17 well (Feng et al., 2003). On the other hand, the top surface of the volcanic rocks also worked as the interface of the weathering and eluviation processes; fractures and dissolutive pores simply developed at this interface, which offered favorable connective conditions for petroleum accumulation on the top of the volcanic rocks. Beside the factors that were discussed above, the number of eruptive cycles also controlled porous development in the volcanic rocks to some degree. Commonly, more eruptive cycles would bring about better developed pores in the volcanic rocks. Furthermore, the development of the perfect reservoir sequences in the intermediate–basic type of volcanic rocks was either related to multi-cycled eruption or related to the alteration in multi-lithology. Relatively speaking, the reservoir sequences were poorly developed in the thick and single layered volcanic rocks. For example, in the Dashen 3 well and the Dashen 4 well, the pyroclastics and the intermediate–basic type of volcanic lava occurred alternately; there were several eruptive cycles with swiftly changing lithology. In addition, in the same lithology, the basic type of volcanic rocks had small thickness and better physical properties. On the other hand, in the Dehui faulted depression, the sole lithology of andesite occurred in the Deshen 7 well. The volcanic rock had a great thickness (2,358 – 3,563 m) and poor physical properties, which has been interpreted as the dry layer.

Tectonic Movement Controlled Reservoir Sequences

Tectonic movement and structural position determined the progression of faults and the developmental level of fractures (Zhou et al., 2008). The occurrence of fractures influenced reservoir sequences in the following three areas: (1) The fractures that were created in the vesicular–amygdaloidal zone would improve the connectivity of air holes and increase the permeability of the rock. More important, the surface freshwater or groundwater would alter the volcanic rocks via the dissolutive process along the fractures. With the establishment of the original air holes, the remaining air holes and the intracrystalline pores of the matrix, a huge number of dissolutive pores (or even karst caves) developed. (2) In the highly concentrated area of fractures, the fractural type of reservoir sequences were created; in addition, under certain conditions, dissolutive pores (or even karst caves) were created. (3) The occurrence of fractures could improve the conditions of flow for underground fluids and increase the dissolution. Therefore, this could advance the dissolution process; for example, observation either of the drilling cores or by examination under the microscope indicated that, as the outcome of dissolution, the dissolutive pores developed along the structural fractures. For example, in the Songliao basin, the fractures in the Yingcheng formation basically can be divided into three phases (Chen et al., 1999). The fractures of the early and middle phases had been totally filled out (or partially filled) by calcite and argillaceous substances. However, the fractures of the late phase had not been filled. The non-clogged fractures were effective fractures that represented 68% of the total fractures and functioned as perfect permeable channels. The concentration level and the opening status of fractures varied in different types of lithology. Fractures were well developed in lava because it was rigid and more brittle. In particular, the structural fractures were well developed in the fine grain rhyolite, welded tuffaceous lava, crystal–fragment tuff and brecciaed lava.

The Latter Alteration Controlled Reservoir Sequences

The types of volcanic petrogenesis included compaction, filling, dissolution and alteration. Because it would diminish the physical properties in reservoir sequences, the filling process was harmful for the development of volcanic reservoir sequences. In addition, the compaction was unfavorable for the creation, preservation and development of volcanic reservoir sequences. In particular, it was harmful to pyroclastics. The common petrogenesis alterations included chloritization, calcite metasomatism and zeolitization. These alterations brought both constructive and negative effects to the volcanic rocks. In general, the air holes would not function as storage spaces directly. Firstly, these spaces were filled out by chlorite, zeolite and calcite. Next, these minerals were dissolved by groundwater and then, in order to create reservoir sequences, these holes had to be connected by fractures.

The large magnitude of fluid activitiy was caused by the volcanic activity, the tectonic movement and the release of hydrocarbon. The direct effect that fluids brought to the volcanic rocks was the replacement of materials in the volcanic rocks, which placed the volcanic rocks in an opened system. The fluids could be divided into thermal fluid and acidic fluid that related with organic substances. The fluid altered the texture of the volcanic rocks, which not only greatly improved the physical properties of the volcanic reservoir sequences, but which also made more complicated types of storage spaces. For example, dissolutive alkali feldspar appeared either in the volcanic rocks in the Yingcheng formation in the Songliao basin or in the Carboniferous volcanic rocks in the Zhungeer basin. The direct result of thermal fluid activity included alteration and dissolution of original minerals. At the same time, the creation of new minerals brought about the processes of secondary cementing and filling. The alteration and dissolution increased the porosity of the volcanic rocks. Conversely, the cementing and filling decreased porosity. In particular, they reduced permeability. For example, the thermal fluid activity that was related to the Permian volcano occurred in the Tarim basin.

The volcanic rocks experienced different degrees of weathering and eluviation during the late stage. For most of the volcanic rocks, the developmental levels of the pores were closely related to the eluviation that not only would break the rock, but that would also clearly alter the chemical composition in the rock, such as the dissolving minerals, oxidizing minerals, hydrating minerals and carbonatization. For example, in the Songliao basin, the purple color, welded, andesitic tuff on the top of the Yingcheng formation occurred in the Shengshen 2 well. Because of the weathering and eluviation, the tuffaceous lava of explosive facies was altered from a dense state to a very loose state. In the drilling cores, the tuffaceous lava looked like loose sands. The porosity of the rock was larger than 15% and the permeability was perfect. Therefore, the eluviation was not only one of the important factors that influenced the storage capacity of the volcanic rocks, but it was also a common geological phenomenon in the volcanic rocks. The space from the top of volcanic rocks to the unconformity surface was an important controlling factor that determined the development of the weathering type of reservoir. The weathering and eluviation would easily produce the fractures and dissolutive pores on top of the volcanic rocks. In a cross section of the field, the dissolutive pores, vertical fractures and joints commonly appeared on the top of volcanic rocks; these pores and fractures were developed by the weathering and eluviation processes. Hence, field observation also proved that the alteration in the late stage was an important factor in the development of favorable reservoir sequences.

Dissolution was a process whereby materials were changed from their original state, which would improve the overall porosity. All kinds of secondary pores were related to dissolution. Dissolution was another important factor that controlled storage capacity in the volcanic rocks. Under the processing of acidic water (either organic or non-organic), the unstable materials in the volcanic rocks were dissolved and the secondary pores were created.

13.2   Control Factors for Oil and Gas Accumulation in the Volcanic Reservoirs

Because the volcanic petroleum reservoirs have unique physical properties, the primary control factors for accumulation, the accumulative pattern and types of reservoir were different when compared with regular petroleum reservoirs. To develop the volcanic petroleum reservoir required the collaboration of volcanic lithosome and other accumulative elements.

13.2.1   *Volcanic Entity Controlled the Development of the Oil and Gas Reservoir*

As a type of reservoir sequences, volcanic rocks have their unique characteristics. Firstly, because they were less influenced by compaction, as the depth increased the physical porosity slowly decreased in the volcanic reservoir sequences; even below 4,000 – 5,000 m, the volcanic rocks still maintained perfect physical properties. Secondly, compared with regular clastic reservoir sequences, the control factors for the development of volcanic reservoir sequences were different. In addition, the developmental zone of the volcanic reservoir sequences also varied. Thirdly, because of poor connectivity between different volcanic lithosomes, the character of the reservoir–seal combination and the types of natural gas reservoir were also different.

The analysis indicated that the volcanic eruption edifice, the volcanic lithofacies and the eruption cycle controlled the development of volcanic reservoir sequences together. In general, the volcanic rocks had great thickness; however, the effective volcanic reservoir sequences had a relatively smaller thickness. The natural gas reservoirs were best developed either in the rhyolite of explosive facies and extrusive facies near the volcanic crater or on the top surface of the multi-cycled extrusive face. The exploration practices demonstrated that the discovered gas wells (that yielded a commercial grade gas flow) occurred either in the crater area or near the crater. Conversely, far away from the crater, the Xushen 16 well is a failed well and the Xushen 22 well is a low production gas well (Fig. 13.7). For the volcanic rocks in the extrusive facies, the primary pores were well developed on the top surface of volcanic rocks. Also, the litho-diversity was distinct and thus weathering and eluviation could easily take their course. For example, in the basalt of the Dashen 3 well and the Dashen 1 well, the volcanic gas reservoirs all developed on the top surface of the volcanic rocks.

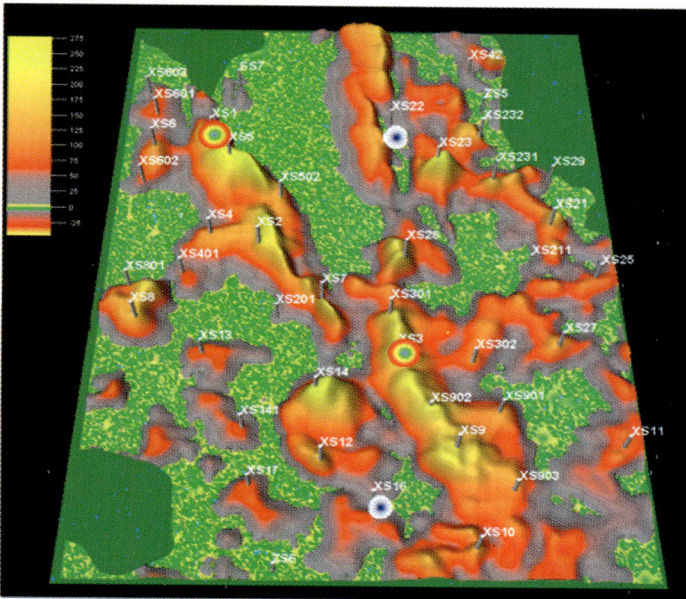

Fig. 13.7. The distribution map of volcanic rocks in the faulted basin in the Xujiaweizi area (according to the Research Institute of Daqing Oil Field)

Because the volcanic eruption contained the characteristics of multi-phases and because the connectivity between the lithosomes was poor, therefore the lithologic gas reservoir was the primary type of volcanic reservoir. The relationship of gas to water was complicated; usually, the gas was positioned on top of the water and the distribution of gas and water were controlled by the volcanic litho-body (Fig. 13.8). The gas reservoirs of different volcanic lithosomes were not connected; multiple sets of gas layers were superimposed vertically and the distribution pattern of the gas reservoirs looked horizonally like a string of pearls. In the volcanic rock, the relationship of gas to water was controlled by volcanic lithosome and volcanic lithology. During the eruption, the volcanic rocks of the same eruptive phase would have very different physical properties if they were located at a different position in the volcano, which brought about the separation of the gas reservoir in the same volcano. Moreover, different volcanic lithosomes also didn't connect. Therefore, several water and gas systems were present, both vertically and horizontally; even a distribution pattern of 'gas on top and water on the bottom' appeared. In the Xushen 13 well, the first member of the Yingcheng formation is a good example (Fig. 13.8); on a horizontal level, the distribution of volcanic gas reservoirs looked like a string of pearls.

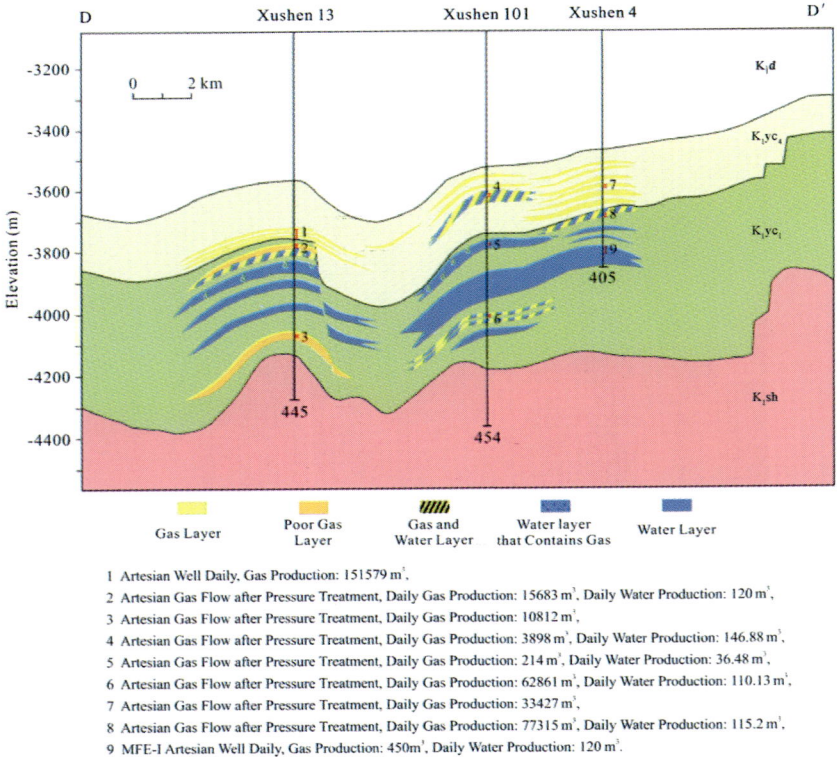

1 Artesian Well Daily, Gas Production: 151579 m³,
2 Artesian Gas Flow after Pressure Treatment, Daily Gas Production: 15683 m³, Daily Water Production: 120 m³,
3 Artesian Gas Flow after Pressure Treatment, Daily Gas Production: 10812 m³,
4 Artesian Gas Flow after Pressure Treatment, Daily Gas Production: 3898 m³, Daily Water Production: 146.88 m³,
5 Artesian Gas Flow after Pressure Treatment, Daily Gas Production: 214 m³, Daily Water Production: 36.48 m³,
6 Artesian Gas Flow after Pressure Treatment, Daily Gas Production: 62861 m³, Daily Water Production: 110.13 m³,
7 Artesian Gas Flow after Pressure Treatment, Daily Gas Production: 33427 m³,
8 Artesian Gas Flow after Pressure Treatment, Daily Gas Production: 77315 m³, Daily Water Production: 115.2 m³,
9 MFE-I Artesian Well Daily, Gas Production: 450m³, Daily Water Production: 120 m³.

Fig. 13.8. Cross section of gas reservoir from the Xushen 13 well–the Xushen 101 well–the Xushen 4 well (according to Research Institute of Daqing Oil Field)

Apart from the volcano controlling the types of volcanic gas reservoirs, faults and regional structures also worked as the major control factors. For example, in the Songliao basin and the Zhungeer basin, the volcanic rocks were distributed along the faults. The types of volcanic gas reservoir include four primary types and six secondary types; the primary types include the volcanic faulted nose (faulted anticline)–lithologic composite type of gas reservoir, the volcanic lithologic type of gas reservoir, the volcanic draped anticline (stratigraphic overlapping)–lithologic composite type of gas reservoir and the volcanic structural type of gas reservoir. The former two types of gas reservoirs mainly developed in the faulted depression; the latter two types of gas reservoirs mainly developed on the uplift area and the slope area. The faulted nose (faulted anticline)–lithologic composite type of gas reservoir had the widest distribution and it was related to the faults (Fig. 13.8).

Foreign generation, self-storage and self-sealing were the important characteristics of the volcanic gas reservoir. Because they were located at different positions, the physical properties of the volcanic rocks varied; therefore, the volcanic rocks could work not only as perfect reservoir sequences, but they could also function as ideal cap rocks. On the other hand, the multiple eruptive phases created many isolated volcanic lithosomes. The isolation offered a favorable seal for the gas reservoir, either horizontally or vertically.Thus, the volcanic rocks directly and perfectly sealed the volcanic gas reservoir. The top surface was the ideal place for developing the gas reservoir. However, the perfect cap rocks were directly developed on the top of the volcanic reservoir. The statistical data of 25 gas wells (that yielded a commercial grade gas flow) indicated that, in 20 gas wells, the direct cap rocks for the gas reservoirs were the volcanic rocks themselves. The possible cause of this phenomenon was that, after they erupted to the surface, the volcanic rocks experienced alteration in weathering and eluviation. The ideal reservoir sequences developed on the top portion of the volcanic rocks. However, in the upper portion of the reservoir sequences, because there were well developed fractures, the influence of the cementing process, filling process and clays formed by weathering were more obvious. Consequently, these effects reduced the physical properties in the topmost portion of the volcanic rocks. Thus, this portion of volcanic rocks directly played the role of cap rock. In the volcanic rocks, the in situ type of accumulative feature lessened the requirement of regional cap rocks for the volcanic gas reservoir. For example, in the Shengping area, although the Deng 2 member did not have well developed mudstone as cap rock, the deep level volcanic gas reservoir was sealed because of the direct cap rock on the top portion of volcanic rock.

13.2.2 *Hydrocarbon Source Rocks Controlled Development of Volcanic Oil and Gas Reservoir*

Because of the independent and isolated nature of volcanic lithosomes, in the volcanic reservoir oil and gas mainly migrated a short distance horizontally. The volcanic petroleum system that contained natural gas was limited within the border of effective source rocks. This phenomenon not only appeared in a rifted basin (such as the Songliao basin), but it also occurred in the Carboniferous volcanic petroleum reservoirs of marine and terrigenous facies in the northern Xinjiang region (Hou and Yang, 2002; Li et al., 2009; Liu et al., 1996; Lou et al., 1998; Hu et al., 1999).

In the deep level of the Songliao basin, there is a faulted basin group that consisted obviously of several speared faulted basins; every faulted basin was an individual petroleum system with its own source area, subsidence center and depositional center.

A large faulted depression contained several depositional centers that controlled the boundary of the gas system. The quality and magnitude of hydrocarbon source rocks determined the scale of the gas reservoir. For example, the Xujiaweizi faulted depression consisted of two depositional centers, which were the Xuxi (western) center and the Xudong (eastern) center (Feng and Liu, 2006; Ni et al., 2009; Yang et al., 2000; Li et al., 1986; 2002). Every depositional center contained a complete depositional system. The coarse grain sediments were deposited near the controlling fault, the hydrocarbon source rocks developed in the central region of the lake and the deposits of delta front facies or the deposits of plain facies occurred on the slope region.

The Xuxi depression was the main contributor to the Xushen 1 natural gas region. Recently, a new well (Xushen 31 well) was drilled on the Xuzhong steep slope region. Even the strata had good physical properties. However, there were no oil and gas shows. This drilling result indicated that the Xuzhong faulted steep slope region was not a primary destination domain for natural gas migration. On the central paleo uplift, the gas testing in the Chang 103 well resulted in a water bed, because this well was far away from the Shahezi hydrocarbon source rock. These exploration results clearly show that the hydrocarbon depression controlled natural gas accumulation (Fig. 13.9). The statistical data indicated that, in the deep level of the Songliao basin, the accumulative location should be near the hydrocarbon source rock of the Shahezi formation and should be near the faults that controlled the depression. These were the basic natural gas accumulative preconditions. Among the 69 wells, 43 wells yielded either a commercial grade gas flow or a low productive gas flow, which represented 62% of the total wells. The distance of these wells from the source rocks was less than 10 km. Secondly, six wells had gas shows, which represented 8.7% of the total wells and the distance of the wells from the source rocks was in the range of 10 – 20 km. Conversely, the wells that were located more than 20 km away from the source rocks did not have any gas shows. On the southern side of the basin, the Changshen 3 well was a failed well because of the long distance from the source rocks. If the distance to the source area was more than 20 km, natural gas could not travel such a long distance to reach the accumulative domain.

In the northern Xinjiang region, the Carboniferous volcanic oil and gas reservoirs were mainly discovered on the Luliang uplift in the Dishuiquan area, on the eastern side of the Zhungeer basin. This area contained two hydrocarbon generating centers: the Dishuiquan hydrocarbon generating center and the

Wucaiwan hydrocarbon generating center. The natural gas reservoirs were distributed along these two hydrocarbon generating centers (Fig. 13.10).

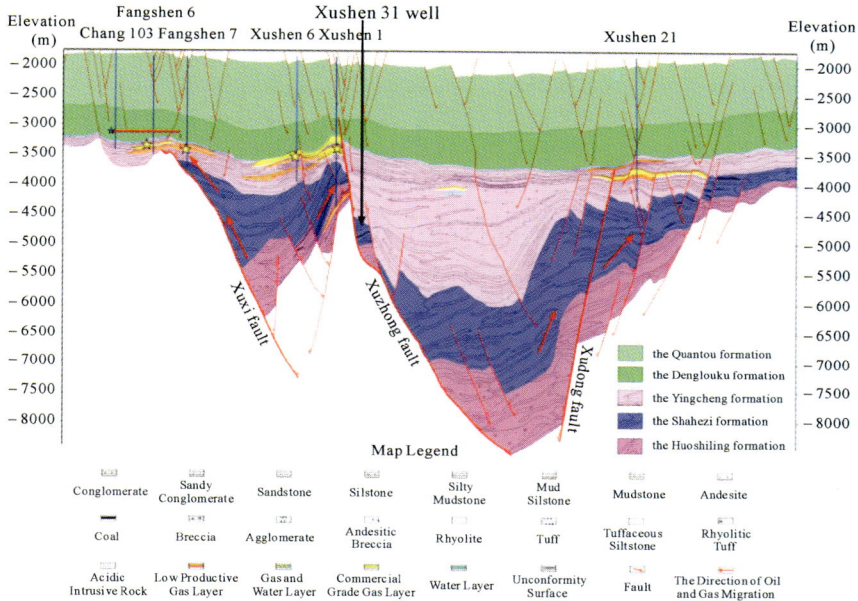

Fig. 13.9.   Cross section of gas reservoir from the Chang 103 well–Xushen 1 well–Xushen 31 well–Xushen 21 well

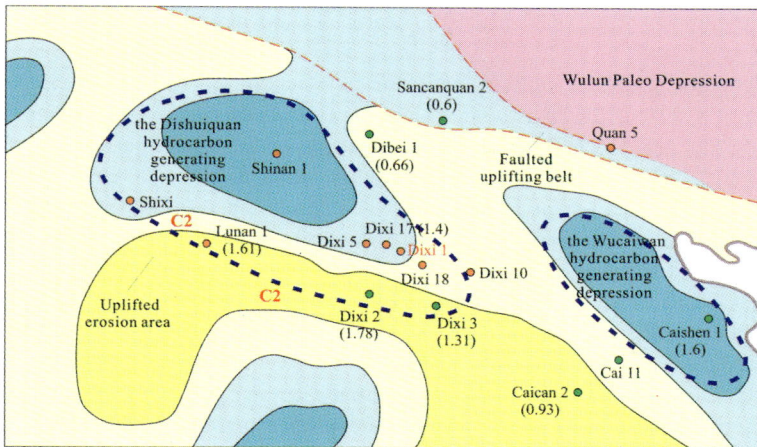

Fig. 13.10.   Distribution map of source rocks and gas reservoir in the Zhungeer basin (according to the Research Institute of Petroleum Exploration and Development of China)

13.2.3 *Faults, Unconformity Surface and Paleo-Uplift Controlled Volcanic Oil and Gas Reservoirs*

The faults controlled the magnitude of the faulted depression. In the Songliao basin, because the faults that controlled the Xujiaweizi depression extended a great distance, had a small dipping angle and a high level of correlation, the faulted depression that was created by these faults covered a large region and was of great thickness. In addition, the hydrocarbon source rocks that developed during the faulted period were of high quality, great gas generating intensity and had abundant resources.

A deeply extended large scale fault controlled the distribution of volcanic rocks. In the Songliao basin, the volcanic rocks in the Yingcheng formation were distributed along the deeply extended large scale faulted belt; in the western faulted depression belt, the volcanic rocks were distributed along the large scale faulted belt at a deep level in an NNE direction. However, in the eastern section, the volcanic rocks in the Xujiaweizi faulted depressed belt were distributed along a large scale fault at a deep level in an NNW direction. In the Xujiaweizi faulted depression, the volcanic rocks were mainly distributed in the three faulted belts of Xuxi, Xuzhong and Xudong, which created three important volcanic accumulative belts for natural gas. Natural gas exploration has already achieved breakthroughs in these volcanic belts. The deeply extended large scale faults worked as migration channels for the fluids at a deep level. Favorable volcanic accumulative belts developed either around deeply extended large scale faults or at the intersection points of multi-direction and multi-phase faults. In addition, the faults benefited the alteration of the volcanic reservoir sequences during the late period. Thus, the discovered natural gas reservoirs of the Xingcheng gas field, Shengping gas field and Xudong gas field (gas reservoir) were all near deeply extended large scale faults.

The intensity of fault activity controlled the accumulative level in the volcanic petroleum reservoir. During the petroleum accumulative period, if the intensity of fault activity was great, the fault could function as the migration channel. On the other hand, after oil and gas accumulated, fault activity could damage the protection of natural gas at a deep level. The combined results of these two kinds of functions stemming from fault activity determined the abundance of oil and gas in the volcanic reservoir.

The unconformity surface controlled the development of the stratigraphic type of oil and gas reservoir. In the Zhungeer basin, the Carboniferous system contained multiple phases of unconformity. The unconformity surface was usually the interface of volcanic rocks–sedimentary rocks; the volcanic (lithologic) stratigraphic type of petroleum reservoir developed under the sedimentary rocks. The multiple sets of favorable reservoir sequences developed under the preconditions of multiple phases of volcanic activity and the weathering and eluviation process. Thus, several sets of unconformity surface related volcanic oil and gas reservoirs developed.

The paleo uplifts that were either in the faulted depression or in the transitional area between two depressions received hydrocarbon supplies from dual directions. These uplifts developed during the early period and they also contained well developed structural traps. Thus, they offered an ideal place for petroleum accumulation, such as in the Songliao basin, the Lefeng low bulge and the paleo central uplift in the Xujiaweizi faulted depression. On the paleo uplifts, all kinds of draped trap were widely developed, which included the stratigraphic overlapping type, the trap of base rock and the structural–lithologic type. These traps offered perfect places for petroleum to accumulate.

13.3   Geochemical Characteristics of Carbon Dioxide in the Volcanic Petroleum Reservoir and the Natural Resources in the Songliao Basin

Carbon dioxide is widely and unevenly distributed in the world. In our current atmosphere, the content of carbon dioxide in the air is 0.037% – 0.038%; however, the content of carbon dioxide in natural gas can be 90% – 99%. A low concentration of carbon dioxide cannot be utilized as a natural resource. According to current technology, the concentration of industrial grade carbon dioxide should not be less than 60%. When the concentration of carbon dioxide is above 90%, the pumped carbon dioxide can be used directly by industry. In 1823, Faraday and Davy in Great Britain were the first scientific team that successfully liquefied carbon dioxide, setting up the beginnings of the commercial exploitation of carbon dioxide. When its concentration reaches a certain level (the content should be larger than 60%), carbon dioxide becomes a precious non-hydrocarbon resource of reasonably high economic value. It is used in crude oil production, the chemical industry, in agricultural, in meteorology, in the environmental industry, health industry and the food and beverage industries (Zhang and Fang, 1996; Zhang et al., 1988). In the USA, Western Europe and Japan, the price of carbon dioxide is occasionally higher than the price of hydrocarbon gas.

In China, carbon dioxide gas fields are mainly distributed in the eastern and southern rifted basins (the latter is mainly located under the sea). At present, there are more than thirty discovered carbon dioxide gas reservoirs (fields), which are located in the Hailaer basin in Inner Mongolia, the Songliao basin, Bohai bay basin, Subei basin, Sanshui basin in Guangdong province, the basin of the Pearl River estuary, Yingge Sea basin, and the Fushan depression in Beibu gulf. From the regional tectonic point of view, these carbon dioxide gas reservoirs are mainly distributed in the faulted belt of the Pacific Rim in eastern China. In these gas reservoirs, all of the carbon dioxide gases have a non-organic originating mechanism. In particular, in recent years pure carbon dioxide gases were discovered in several exploration wells in the Songliao basin.

13.3.1   Geochemical Features of the Carbon Dioxide Gas Reservoir

In the Songliao basin, carbon dioxide gases are widely distributed in the Changling faulted depression, the Dehui faulted depression and the Xujiaweizi faulted depression; the stratigraphic sequences that contain carbon dioxide gas include the strata from the Quaternary to the Yingcheng formation in the Cretaceous system.

The Changling faulted depression is located in the western faulted depression zone in the Songliao basin; several of the carbon dioxide gas reservoirs were discovered in this faulted depression. Carbon dioxide gas reservoirs of high concentration are mainly distributed in the Changling, Qianan and Honggang areas.In the Changling faulted depression, the carbon dioxide gases vertically occur in the strata from a shallow level to a deep level in the Nenjiang formation, Yaojia formation, Qingshankou formation, Quantou formation and Yingcheng formation. In particular, in recent years carbon dioxide gases were discovered in volcanic rocks in the Yingcheng formation in the Changling area; the wells that encountered carbon dioxide gases included the Changshen 2 well, the Changshen 4 well, the Changshen 6 well and the Changshen 7 well. These wells not only contain a high concentration of carbon dioxide gas, but they also have a high production rate; the content of carbon dioxide gas is more than 90%. In the Changshen 6 well, the daily production rate of carbon dioxide gas is 150,646 m^3. Recently, in the Changling faulted depression, the volcanic rocks were found to contain more than $1,500 \times 10^8$ m^3 of carbon dioxide gas within the third order reserve.

In the Dehui faulted depression, the carbon dioxide gas reservoirs are mainly distributed in the Wanjinta area. In Xujiaweizi, carbon dioxide gas reservoirs mainly occur in the volcanic rocks of the Yingcheng formation. In the Changde area, the Dongying formation contains 5.18×10^8 m^3 of the proven carbon dioxide gas reserve. Recently, the highest content (more than 90%) of carbon dioxide gas was discovered in the volcanic rocks in the Xushen 10 well, Xushen 19 well and Xushen 28 well.

Table 13.2 lists major geochemical characteristics of carbon dioxide in the Songliao basin. Because of the $\delta^{13}C_{CO_2} > -8‰$ in the carbon dioxide, and also because the helium isotope was on the heavier side ($R/R_a = 1.9 - 7.2$), this is typical non-organic, mantle originating carbon dioxide. The hydrocarbon gases have a complicated originating mechanism; in addition, the carbon isotope in the hydrocarbon gases is commonly heavier at a deep level (the Denglouku formation and Yingcheng formation), which indicates the non-organic originating mechanism. Conversely, the hydrocarbon gases at a shallow level are associated with a lightweight carbon isotope, which shows that gas is synchronized with the oil layer at a shallow level.

Table 13.2 The geochemical features of CO_2 in high carbon dioxide content wells, Songliao basin

Area	Well number	Lithology	Stratigraphy	Main constituent of gas (%)		Features of Isotope	
				CO_2	CH_4	$\delta^{13}C_{CO_2}(‰)$	R/R_a
	Qianshen 1	Sandstone	K_1q^4	80.09	12.93		
Qianan	Qianshen 8	Sandstone	K_1q^4	85.55	13.39	−3.92	
	Qianshen 10	Sandstone	K_1q^4	80.73	1.95	−3.73	
	Qianshen 11	Sandstone	K_1q^4	95.73	2.18	−5.30	
Gudian	Gu 6	Sandstone	K_1q^4	90.20	4.30		
	Gu 9	Sandstone	K_1q^4	97.05	2.65	−8.44	
Honggang	Hong 81-3-1	Sandstone	K_1f	94.92	2.53	−8.00	
	Hong 77	Sandstone	K_1q^4	96.14	1.71		
	Changshen 4	Volcanic rocks	K_1yc	69.62	22.00	−7.50	2.08
	Changshen 7	Volcanic rocks	K_1yc	77.81	18.56	−5.80	1.90
Changshen	Changshen 2	Volcanic rocks	K_1d	93.98	4.18	−6.70	
	Changshen 2	Volcanic rocks	K_1yc	98.53	0.90	−6.60	4.54
	Changshen 6	Volcanic rocks	K_1yc	98.69	0.41	−6.30	3.78
	Fangshen 9	Volcanic rocks	K_1yc	82.49	15.96	−6.15	
Xujiaweizi	Fangshen 9-1	Volcanic rocks	K_1yc	89.15	9.48	−5.69	
	Xushen 10	Volcanic rocks	K_1yc	90.41	3.76	−4.43	
	Wan 2	Sandstone	K_1q^3	99.02	0.61	−4.04	6.87
	Wan 4	Sandstone	K_1q^3	89.92	9.69	−8.83	
Wujinta	Wan 5	Sandstone	K_1q^3	93.43	3.74	−4.95	6.3
	Wan 5	Sandstone	K_1q^{1+2}	99.48	0.52	−4.60 ~ −6.07	
	Wan 6	Sandstone	K_1q^3	97.77	1.39	−4.31	7.2

The inclusion of the helium isotope in the analysis also indicates that the carbon dioxide gas typically came from the non-organic mantle (Table 13.3).

Table 13.3 The analysis of the helium and argon isotope; samples were from the inclusions

Well number	Depth (m)	Isotope		
		$^3He/^4He\ (\times10^{-6})$	R/R_a	$^{40}Ar/^{36}Ar$
Changshen 1	3,574.7	3.68	2.63	407
Changshen 1-2	3,672	4.34	3.10	1,033
Changshen 103	3,722.5	4.26	3.04	395
Changshen 103	3,725.8	2.92	20.80	1,353
Changshen 3	2,669.5	3.52	2.51	790
Laoshen 1	2,574	1.92	1.37	2,485
Deshen 5	1,718	2.72	1.94	554
Deshen 5	2,580	2.03	1.45	363
Tuoshen 1	2,104	2.46	1.76	2,419
Hong 73	1,873	4.36	3.11	579
Qian 198	2,278.6	7.33	5.24	715
Wan 17	2,095.83	4.45	3.18	1,511
Wan 22	2,168	1.30	0.93	897
Gu 7	1,550	2.96	2.12	501
Gu 9	1,638	5.21	3.72	501

13.3.2   Distribution of the Carbon Dioxide Gas Reservoir

Using a study of the relationship between the discovered carbon dioxide gas reservoirs and secondary order faults, it was noticed that, in the Songliao basin, the carbon dioxide gas reservoirs are mostly distributed in the zone of the deeply extended strike-slip fault lying in an NE-NNE direction. The carbon dioxide that originated in the mantle traveled upward into the formations to form carbon dioxide gas reservoirs. At first, the carbon dioxide gas migrated along the faults in the lithosphere. Secondly, gas moved within the faulted system in the crust. These crust faults worked as connective channels that linked the faults in the lithosphere and the basement. Thirdly, the carbon dioxide gas passed through the fault in the basement. Lastly, the carbon dioxide gas traveled through small faults of the next order in the overlying rock above the basement faults. The gravity and magnetic data indicate that, for the Songliao basin, the faults (that are oriented in an NE-NNE direction) strongly influenced the basement of the basin, revealing that the deeply extended NE-NNE oriented faults were the important channel that connected with the deep mantle. Because the Songliao basin was influenced by the Pacific subduction (that was in an NW direction) during the late period, the faulted belts lying in an NE direction had sinistral strike-slips and connected to the fluids at a deep level. As a result, these faults controlled the distribution of carbon dioxide gas.

In the basin, the discovered carbon dioxide gas reservoirs were mostly distributed along the deeply extended, strike-slip faulted belt that was oriented in an NE-NNE direction, such as in the Changling faulted depression (Fig. 13.10), the Dehui faulted depression and Gulong faulted depression. During the end of the Palaeogene, the Dehui faulted depression went through powerful compressive–shear tectonic movement. Accordingly, an F1 strike-slip fault was created in the core region of the Nongan-Yangdachengzi compressive anticlinal belt with an extended distance of more than 33 km. This strike-slip fault controlled the distribution of carbon dioxide gas reservoirs.

13.3.3   Resource of Non-Organic Carbon Dioxide

In the Songliao basin, the volcanic gas system contained 651.8×10^8 m^3 of the proven carbon dioxide gas reserve and 500×10^8 m^3 of the controlled carbon dioxide gas reserve (Fu et al., 2001).

In the Songliao basin, the estimation of the carbon dioxide gas resource in the volcanic gas system utilized the volcanic rock degassing method and the reserve abundance analogy method.

The Volcanic Rock Degassing Method: As listed in Table 13.4, the major gaseous constituent that volcanic rocks adsorbed was carbon dioxide gas. In particular, the original pores were well developed in the rocks of the volcanic vent facies, the subvolcanic facies and the invasive volcanic facies at a shallow level.

The porous nature of these rocks allowed them to absorb a huge amount of carbon dioxide gas and to create the primary gas reservoir.

Table 13.4   The gas contents in various volcanic rocks and igneous rocks

Type of rock	Composition of gas (%, extrusive rock)				Type of rock	Composition of gas (%, intrusive rock)			
	CO_2	H_2O	CO	CH_4		CO_2	H_2O	CO	CH_4
Basalt	64.6	19.9	13.8	2.5	Gabbro	16.6	77.4	3.7	2.3
Andesite	74.6	13.6	10.6	2.2	Diorite	12.0	80.6	3.2	4.3
Rhyolite	57.1	30.6	12.3		Granite	33.4	59.5	5.2	2.3
					Diabase	44.7	50.7	3.2	1.9

The result of the volcanic degassing test (Table 13.5) showed that every gram of volcanic rock could absorb at least 0.02 ml of carbon dioxide gas. In the Songliao basin, a mere 20,966 km^2 of volcanic rocks occurred in the Yingcheng formation. According to the result of the degassing test, the quantity of carbon dioxide gas that might be absorbed by volcanic rocks could reach $4,000 \times 10^8$ m^3. If considering the additional carbon dioxide gas that might be absorbed by the Huoshiling formation and the volcanic rocks in other strata, in the Songliao basin the carbon dioxide resource is enormous.

Table 13.5   Test results of degassing experiments for volcanic rocks

Stratigraphy	Lithology	T_{CO_2}	Volatile content (ml/g)	CO_2 (ml/g)	CO_2 (W%)
K_1d	Light gray color, acid tuff	2.07	0.0299	0.0218	0.4429
K_1yc	Brownish red color, acid volcanic breccia	2.24	0.0357	0.0286	0.5611
K_1yc	Light green color andesite	2.76	0.0391	0.0357	0.7008
J	Grayish green color, basaltic andesite	2.35	0.0554	0.0493	0.9675
J	Grayish green color, dacite	2.42	0.0513	0.0445	0.8734
J	Basaltic trachyandesite	3.11	0.0484	0.0388	0.7629
J	Grayish green color, andesitic basalt	5.52	0.0789	0.0706	1.3870
J	Grayish black color, andesitic basalt	1.38	0.0462	0.0379	0.7442
J	Brown color tuff	1.21	0.0433	0.0357	0.7009
J_2	Purple red color, altered andesite	2.07	0.0458	0.0396	0.7769
Ks	Multi-colored, volcanic breccia	1.04	0.0364	0.0337	0.6496
Ks	Dark gray color mudstone	1.04	0.0551	0.0485	0.952
Ks	Dark grayish green color tuff	2.07	0.0742	0.0693	1.361
K_1d_1	Dark color mudstone	2.68	0.0542	0.0430	0.844

In the volcanic rocks, the carbon dioxide gas resource per unit area can be calculated by computing the discovered carbon dioxide gas reservoirs by a statistical method. The analogy method is also an effective method for preliminarily working out the quantity of the carbon dioxide gas resource in the volcanic rocks in the Songliao basin. For example, the Changling faulted depression contains thirty-nine volcanic lithosomes that cover 4,000 km^2 of land approximately. Among these volcanic lithosomes, seven of them are connected with the deeply extended faults and they cover 1,184 km^2 of land. According to the numbers of the controlled reserve and the forecast reserve from the Changshen 6 well, and considering the size of the distribution area of the volcanic rocks, the magnitude of the proven carbon dioxide gas reserve abundance is 7.14×10^8 – 16.81×10^8 m^3/km^2. In Changde East, the quantity of the proven carbon dioxide gas reserve is 65.18×10^8 m^3, the gas containing area covers 13.6 km^2 of land, and the scale of the reserve abundance is 4.79×10^8 m^3/km^2. During the calculation of the carbon dioxide gas resource, instead of calculating the resource abundance, reserve abundance is used as an alternative. In the Changling faulted depression only, the quantity of the carbon dioxide gas resource could reach $5,000 \times 10^8$ m^3. In the deep level of the Songliao basin, the volcanic rocks (that were distributed along an NE-NNE oriented, deeply extended strike-slip fault) covered 8,004 km^2 of land. According to the preliminary estimation, in the Songliao basin the enormity of the carbon dioxide gas resource should be $38,339 \times 10^8$ m^3.

References

Chen, J.L., Cai, X.Y., Lin, C.H., et al., 1999. The tectonic feature of rifted basin and its episodic evolution in the northern Songliao basin. Acta Petrolei Sinica, 20(4):14-18.

Deng, J.F., Zhao, G.C., Zhao, H.L., Luo, Z.H., Dai, S.Q., Li, K.M., 2000. The tectonic combination and orogeny of the Yanshanian movement in volcanic rocks–a deep development in eastern China. Geological Review, 46(1): 41-48.

Feng, Z.H., Liu, W., 2006. The type of mechanism of the deep level natural gas in the Xujiaweizi faulted depression. Natural Gas Industry, 26(6):18-20.

Feng, Z.H., Ren, Y.G., Wang, C., et al., 2003. The inclusions in deep level volcanic sequences and related natural gas accumulation in the Songliao basin. Natural Gas Geoscience, 14(6):436-441.

Fu, X.F., Fu, G., Lu, Y.F., 2001. The deep level sealing condition and its control of the natural gas accumulation in the northern Songliao basin. Offshore Oil, 109:42-48.

Guo, Z.Q., 1996. The function of a deep fault for petroleum development in the Songliao Basin. Acta Petrolei Sinica, 17(3):68-74.

Guo, Z.Q., Wang, X.B., Liu, W.L., 1997. The accumulative characteristics of non-organic gas in the Songliao basin. Science in China, Series D, 27(2): 144-148.

Hou, Q.J., Yang, Y.F., 2002. The developmental mechanism of non-organic gas and related exploration in the Songliao basin. Natural Gas Industry, 22(3): 5-10.

Hu, Y.S., Cao, L.Y., Tian, F., 1999. The depositional evolution of the songliao rifted basin group and the model of stratigraphic sequences. Petroleum Explorationist, 4(2):37-39.

Li, C.L., Jin, H., Gao, J.Y., Liu, R.H., Zhu, J., Fan, B.J., 2009. Natural gas accumulation at a deep level in the Songliao basin. Petroleum Geology and

Recovery Efficiency, 16(1):1-4.

Li, J.T., Mo, S.G., He, Z.J., et al., 2004. The occasion of the left-shaped strike-slip movement in the northern section of the greater Xing'an range and its control of tectonic evolution and reconstruction since the Mesozoic era in northeastern China and neighboring area. Earth Science Frontiers, 11(3):157-168.

Li, S.T., Yang, S.G., Wu, C.L., et al., 1987. The late Mesozoic rifted basin and faulted depression in northeastern China. Science in China, Series B, 17(2):185-195.

Lin, Q., Ge, W.C., Sun, D.Y., et al., 1998. The tectonic significance of the Mesozoic volcanic rocks in northeastern China. Chinese Journal of Geology, 33(2):129-139.

Liu, D.L., Chen, F.J., Guan, D.F., et al., 1996. The creation and development of the Songliao basin and the geo-dynamics in the lithosphere. Chinese Journal of Geology, 31(4):397-408.

Liu, X.W., Dang, Y.M., et al., 1998. The depositional characteristics of the deep level in the area east of the Daqing placanticline, in the Northern Songliao Basin. Petroleum Geology & Oilfield Development in Daqing, 17(5):16-18.

Lou, X., Sun, F.J., Shao, M.L., et al., 2003. Deep level coal bed methane and the geochemical characteristics of the source rocks in Songliao basin. Petroleum Exploration & Development, 36(3):339-346.

Ni, Y.Y., Dai, J.X., Zhou, Q.H., et al., 2009. The evidence of non-organic natural gas and estimated reserves in the Xujiaweizi faulted depression. Petroleum Exploration & Development, 36(1):35-45.

Shu, P., Ding, R.X., Ji, X.Y., et al., 2007. The geochronology of volcanic zircon in the reservoir sequences of the Qingshen Gas Field, Songliao basin. Acta Petrologica ET Mineralogica, 32(1):15-20.

Wan, P.J., 2003. The types and characteristics of volcanic lithofacies and the significance of volcanic reservoir sequences. Journal of Jilin University, (4).

Wang, H.Z., Yang, S.N., Li, S.T., 1987. The development of the cenozoic basins in eastern china and neighboring area and the tectonic development of continental margins. Acta Geologica Sinica, 57(3):213-223.

Wang, P.J., Chen, S.M., Liu, W.Z., et al., 2003. The relationship of the volcanic lithofacies and the volcanic reservoir sequences in the Songliao basin. Oil and Gas Geology, 24(1):18-23.

Wang, P.J., Feng, Z.Q., Liu, W.Z., et al., 2008. The Exploration of Lithologic Type of Natural Gas Reservoir in the Volcanic Lithofacies in the Basin. Science Press, Beijing.

Wang, Z.H., Luo, X., Li, J.K., et al., 2008. The distribution forecast of effective hydrocarbon source rocks at a deep level, in the northern Songliao basin. Natural Gas Geoscience, 19(2):204-209.

Yang, Y.F., Zhang, Q., Huang, H.P., et al., 2000. Non-organic gas and its accumulative model in the Xujiaweizi Faulted Depression, in the Songliao Basin. Earth Science Frontiers, 7(4):523-533.

Zhang, Y.Q., Zhao, Y., Dong, S.W., et al., 2004. The characteristics of tectonic evolution in the early cretaceous rifted basin in eastern China and neighboring area. Earth Science Frontiers, 11(3):123-133.

Zhao, D.L., 1996. The formation mechanism of storage space in the volcanic reservoir sequences and the contents of oil and gas. Geological Review, supplementary issue:37-43.

Zhao, H.L., Deng, J.F., Chen, F.J., et al., 1996. The mesozoic volcanic rocks on the southeastern margin of the songliao Basin and the tectonic background of the basin. Earth Science (Journal of China University of Geosciences), 21(4): 421-427.

Zhao, W.Z., Zhou, C.N., Feng, Z.Q., et al., 2008. The geologic features of deep level volcanic gas reservoir in the Songliao basin and related evaluation technique. Petroleum Exploration & Development, 35(2):129-142.

14

Examples of Petroleum Exploration

14.1 Ordovician Petroleum Reservoir in the Tarim Basin

The geological location of the Ordovician buried hill is on the Lunnan low bulge at the Tabei uplift in the Tarim basin (Fig. 9.9). The Lunnan Ordovician buried hill is next to the Caohu depression on the eastern side; it is adjacent to the Halahatang depression on the western side; it utilizes the Luntai fault as its northern boundary and it is next to the Manjiaer depression on the southern side (Liang, 2008; Yang et al., 2008).

The Lunnan Ordovician buried hill occupies a major portion of the Lunnan paleo-uplift, which contains a favorable exploration area of 2,450 km^2. After more than 20 years of drilling exploration in the Ordovician buried hill type of oil reservoirs, the discovered geological oil reserve is more than 1×10^8 t. At the same time, apart from isolated enriched oil reservoirs, the entire buried hill contains oil. Therefore, this area not only has great potential for petroleum exploration and development, but it is also an important domain for in-depth exploration in the Tarim basin.

14.1.1 History of Petroleum Exploration

The exploration history of the Lunnan Ordovician buried hill type reservoir included three phases.

During the early 1990s, the first exploration phase set out to investigate the higher position of the Lunnan buried hill. In 1990, a borehole test for the Ordovician system at large intervals was carried out at the Lunnan 15 well. The well test obtained 1.5 m^3 of oil that could be converted into a daily production rate of 30 m^3/d. After the well was completed, the results of the well test for individual layers indicated a dry well. At same time, the New Star Co. completed the Sha 23 well on the eastern wing of the Tahe #3 structure. The well test also showed a dry well. During 15 – 18 September 1990, a borehole, wall mounted, well completion test was carried out at an interval of 5,224 – 5,478 m in the Sha 23 well; the drawdown pressure was 19.613 MPa and the quantity of obtained fluids (natural gas and a small amount of crude oil) from the test was 0.56 m^3. The report of the well test indicated that "the oil and gas were unevenly distributed in the lithosome of the tested interval that had unsaturated oil content and limited storage capacity."

During the middle of the 1990s, the second exploration phase explored the southern slope of the uplift. The Tahe oil field was discovered. This oil field was a large scale oil and gas field with more than a hundred million tonnes of petroleum reserves, which was discovered on the northern side of the Tarim basin by the New Star Co. during the "Ninth Five-Year Plan". The major portion of this petroleum field consisted of the Ordovician oil and gas reservoirs. In the exploration blocks #3, #4 and #6, the total proven petroleum reserve was 12,950.8×10^4 t of oil equivalent.

The third exploration phase has been in progress from the end of the 1990s to the present; the exploration target is the southwestern slope (the Lungu area). Recently, a 3D seismic survey has been carried out. In addition, based on the structural interpretation, the forecast of reservoir sequences and an estimation of fractures are in progress (Fig. 14.1).

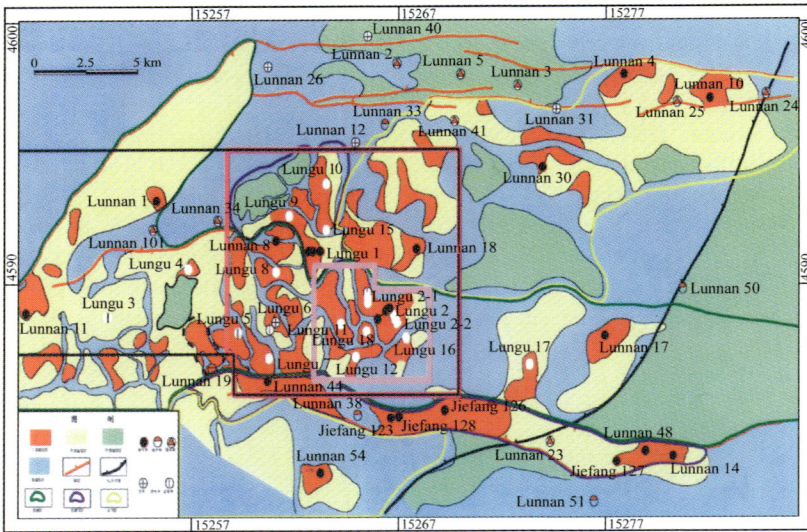

Fig. 14.1.   Evaluation map of the Ordovician buried hill type reservoirs in the Lunnan area

14.1.2   *Condition of the Hydrocarbon Source Rock*

In the Lunnan area, oil and gas in the Ordovician system originated in the marine environment; the petroleum mainly came from the Cambrian–Ordovician systems. In addition, this region was next to the largest hydrocarbon generating center (the

Manjiaer depression) of the Tarim basin (Shi et al., 2005; Zhao et al., 2007). The characteristics of the hydrocarbon generating depression included a longer generating period and multiple phases of oil and gas supply. In the Manjiaer depression, in the Cambrian–Lower Ordovician basins in adjacent areas and in the slope facies, the hydrocarbon source rocks could provide 77.9×10^8 t of resources from the early Hercynian tectonic period and they could offer 53.2×10^8 t of resources from the late Hercynian tectonic period. Although the oil and gas reservoirs that developed during the early Hercynian tectonic period were severely damaged (in this area, the oil and gas might be accumulated on the southern side of the Tahe area that was covered by the Middle–Upper Ordovician strata), the oil and gas reservoirs that developed during the late Hercynian tectonic period and the Himalayan period were well preserved (the oil and gas reservoirs that developed during the late Hercynian tectonic period were only partially damaged). The huge amount of oil and gas resources from these two phases (89.0×10^8 t) offered sufficient resources for the large scale oil and gas reservoirs in this area.

14.1.3 Characteristics of Reservoir Sequences

In the Lunnan area, the unconformity surface was positioned between the Ordovician strata and the Silurian strata. In the Middle–Upper Ordovician series, the lithology of continental shelf facies consisted of sandy mudstone interbedded with limestone. In the Lower Ordovician series, the depositional environment went from the open platform to the platform margin; the lithology was reasonably pure carbonate rocks. At present, the drilling exploration is targeting the limestone that developed in the Yijianfang formation–Yingshan formation in the Lower Ordovician series.

All of the Ordovician reservoir sequences consisted of carbonate rocks that included pelmicrite, grainy pelmicrite, pelmicritic grain stone, sparitic grain stone, dolomitic limestone and karst rock. According to the identification results from the thin sections, the frequency of occurrence for every type of limestone is listed in Table 3.2. This table revealed that, in this area, pelmicrite, grainy pelmicrite, sparitic grainstone and micritic grainstone had occurred more frequently, thus forming the main types of carbonate rocks in this area with an average frequency of 13.8% – 35.7%.

In the Lunnan area, the porosity in the Ordovician reservoir sequences is in the range of 0.04% – 10.82%; among the samples 31.4% contain a porosity of < 0.5%, 81% of the samples contain a porosity of < 1.0%. The permeability is in the range of $(0.001 – 252) \times 10^{-3}$ μm^2 and the average permeability is 0.73×10^{-3} μm^2. Among the samples, 64.3% of the samples have a permeability of < 50.1×10^{-3} μm^2, 92.6% of the samples have a permeability of < 1×10^{-3} μm^2 (Fig. 14.2).

The following are the primary features of the Ordovician carbonate reservoir sequences. (1) Because the matrixes had low porosity and poor permeability, sufficient storage spaces were hard to create. (2) The dissolutive pores, holes and fractures were effective spaces for storage and permeation. (3) The storage capacity had strong heterogeneity in both horizontal or vertical directions. (4) There were several types of reservoir; among these reservoirs, the most important types were the fractural–porous type and the fractural–dissolutive type. (5) The porous type of reservoir with shoals facies developed in an isolated area.

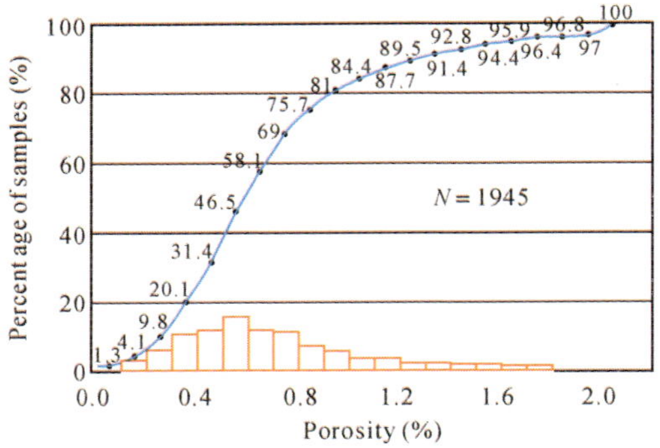

Fig. 14.2. Distribution map of reservoir porosity in the Ordovician system

14.1.4   *Types of Oil and Gas Reservoirs*

Overall, in the Lunnan area, the Ordovician oil and gas reservoir developed on the Akekule uplift; the type of trap was a large scale karst–unconformity trap that was filled with multiple phases of oil and gas with different degrees of maturity. In the Tahe oil field, several factors were related to the seals of the Ordovician oil reservoir; moreover, if combined with a buried hill (or remaining hills) and fault, the composite type of seals and traps might be created (Fig. 14.3).

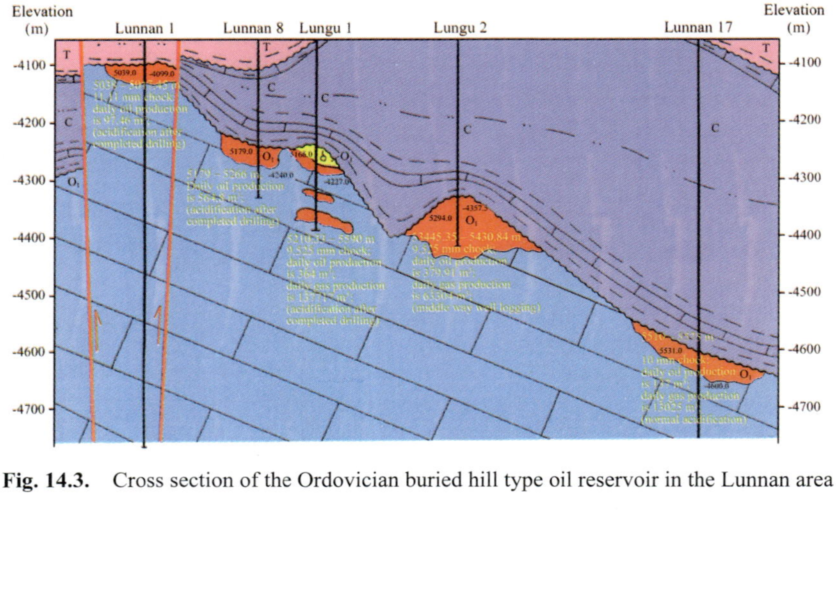

Fig. 14.3.   Cross section of the Ordovician buried hill type oil reservoir in the Lunnan area

14.1.5   Nature of Fluids

The fluid materials in the Ordovician oil and gas reservoir include condensate oil, normal crude oil, heavy oil, natural gas and formation water. In a vertical direction, the nature of the crude oil has a clear distinction and zonation. In the top portion, the density of the crude oil is 0.8186 g/cm^3; this number indicates that the oil is condensate oil. The result of pTV analysis for fluids from wells specifies condensate gas. In the upper portion, the surface density of the oil is 0.8297 g/cm^3, which indicates light crude oil; in the lower portion, the surface density of the oil is 0.853 g/cm^3, which indicates normal crude oil; in the base portion (below 5,545.66 m in the T301 well), the surface density of the oil is 0.966 g/cm^3, which indicates heavy viscous crude oil. In the natural gas, the average content of methane is 84.47%, the content of heavy hydrocarbon is 12.61%, the average densities of both substances are 0.68 g/cm^3, which indicates the transitional characteristic of condensate gas and oil soluble gas.

The diversified accumulative characteristics were closely related to the accumulative phase, preservative condition, the large scale fault and the development level of the reservoir entity.

14.2   Reef Flat Type of Gas Reservoir in the Changxing–Feixianguan Formations in the Longgang Area, Sichuan Basin

The location of the Longgang area is limited to the east of Yilong County, the west of Changping County, the south of Bazhong County and the north of Yingshan County; additionally, the Longgang area covers 7,780 km^2 of land; the topography includes low mountains and hills. The altitude of the mountains is in the range of 500 – 600 m. From the geological point of view, the position of the regional tectonic structure is located at the junction of a gentle folding belt in the central Sichuan basin and a lower flat structural region on the northern side of the Sichuan basin. Overall, it is a monocline that dipped in a northwesterly direction with poorly developed faults. During the Upper Permian–Lower Triassic (when the Changxing–Feixianguan formations were deposited), the Longgang area was located on the western side of "the Kaijiang–Liangping sea trough" (Fig. 14.4). In the Longgang area, the exploration targets include the oolite beach in the Lower Triassic Feixianguan formation and the bio-reef gas reservoir in the Upper Permian Changxing formation. Recently, excellent gas shows have occurred in the Jurassic system at a shallow depth in the Upper Triassic Xujiahe formation and in the Middle Triassic Leikoupo formation.

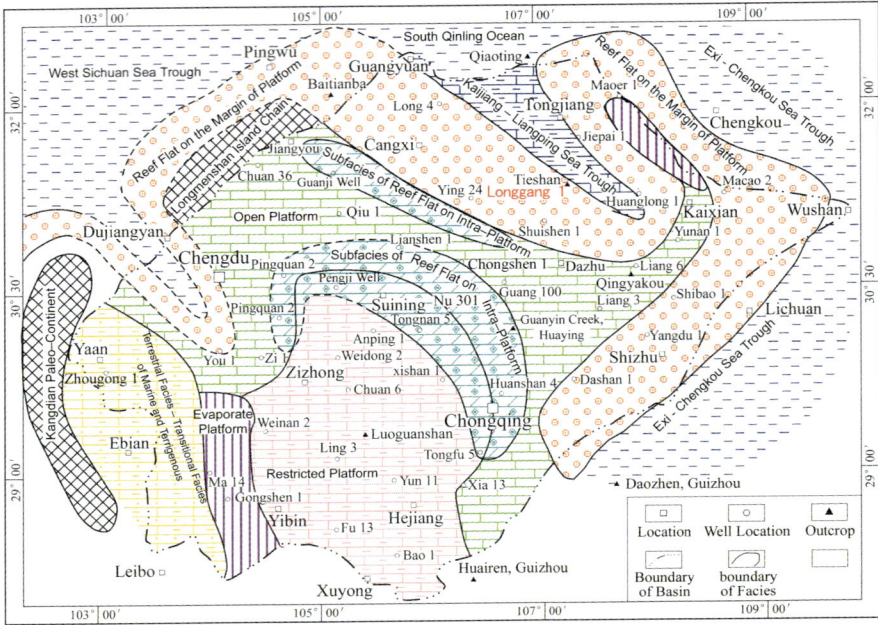

Fig. 14.4.   Distribution map of reef flat reservoir in the Changxing formation–Feixianguan formation, Sichuan basin

14.2.1   Exploration History

In the Longgang area, the initial geological survey was carried out in 1929. After the founding of the People's Republic of China, several companies and research institutes conducted different types of investigations in the Longgang area, which included a petroleum geology survey, comprehensive structural research, a seismic survey and drilling exploration at a shallow depth. Commercial grade gas flows were obtained in the Jurassic Daanzhai formation and the Lianggaoshan formation. However, the overall exploration level was limited in the preliminary stage. Only a few wells were drilled into the Permian strata.

In 2003, in emphasizing the preliminary exploration work, the old seismic data of the Kaijiang–Liangping sea trough was reprocessed. The bio-reef (litho-uplift) and oolitic beach that were distributed around the sea trough were investigated and favorable reservoir sequences in the bio-reef and oolitic beach, which were distributed on the platform margin along the "Kaijiang–Liangping sea trough", were forecast. In addition, the Changxing bio-reef (litho-uplift) in the Longgang area was discovered. This bio-reef is oriented in a northwest–southeast direction; the length of the bio-reef is 40 km and the width is 6 – 4 km. The thickness of the Upper Permian strata is more than 330 m and covers 184.28 km^2 of land. Upwards, the oolitic beach in the Feixianguan formation contained distinctive characteristics and covered 253 km² of land.

In 2006, the Longgang 1 well was drilled, which was an especially risky exploration. In this well, two sets of porous type reservoir sequences that occurred in the bio-reef of the Changxing formation and in the oolitic beach of the Feixianguan formation were discovered. Well tests for both reservoir sequences indicated excellent gas flows with a high production rate. This breakthrough initiated the exploration of the reef flat type of gas reservoir in the Longgang area. In 2007, individual exploration projects were set up. According to the idea of "completed deployment, overall control, step by step implementation and dynamic adjustment", 2,600 km of 2D seismic surveys and 2,600 km² of 3D seismic surveys were completed. In addition, 11 exploration wells were also drilled.

In 2008, in order to understand the size of the reserves in the reef and oolitic beach facies, exploration activity was aimed at the major portion of Longgang. In an effort to obtain new discoveries, the east and west wings were eliminated from the drawing board; 1,000 km² of 3D seismic surveys and 27 exploration wells for two exploration phases were planned. On account of the drilling exploration, the magnitude of the reef flat type of gas reserve in the Longgang area has been further confirmed. Since 2008, exploration activity has been continuing.

14.2.2   Hydrocarbon Source Rocks

In the Longgang area, hydrocarbon source rocks include argillaceous rock and carbonate rock in the Lower Permian series, coal measures and carbonate rocks in the Upper Permian series and carbonate rocks in the Feixianguan formation in the Lower Triassic series. The result of a thermal maturation study of the source rocks revealed that, in the Longgang area, the intensity of hydrocarbon generation in the Permian system was $40 \times 10^8 - 60 \times 10^8$ m³/km². In Longgang and its peripheral areas, the Permian strata generated 31.8×10^{12} m³ of natural gas in total. Thus, the preconditions for the development of a large scale natural gas reservoir were in place.

14.2.3   Reservoir Sequences

The primary reservoir sequences included carbonate rocks of bio-reef facies in the Changxing formation and oolitic beach facies in the Feixianguan formation (Fig. 14.5). In the Longgang area, the bio-reef and oolitic beach appeared on the platform margin and their distributions were controlled by the Kaijiang–Liangping sea trough. Extensive occurrence was the characteristic of the reef and oolitic beach. According to the most recent forecast of reservoir sequences, the thickness of reservoir sequences in the oolitic beach facies is greater than 10 m. These types of reservoir sequences covered 1,133.4 km^2 of land in total. In addition, the thickness of reservoir sequences in the bio-reef facies is also greater than 10 m. These types of reservoir sequences covered 424.66 km^2 of land in total.

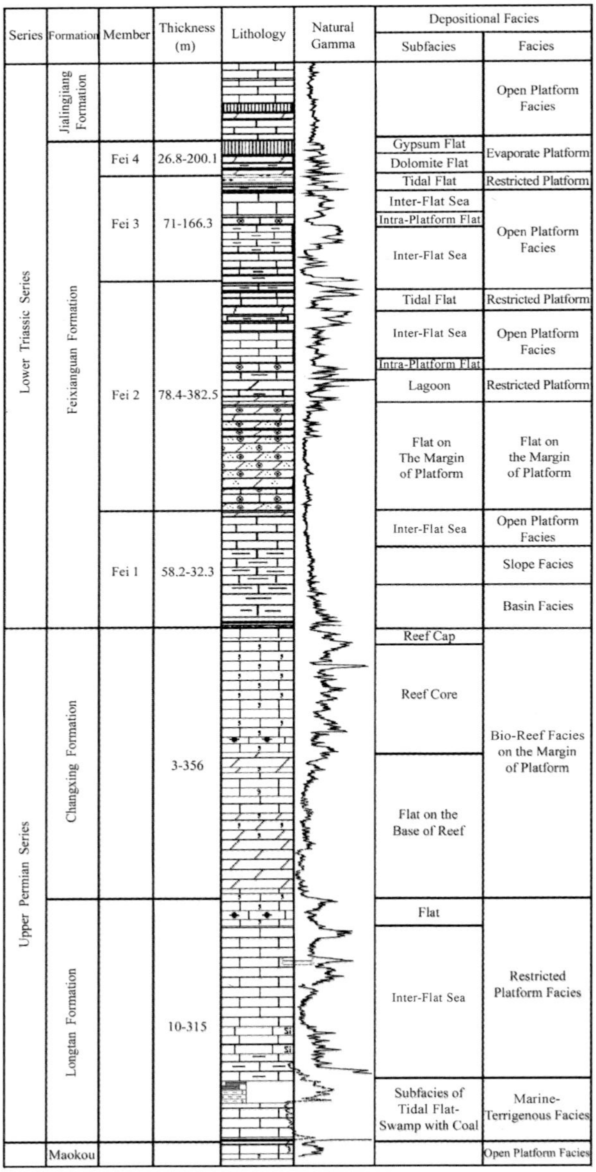

Series	Formation	Member	Thickness (m)	Lithology	Natural Gamma	Depositional Facies	
						Subfacies	Facies
Lower Triassic Series	Jialingjiang Formation						Open Platform Facies
	Feixianguan Formation	Fei 4	26.8-200.1			Gypsum Flat	Evaporate Platform
						Dolomite Flat	
						Tidal Flat	Restricted Platform
		Fei 3	71-166.3			Inter-Flat Sea	Open Platform Facies
						Intra-Platform Flat	
						Inter-Flat Sea	
		Fei 2	78.4-382.5			Tidal Flat	Restricted Platform
						Inter-Flat Sea	Open Platform Facies
						Intra-Platform Flat	
						Lagoon	Restricted Platform
						Flat on The Margin of Platform	Flat on the Margin of Platform
		Fei 1	58.2-32.3			Inter-Flat Sea	Open Platform Facies
							Slope Facies
							Basin Facies
Upper Permian Series	Changxing Formation		3-356			Reef Cap	
						Reef Core	Bio-Reef Facies on the Margin of Platform
						Flat on the Base of Reef	
	Longtan Formation		10-315			Flat	Restricted Platform Facies
						Inter-Flat Sea	
						Subfacies of Tidal Flat-Swamp with Coal	Marine-Terrigenous Facies
	Maokou						Open Platform Facies

Fig. 14.5. Stratigraphic column of the Changxing formation–Feixianguan formation in the Longgang area

In the Longgang area, the completed exploration wells encountered bio-reef and oolitic beach reservoir sequences in various situations. The reservoir sequences of bio-reef facies and oolitic beach facies contained dolomite and dolomitic limestone; the thickness of reservoir sequences and the physical properties varied. In the Changxing formation, the thickness of reservoir sequences is in the range of 5.38 – 67.38 m; the porosity in reservoir sequences is 3.4% – 9.8%. In the Feixianguan formation, the thickness of reservoir sequences is in the range of 5.86 – 57.38 m; the porosity in reservoir sequences is 3.2% – 11.41% (Table 14.1). In the Changxing formation, there were three developmental cycles of the bio-reef. The bio-reef type of reservoir sequences mainly developed in the upper portion of every cycle; the reservoir sequences in the 3rd cycle were the most extensively distributed. High quality dolomitic reservoir sequences mainly occurred on the high land of the platform margin during the early period; on the inner platform, the reservoir sequences primarily contained oolitic limestone.

Table 14.1   Discovered reef flat reservoirs during drilling exploration in the Longgang area

Formation	Well number	Reservoir thickness (m)			Porosity（%）		Major reservoir lithology
		Total thickness	Single layer thickness	Total layers	Range	Mean value	
Feixianguan	Longgang 1	55.6	2.1–43.4	5	2.57–12.1	11.41	Dolomite
	Longgang 2	42.4	1.4–20.8	5	2.54–10.61	9.16	Dolomite
	Longgang 3	14.25	0.8–8.1	7	2.11–14.59	10.32	Dolomite
	Longgang 6	44.1	2.2–20	5	3.18–8.59	5.63	Calcareous Dolomite
	Longgang 7	29.3	1.6–8.88	6	3.31–11.8	4.94	Dolomitic Limestone
	Longgang 8	5.86	0.25–4	4	2.08–8.43	6.47	Dolomitic Limestone
	Longgang 18	20.75	3.38–9.5	4	3.33–4.80	3.81	Dolomitic Limestone
	Longgang 001-1	47.65	1.1–22.1	9	2.65–11.89	8.51	Dolomite
	Jianmen 1	19.25	0.9–15.1	4	3.09–4.68	3.2	Dolomitic Limestone
	Longgang 9	57.38	0.8–6.9	6	3.31–11.8	3.75	Dolomitic Limestone
	Longgang 11	27.63	1.4–9.1	5	2.75–6.21	3.55	Limestone
Changxing	Longgang 1	29.5	0.8–13.4	9	2.11–5.85	4.21	Dolomite
	Longgang 2	67.38	9.4–23.1	7	2.54–6.14	4.9	Dolomite
	Longgang 3	15.86	1.3–10.8	3	4.98–6.98	7.03	Dolomite
	Longgang 6	10.8	1.5–7.6	3	2.49–3.82	4.26	Calcareous Dolomite
	Longgang 7	25.4	0.5–6.75	7	2.06–13.24	9.84	Dolomite
	Longgang 8	30.88	5.5–18	3	3.94–5.85	5.23	Calcareous Dolomite
	Longgang 18	33.25	4.75–10.38	5	2.52–6.21	4.26	Dolomitic Limestone
	Longgang 001-1	16.63	1.75–11.38	3	2.14–3.68	3.4	Dolomitic Limestone
	Jianmen 1	31.75	0.13–7.63	9	2.18–6.17	4.63	Calcareous Dolomite
	Longgang 9	5.38	0.5–4.9	2	2.21–3.54	3.42	Dolomitic Limestone
	Longgang 11	35.5	3.3–10.9	6	2.51–7.90	4.98	Dolomite

The porous nature of the reservoir sequences: the Changxing formation mainly contained intra-crystalline (dissolutive) pores, karst caves and micro fractures with strong heterogeneity. The Feixianguan formation contained three porous types that included intra-granular (intra-crystalline) dissolutive pores in the oolitic dolomite, intra-granular (inter-granular) dissolutive pores in dolomitic oolitic limestone and the casting mode type of pores in the oolitic limestone. Every type of pore could provide primary storage space in any well. Among these types, intra-granular (intra-crystalline) dissolutive pores in the oolitic dolomite and intra-granular (inter-granular) dissolutive pores in dolomitic oolitic limestone occurred on the platform margin. The casting mode type of pores in the oolitic limestone appeared on the inner platform margin. In the reef and oolitic types of reservoir sequences, the dolomite possibly developed in the buried environment. Dolomitization mainly happened at the junction of the reef complex and echinoderm beach.

14.2.4 *Types of Gas Reservoirs*

The Longgang bio-reef and oolitic beach type of gas reservoirs are made of several lithologic traps in the background of a large size monocline; the overall characteristics are multiple gas and water interfaces with multiple pressure systems. Various lithologic traps were not only superimposed vertically, but they could also be connected with each other horizontally. The expanded pattern of the reservoir entity revealed that the Longgang gas reservoir was made of multiple lithologic gas traps. Under the control of reservoir distribution and influenced by the hydrocarbon charging condition and by the structural background, multiple lithologic gas traps were not only superimposed vertically, but they were also connected with each other horizontally. Inside a single gas reservoir, the gas was in the higher position and the water appeared at a lower level.

In the Feixianguan formation, the lower reservoir sequences have six gas reservoirs that are of the oolitic beach type, Longgang 2, Longgang 001-3, Longgang 26, Longgang 6, Longgang 27 and Jianmen 1. The upper reservoir sequences contain two gas reservoirs that also belong to the oolitic beach type, Longgang 001-1 and Longgang 6. In the Changxing formation there are 11 gas reservoirs that belong to the bio-reef type, which are Longgang 8, Longgang 2, Longgang 001-3, Longgang 001-1, Longgang 1, Longgang 28, Longgang 26, Longgang 6, Longgang 27, Jianmen 1 and Longgang 11.

14.2.5 *Nature of Fluids*

In the Longgang area, the natural gas in the bio-reef and oolitic beach type of reservoirs is dry gas; the methane content is more than 90%, the content of H_2S is in the range of 3.99 – 108 g/m^3. Usually, the content of H_2S is in the range of 30 – 70 g/m^3. The quality of the natural gas is better than that in the same kind of gas reservoir in the northeastern Sichuan basin.

In the Longgang area, the hydrocarbon source supplied coal-related gas to the bio-reef and oolitic beach type of gas reservoirs. Also, the natural gas in the bio-reef and oolitic beach type of reservoirs demonstrated the characteristics of coal-related gas. This phenomenon may be associated with the occurrence of the Upper Permian coal layers. In the Longgang 1 well and the Longgang 2 well, the values of the carbon isotopes from methane and ethane are almost the same; the value of the carbon isotope from methane is –29‰ and the value of the carbon isotope from ethane is –23‰. According to the value of the carbon isotope, the natural gas should be of the coal-related gas type (Table 14.2).

Table 14.2 Characteristics of carbon isotopes in natural gas from the Longgang 1 well and Longgang 2 well

Well number	Sample name	Well depth (m)	$\delta^{13}C_{PDB}$ (‰)	Sampling
Longgang 2	Methane	5,989.6 (T_1f)	–29.55	1st time MDT sample
	Ethane		–23.23	
	Carbon Dioxide		–0.13	
	Methane	6,008 (T_1f)	–29.75	
	Ethane		–28.19	
	Methane		–29.79	2nd time MDT sample
	Ethane		–21.91	
	Methane		–29.36	
	Ethane	6011 (T_1f)	–23.88	
	Carbon Dioxide		–5.62	
	Methane		–30.4	
	Ethane		–21.98	
	Carbon Dioxide		–4.84	
Longgang 1	Methane	6,055–6,072 (T_1f)	–29.31	
	Ethane	6,076–6,124 (T_1f)	–23.15	
	Methane	6,202–6,204 (P_2ch)	–29.44	
	Ethane		–22.68	

14.3 Cretaceous Volcanic Gas Field in the Xujiaweizi Faulted Depression, the Songliao Basin

The Xujiaweizi faulted depression is one of the large depressions on the southeastern side of a deeply buried tectonic entity in the northern Songliao basin (Li et al., 2006; Ren et al., 2004). The geographical position of the Xujiaweizi faulted depression is located within the city limits of Daqing, in Heilongjiang province (Fig. 14.6). The surface settings are grassland and farmland with gentle topography; the altitude on the surface is in the range of 143 – 160 m above sea level.

At present, the network of the 3D seismic survey basically covers the main portion of the Xujiaweizi faulted depression, 5,058 km^2 (19 subdivisions) in total. In this area, drilling exploration has reached an advanced level with 129 exploration wells. Commercial grade gas flows have been obtained in 59 wells. Up until 2007, several natural gas fields had been discovered, which were Changde, Shengping, Wangjiatun, and Xushen. Because the total proven reserve of natural gas is 2,457×10^8 m^3, in this area the discovery rate of natural gas was relatively low.

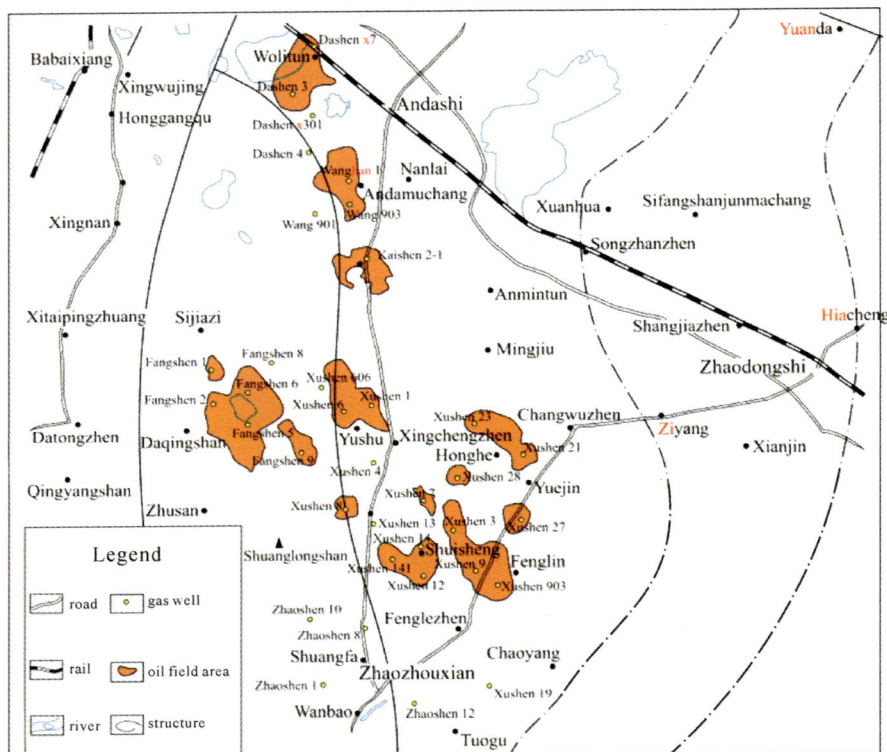

Fig. 14.6. The geographical location of gas reservoirs in the Yingcheng formation, in the Xujiaweizi faulted basin

14.3.1   Exploration History

Before 1988, the exploration in the Xujiaweizi faulted depression emphasized the evaluation of the depression. The drilling exploration mainly focused on structural traps that were located on the paleo uplift, at the edge of the depression. Natural gas reservoirs in the sandstone in the Shengping formation, Changde formation and Denglouku formation were successively discovered. During the period of "the Eighth Five-Year Plan" and "the Ninth Five-Year Plan", by comprehensive study, people further understood the potential of the natural resources in the Xujiaweizi faulted depression. The exploration targets shifted to the inner depression and aimed at the volcanic lithologic trap and the litho-stratigraphic traps in the sandy conglomerate.

In 1988, the Xingcheng nose-like structure (an uplift) was discovered by 2D seismic survey. This structure was the southward extending portion of the Shengping uplift. The extended length of the structure is 30km and the width (in an E-W direction) is 5 – 8 km; the volcanic lithologic traps were well developed in this nose-like structure. This uplift was located inside an ideal hydrocarbon source region. In addition, the overlaying strata were well developed mudstone of the Denglouku formation. Therefore, the best petroleum system was developed in this area. The "uplift in a depression" benefited natural gas accumulation. The Xingcheng nose-like structure had the geological conditions for the development of a large size gas reservoir.

In 2001, based on newer 3D seismic data, based on seismic attribute analysis for the target area, by way of studying seismic facies and seismic data reversion, the exploration target was described in detail. The selected drilling target was decided on the northern side of the Xingcheng nose-like structure and the Xushen 1 well was designed. The well test for volcanic rocks in the Yingcheng formation obtained a highly productive gas flow with a daily production rate of 53.0×10^8 m^3; the well test for the conglomerate obtained a daily production rate of 5.48×10^8 m^3/d. Thus, the Xushen gas field was discovered. Natural gas exploration in the volcanic rocks in the Yingcheng formation has developed into a new phase.

After the discovery of the Xushen gas field, engineers not only designed the network for a 3D seismic survey in the Xingcheng area, but they also planned and drilled a series of exploration wells on the southern and western sides of the Xingcheng structural belt. At same time, exploration wells were drilled in the Shengping structure and the Wangjiatun structure in order to investigate the volcanic rocks in the Yingcheng formation. These exploration wells were successful. In 2005, the total proven gas reserve was $1,018.68 \times 10^8$ m^3, which was discovered in the exploration blocks of Xushen 1, Xushen 8, Xushen 9 and Xushen 2-1.

In 2006, in order to achieve a great exploration victory, synchronized with the evaluation of the proven gas reserve in the Xushen gas field so as to completely estimate the size of the natural gas reserve in the Xujiaweizi faulted depression and in order to uncover the natural gas accumulative pattern of special reservoir sequences in the volcanic rocks and sandy conglomerate, favorable exploration targets, that included the Fengle low bulge on the southern side, the Xudong slope belt on the eastern side and the Anda secondary depression on the northern side, were selected. The preliminary exploration was implemented in promising traps. At the end of 2007, the exploration in the volcanic rocks achieved success. Breakthroughs were achieved in the Xudong, Anda and Fengle areas successively; the region containing a natural gas reserve had been developed. In the Fengle low bulge, the well test for the volcanic reservoir sequences in the Yingcheng formation achieved a commercial grade gas flow with a daily production rate of 51,749 m^3 in the Xushen 12 well. In the Wangjiatun area, the well test for the volcanic reservoir sequences in the Yingcheng formation achieved a daily production rate of 202,190 m^3 in the Wangsheng 1 well. On the Xudong slope belt, the well test for the volcanic reservoir sequences in the Yingcheng formation achieved a highly productive commercial gas flow with a daily production rate of 206,446 m^3 in the Xushen 21 well. In 2007, the Daqing Oil Field Co. Ltd reported that the additional proven gas reserve was $1,015.04 \times 10^8$ m^3, which was discovered in the exploration blocks of Xushen 21, Wangshen 1, Xushen 12 in the Xushen gas field, on the northern side of the Songliao basin.

14.3.2   *Hydrocarbon Source Rock*

In ascending order, the faulted depression in Xujiaweizi contained four sets of hydrocarbon source sequences that occurred in the Huoshiling formation, Shahezi formation, Yingcheng formation and Denglouku formation. Because it developed during the time that the faulted depression was created, the thickness of the Shahezi formation was great and the organic abundance in the strata was high. These preconditions determined that the Shahezi formation was the most important hydrocarbon source rock at a deep level.

The dark colored mudstone of the Shahezi formation was widely distributed. In general, the thickness of the mudstone was more than 400 m. The depositional center consisted of three subsided centers that were oriented in an NNE direction and occurred in an echelon pattern. Strata of great thickness occurred in the areas of Xujiaweizi–Xingshan, south of the Songzhan, and the Zhaozhou region; the greatest thickness could be up to 1,000 m. In the Shengping area and the Shengping–Songzhan area, the Shahezi formation contained coal layers with reasonable thickness; the greatest thickness was 105.5 m. Geochemical analysis showed that, in the Shahezi formation, the primary type of kerogen was type III and there was also type I kerogen and type II kerogen in isolated areas. The abundance of organic material was high; the average TOC is 1.52%, the average S_1+S_2 is 2.02 mg/g, the average chloroform "A" is 0.021%. The test results of vitrinite reflectance revealed that, in the Shahezi formation, the diversity of thermal maturation was evident. The value of R_o was in the range of 1.70% – 3.56%, which indicated that the organic materials were at a highly mature–post mature stage.

14.3.3   *Characteristics of Reservoir Sequences*

In the Xujiaweizi faulted depression, the deeply buried volcanic reservoir sequences were mainly distributed in the upper portion of the Huoshiling formation in the Jurassic system and in the first member of the Yingcheng formation in the Cretaceous system. In some areas, the volcanic reservoir sequences also occurred in the third member of the Yingcheng formation. The lithology of reservoir sequences included rhyolite and andesite of extrusive facies and the volcanic breccia of explosive facies. In the volcanic reservoir sequences, the reservoir physical properties were less influenced by the buried depth. However, they were controlled by the original lithology, litho-facies and tectonic movement during the late period. Therefore, these volcanic rocks functioned as important reservoir sequences at a deep level (Zhao et al., 2006).

● Types of Volcanic Rocks

Based on observation of the drilling core, identification of the thin section and analysis of the chemical constituents, the lithology in the Huoshiling formation was mainly of the basic type, the intermediate type and the acidic type of volcanic rock. The volcanic rock included basaltic andesite, rhyolite, andesite, dacite and andesitic pyroclast. The lithology in the first member of the Yingcheng formation was the intermediate–acidic type of volcanic rock that included rhyolite, dacite and acidic pyroclast. The third member of the Yingcheng formation mainly contained the intermediate–basic type of volcanic rock that included basalt and andesite basalt.

Every area had its own type of volcanic rock; for example, the intermediate–basic type of volcanic rock occurred in the Anda area, rhyolite appeared in the Shengping area, tuff turned up in the Xingcheng area. In the area south of Xingcheng, the types of volcanic rock increased; trachyte and basalt came into view.

● Volcanic Facies

According to the types of volcanic lithology that were discovered in the wells of the Songliao basin, and considering the results from different researchers, in this area the volcanic facies can be categorized into 5 facies and 15 subfacies. Overall, in the northern Songliao basin, most of the volcanoes (that were buried at a deep level) belong to extrusive facies with an intermediate energy level and well developed facies belts. Fig. 14.7 displays the eruption model and the distribution patterns of volcanic rock facies.

● Characteristics of Storage Space in the Volcanic Reservoir

Considering the characteristics of storage space in the volcanic reservoir sequences in the Xujiaweizi faulted depression, and based on the volcanic drilling cores that were recovered from the wells, the storage spaces in the volcanic rocks can be divided into three types, according to their originating mechanism, which are original pores, secondary pores and fractures. The primary pores include air holes, amygdaloidal holes, intra-crystalline pores and the dissolutive pores of phenocryst. The most important fractures include structural fractures and cracked fractures. A good reservoir sequence basically contained these six types of pores and fractures that were discussed above. These pores and fractures were developed in rhyolite, andesite, agglomerate and crystal tuff that occurred in the weathered eluvial belt. When pores and fractures combined together to create storage spaces, the volcanic reservoir sequences were altered subsequently into more sufficient reservoir sequences.

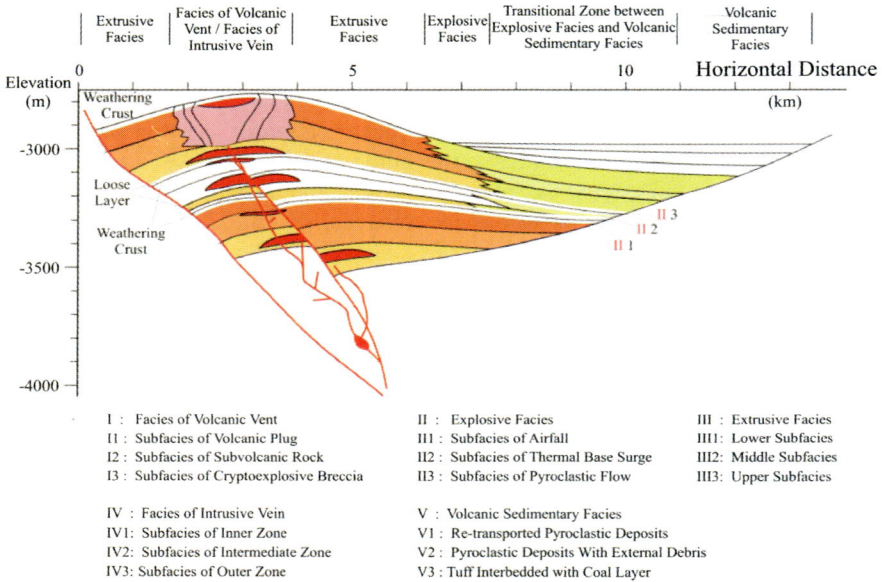

I :	Facies of Volcanic Vent	II :	Explosive Facies	III :	Extrusive Facies
I1 :	Subfacies of Volcanic Plug	II1:	Subfacies of Airfall	III1:	Lower Subfacies
I2 :	Subfacies of Subvolcanic Rock	II2:	Subfacies of Thermal Base Surge	III2:	Middle Subfacies
I3 :	Subfacies of Cryptoexplosive Breccia	II3:	Subfacies of Pyroclastic Flow	III3:	Upper Subfacies
IV :	Facies of Intrusive Vein	V :	Volcanic Sedimentary Facies		
IV1:	Subfacies of Inner Zone	V1 :	Re-transported Pyroclastic Deposits		
IV2:	Subfacies of Intermediate Zone	V2 :	Pyroclastic Deposits With External Debris		
IV3:	Subfacies of Outer Zone	V3 :	Tuff Interbedded with Coal Layer		

Fig. 14.7. Model of volcanic eruption in the Xujiaweizi faulted basin (according to the Research Institute of Daqing Oil Field, 2006)

● Physical Properties of Volcanic Reservoir Sequences

In the Xujiaweizi area, the physical properties of the volcanic reservoir sequences were diversified with strong heterogeneity. The porosity is in the range of 0.5% – 27.5%, the common distribution of porosity is in the range of 7% – 10%. The permeability is in the range of $(0.01–191)\times10^{-3}$ μm^2 and the regular distribution of permeability is in the range of $0.1\times10^{-3}–0.5\times10^{-3}$ μm^2. The reservoir sequences contained well developed fractures that combined with all kinds of pores to create storage spaces and migration channels (Zhou et al., 2001). The lithology of the volcanic rock and related facies was closely associated with physical properties. In general, the permeability in welded tuff is larger than 1×10^{-3} μm^2; conversely, in the agglomerate and tuff, because fractures were underdeveloped, the permeability was commonly smaller than 1×10^{-3} μm^2. Overall, the pyroclast contained the best reservoir physical properties; in addition, the acidic lava contained better reservoir physical properties than those in the intermediate lava and basic lava. Moreover, in the upper subfacies of the extrusive facies and the subfacies of the pyroclastic flow of explosive facies, the porosity and permeability were reasonably good; on the other hand, in the middle subfacies of the extrusive facies and the subfacies of the thermal base surge of the explosive facies, the porosity and permeability were relatively poor.

14.3.4   Types of Natural Gas Reservoirs

In China, the discovered natural gas reserves can be categorized into four large types in the first order classification, and then they can be further divided into six subtypes in the second order classification. In the Xujiaweizi area, there are five different types of natural gas reservoir, which include the volcanic faulted nose (faulted anticline)–lithologic composite gas reservoir, the volcanic lithologic gas reservoir, the volcanic stratigraphic overlapping–lithologic composite gas reservoir, the volcanic draped anticlinal–lithologic composite gas reservoir, the volcanic faulted anticlinal gas reservoir (Fig. 14.8). Among these types of gas reservoir, the volcanic faulted nose (faulted anticline)–lithologic composite gas reservoir is the most common type. For example, in the Shengshen 2-1 gas reservoir (that belongs to the type of volcanic faulted anticlinal–lithologic composite reserve), the volcano erupted many times along the large size, deeply extended, basement fault. A pattern of neo-faulted anticlines developed on the top of the volcanic rocks. In this type of gas reservoir, because the fault controlled one side of the gas reservoir, the volcanic rocks controlled the other side of the reservoir. Thus, the composite gas reservoir developed, which was controlled by the structure and lithology.

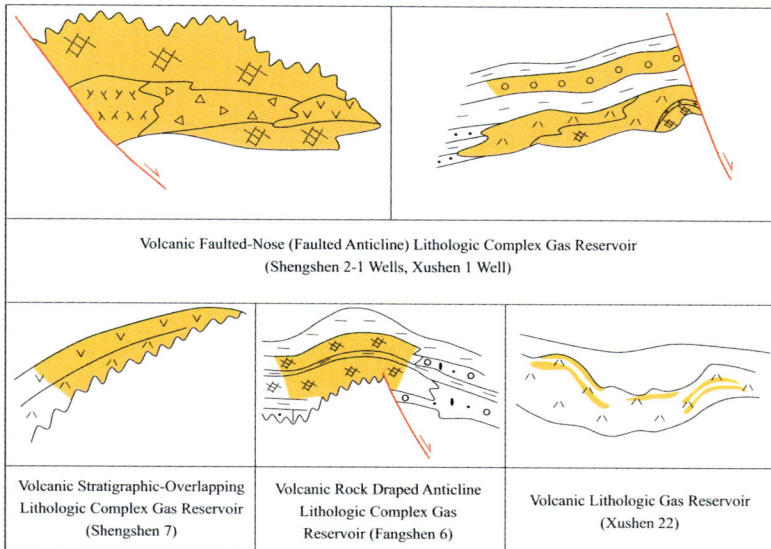

Volcanic Faulted-Nose (Faulted Anticline) Lithologic Complex Gas Reservoir
(Shengshen 2-1 Wells, Xushen 1 Well)

Volcanic Stratigraphic-Overlapping Lithologic Complex Gas Reservoir (Shengshen 7)

Volcanic Rock Draped Anticline Lithologic Complex Gas Reservoir (Fangshen 6)

Volcanic Lithologic Gas Reservoir (Xushen 22)

Fig. 14.8.   Classification of natural gas reservoir in volcanic rocks

Up to now we have discovered 95 volcanic gas reservoirs in the Xujiaweizi faulted depression. Among these gas reservoirs, 40 belong to the faulted nose (faulted anticline)–lithologic composite gas reservoir type. This type of gas reservoir had a broad distribution, which was not only associated with the fault creating mechanism, but which also developed near the special kinds of fault that controlled the depression.

14.3.5   Nature of Fluids

● Constituents of Natural Gas

In the Xingcheng area, because natural gas mainly contains methane (the content of methane is 90% – 95%) with a large aridity coefficient (above 0.98), this is called dry gas. Usually, the contents of C_2–C_5 are in the range of 2.5% – 3% and the content of CO_2 is in the range of 0 – 5%. In addition, there are small amounts of helium gas. In natural gas, the carbon isotopes of methane and ethane are relatively heavy.

In the Wangjiatun–Shengping area, most of the natural gas is dry gas; the content of methane is in the range of 90% – 95% and the content of heavy hydrocarbon of C_{2+} is less than 3%. Moreover, there are small amounts of helium gas and CO_2 gas.

● Geochemical Features of Formation Water

In the Yingcheng formation, the formation water mainly consisted of the type $NaHCO_3$. The salinity of formation water was relatively high, if comparing the salinity in the Quantou formation and the Denglouku formation. The average salinity is 33,709 mg/L. The $CaCl_2$ type of formation water was distributed in the areas of the Fangshen 7 well, Fangshen 9 well, Shengshen 4 well, Songshen 1 well and Songshen 401 well; the distribution of salinity is in the range of 17,566 – 275,106 mg/L and the value of salinity indicates excellent preservation conditions in these areas.

● Formation Pressure

In the Yingcheng formation, the formation pressure is in the range of 16.95 – 40.62 MPa, which is higher than the pressure either in the Denglouku formation or in the Quantou formation. The pressure coefficient is in the range of 0.99 – 1.14, which indicates a normal pressure–high pressure range. Inside the Xujiaweizi faulted depression, several locations have high pressure, which include the area from the Dashen 1 well–Wangshen 1 well, Shengshen 2 well district, Xushen 1 well district, Fangshen 6 well district, Xushen 3 well district, Xushen 9 well district and Zhaoshen 10 well district (Fig. 14.9).

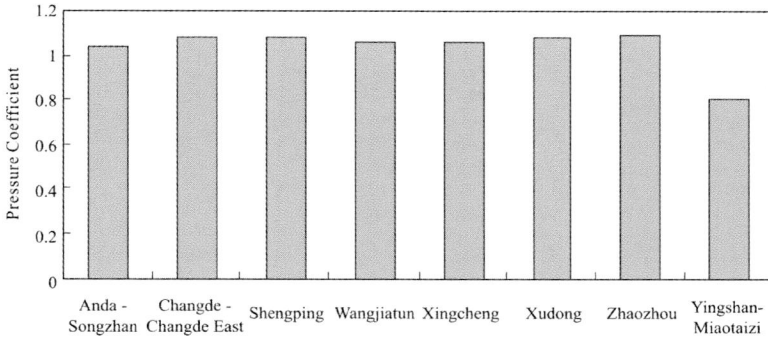

Fig. 14.9.   Diagram of formation pressures in the Yingcheng formation in the northern Songliao basin

14.4   Carboniferous Volcanic Gas Field in the Zhungeer Basin

The Zhungeer basin is one of the large size petroleum basins in China; it is located in the northern Xinjiang region and it covers 130,000 km^2 of land. From the geological point of view, the Zhungeer basin is positioned at the intermediate section of the Kazakhstan plate, the Siberian plate and the Tianshan folding belt. It is a compressed and superimposed basin that experienced a long period of compressive tectonic movement in the late Paleozoic–Mesozoic–Cenozoic eras.

The Zhungeer basin contained the classic type of dual layers. The basement was located beneath the Carboniferous system, the cap rocks developed above the Carboniferous system. The Carboniferous strata occurred immediately above the basement and contained well developed volcanic rocks. Because it was exposed on the surface for a long period, the strata had been altered into sufficient reservoir sequences. In the Ludong–Wucaiwan area, a few oil and gas fields of significant size have been discovered (Jin et al., 2008; Liu et al., 2009). In particular, in recent years exploration of the Carboniferous volcanic rocks has achieved important breakthroughs (Wang et al., 2008; Zhang et al., 2005).

The administrative division of the Ludong–Wucaiwan area is under the supervision of the Altay Prefecture, in the Kazak Autonomous Prefecture of Ili. In the exploration area, the surface landscape is a desert with wide-ranging altitudes and widely distributed sand dunes. The surface altitude is in the range of 340 – 920 m. The tectonic position is limited to the north of the Baijiahai uplift, east of the Mobei uplift, west of Kelameili Mountain and south of the Wulungu depression in the Zhungeer basin. This tectonic position contains the Dinan uplift, Dibei uplift, Dishuiquan depression and Wucaiwan depression; the former three structures are located on the Luliang uplift in the central region and the latter one is located on the Dongbu (or the eastern) uplift (Fig. 14.10). The exploration concentrated on the Dinan uplift where the Kelameili gas field was discovered.

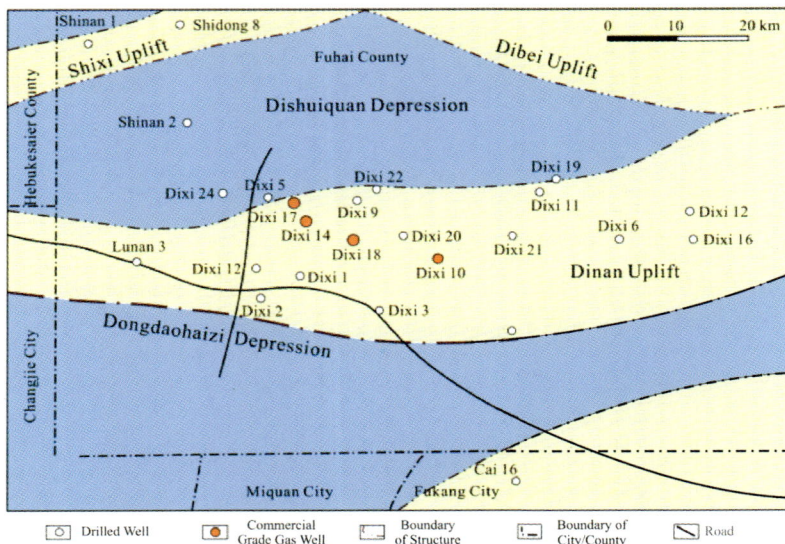

Fig. 14.10.   Location map of the Ludong–Wucaiwan area

14.4.1   Exploration History

The Ludong–Wucaiwan area was one of the earliest exploration targets in the central region. During the 1980s, a preparatory 2D seismic survey had been carried out in this area. In the 1990s, a detailed 2D seismic survey was performed. By the end of 2007, a 2D seismic survey of 15,851.5 km was completed. The density of the seismic network was 1 km×1 km – 7 km×10 km and the 3D seismic network covered 5,480.8 km^2 of land. In this area, 110 exploration wells were drilled in total. Among these wells, 36 of them obtained oil and gas flows. The exploration primarily targeted the strata in the Cretaceous, Jurassic, Triassic, Permian and Carboniferous systems. Oil and gas shows were obtained in all of the targets.

In the early 1990s, drilling exploration was carried out to study the Carboniferous system. In 1993, the drilling of the Dixi 1 well was initiated. In September 1999, for testing the strata of the Carboniferous system, an oil test was performed at an interval of 3,650 – 3,665 m in the Dixi 5 well; the test result for natural gas was 10,740 m^3/d. In addition, the test result for water was 32.03 m^3/d. In April 2004, a production experiment was performed in the Dixi 10 well. During this experiment, the daily production rate of oil was 6.78 m^3/d and the daily production rate of natural gas was 202,390 m^3/d. Thus, a Carboniferous condensate gas reservoir was discovered in the Dixi 10 well district. In 2005, the discovery of condensate natural gas in the Carboniferous system in the Dixi 10 well district revealed a geological gas reserve of 20.30×10^8 m^3.

In 2006, a comprehensive study of natural gas exploration was conducted in the Carboniferous system, on the Dinan uplift. Carboniferous condensate natural gas was discovered in the Dixi 14 well district, Dixi 17 well district and Dixi 18 well district. On the Dinan uplift along the Dishuiquan fault (from west to east) four natural gas reservoirs were discovered in the Carboniferous system. These gas reservoirs were located in the Dixi 17 well district, Dixi 14 well district, Dixi 18 well district and Dixi 10 well district. In addition, isolated, residual oil and gas wells were discovered in the Dixi 20 well and the Dixi 21 well.

In the Wucaiwan depression, the Caican 1 well was located in the middle of the Wucaiwan nose-like structure. A low condensate gas flow was discovered in the Bashan formation in the Carboniferous system; the well test recorded a daily production rate for natural gas of 2,840 m^3/d. The Cai 25 well was located on the Wucaiwan anticline. The well test for the Bashan formation obtained commercial grade oil and gas flows and the daily production rate for oil was 7.5 t/d while the daily production rate for natural gas was 7,156 m^3/d and the daily production rate for water was 5.06 m^3/d. In 2007, Carboniferous gas reservoirs were discovered in the Caishen 1 well and the Cai 53 well. These discoveries indicate that the Carboniferous system had the preconditions for developing the internal episodic type of oil and gas reservoirs. Up to now, four oil and gas fields in the Ludong–Wucaiwan area have been discovered, with a proven natural gas reserve of 51.25×10^8 m^3.

14.4.2   Hydrocarbon Source rocks

In the Dixi and Wucaiwan areas, the primary hydrocarbon source rocks included the Pingdiquan formation in the Permian system and the Dishuiquan formation in the Carboniferous system. The hydrocarbon source rocks in the Permian system contained good types of organic materials with a lower level of thermal mutation. On the other hand, in the Ludong–Wucaiwan area, the Carboniferous hydrocarbon source rocks mainly consisted of dark colored mudstone and the coal measure type of hydrocarbon source rocks also occurred. The abundance of organic materials was high; the average TOC in mudstone was 1.48%, the average TOC in carbonaceous mudstone was 15.53%, and the average TOC in coal measures was 43.78%. The level of maturation was high; the Ro was around 1.5% and most of the organic materials were of type II$_2$–type III. In the study area, these hydrocarbon source rocks were the primary source rocks that produced dry gas.

The Carboniferous hydrocarbon source rocks were widely distributed and they covered 8,000 km^2 of land. Apart from the Kelameili mountain front, almost the entire region contained the same type of source rock (Fig. 14.11). The source rocks were mainly distributed in three locations that were the Dongdaohaizi depressed area, Wucaiwan depressed area and Suosuoquan depressed area. Toward the center of the depression, the thickness of source rock gradually increased. However, toward the uplift, the thickness of source rock decreased. The Dongdaohaizi depressed area is the primary source area for the region. In the most extreme case, the thickness of the source rock represented 90% of the total thickness of the strata. Clearly, this was a major hydrocarbon generating center for the region. Near the location of the Dixi 6 well–Dixi 2 well, the thickness of dark colored mudstone represented more than 90% of the total thickness of strata. Because strong hydrocarbon generation occured in the broad area west of the Dixi 11 well and south of the Sancan 1 well–Lu 26 well, this area became a favorable exploration target. Highly productive gas wells of Dixi 5, Dixi 10 and Dixi 14 were located in this area. Obviously, the source rocks influenced the gas reservoir. Additionally, the area east of the Quan 1 well and the Wucaiwan depression also contained hydrocarbon generating centers; the exploration potential in these areas needs to be further evaluated.

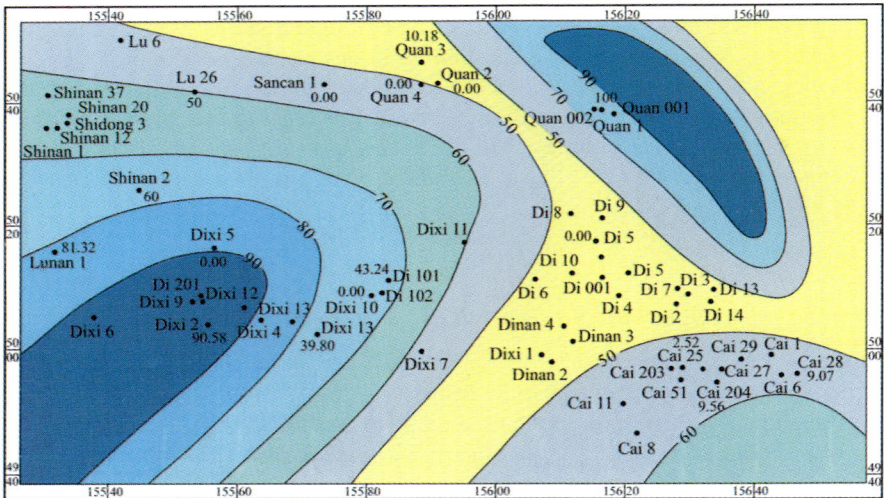

Fig. 14.11.   Distribution map of hydrocarbon generating center in the Ludong–Wucaiwan area (percentage of dark mudstone in the strata)

14.4.3   *Reservoir Sequences*

● **Volcanic Lithology and Related Litho-Facies**

In the Ludong–Wucaiwan area, the Carboniferous volcanic rocks mainly contained pyroclast of explosive facies. The lithology included tuff and volcanic breccias and also extrusive facies, which included andesite, dacite, basalt and rhyolite. In the area of the Wucaiwan–Baijiahai uplift, the most common type of rock was the intermediate type of andesite, followed by the basic type of basalt. The acidic type of rhyolite and dacite was mainly distributed in the Dixi 10 well district.

● **Storage Space and Reservoir Physical Properties**

In the Ludong area, the major types of storage space included micro fractures and secondary pores. The type of storage space was not related to the type of lithology. The main types of storage space included fractures, dissolutive pores in the matrix, partially filled air holes in the breccias (or litho-clast), air holes, dissolutive pores of phenocryst, dissolutive pores of matrix, dissolutive pores of the minerals that developed in the amygdaloidal holes. The Wucaiwan area mainly contained secondary pores and micro fractures and the types of storage space were closely associated with the types of lithology. Andesite and basalt mainly contained fractures, partially filled air holes, air holes, dissolutive pores of phenocryst and dissolutive pores of the minerals that grew in the amygdaloidal holes. Volcanic breccias mainly contained dissolutive pores of matrix and fractures, tuff and clastic lava mainly contained fractures, dissolutive pores of matrix and other types of dissolutive pores. Imaging logging intuitively demonstrated that, in the Wucaiwan area, the pores, holes and fractures were well developed in the volcanic reservoir sequences. The secondary pores were developed by fresh water eluviation under supergenesis. At the high position of a structure, because the eluviation was strong and the filling process was weak, the secondary pores were well developed. Conversely, at the low position of a structure, most of the pores that were created by the dissolution were filled up by chlorite and calcite.

In the Ludong–Wucaiwan area, the Carboniferous volcanic reservoir sequences had strong heterogeneity; porosity and permeability were diversified. The porosity was in the range of 0.2% – 30.1% and the average porosity was 7.8%. Permeability was in the range of 0.001×10^{-3} – 145.69×10^{-3} μm^2 and the average permeability was 3.1×10^{-3} μm^2. The characteristics of physical properties indicated low–intermediate porosity and low permeability. However, andesite, volcanic breccias, dacite and rhyolite had relatively good physical porosity; the average porosity was 9.1%. In comparison, basalt and tuff had poor physical properties; the average porosity was 7.5%. The figures for physical properties evidently indicated that, in the volcanic reservoir sequences, physical properties are closely associated with lithology. The reservoir sequences that either had good or poor physical properties could occur in the explosive facies and the extrusive facies. However, good physical properties appeared in the volcanic breccia that was located near the crater in the explosive facies. In the Cai 27 well, the alternating andesite contained secondary fractures and holes. In the Cai 34 well, andesitic volcanic breccias contained pores and holes. In the Cai 204 well, andesite contained fractures and holes.

14.4.4   Types of Gas Reservoir

The entire Dinan uplift was a natural gas accumulative region that contained three types of traps. In correlation, three types of gas reservoir were also developed in this region, which were the faulted–stratigraphic composite volcanic gas reservoir, volcanic lithologic gas reservoir and the faulted–lithologic composite volcanic gas reservoir (Fig. 14.12).

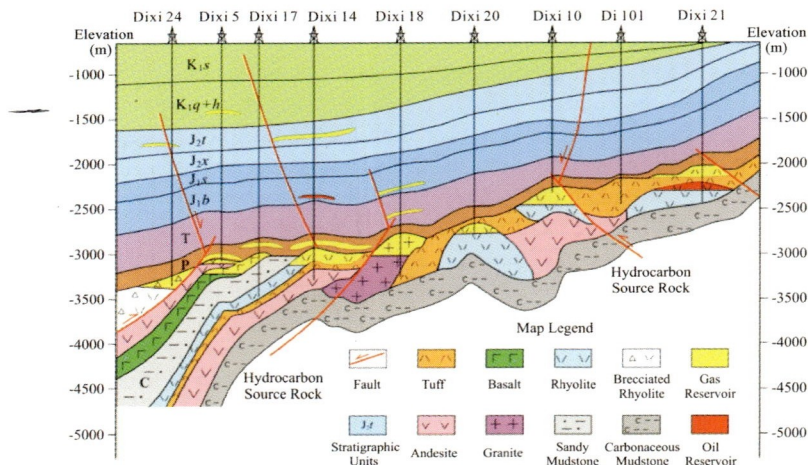

Fig. 14.12.   Types of natural gas reservoir in the Dinan uplift (according to the Research Institute of Xinjiang Oil Field)

● **Structural–Stratigraphic Composite Gas Reservoir in the Volcanic Rocks (the Dixi 17 Gas Field)**

The Dixi 17 well was located at the western end of the Dinan uplift. Basalt was the primary reservoir sequence that was pinched out in an easterly direction. Dark colored mudstone occurred in the direction of the upward-dipping basalt; this mudstone was positioned between the upper and lower volcanic sequences and it combined with the Permian mudstone to create the seals. The southern fault of the Dixi 17 well and the northern fault of the Dixi 17 well passed through the area on the southern and northern sides respectively. This is a faulted–stratigraphic type of gas reservoir.

● **Lithologic Gas Field in the Volcanic Rocks (the Dixi 14 Gas Field)**

The Dixi 14 well is located in the middle section of the Dinan uplift. The lithologies of reservoir sequences were varied, not only including the basic type of basalt and the acidic type of rhyolite of extrusive facies, but also containing tuffaceous breccia of explosive facies. In an easterly direction, pyroclast–lava changed into dark colored mudstone and a coal layer of limnetic facies. This change in facies indicated that, on the eastern side of the Dixi 14 well district, the sealing condition appeared in an upward dipping direction. On the western side of the Dixi 14 well district, the sedimentary rocks occurred between the upper and lower volcanic sequences. The overlaying strata were Permian mudstones. The Dixi 14 well district is a volcanic cone–lithologic composite type of gas reservoir.

- **Faulted–Lithologic Composite Gas Reservoir in the Volcanic Rocks (the Dixi 18 Gas Field)**

In the Dixi 18 well district, dark colored mudstone and coal layers of limnetic facies developed on the western side (these rocks occurred in the Di 404 well); the marine mudstone and the depositional tuff developed on the northern side (these rocks appeared in the Dixi 27 well and the Dixi 181 well); the depositional tuff developed on the eastern side (this rock occurred in the Dixi 20 well district); the Dishuiquan fault was located on the southern side. This is a faulted–lithologic type of gas reservoir.

14.4.5 Nature of Fluids

- **Characteristics of Hydrocarbon Constituents**

In the gas reservoir on the Dinan uplift, the surface density of condensate oil is $0.761 - 0.79$ g/cm^3; at 50 °C, the viscosity is $0.94 - 1.56$ MPa·s; the content of wax is $0.76\% - 3.66\%$; the condensation point is at $-13.2 - 4.2$ °C; the initial boiling point is at $69.2 - 123.5$ °C. In the gas reservoir on the Dinan uplift, the relative density of natural gas is $0.633 - 0.644$; the content of methane is $83.6\% - 87.5\%$; the content of carbon dioxide is $0.1\% - 0.258\%$; the content of nitrogen is $4.08\% - 5.81\%$.

- **Geochemical Features of Formation Water**

In the Dinan uplift, the carboniferous system contained the $CaCl_2$ type of formation water. The distribution of the total salinity is in the range of $8,776.15 - 22,606.56$ mg/L; the content of chlorion is in the range of $6,796.19 - 13,915.4$ mg/L.

● **Pressure of Formation and Related Geothermal Characteristics**

According to the collected data in each well on the Dinan uplift (which included measured formation temperature, pressure data, calculated original formation pressure, the pressure coefficient and the temperature that was registered in the middle portion of the gas reservoir), in every exploration block the formation pressure is in the range of 33.645 – 47.5 MPa; the pressure coefficient is in the range of 1.07 – 1.27; the temperature of the middle portion is in the range of 104.4 – 117.26 °C.

References

Jin, J., Liu, L.F., Yu, X.Y., et al., 2008. The progress of natural gas exploration in the carboniferous volcanic rocks in the Ludong–Wucaiwan area. Natural Gas Industry, 28(5):21-23.

Li, H.G., Lin, X.Y., 2006. The characteristics of deeply extended fault in the Changling Faulted Basin and the potential for natural gas exploration. Oil Geophysical Prospecting, 41(Suppl.):33-36.

Li, J.K., Feng, Z.H., Liu, W., et al., 2006. The study of natural gas accumulative phase at a deep level in the Xujiaweizi faulted basin, the Songliao Basin. Acta Petrolei Sinica, 127(Suppl.):42-45.

Liang, D.G., 2008. The review and forecast of the Ordovician Oil Field in the Lunnan–Tahe area in the Tarim basin. Acta Petrolei Sinica, 29(1):153-158.

Liu, J.T., Zhu, Y.X., Li, Z.G., et al., 2009. The characteristics and main controlling factors of the Carboniferous volcanic oil and gas reservoirs in the Santanghu basin. Lithologic Reservoirs, 21(3):23-27.

Ren, Y.G., Zhu, D.F., Wan, C.B., et al., 2004. The geologic features at a deep level of the northern songliao basin and the exploration direction of natural gas. China Petroleum Exploration, 9(4):12-18.

Shi, H.X., Xu, S.M., Lin, F., et al., 2005. The source rock analysis and exploration forecast in the Lunnan Oil Field in the Tarim basin. Xinjiang Petroleum Geology, 26(6):623-626.

Wang, R.C., Xu, H.M., Shao, Y., et al., 2008. The characteristics of the carboniferous volcanic reservoir sequences in the Ludong area, the Zhungeer basin. Acta Petrolei Sinica, 29(3):350-355.

Yang, N., Lu, X.X., Cheng, M.T., et al., 2008. The characteristics of carbonate reservoir sequences in the Lunnan and Tahe Oil Fields in the Tarim Basin— Using the Sha 107 Well and the Lungu 40 Well as examples. Petroleum Geology & Experiment, 30(3):247-251.

Zhang, C.J., Shi, X., Wu, X.Z., et al., 2005. The requirements of oil and gas accumulation in the carboniferous system and the favorable exploration domain in the Zhungeer basin. China Petroleum Exploration, 15(1):11-15.

Zhao, M.J., Pan, W.Q., Han, J.F., et al., 2007. The accumulative process and accumulative model of the Ordovician buried hill type of oil reservoir in the Lunnan area, the Tarim basin. Chinese Science Bulletin, 52(S.):174-183.

Zhao, Z.H., Guo, Z.J., Huan, B.F., et al., 2006. The geochemical features of the late Paleozoic volcanic rocks and the significance of structural–magmatic evolution in the Santanghu basin, the Xinjiang region. Acta Petrologica Sinica, 22(1):199-214.

Zhou, F.Y., Sun, Y.S., Zhang, S.C., 2001. The course, phases and direction of oil and gas migration in the Lunnan area, in the Tarim basin. Geological Review, 47(3): 329-334.

Index